Advances in
Machine Learning and
Data Mining for Astronomy

Chapman & Hall/CRC
Data Mining and Knowledge Discovery Series

SERIES EDITOR

Vipin Kumar

University of Minnesota
Department of Computer Science and Engineering
Minneapolis, Minnesota, U.S.A

AIMS AND SCOPE

This series aims to capture new developments and applications in data mining and knowledge discovery, while summarizing the computational tools and techniques useful in data analysis. This series encourages the integration of mathematical, statistical, and computational methods and techniques through the publication of a broad range of textbooks, reference works, and handbooks. The inclusion of concrete examples and applications is highly encouraged. The scope of the series includes, but is not limited to, titles in the areas of data mining and knowledge discovery methods and applications, modeling, algorithms, theory and foundations, data and knowledge visualization, data mining systems and tools, and privacy and security issues.

PUBLISHED TITLES

UNDERSTANDING COMPLEX DATASETS:
DATA MINING WITH MATRIX DECOMPOSITIONS
David Skillicorn

COMPUTATIONAL METHODS OF FEATURE SELECTION
Huan Liu and Hiroshi Motoda

CONSTRAINED CLUSTERING: ADVANCES IN
ALGORITHMS, THEORY, AND APPLICATIONS
Sugato Basu, Ian Davidson, and Kiri L. Wagstaff

KNOWLEDGE DISCOVERY FOR COUNTERTERRORISM
AND LAW ENFORCEMENT
David Skillicorn

MULTIMEDIA DATA MINING: A SYSTEMATIC
INTRODUCTION TO CONCEPTS AND THEORY
Zhongfei Zhang and Ruofei Zhang

NEXT GENERATION OF DATA MINING
Hillol Kargupta, Jiawei Han, Philip S. Yu,
Rajeev Motwani, and Vipin Kumar

DATA MINING FOR DESIGN AND MARKETING
Yukio Ohsawa and Katsutoshi Yada

THE TOP TEN ALGORITHMS IN DATA MINING
Xindong Wu and Vipin Kumar

GEOGRAPHIC DATA MINING AND
KNOWLEDGE DISCOVERY, SECOND EDITION
Harvey J. Miller and Jiawei Han

TEXT MINING: CLASSIFICATION, CLUSTERING, AND
APPLICATIONS
Ashok N. Srivastava and Mehran Sahami

BIOLOGICAL DATA MINING
Jake Y. Chen and Stefano Lonardi

INFORMATION DISCOVERY ON ELECTRONIC HEALTH
RECORDS
Vagelis Hristidis

TEMPORAL DATA MINING
Theophano Mitsa

RELATIONAL DATA CLUSTERING: MODELS,
ALGORITHMS, AND APPLICATIONS
Bo Long, Zhongfei Zhang, and Philip S. Yu

KNOWLEDGE DISCOVERY FROM DATA STREAMS
João Gama

STATISTICAL DATA MINING USING SAS APPLICATIONS,
SECOND EDITION
George Fernandez

INTRODUCTION TO PRIVACY-PRESERVING DATA
PUBLISHING: CONCEPTS AND TECHNIQUES
Benjamin C. M. Fung, Ke Wang, Ada Wai-Chee Fu, and
Philip S. Yu

HANDBOOK OF EDUCATIONAL DATA MINING
Cristóbal Romero, Sebastian Ventura,
Mykola Pechenizkiy, and Ryan S.J.d. Baker

DATA MINING WITH R: LEARNING WITH
CASE STUDIES
Luís Torgo

MINING SOFTWARE SPECIFICATIONS: METHODOLOGIES
AND APPLICATIONS
David Lo, Siau-Cheng Khoo, Jiawei Han, and Chao Liu

DATA CLUSTERING IN C++: AN OBJECT-ORIENTED
APPROACH
Guojun Gan

MUSIC DATA MINING
Tao Li, Mitsunori Ogihara, and George Tzanetakis

MACHINE LEARNING AND KNOWLEDGE DISCOVERY FOR
ENGINEERING SYSTEMS HEALTH MANAGEMENT
Ashok N. Srivastava and Jiawei Han

SPECTRAL FEATURE SELECTION FOR DATA MINING
Zheng Alan Zhao and Huan Liu

ADVANCES IN MACHINE LEARNING AND DATA MINING
FOR ASTRONOMY
Michael J. Way, Jeffrey D. Scargle, Kamal M. Ali, and
Ashok N. Srivastava

Advances in Machine Learning and Data Mining for Astronomy

Edited by
Michael J. Way, Jeffrey D. Scargle,
Kamal M. Ali, and Ashok N. Srivastava

CRC Press
Taylor & Francis Group
Boca Raton London New York

CRC Press is an imprint of the
Taylor & Francis Group, an **informa** business

A CHAPMAN & HALL BOOK

CRC Press
Taylor & Francis Group
6000 Broken Sound Parkway NW, Suite 300
Boca Raton, FL 33487-2742

First issued in paperback 2016

© 2012 by Taylor & Francis Group, LLC
CRC Press is an imprint of Taylor & Francis Group, an Informa business

No claim to original U.S. Government works

ISBN 13: 978-1-138-19930-9 (pbk)
ISBN 13: 978-1-4398-4173-0 (hbk)

Library of Congress Cataloging-in-Publication Data

Advances in machine learning and data mining for astronomy / editors, Michael J. Way ... [et al.].
    p. cm. -- (Chapman & Hall/CRC data mining and knowledge discovery series)
    Summary: "This book provides a comprehensive overview of various data mining tools and techniques that are increasingly being used by researchers in the international astronomy community. It explores this new problem domain, discussing how it could lead to the development of entirely new algorithms. Leading contributors introduce data mining methods and then describe how the methods can be implemented into astronomy applications. The last section of the book discusses the Redshift Prediction Competition, which is an astronomy competition in the style of the Netflix Prize"-- Provided by publisher.
    Includes bibliographical references and index.
    ISBN 978-1-4398-4173-0 (hardback : acid-free paper)
    1. Astronomy--Data processing. 2. Data mining. 3. Machine learning. I. Way, Michael J.

QB51.3.E43A386 2012
522'.85631--dc23
2011036610

Visit the Taylor & Francis Web site at
http://www.taylorandfrancis.com

and the CRC Press Web site at
http://www.crcpress.com

Contents

Foreword, ix

Editors, xi

Perspective, xiii

Contributors, xxv

PART I   **Foundational Issues**

CHAPTER 1 ▪ Classification in Astronomy: Past and Present   3

   ERIC FEIGELSON

CHAPTER 2 ▪ Searching the Heavens: Astronomy, Computation,
Statistics, Data Mining, and Philosophy   11

   CLARK GLYMOUR

CHAPTER 3 ▪ Probability and Statistics in Astronomical Machine Learning
and Data Mining   27

   JEFFREY D. SCARGLE

PART II   **Astronomical Applications**
SECTION 1   **Source Identification**

CHAPTER 4 ▪ Automated Science Processing for the Fermi
Large Area Telescope   41

   JAMES CHIANG

CHAPTER 5 ▪ Cosmic Microwave Background Data Analysis   55

   PANIEZ PAYKARI AND JEAN-LUC STARCK

CHAPTER 6 ■ Data Mining and Machine Learning in Time-Domain
Discovery and Classification   89

JOSHUA S. BLOOM AND JOSEPH W. RICHARDS

CHAPTER 7 ■ Cross-Identification of Sources: Theory and Practice   113

TAMÁS BUDAVÁRI

CHAPTER 8 ■ The Sky Pixelization for Cosmic Microwave
Background Mapping   133

O.V. VERKHODANOV AND A.G. DOROSHKEVICH

CHAPTER 9 ■ Future Sky Surveys: New Discovery Frontiers   161

J. ANTHONY TYSON AND KIRK D. BORNE

CHAPTER 10 ■ Poisson Noise Removal in Spherical Multichannel Images:
Application to Fermi Data   183

JÉRÉMY SCHMITT, JEAN-LUC STARCK, JALAL FADILI, AND SETH DIGEL

SECTION 2   **Classification**

CHAPTER 11 ■ Galaxy Zoo: Morphological Classification
and Citizen Science   213

LUCY FORTSON, KAREN MASTERS, ROBERT NICHOL, KIRK D. BORNE,
EDWARD M. EDMONDSON, CHRIS LINTOTT, JORDAN RADDICK,
KEVIN SCHAWINSKI, AND JOHN WALLIN

CHAPTER 12 ■ The Utilization of Classifications in High-Energy
Astrophysics Experiments   237

BILL ATWOOD

CHAPTER 13 ■ Database-Driven Analyses of Astronomical Spectra   267

JAN CAMI

CHAPTER 14 ■ Weak Gravitational Lensing   287

SANDRINE PIRES, JEAN-LUC STARCK, ADRIENNE LEONARD,
AND ALEXANDRE RÉFRÉGIER

Chapter 15 ▪ Photometric Redshifts: 50 Years After   323

     Tamás Budavári

Chapter 16 ▪ Galaxy Clusters   337

     Christopher J. Miller

Section 3   **Signal Processing (Time-Series) Analysis**

Chapter 17 ▪ Planet Detection: The Kepler Mission   355

     Jon M. Jenkins, Jeffrey C. Smith, Peter Tenenbaum, Joseph D. Twicken, and Jeffrey Van Cleve

Chapter 18 ▪ Classification of Variable Objects in Massive Sky Monitoring Surveys   383

     Przemek Woźniak, Łukasz Wyrzykowski, and Vasily Belokurov

Chapter 19 ▪ Gravitational Wave Astronomy   407

     Lee Samuel Finn

Section 4   **The Largest Data Sets**

Chapter 20 ▪ Virtual Observatory and Distributed Data Mining   447

     Kirk D. Borne

Chapter 21 ▪ Multitree Algorithms for Large-Scale Astrostatistics   463

     William B. March, Arkadas Ozakin, Dongryeol Lee, Ryan Riegel, and Alexander G. Gray

Part III   **Machine Learning Methods**

Chapter 22 ▪ Time–Frequency Learning Machines for Nonstationarity Detection Using Surrogates   487

     Pierre Borgnat, Patrick Flandrin, Cédric Richard, André Ferrari, Hassan Amoud, and Paul Honeine

Chapter 23 ▪ Classification   505

     Nikunj Oza

CHAPTER 24 ■ On the Shoulders of Gauss, Bessel, and Poisson: Links, Chunks, Spheres, and Conditional Models   523

WILLIAM D. HEAVLIN

CHAPTER 25 ■ Data Clustering   543

KIRI L. WAGSTAFF

CHAPTER 26 ■ Ensemble Methods: A Review   563

MATTEO RE AND GIORGIO VALENTINI

CHAPTER 27 ■ Parallel and Distributed Data Mining for Astronomy Applications   595

KAMALIKA DAS AND KANISHKA BHADURI

CHAPTER 28 ■ Pattern Recognition in Time Series   617

JESSICA LIN, SHERI WILLIAMSON, KIRK D. BORNE, AND DAVID DEBARR

CHAPTER 29 ■ Randomized Algorithms for Matrices and Data   647

MICHAEL W. MAHONEY

INDEX, 673

Foreword

Like many sciences, astronomy has reaped immense benefit from explosive advances in computational and information technology. The Age of Digital Astronomy is such an extremely data-rich environment so as to defy traditional methodologies and approaches to analyzing and extracting new knowledge from the data. The scientific discovery process is increasingly dependent on the ability to analyze massive amounts of complex data generated by scientific instruments and simulations. Analysis is rapidly becoming the bottleneck, if not choke point, in the process. This situation has motivated needs for innovation and for fostering of collaborations with the computer science and technology community to bring advances in those fields to bear on the science investigations. This book documents many such successful collaborations among computer scientists, statisticians, and astronomers, applying state-of-the-art machine learning and data mining techniques in astronomy. The book itself is the product of such collaboration.

The value and potential impact of this book lies in building upon and extending the underlying interdisciplinary collaborations, along both the science and computer science axes. The massive data-richness and -complexity challenges are not unique to astronomy, but rather are shared with many science disciplines. As a matter of fact there is a broad emerging attention to and emphasis on the area of data-intensive or "big data" science. Some of the challenges are unique, but many are common and lend themselves to general tools that transcend traditional boundaries between the science disciplines.

Prospects are bright for the value and impact interdisciplinary teams of computational scientists, applied mathematicians, and domain scientists can bring to the science endeavor. Bold new ideas and innovations will go beyond improving current techniques to seek breakthrough possibilities. An emerging field of research analytics will push new frontiers for data analysis, not just better hammers. An educational component will evolve along with it for not only training the next generation, but also adapting existing scientists to this collaborative mode of research with computer scientists.

Joe Bredekamp
NASA Science Mission Directorate

Editors

Michael J. Way, PhD, is a research scientist at the NASA Goddard Institute for Space Studies in New York and the NASA Ames Research Center in California. He is also an adjunct professor in the Department of Physics and Astronomy at Hunter College in the City University of New York system. He is the leader of the biennial New York Workshop on Computer, Earth, and Space Science. His background is mainly in the astrophysical sciences where he is working to find new ways to understand the multiscale structure of our universe, modeling the atmospheres of exoplanets, and applying kernel methods to new areas in astronomy. He is also involved in interdisciplinary research with climate scientists, astrobiologists, mathematicians, statisticians, and researchers in machine learning and data mining.

Jeffrey D. Scargle, PhD, is an astrophysicist in the Space Science and Astrobiology Division of the NASA Ames Research Center. For high-energy astrophysics applications he has developed algorithms for time-series analysis, image processing, and density estimation that are in wide use in a number of fields. He has participated in numerous NASA astrophysics missions and is a member of the Fermi Gamma Ray Telescope collaboration. His main interests are: variability of astronomical objects, including the Sun, sources in the Galaxy, and active galactic nuclei; cosmology; plasma astrophysics; planetary detection; data analysis and statistical methods.

Kamal M. Ali, PhD, is a research scientist in machine learning and data mining. His PhD work centered on combining classifiers, ensembles, and voting machines. He carried out research at IBM Almaden, Stanford University, TiVo, Vividence and Yahoo. He currently maintains a consulting practice and is cofounder of the start-up Metric Avenue. He has published dozens of papers in peer-reviewed journals and conferences, and has served as chair at the International Conference of Knowledge Discovery and Data Mining (ACM SIGKDD) and as reviewer for dozens of conferences and journals. His work has ranged from engineering the TiVo Collaborative Filtering Engine, which is used in millions of TiVos worldwide, to diverse research in combining classifiers, search engine quality evaluation, and learning association rules. Recently, at Stanford University, his research focused on design, execution, and improvement of plans executed in a cognitive architecture. His current research is in combining machine learning in conditional random fields with linguistically rich features to make machines better at reading Web pages.

Ashok N. Srivastava, PhD, is the principal scientist for Data Mining and Systems Health Management at NASA Ames Research Center. He was formerly the principal investigator for the Integrated Vehicle Health Management research project at NASA. His current research focuses on the development of data mining algorithms for anomaly detection in massive data streams, kernel methods in machine learning, and text mining algorithms. Dr. Srivastava is also the leader of the Intelligent Data Understanding group at NASA Ames Research Center. The group performs research and development of advanced machine learning and data mining algorithms in support of NASA missions. He performs data mining research in a number of areas in aviation safety and application domains such as earth sciences to study global climate processes and astrophysics to help characterize the large-scale structure of the universe.

Perspective

Michael J. Way
NASA Goddard Institute for Space Studies

Jeffrey D. Scargle
NASA Ames Research Center

Kamal M. Ali
Stanford University

Ashok N. Srivastava
NASA Ames Research Center

CONTENTS

0.1	Introduction	xiii
0.2	The Age of Digital Astronomy	xiv
	0.2.1 Detectors and Data Acquisition Technology	xv
	0.2.2 Telescopes and Related Technology	xv
	0.2.3 Data Accessibility	xvi
	0.2.4 Data Analysis Methodology	xvii
0.3	Organization of This Book	xviii
	0.3.1 Foundational Issues	xviii
	0.3.2 Astronomical Applications	xix
	0.3.2.1 Source Identification	xix
	0.3.2.2 Classification	xx
	0.3.2.3 Signal Processing (Time-Series Analysis)	xxi
	0.3.2.4 The Largest Data Sets	xxii
	0.3.3 Machine Learning Methods	xxii
0.4	About the Cover	xxiii
References		xxiv

0.1 INTRODUCTION

As our solar system is deeply embedded in the disk of our Milky Way Galaxy, it is difficult to obtain a good picture of the structure of our own galaxy and the universe as a whole.[*] In

---

[*] See Wright (1750) for one of the first detailed treatments.

much the same way it is difficult for astronomers today to gain a comprehensive perspective on the revolution that is completely changing the nature of their science. Our aim is to provide some of this perspective.

0.2 THE AGE OF DIGITAL ASTRONOMY

Today astronomers are immersed in an era of rapid growth and changes in scientific methodologies and technologies, leading to astonishing shifts in our world view. Most aspects of this revolution rely on the concept of digital data, hence this era is aptly called the *Age of Digital Astronomy*. One wonders whether the giants who invented the scientific method—Aristotle, Galileo, Newton—had even the slightest inkling of how our scientific world would grow and proliferate from the seeds they planted. Likewise, one can ponder the same (unanswerable) question for more recent giants—Baade, Bessel, Chandrasekhar, Einstein, Hubble, and Struve. These musings may help us to assess our own ability to foresee the future progress of astronomical research.

Recounting a few historical highlights may help set the context of this revolution. Along with botany and biology, astronomy has always been among the sciences most oriented toward data collection. Sometimes criticized as "mere stamp collecting," astronomical data acquisition in prehistory was far from purposeless. The collected "stamps" were effective guides for planting crops and impressing the populace. Found around the world, megalithic sites were arguably the first instruments with predictive power for estimating equinoxes and seasonal sunrises and sunsets. Ancient Babylonians, Egyptians, Chinese, and Greeks gathered enough data in their time to make increasingly accurate predictions of eclipse cycles, precession of the Earth's axis, and planetary motions. The few surviving Mayan codices contain a wealth of data on Venus, the Sun, and the Moon that informed their sophisticated calendric and agricultural technology. Who knows what other riches of the same kind were lost to the invasion of the conquistadors?

These early databases eventually allowed their curators to mine the data about the heavens in search of useful mathematical representations. Several early pre-telescopic source catalogs include those by anonymous Babylonian star-watchers as early as 1500 BCE, the Chinese Dunhuang Map (ca. 700 ACE) derived from observations of Wu Xian (ca. 1000 BCE), and of Gan De and Shi Shen (fourth century BCE)[*] and the perhaps better-known Greek astronomer and trigonometer Hipparchus (ca. 140 BCE). Astronomy has always been at the forefront of categorizing large ensembles of data, both for utilitarian purposes and for understanding the underlying phenomena.

Collected over long periods of time, these data led to mathematical models with much predictive power—evidenced by the most sophisticated device from ancient times discovered to date: The Antikythera mechanism (Evans et al., 2010; Marchant, 2010). From the likes of the Greek astronomers Hipparchos and Ptolemy, then the Islamic era observers, Al-Sufi and Ulugh-Beg to the Renaissance brilliance of Tycho Brahe, the accuracy of ancient astrometric observations progressed steadily. By the time of Brahe's technologically advanced

---

[*] In the Wade–Giles system these three names are: Wu Hsien, Kan Te, and Shih Shen, respectively.

Uranibourg, arguably the first organized research unit in history (Christianson, 2000), observations were accurate enough for his "postdoctoral fellow" Johannes Kepler to discern the elliptical orbit of Mars, therefore, solidifying the Copernican revolution even before the invention of telescopes (1608).

Starting in the 1600s, telescopic observations produced much varied data on everything from comets to nebulae,[*] including brightness variability of stars and supernovae. It became increasingly hard for rudimentary data classification to yield even a crudely realistic picture of the Universe. So our era is not the first time that a technology-driven explosion of observations led to hectic attempts to understand the implications of the data. Even Galileo (the founder of the scientific method) failed to make sense of his observations of Saturn's "ears."[†] He astutely published phenomenological descriptions of this and the other astonishing sights appearing in his telescope, leaving classification and ultimately understanding of the phenomena as later steps in the scientific process. In the modern paradigm, exploration and unsupervised learning yield discoveries and principles that inform subsequent supervised learning. And in fairness to Galileo and other early planetary astronomers, one should note that Saturn has not yet completely surrendered her secrets to our latest technological attacks using highly instrumented probes situated near the planet.

Astronomical science has thus seen a number of technology breakthroughs followed by leaps in fundamental understanding. It is now clear that we are currently embedded in another such episode. Let us quickly review the relevant technological advances in solid–state physics, material sciences, computer science, and statistical science that have contributed to astronomy in the most recent decades.

0.2.1 Detectors and Data Acquisition Technology

One of the authors remembers as a graduate student spending nights at the then huge—but now obsolete, small, and mothballed—telescopes on Mt. Wilson, and then spending days drawing lines through noisy traces cranked out by strip-chart recorders (the old-fashioned way of averaging out observational errors!), then reading numbers off the scale printed on the chart paper. In those days, the big new development was improving photometry by replacing visual and photographic methods with photoelectric detectors and electronic recording devices. Then came multichannel versions of these same devices. Compare these crudities with modern high-quantum efficiency CCD detector systems for photometry and spectroscopy, high-speed electronic recording devices, and sophisticated on-board analysis of highly complex data streams such as those found on the Fermi Gamma Ray Space Telescope.

0.2.2 Telescopes and Related Technology

In the last century, the collecting area of optical telescopes has grown rapidly through a number of technological advances in the lightweight metals, light mirror design, and segmented mirrors to name but a few. Initially, the largest telescopes gave up their status quite

---

[*] Now known to include star-forming regions in our own galaxy and external galaxies such as Andromeda.

[†] Saturn's "ears" were later correctly interpreted by Christiaan Huygens (De causis mirandorum Saturni phænomenôn, 1659).

slowly. For example, the 100 in. (2.5 m) Hooker was the world's largest telescope from 1917 to 1948. In 1949, the Palomar 200 in. (5 m) took over this title, and was superseded in 1975 by the BTA-6 236 in. (6 m) telescope in Russia, and then in 1993 by the segmented mirror telescope Keck 1 (10 m). Today, the European Southern Observatory can be considered the largest optical telescope: advances in interferometry allow the light from their four 8 m telescopes to be combined. The next decade should bring us at least two single-dish telescopes in the 30–40 m range—hopefully one in each hemisphere.

At the end of the Cold War, astronomers leveraged a number of previously classified infrared detector technologies. These now allow ground-based observations through transparent windows in the 1–10 μm wavelength range. Advances in adaptive optics have enabled a number of ground-based observatories to obtain high-resolution imaging of objects in the near-infrared comparable to that obtained by space telescopes operating well above the turbulence of the Earth's atmosphere.

In space, above the absorbing layers of the atmosphere, telescopes range from the microwave regime (COBE, WMAP, Planck) through the optical (HST), the infrared (IRAS, ISO, Spitzer, Herschel), to the x-ray (ROSAT, Chandra, XMM-Newton, Swift), and γ-ray (CGRO, FGST) regimes.[*] These have opened up previously unexplored views of the world. So have ground-based observations of cosmic rays and neutrinos—that is particles, not light. All of these programs have required new developments in one or more areas of detector technology, spacecraft design, data processing, mirror construction, and spectrograph design.

An even bolder leap is in progress, on terra firma, in the form of the search for gravitational waves (cf. Sam Finn's chapter on corresponding data analysis considerations), another mode of observation transcending electromagnetic radiation and possibly representing the last such astronomical frontier. In addition, robotic telescopes and networks of telescopes are undergoing rapid development. From the point of view of information technology, perhaps the most innovative such leap is the beam-forming pipeline of the Low Frequency Array (Mol and Romein, 2011) or LOFAR, a software telescope utilizing real-time supercomputer-based signal processing to bring together data from a pan-European array of tens of thousands of radio antennas.

0.2.3 Data Accessibility

A less touted but arguably even more important element of the Age of Digital Astronomy derives from the radical modes of data promulgation. No longer does one send postcards and letters to colleagues requesting pre- or reprints of articles and tabulated data. For scientific presentations 4 × 6 in. glass lantern slides and the corresponding bulky and hot projectors run by students have given way to more compact media. While there is no denying the impact of modern information technology on communication of scientific, engineering, and technical information, many aspects of busy, regimented, bulletized powerpoint presentations are counterproductive (e.g., Tufte, 2003).

---

[*] COBE: COsmic Microwave Background Explorer; WMAP: Wilkinson Microwave Anisotropy ProbePlanck; HST: Hubble Space Telescope; IRAS: InfraRed Astronomical Satellite; ISO: Infrared Satellite Observatory; ROSAT: ROentgen SATellite; CGRO: Compton Gamma-Ray Observatory; FGST: Fermi Gamma Ray Space Telescope.

The Internet revolution has rather suddenly enabled the average astronomer to obtain all kinds of data in different wavelength regimes from a variety of Web sites—both focused collections from single observatories and organized synoptic collections of related data from disparate observatories. To name a few: HEASARC, MAST, SAF, IPAC, LAMBDA, and the PDS.[*] These can all be accessed via a modest-sized computer connected to the global Internet, a far cry from 30 years ago when it was typical to write institutions for data tapes to be sent via the postal service!

The Internet has also led to tremendous advances in tools for rapid dissemination of time-critical data. Even 20 years ago one might wait for an International Astronomical Union "Astronomical Telegram" to inform one of important, time-sensitive events. Today there is onboard data processing on high-energy satellites—such as the Fermi Gamma Ray Space Telescope and the Swift hard x-ray telescope—and real-time international alert systems. We are seeing that machine readable data are exchanged between software agents and the hardware systems that host them. In fact, autonomous robotic agents are gradually becoming capable of making their own decisions as to what should be observed next and how.

0.2.4 Data Analysis Methodology

Everyone knows of the exponential growth of computation speed, fast-access memory, and permanent data storage capacity—collectively and loosely called Moore's Law. In comparison the role of storage is sometimes overlooked, but the ability to store and rapidly recover huge data sets—including but not limited to observational data (the coming LSST[†] cornucopia is a good example)—is of equal importance. The ability to archive massive amounts of intermediate computations, complex calibration tables, and theoretical simulations has altered scientific computation in major ways.

Most often the raw size of databases is emphasized: so-and-so many tera- or petabytes per year, per day, and so on. But the complexity of databases and their analyses is much more important. Raw size is almost automatically compensated for by increases in data handling and processing capabilities. Indeed, these two aspects feed back into each other to form a kind of mutually regulated system. But data complexity requires more—clever techniques, such as artificial intelligence, machine learning (at least for now still designed by clever humans, not machines), sophisticated data mining procedures, and so on. If there is any facet of the Age of Digital Astronomy where progress is lagging it is in addressing this complexity. The resulting difficulties affect a number of fields, as demonstrated by a recent issue of *Science*[‡] (e.g., Baraniuk, 2011) and an entire book (Hey et al., 2009) to name two recent examples.

How much exponential progress has been made since Karl Friedrich Gauss invented least squares and, less well known, least-absolute-value (robust) solutions of overdetermined

---

[*] HEASARC: High Energy Astrophysics Science Archive Research Center; MAST: Multimission Archive at STSci; SAF: Science Archive Facility for the European Southern Observatory and Space Telescope European Coordinating Facility; IPAC: Infrared Processing and Analysis Center; LAMBDA: Legacy Archive for Microwave Background Data Analysis; PDS: Planetary Data System.

[†] Large Synoptic Survey Telescope.

[‡] http://www.sciencemag.org/site/special/data

linear equation sets? Or since Jean Baptiste Joseph Fourier invented harmonic analysis? From mathematical and statistical scientists the answer is surely "quite a lot." Advanced modeling techniques, harmonic analysis, time-frequency and timescale distributions, wavelet methods, generalized basis pursuit, compressive sensing and other very powerful data analysis methods populate the pure and applied mathematics landscapes. Biostatisticians routinely work with models containing many more parameters than data points.

While all this advanced technology is just now working its way into astronomy, we are not very far past the day when the only thing astronomers did with time-series data (beside plotting them, and maybe poring through many plots until some unusual feature, perhaps a statistical fluke, was uncovered) was to compute the Fourier transform; if the data were unevenly spaced, interpolate! Maybe the next step (computing a power spectrum) could be taken, but often blissfully ignoring the fact that this function is very noisy—perversely noisy in such a way that accumulating more data does little good. More generally, astronomers were locked in a mental box that limited acceptable data analysis methods to a very straight and narrow path. Until recently, one could cogently argue that astronomical application of data analysis methods had not really advanced much in the last century or so. But this activity is undergoing its own slightly tardy revolution. This is the subject of our book.

The above discussion raises many questions. Is the current "revolution" qualitatively different from many previous episodes in the history of science, such as those mentioned above? Are there always synergistic advances in technology and science, and we are now fooling ourselves into placing more importance (as every generation seems to do) on current affairs? What will the state of astronomical and physical knowledge and information technology be like 10 years from now? What about 100 years from now? Will scientists then look back to today and not even see what we are currently declaring a revolution? You will not find answers to these questions in this book, but you will find a practical guide to many of the most important advances in machine learning and data mining in astronomy today. Perhaps in 10 years time a similar volume will be produced documenting the state of the art, allowing some comparisons. It could also be that there will be nothing to compare because machine learning and data mining will have advanced so far that the techniques discussed here will be as pedestrian as least-squares fitting of straight lines is today. We hope that this will be the case!

0.3 ORGANIZATION OF THIS BOOK

The book is divided into three parts: Foundational Issues, Astronomical Applications, and Machine Learning methods.

The cross-cutting nature of many of the topics, concepts and analysis methods considered here inevitably has led to an imperfect chapter classification scheme. Many of the contributions cross over the broadest distinctions, often describing both methods and applications. Nevertheless, we hope that the reader finds the following outline of sections useful.

0.3.1 Foundational Issues

Introductory chapters in Part I provide some context to issues that have confronted the astronomical sciences historically, particularly in probabilistic and statistical aspects of classification and cluster analysis. These issues are important in health and social sciences as well as the physical sciences. Chapter 1, "Classification in Astronomy: Past and Present" by Eric Feigelson, a long-time champion of astrostatistics, provides a historical commentary on developments in the relation between statistics and astronomy over the last century or so.[*] Philosopher-scientist Clark Glymour, in "Searching the Heavens: Astronomy, Computation, Statistics, Data Mining and Philosophy" and astronomer-data analyst Jeff Scargle, in "Probability and Statistics in Astronomical Machine Learning and Data Mining," provide commentaries on some rocky aspects of this relationship.

0.3.2 Astronomical Applications

Part II describes a number of astrophysics case studies that leverage a range of machine learning and data mining technologies. The considerable overlap between the four subsections of this part, and the chapters in them, should be kept in mind by the reader wishing to explore a specific topic. Indeed, many of the chapters in this part could equally well be placed in more than one subsection. For example, the chapter, "Data Mining and Machine Learning in Time Domain Discovery and Classification," could easily be put in any of the four sections. Classes and clusters can overlap!

0.3.2.1 Source Identification

After pipeline processing of raw data, the starting analysis of astronomical surveys is usually source detection. Section 1 of Part II covers this topic of identification and preliminary characterization of sources in several contexts, covering the range from long wavelengths (microwave) to short (gamma rays).

NASA's Fermi Gamma-ray Space Telescope has two kinds of automated data analysis (to be distinguished from routine data reduction). In addition to the classification technology described in the chapter in that section, a top-level automated analysis of the photon event data sent to the ground is also carried out. This is described in James Chiang's chapter "Automated Science Processing for the Fermi Large Area Telescope."

Philosophically, one of the most important developments in the last few decades is the ability to detect structure in the Cosmic Microwave Background (CMB)—a snapshot of our universe at the mere age of 400,000 years. The identification of the structure in this image, as detailed in the chapter "Cosmic Microwave Background Data Analysis" by Paniez Paykari and Jean-Luc Starck, is important because it provides clues to physical conditions at a very early stage in the evolution of the Universe. These clues can be detected and studied only after they are segregated from three other features: a sea of foreground sources (the most important being our Milky Way galaxy), the spatial pattern imprinted on the data by the velocity of the Earth (this essentially absolute motion would surprise Galileo and possibly even Einstein), and the much brighter constant (i.e., isotropic) component of the CMB itself.

---

[*] We write this introduction on the heels of the quinquennial conference Statistical Challenges in Modern Astronomy expertly organized by astronomer Eric Feigelson and his statistician colleague G. Jogesh Babu, where many of these issues were explored in great depth. The interested reader will be richly rewarded by consulting the volume from this and earlier ones in the series of books titled *Statistical Challenges in Modern Astronomy, I–IV, V* on the way.

In "Data Mining and Machine Learning in Time-Domain Discovery and Classification," Joshua Bloom and Joseph Richards discuss the challenges posed by the changing heavens; that is, discovering and characterizing singular events in time-series data. This topic will become increasingly important as more and more photometric survey projects yield huge collections of astronomical time series. An important and surprisingly difficult problem deriving from the proliferation of these large area surveys centers around consolidating source identification. For a very long time astronomical nomenclature has suffered from the confusion of a single object having upwards of dozens of names. A related issue is the identification of incarnations of the same object in catalogs obtained by various instruments at different wavelengths and different spatial resolutions. Tamas Budavari discusses this problem in "Cross-Identification of Sources: Theory and Practice."

A crucially useful tool in dealing with analysis of Cosmic Background and other all-sky data is described in "The Sky Pixelization for Cosmic Microwave Background Mapping," by Oleg Verkhodanov and Andre Doroshkevich.

Probably the most overarching thread in this book centers around the above-mentioned multitude of existing and planned surveys—the latter described in "Future Sky Surveys: New Discovery Frontiers," by Anthony Tyson and Kirk Borne. The most ambitious of these projects will yield multiple reobservations of the same areas of the sky for (1) the study of known variables such as active galactic nuclei, (2) detection of new examples of specific classes of objects, such as supernovae and planetary transits, and (3) detection and characterization of new types of transient and otherwise variable objects.

In the Fermi data pipeline, after the photon events are selected and sent to the ground (the subject of Bill Atwood's chapter in the classification section) various data analysis processes extract information about the sources. The chapter, "Poisson Noise Removal in Spherical Multichannel Images: Application to Fermi Data," by Jérémy Schmitt, Jean-Luc Starck, Jalal Fadili, and Seth Digel outlines a variety of approaches to this task, largely making use of the Poisson nature of the signal.

0.3.2.2 Classification

The process of placing elements from a data set into one of a relatively small number of sets (i.e., classes) is called *classification*. This simple operation is at the heart of the reductionist process in science: such assignment provides simplification and understanding, and enables the derivation of useful quantities. This section covers a large range of wavelengths and problems, exploring everything from gravitational lensing to identifying organic molecules in space. Some people still think of astronomy as "postage stamp collecting," and indeed this is still an important aspect of the field. The Galaxy Zoo project brings this aspect to a completely new level, as the reader will find in "Galaxy Zoo: Morphological Classification and Citizen Science," by Lucy Fortson, Karen Masters, Robert Nichol, Kirk D. Borne, Edward M. Edmondson, Chris Lintott, Jordan Raddick, Kevin Schawinski, and John Wallin.

Let us move on to machine learning in space! NASA's Fermi Gamma-ray Space Telescope has two kinds of automated data analysis (to be distinguished from routine data reduction), one of which—automated science processing—was discussed above with regard to source identification. In addition, because of the nature of the silicon strip detector technology and the cosmic ray background facing the instrument, a sophisticated classification analysis is required to capture true photon detections that reject the large background. This process is described by Bill Atwood in "The Utilization of Classifications in High Energy Astrophysics Experiments."

An important problem area in astronomy centers around the identification of the atomic and molecular composition of a wide range of astronomical sources, often using large spectroscopic databases. In his chapter, "Database-Driven Analyses of Astronomical Spectra," Jan Cami discusses modern approaches to this subject. He focuses on identifying polycyclic aromatic hydrocarbons (PAHs) in the interstellar medium and molecular layers in the circumstellar envelopes of asymptotic giant branch stars.

Another theme of modern astronomy is the exploitation of the gravitational bending of light of distant objects by massive objects lying near the line of sight to the distant object. This topic is detailed in "Weak Gravitational Lensing," by Sandrine Pires, Jean-Luc Starck, Adrienne Leonard, and Alexandre Réfrégier. Although first appearing in the literature in the mid-1980s, this topic has received more attention only in the last decade or so.

The information content of large-scale sky surveys is greatly enhanced by applying machine learning techniques to derive approximate values of the distances to galaxies based purely on their colors. The resulting *photometric redshifts* are less accurate than directly measured *spectroscopic redshifts*, but can be obtained with much fewer observational resources. As a result, the number of galaxies with approximately known distances is dramatically increased. This topic, crucial to existing and future sky surveys, is approached from a new angle in "Photometric Redshifts: 50 Years After," by Tamas Budavari.

A key data analysis task in astronomy is to identify clusters in data of varying dimensionality. This process can be viewed as a variant of classification in which the identities of the classes are defined by the distribution of the data points in the space in which the clustering is being attempted. Christopher Miller discusses this task in the context of the large-scale structure of the Universe in the chapter, "Galaxy Clusters."

0.3.2.3 Signal Processing (Time-Series Analysis)

Until recently, the most significant applications of signal processing techniques in astronomy were analyses of luminosity fluctuations. As the twenty-first century unfolds, time-series analyses at multiple wavelengths and sampling cadences will increase, largely but not exclusively driven by surveys. This subsection shows how in the last 10 years the relevant information technology has expanded into new astronomical domains.

The Kepler mission was designed to detect Earth-sized planets around nearby stars. It looks at time series of hundreds of thousands of stars to disentangle the interesting variable star sources from their planetary-transit objects of interest. In their chapter, "Planet Detection: The Kepler Mission," Jon Jenkins, Jeffrey Smith, Peter Tenenbaum, Joseph Twicken, and Jeffrey Van Cleve describe algorithms to detect weak planetary transit signatures against a background of instrumental noise and intrinsic stellar variability based upon advanced signal processing techniques and Bayesian analysis.

A view of discovery and classification aspects of time-domain astronomy complementary to the Bloom–Richards chapter in the Source Identification section is "Classification of Variable Objects in Massive Sky Monitoring Surveys," by Przemek Woźniak, Łukasz Wyrzykowski, and Vasily Belokurov.

Also relying upon sophisticated time-series analysis is gravitational wave astronomy (GWA). GWA opens up a completely new domain to astronomy that has never been previously explored. The chapter, "Gravitational Wave Astronomy," by Sam Finn discusses the corresponding new detection techniques and the underlying noise distribution against which signals are to be detected.

0.3.2.4 The Largest Data Sets

The last subsection deals with how astronomers and computational scientists deal with the large quantities of complex data now and in the near future. Scientists downloading large datasets to their desktops to run the same software they wrote with their thesis advisor 20 years ago is becoming less and less viable. The sizes of datasets are growing faster than the individual workstation's ability to hold the data in memory or even disk—not to mention I/O limitations. This is truly a paradigm shift, one that many astronomers have not yet comprehended. These and other issues, especially the role of modern machine learning methods in this setting, are discussed in the two chapters in this section: "Virtual Observatory and Distributed Data Mining," by Kirk Borne and "Multitree Algorithms for Large-Scale Astrostatistics," by William March, Arkadas Ozakin, Dongryeol Lee, Ryan Riegel, and Alex Gray.

0.3.3 Machine Learning Methods

The third section of the book has eight contributions from the developers of algorithms and practitioners of machine learning and data mining. Many of the techniques discussed in the Astronomical Applications section are here discussed in more detail. We have asked these authors to consider how their expertise may appeal to astronomers in general. It is fortunate, but not an accident, that several of the authors have collaborated extensively with astronomers. Indeed the labels astronomer, mathematician, and statistician are somewhat pale descriptions of the multidisciplinary training and expertise of many of today's scientists.

The concept of time–frequency distributions (also known to astronomers as spectrograms or dynamic power spectra) is quite old, but has more recently proliferated in the areas of application, mathematical development, and computational implementation. The chapter "Time–Frequency Learning Machines for Nonstationarity Detection Using Surrogates" by Pierre Borgnat, Patrick Flandrin, Cédric Richard, André Ferrari, Hassan Amoud, and Paul Honeine discusses this topic with this book's audience in mind.

The processes of classification and cluster analysis are discussed in appropriately named chapters, providing detailed discussion of current concepts and methodologies: Nikunj Oza's "Classification" and Kiri Wagstaff's "Data Clustering." The chapter, "Ensemble Methods: A Review," by Matteo Re and Giorgio Valentini described cluster analysis techniques that combine results from more than one analysis of a data set, thereby extracting more information and improving reliability.

"On the Shoulders of Gauss, Bessel and Poisson: Links, Chunks, Spheres and Conditional Models," by William Heavlin, offers an innovative discussion inverting the conventional view of the relationship between astronomy and statistical theory. This contribution outlines how astronomers' physical intuition has advanced the science of statistical modeling, and continues to do so.

In "Parallel and Distributed Data Mining for Astronomy Applications," Kamalika Das and Kanishka Bhaduri address a thread underlying the massive computations that are needed in many applications: namely matching the structure of the computational system to the nature of the algorithms and the data analysis problems.

From the discussion regarding applications in Part II, the reader will recognize that the chapter, "Pattern Recognition in Time Series," by Jessica Lin, Sheri Williamson, Kirk Borne, and David DeBarr is currently of great importance in astronomy—and one that will surely grow with time.

Michael Mahoney's contribution, "Randomized Algorithms for Matrices and Data," discusses a computational topic of pervasive importance to the data analyses discussed in this book.

We believe this book will contribute a unique perspective on machine learning and data mining for astronomy today. In comparison, there have been a number of excellent conference proceedings in years past such as *Mining the Sky* (Banday et al., 2001) and *Automated Data Analysis in Astronomy* (Gupta et al., 2002) although these books are both approximately 10 years old. There are also the annual Astronomical Data Analysis Software and Systems conference proceedings (e.g., Evans et al., 2011) but most contributions are quite limited in scope. The Feigelson and Babu *Statistical Challenges in Modern Astronomy* conferences and proceedings have also been an excellent resource (e.g., Feigelson and Babu, 1992, 2007) as mentioned above.

In conclusion, we offer special thanks to Randi Cohen of Taylor & Francis for her patient, professional, and highly competent work, orchestrating the efforts of four editors and multiple authors and reviewers. We are also extremely gratified by the enthusiastic participation of the many world class scientists who wrote the chapters and otherwise helped make this project a reality. We are confident that the resulting volume will be useful to astronomers, computational scientists, and statisticians alike.

0.4 ABOUT THE COVER

The top portion of the cover image depicts a simulated interaction of a single high-energy photon with the Fermi Gamma-ray Space Telescope. An incident photon interacts with one of the tungsten conversion layers producing an electron–positron pair and subsequent particle shower. The onboard data acquisition system processes these events and transmits the resulting detector signals to the ground where a sophisticated reconstruction algorithm analyzes the three-dimensional tracks and computes a large number of parameters characterizing each event.

Further analysis using a classification tree and other algorithms yields scientific results such as the all-sky map, shown in the bottom portion, which is based on 3,387,139 photons with energies of 1 GeV and above. It shows thousands of point sources, ranging from wildly variable active galaxies to presently unknown objects, as well as the diffuse emission in the plane of the Milky Way.

Arrows symbolize the flow of information from the instrument to results and back. Modern astrophysics requires a tight coupling of science goals (derived from new ideas and previous observations) with the design of both instruments and data analysis methods. The techniques described in this book are proving crucial to this interaction.

REFERENCES

Banday, A.J., Zaroubi, S. and Bartelmann, M. 2001, *Mining the Sky: Proceedings of the MPA/ESO/MPE Workshop*, Garching, Germany, July 31–August 4, 2000 (ESO Astrophysics Symposia), ISBN-10: 3540424687.

Baranuik, R.G. 2011, More is less: Signal processing and the data deluge, *Science*, 331, 6018, 717.

Christianson, J.R. 2000, On Tycho's island: Tycho Brahe and his assistants, Cambridge, England: Cambridge University Press, ISBN: 978-0-521-65081-6.

Evans, I.N., Accomazzi, A., Mink, D.J., and Rots, A.H. 2011, *Astronomical Data Analysis Software and Systems XX (ADASSXX)*, Astronomical Society of the Pacific Conference Series, Vol. 442, ISBN: 978-1-58381-764-3.

Evans, J., Carman, C.C., and Thorndike, A.S. 2010, Solar anomaly and planetary displays in the Antikythera mechanism, *Journal for the History of Astronomy*, 41 (1, no. 142), 1–39. http://www.shpltd.co.uk/jhacont2010.html.

Feigelson, E.D. and Babu, G.J. 1992, *Statistical Challenges in Modern Astronomy*, New York, USA: Springer, ISBN-10: 0387979115.

Feigelson, E.D. and Babu, G.J. 2007, *Statistical Challenges in Modern Astronomy IV*, San Francisco, USA: Astronomical Society of the Pacific, ISBN-10: 1583812407.

Gupta, R., Singh, H.P., and Bailer-Jones, C.A.L. 2002, *Automated Data Analysis in Astronomy*, New Delhi, India: Narosa Publishing House, ISBN-10: 0849324106.

Hey, T., Tansley, S., and Tolle, K. 2009, *The Fourth Paradigm: Data-Intensive Scientific Discovery*, Redmond, Washington, USA: Microsoft Research, ISBN-10: 0982544200.

Marchant, J. 2010, *Decoding the Heavens: A 2,000-Year-Old Computer and the Century-Long Search to Discover Its Secrets*, Cambridge, Massachusetts, USA: Da Capo Press, ISBN-10: 9780306818615.

Mol, J.D. and Romein, J.W. 2011, The LOFAR Beam Former: Implementation and performance analysis, http://arxiv.org/abs/1105.0661.

Tufte, E.R. 2003, *The Cognitive Style of PowerPoint: Pitching Out Corrupts Within*, Cheshire, CT: Graphics Press, ISBN: 0961392169.

Wright, T. 1750, *An Original Theory or New Hypothesis of the Universe*, New York: Elsevier, 1971.

Contributors

Hassan Amoud
Institut Charles Delaunay
Université de Technologie de Troyes
Troyes, France

Bill Atwood
Santa Cruz Institute for Particle
    Physics
University of California
Santa Cruz, California

Vasily Belokurov
Institute of Astronomy
Cambridge University
Cambridge, United Kingdom

Kanishka Bhaduri
Intelligent Systems Division
NASA Ames Research Center
Moffett Field, California

Joshua S. Bloom
Department of Astronomy
University of California, Berkeley
Berkeley, California

Pierre Borgnat
Laboratoire de Physique
École Normale Supérieure de Lyon
Lyon, France

Kirk D. Borne
Department of Computational and Data
    Sciences
George Mason University
Fairfax, Virginia

Joe Bredekamp
NASA Science Mission Directorate
Washington, DC

Tamás Budavári
Department of Physics and
    Astronomy
The Johns Hopkins University
Baltimore, Maryland

Jan Cami
Department of Physics and Astronomy
University of Western Ontario
London, Ontario, Canada
and
Department of Physics and Astronomy
SETI Institute
Mountain View, California

James Chiang
SLAC National Accelerator Laboratory
Kavli Institute for Particle Astrophysics
    and Cosmology
Menlo Park, California

Kamalika Das
Intelligent Systems Division
NASA Ames Research Center
Moffett Field, California

David DeBarr
Microsoft Corporation
Seattle, Washington

Seth Digel
SLAC National Accelerator Laboratory
Kavli Institute for Particle Astrophysics
    and Cosmology
Menlo Park, California

A.G. Doroshkevich
Lebedev Physical Institute
Russian Academy of Sciences
Moscow, Russia

Edward M. Edmondson
Institute of Cosmology and
    Gravitation
University of Portsmouth
Portsmouth, United Kingdom

Jalal Fadili
Groupe de Recherche en Informatique,
    Image, Automatique et Instrumentation
    de Caen
CNRS–ENSICAEN–Université de Caen
Caen, France

Eric Feigelson
Department of Astronomy and
    Astrophysics
Pennsylvania State University
University Park, Pennsylvania

André Ferrari
Laboratoire Fizeau
Université de Nice Sophia Antipolis
Nice, France

Lee Samuel Finn
Department of Astronomy and Astrophysics
Pennsylvania State University
University Park, Pennsylvania

Patrick Flandrin
Laboratoire de Physique
École Normale Supérieure de Lyon
Lyon, France

Lucy Fortson
School of Physics and Astronomy
University of Minnesota
Minneapolis, Minnesota

Clark Glymour
Department of Philosophy
Carnegie Mellon University
Pittsburgh, Pennsylvania

Alexander G. Gray
Computational Science and Engineering
    Division
Georgia Institute of Technology
Atlanta, Georgia

William D. Heavlin
Google, Inc.
Mountain View, California

Paul Honeine
Institut Charles Delaunay
Université de Technologie de Troyes
Troyes, France

Jon M. Jenkins
SETI Institute
Mountain View, California
and
Space Science and Astrobiology Division
NASA Ames Research Center
Moffett Field, California

Dongryeol Lee
Computational Science and Engineering
   Division
Georgia Institute of Technology
Atlanta, Georgia

Adrienne Leonard
CEA/DSM–CNRS–Université
   Paris Diderot
Gif-sur-Yvette, France

Jessica Lin
Department of Computer Science
George Mason University
Fairfax, Virginia

Chris Lintott
Department of Physics
University of Oxford
Oxford, United Kingdom
and
Department of Astronomy
Adler Planetarium and Astronomy
   Museum
Chicago, Illinois

Michael W. Mahoney
Department of Mathematics
Stanford University
Stanford, California

William B. March
Computational Science and Engineering
   Division
Georgia Institute of Technology
Atlanta, Georgia

Karen Masters
Institute of Cosmology and
   Gravitation
University of Portsmouth
Portsmouth, United Kingdom

Christopher J. Miller
Department of Astronomy
University of Michigan
Ann Arbor, Michigan

Robert Nichol
Institute of Cosmology and
   Gravitation
University of Portsmouth
Portsmouth, United Kingdom

Nikunj Oza
Intelligent Systems Division
NASA Ames Research Center
Moffett Field, California

Arkadas Ozakin
Computational Science and Engineering
   Division
Georgia Institute of Technology
Atlanta, Georgia

Paniez Paykari
CEA/DSM–CNRS–Université Paris
   Diderot
Gif-sur-Yvette, France

Sandrine Pires
CEA/DSM–CNRS–Université Paris
   Diderot
Gif-sur-Yvette, France

Jordan Raddick
Department of Physics and Astronomy
The Johns Hopkins University
Baltimore, Maryland

Matteo Re
Dipartimento di Scienze
   dell'Informazione
Università degli Studi di Milano
Milan, Italy

Alexandre Réfrégier
CEA/DSM–CNRS–Université Paris
    Diderot
Gif-sur-Yvette, France

Cédric Richard
Laboratoire Fizeau
Université de Nice Sophia Antipolis
Nice, France

Joseph W. Richards
Department of Astronomy
and
Department of Statistics
University of California, Berkeley
Berkeley, California

Ryan Riegel
Computational Science and Engineering
    Division
Georgia Institute of Technology
Atlanta, Georgia

Jeffrey D. Scargle
Space Science and Astrobiology Division
NASA Ames Research Center
Moffett Field, California

Kevin Schawinski
Yale Center for Astronomy
    and Astrophysics
Yale University
New Haven, Connecticut

Jérémy Schmitt
Laboratoire AIM
CEA/DSM–CNRS–Universite Paris
    Diderot
Gif-sur-Yvette, France

Jeffrey C. Smith
SETI Institute
Mountain View, California
and
Space Science and Astrobiology Division
NASA Ames Research Center
Moffett Field, California

Jean-Luc Starck
Laboratoire AIM
CEA/DSM–CNRS–Universite Paris Diderot
Gif-sur-Yvette, France

Peter Tenenbaum
SETI Institute
Mountain View, California
and
Space Science and Astrobiology Division
NASA Ames Research Center
Moffett Field, California

Joseph D. Twicken
SETI Institute
Mountain View, California
and
Programs and Projects Directorate
NASA Ames Research Center
Moffett Field, California

J. Anthony Tyson
Department of Physics
University of California, Davis
Davis, California

Giorgio Valentini
Dipartimento di Scienze
    dell'Informazione
Università degli Studi di Milano
Milan, Italy

Jeffrey Van Cleve
SETI Institute
Mountain View, California
and
Science Directorate
NASA Ames Research Center
Moffett Field, California

O.V. Verkhodanov
Special Astrophysical Observatory
Russian Academy of Sciences
Karachaj-Cherkesia, Russia

Kiri L. Wagstaff
Jet Propulsion Laboratory
California Institute of Technology
Pasadena, California

John Wallin
Department of Physics and Astronomy
Middle Tennessee State University
Murfreesboro, Tennessee

Sheri Williamson
Department of Computer Science
George Mason University
Fairfax, Virginia

Przemek Woźniak
Los Alamos National Laboratory
Los Alamos, New Mexico

Łukasz Wyrzykowski
Institute of Astronomy
Cambridge University
Cambridge, United Kingdom

I

Foundational Issues

CHAPTER 1

Classification in Astronomy

Past and Present

Eric Feigelson
Pennsylvania State University

CONTENTS

1.1   Astronomical Classification in the Past   3
1.2   Unsupervised Clustering in Astronomy   4
1.3   The Rise of Supervised Classification   5
1.4   Machine Learning Classification in Astronomy Today   6
Acknowledgments   9
References   9

1.1 ASTRONOMICAL CLASSIFICATION IN THE PAST

Astronomers have always classified celestial objects. The ancient Greeks distinguished between *asteros*, the fixed stars, and *planetos*, the roving stars. The latter were associated with the Gods and, starting with Plato in his dialog *Timaeus*, provided the first mathematical models of celestial phenomena. Giovanni Hodierna classified nebulous objects, seen with a Galilean refractor telescope in the mid-seventeenth century into three classes: "Luminosae," "Nebulosae," and "Occultae." A century later, Charles Messier compiled a larger list of nebulae, star clusters and galaxies, but did not attempt a classification. Classification of comets was a significant enterprise in the 19th century: Alexander (1850) considered two groups based on orbit sizes, Lardner (1853) proposed three groups of orbits, and Barnard (1891) divided them into two classes based on morphology.

Aside from the segmentation of the bright stars into constellations, most stellar classifications were based on colors and spectral properties. During the 1860s, the pioneering spectroscopist Angelo Secchi classified stars into five classes: white, yellow, orange, carbon stars, and emission line stars. After many debates, the stellar spectral sequence was refined by the group at Harvard into the familiar OBAFGKM spectral types, later found to be a sequence on surface temperature (Cannon 1926). The spectral classification is still being extended with recent additions of O2 hot stars (Walborn et al. 2002) and L and T brown dwarfs (Kirkpatrick 2005). Townley (1913) reviews 30 years of variable star classification, emerging with six classes with five subclasses. The modern classification of variable stars has about 80 (sub)classes, and is still under debate (Samus 2009).

Shortly after his confirmation that some nebulae are external galaxies, Edwin Hubble (1926) proposed his famous bifurcated classification of galaxy morphologies with three classes: ellipticals, spirals, and irregulars. These classes are still used today with many refinements by Gerard de Vaucouleurs and others. Supernovae, nearly all of which are found in external galaxies, have a complicated classification scheme: Type I with subtypes Ia, Ib, Ic, Ib/c pec and Type II with subtypes IIb, IIL, IIP, and IIn (Turatto 2003). The classification is based on elemental abundances in optical spectra and on optical light curve shapes. Tadhunter (2008) presents a three-dimensional classification of active galactic nuclei involving radio power, emission line width, and nuclear luminosity.

These taxonomies have played enormously important roles in the development of astronomy, yet all were developed using heuristic methods. Many are based on qualitative and subjective assessments of spatial, temporal, or spectral properties. A qualitative, morphological approach to astronomical studies was explicitly promoted by Zwicky (1957). Other classifications are based on quantitative criteria, but these criteria were developed by subjective examination of training datasets. For example, starburst galaxies are discriminated from narrow-line Seyfert galaxies by a curved line in a diagram of the ratios of four emission lines (Veilleux and Osterbrock 1987). Class II young stellar objects have been defined by a rectangular region in a mid-infrared color–color diagram (Allen et al. 2004). Short and hard γ-ray bursts are discriminated by a dip in the distribution of burst durations (Kouveliotou et al. 2000). In no case was a statistical or algorithmic procedure used to define the classes.

1.2 UNSUPERVISED CLUSTERING IN ASTRONOMY

One algorithmic procedure for finding distinct clusters gained considerable popularity in astronomy: the "friends-of-friends" or "percolation" algorithm. Its principal use was to identify (hopefully) physically bound groups of galaxies from the two-dimensional distribution of galaxies on the sky without prior information about the group locations or populations. Turner and Gott (1976) drew circles around each galaxy with a radius chosen to find a selected overdensity factor in the galaxy surface density. When the circles of proximate galaxies overlap, they were considered to be members of a group. This algorithmic approach allows the data, rather than the astronomer, to determine whether a cluster is small with few members or large and populous. This was an effort to supersede earlier approaches based on pre-defined cluster richness and size (Abell 1957). The moniker "friends-of-friends" algorithm was coined by Press and Davis (1982), and has since been used in hundreds of papers during the 1990s and 2000s. Most of these studies continued to treat galaxy clustering, but the method spread in the community to finding concentrations in a variety of two- or three-dimensional spatial or multivariate point processes.

It was occasionally recognized that the friends-of-friends method had mathematical relationships to nonparametric hierarchical clustering (Rood 1988) and the pruned minimal spanning tree (Barrow et al. 1985). In statistical parlance, the astronomers' friends-of-friends algorithm is equivalent to single-linkage agglomerative clustering. But there was little effort to learn from extensive experience with hierarchical clustering in other fields.

Unfortunately, single-linkage clustering is universally criticized by experts in unsupervised clustering methodology. It produces elongated, filamentary structure that tend to erroneously merge distinct structures due to interstitial noise or outlying objects; this is called "chaining." Kaufman and Rousseeuw (1990, 2005) review the methodological literature and conclude that "Virtually all authors agreed that single linkage was least successful in their [simulation] studies." Everitt et al. (2001) similarly report that "Single linkage, which has satisfactory mathematical properties and is also easy to program and apply to large data sets, tends to be less satisfactory than other methods because of "chaining." Statisticians instead recommend average-linkage or Ward's minimum variance criterion for most purposes.

For problems where the populations are not sparse and can be reasonably well fit by multivariate normal (MVN) distributions, numerous well-established methods are available. The normal mixture model can be applied, and unsupervised clustering can be pursued using maximum likelihood estimation (McLachlan and Peel 2000). Computations are straightforward using the expectation–maximization (EM) algorithm (McLachlan and Krishnan 2008), and the optimal number of classes can be selected by penalized likelihood measures such as the Bayesian information criterion (Fraley and Raftery 2002). These statistical methods are very widely used in other fields for clustering and other multimodal modeling problems, but they are not often applied in astronomy.

In addition, astronomers rarely seek to validate a clustering scheme found by the percolation algorithm. When multivariate normal (MVN) distributions are assumed to apply, well-established multivariate analysis of variance (MANOVA) tests can be used: Hotelling's T^2, Wilks' Λ, and Pillai's trace compare the within-group variance to the inter-group variance, emerging with well-defined probabilities for the hypothesis that the two groups are distinct (Johnson and Wichern 2002). Normal mixture models and MANOVA tests, together with nonparametric agglomerative clustering algorithms, indicate that three (rather than two) classes of γ-ray bursts are present in an important catalog (Mukherjee et al. 1998). When the MVN assumption does not hold, group structure and membership can be investigated through bootstrap resampling.

1.3 THE RISE OF SUPERVISED CLASSIFICATION

The unsupervised clustering of spatial or multivariate point processes has inherent uncertainties given the vast range of possible structures. But astronomers often study problems where considerable prior knowledge is available. Often a limited sample of nearby, bright objects in a class have been intensively studied. These prototypes can then serve as "training sets" for classification of larger, less well-characterized samples. The use of training sets, together with often-sophisticated algorithms for supervised classification, falls under the rubric of "machine learning." Nilsson (2010, Chapter 29) gives a brief and readable history of machine learning techniques.

Classification is particularly important for the enormous datasets arising from wide-field surveys starting with the 2MASS and Sloan Digital Sky Survey (SDSS) and leading to an alphabet-soup of current and planned surveys: Dark Energy Survey (DES), Carnegie Supernova Project (CSP), Palomar Transient Factory (PTF), Panoramic Survey Telescope and Rapid Response System (Pan-STARRS), Large Sky Area Multi-Object Fibre Spectroscopic

Telescope (LAMOST), Large Synoptic Survey Telescope (LSST), and others (see Chapter 9 by Tyson and Borne, "Future Sky Surveys," in this volume). These projects produce million-to-billion object multivariate samples with combinations of photometric magnitudes, multi epoch variability, and spectra. The survey data can be transformed into multivariate tabulations of astrophysically interesting variables such as color indices, variability properties (amplitude, timescale, periodicity, light curve shape), and spectral properties (temperature, reddening, redshift, emission lines). One can characterize the various classes of stellar and extragalactic objects in the multivariate properties available from the wide-field surveys. The challenge is then to apply machine learning classification techniques to understand the larger surveys.

Although, mathematical foundations such as the linear discriminant analysis (LDA, Fisher 1936) and the perceptron (Rosenblatt 1962) were known for decades, powerful supervised learning methods did not emerge until the 1990s with the development of backpropagation, bootstrapping, bagging and boosting, and other computationally intensive techniques (Duda et al. 2001; Hastie et al. 2009). The important class of support vector machines (SVMs), for example, generalized LDA by finding hypersurfaces that maximally separate the classes in training sets. SVMs are one of the most successful of the machine learning classification algorithms. The concepts underlying SVMs date back to Vapnik and Lerner (1963) and the method matured with the incorporation of kernel smoothers (Wahba 1990) and statistical validation (Shawe-Taylor et al. 1998). SVMs demonstrated high success in speech and handwriting recognition, computer vision, and other fields, often outshining neural networks and other methods.

Figures 1.1 and 1.2 illustrate an application of SVMs to an astrometric photometric survey (Feigelson and Babu 2012). We consider here a test set of 17,000 point sources selected from the SDSS survey (York et al. 2000). The five photometric measurements give a four-dimensional space of color indices: $u - g$, $g - r$, $r - i$, and $i - z$. Three major classes are represented in this dataset: main sequence (and red giant) stars, white dwarfs, and quasars. The top panels of Figure 1.1 show two-dimensional projections of the four-dimensional training sets derived from well-studied samples observed in the SDSS: 5000 main sequence stars from Ivezic et al. (2007), 2000 white dwarfs from Eisenstein et al. (2006), and 2000 quasars from Schneider et al. (2005). The bottom panels show the classification of the test set produced by the function *svm* in the CRAN package *e1071* (Dimitriadou et al. 2010) within the R statistical computing environment (R Development Core Team 2010). The classification is performed with a radial basis kernel. Figure 1.2 shows the high success rate of the classifier when applied to the training dataset. This was the best performance of several machine learning techniques (LDA, k-Nearest-neighbor, Classification and Regression Trees, and SVM) examined by Feigelson and Babu (2012).

1.4 MACHINE LEARNING CLASSIFICATION IN ASTRONOMY TODAY

Table 1.1 gives a rough bibliometric summary of modern classification usage in contemporary astronomy. The search is restricted to four journals that publish most of the astronomical research worldwide, and the terms are listed in order of popularity in Web pages found with *Google*. Except for single-linkage hierarchical clustering, (the "friends-of-friends" algorithm

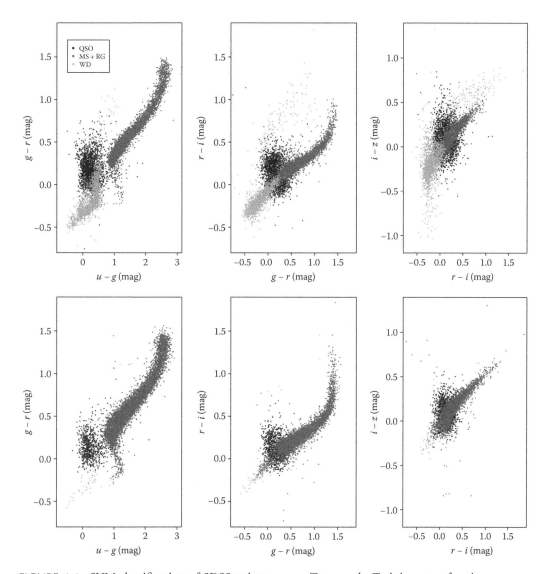

FIGURE 1.1 SVM classification of SDSS point sources. Top panels: Training sets of main sequence (mid-gray), white dwarfs (light-gray), and quasars (black). The three panels show two-dimensional projections of the four-dimensional SDSS color space. Bottom panels: Application of an SVM classifier based on these training sets to a test set of 17,000 SDSS point sources. (From Feigelson, E. D. and Babu, G. J. 2012, *Modern Statistical Methods for Astronomy with R Applications*, Cambridge University Press. With permission.)

discussed above), usage of machine learning and related methodologies in mainstream astronomical studies is quite low. For comparison, the four journals published ∼70,000 papers during the 2001–2010 decade.

The most established of modern classification methods in astronomy are automated neural networks. They are, for example, a component of the successful SExtractor software package for identifying faint sources in images (Bertin and Arnouts 1996). Neural networks have been successfully used in star/galaxy discrimination, photometric redshift estimation,

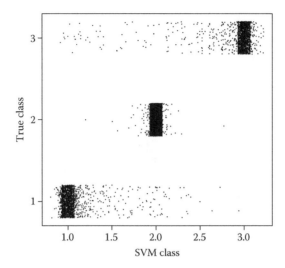

FIGURE 1.2 Validation of the SDSS SVM classifier by application to the training set. Class 1 = quasars, Class 2 = main sequence stars, Class 3 = white dwarfs.

TABLE 1.1   Usage of Classification and Clustering Methods

Method	Google[a]	ADS[b]
Data mining	16.6 million	20
Neural network	7.1 million	135
Machine learning	5.6 million	20
Hierarchical clustering[c]	950 thousand	400
SVM	870 thousand	15
LDA	500 thousand	5
Random forest	260 thousand	3
Bayesian classifier	230 thousand	15
Classification and regression tree	160 thousand	0
k-Nearest neighbor	110 thousand	9
Model-based clustering	56 thousand	2

[a] Google Web page hits, early 2011.
[b] Refereed articles in the four largest astronomical journals (*Astrophysical Journal, Astronomical Journal, Monthly Notices of the Royal Astronomical Society, Astronomy & Astrophysics*) accessed through the NASA/Smithsonian Astrophysics Data System, early 2011. The method must appear in the Title, Abstract, or Keywords to be counted.
[c] Hierarchical clustering includes the astronomers' "friends-of-friends" algorithm.

and the classification of galaxy morphologies, galaxy spectra, and stellar spectra. Bayesian classifiers have been used for classifying kinematic stellar populations, quasar and star spectra, galaxy morphologies, and photometric redshifts. SVMs have classified variable star light curves. Random Forests select supernova light curves and estimate photometric redshifts. k-Nearest-neighbor algorithms assist with photometric selection of quasars. Machine learning classification is also often used in solar and space weather physics.

We see that, although a vanguard of experts is active, the vast majority of astronomers are only vaguely aware of recent advances in machine learning classification techniques and do not incorporate them into their research programs. This volume can thus play an important role in educating astronomers on the value of modern classification methodology and promoting their use throughout the astronomical community.

ACKNOWLEDGMENTS

This review was supported by NSF grant SSI-1047586 (G. J. Babu, principal investigator).

REFERENCES

Abell, G. O. 1957, The distribution of rich clusters of galaxies, *Astrophys. J. Suppl.*, 3, 211–288.

Alexander, S. 1850, On the classification and special points of resemblance of certain of the periodic comets; and the probability of a common origin in the case of some of them, *Astron. J.*, 1, 147–150.

Allen, L. E. et al. 2004, Infrared array camera (IRAC) colors of young stellar objects, *Astrophys. J. Suppl.*, 154, 363–366.

Barnard, E. E. 1891, On a classification of the periodic comets by their physical appearance, *Astron. J.* 11, 46.

Barrow, J. D., Bhavsar, S. P., and Sonoda, D. H. 1985, Minimal spanning trees, filaments and galaxy clustering, *Mon. Not. R. Astron. Soc.*, 216, 17–35.

Bertin, E. and Arnouts, S. 1996, SExtractor: Software for source extraction, *Astron. Astrophys. Suppl.*, 117, 393–404.

Cannon, A. J. 1926, Classifying the stars, in *The Universe of Stars* (H. Shapley and C. H. Payne, eds), Harvard Observatory, Cambridge, MA, p. 101.

Dimitriadou, E., Hornik, K., Leisch, F., Meyer, D., and Weingessel, A. 2010, e1071: Misc functions of the Department of Statistics (e1071), version 1.5-24, TU Wien.

Duda, R. O., Hart, P. E., and Stork, D. G. 2001, *Pattern Classification*, 2nd edition, Wiley, New York, NY.

Everitt, B. S., Landau, S., and Leese, M. 2001, *Cluster Analysis*, 4th edition, Arnold, West Sussex, UK.

Feigelson, E. D. and Babu, G. J. 2012, *Modern Statistical Methods for Astronomy with R Applications*, Cambridge University Press, Cambridge, UK.

Fisher, R. A., 1936, The use of multiple measurements in taxonomic problems, *Ann. Eugenics*, 7, 111–132.

Fraley, C. and Raftery, A. E. 2002, Model-based clustering, discriminant analysis, and density estimation, *J. Am. Stat. Assoc.*, 97, 611–631.

Hastie, T., Tibshirani, R., and Friedman, J. 2009, *The Elements of Statistical Learning: Data Mining, Inference, and Prediction*, 2nd edition, Springer, New York.

Hubble, E. P. 1926, Extragalactic nebulae, *Astrophys. J.* 64, 321–369.

Johnson, R. A. and Wichern, D. W. 2002, *Applied Multivariate Statistical Analysis*, Prentice-Hall, Upper Saddle River, NJ.

Kaufman, L. and Rousseeuw, P. J. 1990. *Finding Groups in Data: An Introduction to Cluster Analysis*, Wiley-Interscience, Hoboken, NJ.

Kaufman, L. and Rousseeuw, P. J. 2005, *Finding Groups in Data: An Introduction to Cluster Analysis*, 2nd edition (paperback), Wiley, Hoboken, NJ.

Kirkpatrick, J. D. 2005, New spectral types L and T, *Annu. Rev. Astron. Astrophys.*, 43, 195–245.

Kouveliotou, C., Meegan, C. A., Fishman, G. J., Bhat, N. P., Briggs, M. S., Koshut, T. M., Paciesas, W. S., and Pendleton, G. N. 1993. Identification of two classes of gamma-ray bursts. *Astrophys. J. Lett.*, 413, L101–L104.

Lardner, D. 1853, On the classification of comets and the distribution of their orbits in space, *MNRAS*, 13, 188–192.

McLachlan, G. and Peel, D. 2000, *Finite Mixture Models*, Wiley, New York, NY.

McLachlan, G. J. and Krishnan, T. 2008, *The EM Algorithm and Extensions*, Wiley, New York, NY.

Mukherjee, S., Feigelson, E. D., Babu, J. G., Murtagh, F., Fraley, C., and Raftery, A. 1998, Three types of gamma-ray bursts, *Astrophys. J.*, 508, 314–327.

Nilsson, N. J. 2010, *The Quest for Artificial Intelligence: A History of Ideas and Achievements*, Cambridge University Press, Cambridge, UK.

Press, W. H. and Davis, M. 1982, How to identify and weigh virialized clusters of galaxies in a complete redshift catalog, *Astrophys. J.*, 259, 449.

R Development Core Team, 2010, *R: A Language and Environment for Statistical Computing*, R Foundation for Statistical Computing, Vienna.

Rood, H. J. 1988, Voids, *Annu. Rev. Astron. Astrophys.*, 26, 245–294.

Rosenblatt, F. 1962, *Principles of Neurodynamics: Perceptrons and the Theory of Brain Mechanisms*, Spartan Books, Washington D.C.

Samus, N. N. 2009, GCVS [General Catalogue of Variable Stars] variability types, http://www.sai.msu.su/gcvs/gcvs/iii/vartype.txt. See also http://www.sai.msu.su/gcvs/future/classif.htm.

Shawe-Taylor, J. et al. 1998, Structural risk minimization over datadependent hierarchies, *IEEE Trans. Inf. Theory*, 44, 1926–1940.

Tadhunter, C. 2008, An introduction to active galactic nuclei: Classification and unification, *New Astron. Rev.*, 52, 227–239.

Townley, S. D. 1913, Classification of variable stars, *Publ. Astron. Soc. Pac.*, 25, 239–243.

Turatto, M. 2003, Classification of supernovae, in *Supernovae and Gamma-Ray Bursts* (K. Weiler, ed.), *Lecture Notes in Physics*, Springer, Berlin, Germany, 598, pp. 21–36.

Turner, E. L. and Gott, J. R. 1976, Groups of galaxies. I. A catalog, *Astrophys. J. Suppl.* 32, 409–427.

Vapnik, V. and Lerner, A. 1963, Pattern recognition using generalized portrait method, *Autom. Remote Control*, 24, 774–780.

Veilleux, S. and Osterbrock, D. E. 1987, Spectral classification of emission line galaxies, *Astrophys. J. Suppl*, 63, 295–310.

Wahba, G. 1990, *Spline Models for Observational Data, SIAM-SIMS Conf. Ser. 59*. SIAM, Philadelphia, PA.

Walborn, N. R. et al. 2002, A new spectral classification system for the earliest O stars: Definition of type O2, *Astron. J.*, 123, 2754–2771.

Zwicky, F. 1957, *Morphological Astronomy*, Springer, Berlin, Germany.

CHAPTER 2

Searching the Heavens

Astronomy, Computation, Statistics, Data Mining, and Philosophy

Clark Glymour
Carnegie Mellon University

CONTENTS

2.1	The Conflicts	11
	2.1.1 Astronomy	11
2.2	Planetary Theory from Data Mining: Ptolemy, Copernicus, Kepler, Newton	12
2.3	Computation and Estimation: From Boscovitch to Legendre	13
2.4	From Philosophical Skepticism to Bayesian Inference: Hume, Bayes, and Price	15
2.5	Early Planet Hunting	17
2.6	Bayesian Statistics, Data Mining, and the Classical Tests of General Relativity	18
2.7	Statisticians Become Philosophers	22
2.8	The Digital Revolution	24
2.9	Statistics and Machine Learning Meet One Another	25
2.10	A Brief Reflection	26

2.1 THE CONFLICTS

2.1.1 Astronomy

Our first and purest science, the mother of scientific methods, sustained by sheer curiosity, searching the heavens we cannot manipulate. From the beginning, astronomy has combined mathematical idealization, technological ingenuity, and indefatigable data collection with procedures to search through assembled data for the processes that govern the cosmos. Astronomers are, and ever have been, data miners, and for that reason astronomical methods (but not astronomical discoveries) have often been despised by statisticians and philosophers. Epithets laced the statistical literature: *Ransacking! Data dredging! Double Counting!* Statistical disdain was usually directed at social scientists and biologists, rarely if ever at astronomers, but the methodological attitudes and goals that many twentieth-century philosophers and statisticians rejected were creations of the astronomical tradition. The philosophical criticisms were earlier and more direct. In the shadow (or in Alexander Pope's phrasing, the light) cast on nature in the eighteenth century by the Newtonian triumph, David Hume revived arguments from the ancient Greeks to challenge the very possibility of coming to know what causes what. His conclusion was endorsed in the twentieth century by many philosophers who found talk of causation unnecessary or unacceptably metaphysical, and absorbed by many statisticians as a general suspicion of causal claims, except possibly when they are founded on experimental manipulation. And yet in the hands of a mathematician, Thomas Bayes, and another mathematician and philosopher, Richard Price, Hume's essays prompted the development of a new kind of statistics, the kind we now call "Bayesian."

The computer and new data acquisition methods have begun to dissolve the antipathy between astronomy, philosophy, and statistics. But the resolution is practical, without much reflection on the arguments or the course of events. So, I offer a largely unoriginal history, substituting rather dry commentary on method for the fuller, livelier history of astronomers' ambitions, politics, and passions.[*]

2.2 PLANETARY THEORY FROM DATA MINING: PTOLEMY, COPERNICUS, KEPLER, NEWTON

By "data mining" I mean inferences about scientific hypotheses based on historical data that was collected for some other purpose or previously used, or inferences to hypotheses prompted by aspects of the very data under scrutiny.

Ptolemy worked from his own observations and historical data, apparently taken from Hipparchus. Unelaborated, the familiar epicycle on deferent device and the geocentric hypothesis left unspecified the order and distances of the planets from the Earth, the radii of the deferents, and radii, centers, and numbers of the epicycles, and the angular velocities of deferents and epicycles—that is, almost everything. Many of these quantities remained indeterminate in Ptolemy's theory, which needed only various ratios to account for planetary motions. The order of the planets remained formally indeterminate as well, although the periods between oppositions, periods of revolutions of longitude, and comparative brightness gave grounds for Ptolemy's ordering of the superior planets. He ordered the inferior planets incorrectly, placing Mercury closer to Earth and farther from the Sun than Venus. In addition to the epicycle on deferent, Ptolemy used an eccentric (moving, for the superior planets), the off-Earth center of the deferents, and an equant, the off-Earth point with respect to which the deferents have constant angular velocities. In principle, we now know thanks to Harold Bohr, Ptolemy could have made do instead with more epicycles, as did some Arab astronomers. All the planetary, solar, and lunar parameters came from mining the data.

Although it is not fashionable in our mathematical times to think of it this way, the astronomer's task is not only to collect data but also to find an *explanation* of a body of

---

[*] My accounts of various episodes in the astronomical tradition are taken from standard sources, especially Otto Neugebauer, *The Exact Sciences in Antiquity*. Princeton: Princeton University Press, 1952; Richard Baum and William Sheehan, *In Search of Planet Vulcan*, Basic Books, 1997; Jeffrey Crelensten, *Einstein's Jury*, Princeton University Press, 2006; and Stephen Stigler, *The Measurement of Uncertainty*, Harvard University Press, 1990. Methodological commentary is mine, not that of these sources.

phenomena, an explanation that is consonant with physical knowledge, and, in the best case, at least approximately true. "Explanation" is a vague notion for which there is no formal theory. Ptolemy is a case in point. Over a long interval of time, each superior planet approximately satisfies a regularity: the number of solar years in that interval is equal to the number of oppositions of the planet in that interval plus the number of revolutions of longitude of the planet (with respect to the fixed stars) in the interval. To accommodate this regularity, Ptolemy required that the line from each superior planet to its epicyclic center on the deferent always remain parallel to the Earth–Sun line. He observed that he had no explanation for this constraint, and thereby expressed a sense—difficult to make precise— that constraints among parameters in a theory should emerge from the essential structure of the theory. That is exactly what heliocentric theory does in this case: the approximate regularity reduces to a necessary truth on the heliocentric model, and that was one of Kepler's arguments for a sun-centered planetary system.

The epitome of data mining was Kepler's exploitation of the data left by Tycho Brahe. We know Kepler built his laws from working over Brahe's data with enormous computational skill and ingenuity, one hypothesis after another. (Kepler gave voice to another idea in advance of his time. Making part of his living as an astrologer, when one of Kepler's forecasts—of weather as I recollect, but do not trust me—went wrong, he complained of his critics that they did not understand the role of chance variations.)

Descriptions of nature's regularities are an efficient form of data compression. When laws such as Kepler's are used to infer or support other generalizations, the data are being mined indirectly. Amongst many things, Newton was a data miner. The argument for universal gravitation in Book III of the Principia uses received data from astronomers on the orbits of the satellites of Jupiter and Saturn and of the moon, Kepler's laws, and simple experiments with pendula.

Data often require a "massage" to reveal the processes that have generated them. Newton did not explicitly use Kepler's first law, although he of course knew it (perhaps because he wished to claim he had "proved" it, whereas Kepler had merely "guessed" it). Instead, in his argument for the law of universal gravitation, Newton assumed that the planets and satellites (except for the moon) move in circular orbits, and then he used the force law to show that the orbits are ellipses.

2.3 COMPUTATION AND ESTIMATION: FROM BOSCOVITCH TO LEGENDRE

The value of data mining depends on the quality of the data and the methods applied to them. Astronomy created the study of both. Near the end of the eighteenth century, Nevil Maskelyne, the fifth Astronomer Royal, dismissed his assistant observer for errors in his timing of transits across an observing wire. Friedrich Bessel soon found by experiment that individuals differed, some estimating transits early, some late. Formulas were developed for pairwise comparisons of differences of observers in timing a transit, but experimental work soon enough developed an equation—the "personal equation"—for absolute errors. So began psychophysics and the study of human factors.

Among the first problems in astronomical data analysis is to find a correct way to estimate a relationship among variables from a data sample, and to find a method to test a hypothesis with free parameters against sample data. Eighteenth-century astronomers sometimes used a nonprobabilistic method to test hypotheses against inconsistent observations, the method now sometimes called "uncertain but bounded error" in engineering texts. The idea is simple: in using a set of measurements either to test hypotheses or to estimate parameters, the investigator believes that the errors in the measurements are within some specific bound. That is, when a value x of variable X is recorded in measurement, the true value of X on that occasion is within ε of x where ε is some positive quantity assumed known. The measured values of variables then determine bounds on the values of parameters, and may exclude a hypothesis when no values of its parameters exist that are consistent with both the measurements and the error bounds assumed.

In 1755, Father Boscovitch analyzed five data points on the length of meridian arc at various latitudes, taken for the purpose of testing the Newtonian hypothesis of an ellipsoidal Earth. Boscovitch had five data points and a linear equation (see Stigler, p. 42) in two unknowns, allowing 10 determinations of the unknown parameters. He computed all 10 values, and also computed the average value of one of the parameters, the ellipticity, and argued that the difference between the individual values of the ellipticity and the average value was too large to be due to measurement error. He concluded that the elliptical hypothesis must be rejected. Boscovitch's argument is valid if we accept his bounds on the errors of measurement. Stigler claims that Boscovitch's error bounds were unreasonable at the time, and using a more reasonable error bound (100 toises—about 600 ft) he finds that the least-squares line is within the error interval for all five observations. But that is only to deny Boscovitch's premise, not the validity of Boscovitch's argument. Boscovitch's procedure is sensible, simple in description, informative about the truth or falsity of hypotheses of interest, and requires only an elementary kind of prior belief that could readily be elicited from scientific practitioners. It corresponds to an interval estimation procedure that is, if the assumptions about error bounds are correct, logically guaranteed to give an interval estimate containing the true value if one exists. Boscovitch himself later abandoned the method for a procedure that minimized the sum of absolute values of differences between estimate and observations, the $L1$ norm as it is now called.

Uncertain, but bounded, errors have advantages. If the error bounds are correct, then the method converges to a correct interval. But the method had the overwhelming disadvantage for scientists before the late twentieth century that, by hand calculation, uncertain but bounded error estimation is intractable except in the simplest of cases. Early in the nineteenth century, uncertain, but bounded, error methods and $L1$ norm methods were rapidly displaced by least squares, a criterion that had been proposed and investigated in the middle of the eighteenth century but was not publicly solved (Carl Gauss, as usual, claimed priority) until the appendix of Adrien Legendre's essay on the orbits of comets in 1805.

Part of the reason for the success of least squares was that Gauss and Pierre Laplace gave least squares what remains the standard justification for its use: the expected value of least-squares estimates is the mean for normally distributed variables, and least squares minimizes the expected squared error of the estimate. The central-limit theorem justified the assumption of a Normal distribution of measurement errors as the limit of the binomial distribution, or more substantively, the Normal results in the limit from summing appropriately small, unrelated causes. But, as Abraham DeMoivre emphasized in *The Doctrine of Chances*, probability had no logical connection with truth. Jacob Bernoulli's theorem, for example, did not say (contrary to proposals of some twentieth-century frequentists) that a binomial probability could be *defined* as the limiting frequency of a sequence of trials; for *that* purpose, the theorem would have been twice circular, requiring that the trials be independent and giving convergence only in probability. In contrast, under assumptions investigators were disposed to make, the method of uncertain but bounded error afforded finite sample guarantees about where the truth lay, assuming only what astronomers thought they knew. But, with Legendre's method, least-squares solutions could be computed by hand, and provided a uniform assumption not dependent on individual judgments of measurement accuracy. For those benefits, astronomers were willing to change what they assumed.

2.4 FROM PHILOSOPHICAL SKEPTICISM TO BAYESIAN INFERENCE: HUME, BAYES, AND PRICE

Plato's *Meno* is the ancient source of skepticism about the very possibility of using experience to learn general truths about the world. After a sequence of conjectures and counterexamples to definitions of "virtue," Meno asks Socrates how they would know they had the right answer if they were to conjecture it. In the second century AD, Sextus Empiricus revived Meno's query in a more direct form: the absence of an available counterexample to a generalization does not entail that there is no counterexample yet to be found; hence, general knowledge cannot be founded on experience.

David Hume received a classical eigtheenth-century British education, and he surely knew Plato's writings and probably some of those of Sextus as well. In 1740, he published *A Treatise of Human Nature*, which sought to trace the basis for claims of scientific knowledge to human psychology, and sought to discover the facts and regularities of psychology by informal introspection, the "Experimental Method" as the full title of his book claimed. No philosopher has been so much cited by scientists, and so little read. Hume distinguished between propositions about "matters of fact" and propositions about "relations of ideas," and he retained a traditional account of mental faculties, including Reason and Imagination. The scope of Reason is confined to necessary relations of ideas, in effect to mathematics and logic. Our senses deliver "impressions" and we have as well internal impressions, as of hunger and pain and pleasure and memory. The mind copies and abstracts impressions into "ideas," which the mind then compounds to form more complex ideas, and associates (by similarity of content, nearness or "contiguity" in time, etc.) to make inferences about as-yet-unobserved sequences of impressions. The intelligibility of language depends on the sources of terminology in ideas or impressions of sensation. To find what an expression such as "cause" or "probability" means, one must reduce its content to a composition of claims about ideas and their relations, and hence ultimately to a source in impressions, whether internal or external. Some of these associations are so frequent that they become habits, which we make without reflection or deliberation, involuntarily as it were. We expect the sun to rise tomorrow from habit, not knowledge. The source of our notion of causality is not in any impressions of sensation, but in the habitual association of ideas by the constancy of one kind of impression succeeded by another closely in space and time. Neither do we perceive causal relations, or "necessary connections" between events, nor, by the arguments of Plato and Sextus (from whom Hume borrows but does not cite), can causal relations be established by Reason. So there are two threads. A psychological argument about meaning: "causality" refers to events of a kind that are contiguous in space, and constantly conjoined in a fixed order. (In one phrase, which never reoccurs in his writings, Hume does suggest that causal claims are about counterfactuals—the effect would not have occurred had the cause not been present.) And a logical argument: constant conjunction of observed events does not entail that as yet unobserved events will satisfy the same regularity. Although frequentist statistical reasoning is drenched in counterfactual reasoning, twentieth-century statisticians embraced the first thread of Hume's argument and overcame the second by appeal to probability.

What of the probability of causal judgments? Hume has the same psychological take: probability *is* merely degree of "opinion," and of course, opinions differ. There is no "objective" content to causal or probabilistic claims, and they cannot be known by Reason or through experience, and hence not at all.

It is not known whether Hume's writing helped to prompt Thomas Bayes to work out his account of probability and inverse inference, but Hume's work certainly influenced the publication in 1784 of Bayes' "Essay towards Solving a Problem in the Doctrine of Chances." Bayes' literary executor was a fellow reverend, mathematician, and Fellow of the Royal Society, Richard Price, who saw to the posthumous publication of two of Bayes' essays. Price also had philosophical interests, and he found in Bayes' work a response to Hume's skepticism about the possibility of knowledge of nature, and, not incidentally, to Hume's skepticism about miracles. Price gave Bayes' essay its title, an introduction, and an appendix directed to Hume without naming him: Price used Bayes' method to calculate the probability that the sun will rise tomorrow given that it has been seen to rise on a first day, and so on.

Bayes formulated his theory of probability in terms of the betting odds one "ought" to hold given a prior state of knowledge. Bayes construes pure ignorance of the probability of an event type as the prior belief that there is an equal chance for any frequency of the event type in a given finite sequence of trials, and from that assumption, infers that the prior probability distribution should be uniform. (George Boole later objected that if the equal chances are for each sequence of possible outcomes in n trials, for every n, then the prior must be exactly $1/2$.) Hume, not Bayes, was the subjectivist about probability, and the stimulus of his skepticism was the last useful contribution that philosophy was to make to probability for more than a century.

Bayes' method had two difficulties, the selection of the prior probability distribution and computability. Posterior probabilities could not be computed analytically except in simple cases. Bayes' theory had a minor following in the nineteenth century, and was dismissed by the twentieth-century giants of statistics until the digital computer, and the approximation algorithms it prompted, changed Bayesian methods from an interesting mathematical/philosophical toy into a practical tool.

2.5 EARLY PLANET HUNTING

Observing is mining the data of our senses. The invention of the telescope soon prompted sky surveys. What was found was sometimes not what was sought. In 1781, using a 7 foot reflector with a 6 and 1/2 inch mirror, of his own construction, in the course of a sky survey of objects down to the 8th magnitude, William Hershel, the most astonishing amateur astronomer ever, found the object we now call Uranus (Hershel had another name in mind, for King George III), and put to rest the ancient assumption of six planets. Hershel thought for some time that he had seen a comet, but the subsequent calculation of a nearly circular orbit settled the matter. Now, besides comets, there were other solar system objects to look for.

Early planet hunting was a multistage data mining process of anomaly detection and classification: survey the sky for an object that does not move with the fixed stars or that has a distinctive disk; estimate its orbit sufficiently to demonstrate that the object is a planet. The search was not random. As early as 1715, David Gregory had noted a progression of planetary distances, repeated in subsequent textbooks. In 1764, in a translation of Charles Bonnet's *Contemplation de la Nature*, Johann Titus elaborated the point, adding that there should be a planet between Jupiter and Mars and calling for a search. In 1772, Johann Bode repeated the point, initially without reference to Titus. In keeping with Stigler's law of eponymy (nothing in science is named after the person who first discovered it), the corresponding power relation has come to be known as "Bode's law." Baron von Zach organized a group of astronomers to look for the missing planet, but Ceres was discovered serendipitously in 1801 by Giuseppe Piazza in collaboration with Gauss, who did not make the observations but calculated the orbit. The distance was closer to the prediction from Bode's law than was that of any of the other known extraterrestrial planets except for Jupiter.

The celebrated story of the discovery of Neptune does not need repetition here, except to note that the process pursued independently by Urban LeVerrier and by John Adams was in interesting ways different from the data mining procedure that found Uranus. In this case, the anomaly was not sought, but found in the motion of Uranus. Adams assumed a single planet, initially located in accord with Bode's law (which Neptune actually substantially violates), calculated the predicted perturbations of Uranus, changed the hypothesis to better account for the residuals, and iterated until a satisfactory solution was found. Adams' solution changed several times, and his method appears to have been one of the reasons that James Challis, the director of the Cambridge Observatory, was reluctant to search for the hypothetical planet. At the nudging of the Astronomer Royal, Challis eventually did a plodding search, and actually observed Neptune, but (like many others before) failed to recognize the planet. LeVerrier had similar problems with French astronomers, and the planet was eventually found at his suggestion by the Berlin Observatory.

It is tempting to speculate (and why resist?) that the episode had a benign influence on astronomical practice. Many astronomers were subsequently eager to use their telescopes to search for effects that might explain the subtle anomalies in the orbit of Mercury.

The discovery of Neptune raised a question about language and knowledge that no philosopher at the time seems to have seriously addressed. (The contemporary philosophers had chiefly become a conservative burden on scientific advances: Georg Hegel, in an incompetent dissertation on Newtonian celestial mechanics, had proposed replacing Bode's law with a "philosophical" series taken from Plato; William Whewell opposed atomic theory and evolution; Auguste Comte announced that the material composition of stars would never be known.) Neptune did not follow the orbits computed either by Adams or by LeVerrier. Benjamin Pierce argued that the discovery of the planet was a "happy accident"—the observed planet was not the computed planet. Indeed, Adams and LeVerrier had the good fortune to calculate Neptune's position near conjunction with Uranus, when perturbations were maximal. John Hershel argued that the essential thing in discovering the planet was not the calculation of an orbit, but the identification of its position, which is what Adams and LeVerrier had almost given (there were errors in their distances). Herschel's point can be viewed as the primacy of ostension—pointing—over description in referring to an object. What counted most for success of reference were not the various descriptions that Adams and LeVerrier gave of the object—"the cause of the perturbations of Uranus' orbit," "the body with such and such orbital elements"—but that they could say for the time being where to look to see it, and that it was a planet. In retrospect, Leverrier's indignation at the "happy accident" account seems more appropriate than either Pierce's criticism or Herschel's response to it: Leverrier's "happy accident," like most good results of data mining, depended on a very unaccidental and thoughtful program of search and inference.

Mining historical data to improve orbital elements for the planets, in 1858 LeVerrier announced an anomalous advance of 38" of arc per century in Mercury's perihelion, which he thought could be explained by several small planets orbiting between the Sun and Mercury. So began a search that lasted for the rest of the century and found lots of sunspots, but no planet. Near the end of the century, LeVerrier's calculation was corrected by Simon Newcomb using better (and, unlike LeVerrier, consistent) planetary masses—and worsening the anomaly. By 1909, when Newcomb reviewed the history, astronomers and physicists were left with no alternatives but to alter the inverse square law, as Asaph Hall had suggested, or to postulate some as-yet-undetectable massive miasma such as Hugo von Seeliger's "zodiacal light."

2.6 BAYESIAN STATISTICS, DATA MINING, AND THE CLASSICAL TESTS OF GENERAL RELATIVITY

In October of 1915, Einstein produced an explanation of Newcomb's 43 second per century anomalous advance of the perihelion of Mercury. The paper provoked controversy at the time—Einstein did not have the field equations he was soon to produce, and used what amounted to a special case (in 1916, Schwarzschild and Droste independently found the exact solution for a spherically symmetrical field); his paper was littered with typographical errors (kept in the standard English translation!) because the paper was printed by the Berlin Academy without review; his derivation was obscure enough that a central point—whether he was using a power series approximation he did not explicitly give or merely analyzing a single-term deviation—are still matters of scholarly discussion. Any number of contrary hypotheses and criticisms were published, including the fact that Einstein's and Schwarzschild's analyses of perihelion advance neglected planetary perturbations of Mercury's orbit amounting to about 5" of arc per century.[*]

William Jeffereys and James Berger[†] have argued that Einstein's explanation of the anomalous advance was a triumph of subjective Bayesian reasoning. A subjective view of probability holds, with Hume, that probability is opinion, but opinion with degrees constrained by the axioms of probability, and that, for rationality, on coming to know or accept some datum, the probability of any claim (including that datum) must be altered to its conditional probability on the datum. The subjective Bayesian interpretation of probability was first given a justification in terms of betting odds by Frank Ramsey in 1926. Since digital computing made computation of posterior probabilities feasible, a considerable number of astronomers have claimed to be "Bayesian" in this sense. Useful as Bayesian statistics now is for astronomy and many other sciences, it does not provide a plausible account of the reasoning in the classical tests of general relativity; instead, the circumstances of the classical tests illustrate fundamental problems for subjective probability as an account of an important aspect of actual scientific reasoning.

Jeffereys and Berger compute posterior comparisons, conditional on the perihelion advance value, for general relativity and for alternative explanations of the perihelion advance at the time which, unlike the relativistic prediction, had free parameters that must be adjusted to produce the advance. As usual with posterior ratios for alternative hypotheses on the same data, the prior probability of the data in the usual expression for conditional probability can be canceled. And that hides a problem.

Bayesians abhor "double counting": starting from a prior opinion or subjective probability distribution D1, considering a datum (or body of data) and in light of it altering the prior opinion to D2, and then computing and announcing a posterior probability distribution D3 by conditioning the altered distribution, D2, on the datum. The procedure is regarded as especially vicious when D2 is announced as the prior. The exact logical error in double counting from a subjectivist viewpoint is never (in my reading anyway) made explicit. Where the final posterior distribution D3 equal to D1 conditional on the datum, ignoring D2, there would be no problem in having peeked at the datum. If the initial probability D1 is consistently altered to obtain D2 by *conditioning* on the datum, then subsequent conditioning again on the datum is harmless and does not change any degrees of belief: in that case D3 must equal D2 because in D2 the datum (conditioned on itself) has probability 1, and conditioning (to form D3) on a proposition (the datum) that already has probability 1 does not change the probabilities of any other propositions. Double counting can be objectionable from a subjective Bayesian point of view only if the intermediate distribution D2 is incoherent (i.e., does not satisfy the axioms of probability) or disingenuous, and in particular if in D2 the datum is *not* given probability 1 although in D2 hypotheses of interest are given their conditional probability (in D1) on the datum.

---

[*] See J. Earman and M. Janssen, "Einstein's explanation of the motion of Mercury's perihelion," in J. Earman, M. Janssen and John Norton, eds. *The Attraction of Gravitation: New Studies in the History of General Relativity.* Birkhauser, Boston, 1993.

[†] J. Berger and W. Jeffereys, The application of robust Bayesian analysis to hypothesis testing and Occam's razor. *Statistical Methods and Applications*, 1, 17–32.

In devising the general theory, Einstein peeked at the perihelion data. In fact, Einstein developed a series of relativistic gravitational theories between 1907 and 1915, and repeatedly used the perihelion advance to test and reject hypotheses. So, from a subjective Bayesian point of view, in regarding the derivation of the perihelion advance from the general theory as a confirmation of the general theory, Einstein must either have been (Bayesian) incoherent, or else have been mistaken that the perihelion advance increased the probability of the general theory of relativity.[*] Even that is not quite right, because at the time of his perihelion advance paper, general relativity was not in the algebra of propositions to which Einstein could have given degrees of belief: neither he, nor anyone else, had yet conceived the theory. More generally, as a Bayesian economist, Edward Leamer[†] seems first to have made vivid, subjective probability gives no account of the rational transition of subjective degrees of belief upon the introduction of novel hypotheses, and of how the logical relations of novel hypotheses with previously accepted data are to be weighted in that transition. That is exactly what was going on in the case of general relativity and the anomalous advance of Mercury's perihelion.

Photography changed astronomy. Photography made possible data mining that was not dependent on whatever the observer happened to record in the telescope's field of view. And it did much more. It made possible long exposures, increasing the resolving power of the instrument; it allowed the recording of spectral lines; and it allowed repeated examination and measurement of images. Only digitalization, which came a century later, was comparably important to astronomical methodology. By the end of the nineteenth century astrophotographers had available a gelatin-coated "dry plate" process that replaced the elaborate "collodion" process that had made astrophotography feasible by the 1850s but required preparation of a glass plate immediately before exposure and development immediately after. Dry plates enormously facilitated photography during eclipse expeditions. And so it was that when Einstein predicted a gravitational deflection of light, astronomers ventured to solar eclipse sites to prove him right or wrong.

Karl Popper regarded Einstein's prediction of the gravitational deflection of light as the paradigm of scientific method: conjecture, predict, test, and revise if refuted. The historical sequence was rather different: conjecture and predict, conjecture and predict differently, test and refute, retain conjecture and prediction, test and refute and test and not refute, test and not refute. Einstein produced a relativistic theory of gravity in 1911 which yielded a "Newtonian" deflection of 0.87" at the limb of the sun. An eclipse expedition from the Argentine National observatory attempted to measure the deflection in 1912 but was rained out. Expeditions from Argentina, Germany, and the Lick Observatory tried again in 1914 but the German expedition became Russian prisoners of war before reaching the observation site and the other two groups of observers were rained out. In 1915, using his new theory, Einstein predicted a deflection of 1.74" of arc. An expedition from the Lick Observatory took eclipse photographs at Goldendale, Washington in 1918. Heber Doust Curtis, the

---

[*] Berger (private correspondence) has acknowledged that Einstein's use of the perihelion advance vitiates the subjective Bayesian account of Einstein's reasoning.

[†] Edward Leamer, *Specification Searches*, Wiley, 1978. A similar argument is given with respect to the perihelion advance in Clark Glymour, *Theory and Evidence*, Princeton University Press, 1980.

astronomer in charge of reducing the observations, and W. W. Campbell, the Director of the observatory, on separate occasions initially announced that the observations were inconsistent with Einstein's second prediction. For reasons to be discussed later, Campbell refused to publish a final account of the 1918 measurements. Late in the summer of 1919, the results of three sets of measurements with three telescopes from two British expeditions, one to Sobral, Brazil, the other to Principe, off the coast of Africa, were announced.

The British observations and data analyses have recently been discussed several times by historians,[*] but the first blush results were these: One instrument in Brazil gave fuzzy images and some rotational distortion but showed a lot of stars; measurement of the glass plates and reduction by least squares gave the "Newtonian" deflection with a substantial standard deviation (0.48"); a second instrument give precise images, a small standard deviation (0.178"), and a deflection (1.98") substantially above the prediction of general relativity, which would have been rejected at about the 0.1 significance level (no such test was reported at the time and no variances calculated—the standard at the time was "probable error"). A third instrument, at Principe, supervised by Arthur Eddington, showed only five stars, not very clearly. By aid of some legerdemain, Eddington calculated a deflection of about 1.65" with a standard deviation (0.44") about equal to that of the first Sobral instrument. Setting up the problem as a forced choice between the Newtonian and the general relativistic values, at a joint meeting of the Royal Astronomical Society of London in the summer of 1919, Frank Dyson, the Astronomer Royal, announced that Einstein's theory had been confirmed. No explicit Bayesian analysis was feasible at the time; modern methods for combining estimates from multiple data sets, or for testing hypotheses—let alone from multiple data sets—were not available. Useful statistical methods were limited to least-squares and probable errors. (I have never seen a modern analysis taking account of the three data sets, their varying numbers of stars, and the three distinct group variances. My vague guess is that a reasonable hypothesis test would reject both the 0.87" and 1.74" hypotheses, and a Bayesian comparison of the two with equal priors would favor the larger deflection. There were any number of possible deflection values in between, but none were considered at the meeting.) The conflicting data from Brazil were explained away post hoc, and Eddington, in his *Mathematical Theory of Relativity*—the first and for a good while, standard reference on the theory in English—pretended the expedition to Brazil had never occurred. Needless to say, a lot of astronomers, including Campbell, were unconvinced. Under Campbell's direction, a Lick Observatory eclipse expedition to Australia in 1922 produced the first unequivocal confirmation of the general relativistic prediction.

At the joint meeting in the summer of 1919, after the presentation of the results of the expeditions and Dyson's verdict, the acerbic and ambitious Ludwig Silberstein, who was understandably not fond of English weather, objected that solar spectra could "even in this

---

[*] See J. Earman and C. Glymour, Relativity and eclipses: The British expeditions of 1919 and their predecessors. *Historical Studies in the Physical Sciences*, 11, 49–85, 1980. A defense of Eddington's and Dyson's interpretation of the measurements is given by B. Almass, in Trust in expert testimony: Eddington's 1919 eclipse expedition and the British response to general relativity. *Studies in History and Philosophy of Science Part B: Studies in History and Philosophy of Modern Physics* 13, 57–67, 2009.

country, be observed many times a year." Silberstein advocated a theory called "Relativity without the Equivalence Principle." Silberstein's point was that measurements of solar spectra to test the gravitational red shift were not going Einstein's way. Even Eddington's endorsement of the theory was initially qualified because of the solar spectra results.

The case of the gravitational red shift presents an example of what might be called "reverse data mining": using a theory to alter the interpretation of previous data that did not support it. Einstein had given heuristic arguments for a gravitational red shift in 1907 and 1911, and in 1916 he gave another derivation from general relativity assuming a spherically symmetric and static gravitational field. The argument was heuristic and theorists had a devilish time keeping straight clock measurement, coordinate time, and proper time. Even the great Max von Laue derived a blue shift! Work in the 1880s had found that solar lines shifted most often toward the red compared with terrestrial sources, varied with the element considered, with the line intensity, and were not proportional to the wavelength. A Doppler shift could not explain them. In the first decade of the twentieth century, solar shift measurements were made by a variety of observers, of course without reference to relativity or the equivalence principle. Between 1915 and 1919, repeated measurements were made to test the gravitational red shift with various lines, notably and persistently by Charles St. John at the Mt. Wilson Observatory. It was not until after the announcement of the British eclipse expeditions that consistent confirmations of the predicted red shift appeared. Some of the confirming measurements were new, but many were by reverse data mining. Old measurements conflicting with the data were reviewed and reanalyzed in accord with the relativistic hypothesis, for example, by claiming that solar lines had previously been mismatched with terrestrial lines. It is hard not to think that the reanalyses were driven with the aim of reconciling the historical measurements with the new theory. (The reworking of old numbers to accord with a new theory was not novel in science. In 1819, for example, Dulong and Petit proposed that specific heats of elements are proportional to their atomic weights. They had measured specific heats calorimetrically, but the only atomic weights available were the tables produced by Berzelius, largely by guesswork, assumptions about molecular formulas, and chemical analogy. Dulong and Petit simply replaced the entries in the Berzelius tables with values that agreed with their hypothesis. Good methods may more often find the truth than bad methods, but truth does not care about how it is found.)

2.7 STATISTICIANS BECOME PHILOSOPHERS

By the early twentieth century, astronomy and physics had established the causes of the motions of the solar system bodies, and were beginning to unravel the cause of the sun's heat, the composition of stars, classifications and laws that allowed distance measurements to stars, the resolution of nebulae into galaxies, a relation between red shifts and distances—Hubble's law—that gave insight into cosmological structure, and much more. The laws, mechanisms, and natural kinds (e.g., star types) of the universe were being revealed by physical theory and observation, connected by serendipity and data mining. Meanwhile, statistics was undergoing a revolution limited by two factors, computation and the suspicion of causality, a suspicion sustained by philosophy.

In the eighteenth and nineteenth centuries, in the hands of Laplace, Bayes, Price, George Boole, George Udny Yule, and many others, probability was the mathematical tool for the discovery of causes. There was no other mathematics of causal relations, no formal representation of causal structures as mathematical objects other than those implied by concrete theories, whether atomic theory or Newtonian celestial dynamics. That division continued with the remarkable development of experimental design by Ronald Fisher, whose analyses always took known potential causal relations as providing the structure within which to apply his computationally tractable statistical methods for their estimation and testing. Among philosophers and many statisticians, however, the twentieth century saw a remarkable change in outlook, signaled by Karl Pearson's announcement in 1911 in *The Grammar of Science* that causation is nothing but correlation and by Bertrand Russell's argument 2 years later that the notion of causation was a harmful relic of a bygone age. Russell's argument focused on the reversibility of Newtonian equations, Pearson's on a kind of subjectivism in the spirit of George Berkeley, the eighteenth-century idealist philosopher. (Pearson, possibly the worst philosopher ever to write in English, claimed that material objects do not exist: they are merely illusions produced by (a material object) *the brain*! Pearson seems not to have noticed the contradiction.) Positivist philosophers of the time such as Rudolf Carnap and Carl Hempel avoided the notion of causation altogether and sought to replace it with deductive relations among sentences. Just as in the seventeenth century, Gottfried Leibniz adopted a bizarre metaphysics apparently because the formal logic he knew could not accommodate relations, the logically minded philosophers of the early twentieth century seemed to have been suspicious of causality because causal phrasings were not easily accommodated by the novel and impressive developments of formal logic that had begun in the 1870s. Fisher's attitude seems to have been that talk of causation makes sense in experiments, but not otherwise. (He seems to have had a similar view about talk of probability.) A lot of astronomy was about finding the causes of observations, but Fisher, whose Bachelor's degree was in astronomy and who had worked initially on the theory of errors for astronomical observations, did not appear to see any tension between that and his commitment to experiment. A general sense that "causation" was an objectionably metaphysical notion became pervasive; "probability" was acceptable, even though no one could provide a tenable explanation of what the term referred to. The chief difference was that probability had a developed, and developing, mathematical theory, and causation did not. A second difference was that with observational data, associations could, it seemed, always be explained by postulating further common causes, whether historical or contemporary with the associated events, and so no direct causal relationships could be established. Fisher used the second argument to attack the conclusion that smoking causes cancer, and even as late as 1998, Judea Pearl, the computer scientist who was later to become the most prominent contemporary champion of causation, used it to reject causal interpretations of nonexperimental data. Causal claims based on observing might make sense, but they were always unwarranted. The counterexamples provided by astronomy and evolution were seldom noted.

Along with an aversion to causation, prompted by the wide influence of statistical hypothesis, testing twentieth-century statistics developed an aversion to search. The principal argument, repeated in scores, perhaps hundreds of sources throughout the century, was by worst case. Given a sample of N independent variables, and, say, a hundred hypotheses of correlation tested on the sample, the chance that at least one null hypothesis of zero correlation would be rejected at a 0.05 alpha level can be much higher than 0.05, depending on the sample size. Statistical methodology produced two responses: the "Bonferroni adjustment," which reduced the rejection region by the reciprocal of the number of tests; and the methodological dogma that a hypothesis to be tested must be specified before the data used in testing it is examined. It seemed not to matter (and, as far as I have read, almost never noted) that the same worst-case argument would apply if 100 independent scientists each prespecified one of the 100 hypotheses and tested it on the sample, or even if each of the 100 independent scientists each had their own samples from the same probability distribution. Neither did it matter that the worst and expected cases might differ. The logical conclusion of the conception of science as implicit or explicit hypothesis testing seemed to be that subsequent testing of other hypotheses on samples from the same distribution, or on the same sample, should reduce the confidence in whatever hypotheses passed earlier tests. The unavoidable implication was that almost the whole of the history of science, and especially astronomy, was a statistical mistake. In an unpublished paper, one prominent economist, Hal White, drew that very conclusion. By the 1960s, conventional statistics was fundamentally in conflict with science, and Bayesian remedies were unconventional and in practice unavailable. Then both science and statistics were changed by a profound technological revolution: the computer and the digitalization of data.

2.8 THE DIGITAL REVOLUTION

The twentieth century saw an expansion of sky surveys with new telescopes, and near mid-century the practical use in astronomy of sky surveys outside the visible spectrum using radio, infrared and, later, x-ray and γ-ray detection. Those remarkable developments were complemented by the development of the digital computer and, around 1970, of the charge-coupled device (CCD) camera and other solid state digital detectors. Early radio telescope instrumentation produced huge amounts of data, printed on page after page. Magnetic recording in the 1960s, followed by digitalization and automated computation of correlations allowed real-time reduction of raw data and the production of images for scientific analysis. Once data were digitalized, all of the convenience, speed, and potential analytical power of the computer could be put to work on the flood of data new instrumentation was producing.

It may be that the digital revolution had a social effect on data-intensive sciences such as astronomy and psychology. While particular female astronomers from the nineteenth century such as Caroline Herschel are historical quasi-celebrities, women were rare in astronomy until past the middle of the twentieth century when their numbers began to grow. Before the digital revolution it was common for technically able, educated women to work as "computers" in observatories. For the most part, the careers of these women are unknown, but there are exceptions. Adelaide Hobe, for example, was an assistant at the Lick Observatory in the first quarter of the twentieth century. She conducted the remeasurement and recalculation of the gravitational deflection from the plates of the Observatory's 1918 eclipse expedition. Her work convinced Campbell that the photographs were not decisive against Einstein, and may have helped to prompt the Lick Observatory expedition to Australia in 1922. Curtis, the astronomer who had done the original measurements and calculations, and who by 1920 had moved to become director of an observatory—the Allegheny, in Pittsburgh—turned out to have been arithmetically challenged. The more capable Hobe suffered the indignities of her time. Graduating from the University of California in physics (although her transcript says "Home Economics") and unwelcome in graduate school because of her sex, she had taken work at the Lick Observatory as a human computer. One astronomer noted that it was a pity she was a woman. The University of California, and other universities, did come to admit women to graduate study in astronomy, but it is difficult to believe that the coincidence of growth of women in the subject and the emergence of technology that removed the need for human computers was entirely a matter of chance.

2.9 STATISTICS AND MACHINE LEARNING MEET ONE ANOTHER

The computer prompted new statistics and rejuvenated old, and in consequence the logical divide between scientific practice and statistical orthodoxy began to be bridged. Bayesian statistics suddenly became a useful tool. The calculation of posterior probabilities became possible through fast-converging asymptotic approximations, notably Schwartz's, and through Monte Carlo methods, notably Gibb's sampler and Metropolis algorithms. Bayesian methods developed for analyzing time series and their breaks, and for fusing disparate data sets. In astronomy, cosmology, and elsewhere, Bayesian statistics succeeded by treating each new problem with model-driven likelihoods and priors generally given by mathematical convenience, dispensing in practice if not always in rhetoric with any pretense that the probabilities represented anyone's, or any ideal agent's, degrees of belief, but yielding scientifically valuable comparisons of hypotheses nonetheless. Bayesian methods made issues of multiple hypothesis testing, for example, in searches for extrasolar planets, largely irrelevant.

The cornucopia of new digital data in genomics, Earth science, and astronomy begged for automated search methods, methods of handling hitherto unthinkable numbers of variables, and data sets varying from small in number of units to, well, astronomical. They were not long in coming. The major developments occurred first in classification, regression, and clustering. Flavors of neural net classifiers abounded, followed by probabilistic decision trees, support vector machines, Markov blanket variable selection algorithms, and more. In many cases, successful classification or regression became possible even when the number of variables in a data set is two orders of magnitude greater than the sample size. New source separation methods developed, notably varieties of independent components algorithms. Statistical and model-based data correction, as in outlier rejection, were supplemented with classifier-based data correction, as with "boosting" and "data polishing." Problems of overfitting, which traditionally would have been addressed only by testing out-of-sample predictions (no peeking!), were addressed through resampling methods. The worst-case Bonferroni correction is coming to be replaced by the more sensible false discovery rate. As the contributions to this volume show, developments such as these are becoming central to the analysis of astronomical data.

In many cases, these new developments were heuristic, but in other cases—Markov blanket variable selection for example—they came with well-developed asymptotic theories in place of more statistically standard but heuristic procedures, and in other cases—support vector machines, for example—with demonstrable finite sample error properties.

Twentieth-century suspicions about causality—especially about causal conclusions drawn by computational procedures applied to observational data—still remain, despite dramatic research developments in the last 20 years in machine learning of graphical causal models. This research fractured model specification searches into a variety of cases of varying informativeness. Correlation, for example, is not causation, but correlation is a second moment quantity: in linear systems with non-Gaussian additive, independent disturbances, higher moments determine direction of influence. Consistent, computable search procedures, generalized to tolerate unknown latent variables and sample selection bias, provided set-valued model estimates for any distribution family for which conditional independence tests are available; algorithms for linear and nonlinear models with additive errors provided finer, in some cases unique, estimates of causal structure; correct algorithms for estimating relations among indirectly measured variables appeared for linear and some nonlinear systems. The new methods were adapted to time series. Most recently, work on machine learning of graphical models has extended traditional factorial designs to reveal enormous possible efficiencies in experimental design and has provided guides to the integration of fragmentary causal models. Some of this work has been put to use in genomics, brain imaging, biology, and economics but as yet has had little if any impact in astronomy.

2.10 A BRIEF REFLECTION

The essays in this volume illustrate some of the many applications of machine learning and novel statistics in astronomy, which need no review here. They illustrate a wave of new inference methodology that new data acquisition methods are forcing. From the seemingly routine task of matching sky catalogs to the excitement of the search for new planetary systems, machine learning, statistics, and astronomy are happily merging. Bad arguments for old shibboleths are ignored, real concerns for reliable inference retained. Philosophers are ignored. All good.

CHAPTER 3

Probability and Statistics in Astronomical Machine Learning and Data Mining

Jeffrey D. Scargle
NASA Ames Research Center

CONTENTS

3.1	Detecting Patterns and Assessing Their Significance	28
3.2	Assessing Causality	30
3.3	Correlation versus Dependence	31
3.4	Probability in Science	32
3.5	Cosmic Variance	33
Appendix: Tales of Statistical Malfeasance		34
References		36

Statistical issues peculiar to astronomy have implications for machine learning and data mining. It should be obvious that statistics lies at the heart of machine learning and data mining. Further it should be no surprise that the passive observational nature of astronomy, the concomitant lack of sampling control, and the uniqueness of its realm (the whole universe!) lead to some special statistical issues and problems.

As described in the Introduction to this volume, data analysis technology is largely keeping up with major advances in astrophysics and cosmology, even driving many of them. And I realize that there are many scientists with good statistical knowledge and instincts, especially in the modern era I like to call the *Age of Digital Astronomy*. Nevertheless, old impediments still lurk, and the aim of this chapter is to elucidate some of them. Many experiences with smart people doing not-so-smart things (cf. the anecdotes collected in the Appendix here) have convinced me that the cautions given here need to be emphasized.

Consider these four points:

1. Data analysis often involves searches of many cases, for example, outcomes of a repeated experiment, for a feature of the data.

2. The feature comprising the goal of such searches may not be defined unambiguously until the search is carried out, or perhaps vaguely even then.

3. The human visual system is very good at recognizing patterns in noisy contexts.

4. People are much easier to convince of something they want to believe, or already believe, as opposed to unpleasant or surprising facts.

One can argue that all four are good things during the initial, exploratory phases of most data analysis. They represent the curiosity and creativity of the scientific process, especially during the exploration of data collections from new observational programs such as all-sky surveys in wavelengths not accessed before or sets of images of a planetary surface not yet explored.

On the other hand, confirmatory scientific studies need to adopt much more circumscribed methods. Each of these issues conceals statistical traps that can lead to grave errors, as we shall now see. These four ideas thread most of the following discussion.

3.1 DETECTING PATTERNS AND ASSESSING THEIR SIGNIFICANCE

The human eye–brain system is very good at recognizing unusual patterns in scientific images. We have an uncanny ability to bring together disparate parts of an image to make up familiar features. But images are commonly corrupted by meaningless pattern fragments due to noise or other quasi-random features. The question is then whether a feature identified by eye is real—that is, an actual structure is the imaged astronomical object and not a noise fluctuation. This decision can be very difficult, especially in the context of exploratory inspection of a large set of images. *Post facto* statistical analysis is difficult without quantitative knowledge of characteristics of the noise and image clutter and—most important—a completely predefined search method (including the conditions under which the search will be terminated) and an unambiguous definition of what would be considered unusual.

Not surprisingly, in practice these conditions are seldom met in any rigorous way. Indeed one can argue that imposing them is likely to stifle effective exploration of new data sets. But without them, meaningful assessment of the significance of any discovery is problematic. Some data manipulation procedures can provide estimates of the frequency with which patterns emerge from noisy data absent any real features. But such *false-positive rates* are of limited value, largely because of the importance of selection effects imposed by the human in the loop and by nature. Assessment of the significance of a feature detected during a search through data is not a simple statistical matter, and in particular cannot be achieved with standard probability arguments alone.

This is a pervasive, important issue not limited to images but applying to any observational or experimental data. Consider an experiment consisting of one or more trials. Many statistical methods have been designed to assess the significance when the experiment consists of

    a. A single trial, defined *prior* to carrying out the experiment, or

    b. Several trials; after the fact the "best" of them is taken as the outcome.

Methods of the first kind are said to be *a priori*, the second *post facto*—according to whether the outcome is defined prior to the experiment or after it has been carried out. Clearly, very different standards must be used to judge statistical significance in the two cases. The same outcome correctly judged as highly significant in setting (a) may be an insignificant statistical fluctuation in setting (b). Failure to realize this point has more than once led to colossal blunders; witness example (2) in the Appendix. Physicists call taking multiple trials into account the look-elsewhere effect—because in effect the experimenter has looked not only at the identified case but also at others.

The natural propensity for humans, even scientists, to believe what they want to believe[*] leads to the related danger of *fiddling with the data*. Here I mean combing through either raw data or the products of an automated pattern perception process—perhaps "cleaning up" the data and rejecting outliers, and with an attitude of ignoring some data and emphasizing others based on a preconceived notion. Often what one is looking for or would consider interesting is not unambiguously specified ahead of time. This search commonly continues until one indeed finds something interesting. The result may be a feature in an image or a structure in some data representation that is truly unusual . . . but again the question is: Is it real or is it a statistical fluctuation?

I am not trying to promote the notion that such fiddling should not be allowed. On the contrary, this activity is essential to progress in science; all scientists great and small do it to some extent, in various circumstances. The point is that in assessing the significance of outcomes one must take into account such fiddling. If p is the probability of a statistical fluctuation being (incorrectly) judged as real at a single trial, then the probability that in a set of N trials, no such misjudgment will be made is $(1-p)^N$ because the probability of a chain of independent events is the product of their individual probabilities. Then the false positive probability, defined to be the chance of finding one or more such misjudgments in the N-fold sample is

$$\text{Prob}(\text{one or more false positives}) = 1 - (1-p)^N$$

This useful and widely applicable formula is only valid for independent trials. It is hard to treat real-world situations in this way because the key question of whether or not the many trials are statistically independent is so difficult.

An alternative is that after something has been discovered in the process of fiddling with the data, then a new independent experiment is designed for testing the hypothesis that the discovery is real.

---

[*] Remember Richard Feynman's oft-quoted remark "The first principle is that you must not fool yourself—and you are the easiest person to fool."

A very common example in high-energy astrophysics is to examine energy spectra of photon-counting data, searching for peaks that may represent line emission. The term "bump hunting" is appropriately used for this exercise. The statistical problem seems deceptively simple: account for the fact that such searches are comprised of many effective trials, since many energy bins were inspected for a bump. Peak hunting in power spectra is essentially the same problem, and is acerbated by the usual practice of considering a fine grid of overlapping bins— "I don't want to miss a peak that lies between the usual set of frequencies (i.e., integer multiples of the fundamental frequency, up to the Nyquist frequency)." The resulting trials are not independent, vitiating simple significance analysis of independent trials. Many of the false detections of periodicities in quasar redshifts derived some of their apparent statistical significance by not adjusting for this kind of "parameter tuning" in power spectra of the redshift distribution. See for example, [1] for a discussion on how to estimate the effective number of independent trials in such situations.

"Many trials" effects can also cloud the assessment of theoretical results, as pointed out by Don Lamb in a seminar at Stanford University on Monday, April 18, 2011. Theoreticians can lose track of how many ideas they pursue and discard before arriving at a satisfactory and publishable result. This makes the significance of the result and its agreement with available data problematic for reasons similar to those we have been discussing.

And finally note that biasing selections can at least seem innocent and not necessarily will-ful fudging. Suppose you are computing the power spectrum of some time series data, trying out various taper functions. There are many such windows: Tukey, Hanning, Hamming, Lanczos, triangular, cosine, and so on. If one of them shows an interesting result, say in accord with your pet theory, what is wrong with choosing that window, trumpeting its virtues, and citing copious research on its properties? [By the way, multitapering [2] is a refreshing alternative to this window-out-of-a-hat approach.] Well, I hope you see what is wrong from the point of view of statistical significance. I think many practicing astronomers can identify other data analysis temptations that can be rationalized to seem statistically innocent.

3.2 ASSESSING CAUSALITY

In addition, the "many trials" aspect of this issue can be provided by Nature as well as by human data miners. Take the case of time-series observations, at two wavelengths, of the luminosity of a randomly varying astronomical object characterized by flares (short outbursts of radiation). The question almost always asked is "Are the variations at these two wavelengths correlated?*" Typically the answer is "yes" if one or more flares occur at more or less the same time. During many talks presenting multiwavelength observations of variable sources I have mumbled this joke to myself: "If the flares occur at the same time, they are correlated; if not, they are correlated with a time-lag. Either way they are correlated!" Unfortunately, this is not entirely a joke. (Of course good evidence for real effects can accumulate if there are many simultaneous flares with no or few counterexamples.) A statistically sound procedure is to consider the results of this kind of data analysis as exploratory only—and then make predictions that are testable on new, independent, data.

---

* Technically this is the wrong question. See Section 3.3.

Most scientists are generally aware of the notion "correlation does not necessarily mean causation"—that is, just because two time series have, say, a strong peak in their cross-correlation function at some lag does not mean that a physical process in the first one causes a process in the second one. In addition to the problem of assessing the statistical significance of such peaks there is the notion that the processes underlying two apparently correlated time series may be driven by a third process for which there is no data. Elucidation of causal relations in time-series data is inherently difficult, but some publications in this topic [3,4] are worth consulting.

In experimental sciences one can address this issue by carrying out repeated experiments with different conditions for the putative cause and observing the behavior of the putative effect. In the observational science of astronomy we do not have such recourse; progress comes from careful combination of observational, theoretical, and data-analytic efforts. It is important to keep clearly in mind the precise meanings of and relations between functions to quantify connections between time series.

3.3 CORRELATION VERSUS DEPENDENCE

The need for a quantitative assessment of the degree to which two different items of data (time series, images, etc.) are related to each other often arises. In statistics textbooks the most common notion is that of the *correlation* or *covariance* (we do not distinguish these two, which differ only by normalization constants) between two random variables X and Y, measured by averaging their product:

$$C(X, Y) = E[XY], \tag{3.1}$$

where E stands for the average or expected value, and it is assumed that the mean values of X and Y have been subtracted. But correlation is a relatively weak measure of how closely X and Y are related. The strongest such characterization is *dependence*. Two variables are independent if their joint probability density is the product of their individual probability densities:

$$P_{XY}(x, y) = P_X(x)P_Y(y) \tag{3.2}$$

Independent processes are always uncorrelated, but the reverse is not true. This distinction is important for two reasons: (a) highly dependent processes can be uncorrelated, and (b) dependence, not correlation, is usually relevant physically.

On the latter point consider two causally disconnected flaring processes. (The flare events do not "talk" to each other.) Their luminosity time series will be independent, not just uncorrelated. The difficulty of quantitative assessment of dependence is undoubtedly responsible for the relative little use made of this concept. See Ref. [5] for discussion on some methods for this. On the other hand, note that the simple scatter plot of values of X versus Y elucidates dependence to a considerable degree.

Modern developments in data mining and information technology may have unwittingly exacerbated some of these problems. The ease of collecting, processing, and visualizing large amounts of data increases the incidence of unreal patterns that are just random fluctuations. Even very effective advanced scientific visualization facilities, such as the NASA

Ames Hyperwall and the portable version of it called Viewpoints [6], can be viewed as engines that crank out so many views of high-dimensional data that statistical fluctuations aplenty are inevitable. Counterbalancing this effect is the contribution of automated, objective data processing and mining methodologies to facilitate significance evaluation techniques such as variance analysis via bootstrap, surrogate (or data scrambling) techniques, and reclassification tests (leave one case out, reclassify based on the remaining cases—and average over the case left out).

3.4 PROBABILITY IN SCIENCE

Was the incident described in Item 1 in the Appendix here a rare anomaly? Perhaps, but I think there is considerable confusion about the role of uncertainty and its quantification, especially in data mining. This confusion arises at least partly from the prevalence of two statistical formalisms—Bayesian and frequentist. This is not the place for a full discussion on these approaches and their relationships. For some interesting insights one might consult [7]. Nor will I discuss various attempts at reconciling or combining the two methodologies (see [8]), but it is worthwhile to quote from Brad Efron's article [9] on this topic:

> The 250-year debate between Bayesians and frequentists is unusual among philosophical arguments in actually having important practical consequences. Whenever noisy data is a major concern, scientists depend on statistical inference to pursue nature's mysteries. 19th Century science was broadly Bayesian in its statistical methodology, while frequentism dominated 20th Century scientific practice. This brings up a pointed question: which philosophy will predominate in the 21st Century? One thing is already clear – statistical inference will pay an increased role in scientific progress as scientists attack bigger, messier problems in biology, medicine, neuroscience, the environment, and other fields that have resisted traditional deterministic analyses. This talk argues that a combination of frequentist and Bayesian thinking will be needed to deal with the massive data sets scientists are now bringing us.

Brad's thinking was apparently influenced by a joint particle physics and statistics conference [10] he attended a year earlier. The same influences would almost certainly arise from considering current and planned astronomical sky surveys, massive computational simulations and other developments described in this book.

I believe that some clarity can be cast onto the Bayesian/frequentist debate by realizing that the meaning of the term "probability" is quite different in the two approaches. In data analysis settings the goal of both is to figure out what can be learned from the data at hand. Bayesians compute probability distributions of random models given the data. Frequentists proceed by quantifying how unusual the actually observed data are in the context of the random distribution of possibly observable data. Same goal, but different view of what is random.

The idea underlying the concept of probability in both cases is a quantitative representation of uncertainty. Because of the many different facets of uncertainty and probability a clear definition of the concept is slippery. Probability is meant to be a quantification of uncertainty about past, present, or future events. But it is also incorporated into general physical laws and descriptions of physical systems, both internally (as in statistical mechanics) and as applied to the physical descriptions themselves (as in quantum mechanics).

Let us start with the most straightforward, physics-based notion of probability. By saying p is the probability of outcome O of experiment E a frequentist means that p is approximately the fraction of times O is the outcome if E is repeated many times, and approaches p in the limit as the number of trials becomes larger. This concept underlies much of the physical sciences including quantum mechanics, where the classical notion of uncertainty is implemented via the concept of probability amplitudes [11]. Physicists must face the impossibility of exactly reproducing the conditions of experiments. Astronomers must face situations where only one observation is possible, coupled with an inability to condition experiments at all. Another technical problem is the difficulty of representing what is known before an observation. In the Bayesian formalism this is necessary in order to correctly quantify what is known after—especially when there is no such prior knowledge or, even if there is, one does not wish to include it in the analysis.

Bayesian analysis addresses the degree to which one should believe in the truth of a hypothesis. This assessment is typically based on empirical data, but the analysis applies to hypotheses not to outcomes of individual experiments. That is to say the subject of Bayesian data analysis is not the Universe, but the state of one's belief about it. The whole game is to assess uncertainties in our view of the world. Some are bothered by philosophical issues, such as the apparent subjectivity of this formalism and tricky issues about the meaning of terms like "belief" and "truth." In a nutshell: The Bayesian formalism is constructed to correctly represent logical analysis of how new data should alter prior beliefs about hypotheses [7].

Science in general and astronomy in particular deal with both statistical analysis of outcomes of experiments (observations in astronomy) and construction and evaluation of hypotheses about nature based on such analysis. So it is natural that astronomical machine learning and data mining practitioners view the world from a sort of combined Bayesian and frequentist perspective and make use of both formalisms. Brad Efron was right. Hence, it is important to keep in mind the differences discussed here.

A simple way to think of all this is this oft-stated description: Bayesians consider models of the natural world as uncertain, to be treated as random processes; on the other hand, data are fixed and taken to be completely known (but with account of observational errors). Frequentists regard the scientific model as fixed but unknown and the data as samples from a random process. The slogan of the Bayesian camp might be "models random, data deterministic" while that of the frequentists might be "models deterministic, data random." Faced with an observation that is unlikely to be drawn from the presumed data probability distribution under a given hypothesis, the frequentist tries to quantify the resulting disbelief in the hypothesis ("rejecting the null hypothesis")—without necessarily saying what they believe should be adopted instead. In the same situation the Bayesian tries to quantify relative belief in a set of two or more hypotheses, taking into account both the data at hand and prior beliefs in the hypotheses.

3.5 COSMIC VARIANCE

Any analysis of astronomical data, automated or otherwise, has to account for one crucial problem resulting from the passive nature of observations. Astronomy, other than in nearby solar system contexts, is not an experimental science. Accordingly in many cases there is only one sample, both in practice and in principle. There is only one Universe.[*] Of course in many other settings there are many samples . . . billions of stars, probably a comparable number of planets, trillions of galaxies, and so on. In these contexts one has the same situation as in psychology—many unique, individual human beings who as a group can be discussed statistically. But here too the nature of the underlying random process that we are trying to understand "statistically" is unclear.

In summary, modern cosmological observations, in particular of the multiscale structure of the Universe and the cosmic microwave background have put us face-to-face with images that contain all of the objects or structures that we as humans will ever be able to observe. Much of the structure appears to be random, hypothesized to be growth from purely random initial conditions in the Big Bang. So this is the grandest of all possible *spatial statistics*. Indeed the billions of years it takes light to reach us from the most remote observable objects it makes this subject actually *spatio-temporal statistics*.

Suppose we notice a feature in the data from this only available snapshot of the Universe that seems odd or unusual in some way—we are tempted to say "nonrandom." Statistically the feature is an outlier in some distribution. Is this feature "real" and due to some identifiable physical effect? [This is the same issue discussed above for large samples, only here in the one-sample-only context.] Or is it "just random," due to a statistical fluctuation? Is this distinction even meaningful? What do you mean by "statistical" in a single sample? What does it mean to say that the initial conditions of the Universe are random? These are all difficult questions.

This issue is connected with a concept called *cosmic variance*. To quote from Wikipedia:

> . . . even if the bit of the universe observed is the result of a statistical process, the observer can only view one realization of that process, so our observation is statistically insignificant for saying much about the model, unless the observer is careful to include the variance. This variance is called the cosmic variance and is separate from other sources of experimental error: a very accurate measurement of only one value drawn from a distribution still leaves considerable uncertainty about the underlying model.

In summary: many of the standard problems generic to astronomical data analysis hold as well, or are perhaps worse, in the context of automated machine learning and data mining methods. Cosmic variance, the one-sample problem, data fiddling, human subjectivity . . . all present challenges to the data miner and machine learning scientists alike.

I am grateful to philosopher-scientist Clark Glymour and Mike Way for numerous helpful suggestions.

APPENDIX: TALES OF STATISTICAL MALFEASANCE

Here are a few examples of actual events that have given me a dim view of both popular and professional knowledge of statistics. These are perhaps amusing anecdotes, but they

---

[*] Pragmatically speaking this statement holds, given current observational technology, despite so-called multi-verse cosmologies which deny it.

signal that this kind of malfeasance cuts across many serious fields, including medicine and politics, as well as the topics of this book.

1. At a professional conference I once listened to a speaker describe a data mining system that, as far as I could tell, was meant to identify outliers. My question about how he evaluated the statistical significance of the identifications led to the all-too-common exchange where neither side seems to understand what the other is saying. This confusion ended abruptly when the speaker averred that his system was not statistical at all! Of course nothing could be closer to the heart of statistical science than assessing whether extreme values are truly anomalous, as opposed to simply lying in the tail of a probability distribution.

2. As redshift measurements of quasars started to accumulate in the early 1970s, some astronomers thought that the measured values were not random, but clustered. In [12], statistical significance of these clusters was estimated as though the relevant redshift ranges had been selected ahead of time, not chosen *post facto* as they were by selecting the densest regions in redshift space. The author of [12] rejected the hypothesis of random (unclustered) redshifts because the most highly populated bins contained more points than would be expected in *a fixed, preselected redshift bin*. But the issue is whether this number is greater than what would be expected in the bins with the highest populations. Details and the absurd conclusion reached are described in Ref. [13].

3. On a radio talk show the president of a major political polling company was taking calls from the public. One caller pointed out one way that telephone surveys can reach a biased selection of the voting public. The president responded that such biases in his polls are eliminated by obtaining larger samples. Of course this only makes things worse by making a biased conclusion seem more significant as the statistical errors decrease. More generally, this issue is the classic and often misunderstood difference between systematic and random errors, a difference that confounds many otherwise savvy scientists. ("We estimate the total error by adding random and systematic errors in quadrature [because we don't know what else to do].")

4. In medical research there is a technique called "meta-analysis" that collects results of clinical trials to increase statistical significance. One worries about *publication bias*: if only those trials with good results are published and make it into the meta-analysis the results will be biased. Almost every worker in this field appeals to the "Rosenthal fail-safe formula." Unfortunately, this formula is completely wrong, as described in Ref. [14]. Even more unfortunately decisions about which drugs are effective and safe are often based on highly biased meta-analysis incorrectly validated by the bogus "fail-safe" test. The reader interested in this topic will no doubt enjoy the article [15].

Much more important than these few anecdotes is the fact that these are just the tip of one iceberg in a large ice field pervading the ocean of modern science.

REFERENCES

1. Horne, J. H. and Baliunas, S. L., 1986. A prescription for period analysis of unevenly sampled time series, *Astrophysical Journal*, 302, 757.
2. Percival, D. B. and Walden, A. T., 1993. *Spectral Analysis for Physical Applications: Multitaper and Conventional Univariate Techniques*. Cambridge: Cambridge University Press.
3. Chu, T., Danks, D., and Glymour, C., 2004. Data driven methods for Granger causality and contemporaneous causality with non-linear corrections: Climate teleconnection mechanisms; http://www.hss.cmu.edu/philosophy/faculty-glymour.php
4. Glymour, C., 2001. Instrumental probability, *The Monist*, 84(2), 284. http://www.hss.cmu.edu/philosophy/glymour/glymour2001.pdf.
5. Gabor, J. S. and Rizzo, M. L., 2009. Brownian distance covariance, *The Annals of Applied Statistics*, 3, 1236–1265.
6. Gazis, P. R., Levit, C., and Way, M. J., 2010. Viewpoints: A high-performance high-dimensional, exploratory data analysis tool, *PASP*, 122, 1518, http://www.nas.nasa.gov/Resources/Visualization/visualization.html; http://astrophysics.arc.nasa.gov/~pgazis/viewpoints.htm
7. Jaynes, E., 2003. *Probability Theory: The Logic of Science*. Cambridge, UK: Cambridge University Press, and see http://bayes.wist.edu/etj/prob/book.pdf
8. Bayarri, M. and Berger, J., 2004. The interplay between Bayesian and frequentist analysis. *Statistical Science*, 19, 58–80. http://www.isds.duke. edu/~berger/papers/interplay.html
9. Efron, B., 2005. Modern science and the Bayesian-frequentist controversy. //www-stat. stanford.edu/brad/papers/NEW-ModSci 2005.pdf.
10. PHYSTAT 2003, a Conference held at SLAC on September 8th–11th, 2003. http://www.slac. stanford.edu/econf/C030908/
11. Feynman, R. P. and Hibbs, A. R., 2005. *Quantum Mechanics and Path Integrals*, Emended Edition, Styer, D. F. Emender (Ed.). Mineola, NY: Dover Publications.
12. Varshni, Y. P., 1976. The red-shift hypotheses for quasars: Is the earth the center of the universe?, *Astrophysics and Space*, 51, 121–124.
13. Weymann, R., Boroson, T., and Scargle, J., 1978. Are quasar redshifts randomly distributed?: Comments on the paper by Varshni "Is the Earth the Center of the Universe," *Astrophysics and Space Science*, 53, 265.
14. Scargle, J., 2000. Publication bias: The "File-Drawer" problem in scientific inference, *Journal of Scientific Exploration*, 14, 91–106.
15. Lehrer, J., 2010. The truth wears off. *New Yorker*, Dec 13.

II

Astronomical Applications

1

Source Identification

CHAPTER 4

Automated Science Processing for the Fermi Large Area Telescope

James Chiang

Kavli Institute for Particle Astrophysics and Cosmology

CONTENTS

4.1 Introduction   41
4.2 Expected Signal Types   43
4.3 ASP Implementation   45
    4.3.1 Known GRBs in LAT Data   45
    4.3.2 Blind Search for GRBs   47
    4.3.3 Monitored and Recurring Flaring Sources   50
4.4 Implementation and Execution   52
References   53

4.1 INTRODUCTION

The Large Area Telescope (LAT) onboard the Fermi γ-ray Space Telescope provides high sensitivity to emission from astronomical sources over a broad energy range (20 MeV to >300 GeV) and has substantially improved spatial, energy, and timing resolution compared with previous observatories at these energies [4]. One of the LAT's most innovative features is that it performs continuous monitoring of the γ-ray sky with all-sky coverage every 3 h. This survey strategy greatly enables the search for transient behavior from both previously known and unknown sources. In addition, the constant accumulation of data allows for increasingly improved measurements of persistent sources. These include the Milky Way Galaxy itself, which produces γ-ray emission as a result from interactions of cosmic rays with gas in the Galaxy, and potential signals from candidate dark matter particles in the Milky Way and its neighboring galaxies.

The automated science processing (ASP) functionality of the Fermi Instrument Science Operations Center (ISOC) is a part of the automated data pipeline that processes the raw data arriving from the spacecraft and puts it into a form amenable to scientific analysis. ASP operates at the end of the pipeline on the processed data and is intended to detect and characterize transient behavior (e.g., short time scale increases or "flares" in the γ-ray flux) from astronomical sources. On detection of a flaring event, ASP will alert other observatories on a timely basis so that they may train their telescopes on the flaring source in order to detect possible correlated activity in other wavelength bands. Since the data from the LAT is archived and publicly available as soon as it is processed, ASP serves mainly to provide triggers for those follow-up observations; its estimates of the properties of the flaring sources (flux, spectral index, location) need not be the best possible, as subsequent off-line analysis can provide more refined information on source properties during a flare.[*]

Several features of the Fermi LAT data constrain the ability to detect transient signals. The main limitation is the detection rate of celestial photons. Even though the LAT is more sensitive than previous telescopes at these energies, the nominal photon detection rate is still significantly <10 counts per second (Hz). Spread over the whole sky in one 3 h interval, this is <2 photons per square degree. Another difficulty arises from the angular dependence of the photon sensitivity with respect to the "boresight" of the instrument. The LAT "effective area"[†] drops off roughly as $\cos \theta$, where θ is the off-axis angle with respect to the instrument z-axis. The resulting field-of-view (FOV) is about 2.4 sr, or about 20% of the whole sky. At the few percent level, there is also a modulation of the effective area with azimuthal angle ϕ about the z-axis. These effects, combined with occultations by the Earth and the rocking motion of the satellite between the northern and southern celestial hemispheres on alternate orbits, means that the exposure to any point on the sky is modulated at the orbital timescale ($\sim$95 min); and this modulation will be strongly imprinted on the count rate from a given location on the sky. In addition, the Fermi orbit precesses on a 53.4-day period and that will also modulate the signal from any given source.

As discussed in Chapter 2a (also see Ref. [4]), the raw, unprocessed data stream transmitted to the ground by the LAT is dominated by charged particle backgrounds whose rates are $\mathcal{O}(10^2)$ higher than the photon rates. Charged particles are rejected via a complex particle reconstruction and pattern recognition analysis of the signals in the various detector elements. However, even after these particle rejection steps, there are still some residual backgrounds. Since the incident charged particle rates depend strongly on the location of the satellite in the Earth's magnetosphere, the backgrounds are also modulated at the orbital timescale. Finally, the Earth's limb is a strong source of γ-rays, which are produced via cosmic ray interactions with the atmosphere. During normal survey pointing, the Earth's limb is kept out of the nominal FOV, but for pointing maneuvers known as autonomous repoints, which are triggered by GRB detections, and for an inertial pointing toward a particular celestial source, the Earth limb can enter the FOV and its emission will be a substantial background. Special data selections are generally needed to handle data streams that contain

---

[*] Exceptions to this are the locations on the sky of gamma-ray bursts (GRBs) which we will discuss in detail below.

[†] The effective area is proportional to the probability that a photon from a given direction on the sky will be detected by the instrument. It is the collecting area of an equivalent perfect detector.

Earth limb emission, and often the best strategy is simply to omit data from time intervals when the sources of interest are near the Earth's limb.

4.2 EXPECTED SIGNAL TYPES

The types of signals that ASP considers are all produced by point sources. This includes GRBs and longer timescale transients such as flares from a class of active galactic nuclei (AGNs) known as "blazars." The temporal properties of the emission from these sources span a large range of timescales, $0.1-\gtrsim 10^6$ s; and the properties of the relevant source types influence the detection strategies that are employed. GRBs are extragalactic, and so they should be distributed isotropically on the sky. This means that they will mostly be seen outside of the Galactic plane, where the Galactic diffuse emission might otherwise make identifying and localizing them difficult. GRBs are unique events in that they are one-time occurrences. This is in contrast to repeating sources such as blazars. Although GRBs are believed to be associated with stellar populations (either as collapsing massive stars or coalescing compact binaries) and would thus be located within a galaxy, the sky density of viable host galaxies is so high that directed monitoring of individual galaxies for GRBs is impractical.

During the earliest epochs of their burst emission, GRBs are the most luminous objects in the sky over many decades in photon energy. In the LAT, they can produce pulses with peak photon count rates as large as 10^2 Hz. These pulses typically last up to a few seconds so that they can stand out above the nominal count rate of a few Hz, but only a few GRBs have had signals this bright in the LAT energy band. Figure 4.1 shows the first two GRBs detected by the LAT. The brighter one, GRB 090816C, was extremely luminous and can be clearly seen in the unfiltered photon data. In contrast, the dimmer object, GRB 080825C, is not at all evident in the unfiltered photon data stream (solid histogram) and is only just evident if one restricts the data to an acceptance cone centered on the GBM sky location (dotted histogram). One of the key goals of ASP is to find these faint bursts and alert the astronomical community on a timely basis.

The other predominant signal type for which ASP searches are flares from the blazars. Blazar emission is believed to be produced by relativistic jets of material that are powered by accretion onto supermassive black holes ($M_{BH} \gtrsim 10^7 M_\odot$). Blazar flares that are detectable in the LAT band typically last from hours to weeks, and on these timescales, they can be the brightest point sources in the GeV sky. Depending on the hardness of the spectrum, this can mean of order 10–100 photons/day. Harder sources, though generally producing fewer photons, are easier to detect because the point spread function (PSF) of the LAT is substantially sharper at higher energies: the 68% containment radius at 1 GeV is $\sim 0.1°$ versus $\sim 3.5°$ at 100 MeV. Like GRBs, blazars are extragalactic and so are distributed isotropically on the sky. Unlike GRBs, blazar activity is a recurrent phenomena and the theoretical models of the underlying physical processes can predict or constrain the spectral and temporal properties of this emission. Therefore, it is a useful strategy to monitor a well-selected, predefined set of sources routinely to look for specific flaring episodes as well as to characterize the longer timescale behavior. The monitored source list for ASP is based on spectral and temporal properties observed at other wavelength bands, the x-ray and TeV bands in particular. Figure 4.2 shows the daily flux estimates (or upper limits) for the blazar 3C 273

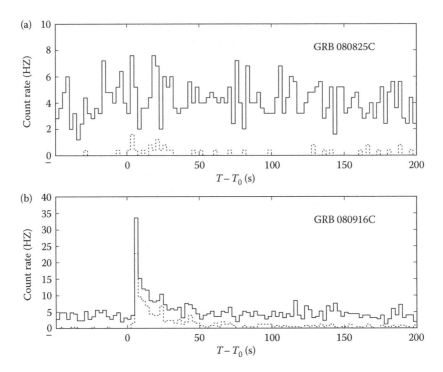

FIGURE 4.1 Example light curves from Fermi-LAT photon event data for a relatively dim burst, GRB 080825C ([3]; a), and a fairly bright burst, GRB 080916C ([2]; b). The solid histograms are all of the LAT events that pass the "transient" class data selection between 30 MeV and 300 GeV. The transient class events have less stringent data selections on the instrument reconstruction variables in order to provide greater sensitivity at lower energies at the expense of somewhat higher residual backgrounds [4]. These data are intended for the analysis of sources that flare on very short timescales (shorter than 10–100 s of seconds), so that even with the increased background levels, the source fluxes are expected to dominate the signal. The dotted histograms are these events within a 15° acceptance cone centered on the GBM Notice position; an additional cut that excludes all photons with reconstructed directions at an angle greater than 105° is also applied in order to minimize contamination from Earth limb photons.

that have been computed by ASP as part of the Fermi-ISOC pipeline. This object was the first extragalactic source seen at >100 MeV energies [15]; it exhibits similar variability as seen in the LAT light curve at all wavelengths and has been extensively studied for many years. It is one of the original Fermi-LAT-monitored sources.

Other sources are expected to produce detectable flaring signals in the LAT. These include lower mass-accreting systems containing a neutron star or stellar mass black hole. Since these sources must be within the Galaxy to be detectable, they will mostly be confined to the Galactic plane region where the diffuse emission from the Galaxy is intense and makes localization more difficult. Finally, there may be some "known unknown" sources whose properties are not well established since they have not yet been observed. These would include emission from the core-collapse of Population III stars; these would be similar to GRBs, but are expected to produce flares on much longer timescales, $\mathcal{O}(10^3–10^4)$ s [16].

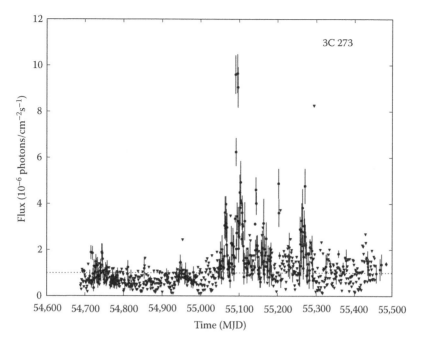

FIGURE 4.2 Daily flux measurements from ASP for the blazar 3C 273 for the energy range 0.1–300 GeV. The black points (with errors) are flux measurements and the downward triangles (points without errors) are 95% confidence level upper limits. Upper limits are reported if the source flux is not significantly different from zero (using a likelihood ratio test) for that epoch. The horizontal dashed line at 10^{-6} photons $cm^{-2} s^{-1}$ indicates the threshold for reporting flares from new sources.

4.3 ASP IMPLEMENTATION

4.3.1 Known GRBs in LAT Data

GRBs are most easily detectable at energies below the LAT range, with their characteristic event energies being between 200 keV–1 MeV. The Gamma-ray Burst Monitor (GBM), the other instrument onboard the Fermi satellite, is designed to detect photons from bursts in this energy range [10]. The GBM is an all-sky monitor, with roughly uniform sensitivity to the entire unocculted sky. Its timing capabilities are sufficient to resolve time structure in the GRB light curve on submillisecond timescales, but it can only localize most bursts to no better than a few degrees. These positional errors are too large for most ground-based follow-ups to find a fading optical counterpart, and so one of the intended goals for LAT detections is that its superior angular resolution at >GeV energies would allow for more precise GRB positions and hence more productive follow-ups for those bursts that have GeV emission.

The ASP GRB refinement task takes as input the precise timing information and rough positional information from the GBM to look for LAT emission. In addition to GBM inputs, other satellites can detect and localize GRBs. The Swift mission [8], in particular, can obtain very precise locations of GRBs using its Burst Alert Telescope (BAT) detector. However, since it is not coaligned with the LAT FOV as the GBM is, most bursts seen by Swift-BAT lie outside of the LAT FOV. Other missions that can also provide input information on bursts for ASP are the INTEGRAL mission [17] and the hard x-ray detector aboard the Suzaku satellite [11]. GRB detections by the various missions are broadcast on the GRBs coordinates network (GCN)* with as short as possible latencies after trigger so that automated telescopes can search for counterparts. The ASP pipeline subscribes to these notices and sets up automated pipeline task instances to search for GRB signals in the LAT data.

Before the ASP GRB refinement processing jobs themselves are started, the trigger times and the position estimates from the various notices that each mission issues for any given burst are stored in the database tables by the processing software. The GRB refinement task then proceeds in several steps. The coordinates used to seed the GRB refinement task are from the position estimate with the smallest error radius. The refinement task extracts the photon event data in a 15° radius acceptance cone centered on this location and in a 200 s time interval centered on the trigger time, T_0. The arrival times of the selected events are then analyzed using the unbinned version of the Bayesian Blocks algorithm [9,13]. The threshold is set to a false-positive rate of 1%. If "changepoints" are found that indicate increases in the source flux from one interval to the next, then a potential burst is deemed to have been detected, and the data are re-extracted using the time window defined by the first and last changepoints. If no changepoints are found, then the burst is flagged as a nondetection and a nominal extraction window of $(T_0, T_0 + 60\,\text{s})$ is used for an upper-limit calculation. Figure 4.3 shows the Bayesian Blocks reconstruction for the LAT data of GRB 080825C.

The newly extracted event data are analyzed to find the best-fit position of the burst; and for this, the energy-dependent LAT PSF is used in an unbinned maximum likelihood analysis. Since the burst signal is expected to dominate any diffuse or residual background emissions, the source model used in the likelihood function consists only of a point source. For the first step, the maximum likelihood estimate (MLE) of the burst position is found via a Nelder–Mead algorithm [12], and the samples of the likelihood surface evaluated during the optimization are used to estimate the error radius under the assumption that the log-likelihood surface is quadratic in the vicinity of the MLE position. Since the log-likelihood function may not be well approximated by a quadratic, more reliable error contours are computed by mapping out the likelihood surface in detail. This is accomplished by fitting a point source on a grid of positions centered on the best-fit location with the scale of the grid set using the error radius estimate from the prior localization step. The error contours for 68%, 90%, and 99% confidence limits correspond to changes of 1.15, 2.30, and 4.60† in the log-likelihood with respect to the MLE value, respectively. The LAT error contours for GRB 080825C are shown in Figure 4.4.

In the case of a LAT detection, the spectral parameters of the GRB emission are derived at the best-fit position, whereas in the case of a nondetection, the seed coordinates from the GCN Notice are used and a flux upper limit is computed. Since upper limits require a model of the sky in the null case, the celestial diffuse emission and residual charged particle backgrounds are included in the model for the unbinned likelihood analysis. The GRB is

---

* http://gcn.gsfc.nasa.gov/
† These are one-half of the $\Delta \chi^2$ values corresponding to those confidence levels for two degrees of freedom.

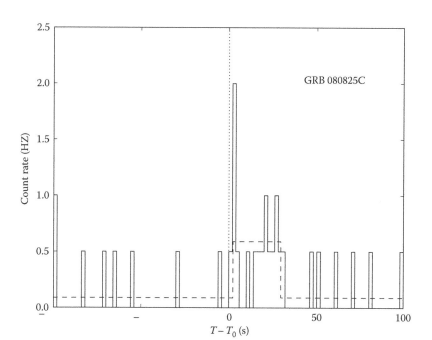

FIGURE 4.3 Histogram of the LAT-detected, transient class events for GRB 080825C (solid). These events were selected to be within 15° of the GBM localization and within 100 s of the GBM trigger time at T_0 (vertical dotted line). The dashed curve is the Bayesian Blocks reconstruction applied to these data. The inferred burst interval for the emission in the LAT band lasts approximately 27 s starting at $T_0 + 2.4$ s.

again modeled as a point source with the photon index and overall normalization as free parameters. The background components comprise a Galactic diffuse component based on the GALPROP model of the Galactic emission [14] and an isotropic component that is meant to account for both the extragalactic diffuse component and the residual instrumental background. As the spectral shape and distribution on the sky of the Galactic component are determined by the GALPROP model, only the overall normalization of that component is allowed to vary.

4.3.2 Blind Search for GRBs

It is possible that a GRB may produce a detectable signal in the LAT data but is not detected by GBM or other instruments at lower energies. Given the current models of GRBs, this is very unlikely, and so a discovery of a GeV-only burst would either have deep implications for GRB physics or would be evidence of a new type of astrophysical source. To find such signals, the LAT data stream is analyzed to find clusters of photons in both time and sky location. Further discussion on the blind search strategy may be found in Ref. [5].

The data are time ordered and partitioned into samples of N events. The appropriate value of N depends on the mean event rate and the typical timescale of interest. For the event selection used for GRB searches in LAT data, the mean data rate is ~4 Hz. The relevant timescales for GRBs are in the range 0.1–100 s, and the typical GRB pulse duration is generally

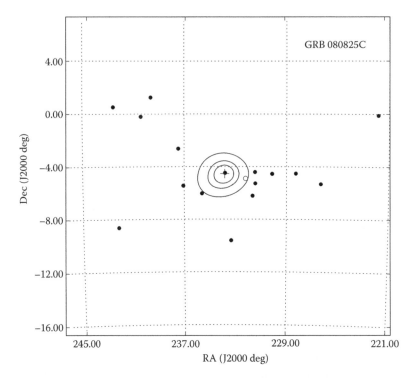

FIGURE 4.4 Error contours for the LAT data of GRB 080825C at the 68%, 90%, and 99% confidence level. The data used in the analysis were the 16 events that arrived within the time window ($T_0 + 2.4$ s, $T_0 + 26.6$ s) found by the Bayesian Blocks analysis (see Figure 4.3). The reconstructed directions of these events are plotted as points. The + sign shows the best-fit position. The event closest to the best-fit position has the highest energy of the sample of 571 MeV. Since the PSF is much narrower at high energies, the location of this event likely dominates the MLE of the GRB coordinates. The GBM location that was used as input to this analysis is plotted as the open circle.

less than a few seconds. We have chosen a partition size of $N = 30$ so that we are sensitive to single pulse variability on timescales $\lesssim 10$ s.

For each partition, we compute a clustering statistic that is based on the probabilities that the events arise from uniform distributions in time and space. The statistic is composed of two parts: the first is the sum of the logarithms of the spatial probabilities, and the second is the sum of the logarithms of the temporal probabilities. For a given candidate source position, the probability of finding an event within θ degrees is

$$p_s = (1 - \cos\theta)/2. \tag{4.1}$$

The spatial part of the clustering statistic is then

$$L_s = -\sum_{i=1}^{N} \log p_s = -\sum_{i=1}^{N} \log((1 - \cos\theta)/2). \tag{4.2}$$

Since $\log p_s \leq 0$, L_s is nonnegative. For the temporal part, we consider the probability that the arrival times of two consecutive events differ by ΔT:

$$p_t = 1 - \exp(-r_t \Delta T), \tag{4.3}$$

where $r_t = N/(t_0 - t_{N-1})$ is the mean rate at which events arrive within a given partition of N events, and t_i labels the arrival time of the ith event in the sample. The temporal part of the clustering statistic is

$$L_t = -\sum_{i=1}^{N-1} \log p_t = -\sum_{i=1}^{N-1} \log(1 - \exp(-r_t \Delta T_i)), \tag{4.4}$$

where $\Delta T_i = t_{i+1} - t_i$. A candidate burst is deemed to have occurred within a given partition if

$$L_p \equiv L_t + L_s > L_{th}. \tag{4.5}$$

The threshold, L_{th}, is found empirically from simulations and from the LAT data itself and set so that the expected false-positive rate is $<5.9 \times 10^{-7}$, which is approximately 5σ.

The estimate of the direction of the burst candidate is found via a crude clustering algorithm: For each event in the partition, the number of events within a 10° acceptance cone centered on the target event is found. The 10° cone size is approximately the 95% containment radius for the PSF at 100 MeV. For the cone with the largest number of events, the mean direction is computed simply by summing the unit vectors for each event. If more than one acceptance cone has the largest number of events, then the first cone that is found is used. Although this algorithm is fairly crude, for small numbers of events it was found to have comparable or better performance than one which fit the event distributions to the energy-dependent PSF in an unbinned maximum likelihood analysis.

The data from the LAT are organized in "runs." The start and stop times of each run correspond to either the time when the spacecraft passes through the ascending node of its orbit[*] or when it enters the South Atlantic Anomaly (SAA).[†] Therefore, run durations are less than or equal to the orbital period or about 95 min. The blind search algorithm is applied to data for each run, and burst candidates are collected. For each candidate burst found via the blind search task, the midpoint time of the sample partition is entered as the trigger time in the GCN database table and the mean direction for that partition found via the above clustering algorithm is used as the position estimate. Since ASP blind search candidates are entered in the database tables the same way as normal GCN notices, the GRB refinement tasks are run on them in the same fashion as for other bursts. Figure 4.5 shows the time history for the data run that contains data for GRB 080916C.

---

[*] The spacecraft's ascending node is the location where it crosses the Earth's equatorial plane, moving South to North.

[†] The SAA is a region of the Earth's magnetosphere located approximately off the eastern coast of South America, just below the equator. In this region, copious amounts of charged particles are trapped and cause detectors such as the LAT to trigger at very high rates. Since the LAT is turned off during SAA passages, the entry and exit points are convenient for defining the run boundaries.

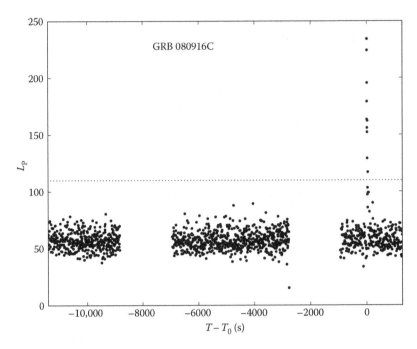

FIGURE 4.5 Time history of log-probability L_p (Equation 4.5) for the GRB blind search for the runs that contain data for GRB 080916C. The time axis is referenced to the trigger time, T_0, of this burst. The dotted horizontal line shows the $\sim 5\sigma$ threshold. There are 1271 intervals considered in these runs. The gaps correspond to SAA passages.

4.3.3 Monitored and Recurring Flaring Sources

Prior to the launch of Fermi, a list of 18 interesting sources was compiled that were to be monitored on daily and weekly bases to determine if any unusual flaring activity is seen. The flux estimates for these integration intervals are posted on a publicly available Web site,[*] and if any source rises above a predetermined threshold, the astrophysics community is notified via an Astronomer's Telegram.[†] The original list of sources was largely (and somewhat subjectively) based on previous detections by energetic gamma-ray experiment telescope (EGRET) or ground-based TeV telescopes. Therefore, in addition to the original list of monitored sources, any new object that flared above the threshold would also have its data publicly released and notifications of the flare activity would be sent out. The task of detecting new flaring sources poses the biggest challenge to ASP.

The strategy for finding new flaring sources is similar to that used to generate the LAT catalog [1]. ASP considers data in integration intervals of 6 h, 1 day, and 1 week. These are spaced roughly logarithmically, but they were chosen more to accommodate human timescales than to reflect the expected underlying variability of the sources. For each integration interval, an all-sky counts map is computed, and this map is analyzed using a continuous wavelet transform. This transformation will smooth out statistical fluctuations

---

[*] http://fermi.gsfc.nasa.gov/ssc/data/access/lat/msl_lc/
[†] http://www.astronomerstelegram.org/

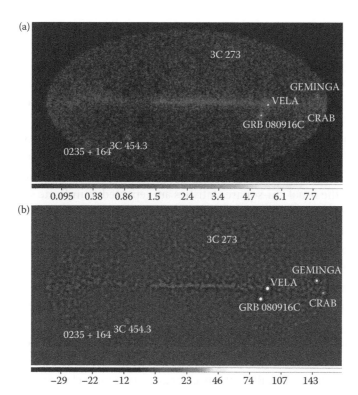

FIGURE 4.6 (**See color insert.**) (a) All-sky counts map of the data for the interval 00:00:00–06:00:00, September 16, 2008. During this interval, GRB 080916C was detected. This map is binned in an Aitoff-Hammer projection with 0.5° × 0.5° pixels at the image center. The map has been smoothed with a 1° width Gaussian kernel. The bar along the bottom edge shows the mapping between counts per pixel and color. (b) The wavelet-transformed map produced by the pgwave program. The sources found by ASP for this interval are indicated in both maps. The bright pulsars, Vela, Geminga, and Crab are easily seen, as is GRB 080916C. 0235 + 164, 3C 454.3, and 3C 273 are blazars. The three remaining sources cannot be sufficiently well localized with these data to provide a firm identification.

while enhancing the signal at the location of point sources. The transformed maps are thresholded and a source finding algorithm is applied. The details of this step of the analysis are presented in Ref. [6] and are implemented in a third-party software tool called "pgwave." See Figure 4.6.

Since the maps used in the wavelet analysis step are somewhat coarse (0.5° per side pixels for the 6 h integrations and 0.25° for the 1 day, and 1 week integrations), an additional point source localization step is required. This step proceeds similarly as the localization used in the GRB refinement analysis, except that instead of an unbinned analysis, the data are binned. In order to take advantage of the better PSF resolution at higher energies, HEALPix[*] binning is employed so that smaller pixels can be used at energies where the spatial resolution is better. The GALPROP-based Galactic diffuse model is also binned on the same scales for the selected range of energies, and a model comprising a point source, the

---

[*] http://healpix.jpl.nasa.gov/

aforementioned Galactic diffuse model and an isotropic component are fit at each candidate point source location. The location of the point source is adjusted using a gradient search. In the process of performing the HEALPix-based localization, a test statistic T_s equal to twice the log-likelihood ratio of the alternative and null hypotheses is also computed for each source. These T_s values are thresholded at approximately the 4-sigma level to determine the final set of point source candidates for a given interval.

In order to search for intra-interval variability, the event data within a $3°$ acceptance cone are histogrammed in six equal sub-bins that cover the full interval, and the exposure in each bin is computed. Under the assumption of an intrinsically constant source, the chi-square is computed, and the usual chi-square statistics are reported to indicate sources that may be varying within the time interval considered.

Finally, given the candidate source list from the wavelet- and HEALPix-based analyses, a full unbinned likelihood analysis is performed. The sources are grouped into "regions-of-interest" (ROIs) based on their proximity to each other on the sky. Within each ROI, all the source parameters, including those of the diffuse components, are fit simultaneously. Since the mean distance between sources can be significantly smaller than size of the PSF, especially at the lower energies and for the week-long intervals, it is important that neighboring sources be fit simultaneously in order to obtain accurate parameter estimates. The ROIs are of order $10–15°$ in radius and are designed to give accurate source parameter estimates for all the sources contained within. The ROIs themselves are found via a clustering algorithm similar to that used by in the first Fermi-LAT catalog [1].

Before proceeding with the ROI-based, unbinned analysis, the candidate source list must be reconciled with the predefined monitored list. This is done simply by finding the candidate source that is closest to the each of the monitored sources, and if the monitored source is within the 90% error radius, that candidate source is replaced by the monitored source using its standard coordinates.

4.4 IMPLEMENTATION AND EXECUTION

The various steps in the ASP tasks, especially for the discovery of the longer-term flaring sources, are based on the procedures that one would follow if doing the analyses by hand. Their implementation and execution in the Fermi ISOC pipeline rely on a combination of low-level analysis tools, Python scripts, batch-job scheduling software, and commercial databases. These components are sewn together using a Java-based pipeline that was developed within the ISOC and which allows for very general task definition and management of a large number of concurrent processes [7]. All the ISOC pipeline tasks, as well as the pipeline itself, were extensively tested prior to the launch of Fermi and exercised by operating on very detailed simulations of the γ-ray sky. The data were fed in under conditions designed to replicate the data flow properties that were expected to occur for the actual flight data-taking conditions. This sort of testing was essential since the final LAT data products, including the results from ASP, are required to be released to the public as soon as possible.*

---

* Typical latencies from the time the photons are detected at the spacecraft to the time they are available from the public Fermi Web site are ∼5 h on average. Considering that hundreds of CPUs on the SLAC compute farm need

A limitation of this development model is that the ASP tasks will probably only find exactly the types of phenomena they were designed and tested to find. ASP does have some sensitivity to flux variability on intermediate timescales of 10^3–10^4 s (as have been proposed for exploding Population III stars), but it is certainly not optimized for these cases. More general methods for detecting flaring sources that are not tuned to specific types of events may have some potential usefulness in an automated analysis setting such as ASP.

One possibility for addressing these sorts of limitations would be to incorporate machine learning into the implementation, where the training input would come from detailed analyses of the LAT data that has been performed by human analysts. As we noted, ASP was never intended to provide definitive characterizations of either the detection significances or the parameter estimates of the flaring sources. Hence, it was planned that human screening that includes a more careful follow-up analysis of the data would be performed using the ASP results as a starting point. For the longer timescale flares, these duties are performed by a group of "Flare Advocates" which comprises members of the LAT team's AGN and Galactic science groups. For the GRB-related follow-ups, they are performed by the joint LAT-GBM GRB science group. In a machine learning context, this sort of training feedback could in principle be useful for reducing false positives, but it is difficult to see how it could result in finding previously unanticipated signal types, since the starting points for the follow-up analysis are the ASP results themselves.

REFERENCES

1. A. A. Abdo, M. Ackermann, M. Ajello, A. Allafort, E. Antolini, W. B. Atwood, M. Axelsson et al. Fermi large area telescope first source catalog. *Astrophysical Journal Supplements*, 188:405–436, 2010.
2. A. A. Abdo, M. Ackermann, M. Arimoto, K. Asano, W. B. Atwood, M. Axelsson, L. Baldini et al. Fermi observations of high-energy gamma-ray emission from GRB 080916C. *Science*, 323:1688–1693, 2009.
3. A. A. Abdo, M. Ackermann, K. Asano, W. B. Atwood, M. Axelsson, L. Baldini, J. Ballet et al. Fermi observations of high-energy gamma-ray emission from GRB 080825C. *Astrophysical Journal*, 707:580–592, 2009.
4. W. B. Atwood, A. A. Abdo, M. Ackermann, W. Althouse, B. Anderson, M. Axelsson, L. Baldini et al. The large area telescope on the Fermi gamma-ray space telescope mission. *Astrophysical Journal*, 697:1071–1102, 2009.
5. D. L. Band, M. Axelsson, L. Baldini, G. Barbiellini, M. G. Baring, D. Bastieri, M. Battelino et al. Prospects for GRB science with the Fermi large area telescope. *Astrophysical Journal*, 701:1673–1694, 2009.
6. F. Damiani, A. Maggio, G. Micela, and S. Sciortino. A method based on wavelet transforms for source detection in photon-counting detector images. I. Theory and general properties. *Astrophysical Journal*, 483:350–369, 1997.
7. D. L. Flath, T. S. Johnson, M. Turri, and K. A. Heidenreich. GLAST (FERMI) Data-processing pipeline. In D. A. Bohlender, D. Durand, and P. Dowler, ed., *Astronomical Data Analysis Software and Systems XVIII*, Volume 411 of Astronomical Society of the Pacific Conference Series, pp. 193–196, 2009.

---

to be running continuously to process the LAT data, this is quite an achievement, of which, ASP component is a relatively small part.

8. N. Gehrels, G. Chincarini, P. Giommi, K. O. Mason, J. A. Nousek, A. A. Wells, N. E. White et al. The Swift gamma-ray burst mission. *Astrophysical Journal*, 611:1005–1020, 2004.

9. B. Jackson, J. D. Scargle, D. Barnes, S. Arabhi, A. Alt, P. Gioumousis, E. Gwin, P. Sangtrakulcharoen, L. Tan, and T. T. Tsai. An algorithm for optimal partitioning of data on an interval. *IEEE Signal Processing Letters*, 12:105–108.

10. C. Meegan, G. Lichti, P. N. Bhat, E. Bissaldi, M. S. Briggs, V. Connaughton, R. Diehl et al. The Fermi gamma-ray burst monitor. *Astrophysical Journal*, 702:791–804, 2009.

11. K. Mitsuda, M. Bautz, H. Inoue, R. L. Kelley, K. Koyama, H. Kunieda, K. Makishima et al. The x-ray observatory Suzaku. *Publication of the Astronomical Society of Japan*, 59:1–7, 2007.

12. J. A. Nelder and R. Mead. A simplex method for function minimization. *Computer Journal*, 7:308–313, 1965.

13. J. D. Scargle. Studies in astronomical time series analysis. V. Bayesian blocks, a new method to analyze structure in photon counting data. *Astrophysical Journal*, 504:405–418, 1998.

14. A. W. Strong, I. V. Moskalenko, and O. Reimer. Diffuse galactic continuum gamma rays: A model compatible with EGRET data and cosmic-ray measurements. *Astrophysical Journal*, 613:962–976, 2004.

15. B. N. Swanenburg, W. Hermsen, K. Bennett, G. F. Bignami, P. Caraveo, G. Kanbach, H. A. Mayer-Hasselwander et al. COS B observation of high-energy gamma radiation from 3C273. *Nature*, 275:298ff., September 1978.

16. K. Toma, T. Sakamoto, and P. Mészáros. Population III GRB afterglows: Constraints on stellar masses and external medium densities. *The Astrophysical Journal*, 731:127 (16pp), 2011.

17. C. Winkler, T. J.-L. Courvoisier, G. Di Cocco, N. Gehrels, A. Giménez, S. Grebenev, W. Hermsen et al. The INTEGRAL mission. *Astronomy and Astrophysics*, 411:L1–L6, 2003.

CHAPTER 5

Cosmic Microwave Background Data Analysis

Paniez Paykari and Jean-Luc Starck

CEA/DSM–CNRS–Université Paris Diderot

CONTENTS

5.1	Introduction to CMB Cosmology		56
	5.1.1	Temperature and Polarization Power Spectrum	58
5.2	The CMB Data Analysis Pipeline		60
	5.2.1	From Raw to Time-Ordered Data	60
	5.2.2	Map-Making	61
	5.2.3	From Multichannel Maps to Cosmological Parameters	61
5.3	Point Sources		63
	5.3.1	Matched Filter (MF)	63
	5.3.2	Mexican Hat Wavelet (MHW)	63
	5.3.3	PowellSnake	63
	5.3.4	SExtractor	64
	5.3.5	SZ Cluster Extraction	64
5.4	Component Separation		65
	5.4.1	Modeling the Sky Emission	65
	5.4.2	Component Separation in the Pixel Domain	66
		5.4.2.1 Template Fitting	66
		5.4.2.2 Internal Linear Combination	67
		5.4.2.3 FastICA	68
		5.4.2.4 Correlated Component Analysis (CCA)	69
	5.4.3	Component Separation in the Spherical Harmonic Domain	69
		5.4.3.1 Maximum Entropy Method (MEM)	69
		5.4.3.2 Spectral Matching ICA	70
	5.4.4	Component Separation in the Wavelet Domain	70
		5.4.4.1 Toward Wavelet-Based Methods	70
		5.4.4.2 Generalized Morphological Component Analysis (GMCA)	71
	5.4.5	Comments	71

5.5   Power Spectrum Estimation   71
      5.5.1   Low-ℓ Codes   72
              5.5.1.1   MADCAP   72
              5.5.1.2   Commander (Gibbs Sampling)   73
      5.5.2   High-ℓ Codes   73
              5.5.2.1   Monte Carlo Apodized Spherical Transform Estimator   73
              5.5.2.2   Xpol   75
              5.5.2.3   Spatially Inhomogeneous Correlation Estimator (SPICE)   76
5.6   Cosmological Parameters   76
5.7   CMB Map Statistical Analysis   78
      5.7.1   Sparse Inpainting as a Solution to the Curse of Missing Data   79
5.8   Polarization   79
Acknowledgment   82
References   82

5.1 INTRODUCTION TO CMB COSMOLOGY

About 400,000 years after the Big Bang the temperature of the Universe fell to about a few thousand degrees. As a result, the previously free electrons and protons combined and the Universe became neutral. This released a radiation which we now observe as the cosmic microwave background (CMB). The tiny fluctuations[*] in the temperature and polarization of the CMB carry a wealth of cosmological information. These so-called temperature anisotropies were predicted as the imprints of the initial density perturbations which gave rise to the present large-scale structures such as galaxies and clusters of galaxies. This relation between the present-day Universe and its initial conditions has made the CMB radiation one of the most preferred tools to understand the history of the Universe. The CMB radiation was discovered by radio astronomers Arno Penzias and Robert Wilson in 1965 [72] and earned them the 1978 Nobel Prize. This discovery was in support of the Big Bang theory and ruled out the only other available theory at that time—the steady-state theory. The crucial observations of the CMB radiation were made by the Far-Infrared Absolute Spectrophotometer (FIRAS) instrument on the Cosmic Background Explorer (COBE) satellite [86]—orbited in 1989–1996. COBE made the most accurate measurements of the CMB frequency spectrum and confirmed it as being a black-body to within experimental limits. This made the CMB spectrum the most precisely measured black-body spectrum in nature. The CMB has a thermal black-body spectrum at a temperature of 2.725 K: the spectrum peaks in the microwave range frequency of 160.2 GHz, corresponding to a 1.9 mm wavelength. The results of COBE inspired a series of ground- and balloon-based experiments, which measured CMB anisotropies on smaller scales over the next decade. During the 1990s, the first acoustic peak of the CMB power spectrum (see Figure 5.1) was measured with increasing sensitivity and by 2000 the BOOMERanG experiment [26] reported that the highest power fluctuations occur at scales of about one degree. A number of ground-based interferometers provided

---

[*] These tiny fluctuations are of the order of $O(10^{-5})$ and $O(10^{-7})$ for temperature and polarization, respectively.

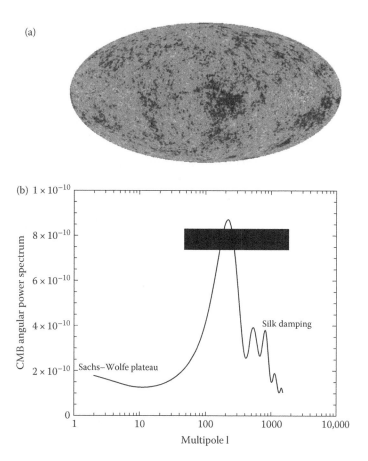

FIGURE 5.1 (**See color insert.**) (a) CMB map seen by WMAP.* (b) CMB angular power spectrum, showing the important scales discussed in the text.

measurements of the fluctuations with higher accuracy over the next three years, including the Very Small Array [16], Degree Angular Scale Interferometer (DASI) [61], and the Cosmic Background Imager (CBI) [78]. DASI was the first to detect the polarization of the CMB and the CBI provided the first E-mode polarization spectrum with compelling evidence that it is out of phase with the T-mode spectrum.

In June 2001, NASA launched its second CMB mission (after COBE), Wilkinson Microwave Anisotropy Explorer (WMAP) [44], to make much more precise measurements of the CMB sky. WMAP measured the differences in the CMB temperature across the sky creating a full-sky map of the CMB in five different frequency bands. The mission also measured the CMB's E-mode and the foreground polarization. As of October 2010, the WMAP spacecraft has ended its mission after nine years of operation. Although WMAP provided very accurate measurements of the large angular-scale fluctuations in the CMB, it did not have the angular resolution to cover the smaller–scale fluctuations that had been observed by previous ground-based interferometers. A third space mission, the Planck Surveyor [1],

---

* http://map.gsfc.nasa.gov/

was launched by ESA[*] in May 2009 to measure the CMB on smaller scales than WMAP, as well as making precise measurements of the polarization of CMB. Planck represents an advance over WMAP in several respects: it observes in higher resolution, hence allowing one to probe the CMB power spectrum to smaller scales; it has a higher sensitivity and observes in nine frequency bands rather than five, hence improving the astrophysical foreground models. The mission has a wide variety of scientific aims, including: (1) detecting the total intensity/polarization of the primordial CMB anisotropies; (2) creating a galaxy-cluster catalogue through the Sunyaev-Zel'dovich (SZ) effect [93]; (3) observing the gravitational lensing of the CMB and the integrated Sachs Wolfe (ISW) effect [82]; (4) observing bright extragalactic radio and infrared sources; (5) observing the local interstellar medium, distributed synchrotron emission, and the galactic magnetic field; (6) studying the local Solar System (planets, asteroids, comets, and the zodiacal light). Planck is expected to yield data on a number of astronomical issues by 2012. It is thought that Planck measurements will mostly be limited by the efficiency of foreground removal, rather than the detector performance or duration of the mission—this is particularly important for the polarization measurements.

Technological developments over the last two decades have accelerated the progress in observational cosmology. The interplay between the new theoretical ideas and new observational data has taken cosmology from a purely theoretical domain into a field of rigorous experimental science and we are now in what is called the precision cosmology era. The CMB measurements have made the inflationary Big Bang theory the standard model of the early Universe. This theory predicts a roughly Gaussian distribution for the initial conditions of the Universe. The power spectrum of these fluctuations agrees well with the observations, although certain observables, such as the overall amplitude of the fluctuations, remain as free parameters of the cosmic inflation model.

5.1.1 Temperature and Polarization Power Spectrum

The observable quantity is the temperature of the CMB, which can be described as

$$T(\hat{p}) = T_{\text{CMB}}[1 + \Theta(\hat{p})], \tag{5.1}$$

where $\Theta(\hat{p})$ is the temperature anisotropy in direction $\hat{p}$. This temperature field is expanded on the spherical harmonic functions with coefficients $a_{\ell m}$

$$\Theta(\hat{p}) = \sum_{\ell m} a_{\ell m} Y_{\ell m}(\hat{p}), \tag{5.2}$$

where ℓ is the multipole moment, which is related to the angular size on the sky via $\ell \sim 180°/\theta$, and m is the phase ranging from $-\ell$ to ℓ. For a Gaussian random field the average and variance carry all the information of the field. In case of $a_{\ell m}$, the average vanishes and the variance is given by

$$\langle a_{\ell m} a^*_{\ell' m'} \rangle = \delta_{\ell \ell'} \delta_{m m'} C_\ell, \tag{5.3}$$

---

[*] http://www.esa.int/SPECIALS/Planck/index.html

where C_ℓ is called the CMB angular power spectrum and it only depends on ℓ due to the isotropy assumption. This spectrum depends on the cosmological parameters through an angular transfer function $T(k, \ell)$ as

$$C_\ell = 4\pi \int \frac{dk}{k} T^2(k, \ell) P(k), \qquad (5.4)$$

where k defines the scale and $P(k)$ is the primordial power spectrum (defining the initial conditions of the Universe). Up to 1996, the angular transfer function was calculated by solving a series of coupled Boltzmann equations simultaneously—a very time–consuming process. In 1996, Seljak and Zaldarriaga [83] devised a new method for the calculation of this transfer function, improving its speed greatly—this is exactly what codes like the CMBFast [83] and CAMB [63] are based on.

The CMB power spectrum is what is used to estimate the cosmological parameters and hence accurate measurements of this spectrum from CMB experiments is the main goal of any CMB data analysis. The shape of the angular power spectrum is related to the physics of the oscillations of photons in the photon–baryon fluid at the time of recombination. The relative height and position of the peaks and troughs of the spectrum are of great importance as they are a direct impact of the cosmological parameters measured. For example, the first peak corresponds to the horizon scale at the time of recombination. It shows how far radiation has traveled since the Big Bang until the time of recombination. At angular scales above $10°$ ($\ell \lesssim 20$) main source of fluctuations is the Sachs–Wolfe effect [82]. This effect is due to the gravitational redshift of the CMB photons causing the CMB spectrum to appear uneven. Also information about the present day galaxies can be obtained by Silk damping on angular scales $\ell \gtrsim 1000$. The Silk scale corresponds to the size of galaxies of the present day. Hence, every aspect of the spectrum carries an important piece of information about cosmology and this reflects on the importance of the accurate measurements of this spectrum from CMB experiments.

Apart from the temperature anisotropy, the CMB radiation is polarized. This polarization is due to the Compton scattering at the time of recombination which thermalizes the CMB radiation. Therefore, there are three types of $a_{\ell m}$; a^E, a^B, and a^T (E-mode, B-mode, and temperature respectively), which can form six types of power spectra

$$\langle a^X_{\ell m} a^Y_{\ell' m'} \rangle = \delta_{\ell \ell'} \delta_{m m'} C^{XY}_\ell, \qquad (5.5)$$

where $X, Y \in \{T, B, E\}$. The C^{TT}_ℓ power spectrum is the temperature power spectrum obtained previously. We expect $C^{BE}_\ell = C^{BT}_\ell = 0$ as any correlation between B and T/E would correspond to parity violation at recombination. The decomposition into E and B modes are particularly helpful because scalar/density fluctuations cannot produce B modes (B modes are only produced by directed quantities such as gravitational waves or lensing). Hence, a B–type detection is a direct signature of the presence of a stochastic background of gravitational waves. This would provide an invaluable information about inflation.

5.2 THE CMB DATA ANALYSIS PIPELINE

As explained previously, the aim of all the CMB experiments is to estimate the cosmological parameters. Figure 5.2 shows a schematic illustration of the steps involved in estimating the cosmological parameters from CMB experiments. Each step involves a compression of information and hence the best techniques at each step are the ones with the least information loss. Below, we will go through the different steps shown in the figure. Needless to say that this review may not do full justice to much of the work that has been done on this topic over the years. Nonetheless, we have tried to cover as much as possible, space permitting, and at least mention all the exciting work even if all the details are not fully covered. You can also refer to Ref. [22] for shorter review on the CMB data analysis.

5.2.1 From Raw to Time-Ordered Data

The raw-measured data from the satellite are preprocessed, cleaned (e.g., by removing glitches such as cosmic ray hits), and checked for any systematic problems. Each detecter's time stream noise correlation is characterized. The result of processing the raw data is the time-ordered data set (TOD), which is simply a list of the positions and temperatures of all the pixels observed, in chronological order. For single-difference experiments, such as WMAP, the TOD consists of pairs of pixel positions and temperature difference. For more general chopping schemes, each temperature in the TOD is some linear combination of the temperature across the sky. In principle, the cosmological parameters can be measured with the smallest error bars possible by performing a brute force likelihood analysis on the TOD. However, in practice, this is numerically unfeasible for large data sets and hence an intermediate step of reducing the TOD to sky maps and then a power spectrum is necessary.

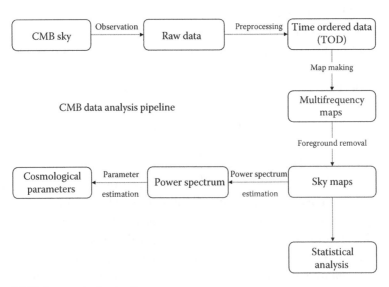

FIGURE 5.2 CMB data analysis pipeline.

5.2.2 Map-Making

The map-making process passes the cosmological information from the TOD into a much smaller data set, the map. The best map-making method should retain all the cosmological information from the TOD so that the parameters can be measured just as accurately from the map(s) as from the full TOD. By linearity, the TOD data can be written as

$$d_t = s_t + n_t = A_{tp}s_p + n_t, \tag{5.6}$$

where s_p is the temporally constant (but spatially varying) CMB signal and n_t is the temporal detector noise. The matrix A_{tp} is the pointing matrix which gives the weight of each pixel p in time sample t. This matrix is typically very sparse with normally only one nonzero entry for a total power temperature observation, two nonzero entries for a differencing temperature observation and three nonzeros for a total power polarization observation; the nonzero values in the rows correspond to the pixels being observed, at the time denoted by the row, and the columns, that typically have many nonzero entries, correspond to all the times a given pixel has been observed. The Gaussian noise likelihood function can be maximized over all possible sky signals to yield the map-making equations

$$N_{pp'}^{-1} = A_{tp}^T N_{tt'}^{-1} A_{t'p'}, \tag{5.7}$$

$$z_p = A_{tp}^T N_{tt'}^{-1} d_{t'}, \tag{5.8}$$

$$d_p = N_{pp'} z_{p'}. \tag{5.9}$$

This results in the sky maps for the different frequencies, where each frequency map is the weighted average of all the maps of the different channels at that frequency. One of the advantages of map-making is that it enables inclusion of maps at additional frequencies from other experiments. Generally, a Planck sky map will have $\sim 10^8$ pixels, while a WMAP map has $\sim 10^7$ pixels.

5.2.3 From Multichannel Maps to Cosmological Parameters

In the CMB experiments, as in many astrophysical observations, signals contain contributions from several components or sources. In efficiently designing experiments one aims to maximize the ability to subtract foregrounds and minimize the susceptibility to systematic errors. For example, incomplete sky coverage increases the sample variance (by a factor of about $1/\sqrt{\text{Covered Area}}$) and smears out features in the power spectrum. Therefore, one needs to choose an area such that the S/N per resolution element is of the order of unity or greater. Also, avoiding regions narrower than a few degrees in the smallest direction is helpful as a cold dark matter spectrum has information on scales of order $\Delta_\ell = 30$.

Generally, the foregrounds are classified into three categories; diffuse galactic emission, extragalactic sources, and solar system emission. The main emission mechanisms from our own galaxy are synchrotron radiation, free–free emission (lower frequencies), and astrophysical dust emissions (higher frequencies). The synchrotron emission is due to the relativistic electrons being accelerated in the galactic-magnetic field. The free–free emission comes from the thermal bremsstrahlung that is caused by the acceleration of hot electrons in the interstellar gas. The observed dust emission is the total emission from all the dust grains along the line of sight. The extragalactic emissions include extragalactic sources, point sources, and clusters of galaxies. To remove the foreground contamination in the CMB data, one can use prior information of the foreground signals to reduce their impact on the data: (1) For all-sky experiments the regions suspected to have significant foreground emissions are masked; for localized components in real space (like the galactic plane) one can discard or down-weigh the polluted regions. The drawbacks, however, are the resulting incomplete sky maps. (2) For ground-based/balloon-borne experiments one would observe in the regions where the contamination is minimal, for example, in directions away from the galactic plane. (3) The CMB itself dominates at a frequency of ~70–100 GHz, but for an efficient component separation a range of channels needs to be observed. Note that ground-based observations are limited by the frequency window permitted by the atmosphere. Based on the observations made at the different frequencies, an estimate of the foreground emissions can be obtained and subtracted from the observations. This procedure is such model dependent but can help to reduce the amount of cut sky.

Component separation consists of estimating a set of parameters which describe the components of interest. For example, it could be parameters describing the statistical properties such as power spectra and spectral indices. However, this is very difficult in practice and there are many methods that have been developed in order to recover the CMB from the multichannel data. The main difficulties are

- The noise is not stationary as different sections of the sky are observed different number of times, according to the scanning strategy adopted.

- The point sources cannot be considered as one template in different frequencies, as each source has its own spectrum.

- The spectral index of the emissions due to dust and synchrotron has a spatial variation.

- The maps at different channels have different resolutions with beams not necessarily isotropic or spatially invariant.

This means to estimate the CMB, first the point sources should be detected and removed (or masked) at each individual channel, then different foreground emissions are removed and a CMB map is recovered using component separation techniques. The CMB power spectrum is then computed from the map, from which the cosmological parameters are estimated. A statistical analysis is also performed on the CMB map, which aims at, for example, determining the Gaussianity of the CMB or testing if it is isotropic as predicted in the theory. Other statistics such as measuring weak gravitational lensing or detecting weak gravitational waves are also part of the analysis.

In subsequent sections, we describe in detail these different steps, and we address the specific case of polarized data.

5.3 POINT SOURCES

5.3.1 Matched Filter (MF)

This is defined as a linear filter that maximizes the amplification of the signal. Given a filter ψ (note that parts of the data that have similar shapes to the filter will be enhanced, therefore, the filter should have a similar shape to the sought signal) the filtered field is

$$w(x) = \int y(q)\psi(q)e^{-iqx}\,dq, \tag{5.10}$$

where $y(q)$ is the Fourier transform of the data $y(x) = s(x) + n(x)$. It can be shown that the field will have the following shape at the position x_0 of the signal

$$w(x_0) = 2\pi \int qs(q)\psi(q)\,dq, \tag{5.11}$$

with variance

$$\sigma_w^2 = 2\pi \int qP(q)\psi^2(q)\,dq, \tag{5.12}$$

where $P(q)$ is the power spectrum of the noise: $\langle n(q)n^*(q')\rangle = P(q)\delta^2(q - q')$, with $n(q)$ being the Fourier transform of the noise $n(x)$. The point sources are recovered by satisfying two conditions: (1) $\langle w(x_0)\rangle = A$, where A is the amplitude of the signal at position x_0; (2) minimizing the variance σ_w^2 with respect to the filter ψ.

Applying this method to first year WMAP data [12] has produced a catalog of 208 extragalactic point sources.

5.3.2 Mexican Hat Wavelet (MHW)

The MHW has been used for the detection of point sources with Gaussian profiles [43]. The MHW is the second derivative of the Gaussian function, which has the following form in Fourier space:

$$\psi(q) \propto (qR^2)\exp\left(-\frac{(qR)^2}{2}\right), \tag{5.13}$$

where R is the scale of MHW. To detect the point sources the wavelet coefficient map at a given R is studied and those wavelet coefficients above a certain threshold are identified as point sources. The reason this works is that in the wavelet coefficient map a large fraction of the background is removed while the amplitudes of the point sources are enhanced. Note that the enhancement depends on the scale R. Each image has an optimal scale that gives the maximum amplification for the point sources and this is determined from the data. The IFCA MHW filter [66] is based on this method.

5.3.3 PowellSnake

PowellSnake [23] is a Bayesian approach. In this method, the likelihood for the parameters characterizing the discrete objects is replaced by an alternative exact form, which makes it much quicker to evaluate. Rather than using Markov chain Monte Carlo (MCMC) methods to search the posterior to detect objects, the local maxima of the posterior are located in the parameter space that parametrizes the object. The maxima are located using a simultaneous multiple minimization code based on Powell's [76]. The method uses a one-dimensional minimization algorithm, which in this case, is an enhanced version of the Brent's method.[*] The end-point of each minimization is a local maximum in the posterior, which gives the optimal parameters for the detected object. A Gaussian approximation to the posterior is then constructed about the peak and the detection is either accepted or rejected based on an evidence criterion; a Bayesian model selection is performed using an approximate evidence value based on a Gaussian approximation to the peak. This Gaussian approximation also provides the covariance matrix for the derived parameters of the objects. If the detection is accepted, then the detected object is subtracted from the map before the next minimization is launched.

5.3.4 SExtractor

One of the most widely used software packages for source detection is SExtractor [15]. Its success is in its ability to deal with very large images and its robustness. The first step in SExtractor is to estimate the background accurately to avoid any biases in the flux estimation. For this, the image is partitioned into blocks, and the local sky level in each block is estimated from its histogram [17,49]. A filter is applied to the background measurements in order to correct for spurious background values. Then, for an optimized detection, the image is convolved with a filter, the shape of which should match the shape of the sought signal. After, the pixels with values larger than a threshold level, T, are considered as significant, that is, belonging to an object. The threshold level is generally chosen as $T = B + N\sigma$, where B is the background estimation at that pixel, σ is the standard deviation of the noise and N is a constant (typically 3–5). The next step is to isolate the blended objects which are connected; sources which are extremely close to each other are deblended if a saddle point is found in the intensity distribution. Spurious detections due to neighboring bright objects are cleaned, and finally the centroids of each source is determined and a photometry in an elliptical Kron aperture [15,57] is performed.

5.3.5 SZ Cluster Extraction

The thermal SZ effect [94] is a spectral distortion of the CMB blackbody spectrum. This is caused by the inverse Compton scattering of the CMB photons by hot electrons in the interstellar gas of a galaxy cluster. There is also the kinetic SZ effect which is due to the radial peculiar velocities of clusters producing secondary anisotropies in the CMB via the Doppler effect. There are techniques for extraction of the kinetic SZ effect. However, as this emission is very weak its extraction is very challenging. Therefore, we focus on the extraction of the thermal SZ effect here.

The techniques for detection of the thermal SZ effect are very similar to the point source extraction techniques. This means that the methods discussed above can easily be adapted to

---

[*] Brent method is an interpolation scheme alternating between parabolic steps and golden sections.

detect the SZ effect, by applying the correct profile of the sought signal. However, the thermal SZ effect has a very specific frequency signature that can be used to detect it—provided multifrequency observations are available. Such a multichannel matched filter was proposed in Ref. [69], and was selected by the Planck collaboration for the release of the Planck Early Sunyaev–Zeldovich Cluster Catalogue [74].

There are also other techniques that can be applied to a CMB map such as biparametric scale-adapter filter (BSAF) [67] and a Bayesian approach [47]. Briefly, BSAF uses extra information apart from just the amplitude of the point sources. In particular, it estimates the number of the maxima due to the background itself and compares it to the number of maxima due to the background plus the point sources by applying the Neyman–Pearson detector. The Bayesian approach is based on the evaluation of the posterior distribution $P(\theta|d)$ of the parameters θ gives the data d. Two different strategies are proposed for the detection of compact sources: (1) an exact approach trying to detect all objects present in the data all at the same time; (2) an iterative approach (McClean algorithm). For both methods a MCMC technique is used to explore the parameters space modeling the objects.

For the Planck Early Release Compact Source Catalogue, PowellSnake and SExtractor were the two algorithms selected as they detected the largest number of sources with high reliability at high Galactic latitude; PowellSnakes for frequencies 30–143 GHz, and SExtractor for frequencies 217–857 GHz [74].

5.4 COMPONENT SEPARATION

5.4.1 Modeling the Sky Emission

All components in a CMB sky observation are assumed to have emissions that can be separated into spatial and spectral parts so that an emission process j is written as

$$x_j(v, p) = a(v)s_j(p),\tag{5.14}$$

and the observation has the form

$$y_i(p) = \sum_j x_j(v_i, p) + n_i(p),\tag{5.15}$$

where i denotes the detector and $n_i(p)$ is the detector's noise contribution. For each component j this takes the form

$$y(p) = As(p) + n(p),\tag{5.16}$$

where A is the mixing matrix with the number of rows and columns representing the number of detectors and number of components, respectively. The problem of component separation involves inverting the mixing matrix A giving a solution such that $\hat{s} = Wy$ is as good an estimator of s as possible. This is an ill–posed problem dubbed the Blind Source

Separation problem (BSS). Possible inversion methods include:

1. A simple inversion in the case of a square and nonsingular A; $W = A^{-1}$. The solution is unbiased, but may be noisy.

2. A pseudo-inverse in case of having more channels than components; $W = [A^{\dagger}A]^{-1}A^{\dagger}$. Here nothing is known about the level of noise and signal. Note that as there is no noise weighting one bad channel could contaminate all the data after inversion. The solution is unbiased.

3. A generalized least-squares solution in case of knowing the noise correlation matrix C_n; $W = [A^{\dagger}C_n^{-1}A]^{-1}A^{\dagger}C_n^{-1}$. This is the best linear solution in the limits of high S/N. The solution is again unbiased.

4. A Wiener Solution in the case where the correlation between the sources, C_s, is known; $W = [A^{\dagger}C_n^{-1}A + C_s^{-1}]^{-1}A^{\dagger}C_n^{-1}$. One is minimizing the variance of a stochastic signal. This solution is biased, but can be debiased by multiplying the Wiener matrix by a diagonal matrix removing, for each mode, the impact of filtering. There is also a second form for the Wiener filter; $W = C_s A^{\dagger}[C_n + AC_s A^{\dagger}]^{-1}$, where in the limits of high S/N it tends to the pseudo-inverse case.

The very first approach to derive the CMB is called "template fitting" which, consists of fitting sky templates to all the non-CMB sky emissions and remove them from the maps. However, more advanced methods have been developed to tackle the BSS problem. For example, the independent component analysis (ICA) methods have been developed which mostly rely on the statistical independence of the sources. Although independence is a strong assumption, it is in many cases physically acceptable, and provides much better solutions than using a simple second-order decorrelation method, such as the principal component analysis. Most of the ICA methods, such as FastICA, assume that sources are statistically independent and non-Gaussian. However, there are also other ones, such as spectral matching ICA (SMICA), which considers the case of mixed stationary Gaussian components and goes further by taking into account the additive instrumental noise. This method works in spherical harmonic domain, which has the advantage of better control of the beams of the instrument. Sparsity was also recently proposed for component separation and several methods have been extended to work in the wavelet domain, or to explicitly use a sparsity criterion. This section reviews a few methods for component separation.

5.4.2 Component Separation in the Pixel Domain

5.4.2.1 Template Fitting

Having the foregrounds as additional components of the microwave sky, one can perform a fit of the template to the data for foreground analysis. For template vector, T, the template-corrected data has the form

$$\tilde{y} = y - \sum_j \beta_j T_j, \tag{5.17}$$

where the best-fit amplitude, β_j, for each foreground template can be obtained by minimizing $\tilde{y}^T C^{-1} \tilde{y}$, where C is the total covariance matrix for the template-corrected data $C = \langle \tilde{y} \tilde{y}^T \rangle$. The minimization leads to

$$\sum_j T_j^T C^{-1} T_j \beta_j = T_j^T C^{-1} y, \tag{5.18}$$

where $T_j^T C^{-1} T_j$ has the information about the cross-correlation between the templates themselves. This equation is equally valid in pixel space or harmonic space. However, in pixel space dealing with incomplete sky coverages is easier and, in addition, the noise is usually a diagonal matrix. On the other hand, in this space the signal covariance matrix is large and not sparse, where in harmonic space and under the assumption of Gaussianity, the signal covariance is diagonal. Although, approximating the noise as uniform and uncorrelated over the sky, one can make the noise covariance diagonal in this space too.

Template cleaning has a number of advantages, the first to be its simplicity. The technique makes full use of the spatial information in the template map, which is important for the nonstationary, highly non-Gaussian emission distribution, typical of Galactic foregrounds. It is also possible to fit multiple template maps to a single–frequency channel, where in pixel-by-pixel techniques at least one frequency channel is required to fit each foreground component. There are also disadvantages to this technique, that is, imperfect models of the templates could add systematics and non-Gaussianities to the data. Refer to Ref. [28] for a more detailed description of template fitting techniques.

Template cleaning of the COBE/FIRAS data reduced a complicated foreground by a factor of 10 by using only three spatial templates [37]. WMAP team used a more complex technique, called the internal linear combination (ILC), explained next, for their template fitting [39].

5.4.2.2 Internal Linear Combination

In this method, very little is assumed about the different components in the signal. The main component is assumed to have the same template in all the frequency bands and the observations are calibrated with respect to this component. Data y have the form

$$y_i(p) = s(p) + f_i(p) + n_i(p), \tag{5.19}$$

where i denotes the frequency channels, $f_i(p)$ and $n_i(p)$ are the foreground and noise contributions in pixel p respectively. One then looks for the solution

$$\hat{s}(p) = \sum_i w_i(p) y_i(p), \tag{5.20}$$

where the weights $w_i(p)$ maximize a certain criterion about the reconstructed estimate $\hat{s}(p)$, while keeping the component of interest unchanged, and satisfy $\sum_i w_i = 1$. The simplest case is assuming the weights are independent of p and try to minimize the variance σ^2 of the estimated map. Hence, having

$$\hat{s}(p) = s(p) + \sum_i w_i f_i(p) + \sum_i w_i n_i(p), \tag{5.21}$$

under the assumption of decorrelation between $s(p)$ and all the foregrounds or noise. The variance of the ILC map is

$$\sigma^2 = w^\dagger C w, \tag{5.22}$$

where $C = \langle yy^\dagger \rangle$ with y and w standing for vectors of elements y_i and w_i. The minimum is obtained using the Lagrange multiplier method, which has as a solution

$$w_i = \frac{\sum_j [C^{-1}]_{ij}}{\sum_{ij} [C^{-1}]_{ij}}. \tag{5.23}$$

Note that the ILC method minimizes the *total* variance of the ILC map which means the weights are strongly constrained by regions close to the galactic plane where most of the foregrounds are constrained. Away from the galactic plane and on small scales, the best linear combination for cleaning the CMB map from foregrounds and noise might be different from regions close to the galactic plane or the large scales. To improve on this, the map is decomposed in several regions and ILC is applied to them independently. The ILC method is useful when no prior information is known about the different components (in this method the only prior knowledge is the CMB behavior). Therefore, it is more of a foreground removal than a component separation technique. Prior information about the different astrophysical components, if present, can be used in efficiently removing their contributions to the CMB sky. In particular their morphology, their localization, their frequency scaling can be used in understanding their emission in the CMB sky.

5.4.2.3 FastICA

FastICA [48] solves a BSS[*] problem. The simplest mixture model takes the form $y = As$ where as before A is the mixing matrix and the entries of s are assumed to be independent random variables. Note that the independent sources can only be recovered up to a permutation and a scaling of the entries of s. Although independence is a strong assumption, it is in many cases physically plausible.

ICA methods were developed to solve the BSS problem. Algorithms for BSS and mixing matrix estimation depend on the model used for the probability distribution of the sources. In a first set of techniques, source separation is achieved in a noise-less setting, based on the non-Gaussianity of all but possibly one of the components. Most mainstream ICA techniques belong to this category, for example, FastICA. In a second set of blind techniques, the components are modeled as Gaussian processes, either stationary or nonstationary and, in a given representation, separation requires that the sources have nonproportional variance profiles. Moving to a Fourier representation, the idea is that colored components can be separated based on the diversity of their power spectra.

The FastICA technique is meant for the analysis of a combination of independent non-Gaussian sources in a noise-less setting. It is a so-called orthogonal ICA method where the

---

[*] BSS is a problem that occurs in multidimensional data processing where the overall goal is to recover unobserved sources s from a mixture of them y, without assuming any forms for the sources.

components are sought by maximizing a measure of non-Gaussianity assuming they are independent. Non-Gaussianity is assessed in FastICA using a contrast function G based on a nonlinear approximation to neg-entropy [48]. In a simple deflation scheme (for spherical data) the directions are found sequentially: a direction r of maximal non-Gaussianity is sought by maximizing

$$J_G(r) = (\mathcal{E}\{G(r^T y_{white})\} - \mathcal{E}\{G(\mu)\})^2, \tag{5.24}$$

where $\mathcal{E}$ is the expectation operator, y_{white} is the renormalized data and μ stands for centered unit variance Gaussian variable, under the constraint that r has unit norm and is orthogonal to the directions found previously. For example, the contrast function G can be $G_0(u) = (1/a)\log(\cosh(au))$, where a is a constant to be determined depending on the application [48]. A complete description of this method can be found in Ref. [48] and references therein.

5.4.2.4 Correlated Component Analysis (CCA)

This method [9] is a semiblind approach that estimates the mixing matrix on subpatches of the sky based on second-order statistics of the data. It makes no assumptions about the independence of the sources. This method adopts the commonly used models for the sources to reduce the number of parameters estimated and exploits the spatial structure of the source maps. The spatial structure of the maps are accounted for through the covariance matrices at different shifts (τ, ψ)

$$C_d(\tau, \psi) = AC_s(\tau, \psi)A^t + C_n(\tau, \psi), \tag{5.25}$$

where $C_d(\tau, \psi)$ is estimated from data and the noise covariance matrix $C_n(\tau, \psi)$ is derived from the map-making noise estimations. Then by minimizing the equation

$$\sum_{\tau, \psi} \|AC_s(\tau, \psi)A^t - [C_d(\tau, \psi) - C_n(\tau, \psi)]\|, \tag{5.26}$$

where the Frobenius norm is used, one can estimate the mixing matrix and the free parameters of the source covariance matrix. Given as estimate of C_s and C_n, the above equation can be inverted and component maps obtained via the standard inversion techniques of Wiener filtering or generalized least-squares inversion. To obtain a continuous distribution of the free parameters of the mixing matrix, CCA is applied to a large number of partially overlapping patches.

5.4.3 Component Separation in the Spherical Harmonic Domain

5.4.3.1 Maximum Entropy Method (MEM)

Having a hypothesis H in which the measured data d is a function of the underlying signal s one can follow Bayes' theorem, which tells us the posterior probability is the product of the likelihood and the prior probability divided by the evidence

$$P(s|d, H) = \frac{P(d|s, H)P(s|H)}{P(d|H)}. \tag{5.27}$$

Then following the maximum entropy principle, one uses a prior distribution which maximizes the entropy, given a set of constraints. Hobson and collaborators [46] argue that for such purposes an appropriate prior for the astrophysical components s is

$$P(s) = \exp[-\alpha S_c(s, m_u, m_v)], \tag{5.28}$$

$$S_c = \sum_j \left\{ [s_j^2 + 4m_{uj}m_{vj}]^{1/2} - m_{uj} - m_{vj} - s_j \ln \left[\frac{\left[s_j^2 + 4m_{uj}m_{vj} \right]^{1/2} + s_j}{2m_{uj}} \right] \right\}, \tag{5.29}$$

where m_u and m_v represent the astrophysical components. The MEM can be implemented in the spherical harmonic domain where the separation is performed mode-by-mode which speeds up the optimization. FastMEM is an algorithm based on this; it is a nonblind method, which means the spectral behavior of the components must be known beforehand. Further details of this method are presented in Ref. [91].

5.4.3.2 Spectral Matching ICA

This technique also solves a BSS problem, but is computationally very different from FastICA. SMICA [27] is based on spectral statistics that are localized in frequency instead of space, which are simply the spectra of the channels. For multichannel maps $y_i(p)$ one computes

$$\hat{R}_\ell = \frac{1}{2\ell + 1} \sum_m y_{\ell m} y^\dagger_{\ell m}, \tag{5.30}$$

for each ℓ and m. One then models the ensemble-average as $R_\ell = \langle \hat{R}_\ell \rangle = \sum_j R^j_\ell$ where the sum is over the components. For each component, R^j_ℓ is a function of a parameter vector θ^j, where the parameterization embodies the prior knowledge about the components. The parameters are determined by minimizing the *spectral mismatch*

$$\min_\theta \sum_\ell (2\ell + 1) K(\hat{R}_\ell | \sum_j R^j_\ell), \tag{5.31}$$

where $K(C_1|C_2)$ is a measure of mismatch between C_1 and C_2.

5.4.4 Component Separation in the Wavelet Domain

5.4.4.1 Toward Wavelet-Based Methods

Working in the pixel space has the advantage of decomposing the sky into patches and processing them independently. Having a mixing matrix per patch, one can have a better grasp of the spatial variation of the total matrix. The smaller the patches are the better the variations are captured, but the higher the noise is, and so there is a trade-off between the two. Working in the spherical harmonic domain makes the spatial variation analysis harder but has the advantage of better modeling of the beam.

Working in the wavelet space is a way to use the best of both worlds. By making use of undecimated wavelet transform, one can transform each channel into the wavelet domain, partition each wavelet band into blocks and perform a component separation per frequency band, per patch. Wavelets tend to grab the informative coherence between pixels while averaging the noise contributions, thus enhancing structures in the data. Hence, a wavelet representation often leads to a more robust noise. Following this principle, ILC and SMICA have been extended to use wavelets [35,71].

5.4.4.2 Generalized Morphological Component Analysis (GMCA)

This method again solves a BSS problem, but goes further by using the sparsity of the components in the wavelet domain [18].

Assume Φ is a signal representation (such as wavelet basis, curvelet frame, etc.) in which each source is assumed to be sparse; $s_j = \Phi\alpha_j$, where α_j are the decomposition coefficients. The sparsity of the sources means that most of the entries of α_j are equal or very close to zero. The multichannel noiseless data y can be written as

$$y = A\alpha\Phi^{\mathrm{T}}, \tag{5.32}$$

where the GMCA seeks an unmixing scheme through the estimation of A, which leads to the sparsest sources s. This is expressed by the following optimization problem written in the augmented Lagrangian form

$$\min \frac{1}{2}\|y - A\alpha\Phi^{\mathrm{T}}\|_{\mathrm{F}}^2 + \lambda \sum_j \|\alpha_j\|_p, \tag{5.33}$$

where typically $p = 0$ (or its relaxed convex version with $p = 1$) and $\|X\|_{\mathrm{F}} = (\mathrm{trace}(X^{\mathrm{T}}X))^{1/2}$ is the Frobenius norm. The details of this method is presented in Ref. [18], where it is shown that the GMCA is very robust to noise.

5.4.5 Comments

As presented above, there are many methods proposed to tackle the difficulties of component separation. They all work differently and depending on what the final scientific goal is one might perform better than the others. For example, one may work better on large scales while another could do a better job on small scales. Therefore, the "best" map could be different depending on what the goal is. A first comparison of methods has been done in Ref. [60]. Also, with the future Planck release in early 2014, we will certainly have a much better understanding of what methods work better to recover the CMB map for Planck.

5.5 POWER SPECTRUM ESTIMATION

As discussed before, if the statistical properties of the CMB fluctuations are isotropic and Gaussian all the cosmological information in a sky map is contained in its power spectrum. This means that all the information from a data set can be reduced to just a few thousand numbers, greatly facilitating parameter estimation. However, a straightforward expansion in spherical harmonics is not the best way to measure the power spectrum: (1) one always has incomplete sky coverage; (2) one wishes to give less weight to noisier pixels in order not to destroy information. Both of these facts spoil the orthogonality of the spherical harmonics. Any quadratic combination of pixels will, when appropriately normalized, measure some weighted average of the power spectrum—the weights are known as the window function. The nonorthogonality simply means that it is impossible to obtain an ideal (Kronecker delta) window function. Instead, the best you can do is to get a window function whose width is about the inverse of the smallest angular map dimension in radians, which is usually adequate for all practical purposes adequate.

To be able to use the power spectrum for estimation of the cosmological parameters, one needs to know the complete likelihood function $P(d|C_\ell(\boldsymbol{\theta}))$, where $\boldsymbol{\theta}$ are the underlying cosmological parameters. Hence, an important output of the power spectrum estimation step is a prescription for evaluation of the model spectra likelihood function. However, this evaluation is not computationally feasible at the full-map resolution and hence there are different methods to calculate the likelihood function at low and high ℓ. The low-ℓ codes use low-resolution maps (e.g., Healpix maps of $N_{\text{side}} = 8, 16$) and determine the properties of the likelihood as a function of the C_ℓ parameters using Bayesian statistics. The high-ℓ codes use unbiased frequentist estimators to form quadratic functions of the data (pixels of the map) and determine $\widehat{C_\ell}$, such that $\langle\widehat{C_\ell}\rangle = C_\ell$. Below we will summarize a few codes.

5.5.1 Low-ℓ Codes

5.5.1.1 MADCAP

MADCAP maximizes the log-likelihood function using a quasi Newton–Raphson (NR). It uses the Fisher matrix

$$F_{\ell\ell'} = \frac{1}{2}\text{trace}\left(C^{-1}\frac{\partial C}{\partial C_\ell}C^{-1}\frac{\partial C}{\partial C_{\ell'}}\right), \tag{5.34}$$

to find the location where

$$\frac{\partial \ln P(d|C_{\ell'})}{\partial C_\ell} = 0. \tag{5.35}$$

The NR iteration step involves

$$\delta C_\ell = \frac{1}{2}\sum_{\ell'}(F^{-1})_{\ell\ell'}\left[d^T C^{-1}\frac{\partial C}{\partial C_\ell}C^{-1}d - \text{trace}\left(C^{-1}\frac{\partial C}{\partial C_{\ell'}}\right)\right]. \tag{5.36}$$

In some cases, it is necessary to bin the C_ℓ to make the calculations computationally feasible. The binned power spectrum satisfies $C_\ell = \sum_{\ell\in b} C_b P_\ell$, where P_ℓ is the shape function usually taking the form $P_\ell \propto (\ell(\ell+1))^{-1}$.

BolPol is a similar quadratic maximum likelihood estimator, which is equivalent to one step of NR iteration. Present, it can handle maps of Healpix resolution $N_{\text{side}} = 32$. BoLike is another similar code which can provide likelihood-based confidence intervals. Refer to Gruppuso et al. [41], where both of these methods have been applied to WMAP data.

5.5.1.2 Commander (Gibbs Sampling)

The Commander [33] algorithm maps out the joint CMB-foreground posterior distribution by sampling. This method is very general and flexible, which means any parametric foreground model can be included in the analysis. It provides the exact joint CMB and foreground posterior distributions, from which the exact marginal CMB power spectrum and sky signal posterior distributions can be obtained.

The idea behind this technique is to draw samples from the joint density $P(C_\ell, s|d)$ and then marginalize over the signal to obtain probability density $P(C_\ell|d)$. This is because the theory of Gibbs sampling states that sampling from the conditional densities $P(s|C_\ell, d)$ and $P(C_\ell|s, d)$ will converge to sampling from the joint density $P(C_\ell, s|d)$ after the initial burn-in period. As the joint distribution is probed one can quantify joint uncertainties if desired. The map sampling process is performed in solving the following two equations simultaneously. The first is to solve for the mean field map x

$$\left[C^{-1} + \left(\sum_i A_i^T C_{n,i}^{-1} A_i \right) \right] x = \sum_i A_i^T C_{n,i}^{-1} d_i, \tag{5.37}$$

$$\left[C^{-1} + \left(\sum_i A_i^T C_{n,i}^{-1} A_i \right) \right] y = C^{-1/2} \omega^0 + \sum_i A_i^T C_{n,i}^{-1/2} \omega_i, \tag{5.38}$$

where d_i is the residual signal map i and ω_i are Gaussian white-noise maps having zero mean and unit variance. Any other component one may wish to include in the analysis will be subtracted from the data so that an actual residual map can be formed, from which the mean field map is computed. The mean field map is a generalized Wiener filtered map, meaning it is biased. For constructing an unbiased sample one must add a fluctuation map having the properties such that the sum of the two fields forms a sample from the distribution of the correct mean and covariance. The second equation above is the appropriate equation for this fluctuation map.

TEASING, a fast approximation scheme, is another method that approximates the low-ℓ likelihood using: (1) parametric models for the conditional and marginal likelihood; (2) a Gaussian copula model to assemble the marginal distribution into a joint distribution. For further information about TEASING refer to Ref. [10].

5.5.2 High-ℓ Codes

5.5.2.1 Monte Carlo Apodized Spherical Transform Estimator

Monte Carlo Apodized Spherical Transform Estimator (MASTER) [45] is a method based on a direct spherical harmonic transform (SHT) of the CMB map and allows one to implement particular properties of a given CMB experiment, such as the survey geometry, instrumental noise behavior, and so on. The unwanted contribution of the instrumental noise, any necessary alteration of either the recorded data stream or the raw map of the sky (introduced during the data analysis) can be calibrated in Monte Carlo (MC) simulations of the modeled observation and can then be removed or corrected for in the estimated power spectrum.

The harmonic mode-mode coupling, which is induced by the incomplete sky coverage, can analytically be corrected for to obtain an unbiased estimated power spectrum.

One can define a pseudo-power spectrum $\tilde{C}_\ell$ as

$$\tilde{C}_\ell = \frac{1}{2\ell + 1} \sum_{m=-\ell}^{\ell} |\tilde{a}_{\ell m}|, \tag{5.39}$$

where $2\ell + 1$ are the number of degrees of freedom and the coefficients $\tilde{a}_{\ell m}$ are defined as

$$\tilde{a}_{\ell m} = \int d\Omega \, \Theta(\Omega) W(\Omega) Y_{\ell m}^* \tag{5.40}$$

$$\approx \Omega_p \sum_p \Theta(p) W(p) Y_{\ell m}^*(p). \tag{5.41}$$

$W(\Omega)$ is a position-dependent weighting function applied to the map. The integral over the whole sky is approximated by a summation over the pixels (with surface area Ω_p) of the CMB map. The ensemble average of this spectrum is related to the full-sky angular spectrum C_ℓ by

$$\langle \tilde{C}_\ell \rangle = \sum_{\ell'} M_{\ell\ell'} F_{\ell'} B_{\ell'}^2 \langle C_{\ell'} \rangle + \langle \tilde{N}_\ell \rangle, \tag{5.42}$$

where $M_{\ell\ell'}$ describes the effect of mode-mode coupling due to the cut sky, B_ℓ is a window function taking care of the smoothing effects of the beam and finite pixel size, F_ℓ is a transfer function modeling the filtering that is applied to the data/maps, and $\langle \tilde{N}_\ell \rangle$ is the average power spectrum of the noise, which can be extracted from the actual data stream.

To reduce the correlations between the C_ℓ, which is induced by the cut sky, and also to reduce the errors on the estimated power spectrum, one can bin the power spectrum in ℓ. The binned power spectrum is $C_b = P_{b\ell} C_\ell$, where P is the binning operator. Therefore, an unbiased estimator for the power spectrum of the whole sky is given by

$$\hat{\mathcal{C}}_b = K_{bb'}^{-1} P_{b'\ell}(\tilde{C}_\ell - \langle \tilde{N}_\ell \rangle_{\text{Monte Carlo}}), \tag{5.43}$$

where $K_{bb'} = P_{b\ell} M_{\ell\ell'} F_{\ell'} B_{\ell'}^2 Q_{\ell'b}$ and $Q_{\ell b} P_{b\ell'} \langle C_{\ell'} \rangle = Q_{\ell b} \langle \mathcal{C}_b \rangle$, with an estimator of the noise having the form

$$\hat{\mathcal{N}}_b = K_{bb'}^{-1} P_{b'\ell} \langle \tilde{N}_\ell \rangle_{\text{Monte Carlo}}. \tag{5.44}$$

This method has been successfully applied to the WMAP map; in this case a hybrid method was used for the power spectrum estimation, where for $\ell \leq 32$ the spectrum is obtained using a Blackwell–Rao estimator applied to a chain of Gibbs samples and for $\ell > 32$ the spectrum is derived from the MASTER pseudo-C_ℓ quadratic estimator [11].

cROMAster is an implementation of the MASTER method extended to polarization [56] and has been applied to BOOMERang data [52,70]. CrossSpect is again a pseudo-C_ℓ [45]

estimator that computes cross–power spectra from two different detectors. XFaster [80] is again a similar method based on MASTER to estimate the pseudo-C_ℓ for temperature and polarization.

5.5.2.2 Xpol

Xpol estimates the angular power spectra by computing the cross-power spectra between different input maps of multiple detectors of the same experiment or from different instruments, calculating analytical error bars for each of the maps. The cross-power spectra are then combined by making use of a Gaussian approximation for the likelihood function. The method also computes an analytical estimate of the cross-correlation matrix from the data, avoiding any MC simulations.

For each power spectra, Xpol can estimate the cross-correlation matrix at different multipoles, from which error bars and the covariance matrix can be deduced for each cross-power spectra. In the limit of large sky coverage [31], one obtains [96]:

$$
\Xi_{\ell\ell'}^{XY,X'Y'} = \mathcal{M}_{\ell\ell_1}^{-1}(XY) \left[\frac{\mathcal{M}_{\ell_1\ell_2}^{(2)}(XX',YY')C_{\ell_1}^{XX'}C_{\ell_2}^{YY'}}{2\ell_2+1} + \frac{\mathcal{M}_{\ell_1\ell_2}^{(2)}(XY',X'Y)C_{\ell_1}^{XY'}C_{\ell_2}^{X'Y}}{2\ell_2+1} \right]
$$
$$
\times \, (\mathcal{M}_{\ell'\ell_2}^{-1}(X'Y'))^t, \tag{5.45}
$$

with

$$
\mathcal{M}_{\ell\ell'}(XY) = E_\ell^X E_\ell^Y M_{\ell\ell'}(XY),
$$
$$
\mathcal{M}_{\ell\ell'}^{(2)}(XX',YY') = E_\ell^X E_\ell^{X'} M_{\ell\ell'}^{(2)}(XX',YY') E_{\ell'}^Y E_{\ell'}^{Y'},
$$

where $X, Y \in \{T, B, E\}$; $M_{\ell\ell'}(XY)$ is the coupling kernel matrix analytically determined from sky masks for X and Y; $M^{(2)}$ is the quadratic coupling matrix determined from the mask product for X and X' correlated with the mask product for Y and Y'; $E_\ell = p_\ell B_\ell \sqrt{F_\ell}$, where p_ℓ is the transfer function describing the smoothing effect induced by the finite pixel size.

To obtain the best estimate of the power spectrum $\tilde{C}_\ell$ by combining the cross–power spectra, one needs to maximize the Gaussian–approximated likelihood function

$$
-2\ln\mathcal{L} = \sum_{XY,X'Y'} [(C_\ell^{XY} - \tilde{C}_\ell)|\Xi^{-1}|_{\ell\ell'}^{XY,X'Y'}(C_{\ell'}^{X'Y'} - \tilde{C}_{\ell'})], \tag{5.46}
$$

and the estimate of the angular power spectrum is (ignoring correlation between adjacent multipoles)

$$
\tilde{C}_\ell = \frac{1}{2} \frac{\sum_{XY,X'Y'}[|\Xi^{-1}|_{\ell\ell}^{XY,X'Y'} C_\ell^{X'Y'} + C_\ell^{XY}|\Xi^{-1}|_{\ell\ell}^{XY,X'Y'}]}{\sum_{XY,X'Y'} |\Xi^{-1}|_{\ell\ell}^{XY,X'Y'}}. \tag{5.47}
$$

This method has been used on Archeops data to estimate the CMB angular power spectrum [96] and the polarized foreground emission at the submillimeter and millimeter wavelength in Ref. [75].

5.5.2.3 Spatially Inhomogeneous Correlation Estimator (SPICE)

This method differs from the previous methods by introducing the angular correlation function of the signal at distance θ

$$\xi(\theta) = \sum_{\ell} \frac{2\ell + 1}{4\pi} C_\ell P_\ell(\theta), \tag{5.48}$$

where $P_\ell(\theta)$ is the Legendre polynomial. The ensemble average of the measured correlation function satisfies $\langle \tilde{\xi}(\theta) \rangle = \xi^W(\theta)\xi(\theta) + \xi^N(\theta)$, where $\xi^W(\theta)$ and $\xi^N(\theta)$ are the weighting and noise correlation functions, respectively [95]. The advantage of this method over MASTER is that the matrix inversion in the MASTER method (Equation 5.42) becomes a division by ξ^W. The full sky C_ℓ is then given by

$$C_\ell \equiv 2\pi \sum_i w_i \xi(\theta_i) P_\ell(\theta_i), \tag{5.49}$$

where w_i are the weights of the Gauss–Legendre quadrature.

5.6 COSMOLOGICAL PARAMETERS

Once computed from the data, the power spectrum can be used to constrain cosmological models. The power spectrum is a complicated-looking function, because it depends on virtually all cosmological parameters. Therefore, we can use an observed power spectrum to measure the cosmological parameters.

A parameter estimation with a simple χ^2 model fit to the observed power spectrum will give virtually the smallest error bars possible. There are several methods for estimating the cosmological parameters. Here we present the most popular method, MCMC simulations. In this method, the idea is to generate a random walk through the parameter space which then converges toward the most likely parameter values. This is an iterative process. At each step a sample of parameters is chosen (MC) from a proposal probability density. The likelihood for that sample is calculated and depending on the criterion (that only depends on the previous likelihood function value) the likelihood is accepted or rejected (Markov chain). This is called the Metropolis–Hastings algorithm. The code CosmoMC [62] is based on this procedure. Below, we will go through the necessary steps involved in an MCMC run for the cosmological parameter estimation.

As explained above, MCMC is a random walk in the parameter space where the probability of picking a set of parameters at any step is proportional to the posterior distribution $P(\theta|d)$:

$$P(\theta|d) = \frac{P(d|\theta)P(\theta)}{\int P(d|\theta)P(\theta)\, d\theta}, \tag{5.50}$$

where $P(d|\theta)$ is the likelihood of the data d given the parameters θ and $P(\theta)$ holds the prior knowledge about the parameters. Here is the necessary steps for each chain:

1. Start with an initial set of cosmological parameters θ_1 to compute the C_ℓ^1 and the likelihood $P_1 = P(d|\theta_1) = P(\hat{C}_\ell|C_\ell^1)$.

2. Take a random step in the parameter space obtaining a new set of parameters θ_2. Compute the C_ℓ^2 and the likelihood P_2 for this new set.

3. If $P_2/P_1 > 1$, save θ_2 as the new set of cosmological parameters and repeat step 2.

4. If $P_2/P_1 < 1$, draw a random number x from a uniform distribution from 0 and 1. If $x > P_2/P_1$, save θ_1 and return to step 2. If $x < P_2/P_1$ save θ_2 as the parameters set, that is, do as in 4.

5. Stop the chains when the convergence criterion is satisfied and the chains have enough points for a reasonable presentation of the posterior distributions.

Here we present an example of a convergence criterion that was used for WMAP first–year data analysis. Assume having M chains, each having $2N$ elements, of which only N is used. Therefore, y_i^j denotes a point in the parameter space with $i = 1, \ldots, N$ and $j = 1, \ldots, M$. The following expressions can be defined:

Mean of the chains

$$\bar{y}^j = \frac{1}{N} \sum_i y_i^j,$$

Mean of the distribution

$$\bar{y} = \frac{1}{NM} \sum_{ij} y_i^j,$$

Variance between chains

$$B_n = \frac{1}{M-1} \sum_j (\bar{y}^j - \bar{y})^2,$$

Variance within a chain

$$W = \frac{1}{M(N-1)} \sum_{ij} (y_i^j - \bar{y}^j)^2.$$

Then the quantity

$$\hat{R} = \frac{(N-1)/NW + B_n(1+1/M)}{W},$$

monitors the convergence by requiring $\hat{R} < 1$ for converged chains. A few other codes for cosmological parameter estimation include, for example, parameters for the impatient cosmologist (PICO) [36], cosmology population Monte Carlo (CosmoPMC) [54], CosmoNet [6]. Further details on MCMC and cosmological parameter estimation can be found in Refs. [42,64,81,97,98].

5.7 CMB MAP STATISTICAL ANALYSIS

The CMB observations so far have confirmed a standard model of the Universe predicting that the primordial fluctuations are a "realization" of a Gaussian random field. This means that the CMB fluctuations can be completely described by a power spectrum. However, there are several inflationary models that predict departures from Gaussianity that are detectable with the current CMB experiments. To measure the amount of non-Gaussianity, the CMB bispectrum (the three point correlation function in harmonic space) is measured.

Indeed, the non-Gaussian signatures in the CMB can be related to every fundamental questions such as the global topology of the Universe [79], topological defects such as cosmic strings [19], and multifield inflation [13], and so on. However, the non-Gaussian signatures can also have noncosmological origins; the SZ effect [93], gravitational lensing by large–scale structures [14] or the reionization of the Universe [5,24]. They may also be due to foreground emission [50] or to non-Gaussian instrumental noise and systematics [8]. All these sources of non-Gaussian signatures might have different origins and thus different statistical and morphological characteristics. Many approaches have been investigated to detect the non-Gaussian signatures: Minkowski functionals and the morphological statistics [84], the bispectrum and trispectrum [58,65], wavelet and curvelet transforms [5,25,38,87]. Describing all these methods are outside the scope of this chapter.

As the component separation cannot be perfect, some level of residual contributions, most significantly in the galactic region and at the locations of strong radio point sources will unavoidably contaminate the estimated spherical CMB maps. Therefore, it is common practice to mask out those parts of the data (e.g., using the mask shown on Figure 5.3, provided by the WMAP team) in order, for instance, to reliably assess the non-Gaussianity of the CMB field through estimated higher–order statistics (e.g., skewness, kurtosis) in various representations (e.g., wavelet, curvelet, etc.) [51] or to estimate the bi-spectrum of the CMB spatial fluctuations. But the gaps in the data thus created need to be handled properly. For most previously mentioned methods, masked data are a nightmare. Indeed, the effect of the mask on a given statistic may be much larger than the weak signal we are trying to detect.

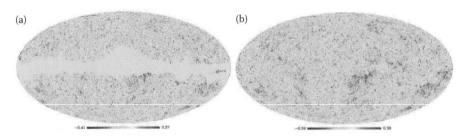

FIGURE 5.3 (**See color insert.**) (a) CMB data map provided by the WMAP team. Areas of significant foreground contamination in the galactic region and at the locations of strong radio point sources have been masked out. (b) Map obtained by applying the MCA-inpainting algorithm on the sphere to the former incomplete WMAP CMB data map.

5.7.1 Sparse Inpainting as a Solution to the Curse of Missing Data

In order to restore the *stationarity* of a partly incomplete CMB map and thus lower the impact of the missing data on the estimated measures of non-Gaussianity or on any other nonlocal statistical test, it was proposed to use an inpainting algorithm on the sphere to fill in and interpolate across the masked regions. The grounds of the inpainting scheme are in the notion of the sparsity of the data set as discussed in the component separation section.

The inpainting problem can be set out as follows: Let X be the ideal complete image, Y be the observed incomplete image, and $\mathcal{M}$ be the binary mask (i.e., $\mathcal{M}_i = 1$ if pixel i has information and $\mathcal{M}_i = 0$ otherwise), hence having $Y = \mathcal{M}X$. Inpainting then aims to recover X, knowing Y, and $\mathcal{M}$. Therefore, one aims to minimize

$$\min_{X} \quad \|\mathcal{S}X\|_0$$
$$\text{s.t.} \quad Y = \mathcal{M}X, \tag{5.51}$$

where $\mathcal{S}$ is the SHT. It was shown in Ref. [32] that this optimization problem can efficiently be solved through an iterative thresholding algorithm called morphological component analysis (MCA):

$$X^{n+1} = \Delta_{\mathcal{S}}^{\lambda_n}(X^n + Y - \mathcal{M}X^n), \tag{5.52}$$

where the nonlinear operator $\Delta_{\mathcal{S}}^{\lambda}(Z)$, (1) decomposes the signal Z onto the spherical harmonics giving the coefficients $\alpha = \mathcal{S}Z$, (2) performs hard/soft thresholding on the coefficients, and (3) reconstructs $\tilde{Z}$ from the thresholded coefficients $\tilde{\alpha}$. In the iteration the threshold parameter λ_n decreases with the iteration number working similar to the cooling parameter of the simulated annealing techniques, that is, it allows the solution to escape from local minima. More details on optimization in inpainting with sparsity can be found in Ref. [89] and the theories behind inpainting on a sphere can be found in Ref. [77].

A simple numerical experiment is shown in Figure 5.3, starting with the full-sky CMB map provided by the WMAP team. This CMB map was partially masked at the pixels where the level of contamination by residual foregrounds is high. Applying the inpainting algorithm, making use of the sparsity of the representation of CMB in the spherical harmonics domain, leads to the map shown on the right of Figure 5.3: the stationarity of the CMB field appears to have been restored. It was shown in Refs. [2,3] that inpainting the CMB map is an interesting approach for analyzing it, especially for non-Gaussianity studies and power spectrum estimation.

The sparse inpainting technique has been used in reconstruction of the CMB lensing [73] and measuring the ISW effect [30].

5.8 POLARIZATION

One of the main goals of the Planck experiment is to constrain the polarization of CMB to a scale never done before. This is of great importance as the data will help break several degeneracies that currently exist. The CMB polarization will also help probe the reionization era as the large–scale polarization, on scales of tens of degrees ($\ell > 10$), is generated at the time reionization. The analysis of the CMB polarization data are very similar to the temperature data. However, some new complications exist in their analysis due to the tensorial nature of the polarization field. In addition, the small amplitude of the CMB polarization means a more careful control of the systematics is necessary.[*] This also makes the polarization data analysis more instrument specific. In spite of the extra complications, the steps presented in the temperature pipeline above can easily be applied to the case of polarization by a simple generalization. A generalization means that the TOD for the polarization data can be written as

$$d_t = s_t + n_t \tag{5.53}$$

$$= A_{tp} s_p + n_t \tag{5.54}$$

$$= A_{tp}(I_p + Q_p \cos 2\psi_t + U_p \sin 2\psi_t) + n_t, \tag{5.55}$$

where A_{tp}, n_t are the pointing matrix and the noise, respectively, ψ_t is the angle between the direction of the detector at time t and the ϕ-direction on the sky. The Stokes parameter I, U, and Q measure the total intensity (I) and the linear polarization (Q and U). For full–sky analysis they are decomposed into spin ± 2 spherical harmonics as

$$Q(\hat{n}) \pm iU(\hat{n}) = \sum_{\ell m} a_{\pm 2, \ell m \mp 2} Y_{(\ell m)}(\hat{n}). \tag{5.56}$$

The spin ± 2 spherical harmonic coefficients, $a_{\pm 2, \ell m}$, are then decomposed into E and B modes as

$$a_{\pm 2, \ell m} = -(a_{\ell m}^E \pm i a_{\ell m}^B), \tag{5.57}$$

with $a_{\ell m}^E = (-1)^m a_{\ell -m'}^{E*}$ and $a_{\ell m}^B = (-1)^m a_{\ell -m'}^{B*}$. This decomposition introduces an extra step in the pipeline presented above, which is the $E-B$-mode separation in the map-making step. The separation can also be done in the power spectrum step, however, a map-level separation is very useful as, for example, analyzing the B-mode would be a useful diagnostic of foreground residuals or unknown systematic effects. The separation amounts to solving for the potentials P_E and P_B in the rank-2 symmetric trace-free tensor, $\mathcal{P}_{ab}$,

$$\mathcal{P}_{ab} = \nabla_{\langle a} \nabla_{b\rangle} P_E + \epsilon_{(a}^c \nabla_{b)} \nabla_c P_B, \tag{5.58}$$

where angle brackets denote the trace-free, symmetric part of the indices, ∇_a is the covariant derivative on the sphere and ϵ_{ab} is the alternating tensor. Note that the maps of the Q and U polarization are not physical quantities (i.e., are basis dependent) and are components of the polarization tensor:

$$\mathcal{P}_{ab}(\hat{n}) = \frac{1}{2} \begin{pmatrix} Q(\hat{n}) & -U(\hat{n}) \sin \theta \\ -U(\hat{n}) \sin \theta & -Q(\hat{n}) \sin^2 \theta \end{pmatrix}, \tag{5.59}$$

---

[*] The CMB radiation has a partial linear polarization with r.m.s of $\sim 6\,\mu K$, compared to $\sim 120\,\mu K$ of the temperature anisotropies.

in spherical polar coordinates. The electric and magnetic parts of $\mathcal{P}_{ab}$ in Equation 5.58 can be decomposed in spherical harmonics as

$$P_E(\hat{\boldsymbol{n}}) = \sum_{\ell m} \sqrt{\frac{(\ell - 2)!}{(\ell + 2)!}} a^E_{\ell m} Y_{\ell m}(\hat{\boldsymbol{n}}), \qquad (5.60)$$

$$P_B(\hat{\boldsymbol{n}}) = \sum_{\ell m} \sqrt{\frac{(\ell - 2)!}{(\ell + 2)!}} a^B_{\ell m} Y_{\ell m}(\hat{\boldsymbol{n}}). \qquad (5.61)$$

The separation can be done trivially for full-sky observations. However, ambiguities arise for cut-sky observations as the orthogonality of the E and B tensor harmonics break. Practical

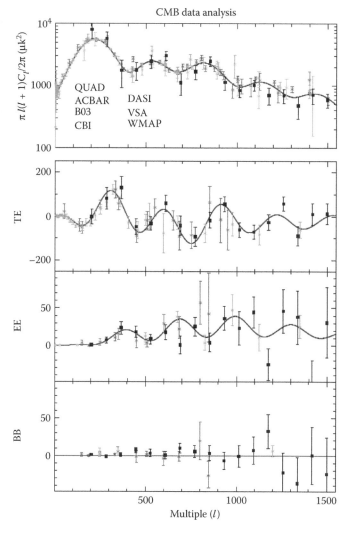

FIGURE 5.4 (**See color insert.**) CMB temperature and polarization power spectra measured by different CMB experiments, each covering different range of scales [4]. (From P. Ade et al. *APJ*, 674: 22–28, 2008. With permission.)

methods for performing the separation can be found in, for example, [20,21,55]. Also [88] have developed multiscale methods, such as polarized wavelets/curvelets, for the statistical analysis of the polarized maps.

Apart from this extra step, the techniques used in each step of the pipeline presented in the temperature case can be applied to the polarization data. For the polarized data the main diffuse polarized foregrounds are the galactic synchrotron (which is linearly polarized in a uniform magnetic field) and the thermal dust emission (which is linearly polarized if non-spherical grains are aligned in a magnetic field). These emissions have been analyzed for the case of Planck by Ref. [34]. At small angular scales extra-galactic radio sources contaminate (equally) to the E- and B-mode power and to remove their effect from the maps one simply excludes the contaminated pixels. Refer to Ref. [68] for the application of the ICA method to the polarization data. There is also the PolEMICA [7], which is the application of SMICA to the polarization data. For the maximum-likelihood component separation [92] methods applied to polarization data refer to Refs. [90] and [53], where the map–making code is called map making through destriping for anisotropy measurments (MADAM). In addition, Ref. [29] use a Metropolis-within-Gibbs MCMC method for component separation.

For the power spectrum estimation, again, the same methods are applied. For example, Ref. [59] apply the Gibbs Sampling method to the polarization data. The advantage of the exact methods such as exact likelihood analysis or Gibbs sampling is that they do not suffer from the so-called E–B coupling that exist in methods such as pseudo-C_ℓ methods [85]; the problem arises due to the unorthogonality of the spherical harmonics on a cut-sky, which may cause leakage from the E-mode power into the B-mode power spectrum. Refer to Ref. [40] for the a detailed explanation of the pseudo-Cl method. Authors of Ref. [55] apply the harmonic ILC to estimate the polarization maps and a quadratic estimator approach to estimate the power spectra.

Figure 5.4 is a summary comparison of power spectrum results for various polarization modes obtained by several observational groups, shown here with permission of the authors of Ref. [4].

ACKNOWLEDGMENT

This work has been supported by the European Research Council grant SparseAstro (ERC-228261).

REFERENCES

1. Planck Collaboration. Planck early results. I. The Planck mission. *Astronomy and Astrophysics*, 536:A1, 2011.
2. P. Abrial, Y. Moudden, J. L. Starck, J. Bobin, M. J. Fadili, B. Afeyan, and M. K. Nguyen. Morphological component analysis and inpainting on the sphere: Application in physics and astrophysics. *Journal of Fourier Analysis and Applications*, 13(6):729–748, 2007.
3. P. Abrial, Y. Moudden, J. L. Starck, M. J. Fadili, J. Delabrouille, and M. Nguyen. CMB data analysis and sparsity. *Statistical Methodology*, 5(4):289–298, 2008.
4. P. Ade, J. Bock, M. Bowden, M. L. Brown, G. Cahill, J. E. Carlstrom, P. G. Castro et al. First season QUaD CMB temperature and polarization power spectra. *APJ*, 674:22–28, 2008.

5. N. Aghanim and O. Forni. Searching for the non-Gaussian signature of the CMB secondary anisotropies. *Astronomy and Astrophysics*, 347:409–418, 1999.

6. T. Auld, M. Bridges, M. P. Hobson, and S. F. Gull. Fast cosmological parameter estimation using neural networks. *MNRAS*, 376:L11–L15, 2007.

7. J. Aumont and J. F. Macías-Pérez. Blind component separation for polarized observations of the cosmic microwave background. *MNRAS*, 376:739–758, 2007.

8. A. J. Banday, S. Zaroubi, and K. M. Górski. On the non-Gaussianity observed in the COBE differential microwave radiometer sky maps. *Astrophysical Journal*, 533:575–587, 2000.

9. L. Bedini, D. Herranz, E. Salerno, C. Baccigalupi, E. E. Kuruouglu, and A. Tonazzini. Separation of correlated astrophysical sources using multiple-lag data covariance matrices. *EURASIP Journal on Applied Signal Processing*, 2005:2400–2412, 2005.

10. K. Benabed, J.-F. Cardoso, S. Prunet, and E. Hivon. TEASING: A fast and accurate approximation for the low multipole likelihood of the cosmic microwave background temperature. *MNRAS*, 400:219–227, 2009.

11. C. L. Bennett, M. Halpern, G. Hinshaw, N. Jarosik, A. Kogut, M. Limon, S. S. Meyer et al. First year Wilkinson microwave anisotropy probe (WMAP) observations: Preliminary maps and basic results. *The Astrophysical Journal*, 148:1, 2003.

12. C. L. Bennett, R. S. Hill, G. Hinshaw, M. R. Nolta, N. Odegard, L. Page, D. N. Spergel et al. First-year Wilkinson microwave anisotropy probe (WMAP) observations: Foreground emission. *APJS*, 148:97–117, 2003.

13. F. Bernardeau and J. Uzan. Non-Gaussianity in multifield inflation. *PRD*, 66:103506–ff., 2002.

14. F. Bernardeau, L. van Waerbeke, and Y. Mellier. Patterns in the weak shear 3-point correlation function. *AAP*, 397:405–414, 2003.

15. E. Bertin and S. Arnouts. SExtractor: Software for source extraction. *Astronomy and Astrophysics, Supplement Series*, 117:393–404, 1996.

16. S. P. Bhavsar and R. J. Splinter. The superiority of the minimal spanning tree in percolation analyses of cosmological data sets. *Monthly Notices of the Royal Astronomical Society*, 282:1461–1466, 1996.

17. A. Bijaoui. Sky background estimation and application. *Astronomy and Astrophysics*, 84:81–84, 1980.

18. J. Bobin, Y. Moudden, J. L. Starck, M. J. Fadili, and N. Aghanim. SZ and CMB reconstruction using generalized morphological component analysis. *Statistical Methodology*, 5(4):307–317, 2008.

19. F. R. Bouchet, D. P. Bennett, and A. Stebbins. Patterns of the cosmic microwave background from evolving string networks. *Nature*, 335:410–414, 1988.

20. E. F. Bunn, M. Zaldarriaga, M. Tegmark, and A. de Oliveira-Costa. E/B decomposition of finite pixelized CMB maps. *PRD*, 67(2):023501–ff., 2003.

21. L. Cao and L.-Z. Fang. A wavelet-Galerkin Algorithm of the E/B decomposition of cosmic microwave background polarization maps. *APJ*, 706:1545–1555, 2009.

22. J.-F. Cardoso. Precision cosmology with the cosmic microwave background. *IEEE Signal Processing Magazine*, 27:55–66, 2010.

23. P. Carvalho, G. Rocha, and M. P. Hobson. A fast Bayesian approach to discrete object detection in astronomical data sets—PowellSnakes I. *MNRAS*, 393:681–702, 2009.

24. P. G. Castro. Bispectrum and the trispectrum of the Ostriker–Vishniac effect. *PRD*, 67:123001–ff., 2003.

25. L. Cayón, J. L. Sanz, E. Martínez-González, A. J. Banday, F. Argüeso, J. E. Gallegos, K. M. Górski, and G. Hinshaw. Spherical Mexican hat wavelet: An application to detect non-Gaussianity in the COBE-DMR maps. *Monthly Notices of the Royal Astronomical Society*, 326:1243–1248, 2001.

26. P. de Bernardis, P. A. R. Ade, J. J. Bock, J. R. Bond, J. Borrill, A. Boscaleri, K. Coble et al. A flat universe from high-resolution maps of the cosmic microwave background radiation. *NAT*, 404:955–959, 2000.

27. J. Delabrouille, J.-F. Cardoso, and G. Patanchon. Multi-detector multi-component spectral matching and applications for CMB data analysis. *Monthly Notices of the Royal Astronomical Society*, 346(4):1089–1102, 2003.

28. J. Dunkley, A. Amblard, C. Baccigalupi, M. Betoule, D. Chuss, A. Cooray, J. Delabrouille et al. Prospects for polarized foreground removal. In S. Dodelson, D. Baumann, A. Cooray, J. Dunkley, A. Fraisse, M. G. Jackson, A. Kogut, L. Krauss, M. Zaldarriaga, and K. Smith, editors, *American Institute of Physics Conference Series*, Vol. 1141, pp. 222–264, June 2009.

29. J. Dunkley, D. N. Spergel, E. Komatsu, G. Hinshaw, D. Larson, M. R. Nolta, N. Odegard et al. Five-year Wilkinson microwave anisotropy probe (WMAP) observations: Bayesian estimation of cosmic microwave background polarization maps. *APJ*, 701:1804–1813, 2009.

30. F. -X. Dupé, A. Rassat, J.-L. Starck, and M. J. Fadili. Measuring the integrated Sachs–Wolfe effect. *AAP*, 534, 2011 (Article number: A51, number of pages: 16).

31. G. Efstathiou. Myths and truths concerning estimation of power spectra: The case for a hybrid estimator. *MNRAS*, 349:603–626, 2004.

32. M. Elad, J.-L. Starck, D. L. Donoho, and P. Querre. Simultaneous cartoon and texture image inpainting using morphological component analysis (MCA). *Applied and Computational Harmonic Analysis*, 19:340–358, 2005.

33. H. K. Eriksen, J. B. Jewell, C. Dickinson, A. J. Banday, K. M. Górski, and C. R. Lawrence. Joint Bayesian component separation and CMB power spectrum estimation. *APJ*, 676:10–32, 2008.

34. L. Fauvet, J. F. Macías-Pérez, J. Aumont, F. X. Désert, T. R. Jaffe, A. J. Banday, M. Tristram, A. H. Waelkens, and D. Santos. Joint 3D modelling of the polarized Galactic synchrotron and thermal dust foreground diffuse emission. *AAP*, 526:A145ff., 2011.

35. G. Faÿ, F. Guilloux, M. Betoule, J.-F. Cardoso, J. Delabrouille, and M. Le Jeune. CMB power spectrum estimation using wavelets. *Physical Review D*, 78(8):083013–ff., 2008.

36. W. A. Fendt and B. D. Wandelt. Pico: Parameters for the impatient cosmologist. *APJ*, 654:2–11, 2007.

37. D. J. Fixsen, E. Dwek, J. C. Mather, C. L. Bennett, and R. A. Shafer. The spectrum of the extragalactic far-infrared background from the COBE FIRAS observations. *APJ*, 508:123–128, 1998.

38. O. Forni and N. Aghanim. Searching for non-Gaussianity: Statistical tests. *Astronomy and Astrophysics, Supplement Series*, 137:553–567, 1999.

39. B. Gold, N. Odegard, J. L. Weiland, R. S. Hill, A. Kogut, C. L. Bennett, G. Hinshaw et al. Seven-year Wilkinson microwave anisotropy probe (WMAP) observations: Galactic foreground emission. *APJS*, 192:15–ff., 2011.

40. J. Grain, M. Tristram, and R. Stompor. Polarized CMB power spectrum estimation using the pure pseudo-cross-spectrum approach. *PRD*, 79(12):123515–ff., 2009.

41. A. Gruppuso, A. de Rosa, P. Cabella, F. Paci, F. Finelli, P. Natoli, G. de Gasperis, and N. Mandolesi. New estimates of the CMB angular power spectra from the WMAP 5 year low-resolution data. *MNRAS*, 400:463–469, 2009.

42. A. Hajian. Efficient cosmological parameter estimation with Hamiltonian MonteCarlo technique. *PRD*, 75(8):083525–ff., 2007.

43. D. Herranz and P. Vielva. Cosmic microwave background images. *IEEE Signal Processing Magazine*, 27:67–75, 2010.

44. G. Hinshaw, J. L. Weiland, R. S. Hill, N. Odegard, D. Larson, C. L. Bennett, J. Dunkley et al. Five-year Wilkinson microwave anisotropy probe observations: Data processing, sky maps, and basic results. *APJS*, 180:225–245, 2009.

45. E. Hivon, K. M. Górski, C. B. Netterfield, B. P. Crill, S. Prunet, and F. Hansen. MASTER of the cosmic microwave background anisotropy power spectrum: A fast method for statistical analysis of large and complex cosmic microwave background data sets. *APJ*, 567:2–17, 2002.

46. M. P. Hobson, A. W. Jones, and A. N. Lasenby. Wavelet analysis and the detection of non-Gaussianity in the cosmic microwave background. *MNRAS*, 309:125–140, 1999.

47. M. P. Hobson and C. McLachlan. A Bayesian approach to discrete object detection in astronomical data sets. *MNRAS*, 338:765–784, 2003.

48. A. Hyvärinen, J. Karhunen, and E. Oja. *Independent Component Analysis*. John Wiley, New York, 2001. 481+xxii pages.

49. M. J. Irwin. Automatic analysis of crowded fields. *Monthly Notices of the Royal Astronomical Society*, 214:575–604, 1985.

50. J. Jewell. A statistical characterization of galactic dust emission as a non-Gaussian foreground of the cosmic microwave background. *Astrophysical Journal*, 557:700–713, 2001.

51. J. Jin, J.-L. Starck, D. L. Donoho, N. Aghanim, and O. Forni. Cosmological non-Gaussian signatures detection: Comparison of statistical tests. *EURASIP Journal*, 15:2470–2485, 2005.

52. W. C. Jones, P. A. R. Ade, J. J. Bock, J. R. Bond, J. Borrill, A. Boscaleri, P. Cabella et al. A measurement of the angular power spectrum of the CMB temperature anisotropy from the 2003 flight of BOOMERANG. *APJ*, 647:823–832, 2006.

53. E. Keihänen, R. Keskitalo, H. Kurki-Suonio, T. Poutanen, and A.-S. Sirviö. Making cosmic microwave background temperature and polarization maps with MADAM. *AAP*, 510:A57ff., 2010.

54. M. Kilbinger, K. Benabed, O. Cappe, J.-F. Cardoso, G. Fort, S. Prunet, C. P. Robert, and D. Wraith. CosmoPMC: Cosmology population Monte Carlo. *ArXiv e-prints*, in review, http://www.citebase.org/abstract?id=oai:arXiv.org:1101.0950

55. J. Kim, P. Naselsky, and P. R. Christensen. CMB polarization map derived from the WMAP 5 year data through the harmonic internal linear combination method. *PRD*, 79(2):023003–ff., 2009.

56. A. Kogut, D. N. Spergel, C. Barnes, C. L. Bennett, M. Halpern, G. Hinshaw, N. Jarosik et al. First-Year Wilkinson microwave anisotropy probe (WMAP) Observations: Temperature-polarization correlation. *APJS*, 148:161–173, 2003.

57. R. G. Kron. Photometry of a complete sample of faint galaxies. *ApJS*, 43:305–325, 1980.

58. M. Kunz, A. J. Banday, P. G. Castro, P. G. Ferreira, and K. M. Górski. The trispectrum of the 4 year COBE DMR data. *Astrophysical Journal Letters*, 563:L99–L102, 2001.

59. D. L. Larson, H. K. Eriksen, B. D. Wandelt, K. M. Górski, G. Huey, J. B. Jewell, and I. J. O'Dwyer. Estimation of polarized power spectra by Gibbs sampling. *APJ*, 656:653–660, 2007.

60. S. M. Leach, J.-F. Cardoso, C. Baccigalupi, R. B. Barreiro, M. Betoule, J. Bobin, A. Bonaldi et al. Component separation methods for the PLANCK mission. *AAP*, 491:597–615, 2008.

61. E. M. Leitch, J. M. Kovac, N. W. Halverson, J. E. Carlstrom, C. Pryke, and M. W. E. Smith. Degree angular scale interferometer 3 year cosmic microwave background polarization results. *APJ*, 624:10–20, 2005.

62. A. Lewis and S. Bridle. Cosmological parameters from CMB and other data: A Monte-Carlo approach. *Physical Review D*, 66:103511, 2002.

63. A. Lewis, A. Challinor, and A. Lasenby. Efficient computation of CMB anisotropies in closed FRW models. *Astrophysical Journal*, 538:473–476, 2000.

64. A. R. Liddle. Statistical methods for cosmological parameter selection and estimation. *Annual Review of Nuclear and Particle Science*, 59:95–114, 2009.

65. M. Liguori, E. Sefusatti, J. R. Fergusson, and E. P. S. Shellard. Primordial non-Gaussianity and bispectrum measurements in the cosmic microwave background and large-scale structure. *Advances in Astronomy*, 2010: 980523 (64 pages), 2010.

66. M. López-Caniego, J. González-Nuevo, D. Herranz, M. Massardi, J. L. Sanz, G. De Zotti, L. Toffolatti, and F. Argüeso. Nonblind catalog of extragalactic point sources from the Wilkinson microwave anisotropy probe (WMAP) First 3 year survey data. *ApJS*, 170:108–125, 2007.

67. M. Lopez-Caniego, D. Herranz, R. B. Barreiro, and J. L. Sanz. Some comments on the note "Some comments on the paper 'Filter design for the detection of compact sources based on the Neyman–Pearson detector' by M. Lopez-Caniego et al., (2005, *MNRAS* 359, 993)" by R. Vio and P. Andreani (astroph/0509394). *ArXiv Astrophysics e-prints*, 2005, http://arxiv.org/abs/astro-ph/0509394.

68. D. Maino, S. Donzelli, A. J. Banday, F. Stivoli, and C. Baccigalupi. Cosmic microwave background signal in *Wilkinson Microwave Anisotropy Probe* three-year data with fastica. *Monthly Notices of the Royal Astronomical Society*, 374(4): 1207–1215, 2007.

69. J.-B. Melin, J. G. Bartlett, and J. Delabrouille. Catalog extraction in SZ cluster surveys: A matched filter approach. *A&A*, 459:341–352, 2006.

70. T. E. Montroy, P. A. R. Ade, J. J. Bock, J. R. Bond, J. Borrill, A. Boscaleri, P. Cabella et al. A measurement of the CMB <EE> spectrum from the 2003 flight of BOOMERANG. *APJ*, 647:813–822, 2006.

71. Y. Moudden, J.-F. Cardoso, J.-L. Starck, and J. Delabrouille. Blind component separation in wavelet space: Application to CMB analysis. *EURASIP Journal on Applied Signal Processing*, 2005:2437–2454, 2005.

72. A. A. Penzias and R. W. Wilson. A measurement of excess antenna temperature at 4080 Mc/s. *APJ*, 142:419–421, 1965.

73. L. Perotto, J. Bobin, S. Plaszczynski, J.-L. Starck, and A. Lavabre. Reconstruction of the CMB lensing for Planck. *Astronomy and Astrophysics*, 519. A4, doi: 10.1051/0004-6361/200912001, http://resolver.caltech.edu/CaltechAUTHORS:20101201-100853519.

74. Planck Collaboration. Planck early results. VII. The early release compact source catalogue. *AAP*, 536:A7(36 pages), 2011.

75. N. Ponthieu, J. F. Macías-Pérez, M. Tristram, P. Ade, A. Amblard, R. Ansari, J. Aumont et al. Temperature and polarization angular power spectra of Galactic dust radiation at 353 GHz as measured by Archeops. *AAP*, 444:327–336, 2005.

76. W. H. Press, S. A. Teukolsky, W. T. Vetterling, and B. P. Flannery. *Numerical Recipes in FORTRAN. The Art of Scientific Computing*, Cambridge University Press. 1992, http://www.nr.com/.

77. H. Rauhut and R. Ward. Sparse Legendre expansions via ℓ_1 minimization. *ArXiv e-prints*, 2010.

78. A. C. S. Readhead, B. S. Mason, C. R. Contaldi, T. J. Pearson, J. R. Bond, S. T. Myers, S. Padin et al. Extended mosaic observations with the cosmic background imager. *APJ*, 609:498–512, 2004.

79. A. Riazuelo, J.-P. Uzan, R. Lehoucq, and J. Weeks. Simulating cosmic microwave background maps in multi-connected spaces. *PRD*, 69(10): 103514 pages, 2004.

80. G. Rocha, C. R. Contaldi, J. R. Bond, and K. M. Gorski. Application of XFaster power spectrum and likelihood estimator to Planck. *MNRAS*, 414:823–846.

81. M. Sato, K. Ichiki, and T. T. Takeuchi. Precise estimation of cosmological parameters using a more accurate likelihood function. *Physical Review Letters*, 105(25):251301–ff., 2010.

82. D. Scott, J. Silk, and M. White. From microwave anisotropies to cosmology. *Science*, 268:829–835, 1995.

83. U. Seljak and M. Zaldarriaga. A line-of-sight integration approach to cosmic microwave background anisotropies. *APJ*, 469:437–ff., 1996.

84. S. F. Shandarin. Testing non-Gaussianity in cosmic microwave background maps by morphological statistics. *Monthly Notices of the Royal Astronomical Society*, 331:865ff., 2002.

85. K. M. Smith. Pure pseudo-C estimators for CMBB—modes. *NAR*, 50:1025–1029, 2006.

86. G. F. Smoot, C. L. Bennett, A. Kogut, E. L. Wright, J. Aymon, N. W. Boggess, E. S. Cheng et al. Structure in the COBE differential microwave radiometer first-year maps. *APJ*, 396:L1–L5, 1992.

87. J.-L. Starck, N. Aghanim, and O. Forni. Detecting cosmological non-Gaussian signatures by multi-scale methods. *Astronomy and Astrophysics*, 416:9–17, 2004.

88. J.-L. Starck, Y. Moudden, and J. Bobin. Polarized wavelets and curvelets on the sphere. *AAP*, 497:931–943, 2009.

89. J.-L. Starck, F. Murtagh, and M. J. Fadili. *Sparse Image and Signal Processing*. Cambridge University Press, Cambridge, 2010.

90. F. Stivoli, J. Grain, S. M. Leach, M. Tristram, C. Baccigalupi, and R. Stompor. Maximum likelihood, parametric component separation and CMB B-mode detection in suborbital experiments. *MNRAS*, 408:2319–2335, 2010.

91. V. Stolyarov, M. P. Hobson, A. N. Lasenby, and R. B. Barreiro. All-sky component separation in the presence of anisotropic noise and dust temperature variations. *MNRAS*, 357:145–155, 2005.

92. R. Stompor, S. Leach, F. Stivoli, and C. Baccigalupi. Maximum likelihood algorithm for parametric component separation in cosmic microwave background experiments. *MNRAS*, 392:216–232, 2009.

93. R. A. Sunyaev and I. B. Zeldovich. Microwave background radiation as a probe of the contemporary structure and history of the universe. *Annual Review of Astronomy and Astrophysics*, 18:537–560, 1980.

94. R. A. Sunyaev and Y. B. Zeldovich. The observations of relic radiation as a test of the nature of x-Ray radiation from the clusters of galaxies. *Comments on Astrophysics and Space Physics*, 4:173–ff., 1972.

95. I. Szapudi, S. Prunet, and S. Colombi. Fast analysis of inhomogeneous megapixel cosmic microwave background maps. *ApJ*, 561:L11–L14, 2001

96. M. Tristram, J. F. Macías-Pérez, C. Renault, and D. Santos. XSPECT, estimation of the angular power spectrum by computing cross-power spectra with analytical error bars. *MNRAS*, 358:833–842, 2005.

97. L. Verde, H. V. Peiris, D. N. Spergel, M. R. Nolta, C. L. Bennett, M. Halpern, G. Hinshaw et al. First-year Wilkinson microwave anisotropy probe (WMAP) observations: Parameter estimation methodology. *APJS*, 148:195–211, 2003.

98. D. Wraith, M. Kilbinger, K. Benabed, O. Cappé, J.-F. Cardoso, G. Fort, S. Prunet, and C. P. Robert. Estimation of cosmological parameters using adaptive importance sampling. *PRD*, 80(2):023507–ff., 2009.

CHAPTER 6

Data Mining and Machine Learning in Time-Domain Discovery and Classification

Joshua S. Bloom and Joseph W. Richards
University of California, Berkeley

CONTENTS

6.1	Introduction	89
6.2	Discovery	91
	6.2.1 Identifying Candidates	92
	6.2.1.1 Pixelated Imaging	92
	6.2.1.2 Radio Interferometry	94
	6.2.2 Detection and Analysis of Variability	94
6.3	Classification	95
	6.3.1 Domain-Based Classification	96
	6.3.2 Feature-Based Classification	99
	6.3.2.1 Feature Creation	100
	6.3.2.2 Supervised Approaches	103
	6.3.2.3 Unsupervised and Semisupervised Approaches	105
6.4	Future Challenges	107
	Acknowledgments	108
	References	108

6.1 INTRODUCTION

The changing heavens have played a central role in the scientific effort of astronomers for centuries. Galileo's synoptic observations of the moons of Jupiter and the phases of Venus starting in 1610, provided strong refutation of Ptolemaic cosmology. These observations came soon after the discovery of Kepler's supernova had challenged the notion of an unchanging firmament. In more modern times, the discovery of a relationship between period and luminosity in some pulsational variable stars [41] led to the inference of the size of the Milky way, the distance scale to the nearest galaxies, and the expansion of the Universe (see Ref. [30] for review). Distant explosions of supernovae were used to uncover the existence of dark energy and provide a precise numerical account of dark matter (e.g., [3]). Repeat observations of pulsars [71] and nearby main-sequence stars revealed the presence of the first extrasolar planets [17,35,44,45]. Indeed, time-domain observations of transient events and variable stars, as a technique, influences a broad diversity of pursuits in the entire astronomy endeavor [68].

While, at a fundamental level, the nature of the scientific pursuit remains unchanged, the advent of astronomy as a *data-driven* discipline presents fundamental challenges to the way in which the scientific process must now be conducted. Digital images (and data cubes) are not only getting larger, there are more of them. On logistical grounds, this taxes storage and transport systems. But it also implies that the intimate connection that astronomers have always enjoyed with their data—from collection to processing to analysis to inference—necessarily must evolve. Figure 6.1 highlights some of the ways that the pathway to scientific inference is now influenced (if not driven by) modern automation processes, computing, data-mining, and machine-learning (ML).

The emerging reliance on computation and ML is a general one—a central theme of this book—but the time-domain aspect of the data and the objects of interest presents some unique challenges. First, any collection, storage, transport, and computational framework for processing the streaming data must be able to keep up with the dataflow. This is not necessarily true, for instance, with static sky science, where metrics of interest can be computed off-line and on a timescale much longer than the time required to obtain the data. Second, many types of transient (one-off) events evolve quickly in time and require more observations to fully understand the nature of the events. This demands that time-changing events are quickly discovered, classified, and broadcast to other follow-up facilities. All of this must happen robustly with, in some cases, very limited data. Last, the process of discovery and classification must be calibrated to the available resources for computation and follow-up. That is, the *precision* of classification must be weighed against the *computational cost* of producing that level of precision. Likewise, the cost of being wrong about the classification of some sorts of sources must be balanced against the scientific gains about being right about the classification of other types of sources. Quantifying these trade-offs, especially in the presence of a limited amount of follow-up resources (such as the availability of larger-telescope observations) is not straightforward and inheres domain-specific imperatives that will, in general, differ from astronomer to astronomer.

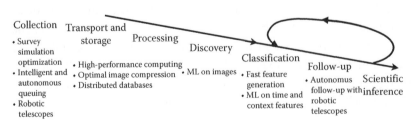

FIGURE 6.1 Data mining, computation, and ML roles in the scientific pathway.

FIGURE 6.2 (**See color insert.**) "The Astronomer," by Johannes Vermeer, c. 1668. In many ways, the epoch of the armchair astronomer is returning to primacy.

This chapter presents an overview of the current directions in ML and data-mining techniques in the context of time-domain astronomy. Ultimately the goal—if not just the necessity given the data rates and the diversity of questions to be answered—is to abstract the traditional role of astronomer in the entire scientific process. In some sense, this takes us full circle from the premodern view of the scientific pursuit presented in Vermeer's "The Astronomer" (Figure 6.2): in broad daylight, he contemplates the nighttime heavens from depictions presented to him on globe, based on observations that others have made. He is an abstract thinker, far removed from data collection and processing; his most visceral connection to the skies is just the feel of the orb under his fingers. Substitute the globe for a plot on a screen generated from an structured query language (SQL) query to a massive public database in the cloud, and we have a picture of the modern astronomer benefitting from the ML and data-mining tools operating on an almost unfathomable amount of raw data.

6.2 DISCOVERY

We take the notion of discovery, in the context of the time domain, as the recognition that data collected (e.g., a series of images of the sky) contains a source which is changing in time in some way. Classification (Section 6.3) is the quantification of the similarity of that source to other known types of variability and, by extension, the inference of *why* that source is changing. The most obvious change to discover is that of brightness or flux. On imaging data, changes in color and position might also be observed.[*] Spectroscopically, changes in emission/absorption properties and apparent velocities might also be sought. Discovery of time-variable behavior is technique specific and, as such, we will review the relevant regimes. Yip et al. [74] discuss variability discovery on spectroscopy line features in the context of active galactic nuclei. Gregory [34] presents ML-based discovery and characterization algorithms for moving source (astrometric- and Doppler-based data) in the context of exoplanets. We focus here on the discovery of brightness/flux variability.

6.2.1 Identifying Candidates

6.2.1.1 Pixelated Imaging

Many new and planned wide-field surveys are drawing attention to the need for data-mining and ML. These surveys will generate repeated images of the sky in some optical or infrared bandpass. These two-dimensional digitized images form the basic input to discovery.[†] The data from such surveys are usually obtained in a "background-limited" regime, meaning that the signal-to-noise on an exposure is dominated by the flux of sources (as the signal) and the background sky brightness (as the dominant noise component). Except in the most crowded images of the plane of the Milky Way, most pixels in the processed images contain only sky flux. Less than a few percent of pixels usually contain significant flux from stars, galaxies, or other astrophysical nebulosities.

There are two broad methods for discovering variability in such images. In one, all sources above some statistical threshold of the background noise are found and the position and flux associated with those sources are extracted to a catalog. There are off-the-shelf code-bases to do this (e.g., [6,31]) but such detection and extraction on images is by no means straightforward nor particularly rigorous, especially near the sky-noise floor. Discovery of variability is found by asking statistical questions (see Section 6.2.2) about the constancy (or otherwise) of the *light curve* produced on a given source, created by cross-correlating sources by their catalog position across different epochs [15]. The other method, called "image differencing," [1,12,66,72] takes a new image and subtracts away a "reference image" of the same portion of the sky; this reference image is generally a sharp, high signal-to-noise composite of many historical images taken with the same instrumental setup and is meant to represent an account of the "static" (unchanging) sky.

Both methods have their relative advantages and drawbacks (see Ref. [24] for a discussion). Since image differencing involves astrometric alignment and image convolution, catalog-based searches are generally considered to be faster. Moreover, catalog searches tend to produce fewer spuriously detected sources because the processed individual images tend to have less "defects" than differenced images. Catalog searches perform poorly, however, in crowded stellar fields (where aperture photometry is difficult) and in regions around

---

[*] Discovery of change in position, especially for fast-moving sources (such as asteroids), inheres its own set of data-mining challenges which we will discuss. See, for example, [38,51].

[†] In each 1 min exposure, for example, the Palomar Transient Factory [39] produces 11 images each from a 2k × 4k CCD array of size 1 sq. arcsecond (0.65 sq. degree per image). Since each pixel is 2 bytes, the amounts to 184 MB of raw data generated per minute. Raw data are preprocessed using calibration data to correct for variable gain and illumination across the arrays; spatially dependent defects in the arrays are flagged and such pixels are excluded from further scrutiny. These initial analysis steps are similar in other synoptic imaging surveys.

galaxies (where new point sources embedded in galaxy light can be easily outshined). Given the intellectual interests in finding variables in crowded fields (e.g., microlensing [9,50]) and transient events (such as supernovae and novae) near galaxies, image-difference-based discovery is considered necessary for modern surveys.

Computational costs aside, one of the primary difficulties with image differencing is the potential for a high ratio of spurious candidate events to truly astrophysical events. A trained human scanner can often discern good and bad subtractions and, for many highly successful projects, human scanners were routinely used for determining promising discovery candidates. The Katzman automatic imaging telescope (KAIT) supernova search [29] makes use of undergraduate scanners to shift through ~1000 images from the previous night. Over 1000 SNe were discovered in 10 years of operations with this methodology [40]. Basic quality/threshold cuts on the metrics about each candidate can be used to present to human scanners a smaller subset of images for inspection; in this way, the Sloan Digital Sky Survey II Supernova Search [33] netted >300 spectroscopically confirmed supernovae discoveries from ~150,000 manually scanned candidates. The Nearby Supernova Factory [2], after years of using threshold cuts, began to use boosted decision trees (Section 6.3.2.2) on metrics from image differences to optimize supernova discovery [4]. Unlike with specific domain-focused discovery surveys (like supernovae searches), many surveys are concerned with discovery and classification of all sorts of variable stars and transients. So unlike in the supernova discovery classifier of Bailey et al. [4] (which was highly tuned to finding transient events near galaxies), discovery techniques must aim to be agnostic to the physical origin of the source of variability. That is, there is an imperative to separate the notion of "discovery" and "physical classification."

In the Palomar Transient Factory, we find at least 100 high-significance bogus candidates for every one real candidate in image differences [8]. With over 1 million candidates produced nightly, the number of images that would have to be vetted by humans is unfeasible. Instead, we produced a training set of human-vetted candidates, each with dozens of measured features (such as FWHM, ellipticity; see Ref. [8]). These candidates are scored on a scale from 0 to 1 based on their inferred likelihood of being bogus or astrophysically "real." We developed a random forest (RF) classifier on the features to predict the 1–0 real-bogus value and saved the result of the ML-classifier on each candidate. These results are used to make discovery decisions in palomar transient factory (PTF). After one year of the survey, we also created training sets of real and bogus candidates by using candidates associated with known/confirmed transients and variables [47]. Figure 6.3 shows the "receiver operating characteristic" (ROC) curve for a RF classifier making use of the year-one training sample.

If all real sources occurred at just one epoch of observation, then ROC curves such as those depicted in Figure 6.3 would directly reflect discovery capabilities: type I error (false-positive rate) would be 1 minus the purity of discovery and type II error (false-negative rate) would be 1 minus the efficiency of discovery. However, most transient events occur over several epochs and bogus candidates often do not recur at precisely the same location. Therefore, turning candidate-level ROC curves to global discovery efficiency/purity quantities is not straightforward. In PTF we require two high-quality ML-score candidates within a 12-day window to qualify a certain position on the sky as a discovery of a true astrophysical

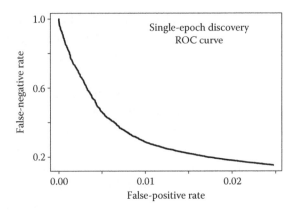

FIGURE 6.3 ROC curve for image-differenced candidates based on a training sample of 50,000 candidates from the Palomar Transient Factory. High efficiency and high purity are on the bottom left of the plot. (From S. Nehagaban et al. Image classification of transients and variables. In preparation. With permission.)

source.* In the first 8 months of producing automatic discoveries with PTF, our codebase independently discovered over 10,000 transients and variable stars.

6.2.1.2 Radio Interferometry

Traditionally, radio images are generated from raw u-v interferometric data using a human-intensive process to iteratively flag and remove spurious baseline data. Phase drift due to the ionosphere, instrumental instability, and terrestrial radio-frequency interference (RFI) are all impediments to automatically producing clean images of the sky. Given the massive data rates soon expected from wide-field surveys (e.g., LOw Frequency ARray: LOFAR; Australian Square Kilometre Array Pathfinder; ASKAP), there is a pressing need to autonomously produce clean images of the radio sky. Algorithmic innovations to speed automatic image creation have been impressive (e.g., [49]). For RFI mitigation, a genetic algorithm approach has produced promising results [32]. Once images are made, sources are detected much the same way as with optical imaging† and catalog searches are used to find transients and variables [11,19].

6.2.2 Detection and Analysis of Variability

For catalog-based searches, variability is determined on the basis of the collection of flux measurements as a function of time for a candidate source. Since variability can be manifested in many ways (such as aperiodic behavior, occasional eclipsing, etc.) one single metric on variability will not suffice in capturing variability [13,23,25,62,64]. A series of statistical questions can be asked as each new epoch is added to a light curve. Are the data consistent with an unchanging flux, in a χ^2 sense? Are there statistically significant deviant data points? How are statistical outlier flux measurements clustered in time? Significant

---

* This discovery is designed to find fast-changing events, of particular interest to the PTF collaboration. We also require at least two observations >45 min separated in time, to help remove moving asteroids from the discovery set.
† Note that McGowan et al. [46] have developed an ML approach to faint source discovery in radio images.

variability of periodic sources may be revealed by direct periodogram analysis ([60]; see also Ref. [5]). In the Poisson detection limit, such as at γ-ray wavebands or with detections of high-energy neutrinos, discovering variability in a source is akin to asking the question of whether there is a statistically significant change in the rate of arrival of individual events; for this, there are sophisticated tools (such as Bayesian blocks) for analysis [59,65]. One of the important real-world considerations is that photometric uncertainty estimates are always just estimates, based on statistical sampling of individual image characteristics. Systematic errors in this uncertainty (either too high or too low) can severely bias variability metrics (cf. [25]). Characterizing efficiency-purity from systematic errors must be done on a survey by survey basis.

6.3 CLASSIFICATION

Determining the physical origin of variability is the basic impetus of classification. But clearly what is *observed* and what is *inferred to cause that which is observed* are not the same, the latter deriving from potentially several interconnected and complex physical processes. A purely physical-based classification schema is then reliant upon subjective and potentially incorrect model interpretation. For instance, to say that the origin of variability is due to an eclipse requires an intuitive leap, however, physically relevant, from observations of a periodic dip in an otherwise constant light curve. A purely observational-based classification scheme, on the other hand, lacks the clarifying simplicity offered by physical classification. For example, how is a periodic light curve "dipping" (from an eclipsing system) different, quantitatively, than an extreme example of periodic brightness changes (say from a pulsational variable)? To this end, existing classification taxonomies tend to rely on an admixture of observational and physical statements: when a variable source is found, the goal is in finding how that source fits within an established taxonomy.

Phenomenological and theoretical taxonomies aside, the overriding conceptual challenge of classification is that no two sources in nature are identical and so the boundaries between classes (and subclasses) are inherently fuzzy: there is no ground truth in classification, regardless of the amount and quality of the data. With finite data, the logistical challenge is in extracting the most relevant information, mapping that onto the quantifiable properties derivable from instances of other variables, and finding an (abstractly construed) distance to other sources. There are several reasons for classification:

1. *Physical Interest*: Understanding the physical processes behind the diversity of variability requires numerous examples across the taxonomy. Studying the power-spectrum of variability in high signal-to-noise light curves can be used to infer the interior structure of stars (astroseismology). Modeling detached eclipsing systems can be used to infer the mass, radius, and temperatures of the binary components.

2. *Utility*: Many classes of variables have direct utility in making astrophysically important measurements that are wholly disconnected from the origin of the variability itself. Mira, RR Lyrae, and Cepheids are used for distance ladder measurements, providing

probes of the structure, and size of the universe. Calibrated standard-candle measurements of Ia and IIP supernovae are cosmographic probes of fundamental parameters. Short-period AM CVn systems serve as a strong source of "noise" for space-based gravity wave detectors; finding and characterizing these systems through optical variability allows the sources to be effectively cleaned out of the detector datastream, allowing more sensitivity searches for gravity waves in the same frequency band.

3. *Demographics*: Accounting for various biases, the demographics from classification of a large number of variable stars can be used to form and understand the evolutionary life-cycle of stars across mass and metallicity. Understanding the various ways in which high-mass stars die can be gleaned from the demographics of supernova subtypes.

4. *Rarities and Anomalies*: Finding extreme examples of objects from known classes or new examples of sparsely populated classes has the potential to inform the understanding of (more mundane) similar objects. The ability to identify anomalous systems and discover new types of variables—either hypothesized theoretically or not—is likewise an important feature of any classification system.

Expert-based (human) classification has been the traditional approach to time-series classification: a light-curve (and colors, and position on the sky, etc.) is examined and a judgement is made about class membership. The preponderance of peculiar outliers of one (historical) class may lead to a consensus that a new sub-class is warranted.[*] Again, with surveys of hundreds of thousands to billions of stars and transients, this traditional role must necessarily be replaced by ML and other data-mining techniques.

6.3.1 Domain-Based Classification

Some of the most fruitful modern approaches to classification involve domain-specific classification: using theoretical and/or empirical models of certain classes of interest to determine membership of new variables in that class. Once a source is identified as variable, its location in color-luminosity space can often provide overwhelming evidence of class membership (Figure 6.4). Hertzsprung–Russell (H–R) diagrams obviously require distance to the source to be known accurately and so, until Gaia [52], it has its utility restricted to those with parallax previously measured by the Hipparcos survey. For some sources, such as RR Lyrae and quasars, location in color–color space suffices to provide probable classification (Figure 6.5). Strict color cuts or more general probabilistic decisions on clustering[†] within a certain color–color space can be performed. (Regardless, reddening and contamination often make such classification both inefficient and impure.)

---

[*] For example, type Ia supernovae, likely due to the explosion of a white dwarf, appear qualitatively similar in their light curves to some core-collapsed supernovae from hydrogen-stripped massive stars (Type Ib/Ic). Yet the presence or absence of silicon in the spectra became the defining observation that led to very different physical inferences for similar phenomenological types of supernovae.

[†] Such classification decisions can make use of the empirical distribution of sources within a class and uncertainties on the data for a given instance [10].

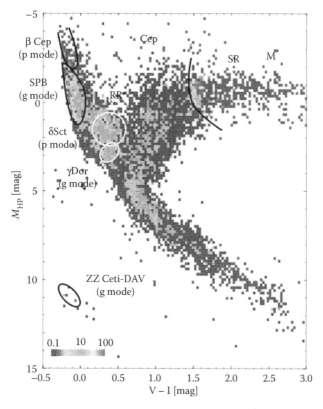

FIGURE 6.4 (**See color insert.**) Fractional variability of stars across the H–R diagram derived from Hipparcos data. Red indicates significant variability and blue low-amplitude variability (10% peak-to-peak). Identification of colors coupled with distances provide a rather clean path to classification. (From L. Eyer and N. Mowlavi. *J. Phys. Conf. Ser.*, 118(1):012010, 2008. With permission.)

Of considerable interest, given that historical and/or simultaneous color information is not always available and that unknown dust can affect color-based classification, is to classify using time-series data alone. For some domains, the light curves tell much of the story. Well before the peak brightness in a microlensing event, for example, an otherwise quiescent star will appear to brighten monotonically with a specific functional form simply described by lensing physics. By continuously fitting the light curve of (apparently) newly variable stars for such a functional form, a statistically rigorous question can be asked about whether that event appears to be microlensing or not. For a sufficiently homogeneous class of variables, an empirical light curve can be fit to the data and those sources with acceptable fits can be admitted to that class. This was done to discover and classify RR Lyrae stars in the sloan digital sky survey (SDSS) Stripe 82 dataset [61]. Such approaches require, implicitly, a threshold of acceptability. However, using cuts based on model probabilities and goodness-of-fit values can be damaging: these metrics are often a poor description of class probabilities due to the overly restricted space of template models under consideration as well as other modeling over-simplifications. A better approach is to use a representative training set of sources with known class to estimate the ROC curve for the model fits, and to then pick the threshold

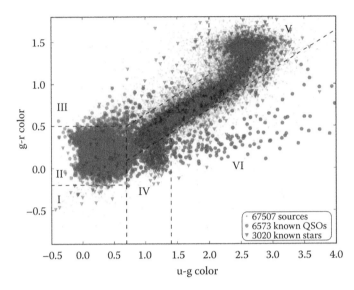

FIGURE 6.5 Color–color plot showing variable sources from Stripe 82. Region II is the traditional quasi-stellar object (quasar) (QSO) locus and Region IV is the region populated by most RR Lyrae. There are clearly many QSOs that fall outside region IV (particularly high redshift QSOs), some of which are in the RR Lyrae region. (From N. R. Butler and J. S. Bloom. *AJ*, 141:93, 2011. With permission.)

value corresponding with the desired efficiency and purity of the sample. If the training set is truly representative, this ensures a statistical guarantee of the class efficiency and purity of samples generated by this approach.

A related, but less strong statement can often be made that the variability has "class-like" variability. For example, there is no one template of a quasar light curve but since quasars are known to vary stochastically like a damped random walk [42] with some characteristic timescale that correlates only mildly with luminosity, it is possible to capture the notion of whether a given light curve is statistically consistent with such behavior. In Butler and Bloom [16] we created a set of features designed to capture how much a variable was "quasar like" and found a high degree of efficiency and purity of quasar identification based on a spectroscopic validation sample (Figure 6.6). Some variable stars, such pulsating super giants and x-ray binaries, also show this QSO-like behavior; so it is clear that such domain-specific statistical features alone cannot entirely separate classes.

There is significant utility in restricting the model fits to a restricted set of classes. Indeed, one of the most active areas of domain-specific classification is in supernova subclassing. By assuming that a source is some sort of supernovae, a large library of well-observed supernova light curves (and photometric colors) can be used to infer the sub-type of a certain instance, especially when quantifying the light curve trajectory through color–color space [53]. Provided that the library of events spans (and samples) sufficiently well the space of possible subclasses (and making use of available redshift information to transform templates appropriately), Bayesian odds ratios can be effectively used to determine membership within calibrated confidence levels (see Ref. [48]).

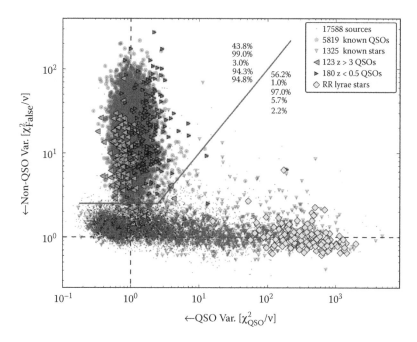

FIGURE 6.6 (**See color insert.**) Variability selection of quasars. Using a Bayesian framework to connect light curves of point sources to damped random walk behavior, statistics that account for uncertainty and covariance can be developed to find QSO-like behavior. This selection (green line) is highly efficient at finding known QSOs (∼99%) and impure at the 3% level. (From N. R. Butler and J. S. Bloom. *AJ*, 141:93, 2011. With permission.)

6.3.2 Feature-Based Classification

An abstraction from domain-specific classification (such as template fitting) is to admit that the totality of the available data encodes and reflects the true classification, irrespective of whether we understand the origin of that variability or have quantified specifically what it means to belong to a certain class. We classify on *features*, metrics derived from time-series and contextual data. There are a number of practical advantages to this transformation of the data. First, feature creation allows heterogeneous data to be mapped to a more homogeneous m-dimensional real number line space. In this space, instances of variable objects collected from different instruments with different cadences and sensitivities can be directly intercompared. This is the sort of space where ML algorithms work well, allowing us to bring to bear the richness of the ML literature to astronomical classification. Second, features may be arbitrary simple (e.g., median of the data) or complex. So in cases with only limited data availability—when, for instance, light curve fitting might fail—we have a subset of metrics that can still be useful in classification. Many ML frameworks have prescriptions for dealing with missing data that do not bias the results. Third, many feature-based classification methods produce class probabilities for each new source, and there are well-prescribed methods in ML both for calibrating the classification results and to avoid overfitting. Last, ML approaches allow us to explicitly encode the notion of loss (or "cost") in the classification process, allowing for a controlled approach to setting the efficiency and purity of the final results.

There is, of course, a huge space of possible features and many will be significantly related to others (e.g., mean and median will strongly correlate). One of the interesting advantages of some ML techniques is the classification robustness both in the face of feature covariance and "useless" features. This is freeing, at some level, allowing us to create many feature generators without worry that too many kitchen sinks will sink the boat. The flip side, however, is that there are always more features on the horizon than those in hand that could be incrementally more informative for a certain classification task.

Methods for feature-based classification of time-varying sources in astronomy come in one of two flavors. The first are *supervised* methods, which use both the features and previously known class labels from a set of training data to learn a mapping from feature to class space. The second are *unsupervised* methods (also called statistical clustering), which do not use class labels and instead seek to unveil clustering of the data in feature space. The end goals of these approaches are different: supervised classification attempts to build an accurate *predictive* model where, for new instances, the true classes (or class probabilities) can be predicted with as few errors as possible, whereas unsupervised classification seeks a characterization of the distribution of features, such as estimating the number of groups and allocation of the data points into those groups. A common technique (e.g., [26]) is to blend the two by first performing unsupervised classification and subsequently analyzing the resultant clusters with respect to a previously known set of class labels.

6.3.2.1 Feature Creation

The two broad classes of features, *time-domain* and *context*, each provide unique value to classification but also inhere unique challenges. The most straightforwardly calculated time-domain features are based on the distribution of detected fluxes, such as the various moments of the data (standard deviation, skewness, kurtosis). Variability metrics, such as χ^2 under an unchanging brightness hypothesis and the so-called Stetson variability quantities [64], are easily derived and make use of photometric uncertainties. Quantile-based measurements (such as the fraction of data observed between certain flux ranges) provide some robustness to outliers and provide a different view of the brightness distribution than moments. Inter-comparisons (e.g., ratios) of these metrics across different filters may themselves be useful metrics.

Time-ordered metrics retain phase information. Frequency analysis—finding significant periodicity in the data—provides powerful input to the classification of variable stars (Figure 6.7). There are significant limitations to frequency-domain features, most obvious of which is that a lot of time-series data is required to make meaningful statements: with three epochs of data, it makes no sense to ask what the period of the source is. Even in the limit that a frequency of interest (f_0) is potentially sampled well in a Nyquist sense (where the total time duration of the light curve is longer than $\sim 2/f_0$), the particular cadence of the observations may strongly alias the analysis, rendering significance measurements on peaks in the periodogram intractable. And unless the sources are regularly sampled (which, in general, they are not) there will be covariance across the power spectrum.

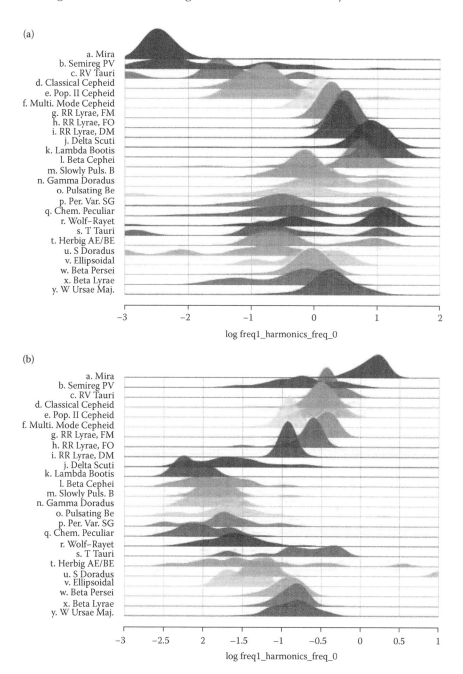

FIGURE 6.7 Distribution of two frequency-domain features derived for 25 classes of variable stars from optical gravitational lensing experiment (OGLE) and Hipparcos photometry: (a) log of the most significant frequency (units of day^{-1}) from a generalized Lomb–Scargle periodogram analysis, and (b) the log of the amplitude of the most significant period, in units of magnitude. Mira variables (top) are long-period, high-amplitude variables, while delta Scuti stars (10th from top) are short-period, low-amplitude variables. Aperiodic sources, such as S Doradus stars (5th from bottom), have a large range in effective dominant period. (From J. W. Richards et al., *ArXiv e-prints*, 1101.1959. *ApJ*, 2011, accepted. With permission.)

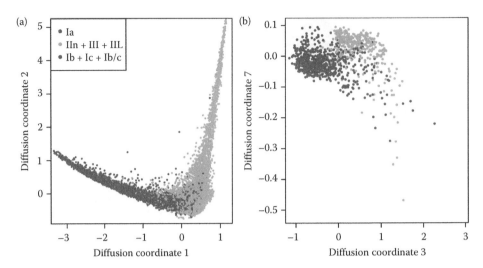

FIGURE 6.8 (**See color insert.**) Light-curve distance measures can be used in conjunction with spectral methods, such as diffusion map, to compute informative features. In this example, a spline-based distance between supernova light curves, designed to capture both shape and color differences, was used. In the first two diffusion map coordinates (a), Type Ia and II SNe are distinguished, whereas higher features (b) reveal some separation between Ia and Ib/c supernovae in the subset of spectroscopic supernovae. (From J. W. Richards et al., *ArXiv e-prints*, 1103.6034, 2011. With permission.)

Finding significant periods can mean fitting a small amount of data over millions of trial frequencies, resulting in frequency-domain features that are computationally expensive.[*] We review techniques and hybrid prescriptions for period finding and analysis in Richards et al. [56].

Other time-ordered features may be extracted using a notion of "distance" between a given instance of a light curve and all others. For instance, to derive features useful for supernova typing, Richards et al. [55] built up a matrix of pairwise distances between each pair of SNe (including both labeled and unlabeled instances) based on interpolating spline fits to the time-series measurements in each photometric band. The pairwise distance matrix was subsequently fed into a diffusion map algorithm that embeds the set of supernovae in an optimal, low-dimensional feature space, separating out the various SN subtypes (Figure 6.8). Other ways of computing distances on light curves have also been proposed. In a variable star analysis, Deb and Singh [20] use the covariance matrix of a set of interpolated, folded light curves to find features using principal components analysis (PCA), while Wachman et al. [67] use cross-correlation as a similarity function for folded light curves. In addition to using distance-based features to capture the time variability of sources, the way in which flux changes in time can be captured by fitting parameters under the assumption that the data are due to a Gaussian process [54,69].

---

[*] One practical approach for data observed with similar cadences is to compute the periodogram at a small number of a fixed set of frequencies and set the power/significance at each of these frequencies to be separate features. Covariance is then implicitly dealt with at the ML level, rather than at the feature-generation level (e.g., Ref. [26]).

We define context-specific features as being all derivable features that are not expected to change in time. The location of the event on the sky, in Galactic or ecliptic coordinates, obviously provides a strong indication of whether the event has occurred in the Galaxy or in the Solar System. Metrics on the distance to the nearest detected galaxy and the parameters of that galaxy (its color, size, inferred redshift, etc.) are crucial features for determining the nature of extragalactic events. Even with very little time-domain data a strong classification statement can be made: for example, an event well off the ecliptic plane that occurs on the apparent outskirts of a red, spiral-less galaxy is almost certainly a type Ia supernova. One of the main challenges with context features is the heterogeneity of the available data. For example, in some places on the sky, particularly in the SDSS footprint, much is known about the stars and galaxies near any given position. Outside such footprints, context information may be much more limited. From a practical standpoint, if context information is stored only in remotely queryable databases, what information is available and the time it takes to retrieve that information may be highly variable in time. This can seriously affect the computation time to produce a classification statement on a given place on the sky.

6.3.2.2 Supervised Approaches

Using a sample of light curves whose true class membership is known (e.g., via spectral confirmation), supervised classification methods learn a statistical model (known as a classifier) to predict the class of each newly observed light curve from its features. These methods are constructed to maximize the predictive accuracy of the classifications of new sources. The goal of these approaches is clear: given a set of previously labeled variables, make the best guess of the label of each new source (and optionally find the sources that do not fit within the given label taxonomy). Many supervised classification methods also predict a vector of class probabilities for each new source. These probabilistic classifiers can be used to compute ROC curves for the selection of objects from a specified science class—such as those in Figure 6.9—from which the optimal probability threshold can be chosen to create samples with desired purity and efficiency.

There are a countless number of classification methods in statistics and ML literature. Our goal here is to review a few methods that are commonly used for supervised classification of time-variable sources in astronomy.

If the class-wise distributions of features were all completely known (along with the class prior proportions), then for a new source we would use Bayes' rule to compute the exact probability that the source is from each class, and classify the source as belonging to the class of maximal probability. This is referred to as *Bayes' classifier*, and is the provable best possible classifier in terms of error rate. In practice, however, we do not know the class-wise feature distributions perfectly. Many methods attempt to estimate the class densities from the training data. In *kernel density estimation* (KDE) classification, the class-wise feature distributions are estimated using a nonparametric kernel smoother. This approach has been used to classify supernova light curves [48]. A pitfall of this technique is the tremendous difficulty in estimating accurate densities in high-dimensional feature spaces via nonparametric methods (this is referred to as the *curse of dimensionality*). To circumvent this problem, *naïve Bayes* performs class-wise KDE on one feature at a time, assuming zero covariance

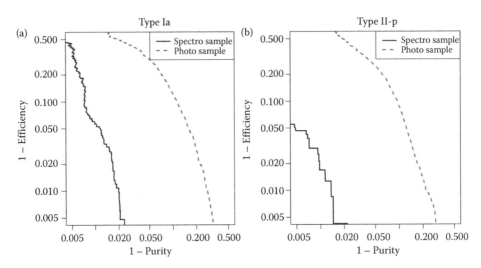

FIGURE 6.9 ROC curves for the selection of supernovae from a RF probabilistic classifier, using data from the SN Photometric Classification Challenge [36]. (a) For classification of Type Ia SNe, in the spectroscopic sample we can achieve 95% efficiency at a 99% purity or >99% efficiency at 98% purity, depending on the threshold. (b) For Type II-P supernovae, the classifier performs even better, with higher efficiency at each given purity level. (From J. W. Richards et al., *ArXiv e-prints*, 1103. 6034, 2011. With permission.)

between features. Although this simplifying assumption is unlikely to be true, naïve Bayes has enjoyed much use, including in time-domain science [43]. A step up from *naïve Bayes* is Bayesian Network classification, which assumes a sparse, graphical conditional dependence structure among the features. This approach was used with considerable success for variable star classification [21,22].

Alternatively, class-wise distributions can be estimated using parametric models. The *Gaussian mixture* classifier assumes that the feature distribution from each class follows a multivariate Gaussian distribution, where the mean and covariance of each distribution are estimated from the training data. This approach is used widely in variable star classification (e.g., [7,21,22]). The advantage of this parametric approach is that it does not suffer from curse of dimensionality. However, if the data do not really follow a mixture of multivariate Gaussian distributions, then predictions may be inaccurate: for example, we showed [56] that using the same set of variable star features, a RF classifier outperforms the Gaussian mixture classifier by a statistically significant margin. Gaussian mixture classifiers are also called *quadratic discriminant analysis* classifiers (or *linear discriminant analysis*, if pooled covariance estimates are used). These names refer to the type of boundaries that are induced between classes, in feature space.

Indeed, many classification methods instead focus on locating the optimal class boundaries. *Support vector machines* (SVMs) find the maximum-margin hyperplane to separate instances of each pair of classes. Kernelization of an SVM can easily be applied to find nonlinear class boundaries. This approach has been used to classify variable stars [21,56,70] and QSOs [37]. The *K-nearest neighbors* (KNN) classifier predicts the class of each object by voting its KNN in feature space, thereby implicitly estimating the class decision boundaries nonparametrically. Another popular method is *Classification trees*, which performs recursive binary partitioning of the feature space to arrive at a set of pure, disjoint regions. Trees are powerful classifiers because they can capture complicated class boundaries, are robust to outliers, are relatively immune to irrelevant features, and easily cope with missing feature values. Their drawback is that due to their hierarchical nature, they tend to have high variance with respect to the training set. Tree ensemble methods, such as *Bagging, Boosting*, and *RF* overcome this limitation by building many classification trees to bootstrapped versions of the training data and averaging their results. Boosting, which has been used by Newling et al. [48] for SN classification and Richards et al. [56] for variable star classification, iteratively reweights the training examples to increasingly focus on difficult-to-classify sources. RF, which was used by multiple entrants in the Supernova Photometric Classification Challenge [36] and by our group [56] for variable star classification, builds de-correlated trees by choosing a different random subset of features for each split in the tree-building process. In Richards et al. [56], we found that RF was the optimal method for a multiclass variable star problem in terms of error rate (Figure 6.10).

In time-domain classification problems, we often have a well-established hierarchical taxonomy of classes, such as the variable star taxonomy in Figure 6.11. Incorporating a known class hierarchy into a classification engine is a research field that has received much recent attention in the ML literature (e.g., [63]). Several attempts for *hierarchical classification* have been made in variable star problems. Debosscher et al. [22] use a 2-stage Gaussian mixture classifier, first classifying binaries versus nonbinaries, while Blomme et al. [7] use a multistage hierarchical taxonomy. In Richards et al. [56], we use two methods for hierarchical classification, both using RF and the taxonomy in Figure 6.11.

Finally, no discussion on supervised classification would be complete without mentioning the hugely popular method *artificial neural networks* (ANN). Although there are several versions of ANN, in their simplest form they are nonlinear regression models that predict class as a nonlinear function of linear combinations of the input features. Drawbacks to ANN are their computational difficulty (e.g., there are many local optima) and lack of interpretability, and for these reasons they have lost popularity in the statistics literature. However, they have enjoyed much success and widespread use in astronomy. In time-domain astronomy, ANNs have been used by for variable star classification [21,28,58] and by one team in the SN Classification Challenge.[*]

6.3.2.3 Unsupervised and Semisupervised Approaches

Unsupervised classification (statistical clustering) methods attempt to find k clusters of sources in feature space. These methods do not rely on any previously known class labels, and instead look for natural groupings in the data. After clusters are detected, labels or other significance can be affixed to them. In time-domain studies, these methods are useful for explorative studies, for instance to discover the number of statistically distinct classes in the data or to discover outliers and anomalous groups. In the absence of confident

---

[*] To be sure, that team's ANN entry fared much worse than their RF entry, using the same set of features.

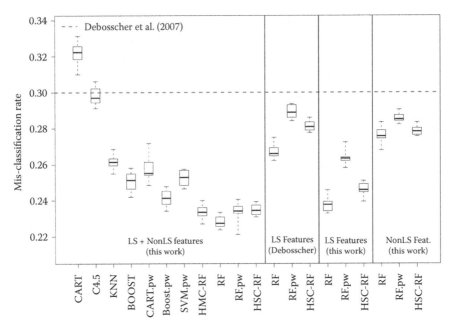

FIGURE 6.10 Distribution of cross-validation error rates for several classifiers on a mixed data set of OGLE and Hipparcos sources (see Ref. [56]). The classifiers are divided based on the features on which they were trained; from left to right: (1) periodic plus nonperiodic features, (2) the Lomb–Scargle features estimated by Debosscher et al. [21], (3) the Lomb–Scargle features estimated by Richards et al. [56], and (4) only nonperiodic features. In terms of mis-classification rate, the RF classifier trained on all of the features perform the best. Classifiers considered are: classification trees (CART and C4.5 variants), KNN, tree boosting (Boost), RF, pairwise versions of CART (CART.pw), RF.pw, and boosting (Boost.pw), pairwise SVM (SVM.pw), and two hierarchical RF classifiers [hierarchical single-label classification (HSC)-RF, hierarchical multi-label classification (HMC)-RF]. All the classifiers plotted, except for single trees, achieve better error rates than the best classifier from Debosscher et al. [21] (dashed line), who considered Bayesian network, Gaussian mixture, Artificial Neural Networks (ANN), and SVM classifiers.

training labels, an unsupervised study is a powerful way to characterize the distributions in the data to ultimately determine labels and build a predictive model using supervised classification.

In time-domain astronomy, the most popular clustering method is *Gaussian Mixture Modeling*. This method fits a parametric mixture of Gaussian distributions to the data by maximum likelihood via the expectation-maximization algorithm. A penalized likelihood or Bayesian approach can be used to estimate the number of clusters present in the data. The `Autoclass` method [18] is a Bayesian mixture model clustering method that was used by Eyer and Blake [26] to cluster variable stars from the All-Sky Automated Survey. Sarro et al. [57] use another variant of Gaussian mixture modeling to cluster a large database of variable stars.

Self-organizing maps (SOM) is another popular unsupervised learning method in time-domain astronomy. This method aims to map the high-dimensional feature vectors down to

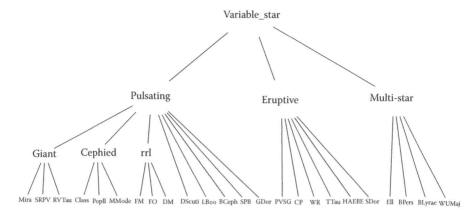

FIGURE 6.11 Variable star classification hierarchy for the problem considered in Ref. [56]. This structure can be used in a hierarchical classifier to yield improved results. The hierarchy is constructed based on knowledge of the physical processes and phenomenology of variable stars. At the top level, the sources split into three major categories: pulsating, eruptive, and multistar systems.

a discretized two-dimensional coordinate plane for easy visualization. SOM is the unsupervised analog of ANN that uses a neighborhood function to preserve the topology of the input feature space. This method has been used previously [14,73] to obtain two-dimensional parametrization of astronomical light curves. In those studies, SOM was performed prior to visual analysis of the labeled sources in this space. This class of approach, where available class labels are ignored to obtain a simple parametrization of the light-curve features and subsequently used in a learning step, is called *semisupervised learning*. The advantage to this technique is that, if the relevant class information is preserved by the unsupervised step, then supervised classification will be easier in the reduced space. Semisupervised classification permeates the time-domain astronomy literature. In addition to the aforementioned SOM studies, other authors have used PCA [20,70] and diffusion map [55] to parametrize time-variable sources prior to classification. Of these studies, only Richards et al. [55] used a statistically rigorous supervised classifier.

6.4 FUTURE CHALLENGES

For any finite collection of photons, our knowledge of the true flux is inherently uncertain. This basic phenomenological uncertainty belies an even greater uncertainty in the physical origin of what we think we are witnessing. As such, any classification scheme of a given variable or transient source must be inherently probabilistic in nature. We have outlined how—with an emerging influence from the ML literature—we can gain traction on the probabilistic classification challenge. Calibrating (and validating) the output probabilities from ML frameworks is still a nascent endeavor.

Feature generation is obviously a key ingredient to classification and we have summarized evidence that RF classifiers are particularly useful at using features that are most relevant to classification and skirting the problem of large covariance between features. On the positive side, this frees us from having to create a small set of perfectly tuned features. However, how do we know when we have exhausted the range of reasonable feature space for classification? Our suspicion is that expert knowledge has already imbued the feature creation process with much of the domain-specific insights implicitly needed for classification: we know for instance that phase offset between the first and second most dominant periods can be a powerful way to distinguish two closely related classes of pulsational variables. There may be information-theoretic (and feature-agnostic) answers to this question, which might be attacked with some genetic programming framework.

On statistical grounds, implicit in the feature generation procedure is that the distribution of features (and their covariances) on the training set will be similar to the set of instances of sources we wish to classify. A gross mismatch of the characteristics of these two sets is likely to be a significant problem for the robustness of the classification statements. No study to date has looked at how we can use the knowledge gleaned from one survey and apply that to classification in another. For instance, if a classifier is blindly trained on one survey to classify objects from another, then it will achieve suboptimal results by not considering differences in feature distribution between the surveys. Ideas from statistics, such as importance sampling and active learning, can be exploited to account for these differences.

As these very basic algorithmic questions are addressed, the computational implications of using events from real surveys will have to be understood. Is feature creation and the application of an existing machine-learned framework fast enough for a given data stream? How can loss functions be embedded in computational choices at the feature and the labeling levels? For streaming surveys, how often should the learning model be updated with newly classified examples from the survey itself? What are the roles of massively parallel hardware (e.g., graphical processing units) in feature generation, learning, and classification?

Astronomical datasets have always presented novel algorithmic, computational, and statistical challenges. With classification based on noisy and sometimes-spurious data, the forefront of all of these endeavors is already being stretched. As astronomers, expanding the ML literature is a means to an end—if not just a way to keep our heads above water—building a vital set of tools for the exploration of the vast and mysterious dynamic universe.

ACKNOWLEDGMENTS

The authors acknowledge the generous support of a CDI grant (#0941742) from the National Science Foundation. We thank Nat Butler and Dan Starr for helpful conversations. J.S.B. thanks those at the Universitat de Barcelona (Spain) for accommodating him in the astronomy department in late 2010, where much of this chapter was written.

REFERENCES

1. C. Alard and R. H. Lupton. A method for optimal image subtraction. *ApJ*, 503:325, 1998.
2. G. Aldering, G. Adam, P. Antilogus, P. Astier, R. Bacon, S. Bongard, C. Bonnaud et al. Overview of the nearby supernova factory. In J. A. Tyson and S. Wolff, editors, *Society of Photo-Optical Instrumentation Engineers (SPIE) Conference Series*, Volume 4836 of *Society of Photo-Optical Instrumentation Engineers (SPIE) Conference Series*, pp. 61–72, December 2002.
3. P. Astier, J. Guy, N. Regnault, R. Pain, E. Aubourg, D. Balam, S. Basa et al. The Supernova legacy survey: Measurement of Ω_M, Ω and w from the first year data set. *A&A*, 447:31–48, 2006.

4. S. Bailey, C. Aragon, R. Romano, R. C. Thomas, B. A. Weaver, and D. Wong. How to find more Supernovae with less work: Object classification techniques for difference imaging. *ApJ*, 665:1246–1253, 2007.

5. T. Barclay, G. Ramsay, P. Hakala, R. Napiwotzki, G. Nelemans, S. Potter, and I. Todd. Stellar variability on time-scales of minutes: Results from the first 5 years of the rapid temporal survey (RATS). *MNRAS*, 413(4): 2696–2708, 2011.

6. E. Bertin and S. Arnouts. SExtractor: Software for source extraction. *A&AS*, 117:393–404, 1996.

7. J. Blomme, J. Debosscher, J. De Ridder, C. Aerts, R. L. Gilliland, J. Christensen-Dalsgaard, H. Kjeldsen et al. Automated classification of variable stars in the asteroseismology program of the Kepler space mission. *ApJ*, 713:L204–L207, 2010.

8. J. S. Bloom, J. W. Richards, P. E. Nugent, R. M. Quimby, M. M. Kasliwal, D. L. Starr, D. Poznanski et al. Automating discovery and classification of transients and variable stars in the synoptic survey era. *PASP*, submitted for publication.

9. I. A. Bond, F. Abe, R. J. Dodd, J. B. Hearnshaw, M. Honda, J. Jugaku, P. M. Kilmartin et al. Real-time difference imaging analysis of MOA Galactic bulge observations during 2000. *MNRAS*, 327:868–880, 2001.

10. J. Bovy, J. F. Hennawi, D. W. Hogg, A. D. Myers, J. A. Kirkpatrick, D. J. Schlegel, N. P. Ross et al. Think outside the color-box: Probabilistic target selection and the SDSS-XDQSO quasar targeting catalog. *ApJ*, 729(2):141, 2011.

11. G. C. Bower, D. Saul, J. S. Bloom, A. Bolatto, A. V. Filippenko, R. J. Foley, and D. Perley. Submillijansky transients in archival radio observations. *ApJ*, 666:346–360, 2007.

12. D. M. Bramich. A new algorithm for difference image analysis. *MNRAS*, 386:L77–L81, 2008.

13. D. M. Bramich, S. Vidrih, L. Wyrzykowski, J. A. Munn, H. Lin, N. W. Evans, M. C. Smith et al. Light and motion in SDSS Stripe 82: The catalogues. *MNRAS*, 386:887–902, 2008.

14. D. R. Brett, R. G. West, and P. J. Wheatley. The automated classification of astronomical light curves using Kohonen self-organizing maps. *MNRAS*, 353:369–376, 2004.

15. T. Budavári, A. S. Szalay, and M. Nieto-Santisteban. Probabilistic cross-identification of astronomical sources. In R. W. Argyle, P. S. Bunclark, and J. R. Lewis, editors, *Astronomical Data Analysis Software and Systems XVII, Astronomical Society of the Pacific Conference Series*, Vol. 394, pp. 165, August 2008.

16. N. R. Butler and J. S. Bloom. Optimal time-series selection of quasars. *AJ*, 141:93, 2011.

17. D. Charbonneau, T. M. Brown, D. W. Latham, and M. Mayor. Detection of planetary transits across a sun-like star. *ApJ*, 529:L45–L48, 2000.

18. P. Cheeseman and J. Stutz. Bayesian classification (autoclass): Theory and results. *Adv. Knowledge Discov. Data Mining*, 180:153–180, 1996.

19. S. Croft, G. C. Bower, R. Ackermann, S. Atkinson, D. Backer, P. Backus, W. C. Barott et al. The Allen telescope array twenty-centimeter survey–A 690 deg^2, 12 Epoch radio data set. I. Catalog and long-duration transient statistics. *ApJ*, 719:45–58, 2010.

20. S. Deb and H. P. Singh. Light curve analysis of variable stars using Fourier decomposition and principal component analysis. *A&A*, 507:1729–1737, 2009.

21. J. Debosscher, L. M. Sarro, C. Aerts, J. Cuypers, B. Vandenbussche, R. Garrido, and E. Solano. Automated supervised classification of variable stars. I. Methodology. *A&A*, 475:1159–1183, 2007.

22. J. Debosscher, L. M. Sarro, M. López, M. Deleuil, C. Aerts, M. Auvergne, A. Baglin et al. Automated supervised classification of variable stars in the CoRoT programme. Method and application to the first four exoplanet fields. *A&A*, 506:519–534, 2009.

23. D. Dimitrov. Data-mining in astrophysics. A search for new variable stars in databases. *Bulg. Astronom. J.*, 12:49, 2009.

24. A. J. Drake, S. G. Djorgovski, A. Mahabal, E. Beshore, S. Larson, M. J. Graham, R. Williams et al. First results from the Catalina real-time transient survey. *ApJ*, 696:870–884, 2009.

25. L. Eyer. Variability analysis: Detection and classification. In C. Turon, K. S. O'Flaherty, and M. A. C. Perryman, editors, *The Three-Dimensional Universe with Gaia, ESA Special Publication*, Vol. 576, pp. 513ff. 2005.

26. L. Eyer and C. Blake. Automated classification of variable stars for all-sky automated survey 1-2 data. *MNRAS*, 358:30–38, 2005.

27. L. Eyer and N. Mowlavi. Variable stars across the observational HR diagram. *J. Phys. Conf. Ser.*, 118(1):012010, 2008.

28. S. M. Feeney, V. Belokurov, N. W. Evans, J. An, P. C. Hewett, M. Bode, M. Darnley et al. Automated detection of classical novae with neural networks. *AJ*, 130:84–94, 2005.

29. A. V. Filippenko, W. D. Li, R. R. Treffers, and M. Modjaz. The Lick observatory Supernova search with the Katzman automatic imaging telescope. In B. Paczynski, W.-P. Chen, and C. Lemme, editors, *IAU Colloq. 183: Small Telescope Astronomy on Global Scales, Astronomical Society of the Pacific Conference Series*, Vol. 246, p. 121, 2001.

30. W. L. Freedman and B. F. Madore. The Hubble constant. *ARA&A*, 48:673–710, 2010.

31. P. E. Freeman, V. Kashyap, R. Rosner, and D. Q. Lamb. A wavelet-based algorithm for the spatial analysis of Poisson data. *ApJS*, 138:185–218, 2002.

32. P. A. Fridman. Radio frequency interference mitigation with phase-only adaptive beam forming. *Radio Sci.*, 40:5, 2005.

33. J. A. Frieman, B. Bassett, A. Becker, C. Choi, D. Cinabro, F. DeJongh, D. L. Depoy et al. The Sloan digital sky survey-II supernova survey: Technical summary. *AJ*, 135:338–347, 2008.

34. P. C. Gregory. Bayesian exoplanet tests of a new method for MCMC sampling in highly correlated model parameter spaces. *MNRAS*, 410:94–110, 2011.

35. G. W. Henry, G. W. Marcy, R. P. Butler, and S. S. Vogt. A transiting "51 Peg-like" planet. *ApJ*, 529:L41–L44, 2000.

36. R. Kessler, B. Bassett, P. Belov, V. Bhatnagar, H. Campbell, A. Conley, J. A. Frieman et al. Results from the supernova photometric classification challenge. *PASP*, 122:1415–1431, 2010.

37. D. W. Kim, P. Protopapas, Y. I. Byun, C. Alcock, and R. Khardon. QSO selection algorithm using time variability and machine learning: Selection of 1,620 QSO candidates from macho lmc database. *ApJ*, 735(2):68, 2011.

38. J. Kubica, L. Denneau, T. Grav, J. Heasley, R. Jedicke, J. Masiero, A. Milani, A. Moore, D. Tholen, and R. J. Wainscoat. Efficient intra- and inter-night linking of asteroid detections using kd-trees. *Icarus*, 189:151–168, 2007.

39. N. M. Law, R. G. Dekany, G. Rahmer, D. Hale, R. Smith, R. Quimby, E. O. Ofek et al. The Palomar transient factory survey camera: First year performance and results. In I. S. McLean, S. K. Ramsay, and H. Takami, editors, *Society of Photo-Optical Instrumentation Engineers (SPIE) Conference Series, Society of Photo-Optical Instrumentation Engineers (SPIE) Conference Series*, Vol. 7735, pp. 77353M-7–7353M-8, 2010.

40. J. Leaman, W. Li, R. Chornock, and A. V. Filippenko. Nearby Supernova rates from the Lick observatory supernova search. I. The methods and database. *MNRAS*, 412(3):1419–1440, 2011.

41. H. S. Leavitt. 1777 Variables in the Magellanic Clouds. *Ann. Harvard Coll. Observ.*, 60:87–108, 1908.

42. C. L. MacLeod, Ž. Ivezić, C. S. Kochanek, S. Kozłowski, B. Kelly, E. Bullock, A. Kimball et al. Modeling the time variability of SDSS stripe 82 quasars as a damped random walk. *ApJ*, 721:1014–1033, 2010.

43. A. Mahabal, S. G. Djorgovski, R. Williams, A. Drake, C. Donalek, M. Graham, B. Moghaddam et al. Towards real-time classification of astronomical transients. In C. A. L. Bailer-Jones, editor, *International Conference on Classification and Discovery in Large Astronomical Surveys*,

Ringberg Castle, Germany, 14–17 October 2008. American Institute of Physics Conference Series, Vol 1082. American Institute of Physics, Melville, NY, 287–293, 2008.

44. G. W. Marcy and R. P. Butler. Detection of extrasolar giant planets. *ARA&A*, 36:57–98, 1998.

45. M. Mayor and D. Queloz. A Jupiter-mass companion to a solar-type star. *Nature*, 378:355–359, 1995.

46. K. E. McGowan, W. Junor, and T. J. W. Lazio. Detecting radio sources with machine learning. In N. Kassim, M. Perez, W. Junor, and P. Henning, editors, *From Clark Lake to the Long Wavelength Array: Bill Erickson's Radio Science*, Astronomical Society of the Pacific Conference Series, Vol. 345, pp. 362, 2005.

47. S. Nehagaban et al. Image classification of transients and variables. In preparation.

48. J. Newling, M. Varughese, B. A. Bassett, H. Campbell, R. Hlozek, M. Kunz, H. Lampeitl et al. Statistical classification techniques for photometric Supernova typing. *MNRAS*, 414(3):1987–2004, 2011.

49. J. E. Noordam. LOFAR calibration challenges. In J. M. Oschmann Jr., editor, *Society of Photo-Optical Instrumentation Engineers (SPIE) Conference Series*, Vol. 5489, pp. 817–825, 2004.

50. B. Paczynski. Gravitational microlensing by the galactic halo. *ApJ*, 304:1–5, 1986.

51. A. H. Parker and J. J. Kavelaars. Pencil-beam surveys for trans-Neptunian objects: Novel methods for optimization and characterization. *PASP*, 122:549–559, 2010.

52. M. A. C. Perryman. GAIA: An astrometric and photometric survey of our galaxy. *Ap&SS*, 280:1–10, 2002.

53. D. Poznanski, A. Gal-Yam, D. Maoz, A. V. Filippenko, D. C. Leonard, and T. Matheson. Not color-blind: Using multiband photometry to classify supernovae. *PASP*, 114:833–845, 2002.

54. C. E. Rasmussen. Gaussian processes in machine learning. In O. Bousquet, U. von Luxburg, and G. Rätsch, editors, *Advanced Lectures on Machine Learning*, Lecture Notes in Computer Science, Vol. 3176, pp. 63–71. Springer, Berlin, 2004. doi:10.1007/978-3-540-28650-9_4.

55. J. W. Richards, D. Homrighausen, P. E. Freeman, C. M. Schafer, and D. Poznanski. Semi-supervised learning for supernova photometric classification. *ArXiv e-prints*, 1103.6034, 2011.

56. J. W. Richards, D. L. Starr, N. R. Butler, J. S. Bloom, J. M. Brewer, A. Crellin-Quick, J. Higgins, R. Kennedy, and M. Rischard. On machine-learned classification of variable stars with sparse and noisy time-series data. *ApJ*, 733(1):10, 2011.

57. L. M. Sarro, J. Debosscher, M. López, and C. Aerts. Automated supervised classification of variable stars. II. Application to the OGLE database. *A&A*, 494:739–768, 2009.

58. L. M. Sarro, C. Sánchez-Fernández, and Á. Giménez. Automatic classification of eclipsing binaries light curves using neural networks. *A&A*, 446:395–402, 2006.

59. J. D. Scargle. Studies in astronomical time series analysis. V. Bayesian blocks, a new method to analyze structure in photon counting data. *ApJ*, 504:405, 1998.

60. A. Schwarzenberg-Czerny. On the advantage of using analysis of variance for period search. *MNRAS*, 241:153–165, 1989.

61. B. Sesar, Ž. Ivezić, S. H. Grammer, D. P. Morgan, A. C. Becker, M. Jurić, N. De Lee et al. Light curve templates and galactic distribution of RR Lyrae stars from Sloan digital sky survey stripe 82. *ApJ*, 708:717–741, 2010.

62. M.-S. Shin, M. Sekora, and Y.-I. Byun. Detecting variability in massive astronomical time series data—I. Application of an infinite Gaussian mixture model. *MNRAS*, 400:1897–1910, 2009.

63. C. N. Silla and A. A. Freitas. A survey of hierarchical classification across different application domains. *Data Mining Knowledge Discov.*, 22(1–2):31–72, 2011.

64. P. B. Stetson. On the automatic determination of light-curve parameters for cepheid variables. *PASP*, 108:851, 1996.

65. P. A. Sturrock and J. D. Scargle. Histogram analysis of GALLEX, GNO, and SAGE neutrino data: Further evidence for variability of the solar neutrino flux. *ApJ*, 550:L101–L104, 2001.

66. A. B. Tomaney and A. P. S. Crotts. Expanding the realm of microlensing surveys with difference image photometry. *AJ*, 112:2872, 1996.

67. G. Wachman, R. Khardon, P. Protopapas, and C. Alcock. Kernels for periodic time series arising in astronomy. In *Proceedings of the European Conference on Machine Learning and Knowledge Discovery in Databases: Part II*. Springer-Verlag, Berlin, Heidelberg, pp. 489–505, 2009.

68. L. M. Walkowicz, A. C. Becker, S. F. Anderson, J. S. Bloom, L. Georgiev, J. Grindlay, S. Howell et al. The impact of the Astro2010 recommendations on variable star science. In *Astro2010: The Astronomy and Astrophysics Decadal Survey, ArXiv Astrophys. e-prints*, Vol. 2010, p. 307, 2009.

69. Y. Wang, R. Khardon, and P. Protopapas. Shift-invariant grouped multi-task learning for Gaussian processes. In *Proceedings of the 2010 European Conference on Machine Learning and Knowledge Discovery in Databases: Part III*, Springer-Verlag, Berlin, Heidelberg, pp. 418–434, 2010.

70. P. G. Willemsen and L. Eyer. A study of supervised classification of Hipparcos variable stars using PCA and support vector machines. *ArXiv e-print* 0712.2898, 2007.

71. A. Wolszczan and D. A. Frail. A planetary system around the millisecond pulsar PSR1257 + 12. *Nature*, 355:145–147, 1992.

72. P. R. Wozniak. Difference image analysis of the OGLE-II Bulge data. I. The method. *ACTAA*, 50:421–450, 2000.

73. Ł. Wyrzykowski and V. Belokurov. Self-organizing maps in application to the OGLE data and Gaia Science alerts. In C. A. L. Bailer-Jones, editor, *American Institute of Physics Conference Series, American Institute of Physics Conference Series*, Vol. 1082, pp. 201–206, 2008.

74. C. W. Yip, A. J. Connolly, D. E. Vanden Berk, R. Scranton, S. Krughoff, A. S. Szalay, L. Dobos et al. Probing spectroscopic variability of galaxies and narrow-line active galactic nuclei in the Sloan digital sky survey. *AJ*, 137:5120–5133, 2009.

Cross-Identification of Sources

Theory and Practice

Tamás Budavári

The Johns Hopkins University

CONTENTS

7.1	Introduction		113
7.2	The Bayesian Approach		115
	7.2.1	Observational Evidence	116
	7.2.2	Probability of an Association	118
	7.2.3	Limited Sky Coverage	120
	7.2.4	How Is This Better?	121
7.3	Advanced Modeling		123
	7.3.1	Folding in the Photometry	123
	7.3.2	Varying Prior over the Sky	124
	7.3.3	Stars with Unknown Proper Motion	124
7.4	Efficient Crossmatching		127
	7.4.1	Recursive Evaluation	127
	7.4.2	Locality and Storage	128
	7.4.3	Matching in Databases	128
	7.4.4	Parallel on Graphical Processing Units	129
7.5	Concluding Remarks		130
References			132

7.1 INTRODUCTION

Astronomy has entered the era of surveys. Tenaciously observing the night sky, dedicated telescopes provide a vast amount of data daily. These datasets are arguably the primary source of our scientific analyses today, and will be even more so in the near future. Even within a single study we often employ several instruments to make specialized observations.

Sometimes we want to see in multiple wavelength ranges, and at other times, to look for changes on various timescales. The result is a set of independent views of the Universe that we want to contrast and combine into a single, consistent picture. Our goal in this chapter is to understand how this can be done in a statistically correct manner.

The problem of cross-identification is as old as astronomy itself. Every time we revisit an object on the sky we match it against previous observations. In general, and we would like to find the common detections in multiple observations, so we can study them as joint measurements of the same individual objects.

Independent observations can be completely different from one another in terms of their measured properties. The sky, however, provides a fixed reference frame to navigate, and the positions of the sources serve as adequate handles to their identities when measured accurately, cf. IAU naming convention. Thus, it comes as no surprise that the cross-identification methods primarily rely on the positions of the detections. In the simplest case, one could even be tempted to associate the closest detections. When using such pragmatic approaches, various conceptual problems emerge. For example, what to do if the astrometric uncertainty varies from object to object? How to include other measurements in addition to the positions? Or how to generalize to more than just two datasets? Instead of being a problem, these questions provide great insights into the opportunity that our data provide.

In Figure 7.1, we plot detections from three surveys: the Sloan Digital Sky Survey (SDSS [1]), the Galaxy Evolution Explorer (GALEX [2]), and the Two Micron All Sky Survey (2MASS [3]). The size of the circles correspond to the approximate astrometric uncertainties, which are approximated by $\sigma = 0.1''$, $0.5''$, and $0.8''$ (arcseconds)[*] for the SDSS, GALEX, and 2MASS, respectively. In fact, we plot 5σ contours for better visibility. This random patch of the sky is a good illustration of the kind of problems we are facing. We see all sorts of combinations: detection by themselves or paired up with others in separate or even in the same datasets. The large (red) 2MASS circles most often contain tiny (black) SDSS dots right in the middle, but not always. The intermediate (blue) GALEX circles sometimes coincide with SDSS sources, but not nearly as frequently as the 2MASS ones, and we see that 3-way SDSS–GALEX–2MASS matches are fairly rare. Some detections are exactly on top of each other, others are farther away. There is a continuum of differences. We see many alignments that are not easy to interpret. The traditional solution in astronomy is to look at the images. The problem is that it is impossible to look through all the images of modern surveys, and one really has to rely purely on statistics to find the right associations. This is *not* a bad thing! On the contrary. By automating the cross-identification we gain consistency.

Let us now focus on the 3-way matches only. Similarly, we could also consider any one of the 2-way combinations. We need to find all the possible combinations within some limits. First of all, it is not clear as to what those limits should be. For every candidate association we want a reliable measure of quality that automatically works in all scenarios. We need a method that yields all potential matches with meaningful results even in the presence of significant variations of the uncertainties, and is extendable to other types of measurements through which we can fold in the expert astronomer's opinion. How? This is

---

[*] One degree is 60 arcminutes or 3600 arcseconds, $1° = 60' = 3600''$.

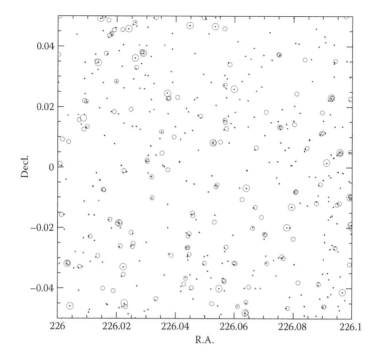

FIGURE 7.1 (**See color insert.**) Select SDSS, GALEX, and 2MASS sources are shown on a random patch of the sky in black, blue, and red, respectively. The size of the radii of the circles correspond to the astrometric uncertainties of $\sigma = 0.1, 0.5$, and 0.8 arc seconds in the three surveys. Here we plot their 5σ contours for better visibility. Cross-identification is about finding the detections in multiple datasets that belong to the same astronomical object.

a difficult problem. Part of the reason is that we do not pose the right question. One would ask whether the angular separation of the detections is small or if the likelihood for a joint position is large, and soon, but the real question is simpler. Do all detections belong to the same astronomical object or not?

In Section 7.2 a Bayesian approach is introduced to sidestep the pragmatic meta-questions, and to directly examine whether a set of given detections belong to the same object. Section 7.3 extends the basic formalism to include physical models, and Section 7.4 discusses efficient strategies for performing associations of the largest datasets. Section 7.5 offers some concluding remarks.

7.2 THE BAYESIAN APPROACH

The question of whether a set of detections belong to the same astronomical object is ill posed. The answer will vary based on what measurements are available, for example, positions, morphological classifications, or color information. The fundamental question is what the different measurements really tell us about each candidate association, in other words what the observational evidence is for an association being correct, and how that translates to probabilities that we can relate to.

7.2.1 Observational Evidence

Same or not? Bayesian hypothesis testing provides a formal, yet simple, tool to decide, which one of these two competing scenarios our data prefer. The hypothesis H that claims that all detections belong to the same object is compared to its complement $\bar{H}$ using an odd called the Bayes factor, which is the ratio of the likelihoods of the two hypotheses,

$$B = \frac{p(D|H)}{p(D|\bar{H})} \tag{7.1}$$

where D represents the data. We can calculate the likelihood of a hypothesis by parameter-izing it, and integrating its prior and likelihood function over the entire parameter space. In case of positional matching, our data are the measured locations, $D = \{x_i\}$ unit vectors. If all detections are from the same object, the model parameter is the common (unknown) true position m, and we write

$$p(D|H) = \int dm\, p(m|H)\, p(D|m, H) \tag{7.2}$$

where $p(m|H)$ is a known prior, for example, the whole celestial sphere; see more on this later. The second term is the likelihood of the parameter m that is, in turn, a product of the probabilities of the independent measurements,

$$p(D|m, H) = \prod_{i=1}^{n} f_i(x_i; m) \tag{7.3}$$

and $f_i(x_i; m)$ is the astrometric uncertainty of the ith detection, which is often assumed to be a normal distribution. Catalogs would simply list the 95% or 1σ limits, for example, for SDSS, $\sigma = 0.1''$.

On the surface of the sphere, the Fisher distribution [4] is the analog of the Gaussian distribution. Considering that for unit vectors $(x - m)^2 = 2 - 2\,xm$, the argument of its exponential cannot be too surprising,

$$F(x; w, m) = \frac{w\, \delta(|x| - 1)}{4\pi \sin\, h\, w} \exp\, (w\, mx) \tag{7.4}$$

where Dirac's δ-symbol in the numerator ensures that the function is only nonzero on the surface of the unit sphere. For high accuracies we indeed get back the Gauss distribution in the tangent plane. In this flat-sky approximation, the precision parameter is related to the width of the normal distribution, $w = 1/\sigma^2$. The Fisher formula, however, is valid even for the largest uncertainties. The value of $w = 0$ corresponds to no positional constraint, a uniform distribution on the whole sky. This means that the $f_i(x_i; m) = F(x_i; m, w_i)$ distri-butions can be good approximations not only for optical catalogs, but also for less accurate observations, such as gamma-ray bursts (GRBs) [5].

The complement hypothesis $\bar{H}$ allows for any one of the detections to come from a different object, and hence its parameters are separate $\{m_i\}$ locations with the same priors and astrometric errors as previously,

$$p(D|\bar{H}) = \prod_{i=1}^{n} \int \mathrm{d}m_i \, p(m_i|\bar{H}) f_i(x_i; m) \tag{7.5}$$

For the aforementioned Fisher distribution and an all-sky prior that is similarly expressed with the Dirac δ as

$$p(m|\cdot) = \frac{1}{4\pi} \delta(|m| - 1) \tag{7.6}$$

the Bayes factor is analytically calculated [6] to be a simple formula of the precision parameters w_i and the observed positions x_i.

$$B = \frac{\sinh w}{w} \prod_{i=1}^{n} \frac{w_i}{\sinh w_i} \quad \text{with } w = \left| \sum_{i=1}^{n} w_i x_i \right| \tag{7.7}$$

If we assume that all positional measurements are highly accurate ($w_i \gg 1$), we can approximate the $\sinh(\cdot)$ function with a single $\exp(\cdot)$, and write

$$B = 2^{n-1} \frac{\prod w_i}{\sum w_i} \exp\left\{ -\frac{\sum_{i<j} w_i w_j \psi_{ij}^2}{2 \sum w_i} \right\} \tag{7.8}$$

where ψ_{ij} represents the angular separation of x_i and x_j. In the 2-way case, the dimensionless Bayes factor simplifies to

$$B = \frac{2}{\sigma_1^2 + \sigma_2^2} \exp\left\{ -\frac{\psi^2}{2(\sigma_1^2 + \sigma_2^2)} \right\} \tag{7.9}$$

where all quantities are in radians, and $\sigma_i^2 = 1/w_i$ as before. The top panel of Figure 7.2 illustrates Equation 7.9 on a logarithmic scale for fixed 0.1" and 0.5" uncertainties that roughly correspond to the precision of the SDSS and GALEX detections. We note that for constant accuracies that do not vary from object to object, a threshold on the Bayes factor $B = B(\psi; \sigma_1, \sigma_2)$ is equivalent to a cut on the angular separation.

When the value of B is much larger than 1, the data suggest a good match, and when B is close to 0, the evidence points to separate objects. While in practice these extrema certainly occur, the interesting regime is the intermediate, gray area. What the measured values really correspond to is difficult to see at first. Using the Bayes factor, however, we can derive the probability for each association and that is a number we can relate to.

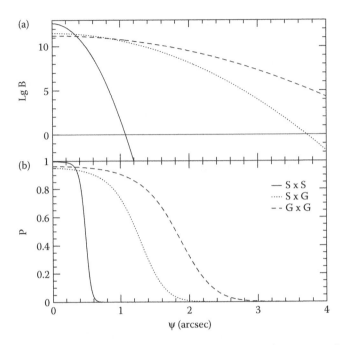

FIGURE 7.2 The panel (a) shows the logarithm of the Bayes factor to base 10 as a function of angular separation. The solid line represents the case of matching detections with $0.1''$ errors, which is the precision of the SDSS. The *dashed* line is for $0.5''$ that approximately matches the typical accuracy of the GALEX observations, and the *dotted* line is their crossmatch. The panel (b) plots the probabilities with identical line styles. The densities are assumed to be 25,000 per square degree for SDSS, half of that for GALEX and 75% for their crossmatch.

7.2.2 Probability of an Association

While for a set of measured positions we can directly calculate the Bayes factor using their uncertainties, it is not possible to assign a probability to their association just based on that information. The reason is that their positional similarity could be purely by chance. For example, in catalogs of nearby stars and high-redshift quasars one could accidentally find objects that are nearby on the sky, but we know that the datasets have no common objects. This is a prior information that has to enter the equation in a formal way. This is where the Bayes factor comes in. By definition, it provides the connection between the prior $P(H)$ and posterior $P(H|D)$ probabilities,

$$P(H|D) = \left[1 + \frac{1 - P(H)}{B\,P(H)}\right]^{-1} \tag{7.10}$$

One can verify this by writing down Bayes' rule for both $P(H|D)$ and $P(\bar{H}|D)$, and recognizing the Bayes factor in their ratio. Since the competing hypotheses are complementary, $P(H) + P(\bar{H}) = 1$ and $P(H|D) + P(\bar{H}|D) = 1$, we get the above dependence.

In order to get to the posterior, we first have to ask what the probability of a match is even before considering their positions (and/or other measured properties). This can only be done by studying the ensemble statistics of the observations. For now, let us consider full sky coverage, and we will deal with the problem of limited field of view later. In a gedanken experiment, we can have two catalogs with identical selection functions [same photometric band and charge-coupled device (CCD) sensitivity]. These two datasets contain the same number of sources, N. The probability of randomly picking the same object from the two sets is $P(H) = N^{-1}$, because whichever object we pick from the first catalog, we have N to choose from the second time, and only one is the previously selected one. Similarly, for k datasets, we have $P(H) = N^{k-1}$. The more general formula is in fact

$$P(H) = \frac{N_\star}{\prod N_i} \tag{7.11}$$

where $\{N_i\}$ are the number of detections in the observed datasets, and $N_\star$ is the number of objects that are common in all. Interestingly all the subtle differences in the selection functions of the observations are encoded into a single number $N_\star$. A number that we do not know, yet.

One thing to notice is the relative insensitivity of large posteriors to the value of the prior. If we overestimated the prior by a factor of 2, a true posterior of 0.95 would only be increased by about 2.4%. Those who only care about a rough estimate of the probabilities could just make an educated guess based on domain knowledge or insight from previous experiments. For example, the bottom panel of Figure 7.2 shows the probability as a function of the separation, same as on top where the observations have astrometric uncertainties 0.1″ and 0.5″. Here we assumed that one of the datasets has 25,000 detections per square degree, the less accurate one is half that density, and the overlap contains 75% of the smaller set. These numbers roughly correspond to the typical statistics of SDSS and GALEX observations, and hence the letters "S" and "G" of the legend. The solid line illustrates the scenario of matching high-precision detections, for example, repeated observations. We see that pairs with less than 0.2″ are practically guaranteed to be the same object, and the posterior is still over 90% at 0.4″ separations. The curve, however, quickly drops, and the probability becomes essentially negligible already at 0.6″. The dashed curve shows the repeated measurements of the less accurate observations. It reaches just over 90% for even the smallest distances but decreases much more slowly. Their crossmatch starts from an even lower maximum but drops faster due to the higher precision.

It is also possible to take out the guesswork completely. A better approach than speculating about the overlap of the selection functions is to determine the prior from the given datasets. We can do this self-consistently by solving the following system:

$$\sum P(H) = N_\star \tag{7.12}$$

$$\sum P(H|D) = N_\star \tag{7.13}$$

where the summation is over all possible associations. For a constant $P(H)$ the first equation is the same as Equation 7.11, because $\prod N_i$ is exactly the number of possible associations to consider. The second equation, however, using the posterior that depends on the prior and all the Bayes factors, is an equation for $N_\star$. In practice, we solve this by initializing the value of $N_\star$ to a reasonable value, such as the minimum of $\{N_i\}$, which is an upper bound, and iterate until convergence. The actual summation can safely be carried out on the $B > 1$ candidates due to the expectedly low value of the prior and the strong dependence of the posterior on B. The convergence is very fast, and hence the probability calculation is extremely efficient in comparison with finding the candidates and calculating the Bayes factors in the first place.

7.2.3 Limited Sky Coverage

So far our discussion has been limited to datasets that cover the entire sky, which is never really the case. While we feel that the results of the cross-identification should not change with the areal coverage as along as the observations are "far" from the edge of the geometry, it is not obvious what dependence the Bayesian approach would actually yield. There are two components to the cross-identification problem—the Bayes factor and the prior probability of the matches—and we have to examine the effect on both.

Previously we calculated the Bayes factor using an all-sky prior. When the field of view is limited, one needs a different prior. The window function $1_A(x)$ describes the geometry: it takes the value of 1 for positions that are within the coverage, and 0 otherwise. This yields an independent constraint on our model parameters, the possible true positions. In general, the window function is convolved with the astrometric precision, but the approximate formula is much simpler when the observations are far from the edge. The overall effect inside the footprint is a boost of the prior when compared with the all-sky case in Equation 7.6

$$p(m|A) = \frac{1_A(m)}{A} \delta(|m| - 1) \tag{7.14}$$

where A is the area in steradians. In essence, the new prior simply limits the integral domain to the observed part of the sky. Within the geometry, the integrals of the Bayes factor will be the same except for the scaling that appears once in the numerator, and n times in the denominator. Its overall effect is the dependence

$$B_A = B \cdot \left(\frac{4\pi}{A}\right)^{1-n} \tag{7.15}$$

where B is the all-sky Bayes factor.

Turning to the probability, we immediately notice that the limited area coverage strongly affects the prior through the number of sources. For homogeneous distributions, the size of a dataset will scale linearly with the area. If ρ represents the number of sources within the limited field of view, the entire sky would contain $N = 4\pi\rho/A$. Hence, the prior scales with the area as

$$P_A(H) = \frac{\rho_\star}{\prod \rho_i} = P(H) \cdot \left(\frac{A}{4\pi}\right)^{1-n} \tag{7.16}$$

The larger the area, the smaller the prior ($n \geq 2$), which is in accord with our expectations.

In practice, the prior is always negligibly small compared to 1, because the datasets are large. Thus, the formula for the posterior probability in Equation 7.10 simplifies to the ratio of

$$P_A(H|D) \simeq \frac{B_A P_A(H)}{1 + B_A P_A(H)} \tag{7.17}$$

which only depends on the product of the Bayes factor and the prior. Due to the scaling of these properties with the area, we see that the effects cancel out, $B_A P_A(H) = B P(H)$, and the posterior is independent of the sky coverage, $P_A(H|D) = P(H|D)$. This is good news! This simplicity enables us to incrementally cross-identify large catalogs as we accumulate new observations.

Sometimes the sky coverage can be important for individual sources that are being studied in great detail. One can formally describe the precise shape of the sky coverage using geometric primitives [7] or approximate it by pixelized maps, for example, using HEALPix [8] or hierarchical triangular mesh (HTM) [9]. In both cases the likelihood integrals are evaluated numerically, but the prior is still simply a function of the area.

7.2.4 How Is This Better?

How is the probabilistic approach better than just, say, associating detections based on angular separations? First of all, it automatically works for more than two datasets, while using a single probability limit to define the threshold. Using separations or the cuts like it would require nontrivial extensions, perhaps an increasingly large set of thresholds for the distances between all the datasets. Lots of decisions to make for an unnecessary short-cut. Second, the method is more meaningful. In the simplest 2-way case, when using fix astrometric errors, a posterior cut will correspond to a distance cut. Although they yield identical associations, the interpretation of a threshold is not obvious. We cannot directly make statistical statements based on that. On the other hand, we know what probabilities mean. And most importantly, the Bayesian approach performs better. Even in a simple 2-way case, when the errors are different for each detection. Here we investigate the quality of the associations of simulations to assess the performance in the case of SDSS and GALEX.

We create a toy universe populated with mock objects, whose clustering matches the measured 2-point angular correlation function of the real data. A uniform random number between 0 and 1, U_{01}, is assigned to each object that encodes its properties. The selection function of a dataset is emulated by selecting objects, whose randomly chosen U_{01} property is within a given interval. The length of the interval will set the surface density of the sources on the sky that can be adjusted to match the real observations, that is, N. We can do this for the SDSS- and GALEX-simulated datasets independently. By varying the overlap of their intervals, we can tune the difference in their selection function, and set a known $N_\star$. Once the selections are done, we simulate the observations by randomizing the positions with realistic errors drawn from the distribution of measured uncertainties. Such simulated catalogs will have exactly the properties of real data, as far as positional cross-identification is concerned [10].

The Bayes factor is directly calculated from the data, and then we solve for the prior just as we would do for the real datasets. The actual prior is known from the selection intervals

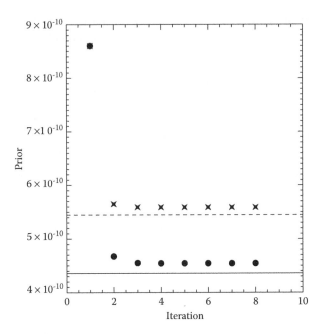

FIGURE 7.3 Iterative solution for the prior converge very fast; see text.

of U_{01}, and so we can compare the result to the truth. Figure 7.3 shows the iteration of Equation 7.13 starting from the upper bound estimate of $\hat{N}_\star = \min(N_{\text{SDSS}}, N_{\text{GALEX}})$ for two matching scenarios. The solid circles illustrate SDSS versus GALEX, while the crosses are limited to good-quality GALEX detections with $S/N > 3$. The speed of the convergence is striking; the estimates are obtained in just a few steps. As expected, the brighter GALEX subset has a larger prior, because it has less objects. We notice that the results are slightly higher than the true priors illustrated by the horizontal lines across the figure. This is because of multiple matches, some of which are false associations due to positional coincidences. We also see that the estimate for a brighter sample is relatively much closer to the truth for its higher accuracy, lower density, and overall less confusion.

Small differences in the prior have tiny effects on large posteriors, as we pointed out earlier, and so this will only result in a minor overestimation of the posterior. For crowded fields the effect can get larger and should be assessed in a statistical way on a case-by-case basis. If the discrepancy is not reasonable, one can perform a simulation in the spirit of our previous exercise to calibrate how much the estimated prior overshoots.

Since we know the truth about each simulated association, we can look at the completeness and rate of false positives as a function of the threshold parameter. The left panel of Figure 7.4 shows results from using angular separations. As we increase the distance, the completeness, shown in solid lines, rises toward 1. The bright sample is shown with thin lines. At the same time, the number of false associations starts to grow. The right panel shows the performance of the probabilistic approach. As the threshold on the posterior probability decreases, the completeness quickly approaches 100%, while keeping the false positives at a lower rate. Noteworthy is the difference between the bright and full GALEX

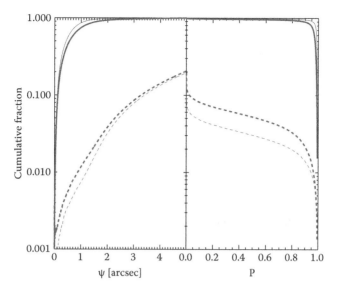

FIGURE 7.4 Performance comparison of mock associations favors the probabilistic method; see text.

samples for which the angular separation performs very similarly, but the Bayesian approach limits the false associations better. Where to cut the matches will depend on the subsequent analysis, but the probability will deliver higher completeness and less false associations everywhere [10].

7.3 ADVANCED MODELING

The Bayesian approach introduced in the previous section not only provides a simple way to quantify and interpret the goodness of the associations, but also serves as a general framework that accommodates other types of observations and more detailed physical modeling. So far we considered positional data, and assumed a constant surface density of static objects. In this section, we briefly look at natural extensions of the method to more complicated cross-identification problems that typically cannot even be addressed in the traditional settings.

7.3.1 Folding in the Photometry

Same or not? We can ask the exact question when facing nonpositional data, for example, photometric measurements. The hypotheses remain the same but their parametrization changes. In addition to a Bayes factor based on positional information, B_x, we can calculate one for the detected fluxes, B_f. Following from its definition, the overall Bayes factor is just their product

$$B = B_x \cdot B_f \tag{7.18}$$

To use the fluxes, one needs a model that can predict the spectral energy distributions (SEDs) of the sources. The usual suspects are templates, empirical [11], or theoretical [12], used also for estimating the physical properties of the sources, for example, photometric redshifts of galaxies. The only difference is that instead of fitting the models to find the maximum likelihood estimates, here we have to integrate over all possibilities,

$$B_f = \frac{\int d\theta \, p(\theta|H) \prod p(f_i|\theta, H)}{\prod \int d\theta_i \, p(\theta_i|\bar{H}) \, p(f_i|\theta_i, \bar{H})} \tag{7.19}$$

where θ represents the parameters of the model, such as redshift, type, luminosity, and so on. Considering that these models are not analytic, the integration would be done numerically in general. The likelihoods can use the photometric uncertainties, for example, Gaussian flux errors.

In practice, one only has to evaluate B_f for the candidate associations, whose positional evidence is already significant. In other words, first a positional cross-identification can be performed keeping a generous pool of candidates for which the photometric information is considered subsequently. The lists of associations can report several Bayes factors of different measurements to enable astronomers to customize the cross-identification to their specific science applications.

7.3.2 Varying Prior over the Sky

In Section 7.2.3, we discussed how the posterior of associations inside the observational footprint is independent of its actual area. The determination of the probability is shown to depend only on the density of sources via the prior. Naturally, there are populations of sources whose density is not constant on the sky. Stars, for example, are more numerous in the Galactic Plane than above and below. Their density changes slowly but significantly as a function of position. In such cases, a constant prior is obviously not a good model.

The solution is to determine the local density. One option is to approximate the changes with an analytic formula. The input datasets can be modeled with some functions $\rho_i = \rho_i(x)$, whose parameters can be statistically determined, but solving for $\rho_\star$ might prove more difficult. However, this may be the only way for small datasets. Another option is to solve for the prior in smaller areas on the sky, where the density is approximately constant. Using some pixelization scheme, for example, HEALPix or HTM, is one way to do the subdivision. Divide and conquer! Every one of the pixels can be considered on their own as independent datasets, and we can solve for a constant local prior the same way as in the case of limited field of view. A combination of the parametric and pixel methods is to find the gradient direction, and use constant bands perpendicularly to that, for example, in galactic latitude for the stars. Our choice will have to be based on the given datasets in question.

7.3.3 Stars with Unknown Proper Motion

Stars pose another challenge for cross-identification methods also because they move. How could one possibly find the same star at different epochs without knowing its velocity? It turns out, we can use the probabilistic framework.

Same or not? We ask the question all over again to evaluate the Bayes factor for H and $\bar{H}$, but this time we model the positions of objects as a function of time. The new parameter is the proper motion μ and in addition to the positional measurements with also keep track of time. If m is the true position of an object at our reference epoch the probability density of finding it at some x will also depend on the time difference Δt. Formally, we can write the likelihoods of the hypotheses in the Bayes factor as before but now using all the parameters of the new model

$$p(D|H) = \int dm \int d\mu\, p(m, \mu|H) \prod_{i=1}^{n} p_i(x_i|\Delta t_i, m, \mu, H) \tag{7.20}$$

$$p(D|\bar{H}) = \prod_{i=1}^{n} \int dm_i \int d\mu_i\, p(m_i, \mu_i|\bar{H})\, p_i(x_i|\Delta t_i, m_i, \mu_i, \bar{H}) \tag{7.21}$$

where $p(m, \mu|\cdot)$ is the prior density of the parameters. If we consider small displacement on the sky between epochs, the true position after Δt time is

$$m' = \frac{m + \mu\Delta t}{|m + \mu\Delta t|} \tag{7.22}$$

which we can directly plug into the astrometric uncertainty, for example, the Fisher distribution.

The devil is, admittedly, in the details. The integrals can only be evaluated with a known prior in hand, which has now two parameters. The joint density can be written as

$$p(m, \mu) = p(m)\, p(\mu|m) \tag{7.23}$$

where the first part is the prior on the position as before, and the second term describes the possible proper motions as a function of the position on the sky. The simplest assumption is a uniform distribution up to a maximum of μ_{max} independently from the position,

$$p(\mu|m) = \begin{cases} 1/\pi\mu_{max}^2 & \text{if } |\mu| < \mu_{max} \\ 0 & \text{otherwise} \end{cases} \tag{7.24}$$

Alternatively one can also use observations to build up more realistic priors that depends on the position [13] or other measured properties of the stars. Next, we compare the performance of the analytic static results to the proper motion model using both of these priors.

We select a small sample of SDSS stars with known proper motions from USNO-B1.0 [14] to study the performance of their identification from multiple observations in Stripe 82. The stars span a wide range of proper motions from 10 to 550 mas/year. Figure 7.5 shows the posterior probability of the star associations as a function of the known proper motion. The three panels illustrate the 2-way matching of observations at different epochs. The time differences are approximately the same within each panel but increase significantly from left to right: 2, 4.5, and 6.5 years. Crosses show results from the analytic static case. We see that the

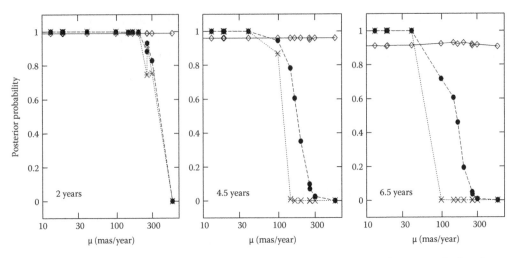

FIGURE 7.5 Probability of 2-way associations as a function of proper motion. As we look at longer time differences between the epochs, our models yield increasing different answers.

probabilities start the plummet as we increase the time between measurements and at around 6.5 years all stars are lost except for the slowest ones below 50 mas/year. Open diamonds illustrate the results from the uniform proper motion prior. Using $\mu_{max} = 600$ mas/year, the stars all have similar probabilities that, however, their value significantly decreases between different panels as the time difference grows. The empirical prior yields the solid circles that first follow the static model but drop much slower as the elapsed time increases. They provide finite probabilities, even if not large enough to make any reasonable selection cut.

Perhaps even more interesting is the next set of panels in Figure 7.6. The symbols refer to the same models as previously, but the panels now differ in the number of epochs. They

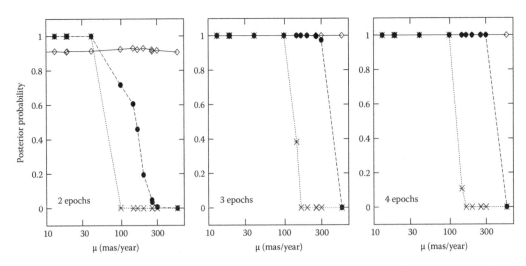

FIGURE 7.6 Probability of 2-, 3-, and 4-way star associations as a function of proper motion. Introducing a 3rd intermediate epoch improves significantly the cross-identification. The effect of the 4th measurement is not so dramatic.

all have the same 6.5-year baseline, but we introduce intermediate points. The left panel is identical to the right-most panel of Figure 7.5, the middle has three epochs, and the right has four. We see that both the constant proper motion and empirical priors do very well at associating the large velocity stars as soon as an additional observation are added. The intermediate points boost even the static models for the lower proper motion stars on the threshold. We offer a simple geometric argument as an explanation for this dramatical change: it is always possible to accurately fit a great circle to two points, but not to three or more. Multiple positions that match the proper motion model are automatically awarded larger likelihoods and yield increased Bayes factors and posterior probabilities.

7.4 EFFICIENT CROSSMATCHING

In practice, one starts with the static model, which will, in fact, find most of the associations at a low processing cost. The more compute-intensive, advanced methods are then to be only applied to the remainder of the sources. Making the searches as efficient as possible, is an active research topic that is especially timely because of the upcoming time-domain surveys. In this section, we discuss some of the practicalities of the problem and detail implementation strategies.

7.4.1 Recursive Evaluation

Performing n-way matching is computationally expensive. In theory, one is to consider all possible associations, whose number is potentially huge:

$$N_1 \times N_2 \times \cdots \times N_n$$

Clearly, one does not need to calculate the Bayes factor for them all. If two observations are very far away, they can never be part of any probable association. This is the motivation behind the following recursive approach, where the datasets are added to the associations one after another. In every step, the list is pruned significantly. Using the spherical normal distribution in the limit of high accuracies, Equation 7.8 can be rewritten to be evaluated incrementally [6] as

$$\ln B = \ln \left(2^{n-1} \frac{\prod w_i}{\sum w_i} \right) - \frac{1}{2} \sum_{k=2}^{n} \frac{a_{k-1}}{a_k} w_k \Delta_k^2 \tag{7.25}$$

with

$$a_k = \sum_{i=1}^{k} w_i \quad \text{and} \quad \Delta_k = x_k - c_{k-1} \tag{7.26}$$

where c_k is the unit vector of the best position for the current partial match,

$$c_k = \left(\sum_{i=1}^{k} w_i x_i \right) \Big/ \left| \sum_{i=1}^{k} w_i x_i \right| \tag{7.27}$$

In the kth step, the maximum search radius is computed from Equation 7.25 such that it reaches a given Bayes factor threshold. In the calculation, one can assume optimal subsequent matches with $\Delta_k^2 = 0$ contributions. We assign, every source within that radius to each k-tuple submatch, and go to the next catalog. From catalog-to-catalog we propagate the quantities that are necessary to calculate the Bayes factor. The recursion formulas are given by the following expressions:

$$a_k = a_{k-1} + w_k \tag{7.28}$$

$$q_k = q_{k-1} + \frac{a_{k-1}}{a_k} w_k \Delta_k^2 \tag{7.29}$$

$$c_k = \left(c_{k-1} + \frac{w_k}{a_k} \Delta_k \right) \bigg/ \left| c_{k-1} + \frac{w_k}{a_k} \Delta_k \right| \tag{7.30}$$

and the initialization is $a_1 = w_1$, $q_1 = 0$ and $c_1 = x_1$. This iterative approach allows us to use fast 2-way crossmatching tools that are the subject of the following paragraphs.

7.4.2 Locality and Storage

Data reside in a variety of storage solutions. An efficient analysis has to deal with the whole hierarchy of stores from hard-drives through the main memory to fast caches. The key to implementing fast cross-identifications is in organizing the data ahead of time, so that detections nearby on the sky are also close on the disk and even in memory. Both HEALPix and HTM pixelizations have been used before for crossmatching with great success. Their speedup comes from the fact that sources within a pixel are known to be nearby. When matching two datasets, we can consider each cell independently and look for matches withing the same cell and its neighbors. Depending on the search radius and the pixel size, one might have to visit farther cells that are not always colocated on disk and/or in memory but in general the algorithms can leverage the caches of modern architectures.

For large surveys, one exceptionally successful method has been the so-called zones algorithm [15]. It is a hybrid solution, whose strength is in its simplicity. The detections are grouped into constant declination zones, within which they are sorted by right ascension. Similar to pixel-based subdivisions, we can process the parts independently but a zone only has two neighbors, as opposed to HEALPix with 8, or HTM with 12 neighboring pixels. For a given search radius, one also knows how far to look in R.A. in each zone. The optimized index essentially yields a continuously sliding window, which is exactly how machines can stream the datasets to the processing units for computation. Just like the size of the pixels in other schemes, the height of the zones is an important parameter to be tuned for the typical search radii, the size/density of the datasets and the limitations of the hardware.

7.4.3 Matching in Databases

For their simplicity, these methods are straightforward to implement in most settings. Databases are the obvious choice, where most of astronomical data reside today. The zones map very well on relational systems. If a zone is identified by its `ZoneID`, we can create a `Link` table that contains pairs of neighboring zones to be considered for crossmatching.

A naive query that joins two tables on the Euclidean distance would run for weeks on the SDSS and GALEX catalogs,

```
SELECT Tab1.ObjID AS ObjID1, ...
       Tab2.ObjID AS ObjID2, ...
FROM Tab1, Tab2
WHERE (Tab1.Cx-Tab2.Cx) * (Tab1.Cx-Tab2.Cx)
    + (Tab1.Cy-Tab2.Cy) * (Tab1.Cy-Tab2.Cy)
    + (Tab1.Cz-Tab2.Cz) * (Tab1.Cz-Tab2.Cz) < @dist2
```

where `@dist2` is the square of the distances threshold. When rewritten to leverage the (`ZoneID, RA`) indices on the source tables and the `Link` table, it completes the job in 15 min on a single machine,

```
SELECT Tab1.ObjID AS ObjID1, ...
       Tab2.ObjID AS ObjID2, ...
FROM Tab1, Tab2, Link
WHERE Tab1.ZoneID = Link.ZoneID1
  AND Tab2.ZoneID = Link.ZoneID2
  AND Tab2.RA BETWEEN Tab1.RA-Link.dRA AND Tab1.RA+Link.dRA
  AND Tab2.Dec BETWEEN Tab1.Dec-@theta AND Tab1.Dec+@theta
  AND (Tab1.Cx-Tab2.Cx) * (Tab1.Cx-Tab2.Cx)
    + (Tab1.Cy-Tab2.Cy) * (Tab1.Cy-Tab2.Cy)
    + (Tab1.Cz-Tab2.Cz) * (Tab1.Cz-Tab2.Cz) < @dist2
```

where `@theta` is the matching radius in degrees and `Link.dRA` is the R.A. limit for the search in each zone. The second query will only consider pairs from zones that are in the link table and within them it would use the index. In addition, a declination comparison is also performed before the expensive distance calculation finally gets done for the filtered candidates. For the sake of simplicity, here we left out the discussion and the constraints to deal with the wraparound problem in R.A., which has no effect on the performance of this query and is only a minor modification of this structured query language (SQL) command.

The last code snippet and implementations of similar approaches can optimize data access and limit the candidates to consider, but the performance is still going to be processing limited. The faster we can do simple arithmetics the better we can perform large-scale cross-identifications.

7.4.4 Parallel on Graphical Processing Units

Computer architectures evolve fast. Their processing power has been increasing exponentially for the last half-century. Scientists and programmers are used to changes that simply make the codes run faster but this might prove untenable in the near future. Following Moore's law, the number of transistors on a single chip, such as our microprocessors, is still growing today. The design patterns, however, have dramatically changed recently. The reason is power consumption. Processors are not speeding up any more, if anything their clocks are being lowered. If applications are to gain performance, they need to make use of the new multicore technologies. For the last decades new CPUs have been always compatible with previous generations at the cost of getting more complicated and inefficient. This will arguably not be the case for long.

The most promising parallel technology today is originated in video processing. The graphical processing units, or GPUs for short, on the most recent video cards are capable of not only rendering millions of pixels on our screens in a fraction of a second but can also perform general-purpose computations. The GPU architecture is significantly different from CPUs. In many ways, it is much simpler, which allows for new opportunities in taking the multicore paradigm to its extreme. A modern GPU can have 512 cores that run close to 25,000 parallel threads at any given moment.

We use C++ and NVIDIA's C for CUDA (Compute Unified Device Architecture) programming languages to prototype a solution of the cross-identification problem [16]. Without sophisticated I/O operations, the limiting factor for the dataset size is the memory. A GeForce GTX 480 card has 1.5 GB memory that can hold about 2×30 million detections with identifiers and positions. For testing, we use identical subsets of the SDSS database. Algorithmic primitives provided by the Thrust library allow managing and sorting them according to the zone algorithm. Using $5''$ zones, we sort each dataset in just 2 s. This is a preprocessing step that could be even be spared in most cases. The cross-identification is implemented as a custom CUDA kernel that is launched for all possible combinations of zones with a matching search radius of $5''$, that is, only neighboring zones are considered. The implementation is admittedly not trivial. Finding the matches is one thing, returning them efficiently is another. The details are outside the scope of this section, but we mention that the key element is to write the results in parallel using accelerated hardware optimization. Our code performs the search in just under 11 s on a single video card. This is over 10 times faster than the performance of a 16-core top-of-the-line Intel CPU. The problem is pleasantly parallel and can be run on multiple GPUs contained in the same chassis. Various optimizations are still to be done, which are expected to increase the speed even further.

7.5 CONCLUDING REMARKS

The probabilistic approach puts the cross-identification problem on a firm statistical foundation. Using hypothesis testing, we can directly calculate the Bayes factor from our data and their uncertainties. We have seen that in the case of the spherical normal distribution named after Fisher, the calculations can be done analytically to perform positional matching. In the limits of constant errors and priors, we made the connection to the simpler approaches that this framework extends. We can naturally handle multiple datasets at the same time even with varying astrometric errors. Using simulated catalogs of SDSS and GALEX, we saw that the probabilistic approach outperforms a simple angular separation threshold. In the Bayes factor, we can naturally introduce other distribution functions to characterize the astrometric uncertainty. One important example is the Kent distribution [17] that can represent "elliptical" errors on the sphere. For completeness, we note that its formula is

$$K(x; w, m, \beta, a, b) = \frac{\delta(|x| - 1)}{c(w, \beta)} \exp\{w\, mx + \beta[(ax)^2 - (bx)^2]\} \qquad (7.31)$$

where β measures the ovalness, a and b are the major and minor axes, and $c(w, \beta)$ is the normalization factor.

In addition to positional information, one can naturally consider other types of data, such as photometric measurements. Using SED modeling or empirical distributions, one can calculate the corresponding Bayes factors and the separate ratios of different data are simply multiplied. In practice, this has important implications. We can enumerate several independent Bayes factors for all possible matching candidates. Instead of just deriving a global set of cross-identifications, we can be flexible and tailor the matching for specific science goals. For example, we can find two sets of associations: one based purely on positions and another based on both positions and photometry. Associations in the former but not in the latter are potentially true but unusual objects, whose photometry does not follow the applied SED model. They could be just artifacts or sources of new discoveries.

Detailed modeling of the proper motion of the stars enables us to perform matching across different epochs even without knowing the velocity of the stars. The numerical integration takes longer but only have to be performed on a small subset of the datasets that could not be associated the static way. During the calculation, one also has the opportunity to solve for the proper motion of each star based on their combined measurements.

The probability of a match is not as straightforward to calculate as the Bayes factor. We looked at how it can be obtained from studying the ensemble properties of the datasets, and why the local density ultimately affects the prior. By definition the posterior is a simple and known analytic function of the prior and the Bayes factor. Clearly spelling out all the assumptions of the algorithm guarantees the consistency of our cross-identifications, so we can better interpret the results. Scientific analyses of combined datasets are intimately intertwined with their cross-identification method. Any set of associations can only be an approximation that depends on our understanding of the astronomical objects. With high-precision data, the approximation is very good, and when it is not, we will know why. Observations typically have a mixture of galaxies and stars. Stars are not isotropically distributed on the sky. Their density increases as we approach the Galactic Plane. Is a constant prior a good model? That depends. If the change in the density is significant within the field of view, one has to classify the detections to separate the stars from the extragalactic sources. The cross-identification will, hence, depend on the classification. If the classification requires all datasets, for example, for SED modeling, to discriminate between the different types of objects, the problem becomes circular but still consistently solvable.

Cross-identification is a difficult problem computationally. A naive approach to evaluating the candidates would finish only in polynomial time. We have looked at faster options that divide the sky to conquer the problem. Using these algorithmic analogs of pixel coaddition, we can obtain results in practically linear timescaling with the area. We recently saw major technological advancements that will be undoubtedly relevant for crossmatching. Computation on new architectures, such as GPUs, need much less power than traditional CPUs. The code that is fast enough to keep up with the upcoming survey telescopes is, however, yet to be written.

In the new era of surveys, automated tools and tests will take the lead in several aspects of the scientific data analysis. Using proper statistical methods is the key to meaningful results. Probabilistic cross-identification is one of such fundamental methods.

REFERENCES

1. D. G. York, J. Adelman, J. E. Anderson, Jr., S. F. Anderson, J. Annis, N. A. Bahcall, J. A. Bakken et al. The Sloan digital sky survey: Technical summary. *Astronomical Journal*, 120:1579–1587, 2000.

2. D. C. Martin, J. Fanson, D. Schiminovich, P. Morrissey, P. G. Friedman, T. A. Barlow, T. Conrow et al. The Galaxy evolution explorer: A space ultraviolet survey mission. *Astrophysical Journal*, 619:L1–L6, 2005.

3. M. F. Skrutskie, R. M. Cutri, R. Stiening, M. D. Weinberg, S. Schneider, J. M. Carpenter, C. Beichman et al. The two micron all sky survey (2MASS). *Astronomical Journal*, 131:1163–1183, 2006.

4. R. Fisher. Dispersion on a sphere. *Royal Society of London Proceedings Series A*, 217, pp. 295–305, May 1953.

5. S. Luo, T. Loredo, and I. Wasserman. Likelihood analysis of GRB repetition. In C. Kouveliotou, M. F. Briggs, and G. J. Fishman, ed., *American Institute of Physics Conference Series*, Vol. 384, pp. 477–481, August 1996.

6. T. Budavári and A. S. Szalay. Probabilistic cross-identification of astronomical sources. *Astrophysical Journal*, 679:301–309, 2008.

7. T. Budavári, A. S. Szalay, and G. Fekete. Searchable sky coverage of astronomical observations: Footprints and exposures. *Publications of the Astronomical Society of the Pacific*, 122:1375–1388, 2010.

8. K. M. Górski, E. Hivon, A. J. Banday, B. D. Wandelt, F. K. Hansen, M. Reinecke, and M. Bartelmann. HEALPix: A framework for high-resolution discretization and fast analysis of data distributed on the Sphere. *Astrophysical Journal*, 622:759–771, 2005.

9. P. Z. Kunszt, A. S. Szalay, and A. R. Thakar. The hierarchical triangular mesh. In A. J. Banday, S. Zaroubi, and M. Bartelmann, ed., *Mining the Sky*, Berlin, Heidelberg, Germany: Springer-Verlag, pp. 631–637, 2001, http://goo.gl/W6JaP

10. S. Heinis, T. Budavári, and A. S. Szalay. Cross-identification performance from simulated detections: Galex and SDSS. *Astrophysical Journal*, 705:739–745, 2009.

11. G. D. Coleman, C.-C. Wu, and D. W. Weedman. Colors and magnitudes predicted for high redshift galaxies. *Astrophysical Journal Supplement*, 43:393–416, 1980.

12. G. Bruzual and S. Charlot. Stellar population synthesis at the resolution of 2003. *Monthly Notices of the Royal Astronomical Society*, 344:1000–1028, 2003.

13. G. Kerekes, T. Budavári, I. Csabai, A. J. Connolly, and A. S. Szalay. Cross identification of stars with unknown proper motions. *Astrophysical Journal*, 719:59–66, 2010.

14. J. A. Munn, D. G. Monet, S. E. Levine, B. Canzian, J. R. Pier, H. C. Harris, R. H. Lupton et al. An improved proper-motion catalog combining USNO-B and the Sloan digital sky survey. *Astronomical Journal*, 127:3034–3042, 2004.

15. J. Gray, M. A. Nieto-Santisteban, and A. S. Szalay. The zones algorithm for finding points-near-a-point or cross-matching spatial datasets. *ArXiv Computer Science e-prints*, 2007, MSR-TR-2006-52: Microsoft Research Technical Report, http://research.microsoft.com/apps/pubs/default.aspx?id=64524.

16. T. Budavári. Cross-identification of astronomical objects: Playing with dice. In *Astronomical Data Analysis Software and Systems (ADASS) XX, Astronomical Society of the Pacific Conference Series*, Vol. 442, San Francisco: Astronomical Society of the Pacific, pp. 79–ff., 2011, http://adsabs.harvard.edu/abs/2011ASPC..442...79B.

17. J. T. Kent. The Fisher–Bingham distribution on the sphere. *Journal of the Royal Statistical Society B*, 44:71–80, 1982.

CHAPTER 8

The Sky Pixelization for Cosmic Microwave Background Mapping

O.V. Verkhodanov
Russian Academy of Sciences

A.G. Doroshkevich
Russian Academy of Sciences

CONTENTS

8.1	Introduction	133
8.2	Introduction to the Map-Making Problem	134
8.3	Pixelization Schemes	136
	8.3.1 Cube Pixelization	137
	8.3.2 Icosahedron Pixelization	138
	8.3.3 Igloo	139
	8.3.4 HEALPix	140
8.4	GLESP	142
	8.4.1 Main Ideas and Basic Relations	143
	8.4.2 Properties of GLESP	146
	8.4.3 Repixelization Problem	148
8.5	Summary	149
	Acknowledgments	150
	Appendix A. Normalized Associated Legendre Polynomials	150
	Appendix B. GLESP Pixel Window Function	151
	Appendix C. The GLESP Package	153
	References	157

8.1 INTRODUCTION

The last decade of research in cosmology was connected with the ambitious experiments including space and ground base observations. Among the most impressive results of these investigations are the measurements of the cosmic microwave background (CMB) radiation like WMAP[*] and Planck.[†] Exactly from the CMB studies, we have started the epoch of the precision cosmology when generally the values of cosmological parameters have been known and present research is devoted to improvement of the precision. These achievements are connected with both the creation of the new facilities in millimeter and submillimeter astronomy (e.g., satellites, receivers, antennas, computers) and development of the methods for the CMB data analysis.

Actually, the process of data analysis contains several technical stages including

1. Registration of time-ordered data (TOD)

2. Pixelization of the CMB data—map preparation

3. Component separation

4. Map statistics analysis

5. Map—spherical harmonics transformation

6. $C(\ell)$–spectrum calculation and spectrum statistics analysis

7. Cosmological parameters estimation

Starting from the cosmic background explorer (COBE) experiment using the so-called Quadrilateralized Sky Cube Projection (see [1–3]), the problem of the whole sky CMB pixelization has attracted great interest and many such schemes were developed. Let us note however that accurate pixelization of the CMB data on the sphere is very important but not the final step of analysis. Usually, the next step implies the determination of the coefficients of the spherical harmonic decomposition of the CMB signal for both anisotropy and polarization. This means that some of the pixelization schemes provide a very accurate map but are inconvenient for further decomposition. This also means that the choice of suitable pixelization schemes depends upon the general goals of the investigation.

In this review, we consider several of the most popular sky map pixelization schemes and link them with the spherical harmonic decomposition of the CMB signal. We start from a short presentation in Section 8.2 of the map-making procedure. In Section 8.3, we briefly describe a few different pixelization schemes, and in Section 8.4, we consider in more details the Gauss–Legendre sky pixelization (GLESP) scheme which is focused on the most effective decomposition of the temperature and polarization maps and the preparation of spectral characteristics of the CMB. A short summary can be found in Section 8.5. Some technical details are given in three appendices.

8.2 INTRODUCTION TO THE MAP-MAKING PROBLEM

In this section, we follow the approach of Natoli et al. [4], who considered the map-making algorithm for the Planck mission and its assumptions.

---

[*] http://lambda.gsfc.nasa.gov
[†] http://sci.esa.int/science-e/www/area/index.cfm?fareaid=17

The primary output of a CMB experiment is stored as the TOD. Let $\mathbf{d}$ be the array of TOD, which consist of N_d sky observations made with a given scanning strategy and at a given sampling rate, for example, three points per FWHM (full width half maximum of the beam). Let a map, $\mathbf{m}$, be a vector containing N_p temperature values, associated with sky pixels of dimension $\sim$FWHM/3. Following Ref. [5], one can assume that the TOD depend linearly on the map:

$$\mathbf{d} = \mathbf{Pm} + \mathbf{n}, \tag{8.1}$$

where $\mathbf{n}$ is a vector of random noise and $\mathbf{P}$ is some known matrix.

The rectangular, $N_d \times N_p$ matrix $\mathbf{P}$ is dubbed a pointing matrix. That is, applying $\mathbf{P}$ on a map "unrolls" the latter on a TOD according to a given scanning strategy. Conversely, applying $\mathbf{P}^T$ on the TOD "sums" them into a map.* The structure of $\mathbf{P}$ depends on what we assume for $\mathbf{m}$. If $\mathbf{m}$ contains a pixelized, but unsmeared, image of the sky, then $\mathbf{P}$ must account for beam smearing. This is the most general assumption and allows one to properly treat, for instance, an asymmetric beam profile. In this case, applying $\mathbf{P}$ to $\mathbf{m}$ implies both convolving the sky pattern with the detector beam response and unrolling $\mathbf{m}$ into a "signal-only" time stream. If, on the other hand, the beam is "at least approximately" symmetric, then it is possible and certainly more convenient to consider $\mathbf{m}$ as the beam-smeared pixelized sky. The structure of $\mathbf{P}$ for a one-horned experiment would then be very simple. Only one element per row would be different from zero, the one connecting the observation of jth pixel to the ith element of the TOD. Hereafter, we will restrict ourselves to this case.

Many methods have been proposed to estimate $\mathbf{m}$ in Equation 8.1 (for a review, see e.g., Ref. [6]). Since the problem is linear in $\mathbf{m}$, the use of a generalized least squares (GLS) method appears well suited. This involves the minimization of the quantity

$$\chi^2 = \mathbf{n}^T \mathbf{Vn} = (\mathbf{d}^T - \mathbf{m}^T)\mathbf{V}(\mathbf{d} - \mathbf{Pm})$$

for some nonsingular, symmetric matrix $\mathbf{V}$. Then, one can construct an estimator $\tilde{\mathbf{m}}$ for the map:

$$\tilde{\mathbf{m}} = (\mathbf{P}^T \mathbf{VP})^{-1} \mathbf{P}^T \mathbf{Vd}. \tag{8.2}$$

The proof that this estimator is unbiased is straightforward. Just note that

$$\tilde{\mathbf{m}} - m = (\mathbf{P}^T \mathbf{VP})^{-1} \mathbf{P}^T \mathbf{Vn}.$$

So, provided that $\langle \mathbf{n} \rangle = 0$, one has $\langle \tilde{\mathbf{m}} \rangle = \mathbf{m}$ (as usual, the symbol $\langle \cdot \rangle$ indicates an average over the ensemble). The map covariance matrix is, then

$$\Sigma^{-1} \equiv \langle (\mathbf{m} - \bar{\mathbf{m}})(\mathbf{m}^T - \bar{\mathbf{m}}^T) \rangle = (\mathbf{P}^T \mathbf{VP})^{-1} \mathbf{P}^T \mathbf{V} \langle \mathbf{nn}^T \rangle \mathbf{VP}(\mathbf{P}^T \mathbf{VP})^{-1}$$

---

* The value of a pixel of this map is the sum of all the observations of that pixel made at different times according to a given scanning strategy.

In order to have a "low noise" estimator, one has to find $\mathbf{V}$ that minimizes the variance of $\bar{\mathbf{m}}$. This is attained if we take $\mathbf{V}$ to be the noise inverse covariance matrix, that is $\mathbf{V}^{-1} = \mathbf{N} \equiv \langle \mathbf{nn}^T \rangle$. Then, $\bar{\mathbf{m}}$ has the very nice property of being, among all linear and unbiased estimators, the one of minimum variance.[*] Then, the GLS solution to the map-making problem is

$$\bar{\mathbf{m}} = \Sigma^{-1} \mathbf{P}^T \mathbf{N}^{-1} \mathbf{d}, \tag{8.3}$$

where

$$\Sigma = \mathbf{P}^T \mathbf{N}^{-1} \mathbf{P}.$$

The statistical properties of detector noise are, as usual, described by a multivariate Gaussian distribution. This fact has two consequences. The first one is rather obvious: if the noise distribution is Gaussian, $\mathbf{m}$ is indeed the maximum likelihood estimator. In fact, in the Gaussian case, the likelihood of the data time stream given the (true) map is

$$L(\mathbf{d}|\mathbf{m}) = (2\pi)^{N_p/2} \times \exp\left\{ -\frac{1}{2}[(\mathbf{d}^T - \mathbf{m}^T \mathbf{P}^T)\mathbf{N}^{-1}(\mathbf{d} - \mathbf{Pm}) + \mathrm{Tr}(\ln \mathbf{N})] \right\}. \tag{8.4}$$

Solving for the maximum would clearly give Equations 8.3 and 8.4 again. The second remark has to do with the notion of a loss-less map, that is, a map which contains all the relevant cosmological information contained in the TOD. A good way to prove that a method is loss-less is to show that the TOD and the related map have the same Fisher information matrix. The estimator defined in Equation 8.3 leads to a loss-less map [6].

8.3 PIXELIZATION SCHEMES

The problem of restoration of the initial CMB distribution is connected with the sky grid selection, since integration in pixel of TOD is required. First, we should note that the problem of integration on a sphere was discussed in the 1970s [7–10]. In the real physical analysis, the problem turns to real digital technologies realized in the celestial grid. To follow this, we should select a pixelization scheme. The simplest one uses equal divisions in latitude and longitude (θ, ϕ). This has been called the equidistant cylindrical projection (ECP). It has the advantages of being both azimuthal and simply hierarchical, in that the data can be easily coarse grained by combining neighboring pixels. The azimuthal symmetry allows for fast spherical harmonic transforms, speeding many operations such as map simulation and inversion [11,12]. The biggest failure of the ECP pixelization is that the pixels near the poles are small and very distorted. In a real experiment, they could be very noisy or even contain no data at all. The ECP scheme can be improved upon by grouping more and more pixels together as one approaches the pole.

---

[*] Any linear estimator can be written in the form $\bar{\mathbf{m}} = \mathbf{Ad} = \mathbf{APm} + \mathbf{An}$. Then, the condition $\mathbf{AP} = \mathbf{I}$ must hold for it to be unbiased. To prove that the variance of $\mathbf{m}$ cannot be smaller than the variance of $\bar{\mathbf{m}}$, one can write

$$\mathrm{Var}(\hat{\mathbf{m}}) = \mathrm{Var}(\bar{\mathbf{m}}) + \mathrm{Var}(\hat{\mathbf{m}} - \bar{\mathbf{m}}) + 2\mathrm{Cov}(\bar{\mathbf{m}}, \hat{\mathbf{m}} - \bar{\mathbf{m}})$$

By substitution, one obtains $\mathrm{Cov}(\bar{\mathbf{m}}, \hat{\mathbf{m}} - \bar{\mathbf{m}}) = (\mathbf{TP} - \mathbf{I})(\mathbf{P}^T \mathbf{N}^{-1} \mathbf{P})^{-1} = 0$. Thus, the variance of $\bar{\mathbf{m}}$ is equal to the variance of $\bar{\mathbf{m}}$ plus a nonnegative quantity.

Also, we should mention the hierarchical triangular mesh[*] (HTM) [13] developed for the Sloan Digital Sky Survey. The HTM sphere pixelization scheme uses triangular pixels which can recursively be subdivided into four pixels. The base pixels are 8, 4 for each hemisphere. They are obtained by the intersection on the sphere of three major big circles. On Earth, they can be represented by the equator and two meridians passing at longitudes 0 and 90 degrees. All these base spherical triangles have the same area. Each of them can then be further divided into four spherical triangles, or trixels, by connecting the middle points of three sides using great circle segments.

The first all sky CMB maps, produced by the COBE satellite, used a pixelization based on the Quadrilateralized Sky Cube Projection, or "quad cube" [3]. The edges of a cube are projected onto a sphere, dividing the sky into six equal areas. These are subdivided into a roughly square, hierarchical lattice. The main drawback of the resulting pixelizations is their lack of azimuthal symmetry, making spherical harmonic transforms time consuming.

At least four methods of the CMB celestial sphere pixelization have been proposed and implemented after the COBE pixelization scheme: the icosahedron pixelizing by Tegmark [14], the igloo pixelization by Crittenden and Turok [15], the Hierarchical Equal Area iso-Latitude Pixelisation (HEALPix)[†] method by Górski et al. [16,17], and GLESP[‡] [18,19].

Two important questions mentioned already by Tegmark in 1996 [14] are now under discussion:

1. What is the optimal method for the choice of the N_{pix} positions of pixel centers, shapes, and sizes to provide (as good as possible) the compact uniform coverage of the sky by pixels with equal areas?

2. What is the best way to approximate any convolutions of the maps by sums using pixels?

The first three above-mentioned pixelization schemes were devoted to solving the first problem as accurate as possible, and the answer to the second question usually follows for the chosen pixelization scheme.

8.3.1 Cube Pixelization
The COBE sky cube pixelization scheme, the first scheme actually used for total sky CMB map, is illustrated in Figure 8.1, and consists of the following steps [14]:

1. The sphere is inscribed in a cube, whose faces are pixelized with a regular square grid.

2. The points are mapped radially onto the sphere.

3. The points are shifted around slightly, to give all pixels approximately equal area.

---

[*] http://skyserver.org/htm
[†] currently http://healpix.jpl.nasa.gov
[‡] http://www.glesp.nbi.dk

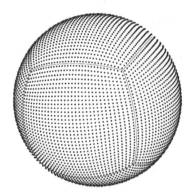

FIGURE 8.1 The cube-based pixelization. (Reproduced from Tegmark, 1996, *ApJ*, 470, L81. With permission.)

A pixel (the area which is closer to a given point than to all other points) is thus approximately square, with a side of length $\sqrt{4\pi/N}$. The points furthest from the pixels lie at the corners of these squares, so $d_{\text{cube}} \approx \sqrt{2\pi/N}$. For a honeycomb grid, a pixel is hexagonal, and one readily computes that $d_{\text{icosa}} \approx \sqrt{8\pi/(3\sqrt{3}N)}$, a value which is about 12% smaller than that for the square grid case. To take advantage of this, one could thus replace the sky cube by a platonic solid with triangular faces, that is, by the tetrahedron, the octahedron, or the icosahedron. The area-equalization is carried out for the sake of the second criterion (see (2.) in the previous section), loosely speaking so that the equal weights that the pixels get when summed over correspond to equal weights $d\Omega$ in an integral. Since the pixels were originally on a rectangular grid on the cube faces, the amount of "stretching" required increases toward the edges of the faces. Both this and the radial projection make the pixels slightly deformed, so that the further out on a face one goes, the more the corresponding pixels on the sphere depart from a regular grid. Because of this, it is clearly desirable to use faces as small as possible, so that the corresponding regions of the sphere are as flat as possible. The platonic solid with the smallest faces is the one with the largest number of faces: the icosahedron, whose faces are 20 triangles (Figure 8.1). Not only does it have the advantage of having more than three times as many faces as the cube, which one would expect to help with criterion (2.), but since the faces are triangles rather than squares, it is better according to criterion (1.) as well [14].

This method allows one to make an implementation of the icosahedron scheme with an area-equalization.

8.3.2 Icosahedron Pixelization

The icosahedron pixelization scheme is illustrated in Figure 8.2, and is akin in spirit to the COBE sky cube method [14]:

1. The sphere is inscribed in an icosahedron, whose faces are pixelized with a regular triangular grid.

2. The points are mapped radially onto the sphere.

3. The points are shifted around slightly, to give all pixels approximately equal area.

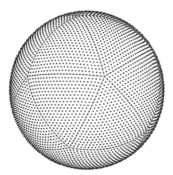

FIGURE 8.2   The icosahedron-based pixelization. (Reproduced from Tegmark, 1996, *ApJ*, 470, L81. With permission.)

Tegmark considers the 3D aspects of the problem as computationally trivial, since any of the 20 icosahedron faces can be rotated to lie in the $Z = 1$ plane by multiplication by an appropriate rotation matrix, and all these rotation matrices can be precomputed once and for all. The mapping between the $Z = 1$ tangent plane and the surface of the unit sphere preserves the direction of a vector and simply changes its length appropriately, either to be unity (on the sphere) or to have $Z = 1$. The straight lines on the tangent plane correspond to great circles on the sphere. Thus, each icosahedron face gets mapped onto a region on the sphere bounded by three great circles.

8.3.3 Igloo

Another suggestion for pixelization is the igloo pixelization [15] which divides a sphere into rows with edges of constant latitude and where each row is divided into identical pixels by lines of constant longitude. The pixels are roughly trapezoidal shaped, becoming nearly rectangular away from the poles. For simplicity and to speed up calculations, the northern and southern hemispheres are tiled identically. There are some advantages of igloo tilings. First, they are quite simple. They are also naturally azimuthal and can be easily made equal area, with most pixels nearly square. But perhaps their biggest advantage is that the pixel edges are defined along constant lines of the spherical polar coordinates θ and φ, allowing for an exact, fast integration of spherical harmonics over the pixels. This is essential in constructing exact simulated skies, and in optimally recovering the sky power spectrum from real data. The simplest example of an igloo pixelization is the ECP pixelization itself (see above). In it, every row has the same number of pixels, and the pixels become quite narrow near the poles. The other models have pixels which are more nearly square and nearly (or exactly) equal area. To do this the number of pixels in each row must decrease as one approaches the poles (Figure 8.3). One can construct igloo pixelizations with either rows equally spaced in latitude, like the ECP, or with pixels of uniform area. The advantage of equal latitude spacing is that the pixelization can be created by a simple rebinning of an ECP pixelization, provided the latter is chosen to have an appropriate number of pixels. In addition, by letting the pixel areas vary, one can make them less distorted. An equal-area pixelization will not be exactly equally spaced in latitude, but has the advantage that all of the pixels will have the same statistical weight.

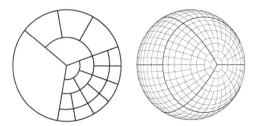

FIGURE 8.3 *Left*: Picture of the polar cap region in the igloo schemes, showing three levels of subdivision to higher-resolution pixels. *Right*: Picture of the 3:6:3 equal area pixelization, which divides the sky into 12 base patches, three at either cap and six $60° \times 60°$ patches at the equator. Here, each base pixel is broken up into 64 smaller pixels. (From R. G. Crittenden and N. G. Turok, 1998, Exactly azimuthal pixelizations of the sky, Report-no: DAMTP-1998-78, astro-ph/9806374. With permission.)

Igloo pixelizations can also be made hierarchical. To do this, one first divides the sphere into a base pixelization with relatively few pixels, optimized to minimize the pixel distortion. Each of these large pixels is then divided into four by bisecting it in longitude and latitude. The latter division is chosen either to keep the pixels the same area or to maintain a constant latitude spacing of the rows. Thus, one creates a finer-grained pixelization with four times the number of pixels. This procedure can be repeated until one reaches the required resolution. Reducing the number of base pixels tends to increase the level of pixel distortions, so some compromise must be found. While there are clear advantages to having fewer base pixels, it is not obvious as to how they should be weighed against the advantages of having more uniform pixels.

The hierarchical division causes some of the subpixels to become more distorted than the coarser pixels, especially near the poles. Away from the poles, there is a limit to how distorted the pixels become, even at the highest resolutions. However, if the polar cap were simply bisected in θ and ϕ, the pixels would become more and more distorted, as occurs in the ECP. Thus, one must use another method for dividing the polar regions. In igloo pixelization, Crittenden and Turok [15] have chosen to initially divide the cap of each pole into three equal wedges. Higher-resolution pixelizations are found by dividing each wedge into four pieces, one central wedge and three pieces surrounding it (see Figure 8.3). This process is iterated, with the interior wedge always being divided in this way and the outer pieces being divided by lines of constant θ and ϕ. This prescription is designed in order to minimize the distortions in the highest-resolution pixels. Crittenden and Turok [15] consider four possible igloo pixelizations: (a) an ECP pixelization, (b) a 12 pixel scheme (3:6:3) divided with equal area, (c) a 12 pixel scheme (3:6:3) divided with equal latitude spacing, and (d) an equal-area scheme with 12,116 base pixels. Figure 8.3 shows the 12 pixel scheme, where the base pixels are shown by bold lines with three pixels at each cap and six around the center. The divisions between the layers lie at $\theta \pm 30°$.

8.3.4 HEALPix

In fact, the first widespread pixelization scheme with an accompanying package was the HEALPix [16,17]. Górski et al. [16] have prepared some requirements to the mathematical structure of discrete whole sky maps and unified them in three items:

1. "Hierarchical structure of the database. This is recognized as essential for very large databases, and was indeed postulated already in the construction of the quadrilateralized spherical cube (or QuadCube, see [29] and http://lambda. gsfc.nasa.gov/product/cobe/skymap info new.cfm), which was used for all the COBE data. A simple argument in favor of this states that the data elements which are nearby in a multi-dimensional configuration space (here, on the surface of a sphere), are also nearby in the tree structure of the data base. This property facilitates various topo-logical methods of analysis, and allows for easy construction of wavelet transforms on triangular and quadrilateral grids through fast look-up of nearest neighbors."

2. "Equal areas of discrete elements of partition. This is advantageous because white noise at the sampling frequency of the instrument gets integrated exactly into white noise in the pixel space, and sky signals are sampled without regional dependence (although care must be taken to choose a pixel size sufficiently small compared to the instrumental resolution to avoid excessive, and pixel shape-dependent, signal smoothing)."

3. "Iso-latitude distribution of discrete area elements on a sphere. This property is essential for computational speed in all operations involving evaluations of spher-ical harmonics. Since the associated Legendre polynomials are evaluated via slow recursions, any sampling grid deviations from an iso-latitude distribution result in a prohibitive loss of computational performance with the growing number of sampling points."

Using these requirements, the HEALPix tessellation of the sphere has been prepared (Figure 8.4).

The base resolution has 12 pixels (or faces) in three rings around the poles and equa-tor. The next level of hierarchy is formed from the previous one by dividing each pixel of the previous level into four equal pixels. Thus, the resolution of the grid is expressed by parameter N_{side} which defines the number of divisions along the side of a base-resolution pixel that is needed to reach a desired high-resolution partition. In HEALPix, all pixel centers are placed on rings of constant latitude, and are equidistant in azimuth on each ring (see Figure 8.4). All iso-latitude rings are located between the upper and lower cor-ners of the equatorial faces (i.e., $-2/3 < \cos\theta_* < 2/3$) or in the equatorial zone. After division into the same number of pixels, we get $N_{eq} = 4 \times N_{side}$. The remaining rings are located within the polar cap regions ($|\cos\theta_*| > 2/3$) and contain a varying number of pixels, increasing from ring to ring, with increasing distance from the poles, by one pixel within each quadrant. So, a HEALPix map has $N_{pix} = 12N_{side}^2$ pixels of the same area $\Omega_{pix} = \frac{\pi}{3N_{side}^2}$.

The HEALPix software unifies programs and libraries. It contains procedures of harmonic decomposition for temperature and polarization ansotropy (most useful ones: "synfast" and "anafast"), estimation of the power spectrum, simulations of models of the primary and foreground sky signal, graphics, and others.

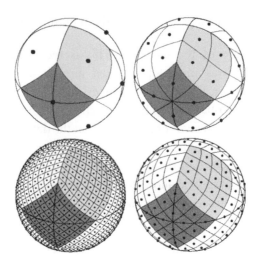

FIGURE 8.4 Orthographic view of the HEALPix partition of the sphere. Overplot of equator and meridians illustrates the octahedral symmetry of HEALPix. Light-gray shading shows one of the eight (four north, and four south) identical polar base-resolution pixels. Dark-gray shading shows one of the four identical equatorial base-resolution pixels. Moving clockwise from the upper left panel the grid is hierarchically subdivided with the grid resolution parameter equal to $N_{side} = 1, 2, 4, 8$, and the total number of pixels equal to $N_{pix} = 12 \times N_{side}^2 = 12, 48, 192, 768$. All pixel centers are located on $N_{ring} = 4 \times N_{side} - 1$ rings of constant latitude. Within each panel the areas of all pixels are identical. (From K. M. Górski et al. 2005, *ApJ*, 622, 759. With permission.)

8.4 GLESP

In this section devoted to the GLESP approach, we shall focus on the problem of processing on the sphere and then determine the scheme of pixelization. As was noted above, the pixelization of the CMB data on the sphere is only some part of the general problem, which is the determination of the coefficients of the spherical harmonic decomposition of the CMB signal for both anisotropy and polarization. These coefficients, which we call $a_{\ell m}$, are used in subsequent steps in the analysis of the measured signal, and in particular, in the determination of the power spectra, $C(\ell)$, of the anisotropy and polarization (see review in Ref. [21]), in some special methods for components separation [22,23] and phase analysis [23–26].

Below we describe a specific method to calculate the coefficients $a_{\ell m}$. It is based on the so-called Gaussian quadratures and is presented in the following section. In this specific pixelization scheme corresponds to the position of pixel centers along the θ-coordinate to the so-called Gauss–Legendre quadrature zeros and it will be shown that this method increases the accuracy of calculations essentially.

Thus, the method of calculation of the coefficients a_{lm} dictates the method of the pixelization. We call our method GLESP, the Gauss–Legendre Sky Pixelization [18,19,27]. We have developed a special code for the GLESP approach and a package of codes which are necessary for the whole investigation of the CMB data, including the determination of anisotropy and polarization power spectra, C_ℓ, the Minkowski functionals, and other statistics.

This section is devoted to description of the main idea of the GLESP method, the estimation of the accuracy of the different steps, and of the final results, the description of the GLESP code and its testing. We do not discuss the problem of how to make integration over a finite pixel size for the TOD. The simplest scheme of integration over pixel area is to use equivalent weight relatively to the center of the pixel. The GLESP code uses this method as, for example, HEALPix and igloo do.

8.4.1 Main Ideas and Basic Relations

The standard decomposition of the measured temperature variations on the sky, $\Delta T(\theta, \phi)$, in spherical harmonics is

$$\Delta T(\theta, \phi) = \sum_{\ell=2}^{\infty} \sum_{m=-\ell}^{m=\ell} a_{\ell m} Y_{\ell m}(\theta, \phi), \tag{8.5}$$

$$Y_{\ell m}(\theta, \phi) = \sqrt{\frac{(2\ell + 1)}{4\pi} \frac{(\ell - m)!}{(\ell + m)!}} P_{\ell}^{m}(x) e^{im\phi}, \quad x = \cos\theta, \tag{8.6}$$

where $P_{\ell}^{m}(x)$ are the associated Legendre polynomials. For a continuous $\Delta T(x, \phi)$ function, the coefficients of decomposition, $a_{\ell m}$, are

$$a_{\ell m} = \int_{-1}^{1} dx \int_{0}^{2\pi} d\phi \Delta T(x, \phi) Y_{\ell m}^{*}(x, \phi), \tag{8.7}$$

where $Y_{\ell m}^{*}$ denotes complex conjugation of $Y_{\ell m}$. For numerical evaluation of the integral equation 8.7, we will use the Gaussian quadratures, a method which was proposed by Gauss in 1814, and developed later by Christoffel in 1877. As the integral over x in Equation 8.7 is an integral over a polynomial of x, we may use the following equality [28]:

$$\int_{-1}^{1} dx \Delta T(x, \phi) Y_{\ell m}^{*}(x, \phi) \equiv \sum_{j=1}^{N} w_j \Delta T(x_j, \phi) Y_{\ell m}^{*}(x_j, \phi), \tag{8.8}$$

where w_j is a proper Gaussian quadrature weighting function. Here, the weighting function $w_j = w(x_j)$ and $\Delta T(x_j, \phi) Y_{\ell m}^{*}(x_j, \phi)$ are taken at points x_j which are the net of roots of the Legendre polynomial

$$P_N(x_j) = 0, \tag{8.9}$$

where N is the maximal rank of the polynomial under consideration. It is well known that Equation 8.9 has N number of zeros in interval $-1 \le x \le 1$. For the Gaussian–Legendre method, Equation 8.8, the weighting coefficients are [28]

$$w_j = \frac{2}{1 - x_j^2} [P_N'(x_j)]^{-2}, \tag{8.10}$$

where ′ denotes a derivative. They can be calculated together with the set of x_j with the *gauleg* code [28].

In the GLESP approach, there are the trapezoidal pixels bordered by θ and ϕ coordinate lines with the pixel centers (in the θ direction) situated at points with $x_j = \cos\theta_j$. Thus, the interval $-1 \leq x \leq 1$ is covered by N rings of the pixels. The angular resolution achieved in the measurement of the CMB data determines the upper limit of summation in Equation 8.5, $\ell \leq \ell_{max}$. To avoid the Nyquist restrictions we use a number of pixel rings, $N \geq 2\ell_{max}$. In order to make the pixels in the equatorial ring (along the ϕ coordinate) nearly squared, the number of pixels in this direction should be $N_\phi^{max} \approx 2N$. The number of pixels in other rings, N_ϕ^j, must be determined from the condition of making the pixel sizes as equal as possible with the equatorial ring of pixels.

Figure 8.5 shows the weighting coefficients, w_j, and the position of pixel centers for the case $N = 31$. Figure 8.6 compares some features of the pixelization schemes used in HEALPix and GLESP. Figure 8.7 compares pixel shapes and distribution on a sphere in a full sky Mollweide projections of HEALPix and GLESP maps.

In definition (8.5) are the coefficients $a_{\ell m}$ complex quantities while ΔT is real. In the GLESP code started from definition (8.5), we use the following representation of the ΔT:

$$\Delta T(\theta, \phi) = \sum_{\ell=2}^{\ell_{max}} a_{\ell 0} Y_{\ell 0}(\theta, \phi) + \sum_{\ell=2}^{\ell_{max}} \sum_{m=1}^{\ell} (a_{\ell m} Y_{\ell m}(\theta, \phi) + a_{l,-m} Y_{\ell,-m}(\theta, \phi)), \quad (8.11)$$

where

$$Y_{l,-m}(\theta, \phi) = (-1)^m Y_{l,m}^*(\theta, \phi), a_{\ell m} = (-1)^m a_{l,-m}^*. \quad (8.12)$$

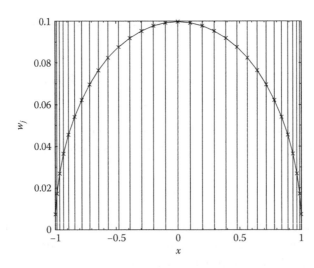

FIGURE 8.5 Gauss–Legendre weighting coefficients (w_j) versus Legendre polynomial zeros ($x_j = \cos\theta_j$) being centers of rings used in GLESP for the case of $N = 31$. Positions of zeros are plotted by vertical lines.

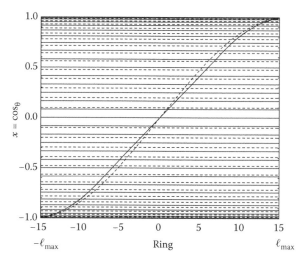

FIGURE 8.6 Ring center position ($x_j = \cos\theta_j$) versus ring number for two pixelization schemes, HEALPix (solid) and GLESP (dashed). Figure demonstrates the case of $N = 31$.

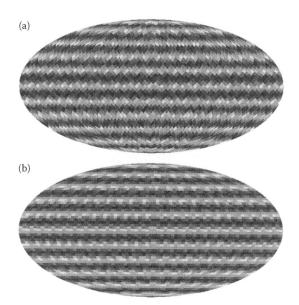

FIGURE 8.7 (**See color insert.**) Schematic representation of two types of pixelization on sphere: HEALPix (a) and GLESP (b). Various colors of pixels are used to show their shape.

Thus,

$$\Delta T(\theta, \phi) = \frac{1}{\sqrt{2\pi}} \sum_{\ell=2}^{\ell_{max}} \mathrm{Re}(a_{\ell,0}) P_\ell^0(\cos\theta) + \sqrt{\frac{2}{\pi}} \sum_{\ell=2}^{\ell_{max}} \sum_{m=1}^{l} \sqrt{\frac{2\ell+1}{2} \frac{(\ell-m)!}{(\ell+m)!}} P_\ell^m(\cos\theta)$$

$$\times [\mathrm{Re}(a_{\ell m}) \cos(m\phi) - \mathrm{Im}(a_{\ell m}) \sin(m\phi)], \tag{8.13}$$

where $P_\ell^m(\cos\theta)$ are the well-known associated Legendre polynomials (see Ref. [29]). In the GLESP code, we use normalized associated Legendre polynomials whose recurrence relation is given in Appendix A.

8.4.2 Properties of GLESP

Following the previous discussion we define the new pixelization scheme GLESP as follows:

- In the polar direction $x = \cos\theta$, we define $x_j, j = 1, 2, \ldots, N$, as the net of roots of Equation 8.9.

- Each root x_j determines the position of a ring with N_ϕ^j pixel centers with ϕ-coordinates ϕ_i.

- All the pixels have nearly equal area.

- Each pixel has weight w_j (see Equation 8.10).

In our numerical code which realizes the GLESP pixelization scheme, we use the following conditions:

- Borders of all pixels are along the coordinate lines of θ and ϕ. Thus, with a reasonable accuracy they are trapezoidal.

- The number of pixels along the azimuthal direction ϕ depends on the ring number. The code allows an arbitrary number of these pixels to be chosen. The number of pixels depends on the ℓ_{max} accepted for the CMB data reduction.

- To satisfy the Nyquist's theorem, the number N of the ring along the $x = \cos(\theta)$ axis must be taken as $N \geq 2\ell_{max} + 1$.

- To make equatorial pixels roughly square, the number of pixels along the azimuthal axis, ϕ, is taken as $N_\phi^{max} = \mathrm{int}(2\pi/d\theta_k + 0.5)$, where $k = \mathrm{int}(N/2 + 0.5)$, and $d\theta_k = 0.5(\theta_{k+1} - \theta_{k-1})$.

- The nominal size of each pixel is defined as $S_{pixel} = d\theta_k \times d\phi$, where $d\theta_k$ is the value on the equatorial ring and $d\phi = 2\pi/N_\phi^{mx}$ on equator.

- The number N_ϕ^j of pixels in the jth ring at $x = x_j$ is calculated as $N_\phi^j = \mathrm{int}(2\pi\sqrt{1 - x_j^2}/S_{pixel} + 0.5)$.

- Polar pixels are triangular.

- Because the number N_ϕ^j differs from 2^k, where k is an integer, we use for the fast Fourier transformation along the azimuthal direction the fastest Fourier transform in the west (FFTW) code (Frigo and Johnson 1997). This code permits one to use not only 2^n approach, but other base numbers too, and provide even higher speed.

With this scheme, the pixel sizes are equal inside each ring, and with a maximum deviation between the different rings of $\sim 1.5\%$ close to the poles (Figure 8.8). Increasing resolution

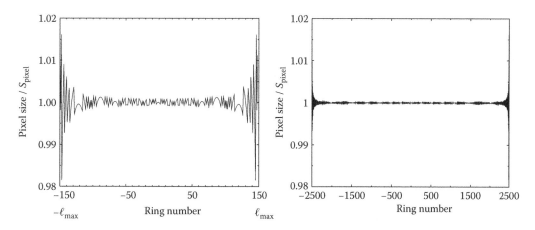

FIGURE 8.8 Pixel size/equator pixel area versus ring number for GLESP for number of rings $N = 300$ and $N = 5000$.

decreases an absolute error of an area because of the in-equivalence of polar and equator pixels proportionally to N^{-2}.

Figure 8.9 shows that this pixelization scheme for high-resolution maps (e.g., $\ell_{max} > 500$) produces nearly equal thickness $d\theta$ for most rings.

GLESP does not have the hierarchical structure, but the problem of the closest pixel selection is on the software level. Despite GLESP being close to the igloo pixelization scheme in the azimuthal approach, there is a difference between the two schemes in connection with the θ-angle (latitude) pixel step selection. Therefore, we cannot unify these two pixelizations. The igloo scheme applied to the GLESP latitude step will give very different pixel areas. The pixels will be neither equally spaced in latitude nor of uniform area, like an igloo requires.

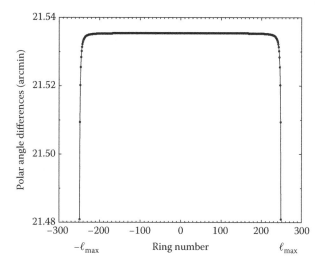

FIGURE 8.9 Pixel size along polar angle ($\ell_{max} = 250$).

8.4.3 Repixelization Problem

Unifying ideas of the nonhierarchical structure of the pixelization, but orienting onto the accuracy of $a_{\ell m}$ calculation, we prepared the GLESP package at http://www.glesp.nbi.dk (see also Appendix C) giving accuracy limited only by precision of the stored data format. However, there are natural limits of the CMB map accuracy connected with

- Level of the system noise

- Level of knowledge about local, galactic, and extragalactic foregrounds to be removed

- Instability of reconstruction methods

- Superimposed systematics errors

The last one connected with some operations often used for map analysis like map rotation in pixel domain and repixelization which could be used for map conversion between different pixelization systems.

To transfer a sky distribution map from one pixel grid (e.g., from HEALPix to GLESP) to another one, we should use one of two ways:

1. To calculate $a_{\ell m}$ coefficients and after that to restore a map in a new grid

2. To use repixelization procedures on the current brightness distribution

Any repixelization procedure will cause loss of information and thereby introduce uncertainties and errors. The GLESP code has procedures for map repixelization based on two different methods in the $\Delta T(\theta, \phi)$-domain: the first one consists in averaging input values in the corresponding pixel, and the second one is connected with spline interpolation inside the pixel grid.

In the first method, we consider input pixels which fell in our pixel with values $\Delta T(\theta_i, \phi_i)$ to be averaged with a weighting function. The realized weighting function is a function of simple averaging with equal weights. This method is widely used in appropriation of given values to the corresponding pixel number.

In the second method of repixelization, we use a spline interpolation approach. If we have a map $\Delta T(\theta_i, \phi_i)$ recorded in the knots different from the Gauss–Legendre grid, it is possible to repixelize it to our grid $\Delta T(\theta'_i, \phi'_i)$ using approximately the same number of pixels and the standard interpolation scheme based on the cubic spline approach for the map repixelization. This approach is sufficiently fast because the spline is calculated once for one vector of the tabulated data (e.g., in one ring), and values of interpolated function for any input argument are obtained by one call of separate routine (see routines "*spline*" to calculate second derivatives of interpolating function and *splint* to return a cubic spline interpolated value in Ref. [28].

Our spline interpolation consists of three steps:

- We set equidistant knots by the ϕ axis to reproduce an equidistant grid.

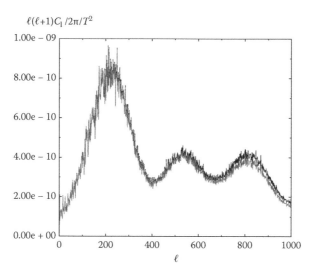

FIGURE 8.10 Power spectra calculated for the initial HEALPix map (upper curve) with $\ell_{max} = 1000$, $N_{side} = 1024$, pixel size $= 11.8026\ \Box'$, and $N_{tot} = 12,582,912$, and for resulting repixelized GLESP map (lower curve) with the closest possible pixel size $= 11.8038\ \Box'$, $N_{tot} = 12,581,579$. Deviations of the power spectra at high ℓ illustrate the ratio of the HEALPix and GLESP window functions.

- We change the grid by $x = \cos(\theta)$ axis to the required GLESP grid.

- Thereafter, we recalculate ϕ knots to the rings corresponding to the GLESP x points.

Figure 8.10 demonstrates the deviation of accuracy of the power spectrum in a case of repixelization from a HEALPix map to a GLESP map with the same resolution. As one can see, for the range $\ell \leq \ell_{max}/2$, repixelization reproduces correctly all properties of the power spectra. For $\ell \geq \ell_{max}/2$, some additional investigations need to be done to take into account the pixel-window function (see Appendix B).

The most standard accurate way is to produce maps just from the time-ordered data or to use transformation from a map of one pixelization to $a_{\ell m}$ coefficients and converse them to a map of another pixelization system.

8.5 SUMMARY

We have considered several pixelization schemes for CMB mapping on the whole sky. Two of them, HEALPix and GLESP, have the accompanying software. Actually, the selection of the pixelization scheme is now a matter of taste. The speed of calculation with modern laptops in both packages cannot show the fundamental difference (e.g., few seconds for map conversion up to $\ell_{max} = 1000$). The last versions of HEALPix (the present one is #2.15) have approximately the same accuracy as GLESP after four iterations which are calculated with approximately the same speed at the single processor. There should be noted that HEALPix calculational procedures are developed to support the MPI/OpenMP procedures that accelerate harmonics conversions, while only the Version 1 of the GLESP package has similar possibilities prepared by Vlad Stolyarov. The GLESP 2 package is mostly oriented to the laptop calculation effectively working up to $\ell_{max} = 4000$ with 2 Gb memory (however, MP procedures could be added in future too).

So, the present way of development of packages is to prepare additional supervising programs of higher-level analysis, making the convenient environment and organization of map-making and correlation procedures for various wavelength ranges. We develop the GLESP in these direction keeping also the main advantages of this scheme:

- It is focused on high-accuracy calculations of $a_{\ell m}$.

- An optimal selection of resolution for a given beam size, which means an optimal number of pixels and a pixel size.

In other cases, using accurately calculated $a_{\ell m}$, one can reproduce any pixelization scheme by the given pixel centers: GLESP, HEALPix, igloo, or icosahedron.

ACKNOWLEDGMENTS

The authors are thankful to Michael Way for useful remarks on the text. Some of the results in this chapter have been derived using the HEALPix [17] package. This work made use of the GLESP* [18,19] package for the further analysis of the CMB data on the sphere. Authors are very grateful to their collegues Pavel Naselsky, Viktor Turchaninov, Per Rex Christensen, Igor Novikov, and Lung-Yih Chiang for their contribution and testing the GLESP package. The study in this field was supported in part by Russian Foundation of Basic Research grant Nr.08-02-00159 and 09-02-00298 and Federal Program "Scientific and Pedagogical Personal Innovative of Russia" Nr. 1336. O.V.V. also acknowledges partial support from the "Dynasty" Foundation.

APPENDIX A. NORMALIZED ASSOCIATED LEGENDRE POLYNOMIALS

In the GLESP code, we use the normalized associated Legendre polynomials f_ℓ^m:

$$f_\ell^m(x) = \sqrt{\frac{2\ell + 1}{2} \frac{(\ell - m)!}{(\ell + m)!}} P_\ell^m(x), \tag{8.14}$$

where $x = \cos\theta$, and θ is the polar angle. These polynomials, $f_\ell^m(x)$, can be calculated using two well-known recurrent relations. The first of them gives $f_\ell^m(x)$ for a given m and all $\ell > m$:

$$f_\ell^m(x) = x\sqrt{\frac{4\ell^2 - 1}{\ell^2 - m^2}} f_{\ell-1}^m - \sqrt{\frac{2\ell + 1}{2\ell - 3} \frac{(\ell - 1)^2 - m^2}{\ell^2 - m^2}} f_{\ell-2}^m. \tag{8.15}$$

This relation starts with

$$f_m^m(x) = \frac{(-1)^m}{\sqrt{2}} \sqrt{\frac{(2m + 1)!!}{(2m - 1)!!}} (1 - x^2)^{m/2},$$

$$f_{m+1}^m = x\sqrt{2m + 3} f_m^m.$$

---

* http://www.glesp.nbi.dk

The second recurrent relation gives $f_\ell^m(x)$ for a given ℓ and all $m \leq l$:

$$\sqrt{(\ell - m - 1)(\ell + m + 2)}f_\ell^{m+2}(x) + \frac{2x(m+1)}{\sqrt{1 - x^2}}f_\ell^{m+1}(x)$$

$$+ \sqrt{(\ell - m)(\ell + m + 1)}f_\ell^m(x) = 0. \tag{8.16}$$

This relation is started with the same $f_\ell^\ell(x)$ and $f_\ell^0(x)$ which must be found with (8.15).

As is discussed in Press et al. (1992, Section 5.5), the first recurrence relation (8.15) is formally unstable if the number of iteration tends to infinity. Unfortunately, there are no theoretical recommendations with regard to the maximum iteration one can use in the quasi-stability area. However, it can be used because we are interested in the so-called *dominant* solution (Press et al. 1992, Section 5.5), which is approximately stable. The second recurrence relation (8.16) is stable for all ℓ and m.

APPENDIX B. GLESP PIXEL WINDOW FUNCTION

For application of the GLESP scheme, we have to take into account the influence of the pixel size, shape, and its location on the sphere on the signal in the pixel and its contribution to the power spectrum $C(\ell)$. The temperature in a pixel is [15,16]

$$\Delta T_p = \int_{\Delta\Omega_p} W_p(\theta, \phi)\Delta T(\theta, \phi)\,d\Omega, \tag{8.17}$$

where $W_p(\theta, \phi)$ is the window function for the pth pixel with the area $\Delta\Omega_p$. For the window function $W_p(\theta, \phi) = 1$ inside the pixel and $W_p(\theta, \phi) = 0$ outside [16], we have from Equations 8.5 and 8.17:

$$\Delta T_p = \sum_{\ell,m} a_{\ell m}W_p(\ell, m),$$

where

$$W_p(\ell, m) = \int d\Omega\, W_p(\theta, \phi)Y_{\ell m}(\theta, \phi)$$

and

$$W_p(\theta, \phi) = \sum_{\ell,m} W_p(\ell, m)Y_{\ell m}^*(\theta, \phi). \tag{8.18}$$

The corresponding correlation function [15] for the pixelized signal is

$$\langle \Delta T_p \Delta T_q \rangle = \sum_{\ell,m} C(\ell)W_p(\ell, m)W_q^*(\ell, m). \tag{8.19}$$

The discreteness of the pixelized map determines the properties of the signal for any pixels and restricts the precision achieved in any pixelization scheme. To estimate this precision, we can use the expansion [15]

$$\Delta T^{\text{map}}(\theta, \phi) = \sum_p S_p \Delta T_p W_p(\theta, \phi) = \sum_{\ell, m} a_{\ell m}^{\text{map}} Y_{\ell m}(\theta, \phi), \qquad (8.20)$$

$$a_{\ell m}^{\text{map}} = \int d\Omega \Delta T^{\text{map}}(\theta, \phi) Y_{\ell m}^*(\theta, \phi) = \sum_p S_p \Delta T_p W_p^*(\ell, m), \qquad (8.21)$$

where S_p is the area of the pth pixel. These relations generalize Equation 8.6, taking properties of the window function into account. The GLESP scheme uses the properties of Gauss–Legendre integration in the polar direction while Azimuthal pixelization for each ring is similar to the igloo scheme, and we get (see Equation 8.4)

$$W_p(\ell, m) = \frac{w_p}{\sqrt{2\pi}\Delta x_p} \exp\left(\frac{im\pi}{N_\phi^p}\right) \frac{\sin(\pi m/N_\phi^p)}{(\pi m/N_\phi^p)} \times \int_{x_p - 0.5\Delta x_p}^{x_p + 0.5\Delta x_p} dx f_\ell^m(x), \qquad (8.22)$$

where $\Delta x_p = (x_{p+1} - x_{p-1})/2$ with x_p the pth Gauss–Legendre knot and N_ϕ^p the number of pixels in the azimuthal direction. This integral can be rewritten as follows:

$$\int_{x_p - 0.5\Delta x_p}^{x_p + 0.5\Delta x_p} dx f_\ell^m(x) \simeq \sum_{k=0}^{\infty} \frac{1 + (-1)^k}{(k+1)!} f^{(k)}{}_\ell^m(x_p) \left(\frac{\Delta x_p}{2}\right)^{k+1}, \qquad (8.23)$$

where $f^{(k)}{}_\ell^m(x_p)$ denotes the kth derivatives at $x = x_p$. So, for $\Delta x_p \ll 1$ we get the expansion of (8.22):

$$W_p^{(2)}(\ell, m) = W_p^{(0)}(\ell, m) \left(1 + \frac{f_\ell^{(2)m}(\Delta x_p)^2}{24 f_\ell^m}\right), \qquad (8.24)$$

where

$$W_p^{(0)}(\ell, m) \simeq \frac{w_p}{\sqrt{2\pi}} \exp\left(\frac{im\pi}{N_\phi^p}\right) \frac{\sin(\pi m/N_\phi^p)}{(\pi m/N_\phi^p)} f_\ell^m(x_p), \qquad (8.25)$$

and $W_p^0(\ell, m)$ is independent of Δx_p. For the accuracy of this estimate we get

$$\frac{\delta W_p(\ell, m)}{W_p(\ell, m)} = \frac{W_p^{(2)}(\ell, m) - W_p^{(0)}(\ell, m)}{W_p^{(0)}(\ell, m)} \simeq \left|\frac{(f'')_\ell^m(\Delta x_p)^2}{24 f_\ell^m}\right|. \qquad (8.26)$$

According to the last version of the HEALPix package, an accuracy of the window function reproduction is about 10^{-3}. To obtain the same accuracy for the $W_p(\ell, m)$, we need to have

$$\Delta x_p \leq 0.15 \left|\frac{f_\ell^m}{(f'')_\ell^m}\right|^{\frac{1}{2}}\Bigg|_{x=x_p}. \qquad (8.27)$$

Using the approximate link between Legendre and Bessel functions for large ℓ [29] $f_\ell^m \propto J_m(\ell x)$ we get

$$\Delta x_p \leq 0.15 x_p / \sqrt{m(m+1)}, \tag{8.28}$$

and for $\Delta x_p \sim \pi/N$ we have from Equation 8.28

$$\frac{\delta W_p(\ell, m)}{W_p(\ell, m)} \geq 10^{-2} \cdot \left(\frac{\ell_{\max}}{N} \right)^2. \tag{8.29}$$

For example, for $N = 2\ell_{\max}$, we obtain $\delta W_p(\ell, m)/W_p(\ell, m) \simeq 2.3 \cdot 10^{-3}$, what is a quite reasonable accuracy for $\ell_{\max} \sim 3000$–6000.

APPENDIX C. THE GLESP PACKAGE

Structure of the GLESP Code

The code is developed in two levels of organization. The first one, which unifies F77 FOR-TRAN and C functions, subroutines and wrappers for C routines to be used for FORTRAN calls, consists of the main procedures: "*signal,*" which transforms given values of $a_{\ell m}$ to a map, "*alm,*" which transforms a map to $a_{\ell m}$, "*cl2alm,*" which creates a sample of $a_{\ell m}$ coefficients for a given C_ℓ and "*alm2cl,*" which calculates C_ℓ for $a_{\ell m}$. Procedures for code testing, parameters control, Kolmogorov-Smirnov analysis for Gaussianity of $a_{\ell m}$ and homogeneity of phase distribution, and others, are also included. Operation of these routines is based on a block of procedures calculating the Gauss–Legendre pixelization for a given resolution parameter, transformation of angles to pixel numbers and back.

The second level of the package contains the programs which are convenient for the utilization of the first-level routines. In addition to the straight use of the already-mentioned four main procedures, they also provide means to calculate map patterns generated by the Y_{20}, Y_{21}, and Y_{22} spherical functions, to compare two sets of $a_{\ell m}$ coefficients, to convert a GLESP map to a HEALPix map, and to convert a HEALPix map, or other maps, to a GLESP map.

Figure 8.11 outlines the GLESP package. The circle defines the zone of the GLESP influence based on the pixelization library. It can include several subroutines and operating programs. The basic program "*cl2map*" of the second level, shown as a big rectangle, interacts with the first-level subroutines. These subroutines are shown by small rectangles and call external libraries for the Fourier transform and Legendre polynomial calculations. The package reads and writes data both in ASCII table and FITS formats. More than 10 programs of the GLESP package operate in the GLESP zone.

The package satisfies the following principles:

- Each program is designed to be easily joined with other modules of a package. It operates both with a given file and standard output.

- Each program can operate separately.

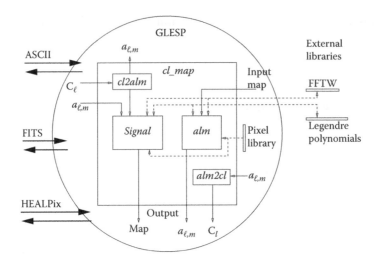

FIGURE 8.11   Structure of the GLESP package.

- Each program is accessible in a command string with external parameters. It has a dialogue mode and could be tuned with a resource file in some cases.

- Output format of resulting data is organized in the standard way and is prepared in the FITS format or ASCII table accessible for other packages.

- The package programs can interact with other FADPS procedures and CATS database (http://cats.sao.ru).

Main Operations

There are four types of operations accessible in the GLESP package:

- Operations related to maps:
    1. Spherical harmonic decomposition of a temperature anisotropy map into $a_{\ell m}$ (*cl2map*)

    2. Spherical harmonic decomposition of a temperature anisotropy and Q,U polarization maps into $a_{\ell m}$ and $e, b_{\ell m}$ coefficients (*polalm*)

    3. Smooth a map with a Gaussian beam (*cl2map*)

    4. Sum/difference/averaging between maps (*difmap*)

    5. Scalar multiplication/division (*difmap*)

    6. Map rotation (*difmap*)

    7. Conversion from Galactic to equatorial coordinates (*difmap*)

    8. Cut temperature values in a map (*mapcut*)

    9. Cut a zone in/from a map (*mapcut*)

    10. Cut out cross sections from a map (*mapcut*)

11. Produce simple patterns (*mappat*)

12. Read ASCII into binary (*mappat*)

13. Read point sources to binary map (*mappat*)

14. Print values in map (*mapcut*)

15. Find min/max values in map sample per pixel (*difmap*)

16. Simple statistic on a map (*difmap*)

17. Correlation coefficients of two maps (*difmap*)

18. Pixel size on a map (*ntot*)

19. Plot figures (*f2fig*)

- Operations related to $a_{\ell m}$:

    1. Synthesize the temperature anisotropy map from given $a_{\ell m}$ (*cl2map*)

    2. Synthesize the temperature anisotropy and Q,U polarization maps from given $a_{\ell m}$ and $e, b_{\ell m}$ coefficients (*polmap*)

    3. Sum/difference (*difalm*)

    4. Scalar multiplication/division (*difalm*)

    5. Vector multiplication/division (*difalm*)

    6. Add phase to all harmonics (*difalm*)

    7. Map rotation in harmonics (*difalm*)

    8. Cut out given mode of harmonics (*difalm*)

    9. Calculate angular power spectrum C_ℓ (*alm2dl*)

    10. Calculate phases (*alm2dl*)

    11. Select the harmonics with a given phase (*alm2dl*)

    12. Compare two $a_{\ell m}$ samples (*checkalm*)

    13. Produce $a_{\ell m}$ of map derivatives (*dalm*)

- Operations related to angular power spectrum C_ℓ:

    1. Calculate power spectrum C_ℓ (*alm2dl*)

    2. Simulate a map by a given C_ℓ (*cl2map*)

    3. Simulate $a_{\ell m}$ by C_ℓ (*createalm*)

- Operations related to phases $\phi_{\ell m}$ and amplitudes $|a_{\ell m}|$:

  1. Calculate phases $\phi_{\ell m}$ (*alm2dl*)

  2. Calculate amplitudes $|a_{\ell m}|$ (*alm2dl*)

  3. Simulate $a_{\ell m}$ by phases (*createalm*)

  4. Select harmonics with a given phase (*alm2dl*)

  5. Add a phase to all harmonics (*difalm*)

Main Programs

The following procedures organized as separate programs in the pixel and harmonics domain are realized now:

alm2dl calculates spectra and phases by $a_{\ell m}$ coefficients

checkalm compares different $a_{\ell m}$ samples

cmap converts HEALPix format maps to the GLESP package format

cl2map converts a map to $a_{\ell m}$ coefficients and $a_{\ell m}$ coefficients to a map, simulates a map by a given C_ℓ spectrum

createalm creates $a_{\ell m}$ coefficients by phases, amplitudes or/and C_ℓ spectrum

dalm calculates the 1st and 2nd derivatives by $a_{\ell m}$ coefficients

difalm calculates arithmetic operations over $a_{\ell m}$ samples

difmap calculates arithmetic operations over maps, produces coordinates transformations

f2fig produces color pictures in GIF images

f2map converts a GLESP map to a HEALPix format map

mapcut cuts amplitude and coordinates in a GLESP map, produces one-dimensional cross sections

mappat produces standard map patterns, reads ASCII data to produce a map, reads point sources position from ASCII files.

polalm converts temperature and Q,U polarization anisotropy maps to $a_{\ell m}$ and $e, b_{\ell m}$ coefficients

polmap converts $a_{\ell m}$ and $e, b_{\ell m}$ coefficients tp temperature and Q,U polarization anisotropy maps

Data Format

The GLESP data are represented in two formats describing $a_{\ell m}$ coefficients and maps.

$a_{\ell m}$ coefficients data contain index describing number of ℓ and m modes corresponding to the HEALPix, real and imaginary parts of $a_{\ell m}$. These three parameters are described by three-fields records of the binary table FITS format.

Map data are described by the three-fields binary table FITS format containing a vector of $x_i = \cos\theta$ positions, a vector of numbers of pixels per each layer N_{ϕ_i}, and set of temperature values in each pixel recorded by layers from the North Pole.

Test and Precision of the GLESP Code

Here, we give some examples of tests to check the code. The first of them is from the analytical maps:

$$Y_{2,0} = \sqrt{\frac{5}{16\pi}}(3x^2 - 1),$$

$$Y_{2,1} = -\sqrt{\frac{15}{8\pi}}x\sqrt{1 - x^2}\cos\phi,$$

$$Y_{2,-1} = -\sqrt{\frac{15}{8\pi}}x\sqrt{1 - x^2}\sin\phi,$$

$$Y_{2,2} = \sqrt{\frac{15}{32\pi}}(1 - x^2)\cos(2\phi),$$

$$Y_{2,-2} = -\sqrt{\frac{15}{32\pi}}(1 - x^2)\sin(2\phi)$$

which are used to calculate and check $a_{\ell m}$. The code reproduces the theoretical $a_{\ell m}$ better than 10^{-7}.

The second test is to reproduce an analytical map $\Delta T(x, \phi) = Y_{\ell m}(x, \phi)$ from a given $a_{\ell m}$. These tests check the calculations of the map and spherical coefficients independently.

The third test is the reconstruction of $a_{\ell m}$ after the calculations of the map, $\Delta T(x, \phi)$, and back. This test allows one to check orthogonality. If the transformation is based on really orthogonal functions it has to return after forward and backward calculation the same $a_{\ell m}$ values.

Precision of the code can be estimated by introduction of a set of $a_{\ell m} = 1$ and reconstruction of them. This test showed that using relation (8.15) we can reconstruct the introduced $a_{\ell m}$ with a precision $\sim 10^{-7}$ limited only by single precision of float point data recording and with a precision $\sim 10^{-5}$ for relation (8.16).

REFERENCES

1. F. K. Chan and E. M. O'Neil, 1976, Feasibility study of a quadrilateralized spherical cube Earth data base, Computer Sciences Corp., EPRF Technical Report.
2. E. M. O'Neill and R. E. Laubscher, 1976, Extended studies of a quadrilateralized spherical cube Earth data base, Computer Sciences Corp, EPRF Technical Report.
3. E. W. Greisen and M. Calabretta, 1993, Representations of celestial coordinates in FITS, *Bull. Am. Astron. Soc.* **182**, 808.
4. P. Natoli, G. de Gasperis, C. Gheller, and N. Vittorio, 2001, A map-making algorithm for the Planck surveyor, *Astron. Astrophys.* **372**, 346, astro-ph/0101252.
5. E. L. Wright, 1996, Paper presented at the IAS CMB Data Analysis Workshop, astro-ph/9612006.
6. M. Tegmark, 1996, An icosahedron-based method for pixelizing the celestial sphere, *ApJ*, **470**, L81.
7. A. H. Stroud, *Approximate Calculation of Multiple Integrals* (Englewood Cliffs: Prentice-Hall), 1971.
8. S. L. Sobolev, *Introduction to the Theory of Cubature Formulae* (Moscow: NAUKA), 1974.

9. I. P. Mysovskikh, The approximation of multiple integrals by using interpolatory cubature formulae, in *Quantitative Approximation*, R. A. Devore and K. Scherer (eds.) (New York: Academic Press), 1976.

10. S. I. Konyaev, 1979, Quadratures of Gaussian type for a sphere invariant under the icosahedral group with inversion (title from the English translation in Mathematical Notes, vol. 25, pp. 326–329, Springer), *Mat. Zametki*, **25**, 629.

11. J. R. Driscoll and D. M. Healy, 1994, Computing Fourier transforms and convolutions on the 2-sphere, *Adv. Appl. Math.*, **15**, 202.

12. P. F. Muciaccia, P. Natoli, and N. Vittorio, 1997, Fast spherical harmonic analysis: A quick algorithm for generating and/or inverting full-sky, high-resolution cosmic microwave background anisotropy maps, *ApJ*, **488**, L63.

13. P. Z. Kunszt, A. S. Szalay, I. Csabai, and A. R. Thakar, 2000, The indexing of the SDSS science archive, *Proc. Astronomical Data Analysis Software & Systems IX*, Astronomical Society of the Pacific Conference Series, **216**. N. Manset, C. Veillet, and D. Crabtree (eds.), 141.

14. M. Tegmark, 1997, How to make maps from cosmic microwave background data without losing information, *ApJ*, 480, L87, astro-ph/9611130.

15. R. G. Crittenden and N. G. Turok, 1998, Exactly azimuthal pixelizations of the sky, Report-no: DAMTP-1998-78, astro-ph/9806374.

16. K. M. Górski, E. Hivon, and B. D. Wandelt, 1999, Analysis issues for large CMB data sets, in *Evolution of Large-Scale Structure: From Recombination to Garching*, A. J. Banday, R. K. Sheth, L. N. da Costa (eds.) (Amsterdam, Netherlands: Enschede : Print Partners Ipskamp).

17. K. M. Górski, E. Hivon, A. J. Banday, B. D. Wandelt, F. K. Hansen, M. Reinecke, and M. Bartelmann, 2005, HEALPix: A framework for high-resolution discretization and fast analysis of data distributed on the sphere, *ApJ*, **622**, 759.

18. A. G. Doroshkevich, P. D. Naselsky, O. V. Verkhodanov, D. I. Novikov, V. I. Turchaninov, I. D. Novikov, P. R. Christensen, and L. -Y. Chiang, 2005, Gauss–Legendre sky pixelization (GLESP) for CMB maps, *Int. J. Mod. Phys.*, 14, 275, astro-ph/0305537.

19. A. G. Doroshkevich, O. V. Verkhodanov, P. D. Naselsky, Ja.Kim, D. I. Novikov, V. I. Turchaninov, I. D. Novikov, L.-Y. Chiang, and M. Hansen, 2011, The Gauss–Legendre sky pixelization for the CMB polarization (GLESP-pol). Errors due to pixelization of the CMB sky, *Int. J. Mod. Phys. D.*, 20, 1053.

20. R. A. White and S. W. Stemwedel, 1992, The quadrilateralized spherical cube and quad-tree for all sky data, *Astronomical Data Analysis, Software and Systems*, ASP Conf. Ser. **25**, D. M. Worrall, C. Biemesderfer, and J. Barnes (eds.) (San Francisco:ASP), 379.

21. E. Hivon, K. M. Górski, C. B. Netterfield, B. P. Crill, S. Prunet, and F. Hansen, 2002, MASTER of the cosmic microwave background anisotropy power spectrum: A fast method for statistical analysis of large and complex cosmic microwave background data sets, *ApJ*, **567**, 2.

22. V. Stolyarov, M. P. Hobson, M. A. J. Ashdown, and A. N. Lasenby, 2002, All-sky component separation for the Planck mission, *MNRAS*, **336**, 97.

23. P. D. Naselsky, A. G. Doroshkevich, and O. V. Verkhodanov, 2003, Phase cross-correlation of the WMAP ILC map and foregrounds, *ApJ*, **599**, L53, astro-ph/0310542.

24. L.-Y. Chiang, P. D. Naselsky, O. V. Verkhodanov, and M. J. Way, 2003, Non-Gaussianity of the derived maps from the first-year Wilkinson microwave anisotropy probe data, *ApJ*, 590, L65, astro-ph/0303643.

25. P. D. Naselsky, A. G. Doroshkevich, and O. V. Verkhodanov, 2004, Cross-correlation of the phases of the CMB and foregrounds derived from the WMAP data, *MNRAS*, **349**, 695, astro-ph/0310601.

26. P. Coles, P. Dineen, J. Earl, and D. Wright, 2004, Phase correlations in cosmic microwave background temperature map, *MNRAS*, **350**, 989, astro-ph/0310252.

27. O. V. Verkhodanov, A. G. Doroshkevich, P. D. Naselsky, D. I. Novikov, V. I. Turchaninov, I. D. Novikov, P. R. Christensen, and L.-Y. Chiang, 2005, GLESP package for full sky CMB maps data analysis and its realization in the FADPS data processing system, *Bull. SAO*, 58, 40.

28. W. H. Press, S. A. Teukolsky, W. T. Vetterling, and B. P. Flannery, 1992, *Numerical Recipes in FORTRAN*, 2nd edition, (New York: Cambridge University Press) (http://www.nr.com).

29. I. S. Gradsteyn and I. M. Ryzhik, 2000, *Tables of Integrals, Series and Products*, 6th Edition (San Diego: Academic Press).

CHAPTER 9

Future Sky Surveys

New Discovery Frontiers

J. Anthony Tyson
University of California, Davis

Kirk D. Borne
George Mason University

CONTENTS

9.1	Introduction	162
9.2	Evolving Research Frontier	163
	9.2.1 Surveys and Moore's Law	163
	9.2.2 Astroinformatics and the Survey Astronomer	164
9.3	New Digital Sky Surveys	164
	9.3.1 SkyMapper	165
	9.3.2 VISTA	166
	9.3.3 Pan-STARRS	166
	9.3.4 LSST	167
	9.3.5 WISE	167
	9.3.6 Kepler	167
	9.3.7 DES	168
	9.3.8 GAIA	168
	9.3.9 WFIRST	168
	9.3.10 Wide-Area Radio Surveys: ALMA, SKA, ASKAP, PiGSS	168
	9.3.11 Étendue Comparison of Various Optical Surveys	169
9.4	The Time Dimension	170
	9.4.1 Early Characterization Required	170
	9.4.2 Accurate Classification Needed	171
9.5	Science Opportunities and Survey Strategy	171
	9.5.1 Inventory of the Solar System	171
	9.5.2 Mapping the Milky Way	172
	9.5.3 Dark Energy	172
	9.5.4 Automated Data Quality Assessment	173

	9.5.4.1	Adaptive Retuning of Algorithm Behavior	175
	9.5.4.2	Data Complexity	175
	9.5.4.3	Automated Discovery	176
	9.5.4.4	Achieving Acceptably Low False Transient Alert Rate	176
9.5.5	Data Mining Research		178
9.6	Scientific Knowledge Discovery from Sky Survey Databases		178
9.7	Petascale Data-to-Knowledge Challenge		179
References			180

9.1 INTRODUCTION

Driven by the availability of new instrumentation, there has been an evolution in astronomical science toward comprehensive investigations of new phenomena. Major advances in our understanding of the Universe over the history of astronomy have often arisen from dramatic improvements in our capability to observe the sky to greater depth, in previously unexplored wavebands, with higher precision, or with improved spatial, spectral, or temporal resolution. Substantial progress in the important scientific problems of the next decade (determining the nature of dark energy and dark matter, studying the evolution of galaxies and the structure of our own Milky Way, opening up the time domain to discover faint variable objects, and mapping both the inner and outer Solar System) can be achieved through the application of advanced data mining methods and machine learning algorithms operating on the numerous large astronomical databases that will be generated from a variety of revolutionary future sky surveys.

Over the next decade, astronomy will irrevocably enter the era of big surveys and of really big telescopes. New sky surveys (some of which will produce petabyte-scale data collections) will begin their operations, and one or more very large telescopes (ELTs = Extremely Large Telescopes) will enter the construction phase. These programs and facilities will generate a remarkable wealth of data of high complexity, endowed with enormous scientific knowledge discovery potential. New parameter spaces will be opened, in multiple wavelength domains as well as the time domain, across wide areas of the sky, and down to unprecedented faint source flux limits. The synergies of grand facilities, massive data collections, and advanced machine learning algorithms will come together to enable discoveries within most areas of astronomical science, including Solar System, exo-planets, star formation, stellar populations, stellar death, galaxy assembly, galaxy evolution, quasar evolution, and cosmology.

Current and future sky surveys, comprising an alphabet soup of project names (e.g., Pan-STARRS, WISE, Kepler, DES, VST, VISTA, GAIA, EUCLID, SKA, LSST, and WFIRST; some of which are discussed in Chapters 17, 18, and 20), will contribute to the exponential explosion of complex data in astronomy. The scientific goals of these projects are as monumental as the programs themselves. The core scientific output of all of these will be their scientific data collection. Consequently, data mining and machine learning algorithms and specialists will become a common component of future astronomical research with these facilities. This synergistic combination and collaboration among multiple disciplines are essential in order to maximize the scientific discovery potential, the science output, the research efficiency, and the success of these projects.

9.2 EVOLVING RESEARCH FRONTIER

Large-scale sky surveys, such as SDSS, 2MASS, GALEX, and many others have proven the power of large data sets for answering fundamental astrophysical questions. This observational progress, based on advances in telescope construction, detectors, and information technology, has had a dramatic impact on nearly all fields of astronomy, and areas of fundamental physics. A new generation of digital sky surveys described here builds on the experience of these surveys and addresses the broad scientific goals of the coming decade.

Astronomy survey science tends to fall into several broad categories: (1) Statistical astronomy—large data sets of uniformly selected objects are used to determine distributions of various physical or observational characteristics; (2) Searches for rare and unanticipated objects—every major survey that has broken new ground in sensitivity, sky coverage, or wavelength has made important serendipitous discoveries, and surveys should be designed to optimize the chances of finding the unexpected; and (3) Surveys of large areas of sky— these become a legacy archive for future generations, allowing astronomers interested in a given area of sky to ask what is already known about the objects there, to calibrate a field photometrically or astrometrically, or to select a sample of objects with some specific properties.

9.2.1 Surveys and Moore's Law

Survey telescopes have been an engine for discoveries throughout the modern history of astronomy, and have been among the most highly cited and scientifically productive observing facilities in recent years. This observational progress has been based on advances in telescope construction, detectors, and above all, information technology.

Aided by rapid progress in microelectronics, current sky surveys are again changing the way we view and study the Universe. The next-generation instruments, and the surveys that will be made with them, will maintain this revolutionary progress. Figure 9.1 charts the trend in optical sky surveys over 50 years. The effect of technology is clear. While the total light-collecting area of telescopes has remained comparatively constant, the information content of sky surveys (using the number of galaxies measured per year to a given signal-to-noise S/N ratio as a proxy) has risen exponentially. This is in large part due to high-efficiency imaging arrays growing to fill the available focal plane, and to the increase in processing power to handle the data flood. Development of software to analyze these digital data in new ways has resulted in a corresponding increase in science output. Photographic surveys had the advantage of large focal plane area early on, but have been eclipsed by more sensitive CCD surveys, driven by the exponential rise in pixel count and computer processing capability—both enabled by the microelectronics "Moore's Law." Plotted versus time is the sum of all CCD pixels on the sky, as well as the number of transistors in a typical CPU. Processing capability keeps up with the data rate. Also plotted is one result of CCD surveys—the number of galaxies photometered per unit time—ranging from a survey using a single 160 Kpixel CCD on a 4-m telescope to a 3.2-Gpixel camera on an 8.4-m telescope.

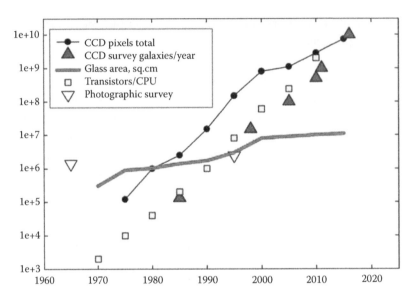

FIGURE 9.1 Data trends in optical surveys of the sky. While photographic surveys covered large area, the data were not as usable as digital data and did not go as faint. Information content (in galaxies surveyed per unit time to a given S/N ratio) in CCD digital surveys roughly follows Moore's law. Processing capability has kept up with pixel count. The most recent survey will scan the sky 100 times faster than the 2000 era survey. These next-generation wide-fast-deep surveys will open the time window on the universe.

9.2.2 Astroinformatics and the Survey Astronomer

The exponential increase in survey data and the resulting science opportunities have resulted in the development of a new breed of scientist, the "survey astronomer," and a new branch of astronomical research, "astroinformatics" (i.e., data-oriented astronomy research[1]). The massive astronomical databases from sky surveys therefore represent fundamental cyberinfrastructure for cyber-enabled scientific discovery by survey astronomers and astroinformatics practitioners. The related computational and data challenges and the exciting corresponding science opportunities are also attracting scientists from high-energy physics, statistics, and computer science. All these research groups are motivated to explore the massive data collections from large sky surveys. The history of astronomy has taught us repeatedly that there are unanticipated surprises whenever we view the sky in a new way. Complete, unbiased surveys are the best technique we have both for discovering new and unexpected phenomena, and for deriving the intrinsic properties of source classes so that their underlying physics can be deduced.

9.3 NEW DIGITAL SKY SURVEYS

Today there is a new crop of exciting sky survey facilities just starting operation or under construction. The success of the Sloan Digital Sky Survey (SDSS), the 2-Micron All-Sky Survey (2MASS), and other surveys, along with the important scientific questions facing us today, have all motivated astronomers to plan the next generation of major surveys. In addition to several ground-based projects, there are proposals for complementary space-based surveys at wavelengths unavailable from the ground (WISE is a currently operating example; and WFIRST is a possible space-based mission for the 2020s—these are both infrared-sensitive facilities). In the optical, wide-field cameras are being built for a variety of telescopes. Some of these will share the telescope with other instruments, limiting the time available for surveys. Others are specifically built wide-field telescopes and cameras dedicated to survey work. Some of these surveys will go appreciably deeper than the SDSS and with better image quality, and will open the time domain to study the variable universe. This chapter further discusses a short summary of four new dedicated survey facilities, followed by brief descriptions of other large data-producing astronomical projects that are either currently operating or that will be operational in the coming one to two decades. All of these facilities will not only contribute to the astronomical data flood, but will also contribute to the enormous scientific knowledge discovery potential from massive data collections at the frontiers of astronomical science.

A comparison of the first four surveys is made through each survey's étendue. The rate which a given facility can survey a volume of the universe is a product of its light-gathering power (effective aperture in square meters) and the area of the sky imaged in a single exposure (in square degrees). This product is the étendue.

9.3.1 SkyMapper

SkyMapper (Figure 9.2) is a 1.35 m modified Cassegrain telescope with $f/4.8$ optics, sited at Siding Spring Observatory near Coonabarabran, NSW Australia.[2] The telescope with an effective 5.2 deg^2 field of view 268 Mpixel CCD camera will survey the southern sky: about one billion stars and galaxies. Its 1-min exposures will tile each part of the observable sky 36 times in multiple optical bands, ultimately creating 500 TB of image data. This will be the first comprehensive digital survey of the entire southern sky. A distilled version of the

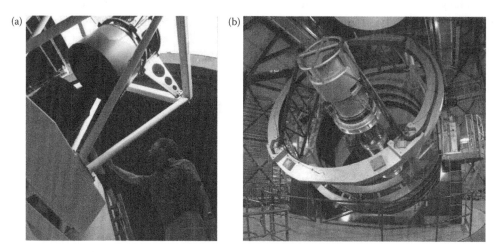

FIGURE 9.2 (a) SkyMapper 1.4 m optical survey telescope with 268 Mpixel CCD camera in Australia. (b) VISTA 4 m near-IR survey telescope and 67 Mpixel HgCdTe hybrid CMOS imaging array in Chile. (Adapted from Emerson, J. and Sutherland, W, *Proc. SPIE*, 4836, 35, 2002. http://apm26.ast.cam.ac.uk/vistasp/)

SkyMapper Survey will be made publicly available and will include a set of images of all the stars, galaxies, and nebulae, as well as a database containing the accurate color, position, brightness, variability, and shape of over a billion objects in the southern sky. SkyMapper has an étendue of 5.2 m^2 deg^2 and is now taking commissioning data.

9.3.2 VISTA

VISTA (Figure 9.2) is a 4-m wide-field near-IR survey telescope for the southern hemisphere, located at ESO's Cerro Paranal Observatory in Chile.[3] It is equipped with a near-infrared camera with 16 hybrid CMOS HgCdTe arrays totaling 67 million pixels of size 0.34 arcsec (0.6 square degree coverage per exposure) and five broad-band filters covering 0.85–2.3 μm, and a narrow-band filter at 1.18 μm. The telescope has an alt-azimuth mount and R–C (Ritchey–Chrétien) optics with a fast $f/1$ primary mirror giving an $f/3.25$ focus to the instrument at Cassegrain. The site, telescope aperture, wide field, and high quantum efficiency detectors will make VISTA the world's premier ground-based near-IR survey instrument. VISTA has an étendue of 6.8 m^2 deg^2 and is now taking science data.

9.3.3 Pan-STARRS

The basis of Pan-STARRS (Figure 9.3) is a 1.8 m $f/4$ R-C telescope and 1.4 Gpixel CCD camera in Hawaii.[4] The camera focal plane contains a 64×64 array of "orthogonal transfer" CCDs, each containing approximately 600×600 pixels, for a total of about 1.4 Gpixels. The planned surveys will use five wide bands spanning 0.4–1.0 μm. PS1 has an étendue of 13 m^2 deg^2. The primary science driver is a 30,000 deg^2 northern hemisphere survey of near-Earth objects of 1 km and even smaller size. However, many other science applications become possible with such a multiband optical survey of the sky. The data will remain proprietary for some time. Ultimately, the plan is to build four such 1.8-m facilities. The pilot system PS1 has been built and is now taking survey data.

(a)

(b)

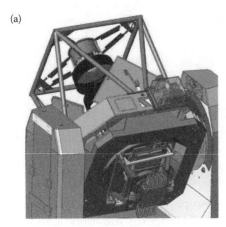

FIGURE 9.3 (a) Pan-STARRS PS1 1.8-m optical survey telescope with 1.4 Gpixel camera in Hawaii. (b) LSST 8.4-m optical survey telescope with 3.2 Gpixel camera planned for Chile.

9.3.4 LSST

With an étendue of 319 m^2 deg^2 the 3-mirror Large Synoptic Survey Telescope[5] (LSST; Figure 9.3) will produce a 6-band (0.3–1.1 μm) wide-field deep astronomical survey of over 30,000 deg^2 of the sky using an 8.4-m (effective aperture 6.7 m) active optics telescope and 3.2-Gpixel camera in Chile. Each patch of sky will be visited 1000 times (2 × 15 s exposures each time) in 10 years. Twenty trillion photometric and temporal measurements covering 20 billion detected objects will be obtained. The ∼15–30 terabytes of pipeline processed data obtained each night will yield a wide-fast survey of the deep optical universe for variability and motion, including asteroids. The deep coverage of 10 billion galaxies will provide unique capabilities for cosmology. Astrometry, 6-band photometry, and time domain data on 10 billion stars will enable studies of galactic structure. The goal is open-data open-source. About 90% of the observing time will be devoted to a uniform deep-wide-fast (main) survey mode. The main deep-wide-fast survey mode will observe a 20,000 deg^2 uniformly deep region of the southern sky up to +15 deg declination about 1000 times (summed over all six bands) during the anticipated 10 years of operations, resulting in a coadded map with 5σ point source limiting r (650 nm) magnitude of 27.7. The remaining 10% of the observing time is likely to be allocated to special programs such as a Very Deep + Fast time-domain survey (the so-called "Deep Drilling" fields). Over 100 science programs are discussed in the LSST Science Book.[5] For transient and variable phenomena LSST will extend time–volume discovery space a thousand times over current surveys. In Section 9.5, we briefly describe some of the major areas scientific discovery has anticipated for the LSST program.

9.3.5 WISE

NASA's space-based Wide-field Infrared Survey Explorer (WISE) mapped the sky in four infrared wavebands: 3.4, 4.6, 12, and 22 μm. WISE achieved 5-sigma point source sensitivities of 0.08, 0.11, 1, and 6 mJy (milli-Janky) in the ecliptic plane of the Solar System, with improved sensitivities toward the ecliptic poles. The astrometric precision of high signal-to-noise (S/N) sources is better than 0.15 arcsec, though the angular resolution is 6–12 arcsec across the four bands. The WISE first data release occurs in 2011, covering nearly 60% of the sky (∼23,600 square degrees). This preliminary data release includes an image atlas (in the four bands), a source catalog (with positions and 4-band photometry) for over 200 million subjects, and an Explanatory Supplement.

9.3.6 Kepler

NASA's space-based Kepler mission includes a telescope of 1.4-m diameter and a 95 Mpixel camera, which stares at a single patch of sky (105 square degrees), obtaining rapidly repeating images in a single broad bandpass (430–890 nm). The field of view encompasses ∼220,000 stars. These stars' temporal variations are cataloged with ultra-high photometric precision (better than 100 micromag). The key science driver is the detection of transiting exo-planets that pass in front of their host stars, but numerous other time-domain astronomy research use cases are also possible and encouraged. The Kepler project team is releasing its data in increments, consistent with its repeated sequential time observations. A Kepler data release in early 2011 included light curves for 165,000 stars, sampled at 30-min cadence, covering 3 months of observation.

9.3.7 DES

The Dark Energy Survey will use an existing telescope (the 4-m Blanco telescope in Chile) outfitted with a new wide-field camera: the DECam. DECam will comprise 520 Mpixels in 62 2K × 2K pixel CCDs, which will image 3 square degrees in its field of view. The survey is planned to cover 5000 square degrees in five SDSS-like optical-band filters: g, r, i, z, and y to an effective limiting magnitude of i ~ 25AB (about 2 magnitudes brighter than LSST) after 5 years. Survey observations will use 30% of the time on the Blanco telescope for 5 years and will chart 300,000 galaxies. DES will be supplemented by IR coverage from VISTA and SZ catalogs of galaxy clusters from the South Pole Telescope. The survey has not yet started. When it does get underway, two public data releases are planned: one approximately half-way through the survey and a final data release after the end of the survey.

9.3.8 GAIA

ESA's space-based GAIA mission will be primarily an astrometric mission, aimed at obtaining extremely accurate and precise positions (to 20 micro-arcsec) for stars down to 20 mag. The survey will require 5 years, using 106 CCDs of 4.5K × 2K each. Because each focal plane observation comprises 1 Gpixel (hence, several gigabits of data) and yet the telemetry link from the satellite is limited to an average rate of 1 Mbit/s, the GAIA facility will perform object detection, extraction, and compression on-board, prior to data transmission. Hence, the main data product from this survey will be the source position and photometry catalogs (not the images). There are actually three instruments onboard: the astrometry instrument, plus a photometer and a high-resolution spectrometer. The addition of spectral and photometric data with the astrometric data will provide rich opportunities for scientific analysis and discovery.

9.3.9 WFIRST

NASA's space-based WFIRST (Wide Field Infrared Survey Telescope) is proposed for launch in the following decade (2020s). The 1.5-m telescope would operate in the near-IR, with an operating mission length of 3 years. The facility would be equipped with a single instrument: a 144-Mpixel focal-plane array with a pixel size of 200 milliarcsec. WFIRST would operate in survey mode, obtaining near-IR photometric catalogs that can be joined with LSST catalogs for increased scientific discovery and analysis of hundreds of millions of sources across the sky.

9.3.10 Wide-Area Radio Surveys: ALMA, SKA, ASKAP, PiGSS

All of the aforementioned surveys make use of optical or near-IR instrumentation. It is natural to think of the data products from these surveys as images, source detections, and science object catalogs. In the field of radio astronomy, several large international facilities are coming online in the years ahead: the Atacama Large Millimeter Array (ALMA); the Square Kilometer Array (SKA); the Australian SKA Pathfinder (ASKAP); and the Allen

Telescope Array, which will conduct the Pi GHz Sky Survey (PiGSS), operating at 3.1 GHz (hence, Pi). For example, the PiGSS project will observe ~250,000 sources over a 2.5-year campaign in a 10,000-square degree sky region, with sensitivity of ~1 mJy. Some repeat observations of subregions of the survey will also be obtained.

The radio phase data (not maps) are not easily searched or viewed or integrated with other data sets. However, additional data products are planned for these new radio survey data sets. Like the FIRST (Faint Images of the Radio Sky at Twenty cm) survey of the early 1990s, these new radio surveys will generate more easily accessible, integrable, and searchable data products from the radio phase data, including digitized maps and source catalogs. One especially promising area for data mining research in this domain will be in automated pattern recognition and feature extraction within hundreds of thousands of radio maps, thereby generating catalogs of image features. Such morphological catalogs, along with source flux catalogs, will put the radio sky survey data collections on equal footing ("first class citizens") with optical/infrared sky survey databases in terms of data usability and data mining ease by the nonexpert.

9.3.11 Étendue Comparison of Various Optical Surveys

To survey a statistically significant portion of the universe, a facility must image deeply in a short time and tile a broad swath of sky many times. The rate which a given facility can survey a volume of the universe is proportional to the étendue. Figure 9.4 compares the étendue of the modern wide-field synoptic digital sky surveys, on a log scale. For good pixel sampling of the PSF, the étendue is also a measure of the data rate. Effective obscured aperture and active imaging field of view are used to calculate the étendue. For an all available sky survey

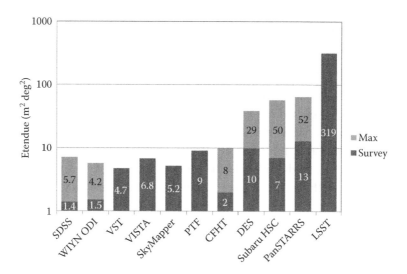

FIGURE 9.4 Étendue of current and planned optical survey telescopes and cameras. Some are dedicated 100% to surveys ("*Survey*"). Others could have higher effective étendue if used 100% in imaging survey mode or if duplicated as in the case of PanSTARRS ("*Max*"). Above 200–300 m² deg² it is possible to undertake a single comprehensive multiband survey of the entire visible sky serving most of the science opportunities, rather than multiple special surveys in series.

lasting a given number of years, higher étendue translates into more revisits to the same patch of sky and greater depth in the coadded image. Obviously, the fraction of nights a facility can spend on one survey is an issue, and this on-sky efficiency multiplies the étendue. Six of the facilities are dedicated to imaging surveys. The remaining share their wide-field cameras with other instruments (estimated fraction). Even so, it can be seen that the Hyper Suprime camera on the 8-m Subaru telescope, with its superb image quality, could be quite effective. For studies of galaxies, faint limiting surface brightness is key. This is enabled by high étendue. Image quality is also important. The PSF width of the facilities in Figure 9.4 spans a range of 5 to 1, and the rate of revisit to a sky patch spans across a 1000-to-1 range.

9.4 THE TIME DIMENSION

The time dimension is a new avenue for exploration, enabled by high étendue. Exploration of the variable optical sky is one of the new observational frontiers in astrophysics. Previous survey catalogs are primarily of the static sky. The Palomar Transient Factory (PTF)[6] is making discoveries today. No optical telescope to date has had the capability to search for transient phenomena at the ultra faint levels over enough of the sky needed to fully characterize phenomena. Variable and transient phenomena have historically led to fundamental insights into subjects ranging from the structure of stars, to the most energetic explosions in the universe, to cosmology. Existing surveys leave large volumes of discovery parameter space (in wavelength, depth, and cadence) unexplored. MACHO and OGLE were pathfinder surveys in time-domain astronomy, mainly focused on gravitational microlensing of stars (see chapter in this book by Wozniac et al.) The exponential increase in survey power shown in Figure 9.1 translates into an exponential increase in time sampling of the variable universe, adding the time dimension to our view of the cosmos.

9.4.1 Early Characterization Required

Because of the wide coverage and broad time sampling, some of the transient object science will be accomplished largely from the databases resulting from the wide-fast-deep surveys, combined where appropriate with multi-wavelength databases from other facilities, space, and ground. For fast or repeating transients, the LSST Deep Drilling subsurvey of ~50 selected 10 square degree fields will yield the highest quality data with excellent time sampling. However, much of the transient science enabled by these new surveys will depend heavily upon additional observations of selected transient objects based on their characterization using the primary survey photometric data. Early characterization of the temporal behavior of a transient object is essential in order for the additional follow-up observations to be judiciously prioritized and scientifically justified. This is especially critical for the LSST survey, which may produce up to 100,000 transient event detections each and every night for 10 years.

Characterization of transients may include measurements of several important parameters, such as: the change in brightness, the rate of rise for a brightening object (or the rate of decline for a dimming object), the color change, the power spectrum of variability, the principal components of variability, higher-order harmonics, periodicity, and monotonicity.

Associated context information will also inform the decision to follow-up on a specific transient, such as: the proximity of the source to another object (such as a galaxy, or a galaxy's nucleus), presence of companions (stars or galaxies), and prior evidence of transient behavior in the source. In most cases for time-domain sky surveys, the early characterization information will come directly from that survey's own data, while the associated information is more likely to be extracted automatically in near real time from external databases (which are accessible through Virtual Observatory search and retrieval protocols). The quality of the prioritization and scientific justification assessment for follow-up observations will depend on the quality of the predictive analytics workflow that is triggered at the onset of each transient event, including accurate probabilistic classification in multiple dimensions. Similarly, the timeliness of the follow-up observations will depend on the efficiency of the characterization workflow, including feature selection, characterization extraction, remote database access, data integration, and efficient probabilistic classification.

Consider the LSST example: every night LSST is expected to deliver data for tens of thousands of astrophysical transients. The majority of these will be moving objects or variable stars. Accurate event classification can be achieved by real-time access to the required context information: multi-color time-resolved photometry and host galaxy information from the survey itself, combined with broad-band spectral properties from external catalogs and alert feeds from other instruments. For photometry, LSST will provide sparsely time-sampled follow-up on hours–days timescales. Because we expect many such transients per field of view, efficient spectroscopic follow-up would best be carried out with multislit or multi-integral field spectroscopic systems.

9.4.2 Accurate Classification Needed

Efficient follow-up will depend on focusing limited resources on the interesting transients. There is always a trade-off between accuracy and completeness. After the first year of operation LSST will be able to produce enough archival and current transient information to enable accurate event *classification*. Combining the characterizations provided by the optical transient data with survey or archival data at other wavelengths will be routine through the Virtual Observatory. We expect that the community will make significant progress on classification of transients before LSST operations begin, given lessons learned from PTF and PS1. Classification accuracy is driven by precision deep photometry and many repeat measurements.

9.5 SCIENCE OPPORTUNITIES AND SURVEY STRATEGY

We describe here some of the key science programs and opportunities anticipated for the LSST project (Section 9.3.4). These data-driven discoveries are enabled by the deep-wide-fast characteristics of the 10-year LSST main sky survey.

9.5.1 Inventory of the Solar System

The small bodies of the Solar System offer a unique insight into its early stages. The planned cadence will result in orbital parameters for several million moving objects; these will be dominated by main-belt asteroids, with light curves and colorimetry for a substantial fraction of detected objects. This represents an increase of factors of 10–100 over the numbers of objects with documented orbits, colors, and variability information. Ground-based optical surveys are the most cost-effective tools for comprehensive near-Earth object (NEO) detection, determination of their orbits, and subsequent tracking. A survey capable of extending these tasks to Potentially Hazardous Objects (PHAs) with diameters as small as 100 m requires a large telescope, a large field of view, and sophisticated data acquisition, processing, and dissemination system. A 10-m class telescope is required to achieve faint detection limits quickly, and together with a large field of view (~10 square degrees), to enable frequently repeated observations of a significant sky fraction—producing tens of terabytes of imaging data per night. In order to recognize PHAs, determine their orbits, and disseminate the results to the interested communities in a timely manner, a powerful and fully automated data system is mandatory. A recent NRC study found that a ground-based telescope with an étendue above 300 m^2 deg^2 will be the most cost-effective way of achieving the Congressional mandate of detecting and determining the orbits of 90% of potentially hazardous asteroids larger than 140 m.[7] An infrared satellite near Venus could also do the job, but at far higher costs.[8]

9.5.2 Mapping the Milky Way

The new generation sky surveys are ideally suited to answering two basic questions about the Milky Way Galaxy: What is the structure and accretion history of the Milky Way? What are the fundamental properties of all the stars within 300 parsecs of the Sun? LSST will enable studies of the distribution of numerous stars beyond the presumed edge of the Galaxy's halo, their metallicity distribution throughout most of the halo, and their kinematics beyond the thick disk/halo boundary, and will obtain direct distance measurements below the hydrogen-burning limit for a representative thin-disk sample of fainter stars. For example, LSST will detect of the order of 10^{10} stars, with sufficient signal-to-noise ratio to enable accurate light curves, geometric parallax, and proper motion measurements for about a billion stars. Accurate multicolor photometry can be used for source classification and measurement of detailed stellar properties such as effective temperature and metallicity.

9.5.3 Dark Energy

Since dark energy changes the expansion rate, it affects two things: the distance–redshift relationship and the growth of mass structure. We would like to measure the way the expansion of our universe and the mass structure changes with cosmic time. Since the expansion of the universe has been accelerating, the development of mass structures via ordinary gravitational in-fall will be impeded. Measuring how dark matter structures and ratios of distances grow with cosmic time—via weak gravitational lensing observations—will provide clues to the nature of dark energy. A key strength of the next-generation surveys is the ability to survey huge volumes of the universe. Such a probe will be a natural part of the all-sky imaging survey: billions of distant galaxies will have their shapes and colors measured. Sufficient color data will be obtained for an estimate of the distance to each galaxy by using photometric redshifts derived from the multiband photometry. Due to its wide coverage of the sky,

LSST is uniquely capable of detecting any variation in the dark energy with direction.[9] By combining with dynamical data, we can test whether the "dark energy" is due to a breakdown of General Relativity on large scales.[10]

The LSST will enable scientists to study the dark energy in four different and complementary ways (Ref. 5, Chapters 13 through 15):

1. The telescope will tomographically image dark matter over cosmic time, via a "gravitational mirage." All the galaxies behind a clump of dark matter are deflected to a new place in the sky, causing their images to be distorted. This is effectively 3-D mass tomography of the universe.

2. Galaxies clump in a nonrandom way, guided by the natural scale that was imprinted in the fireball of the Big Bang. This angular scale (the so-called "baryonic acoustic oscillations" BAO) will be measured over cosmic time, yielding valuable information on the changing Hubble expansion.

3. The numbers of huge clusters of dark matter are a diagnostic of the underlying cosmology. Charting the numbers of these (via their gravitational mirage: weak lensing (WL) over cosmic time, will place another sensitive constraint on the physics of dark energy (see Chapter 14 Pires et al.)).

4. Finally, a million supernovae will be monitored, giving yet another complementary view of the history of the Hubble expansion.

Dark energy affects the cosmic history of the Hubble expansion as well as the cosmic history of mass clustering (which is suppressed at epochs when dark energy dominates). If combined, different types of probes of the expansion history (via distance measures) and dark matter structure history can lead to percent level precision in dark energy parameters. Using the cosmic microwave background as normalization, the combination of these deep probes over wide area will yield the needed precision to distinguish between models of dark energy, with cross checks to control systematic error. Hence, future surveys must measure both the distance and growth of dark matter structure, so that different theoretical models can be easily and uniformly confronted with the data. A 20,000 deg^2 very deep BAO + WL survey can achieve $\sim$0.5% precision on the distance and $\sim$2% on the structure growth factor over a wide range of redshift. It is important that BAO and WL surveys be combined—these surveys by themselves have far smaller cosmological constraining power. Such measurements can test the consistency of dark energy or modified gravity models.[11] For example, some dynamical dark energy models allow large changes or oscillations in the distance–redshift relation over this range of redshifts—these models can then be tested.

9.5.4 Automated Data Quality Assessment

Science breakthroughs enabled by this new generation of surveys, particularly LSST, will be to a large extent based on the precision of the measurements obtained over the duration of the survey. The science data quality assessment (DQA) needs are thus different in kind and level of effort compared with current standards—a direct result of the new regimes that LSST opens. There are three novel aspects of DQA for the LSST survey compared with current astronomy surveys and "big data" science generally:

1. Vastly larger data volumes (1000×) as well as data space (number of dimensions). While this complexity creates its own unique challenges for DQA, it leads to another novel aspect and challenge.

2. Vastly increased precision for cutting edge statistical investigations—key LSST science—made possible by the large size of the database. The main challenge is complexity in the analysis.

3. Unprecedented coverage of the faint time-domain opens the possibility of detection of the unexpected. Pursuit of this exciting science opportunity hinges on innovating DQA in a new dimension.

In scientific investigations there is always a trade-off between completeness and accuracy. Different investigations usually favor different places along this relation. Indeed the broad range of LSST science recounted in the Science Book spans the extremes. Some science programs rely more on completeness, some rely more on precision. Thus, it will be necessary to characterize as accurately as possible this completeness–precision space for the project's major data releases as well as in real time. The current generation surveys have often addressed systematic errors after the fact, during data analysis, and post data release. It is hoped that for LSST much of this understanding will emerge from explorations of the end-to-end simulations during R&D and construction, and that during operations we will have to contend with the residuals and the unknown unknowns. Understanding this to unprecedented levels prior to data release is for LSST an imperative. Finally, there is the inevitable flood of transient object detections and alerts, which calls for accompanying metadata capable of generating event classifications: this must be monitored and the tails of the distributions understood. All this calls for a radically new approach to DQA, with the following characteristics:

- Automated DQA (ADQA) (at least for known tasks)—summaries, continuous monitoring of statistics, distributions, and trends

- Expert human monitoring and interpretation of these ADQA products

- Metadata-Data exploration tools—visualization, and rapid efficient drill-down query capability

- Selected projects that use the data to push the envelope, thus exploring and uncovering residual systematic errors

- Continuous calibration of completeness-precision—injection of artificial signals at all levels

The optimal algorithms that will be designed for processing LSST data will themselves be sensitive to system efficiency, robustness, and changing observing conditions. ADQA must monitor these parameters and their statistical distributions in order to assure that the synoptic aspect of the LSST data is of the highest quality. Examples include subtle variations across the sky in signal to noise, and variations in efficiency in the time domain. Here are three examples:

9.5.4.1 Adaptive Retuning of Algorithm Behavior

Several key algorithms employed in the LSST application pipelines are complex, containing many data-dependent decisions and a large number of tuning parameters that affect their behavior. As observing conditions change, an algorithm may begin to fail for a particular choice of tuning parameters. Petascale data volumes make human intervention in such cases impractical, but it is essential that the pipelines continue to function successfully.

9.5.4.2 Data Complexity

The data challenges posed by current and future digital sky surveys include both data volume and data complexity. The challenges of "big data" are well documented, while the challenges of data complexity are infrequently addressed and consequently less well developed. One of the fundamental features of digital sky surveys that adds to data complexity is the high dimensionality of the data products. For example, the object catalogs frequently have many hundreds of dimensions (i.e., astrophysical parameters derived from the original data): in SDSS there are approximately 420 parameters in the object science table, while in 2MASS there are roughly 450 parameters in its science table. Similarly, LSST will have over 200 measured scientific parameters in its object table. It is easy to see that the intercomparison of multiple catalogs from several sky surveys can rapidly increase the dimensionality of the explored parameter set to more than 1000 dimensions. In such cases, the "curse of dimensionality" needs to be addressed through dimension reduction techniques, in order to reduce the combinatorial explosion of possibilities of parameter combinations that need to be explored. By way of example, consider the classic example of the fundamental plane of elliptical galaxies.[12] In this case, it is found that the majority of elliptical galaxies lie on a 2-dimensional hyperplane within the 3-dimensional parameter space consisting of velocity dispersion, radius, and surface brightness. It would be useful (and hopefully scientifically productive) to explore the existence of similar dimensional reductions in higher dimensions (e.g., the confinement of different classes of astronomical objects or phenomena to an M-dimensional hyperplane in an N-dimensional parameter space, where both M and N are large, and $M \ll N$). This type of analysis addresses the basic data complexity question of "what is the minimum parametric description of an object class?"

The search for (and meaning of) outliers, principal components, parameter clusters, pattern detection, object classification, and even the determination of classification boundaries between different object classes all migrate to a new level of difficulty—all of these data exploration activities will demand the application of novel, sophisticated, and rigorous statistical and machine learning algorithms. Dimension reduction in high-dimensional parameter spaces is therefore essential for arriving at sensible scientific meaning from the enormous complexity of multiparameter dependencies across multiple, large sky surveys, or even within a single sky survey's science catalog. Additional complications and challenges in the analysis of complex data arise from uncertainties in instrument calibration, from observational selection effects, from measurement errors, and in propagating errors to science model constraints. Classification, regression, and density estimations of massive data sets are all affected when the data are truncated in complex ways by observational selection and are contaminated with measurement errors and outliers. Designing efficient statistical emulators that are able to approximate outputs faithfully from complex astrophysical model simulations will thus make statistical inference from large complex data tractable.

To address the data complexity and "curse of dimensionality" challenges presented by the new large digital sky surveys, novel synergistic research is needed across multiple disciplines: machine learning algorithms, statistics methodologies, and astronomical data exploration. The application of new algorithms in all of these research domains will consequently enable and enhance the automated scientific knowledge discovery potential from the LSST (and other large sky surveys') data firehose. Specifically, the recent expansion of research in the field of astrostatistics is a welcome and much needed advancement.[13]

9.5.4.3 Automated Discovery

Spatio-temporal anomaly detection, hyperspace cluster detection, and automated data quality assessment will be essential components of the LSST data processing pipeline. LSST will produce large volumes of science data. The Data Management System will produce derived products for scientific use both during observing (i.e., alerts and supporting image and source data) and in daily and periodic reprocessing (the periodic reprocessing results in released science products). Analysis of the nightly data will also provide insight into the health of the telescope/camera system. An automated data quality assessment system must be developed, which efficiently searches for outliers in image data and unusual correlations of attributes in the database. This will involve aspects of machine learning, both unsupervised (finding the unknown unknowns) and supervised (classifying detected anomalies into known classes of behavior)—see Figures 9.5 and 9.6.

9.5.4.4 Achieving Acceptably Low False Transient Alert Rate

The science mission places high demand on the LSST's ability to accomplish two often competing goals: to rapidly and accurately detect and classify ~100% of the varying and transient objects, and to achieve a low false alarm rate. Given the very high data volume produced by the LSST, the correspondingly large number of detections in each image (up to 1 million objects detected per image), as well as the likelihood of entirely new classes of transients, the LSST will not be able to rely on traditional labor-intensive validation of detections, classifications, and alerts. To achieve the levels of accuracy required, new algorithms for surprise (anomaly, outlier, novelty) detection, characterization, and transient event classification must be created, as well as innovative automated techniques for alert filtering and validation. While not currently planned as part of the LSST data releases, the

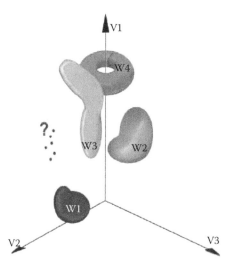

FIGURE 9.5 Unsupervised learning algorithm provides a "simplest" specification of the data. Similar objects cluster in complex regions of *n*-dimensional parameter space. Some anomalies will be instrumentation errors: automating their identification will allow the system to "bootstrap" its knowledge and improve the response. But some of the anomalies will be rare astrophysical objects that are difficult to find any other way. (Courtesy Tom Vestrand.)

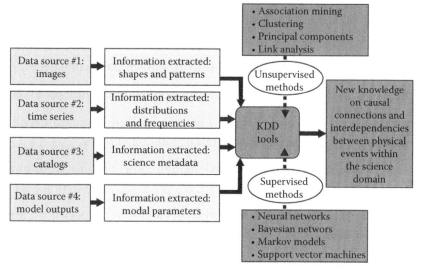

FIGURE 9.6 Schematic illustration of scientific knowledge discovery in databases (*KDD*). Numerous astronomical databases, generated from large sky surveys, provide a wealth of data and a variety of data types for mining. Data mining proceeds first with data discovery and access from multiple distributed data repositories, then continues with the information extraction step, and then is enriched through the application of machine learning algorithms and data mining methods for scientific knowledge discovery. Finally, the scientist evaluates, synthesizes, and interprets the newly discovered knowledge in order to arrive at a greater scientific understanding of the Universe and its constituent parts, processes, and phenomena. *Supervised methods* are those based on training data—these are primarily classification methods. *Unsupervised methods* are those that do not rely on training data or pre-known classes—these include traditional clustering methods.

LSST project, working together with its science collaboration teams, must push development of high-accuracy automated detection, characterization, and classification algorithms during the R&D phase of the project to "proof of principle." During survey operations the statistics of object classifications (and reclassifications) will be an important data quality diagnostic.

9.5.5 Data Mining Research

The LSST project is expected to produce ∼15–30 TB of data per night of observation for 10 years. The final image archive will be ∼100 PB, and the final LSST astronomical object catalog (object-attribute database) is expected to be ∼50 PB, comprising over 200 attributes for 50 billion objects and ∼20 trillion source observations. We anticipate many data mining research opportunities and use cases with the LSST survey database and image archive, including:

- Provide rapid characterizations and probabilistic classifications for one million LSST events nightly.

- Find new correlations and associations of all kinds from the 200 + science attributes.

- Discover voids in multidimensional parameter spaces (e.g., period gaps).

- Discover new and exotic classes of astronomical objects, or new properties of known classes.

- Discover new and improved rules for classifying known classes of objects.

- Identify novel, unexpected behavior in the time domain from time series data.

- Hypothesis testing—verify existing (or generate new) astronomical hypotheses with strong statistical confidence, using millions of training samples.

- Serendipity—discover the rare one-in-a-billion type of objects through outlier detection.

- Science data quality assurance—identify glitches and image processing errors through deviation detection.

9.6 SCIENTIFIC KNOWLEDGE DISCOVERY FROM SKY SURVEY DATABASES

A wide-fast-deep survey of a large fraction of the sky in multiple optical bands (such as that planned for LSST) is required in order to explore many of the exciting science opportunities of the next decade. As was the case with SDSS, we expect the scientific community will produce a rich harvest of discoveries. What can one do with the combined power of all of the new and existing sky surveys (some of which we describe in Section 9.3)? Figure 9.6 schematically illustrates the emerging multimodal knowledge discovery paradigm. The new sky surveys will produce not only the traditional images and catalogs, but they will produce many ancillary data products (some of which will be generated by external entities). These multimodal data products may include spectroscopic data, time series, phase data (e.g., from radio-wavelength surveys), and numerical simulation outputs (e.g., used to model and interpret large complex systems and their observed properties, such as planetary systems, star systems, colliding galaxies, clusters of galaxies, large-scale structure, or the entire cosmos).

For all this to be possible, certain data policies and processes must be in place. For example, data processing, data products, and data access should enable efficient science analysis without a significant impact on the final uncertainties. To enable a fast and efficient response to transient sources, the processing latency for objects that change should be less than a minute after the close of the shutter, with a robust and accurate preliminary classification of reported transients. Likewise, the public availability and accessibility of the data will enable the broader astronomical community to generate more science than the survey collaborations could alone. An open-source open-data policy benefits all and ensures innovation in applications software. Therefore, a survey should plan to make its data and software pipelines public and properly documented in a form that allows the full scientific community to use them.[14]

9.7 PETASCALE DATA-TO-KNOWLEDGE CHALLENGE

Petabyte databases are rare today, but will soon be commonplace. Data sets are growing exponentially, a result of the exponential growth of the underlying technologies which allow us to gather the data. An additional issue is the increasing degree of complexity of the data, requiring dimension reduction for efficient analysis. Naively, one might think that these datasets could be processed and analyzed in constant time, given the same exponential growth of computing power. However, this is true only for linear algorithms and only to the extent that all processing and analysis can be automated. Moreover, most databases are collected for a single purpose. Until now, analysis of these databases has focused on discovery of predetermined features. This is true over a broad spectrum of applications, from high-energy physics, to astronomy, to geosciences. Optimal filters are built to detect given events or patterns. Most often the database itself is prefiltered and tuned to a well-defined application or result, and automated search algorithms are designed which detect and produce statistics on these predetermined features. However, paradigm-shifting discoveries in science often are the result of a different process: on close examination of the data a scientist finds a totally unexpected event or correlation.

Large synoptic imaging projects are a perfect example. A time sequence of images is searched for given features. The unexpected goes undetected. One solution to this failure to detect the unexpected rare event is to build visualization tools which keep the human partially in the loop. This still misses most rare events in huge databases, or the unexpected statistical correlation. In the near future, Gpixel cameras will flood us with data which must be processed and analyzed immediately. Automated software pipelines will relentlessly search for changes. Most any kind of change at bright levels will be detected. The problem occurs at the faint end where most of the volume is.

There, we will have matched filters to detect *known* classes of objects (variable stars, supernovae, moving objects). But we cannot build a matched filter for something that is beyond our current imagination. The data space is position, time, intensity, colors, and motion. One challenge is to efficiently find rare events in large databases and gauge their significance. Another challenge is finding unexpected statistical correlations. There will be a need for developing statistically efficient estimation schemes and methods for assessing estimation error. Combining statistical efficiency with computational efficiency will be a constant challenge, since the more statistically accurate estimation methods will often be the most computationally intensive. Such automated data search algorithms will serve multiple purposes, from revealing systematic errors in the data for known features to discovery of unexpected correlations in nature. This approach is pointless unless we couple with advanced visualization. By combining automated search algorithms with multidimensional visualization we can speed the process of discovery by harnessing simultaneously the best capabilities of the machine and mind.

It will be particularly effective to have data mining and machine learning innovations in place when the next generation surveys are online. Achieving this requires parallel efforts in informatics, statistics, and computer science. Informatics research includes science data indexing and organization, data mining, machine learning, semantics, information retrieval, and visualization. Visualization techniques will specifically aid the development of automated machine learning algorithms. The collaboration will include astronomers and physicists involved in current and upcoming ultra-large surveys, alongside experts in statistics, algorithm development, computer science, data mining, and visualization.

REFERENCES

1. Borne, K., Astroinformatics: Data-oriented astronomy research and education, *JESI*, 3, 5–17, 2010.
2. Keller, S. C. et al., The SkyMapper telescope and the southern sky survey, *PASA*, 24, 1–12, 2007.
3. Emerson, J. and Sutherland, W., Visible and infrared survey telescope for astronomy: Overview, *Proc. SPIE*, 4836, 35, 2002. http://apm26.ast.cam.ac.uk/vistasp/
4. Kaiser, N., Pan-STARRS: A wide-field optical survey telescope array, *SPIE*, 5489, 11–22, 2004. http://www.pan-starrs.ifa.hawaii.edu/
5. LSST Science Collaborations and LSST Project, 2009. LSST Science Book, v2.0, arXiv: 0912.0201, http://www.lsst.org/lsst/scibook
6. Law, N. M. et al. The Palomar transient factory: System overview, performance, and first results, *PASP*, 121, 1395–1408, 2009. http://www.astro.caltech.edu/ptf/
7. Shapiro, I. I. et al. *Defending Planet Earth: Near-Earth Object Surveys and Hazard Mitigation Strategies: Final Report*, National Academies Press, Washington, D.C., 2010.
8. NASA Report to Congress, Near-earth object survey and detection analysis of alternatives, http://www.nasa.gov/pdf/171331main_NEO_report_march07.pdf, 2007.
9. Scranton, R. et al., The case for deep, wide-field cosmology, arXiv:0902.2590, 2009.
10. Zhang, P., Liguori, M., Bean, R., and Dodelson, S., Probing gravity at cosmological scales by measurements which test the relationship between gravitational lensing and matter overdensity, *Phys. Rev. Lett.*, 99, 141,302, 2007.
11. Zhan, H., Knox, L., and Tyson, J. A., Distance, growth factor, and dark energy constraints from photometric Baryon acoustic oscillation and weak lensing measurements, *Ap. J.*, 690, 923–936, 2009.

12. Djorgovski, S. G. and Davis, M., Fundamental properties of elliptical galaxies, *Ap. J.*, 313, 59, 1987.

13. Feigelson, E. D. and Babu, G. J., Statistical challenges in modern astronomy, arXiv:astro-ph/0401404v1, 2004.

14. Kleppner, D., Sharp, P. et al., *Ensuring the Integrity, Accessibility and Stewardship of Research Data in the Digital Age*, National Academies Press, Washington, D.C., 2009.

CHAPTER 10

Poisson Noise Removal in Spherical Multichannel Images

Application to Fermi Data

Jérémy Schmitt and Jean-Luc Starck
CEA/DSM–CNRS–Université Paris Diderot

Jalal Fadili
CNRS–ENSICAEN–Université de Caen

Seth Digel
Kavli Institute for Particle Astrophysics and Cosmology

CONTENTS

10.1	Introduction	184
10.2	Wavelets and Curvelets on the Sphere	186
	10.2.1 The HEALPix Pixellization for Spherical Data	186
	10.2.2 IUWT on the Sphere	186
	10.2.3 Curvelet Transform on the Sphere	187
	10.2.4 Application to Gaussian Denoising on the Sphere	187
10.3	Multiscale Transforms on the Sphere and Poisson Noise	189
	10.3.1 Principle of the Multiscale MS-VSTS	189
	10.3.1.1 VST of a Poisson Process	189
	10.3.1.2 VST of a Filtered Poisson Process	189
	10.3.2 Wavelets and Poisson Noise	190
	10.3.3 Curvelets and Poisson Noise	192
10.4	Application to Poisson Denoising on the Sphere	193
	10.4.1 MS-VSTS + IUWT	193
	10.4.2 Multiresolution Support Adaptation	194
	10.4.3 MS-VSTS + Curvelets	195
	10.4.4 Experiments	196

10.5 Application to Inpainting and Source Extraction   197
    10.5.1 Milky Way Diffuse Background Study: Denoising and Inpainting   197
        10.5.1.1 Experiment   199
    10.5.2 Source Detection: Denoising and Background Modeling   199
        10.5.2.1 Method   199
        10.5.2.2 Experiment   201
10.6 Extension to Multichannel Data   202
    10.6.1 Gaussian Noise   202
        10.6.1.1 2D–1D Wavelet Transform on the Sphere   202
        10.6.1.2 Fast Undecimated 2D–1D Decomposition/Reconstruction   205
        10.6.1.3 Multichannel Gaussian Denoising   206
    10.6.2 Poisson Noise   207
        10.6.2.1 Multiscale Variance Stabilizing Transform   207
        10.6.2.2 Detection–Reconstruction   207
10.7 Conclusion   209
Acknowledgment   209
References   209

10.1 INTRODUCTION

The Fermi Gamma-ray Space Telescope, which was launched by NASA in June 2008, is a powerful space observatory which studies the high-energy γ-ray sky [5]. Fermi's main instrument, the Large Area Telescope (LAT), detects photons in an energy range between 20 MeV and >300 GeV. The LAT is much more sensitive than its predecessor, the energetic gamma ray experiment telescope (EGRET) telescope on the Compton Gamma-ray Observatory, and is expected to find several thousand γ-ray point sources, which is an order of magnitude more than its predecessor EGRET [13].

Even with its relatively large acceptance (∼2 m^2 sr), the number of photons detected by the LAT outside the Galactic plane and away from intense sources is relatively low and the sky overall has a diffuse glow from cosmic-ray interactions with interstellar gas and low-energy photons that makes a background against which point sources need to be detected. In addition, the per-photon angular resolution of the LAT is relatively poor and strongly energy dependent, ranging from >10° at 20 MeV to ∼0.1° above 100 GeV. Consequently, the spherical photon count images obtained by Fermi are degraded by the fluctuations on the number of detected photons. This kind of noise is strongly signal dependent: on the brightest parts of the image like the galactic plane or the brightest sources, we have a lot of photons per pixel, and so the photon noise is low. Outside the galactic plane, the number of photons per pixel is low, which means that the photon noise is high. Such a signal-dependent noise cannot be accurately modeled by a Gaussian distribution. The basic photon-imaging model assumes that the number of detected photons at each pixel location is Poisson distributed.

More specifically, the image is considered as a realization of an inhomogeneous Poisson process. This statistical noise makes the source detection more difficult, consequently it is highly desirable to have an efficient denoising method for spherical Poisson data.

Several techniques have been proposed in the literature to estimate Poisson intensity in 2-dimensional (2D). A major class of methods adopt a multiscale Bayesian framework specifically tailored for Poisson data [18], independently initiated by Timmerman and Nowak [23] and Kolaczyk [14]. Lefkimmiaits et al. [15] proposed an improved Bayesian framework for analyzing Poisson processes, based on a multiscale representation of the Poisson process in which the ratios of the underlying Poisson intensities in adjacent scales are modeled as mixtures of conjugate parametric distributions. Another approach includes preprocessing the count data by a variance stabilizing transform (VST) such as the Anscombe [4] and the Fisz [10] transforms, applied respectively in the spatial [8] or in the wavelet domain [11]. The transform reforms the data so that the noise approximately becomes Gaussian with a constant variance. Standard techniques for independent identically distributed Gaussian noise are then used for denoising. Zhang et al. [25] proposed a powerful method called multiscale (MS-VST). It consists in combining a VST with a multiscale transform (wavelets, ridgelets, or curvelets), yielding asymptotically normally distributed coefficients with known variances. The interest of using a multiscale method is to exploit the sparsity properties of the data : the data are transformed into a domain in which it is sparse, and, as the noise is not sparse in any transform domain, it is easy to separate it from the signal. When the noise is Gaussian of known variance, it is easy to remove it with a high thresholding in the wavelet domain. The choice of the multiscale transform depends on the morphology of the data. Wavelets represent more efficiently regular structures and isotropic singularities, whereas ridgelets are designed to represent global lines in an image, and curvelets represent efficiently curvilinear contours. Significant coefficients are then detected with binary hypothesis testing, and the final estimate is reconstructed with an iterative scheme. In Ref. [21], it was shown that sources can be detected in 3-dimensional (3D) LAT data (2D + time or 2D + energy) using a specific 3D extension of the MS-VST.

To denoise Fermi maps, we need a method for Poisson intensity estimation on spherical data. It is possible to decompose the spherical data into several 2D projections, denoise each projection and reconstitute the denoised spherical data, but the projection induces some caveats like visual artifacts on the borders or deformation of the sources.

In the scope of the Fermi mission, two of the main scientific objectives are in a sense complementary:

- Detection of point sources to build the catalog of γ-ray sources
- Study of the Milky Way diffuse background

The first objective implies the extraction of the Galactic diffuse background. Consequently, we want a method to suppress Poisson noise while extracting a model of the diffuse background. The second objective implies the suppression of the point sources: we want to apply a binary mask on the data (equal to 0 on point sources, and to 1 everywhere else) and to denoise the data while interpolating the missing part. Both objectives are linked: a better knowledge of the Milky Way diffuse background enables us to improve our background model, which leads to a better source detection, while the detected sources are masked to study the diffuse background.

The aim of this chapter is to present a multiscale representation for spherical data with Poisson noise called MS-VST on the sphere (MS-VSTS) [19], combining the MS-VST [25] with various multiscale transforms on the sphere (wavelets and curvelets) [2,3,22]. Section 10.2 presents some multiscale transforms on the sphere. Section 10.3 introduces a new multiscale representation for data with Poisson noise called MS-VSTS. Section 10.4 applies this representation to Poisson noise removal on Fermi data. Section 10.5 presents applications to missing data interpolation and source extraction. Section 10.6 extends the method to multichannel data.

All experiments were performed on HEALPix maps with *nside* = 128 [12], which corresponds to a good pixellization choice for the Gamma ray Large Area Space Telescope (former name of Fermi Gamma Ray Space Telescope) (GLAST)/FERMI resolution.

10.2 WAVELETS AND CURVELETS ON THE SPHERE

New multiscale transforms on the sphere were developed by Starck et al. [22]. These transforms can be inverted and are easy to compute with the HEALPix pixellization, and were used for denoising, deconvolution, Morphological Component Analysis (MCA), and inpainting applications [2]. In this chapter, here we use the isotropic undecimated wavelet transform (IUWT) and the curvelet transform.

10.2.1 The HEALPix Pixellization for Spherical Data

Different kinds of pixellization scheme exist for data on the sphere. For Fermi data, we use the HEALPix representation (Hierarchical Equal Area isoLatitude Pixellization of a sphere) [12], a curvilinear hierarchical partition of the sphere into quadrilateral pixels of exactly equal area but with varying shape. The base resolution divides the sphere into 12 quadrilateral faces of equal area placed on three rings around the poles and equator. Each face is subsequently divided into $nside^2$ pixels following a quadrilateral multiscale tree structure (Figure 10.1). The pixel centers are located on iso-latitude rings, and pixels from the same ring are equispaced in Azimuth. This is critical for computational speed of all operations involving the evaluation of spherical harmonic transforms, including standard numerical analysis operations such as convolution, power spectrum estimation, and so on. HEALPix is a standard pixellization scheme in astronomy.

10.2.2 IUWT on the Sphere

The IUWT on the sphere is a wavelet transform on the sphere based on the spherical harmonics transform and with a very simple reconstruction algorithm. At scale j, we denote $a_j(\theta, \varphi)$ the scale coefficients, and $d_j(\theta, \varphi)$ the wavelet coefficients, with θ denoting the longitude and φ the latitude. Given a scale coefficient a_j, the smooth coefficient a_{j+1} is obtained by a convolution with a low-pass filter $h_j : a_{j+1} = a_j * h_j$. The wavelet coefficients are defined by the difference between two consecutive resolutions : $d_{j+1} = a_j - a_{j+1}$. A straightforward reconstruction is then given by

$$a_0(\theta, \varphi) = a_J(\theta, \varphi) + \sum_{j=1}^{J} d_j(\theta, \varphi). \tag{10.1}$$

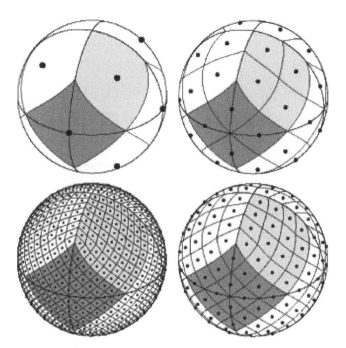

FIGURE 10.1 The HEALPix sampling grid for four different resolutions.

Since this transform is redundant, the procedure for reconstructing an image from its coefficients is not unique and this can be profitably used to impose additional constraints on the synthesis functions (e.g., smoothness, positivity). A reconstruction algorithm based on a variety of filter banks is described in Ref. [22]. Figure 10.2 shows the result of the IUWT on Wilkinson microwave anisotropy probe (WMAP) data (Cosmic Microwave Background).

10.2.3 Curvelet Transform on the Sphere

The curvelet transform enables the directional analysis of an image in different scales. The data undergo an IUWT on the sphere. Each scale j is then decomposed into smoothly overlapping blocks of side-length B_j in such a way that the overlap between two vertically adjacent blocks is a rectangular array of size $B_j \times B_j/2$, using the HEALPix pixellization. Finally, the ridgelet transform [7] is applied on each individual block. The method is best for the detection of anisotropic structures and smooth curves and edges of different lengths. The principle of the curvelet transform is schematized in Figure 10.3. More details can be found in Ref. [22].

10.2.4 Application to Gaussian Denoising on the Sphere

Multiscale transforms on the sphere have been used successfully for Gaussian denoising via nonlinear filtering or thresholding methods. Hard thresholding, for instance, consists of setting all insignificant coefficients (i.e., coefficients with an absolute value below a given threshold) to zero. In practice, we need to estimate the noise standard deviation σ_j in each band j and a coefficient w_j is significant if $|w_j| > \kappa\sigma_j$, where κ is a parameter typically chosen between 3 and 5. Denoting $\mathbf{Y}$ the noisy data and HT_λ the thresholding operator, the filtered

FIGURE 10.2 WMAP data and its wavelet transform on the sphere using five resolution levels (four wavelet scales and the coarse scale). The sum of these five maps reproduces exactly the original data (top left). *Top*: original data and the first wavelet scale. *Middle*: the second and third wavelet scales. *Bottom*: the fourth wavelet scale and the last smoothed array.

data **X** are obtained by

$$\mathbf{X} = \mathbf{\Phi}\mathrm{HT}_\lambda(\mathbf{\Phi}^{\mathrm{T}}\mathbf{Y}), \tag{10.2}$$

where $\mathbf{\Phi}^{\mathrm{T}}$ is the multiscale transform (IUWT or curvelet) and $\mathbf{\Phi}$ is the multiscale reconstruction. λ is a vector which has the size of the number of bands in the used multiscale

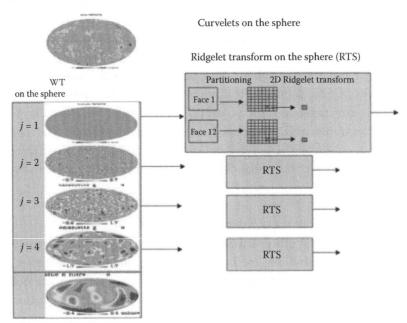

FIGURE 10.3 (**See color insert.**) Principle of curvelets transform on the sphere.

transform. The thresholding operation thresholds all coefficients in band j with the threshold $\lambda_j = \kappa\sigma_j$.

10.3 MULTISCALE TRANSFORMS ON THE SPHERE AND POISSON NOISE

10.3.1 Principle of the Multiscale MS-VSTS

In this section, we propose a multiscale representation designed for data with Poisson noise. The idea is to combine the spherical multiscale transforms with a VST, in order to have a multiscale representation of the data where the noise on multiscale coefficients behaves like Gaussian noise of known variance. With this representation, it is easy to denoise the data using standard Gaussian denoising methods.

10.3.1.1 VST of a Poisson Process

Given Poisson data $\mathbf{Y} := (Y_i)_i$, each sample $Y_i \sim \mathcal{P}(\lambda_i)$ has a variance $\mathrm{Var}[Y_i] = \lambda_i$. Thus, the variance of $\mathbf{Y}$ is signal dependent. The aim of a VST $\mathbf{T}$ is to stabilize the data such that each coefficient of $\mathbf{T}(\mathbf{Y})$ has an (asymptotically) constant variance, say 1, irrespective of the value of λ_i. In addition, for the VST used in this study, $T(\mathbf{Y})$ is asymptotically normally distributed. Thus, the VST-transformed data are asymptotically stationary and Gaussian.

The Anscombe [4] transform is a widely used VST that has a simple square-root form

$$\mathbf{T}(Y) := 2\sqrt{Y + 3/8}. \tag{10.3}$$

We can show that $\mathbf{T}(Y)$ is asymptotically normal as the intensity increases.

$$\mathbf{T}(Y) - 2\sqrt{\lambda} \xrightarrow[\lambda \to +\infty]{\mathcal{D}} \mathcal{N}(0, 1). \tag{10.4}$$

It can be shown that the Anscombe VST requires a high underlying intensity to well stabilize the data (typically for $\lambda \geqslant 10$) [25].

10.3.1.2 VST of a Filtered Poisson Process

Let $Z_j := \sum_i h[i] Y_{j-i}$ be the filtered process obtained by convolving $(Y_i)_i$ with a discrete filter h. We will use Z to denote any of the Z_j's. Let us define $\tau_k := \sum_i (h[i])^k$ for $k = 1, 2, \ldots$. In addition, we adopt a local homogeneity assumption stating that $\lambda_{j-i} = \lambda$ for all i within the support of h.

We define the square-root transform T as follows:

$$T(Z) := b \cdot \mathrm{sign}(Z + c)|Z + c|^{1/2}, \tag{10.5}$$

where b is a normalizing factor. It is proven in Zhang et al. [25] that T is a VST for a filtered Poisson process (with a nonzero-mean filter) in that $T(Y)$ is asymptotically normally distributed with a stabilized variance as λ becomes large.

The MS-VSTS consists in combining the square-root VST with a spherical multiscale transform (wavelets, curvelets, etc.).

10.3.2 Wavelets and Poisson Noise

This subsection describes the MS-VSTS + IUWT, which is a combination of a square-root VST with the IUWT. The recursive scheme is

$$\text{IUWT} \begin{cases} a_j &= h_{j-1} * a_{j-1} \\ d_j &= a_{j-1} - a_j \end{cases}$$

$$\xrightarrow{\text{MS-VSTS} \atop + \text{IUWT}} \begin{cases} a_j &= h_{j-1} * a_{j-1} \\ d_j &= T_{j-1}(a_{j-1}) - T_j(a_j) \end{cases}. \tag{10.6}$$

In Equation 10.6, the filtering on a_{j-1} can be rewritten as a filtering on $a_0 := \mathbf{Y}$, that is, $a_j = h^{(j)} * a_0$, where $h^{(j)} = h_{j-1} * \cdots * h_1 * h_0$ for $j \geqslant 1$ and $h^{(0)} = \delta$, where δ is the Dirac pulse ($\delta = 1$ on a single pixel and 0 everywhere else). T_j is the VST operator at scale j:

$$T_j(a_j) = b^{(j)} \text{sign}(a_j + c^{(j)}) \sqrt{|a_j + c^{(j)}|}. \tag{10.7}$$

Let us define $\tau_k^{(j)} := \sum_i (h^{(j)}[i])^k$. In Zhang et al. [25], it has been shown that, to have an optimal convergence rate for the VST, the constant $c^{(j)}$ associated with $h^{(j)}$ should be set to

$$c^{(j)} := \frac{7\tau_2^{(j)}}{8\tau_1^{(j)}} - \frac{\tau_3^{(j)}}{2\tau_2^{(j)}}. \tag{10.8}$$

The MS-VSTS + IUWT procedure is directly invertible as we have

$$a_0(\theta, \varphi) = T_0^{-1} \left[T_J(a_J) + \sum_{j=1}^{J} d_j \right] (\theta, \varphi). \tag{10.9}$$

Setting $b^{(j)} := \text{sign}(\tau_1^{(j)}) / \sqrt{|\tau_1^{(j)}|}$, if λ is constant within the support of the filter $h^{(j)}$, then we have [25]:

$$d_j(\theta, \varphi) \xrightarrow[\lambda \to +\infty]{\mathcal{D}} \mathcal{N} \left(0, \frac{\tau_2^{(j-1)}}{4\tau_1^{(j-1)2}} + \frac{\tau_2^{(j)}}{4\tau_1^{(j)2}} - \frac{\langle h^{(j-1)}, h^{(j)} \rangle}{2\tau_1^{(j-1)} \tau_1^{(j)}} \right), \tag{10.10}$$

where $\langle ., . \rangle$ denotes inner product.

This means that the detail coefficients issued from locally homogeneous parts of the signal follow asymptotically a central normal distribution with an intensity-independent variance which relies solely on the filter h and the current scale for a given filter h. Let us define $\sigma_{(j)}^2$ the stabilized variance at scale j for a locally homogeneous part of the signal:

$$\sigma_{(j)}^2 = \frac{\tau_2^{(j-1)}}{4\tau_1^{(j-1)2}} + \frac{\tau_2^{(j)}}{4\tau_1^{(j)2}} - \frac{\langle h^{(j-1)}, h^{(j)} \rangle}{2\tau_1^{(j-1)} \tau_1^{(j)}}. \tag{10.11}$$

TABLE 10.1   Precomputed Values of the
Variances σ_j of the Wavelet Coefficients

Wavelet Scale j	Value of σ_j
1	0.484704
2	0.0552595
3	0.0236458
4	0.0114056
5	0.00567026

To compute the $\sigma_{(j)}, b^{(j)}, c^{(j)}, \tau_k^{(j)}$, we only have to know the filters $h^{(j)}$. We compute these filters thanks to the formula $a_j = h^{(j)} * a_0$, by applying the IUWT to a Dirac pulse $a_0 = \delta$. Then, the $h^{(j)}$ are the scaling coefficients of the IUWT. The $\sigma_{(j)}$ have been precomputed for a six-scaled IUWT (Table 10.1).

We have simulated Poisson images of different constant intensities λ, computed the IUWT with MS-VSTS on each image and observed the variation of the normalized value of $\sigma_{(j)}$

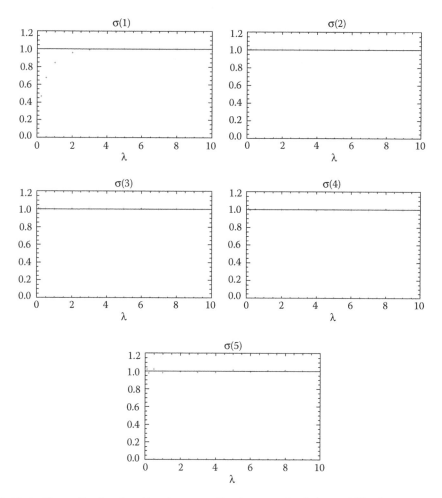

FIGURE 10.4 Normalized value ($(\sigma_{(j)})_{\text{simulated}}/(\sigma_{(j)})_{\text{theoretical}}$) of the stabilized variances at each scale j as a function of λ.

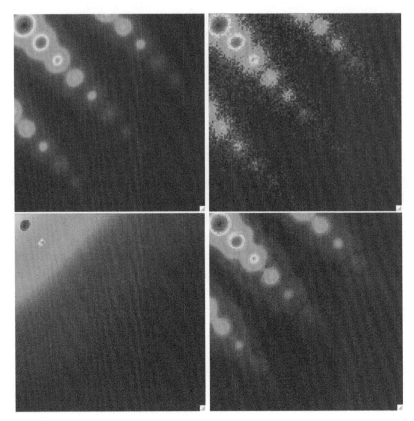

FIGURE 10.5 (**See color insert.**) Comparison of MS-VSTS with Anscombe + wavelet shrinkage on a single face of the first scale of the HEALPix pixellization (angular extent: $\pi/3$ sr). *Top left*: Sources of varying intensity. *Top right*: Sources of varying intensity with Poisson noise. *Bottom left*: Poisson sources of varying intensity reconstructed with Anscombe + wavelet shrinkage. *Bottom right*: Poisson sources of varying intensity reconstructed with MS-VSTS.

$((\sigma_{(j)})_{\text{simulated}}/(\sigma_{(j)})_{\text{theoretical}})$ as a function of λ for each scale j (Figure 10.4). We see that the wavelet coefficients are stabilized when $\lambda \gtrsim 0.1$ except for the first wavelet scale, which is largely noise. In Figure 10.5, we compare the result of MS-VSTS with Anscombe + wavelet shrinkage, on sources of varying intensities. We see that MS-VSTS works well on sources of very low intensities, whereas Anscombe does not work when the intensity is too low.

10.3.3 Curvelets and Poisson Noise

As the first step of the algorithm is an IUWT, we can stabilize each resolution level as in Equation 10.6. We then apply the local ridgelet transform on each stabilized wavelet band.

It is not as straightforward as with the IUWT to derive the asymptotic noise variance in the stabilized curvelet domain. In our experiments, we derived them using simulated Poisson data of stationary intensity level λ. After having checked that the standard deviation in the curvelet bands becomes stabilized as the intensity level increases (which means that the stabilization is working properly), we stored the standard deviation $\sigma_{j,l}$ for each wavelet scale j and each ridgelet band l (Table 10.2).

TABLE 10.2   Asymptotic Values of the Variances $\sigma_{j,k}$ of the Curvelet Coefficients

j	$l = 1$	$l = 2$	$l = 3$	$l = 4$
1	1.74550	0.348175		
2	0.230621	0.248233	0.196981	
3	0.0548140	0.0989918	0.219056	
4	0.0212912	0.0417454	0.0875663	0.20375
5	0.00989616	0.0158273	0.0352021	0.163248

10.4 APPLICATION TO POISSON DENOISING ON THE SPHERE

10.4.1 MS-VSTS + IUWT

Under the hypothesis of homogeneous Poisson intensity, the stabilized wavelet coefficients d_j behave like centered Gaussian variables of standard deviation $\sigma_{(j)}$. We can detect significant coefficients with binary hypothesis testing as in Gaussian denoising.

Under the null hypothesis $\mathcal{H}_0$ of homogeneous Poisson intensity, the distribution of the stabilized wavelet coefficient $d_j[k]$ at scale j and location index k can be written as

$$p(d_j[k]) = \frac{1}{\sqrt{2\pi}\sigma_j} \exp(-d_j[k]^2/2\sigma_j^2). \qquad (10.12)$$

The rejection of the hypothesis $\mathcal{H}_0$ depends on the double-sided p-value:

$$p_j[k] = 2\frac{1}{\sqrt{2\pi}\sigma_j} \int\limits_{|d_j[k]|}^{+\infty} \exp(-x^2/2\sigma_j^2)\,dx. \qquad (10.13)$$

Consequently, to accept or reject $\mathcal{H}_0$, we compare each $|d_j[k]|$ with a critical threshold $\kappa\sigma_j$, $\kappa = 3, 4$, or 5 corresponding respectively to significance levels. This amounts to deciding that:

- If $|d_j[k]| \geqslant \kappa\sigma_j$, $d_j[k]$ is significant

- If $|d_j[k]| < \kappa\sigma_j$, $d_j[k]$ is not significant

Then we have to invert the MS-VSTS scheme to reconstruct the estimate. However, although the direct inversion is possible, it cannot guarantee a positive intensity estimate, while the Poisson intensity is always nonnegative. A positivity projection can be applied, but important structures could be lost in the estimate. To tackle this problem, we reformulate the reconstruction as a convex optimization problem and solve it iteratively with an algorithm based on Hybrid Steepest Descent (HSD) [24].

We define the multiresolution support $\mathcal{M}$, which is determined by the set of detected significant coefficients after hypothesis testing:

$$\mathcal{M} := \{(j, k) | \text{if } d_j[k] \text{ is declared significant}\}. \qquad (10.14)$$

We formulate the reconstruction problem as a convex constrained minimization problem:

$$\text{Arg} \min_{\mathbf{X}} \|\mathbf{\Phi}^T\mathbf{X}\|_1, \text{s.t.}$$

$$\begin{cases} \mathbf{X} \geqslant 0, \\ \forall (j, k) \in \mathcal{M}, (\mathbf{\Phi}^T\mathbf{X})_j[k] = (\mathbf{\Phi}^T\mathbf{Y})_j[k], \end{cases} \tag{10.15}$$

where $\mathbf{\Phi}$ denotes the IUWT synthesis operator.

This problem is solved with the following iterative scheme: the image is initialized by $\mathbf{X}^{(0)} = 0$, and the iteration scheme is, for $n = 0$ to $N_{\max} - 1$:

$$\tilde{\mathbf{X}} = P_+[\mathbf{X}^{(n)} + \mathbf{\Phi}P_{\mathcal{M}}\mathbf{\Phi}^T(\mathbf{Y} - \mathbf{X}^{(n)})], \tag{10.16}$$

$$\mathbf{X}^{(n+1)} = \mathbf{\Phi}\text{ST}_{\lambda_n}[\mathbf{\Phi}^T\tilde{\mathbf{X}}]. \tag{10.17}$$

where P_+ denotes the projection on the positive orthant, $P_{\mathcal{M}}$ denotes the projection on the multiresolution support $\mathcal{M}$:

$$P_{\mathcal{M}}d_j[k] = \begin{cases} d_j[k] & \text{if } (j, k) \in \mathcal{M}, \\ 0 & \text{otherwise} \end{cases}. \tag{10.18}$$

and ST_{λ_n} the soft thresholding with threshold λ_n:

$$\text{ST}_{\lambda_n}[d] = \begin{cases} \text{sign}(d)(|d| - \lambda_n) & \text{if } |d| \geqslant \lambda_n, \\ 0 & \text{otherwise} \end{cases}. \tag{10.19}$$

We chose a decreasing threshold $\lambda_n = N_{\max} - n/N_{\max} - 1, n = 1, 2, \ldots, N_{\max}$.

The final estimate of the Poisson intensity is: $\hat{\mathbf{\Lambda}} = \mathbf{X}^{(N_{\max})}$. Algorithm 10.1 summarizes the main steps of the MS-VSTS + IUWT denoising algorithm.

10.4.2 Multiresolution Support Adaptation

When two sources are too close, the less intense source may not be detected because of the negative wavelet coefficients of the brightest source. To avoid such a drawback, we may update the multiresolution support at each iteration. The idea is to withdraw the detected sources and to make a detection on the remaining residual, so as to detect the sources which may have been missed at the first detection.

At each iteration n, we compute the MS-VSTS of $\mathbf{X}^{(n)}$. We denote $d_j^{(n)}[k]$ the stabilized coefficients of $\mathbf{X}^{(n)}$. We make a hard thresholding on $(d_j[k] - d_j^{(n)}[k])$ with the same thresholds as in the detection step. Significant coefficients are added to the multiresolution support $\mathcal{M}$.

The main steps of the algorithm are summarized in Algorithm 10.2. In practice, we use Algorithm 10.2 instead of Algorithm 10.1 in our experiments.

Algorithm 10.1 MS-VSTS + IUWT Denoising

Require: data $a_0 := \mathbf{Y}$, number of iterations $N_{\max}$, threshold κ

    __Detection__

1: **for** $j = 1$ to J **do**

2:     Compute a_j and d_j using Equation 10.6

3:     Hard threshold $|d_j[k]|$ with threshold $\kappa\sigma_j$ and update $\mathcal{M}$

4: **end for**

    __Estimation__

5: Initialize $\mathbf{X}^{(0)} = 0, \lambda_0 = 1$

6: **for** $n = 0$ to $N_{\max} - 1$ **do**

7:     $\tilde{\mathbf{X}} = P_+[\mathbf{X}^{(n)} + \mathbf{\Phi} P_{\mathcal{M}} \mathbf{\Phi}^{\mathsf{T}}(\mathbf{Y} - \mathbf{X}^{(n)})]$

8:     $\mathbf{X}^{(n+1)} = \mathbf{\Phi}\mathrm{ST}_{\lambda_n}[\mathbf{\Phi}^{\mathsf{T}}\tilde{\mathbf{X}}]$

9:     $\lambda_{n+1} = \frac{N_{\max} - (n+1)}{N_{\max} - 1}$

10: **end for**

11: Get the estimate $\hat{\mathbf{\Lambda}} = \mathbf{X}^{(N_{\max})}$

Algorithm 10.2 MS-VSTS + IUWT Denoising + Multiresolution Support Adaptation

---

Require: data $a_0 := \mathbf{Y}$, number of iterations $N_{\max}$, threshold κ

    __Detection__

1: **for** $j = 1$ to J **do**

2:     Compute a_j and d_j using Equation 10.6

3:     Hard threshold $|d_j[k]|$ with threshold $\kappa\sigma_j$ and update $\mathcal{M}$

4: **end for**

    __Estimation__

5: Initialize $\mathbf{X}^{(0)} = 0, \lambda_0 = 1$

6: **for** $n = 0$ to $N_{\max} - 1$ **do**

7:     $\tilde{\mathbf{X}} = P_+[\mathbf{X}^{(n)} + \mathbf{\Phi} P_{\mathcal{M}} \mathbf{\Phi}^{\mathsf{T}}(\mathbf{Y} - \mathbf{X}^{(n)})]$

8:     $\mathbf{X}^{(n+1)} = \mathbf{\Phi}\mathrm{ST}_{\lambda_n}[\mathbf{\Phi}^{\mathsf{T}}\tilde{\mathbf{X}}]$

9:     Compute the MS-VSTS on $\mathbf{X}^{(n)}$ to get the stabilized coefficients $d_j^{(n)}$

10:     Hard threshold $|d_j[k] - d_j^{(n)}[k]|$ and update $\mathcal{M}$

11:     $\lambda_{n+1} = \frac{N_{\max} - (n+1)}{N_{\max} - 1}$

12: **end for**

13: Get the estimate $\hat{\mathbf{\Lambda}} = \mathbf{X}^{(N_{\max})}$

10.4.3 MS-VSTS + Curvelets

Insignificant coefficients are zeroed by using the same hypothesis testing framework as in the wavelet scale. At each wavelet scale j and ridgelet band k, we make a hard thresholding on curvelet coefficients with threshold $\kappa\sigma_{j,k}$, $\kappa = 3, 4$, or 5. Finally, a direct reconstruction can be performed by first inverting the local ridgelet transforms and then inverting the MS-VST + IUWT (Equation 10.9). An iterative reconstruction may also be performed.

Algorithm 10.3 MS-VSTS + Curvelets Denoising

1: Apply the MS-VST + IUWT with J scales to get the stabilized wavelet subbands d_j

2: Set $B_1 = B_{min}$

3: **for** $j = 1$ to J **do**

4:   Partition the subband d_j with blocks of side-length B_j and apply the digital ridgelet transform to each block to obtain the stabilized curvelet coefficients

5:   **if** j modulo $2 = 1$ **then**

6:     $B_{j+1} = 2B_j$

7:   **else**

8:     $B_{j+1} = B_j$

9:   **end if**

10:   HTs on the stabilized curvelet coefficients

11: **end for**

12: Invert the ridgelet transform in each block before inverting the MS-VST + IUWT

Algorithm 10.3 summarizes the main steps of the MS-VSTS + Curvelets denoising algorithm.

10.4.4 Experiments

The method was tested on simulated Fermi data. The simulated data are the sum of a Milky Way diffuse background model and 1000 γ-ray point sources. We based our Galactic diffuse emission model intensity on the model *gll_iem_v02* obtained at the Fermi Science Support Center [16]. This model results from a fit of the LAT photons with various gas templates as well as inverse Compton in several energy bands. We used a realistic point-spread function for the sources, based on Monte Carlo simulations of the LAT and accelerator tests, which scale with energy approximately as $0.8(E/1\,\mathrm{GeV})^{-0.8}$ degrees (68% containment angle). The positions of the 205 brightest sources were taken from the Fermi 3-month source list [1]. The positions of the 795 remaining sources follow the LAT 1-year point source catalog [17] source distribution: each simulated source was randomly sorted in a box in Galactic coordinates of $\Delta l = 5°$ and $\Delta b = 1°$ around a LAT 1-year catalog source. We simulated each source assuming a power-law dependence with its spectral index given by the 3-month source list and the first-year catalog. We used an exposure of $3.10^{10}\,\mathrm{s\,cm^2}$ corresponding approximatively to 1 year of Fermi all-sky survey around 1 GeV. The simulated counts map shown in this section correspond to photons energy from 150 MeV to 20 GeV.

Figure 10.6 compares the result of denoising with MS-VST + IUWT (Algorithm 10.1), MS-VST + curvelets (Algorithm 10.3), and Anscombe VST + wavelet, shrinkage on a simulated Fermi map. Figure 10.7 shows the results on one single face of the first scale of the HEALPix pixellization (angular extent: $\pi/3\,\mathrm{sr}$). As expected from theory, the Anscombe method produces poor results to denoise Fermi data, because the underlying intensity is too weak. Both wavelet and curvelet denoising on the sphere perform much better. For this application, wavelets are slightly better than curvelets ($SNR_{\mathrm{wavelets}} = 65.8\,\mathrm{dB}$,

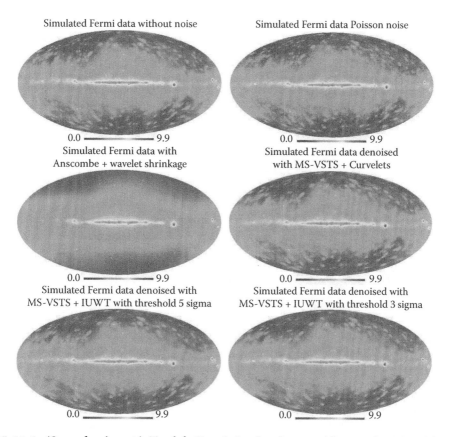

FIGURE 10.6 (**See color insert.**) *Top left*: Fermi simulated map without noise. *Top right*: Fermi simulated map with Poisson noise. *Middle left*: Fermi simulated map denoised with Anscombe VST + wavelet shrinkage. *Middle right*: Fermi simulated map denoised with MS-VSTS + curvelets (Algorithm 10.3). *Bottom left*: Fermi simulated map denoised with MS-VSTS + IUWT (Algorithm 10.1) with threshold $5\sigma_j$. *Bottom right*: Fermi simulated map denoised with MS-VSTS + IUWT (Algorithm 10.1) with threshold $3\sigma_j$. Pictures are in logarithmic scale.

$SNR_{\text{curvelets}} = 37.3$ dB, $SNR(\text{dB}) = 20\log(\sigma_{\text{signal}}/\sigma_{\text{noise}})$. As this image contains many point sources, this result is expected. Indeed wavelets are better than curvelets to represent isotropic objects.

10.5 APPLICATION TO INPAINTING AND SOURCE EXTRACTION

10.5.1 Milky Way Diffuse Background Study: Denoising and Inpainting

In order to extract from the Fermi photon maps, the Galactic diffuse emission, we want to remove the point sources from the Fermi image. As our HSD algorithm is very close to the MCA algorithm [20], an idea is to mask the most intense sources and to modify our algorithm in order to interpolate through the gaps exactly as in the MCA-Inpainting algorithm [2]. This modified algorithm can be called MS-VSTS-Inpainting algorithm. What we want to do is to remove the information due to point sources from the maps, in order to keep only the information due to the galactic background. The MS-VSTS-Inpainting

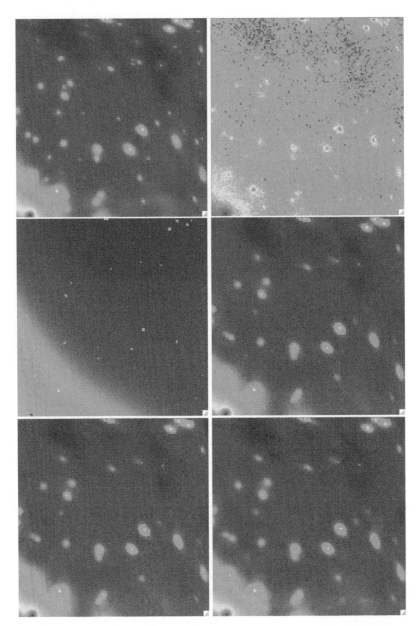

FIGURE 10.7 (**See color insert.**) View of a single HEALPix face (angular extent: $\pi/3$ sr) from the results of Figure 10.6. *Top left*: Fermi simulated map without noise. *Top right*: Fermi simulated map with Poisson noise. *Middle left*: Fermi simulated map denoised with Anscombe VST + wavelet shrinkage. *Middle right*: Fermi simulated map denoised with MS-VSTS + curvelets (Algorithm 10.3). *Bottom left*: Fermi simulated map denoised with MS-VSTS + IUWT (Algorithm 10.1) with threshold $5\sigma_j$. *Bottom right*: Fermi simulated map denoised with MS-VSTS + IUWT (Algorithm 10.1) with threshold $3\sigma_j$. Pictures are in logarithmic scale.

algorithm interpolates the missing data to reconstruct a map of the galactic background, which can now be fitted by a theoretical model. The interpolation uses the sparsity of the data in the wavelet domain : the gaps are filled so that the result is the sparsest possible in the wavelet domain.

The problem can be reformulated as a convex constrained minimization problem:

$$\text{Arg} \min_{\mathbf{X}} \| \mathbf{\Phi}^{\mathrm{T}} \mathbf{X} \|_1, \text{s.t.}$$

$$\begin{cases} \mathbf{X} \geqslant 0, \\ \forall (j, k) \in \mathcal{M}, (\mathbf{\Phi}^{\mathrm{T}} \Pi \mathbf{X})_j[k] = (\mathbf{\Phi}^{\mathrm{T}} \mathbf{Y})_j[k], \end{cases} \tag{10.20}$$

where Π is a binary mask (1 on valid data and 0 on invalid data).

The iterative scheme can be adapted to cope with a binary mask, which gives

$$\tilde{\mathbf{X}} = P_+[\mathbf{X}^{(n)} + \mathbf{\Phi} P_{\mathcal{M}} \mathbf{\Phi}^{\mathrm{T}} \Pi (\mathbf{Y} - \mathbf{X}^{(n)})], \tag{10.21}$$

$$\mathbf{X}^{(n+1)} = \mathbf{\Phi} \text{ST}_{\lambda_n}[\mathbf{\Phi} \tilde{\mathbf{X}}]. \tag{10.22}$$

The thresholding strategy has to be adapted. Indeed, for the inpainting task we need to have a very large initial threshold in order to have a very smooth image in the beginning and to refine the details progressively. We chose an exponentially decreasing threshold:

$$\lambda_n = \lambda_{\max}(2^{(\frac{N_{\max}-n}{N_{\max}-1})} - 1), \quad n = 1, 2, \ldots, N_{\max}, \tag{10.23}$$

where $\lambda_{\max} = \max(\mathbf{\Phi}^{\mathrm{T}} \mathbf{X})$.

10.5.1.1 Experiment
We applied this method on simulated Fermi data where we masked the **500** most luminous sources (with the highest photon per pixel flux). The other sources are not intense enough to be differenciated from the background.

The results are in Figure 10.8. The MS-VST + IUWT + Inpainting method (Algorithm 10.4) interpolates the missing data very well. Indeed, the missing part cannot be seen anymore in the inpainted map, which shows that the diffuse emission component has been correctly reconstructed.

10.5.2 Source Detection: Denoising and Background Modeling
10.5.2.1 Method
In the case of Fermi data, the diffuse γ-ray emission from the Milky Way, due to interaction between cosmic rays and interstellar gas and radiation, makes a relatively intense background. We have to extract this background in order to detect point sources. This diffuse interstellar emission can be modeled by a linear combination of gas templates and inverse compton map. We can use such a background model and incorporate a background removal in our denoising algorithm.

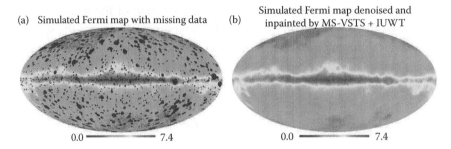

(a) Simulated Fermi map with missing data   (b) Simulated Fermi map denoised and inpainted by MS-VSTS + IUWT

FIGURE 10.8 (**See color insert.**) MS-VSTS-Inpainting. (a) Fermi simulated map with Poisson noise and the most luminous sources masked. (b) Fermi simulated map denoised and inpainted with wavelets (Algorithm 10.4). Pictures are in logarithmic scale.

Algorithm 10.4 MS-VST + IUWT Denoising + Inpainting

Require: data $a_0 := \mathbf{Y}$, mask Π, number of iterations $N_{\max}$, threshold κ

   Detection
1: **for** $j = 1$ to J **do**
2:    Compute a_j and d_j using Equation 10.6
3:    Hard threshold $|d_j[k]|$ with threshold $\kappa\sigma_j$ and update $\mathcal{M}$
4: **end for**
   Estimation
5: Initialize $\mathbf{X}^{(0)} = 0$, $\lambda_0 = \lambda_{\max}$
6: **for** $n = 0$ to $N_{\max} - 1$ **do**
7:    $\tilde{\mathbf{X}} = P_+[\mathbf{X}^{(n)} + \mathbf{\Phi} P_{\mathcal{M}} \mathbf{\Phi}^T \Pi(\mathbf{Y} - \mathbf{X}^{(n)})]$
8:    $\mathbf{X}^{(n+1)} = \mathbf{\Phi}_{\lambda_n}^{ST}[\mathbf{\Phi}^T \tilde{\mathbf{X}}]$
9:    $\lambda_{n+1} = \lambda_{\max}(2^{(\frac{N_{\max} - (n+1)}{N_{\max} - 1})} - 1)$
10: **end for**
11: Get the estimate $\hat{\mathbf{\Lambda}} = \mathbf{X}^{(N_{\max})}$

We denote $\mathbf{Y}$ the data, $\mathbf{B}$ the background we want to remove, and $d_j^{(b)}[k]$ the MS-VSTS coefficients of $\mathbf{B}$ at scale j and position k. We determine the multiresolution support by comparing $|d_j[k] - d_j^{(b)}[k]|$ with $\kappa\sigma_j$.

We formulate the reconstruction problem as a convex constrained minimization problem:

$$\operatorname*{Arg\,min}_{\mathbf{X}} \|\mathbf{\Phi}^T\mathbf{X}\|_1, \text{s.t}$$

$$\begin{cases} \mathbf{X} \geqslant 0, \\ \forall (j,k) \in \mathcal{M}, (\mathbf{\Phi}^T\mathbf{X})_j[k] = (\mathbf{\Phi}^T(\mathbf{Y} - \mathbf{B}))_j[k] \end{cases} \tag{10.24}$$

Then, the reconstruction algorithm scheme becomes

$$\tilde{\mathbf{X}} = P_+[\mathbf{X}^{(n)} + \mathbf{\Phi} P_{\mathcal{M}} \mathbf{\Phi}^T(\mathbf{Y} - \mathbf{B} - \mathbf{X}^{(n)})], \tag{10.25}$$

$$\mathbf{X}^{(n+1)} = \mathbf{\Phi} ST_{\lambda_n}[\mathbf{\Phi}^T\tilde{\mathbf{X}}] \tag{10.26}$$

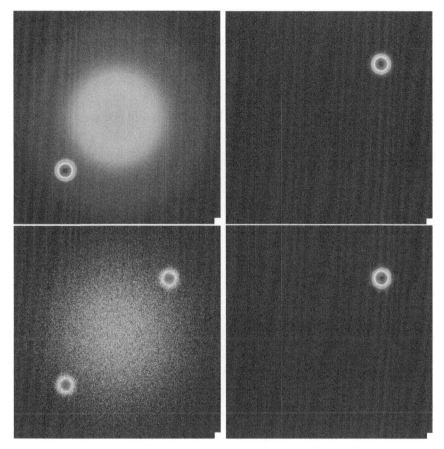

FIGURE 10.9 (**See color insert.**) Theoretical testing for MS-VSTS + IUWT denoising + background removal algorithm (Algorithm 10.5). View on a single HEALPix face. *Top left*: Simulated background : sum of two Gaussians of standard deviation equal to 0.1 and 0.01, respectively. *Top right*: Simulated source: Gaussian of standard deviation equal to 0.01. *Bottom left*: Simulated poisson data. *Bottom right*: Image denoised with MS-VSTS + IUWT and background removal.

The algorithm is illustrated by the theoretical study in Figure 10.9. We denoise Poisson data while separating a single source, which is a Gaussian of standard deviation equal to 0.01, from a background, which is a sum of two Gaussians of standard deviation equal to 0.1 and 0.01, respectively.

Like Algorithm 10.1, Algorithm 10.5 can be adapted to make multiresolution support adaptation.

10.5.2.2 Experiment

We applied Algorithms 10.5 on simulated Fermi data. To test the efficiency of our method, we detect the sources with the SExtractor routine [6], and compare the detected sources with the input source list to get the number of true and false detections. Results are shown in Figures 10.10 and 10.11. The SExtractor method was applied on the first wavelet scale of the reconstructed map, with a detection threshold equal to 1. It has been chosen to optimise

---

Algorithm 10.5 MS-VSTS + IUWT Denoising + Background extraction

---

Require: data $a_0 := \mathbf{Y}$, background B, number of iterations $N_{\max}$, threshold κ

   Detection

1: **for** $j = 1$ to J **do**

2:     Compute a_j and d_j using Equation 10.6

3:     Hard threshold $(d_j[k] - d_j^{(b)}[k])$ with threshold $\kappa\sigma_j$ and update $\mathcal{M}$

4: **end for**

   Estimation

5: Initialize $\mathbf{X}^{(0)} = 0, \lambda_0 = 1$

6: **for** $n = 0$ to $N_{\max} - 1$ **do**

7:     $\tilde{\mathbf{X}} = P_+[\mathbf{X}^{(n)} + \mathbf{\Phi} P_{\mathcal{M}} \mathbf{\Phi}^T(\mathbf{Y} - \mathbf{B} - \mathbf{X}^{(n)})]$

8:     $\mathbf{X}^{(n+1)} = \mathbf{\Phi} ST_{\lambda_n}[\mathbf{\Phi}^T\tilde{\mathbf{X}}]$

9:     $\lambda_{n+1} = \frac{N_{\max} - (n+1)}{N_{\max} - 1}$

10: **end for**

11: Get the estimate $\hat{\mathbf{\Lambda}} = \mathbf{X}^{(N_{\max})}$

---

the number of true detections. SExtractor makes 593 true detections and 71 false detections on the Fermi simulated map restored with Algorithm 10.2 among the 1000 sources of the simulation. On noisy data, many fluctuations due to Poisson noise are detected as sources by SExtractor, which leads to a big number of false detections (more than 2000 in the case of Fermi data).

10.5.2.2.1 Sensitivity to Model Errors.   As it is difficult to model the background precisely, it is important to study the sensitivity of the method to model errors. We add a stationary Gaussian noise to the background model, we compute the MS-VSTS + IUWT with threshold $3\sigma_j$ on the simulated Fermi Poisson data with extraction of the noisy background, and we study the percent of true and false detections with respect to the total number of sources of the simulation and the signal–noise ratio $(\mathrm{SNR(dB)} = 20\log(\sigma_{\mathrm{signal}}/\sigma_{\mathrm{noise}}))$ versus the standard deviation of the Gaussian perturbation. Table 10.3 shows that, when the standard deviation of the noise on the background model becomes of the same range as the mean of the Poisson intensity distribution ($\lambda_{\mathrm{mean}} = 68.764$), the number of false detections increases, the number of true detections decreases, and the SNR decreases. While the perturbation is not too strong (standard deviation <10), the effect of the model error remains low.

10.6 EXTENSION TO MULTICHANNEL DATA

10.6.1 Gaussian Noise

10.6.1.1 2D–1D Wavelet Transform on the Sphere

We propose a denoising method for 2D–1D data on the sphere, where the two first dimensions are spatial (longitude and latitude) and the third dimension is either the time or the energy. We need to analyze the data with a nonisotropic wavelet, where the time–or energy

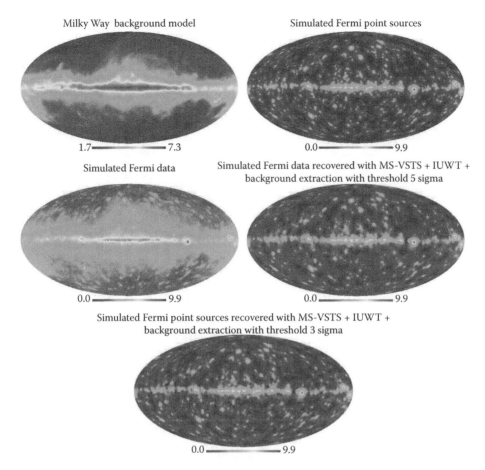

FIGURE 10.10 (**See color insert.**) *Top left*: Simulated background model. *Top right*: Simulated γ-ray sources. *Middle left*: Simulated Fermi data with Poisson noise. *Middle right*: Reconstructed γ-ray sources with MS-VSTS + IUWT + background removal (Algorithm 10.5) with threshold $5\sigma_j$. *Bottom*: Reconstructed γ-ray sources with MS-VSTS + IUWT + background removal (Algorithm 10.5) with threshold $3\sigma_j$. Pictures are in logarithmic scale.

scale is not connected to the spatial scale. An ideal wavelet function would be defined by

$$\psi(\theta, \varphi, t) = \psi^{(\theta,\varphi)}(\theta, \varphi)\psi^{(t)}(t), \tag{10.27}$$

where $\psi^{(\theta,\varphi)}$ is the spatial wavelet and $\psi^{(t)}$ is the temporal (or energy) wavelet. In the following, we will consider only isotropic and dyadic spatial scales, and we denote j_1 the spatial resolution index (i.e., scale 2^{j_1}), j_2 the time (or energy), resolution index. We thus define the scaled spatial and temporal (or energy) wavelets

$$\psi_{j_1}^{(\theta,\varphi)}(\theta, \varphi) = \frac{1}{2^{j_1}}\psi^{(\theta,\varphi)}\left(\frac{\theta}{2^{j_1}}, \frac{\varphi}{2^{j_1}}\right) \quad \text{and} \quad \psi_{j_1}^{(t)} = \frac{1}{2^{j_2}}\psi^{(t)}\left(\frac{t}{2^{j_1}}\right).$$

Hence, we derive the wavelet coefficients $w_{j_1,j_2}[k_\theta, k_\varphi, k_t]$ from a given data set D (k_θ and k_φ are spatial index and k_z a time (or energy) index. In continuous coordinates, this amounts

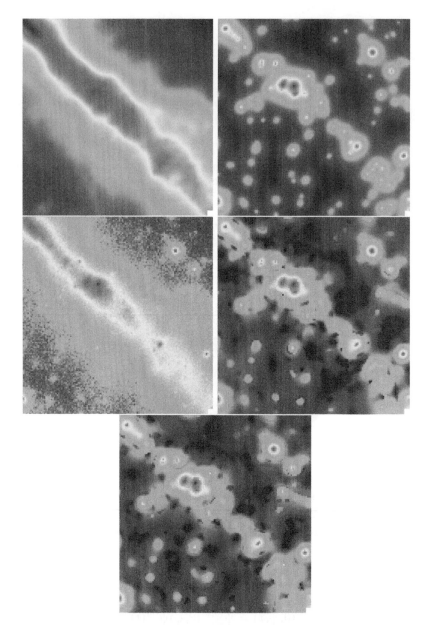

FIGURE 10.11 (**See color insert.**) View of a single HEALPix face (angular extent: $\pi/3$ sr) from the results of Figure 10.10. *Top left*: Simulated background model. *Top right*: Simulated γ-ray sources. *Middle left*: Simulated Fermi data with Poisson noise. *Middle right*: Reconstructed γ-ray sources with MS-VSTS + IUWT + background removal (Algorithm 10.5) with threshold $5\sigma_j$. *Bottom*: Reconstructed γ-ray sources with MS-VSTS + IUWT + background removal (Algorithm 10.5) with threshold $3\sigma_j$. Pictures are in logarithmic scale.

TABLE 10.3   Percent of True and False Detection and SNR versus the
Standard Deviation of the Gaussian Noise on the Background Model

Model Error Std Dev	% of True Detect	% of False Detect	SNR (dB)
0	59.3	7.1	23.8
10	57.0	11.0	23.2
20	53.2	18.9	22.6
30	49.1	43.5	21.7
40	42.3	44.3	21.0
50	34.9	39.0	20.3
60	30.3	37.5	19.5
70	25.0	34.6	18.9
80	23.0	28.5	18.7
90	23.6	27.1	18.3

to the formula

$$
w_{j_1,j_2}[k_\theta, k_\varphi, k_t] = \frac{1}{2^{j_1}} \frac{1}{\sqrt{2^{j_2}}} \iint \int_{-\infty}^{+\infty} D(\theta, \varphi, t)
$$

$$
\times \psi^{(\theta,\varphi)}\left(\frac{\theta - k_\theta}{2^{j_1}}, \frac{\varphi - k_\varphi}{2^{j_1}}\right) \psi^{(t)}\left(\frac{t - k_t}{2^{j_2}}\right) \, \mathrm{d}x \, \mathrm{d}y \, \mathrm{d}z = D * \bar\psi_{j_1}^{(\theta,\varphi)} * \bar\psi_{j_2}^{(t)}(\theta, \varphi, t)
\tag{10.28}
$$

where $*$ is the convolution and $\bar\psi(t) = \psi(-t)$.

10.6.1.2 Fast Undecimated 2D–1D Decomposition/Reconstruction

In order to have a fast algorithm for discrete data, we use wavelet functions associated to filter banks. Hence, our wavelet decomposition consists in applying first an IUWT on the sphere for each frame k_z. Using the spherical IUWT, we have the reconstruction formula

$$
D[k_\theta, k_\varphi, k_t] = a_{J_1}[k_\theta, k_\varphi] + \sum_{j_1=1}^{J_1} w_{j_1}[k_\theta, k_\varphi, k_t], \quad \forall k_t,
\tag{10.29}
$$

where J_1 is the number of spatial scales. To have simpler notations, we replace the two spatial indexes by a single index k_r which corresponds to the pixel index:

$$
D[k_r, k_t] = a_{J_1}[k_r] + \sum_{j_1=1}^{J_1} w_{j_1}[k_r, k_t], \quad \forall k_t,
\tag{10.30}
$$

Then, for each spatial location k_r and for each 2D wavelet scale j_1, we apply a 1D wavelet transform along t on the spatial wavelet coefficients $w_{j_1}[k_r, k_t]$ such that

$$
w_{j_1}[k_r, k_t] = w_{j_1,J_2}[k_r, k_t] + \sum_{j_2=1}^{J_2} w_{j_1,j_2}[k_r, k_t], \quad \forall(k_r, k_t),
\tag{10.31}
$$

where j_2 is the number of scales along t. The same processing is also applied on the coarse spatial scale $a_{J_1}[k_r, k_t]$ and we have

$$a_{J_1}[k_r, k_t] = a_{J_1, J_2}[k_r, k_t] + \sum_{j_2=1}^{J_2} w_{J_1, j_2}[k_r, k_t], \quad \forall (k_r, k_t). \tag{10.32}$$

Hence, we have a 2D–1D spherical undecimated wavelet representation of the input data D:

$$D[k_r, k_t] = a_{J_1, J_2}[k_r, k_t] + \sum_{j_1=1}^{J_1} w_{j_1, J_2}[k_r, k_t] + \sum_{j_2=1}^{J_2} w_{J_1, j_2}[k_r, k_t] + \sum_{j_1=1}^{J_1} \sum_{j_2=1}^{J_2} w_{j_1, j_2}[k_r, k_t]. \tag{10.33}$$

From this expression, we distinguish four kinds of coefficients:

- Detail–detail coefficients ($j_1 \leqslant J_1$ and $j_2 \leqslant J_2$):

$$w_{j_1, j_2}[k_r, k_t] = (\delta - \bar{h}_{1D}) \star (\bar{h}_{1D}^{(j_2-1)} \star a_{j_1-1}[k_r, \cdot] - h_{1D}^{(j_2-1)} \star a_{j_1}[k_r, \cdot]). \tag{10.34}$$

- Approximation–detail coefficients ($j_1 = J_1$ and $j_2 \leqslant J_2$):

$$w_{J_1, j_2}[k_r, k_t] = h_{1D}^{(j_2-1)} \star a_{J_1}[k_r, \cdot] - h_{1D}^{(j_2)} \star a_{J_1}[k_r, \cdot]. \tag{10.35}$$

- Detail–approximation coefficients ($j_1 \leqslant J_1$ and $j_2 = J_2$):

$$w_{j_1, J_2}[k_r, k_t] = h_{1D}^{(J_2)} \star a_{j_1-1}[k_r, \cdot] - h_{1D}^{(J_2)} \star a_{j_1}[k_r, \cdot]. \tag{10.36}$$

- Approximation–approximation coefficients ($j_1 = J_1$ and $j_2 = J_2$):

$$a_{J_1, J_2}[k_r, k_t] = h_{1D}^{(J_2)} \star a_{J_1}[k_r, \cdot]. \tag{10.37}$$

10.6.1.3 Multichannel Gaussian Denoising

As the spherical 2D–1D undecimated wavelet transform just described is fully linear, a Gaussian noise remains Gaussian after transformation. Therefore, all thresholding strategies that have been developed for wavelet Gaussian denoising are still valid with the spherical 2D–1D wavelet transform. Denoting TH the thresholding operator, the denoised cube in the case of additive white Gaussian noise is obtained by

$$\tilde{D}[k_r, k_t] = a_{J_1, J_2}[k_r, k_t] + \sum_{j_1=1}^{J_1} \text{TH}(w_{j_1, J_2}[k_r, k_t]) + \sum_{j_2=1}^{J_2} \text{TH}(w_{J_1, j_2}[k_r, k_t])$$

$$+ \sum_{j_1=1}^{J_1} \sum_{j_2=1}^{J_2} \text{TH}(w_{j_1, j_2}[k_r, k_t]). \tag{10.38}$$

A typical choice of TH is the hard thresholding operator, that is,

$$\text{TH}(x) = \begin{cases} 0 & \text{if } |x| < \tau \\ x & \text{if } |x| \geqslant \tau \end{cases}. \tag{10.39}$$

The threshold τ is generally chosen between 3 and 5 times the noise standard deviation.

10.6.2 Poisson Noise

10.6.2.1 Multiscale Variance Stabilizing Transform

To perform a Poisson denoising, we have to plug the MS-VST into the spherical 2D–1D undecimated wavelet transform. Again, we distinguish four kinds of coefficients that take the following forms:

- Detail–detail coefficients ($j_1 \leqslant J_1$ and $j_2 \leqslant J_2$):

$$w_{j_1,j_2}[k_r, k_t] = (\delta - \bar{h}_{1D}) \star (T_{j_1-1,j_2-1}[\bar{h}_{1D}^{(j_2-1)} \star a_{j_1-1}[k_r, \cdot]]$$
$$- T_{j_1,j_2-1}[h_{1D}^{(j_2-1)} \star a_{j_1}[k_r, \cdot]]). \tag{10.40}$$

- Approximation–detail coefficients ($j_1 = J_1$ and $j_2 \leqslant J_2$):

$$w_{J_1,j_2}[k_r, k_t] = T_{J_1,j_2-1}[h_{1D}^{(j_2-1)} \star a_{J_1}[k_r, \cdot]] - T_{J_1,j_2}[h_{1D}^{(j_2)} \star a_{J_1}[k_r, \cdot]]. \tag{10.41}$$

- Detail–approximation coefficients ($j_1 \leqslant J_1$ and $j_2 = J_2$):

$$w_{j_1,J_2}[k_r, k_t] = T_{j_1-1,J_2}[h_{1D}^{(J_2)} \star a_{j_1-1}[k_r, \cdot]] - T_{j_1,J_2}[h_{1D}^{(J_2)} \star a_{j_1}[k_r, \cdot]]. \tag{10.42}$$

- Approximation–approximation coefficients ($j_1 = J_1$ and $j_2 = J_2$):

$$a_{J_1,J_2}[k_r, k_t] = h_{1D}^{(J_2)} \star a_{J_1}[k_r, \cdot]. \tag{10.43}$$

Hence, all 2D–1D wavelet coefficients w_{j1,j_2} are now stabilized, and the noise on all these wavelet coefficients is Gaussian with known scale-dependent variance that depends solely on h. Denoising is however not straightforward because there is no explicit reconstruction formula available because of the form of the stabilization equations above. Formally, the stabilizing operators T_{j_1,j_2} and the convolution operators along the spatial and temporal dimensions do not commute, even though the filter bank satisfies the exact reconstruction formula. To circumvent this difficulty, we propose to solve this reconstruction problem by using an iterative reconstruction scheme.

10.6.2.2 Detection–Reconstruction

As the noise on the stabilized coefficients is Gaussian, and without loss of generality, we let its standard deviation equal to 1, we consider that a wavelet coefficient $w_{j_1,j_2}[k_r, k_t]$ is significant, that is, not due to noise, if its absolute value is larger than a critical threshold τ, where τ is typically between 3 and 5.

The multiresolution support will be obtained by detecting at each scale the significant coefficients. The multiresolution support for $j_1 \leqslant J_1$ and $j_2 \leqslant J_2$ is defined as

$$\mathcal{M}_{j_1,j_2}[k_r, k_t] = \begin{cases} 1 & \text{if } w_{j_1,j_2}[k_r, k_t] \text{ is significant} \\ 0 & \text{otherwise} \end{cases}. \tag{10.44}$$

We denote $\mathcal{W}$ the spherical 2D–1D undecimated wavelet transform described above, and $\mathcal{R}$ the inverse wavelet transform. We want our solution X to preserve the significant structures of the original data by reproducing exactly the same coefficients as the wavelet coefficients of the input data Y, but only at scales and positions where significant signal has been detected. At other scales and positions, we want the smoothest solution with the lowest budget in terms of wavelet coefficients. Furthermore, as Poisson intensity functions are positive by nature, a positivity constraint is imposed on the solution. It is clear that there are many solutions satisfying the positivity and multiresolution support consistency requirements, for example, Y itself. Thus, our reconstruction problem based solely on these constraints is an ill-posed inverse problem that must be regularized. Typically, the solution in which we are interested must be sparse by involving the lowest budget of wavelet coefficients. Therefore, our reconstruction is formulated as a constrained sparsity-promoting minimization problem that can be written as follows:

$$\min_{\mathbf{X}} \|\mathcal{W}\mathbf{X}\|_1 \text{ subject to } \begin{cases} \mathcal{M}\mathcal{W}\mathbf{X} = \mathcal{M}\mathcal{W}\mathbf{Y} \\ \mathbf{X} \geqslant 0 \end{cases}, \tag{10.45}$$

where $\|\cdot\|$ is the L_1-norm playing the role of regularization and is well known to promote sparsity[9]. This problem can be solved efficiently using the HSD algorithm [24,25], and requires about 10 iterations in practice. Transposed into our context, its main steps can be summarized as follows:

Require: Input noisy data $\mathbf{Y}$, a low-pass filter h, multiresolution support $\mathcal{M}$ from the detection step, number of iterations N_{max}
1: Initialize $\mathbf{X}^{(0)} = \mathcal{M}\mathcal{W}\mathbf{Y} = \mathcal{M}w_Y$
2: **for** $n = 1$ to N_{max} **do**
3:    $\tilde{d} = \mathcal{M}w_Y + (1 - \mathcal{M})\mathcal{W}\mathbf{X}^{(n-1)}$
4:    $\mathbf{X}^{(n)} = P_+(\mathcal{R}\mathrm{ST}_{\beta_n}[\tilde{d}])$
5:    Update the step $\beta_n = (N_{max} - n)/(N_{max} - 1)$
6: **end for**

---

where P_+ is the projector onto the positive orthant, that is, $P_+(x) = \max(x, 0)$, ST_{β_n} is the soft-thresholding operator with threshold β_n, that is, $\mathrm{ST}_{\beta_n}[x] = x - \beta_n \mathrm{sign}(x)$ if $|x| \geqslant \beta_n$, and 0 otherwise.

The final spherical MSVST 2D–1D wavelet denoising algorithm is the following:

Require: Input noisy data **Y**, a low-pass filter h, threshold level τ

1: *Spherical 2D–1D MSVST*: Apply the spherical 2D–1D-MSVST to the data using Equations 10.40 through 10.43.

2: *Detection*: Detect the significant wavelet coefficients that are above τ, and compute the multiresolution support $\mathcal{M}$.

3: *Reconstruction*: Reconstruct the denoised data using the algorithm above.

10.7 CONCLUSION

This chapter presented new methods for the restoration of spherical data with noise following a Poisson distribution. A denoising method was proposed, which used a variance stabilization method and multiscale transforms on the sphere. Experiments have shown that it is very efficient for Fermi data denoising. Two spherical multiscale transforms, the wavelet and the curvelets, were used. Then, we have proposed an extension of the denoising method in order to take into account missing data, and we have shown that this inpainting method could be a useful tool to estimate the diffuse emission. Then, we have introduced a new denoising method on the sphere which takes into account a background model. The simulated data have shown that it is relatively robust to errors in the model, and can therefore be used for Fermi diffuse background modeling and source detection. Finally, we introduced an extension for multichannel data.

ACKNOWLEDGMENT

This work was partially supported by the European Research Council grant ERC-228261.

REFERENCES

1. A. A. Abdo, M. Ackermann, M. Ajello, W. B. Atwood, M. Axelsson, L. Baldini, J. Ballet et al. Fermi/large area telescope bright gamma-ray source list. *Astrophysical Journal Supplement*, 183:46–66, 2009.

2. P. Abrial, Y. Moudden, J. L. Starck, J. Bobin, M. J. Fadili, B. Afeyan, and M. K. Nguyen. Morphological component analysis and inpainting on the sphere: Application in physics and astrophysics. *Journal of Fourier Analysis and Applications*, 13(6):729–748, 2007.

3. P. Abrial, Y. Moudden, J. L. Starck, M. J. Fadili, J. Delabrouille, and M. Nguyen. CMB data analysis and sparsity. *Statistical Methodology*, 5(4):289–298, 2008.

4. F. J. Anscombe. The transformation of Poisson, binomial and negative-binomial data. *Biometrika*, 35(3):246–254, 1948.

5. W. B. Atwood, A. A. Abdo, M. Ackermann, W. Althouse, B. Anderson, M. Axelsson, L. Baldini et al. The large area telescope on the Fermi gamma-ray space telescope mission. *Astrophysical Journal*, 697:1071–1102, 2009.

6. E. Bertin and S. Arnouts. SExtractor: Software for source extraction. *Astronomy and Astrophysics Supplement*, 117:393–404, 1996.

7. E. Candes and D. L. Donoho. Ridgelets: The key to high dimensional intermittency? *Philosophical Transactions of the Royal Society of London*, Series A, 357:2495, 1999.

8. D. L. Donoho. Nonlinear wavelet methods for recovery of signals, densities, and spectra from indirect and noisy data. *Proceedings of the Symposia in Applied Mathematics*, 47:173–205, 1993.

9. D. L. Donoho. For most large underdetermined systems of linear equations, the minimal ℓ^1-norm near-solution approximates the sparsest near-solution. Technical report, Department of Statistics of Stanford University, 2004.

10. M. Fisz. The limiting distribution of a function of two independant random variables and its statistical application. *Colloquium Mathematicum*, 3:138–146, 1955.

11. P. Fryźlewicz and G. P. Nason. A Haar-Fisz algorithm for Poisson intensity estimation. *Journal of Computational and Graphical Statistics*, 13:621–638, 2004.

12. K. M. Górski, E. Hivon, A. J. Banday, B. D. Wandelt, F. K. Hansen, M. Reinecke, and M. Bartelmann. HEALPix: A framework for high-resolution discretization and fast analysis of data distributed on the sphere. *Astrophysical Journal*, 622:759–771, 2005.

13. R. C. Hartman, D. L. Bertsch, S. D. Bloom, A. W. Chen, P. Deines-Jones, J. A. Esposito, C. E. Fichtel et al., Third EGRET catalog (3EG) (Hartman+, 1999). *VizieR Online Data Catalog*, 212:30079–ff, 1999.

14. E. Kolaczyk. Bayesian multiscale models for Poisson processes. *Journal of the American Statistical Association*, 94(447):920–933, 1999.

15. S. Lefkimmiaits, P. Maragos, and G. Papandreou. Bayesian inference on multiscale models for Poisson intensity estimation: Applications to photon-limited image denoising. *IEEE Transactions on Image Processing*, 18(8):1724–1741, 2009.

16. J. D. Myers. LAT background models, 2009. http://fermi.gsfc.nasa.gov/ssc/data/access/lat/BackgroundModels.html.

17. J. D. Myers. LAT 1-year point source catalog, 2010. http://fermi.gsfc.nasa.gov/ssc/data/access/lat/1yr_catalog/.

18. R. Nowak and E. Kolaczyk. A statistical multiscale framework for Poisson inverse problems. *IEEE Transactions Information Theory*, 45(5):1811–1825, 2000.

19. J. Schmitt, J. L. Starck, J. M. Casandjian, J. Fadili, and I. Grenier. Poisson denoising on the sphere: Application to the Fermi gamma ray space telescope. *Astronomy and Astrophysics*, 517, Article number A26, 2010.

20. J.-L Starck, M. Elad, and D. L. Donoho. Redundant multiscale transforms and their application for morphological component analysis. *Advances in Imaging and Electron Physics*, 132:287–348, 2004.

21. J.-L. Starck, J. M. Fadili, S. Digel, B. Zhang, and J. Chiang. Source detection using a 3D sparse representation: Application to the Fermi gamma-ray space telescope. *Astronomy and Astrophysics*, 504:641–652, 2009.

22. J.-L Starck, Y. Moudden, P. Abrial, and M. Nguyen. Wavelets, ridgelets and curvelets on the sphere. *Astronomy and Astrophysics*, 446:1191–1204, 2006.

23. K. Timmerman and R. Nowak. Multiscale modeling and estimation of Poisson processes with application to photon-limited imaging. *IEEE Transactions on Information Theory*, 45(3): 846–862, 1999.

24. I. Yamada. The hybrid steepest descent method for the variational inequality problem over the intersection of fixed point sets of nonexpansive mappings. In D. Butnariu, Y. Censor, and S. Reich, editors, *Inherently Parallel Algorithm in Feasibility and Optimization and their Applications*, pp. 473–504. Elsevier: Amsterdam, 2001.

25. B. Zhang, J. Fadili, and J.-L. Starck. Wavelets, ridgelets and curvelets for Poisson noise removal. *IEEE Transactions on Image Processing*, 11(6):1093–1108, 2008.

2

Classification

CHAPTER 11

Galaxy Zoo

Morphological Classification and Citizen Science

Lucy Fortson
University of Minnesota

Karen Masters and Robert Nichol
University of Portsmouth

Kirk D. Borne
George Mason University

Edward M. Edmondson
University of Portsmouth

Chris Lintott
University of Oxford

Jordan Raddick
The Johns Hopkins University

Kevin Schawinski
Yale University

John Wallin
Middle Tennessee State University

CONTENTS

11.1 Summary 214
11.2 A Brief History of Galaxy Morphology 214
11.3 Genesis of Galaxy Zoo 217
11.4 Galaxy Zoo 1 219
    11.4.1 From Clicks to Classifications 219
        11.4.1.1 Classification Biases 220
        11.4.1.2 Comparison with Other Classifications 220
    11.4.2 Science Results from Galaxy Zoo 1 221
        11.4.2.1 Color and Morphology 222

11.4.2.2 Spiral Arm Directions 222
11.4.2.3 Merging Galaxies 222
11.4.2.4 Active Galaxies 222
11.4.2.5 Rare and Unusual Objects 223
11.5 Evolution of Galaxy Zoo 223
11.5.1 Galaxy Zoo 2 and Hubble Zoo 223
11.5.2 The Citizen Scientists: Motivation and Unexpected Outcomes 224
11.6 The Zooniverse 226
11.6.1 Tasks Suitable for the Zooniverse 228
11.6.2 Data Mining the Zooniverse Results 230
11.6.3 Future Citizen Science Projects 230
11.7 Galaxy Zoo in the Context of Other Citizen Science Projects 232
11.8 Conclusions 233
References 233

11.1 SUMMARY

We provide a brief overview of the Galaxy Zoo and Zooniverse projects, including a short discussion on the history of, and motivation for, these projects as well as reviewing the science these innovative Internet-based citizen science projects have produced so far. We briefly describe the method of applying en-masse human pattern recognition capabilities to complex data in data-intensive research. We also provide a discussion on the lessons learned from developing and running these community-based projects including thoughts on future applications of this methodology. This review is intended to give the reader a quick and simple introduction to the Zooniverse.

11.2 A BRIEF HISTORY OF GALAXY MORPHOLOGY

One of the fundamental facts of the Universe is that most large galaxies* come in two basic shapes which astronomers call "Spirals" and "Ellipticals." The exact details of why this is the case, and how the two types of galaxies relate to each other, remain a major mystery for astronomers. It is central to our understanding of how the creation and evolution of galaxies proceeds with cosmic time and depends on their cosmic location. Significant effort has been spent over the last few decades trying to address these questions.

Edwin Hubble was one of the first astronomers to attempt to systematically address the origin of the shape, or morphology, of galaxies using his famous "Hubble Sequence" or "tuning fork" diagram (Hubble, 1926) which is still in use today (see Figure 11.1). Starting on the left, Hubble classified the elliptical galaxies using the observed ellipticity of the galaxy projected on the sky, giving them a numerical value associated with how round they appeared on the sky. In three dimensions, ellipticals can be triaxial objects, taking a range of morphologies from purely spherical systems through to flattened rugby ball-shaped galaxies.

---

* A galaxy is just a collection of stars; typically billions for a large galaxy.

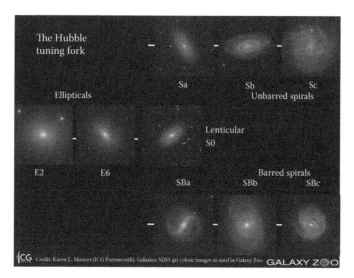

FIGURE 11.1 The Hubble tuning fork illustrated with the type of SDSS color images used in Galaxy Zoo. (Credit: Karen Masters and The Sloan Digital Sky Survey (SDSS) Collaboration, www.sdss.org.)

On the right side of the tuning-fork, Hubble placed spiral or disk galaxies. These galaxies have a central "bulge" of stars, that resemble elliptical galaxies in some ways, embedded in a thin disk of stars that show a range of spiral patterns or "arms." Hubble ordered disk galaxies based on the tightness of these spiral arms and the size of the central bulge. He had two distinct populations of disk galaxies, namely with and without a central bar-like (or linear) structure. At the point where these different classifications met (for spirals with the largest bulges, and tightest wound arms), Hubble placed "lenticular" galaxies which at the time were hypothetical disk galaxies with very large bulges and no spiral arms—they have since been found.

It is a common misconception that Hubble believed the "tuning fork" diagram was an evolutionary sequence, with elliptical galaxies on the left evolving along the sequence to form disk galaxies. In fact, Hubble advised that "temporal connotations are made at one's peril" in an early defense of the classification sequence (Hubble, 1927), going on to say that he set up the classification "without prejudice to theories of [galaxy] evolution" (p. 277). This misconception about Hubble's beliefs probably arose due to his suggestion of the use of "early" and "late" types to describe the progression toward the right along the sequence (although he discussed that this nomenclature was simply for convenience and borrowed terminology commonly used for stellar classification (Hubble, 1926)). Astronomers still call elliptical galaxies "early" types and disk galaxies "late" type galaxies, although we now know that most "late"-type galaxies have much younger stellar populations (ironically more "early type" stars) than most "early"-type galaxies.

Since Hubble, there have been several updates to his classification scheme (for a recent review see Buta, 2011) but key features have remained unchanged. What has changed dramatically is the number of galaxies cataloged and requiring classification. Before the advent of digital detectors in astronomy, astronomers could just visually classify the galaxies they saw via their telescopes and/or on photographic plates. New astronomers were trained to follow the classification rules and provided detailed morphologies for thousands of galaxies. Several large catalogues of nearby galaxies with such classifications exist (e.g., The Hubble Atlas of Galaxies (Sandage, 1961), or the Third Reference Catalogue of Bright Galaxies (RC3) (de Vaucouleurs et al., 1991)), and many of these classifications are collected in the NASA/IPAC Extragalactic Database.[*]

The expert classifier approach quickly became inappropriate with the digital surveys because of the size of the galaxy samples available; for example, the Main Galaxy Sample of the Sloan Digital Sky Survey (SDSS; Strauss et al., 2002; York et al., 2000) is over a million galaxies and simply cannot be visually inspected by any one astronomer (or even all the astronomers in the world working together). It became clear that some automatic method for classifying galaxies was needed, but programming a computer to recognize the complexities of galaxy shapes (spiral arms, bars, disk plus bulges) is very challenging.

The first attempts at an automated classification scheme included the use of artificial neural networks, for example, Lahav et al. (1995) began this work by comparing the classifications of 830 galaxies from a set of six independent experts (R. Buta, H. Corwin, G. de Vaucouleurs, A. Dressler, J. Huchra, and S. van den Bergh). These experts were unanimous in their classification in only 1% of objects, while an agreement of 80% between the experts could only be achieved within a spread of two T-types.[†] The conclusion was that the visual classifications depended on the color, size, and quality of the image used, and that artificial neural networks could be developed to agree to almost the same degree as any pair of expert classifiers. This approach was implemented on the SDSS sample by Ball et al. (2004) with the same basic conclusion as Lahav et al. (1995), that is, that a neural network could reproduce visual morphologies within about 1 T-type.

An alternative approach to developing methods to replicate human classification is to design computational algorithms that attempt to capture the same information. Examples of this approach include the CAS (concentration, asymmetry, clumpiness) structural system of Conselice (2006), which uses a principal component analysis (PCA) to study the diversity of internal structures of a sample of galaxies, and the Zurich Estimator of Structural Types algorithm of Scarlata et al. (2007), which uses a combination of diagnostics of the galaxy shape (that can be measured directly from the galaxy images) and the more traditional Sersic index from the fit to the two-dimensional surface brightness distribution of the galaxy. Such data-oriented methods are very successful at capturing the complexity of galaxy shapes in two-dimensional images but remain hard to translate in terms of the more traditional, established morphological classes discussed above.

In recent years, there has been significant interest in the development of model-based morphologies of galaxies that use established parametric models for the light distribution of galaxies to fit to the two-dimensional images of galaxies. Such methods include Galaxy IMage 2D (GIM2D) (Simard et al., 2002) and Galfit (Peng et al., 2002) which both attempt to fit galaxy images with a combination of a disk and bulge model. These can be used to construct

---

[*] http://nedwww.ipac.caltech.edu/
[†] A T-type is a numerical coding of the Hubble Sequence increasing from negative numbers for ellipticals and lenticulars, through $0 = S0/a$, $1 = Sa$, $3 = Sb$, $5 = Sc$, and so on.

an objective classification scheme such as the bulge-to-disk ratio of galaxies. Unfortunately, these model-fitting techniques are computationally intense and subject to local minima as they search the high-dimensional parameter space for the best–fitting model (in some cases, it can be a 12-parameter model being fit to the galaxy images).

A quicker way to solve the classification problem is to use a proxy for the galaxy class. The most common such proxy is the color of a galaxy as most ellipticals are "red," with their light dominated by older stars, while most spirals are "blue," as they contain areas of active star formation that include luminous blue stars. However, relying on galaxy classification via colors as a proxy misses an important piece of the galaxy evolution story. The color of a galaxy is driven by the stellar (and gas and dust) content of the galaxy, while the shape or morphology of a galaxy reflects its dynamical history that could be very different (and have a different timescale). Therefore, one of the central motivations for the original Galaxy Zoo project was to construct a large sample of early- and late-type galaxy classifications that were independent of color.

11.3 GENESIS OF GALAXY ZOO

The Galaxy Zoo project was inspired by discussions on the limitations of a sample of early-type galaxies produced by Bernardi et al. (2003) from the initial SDSS[*] data (York et al., 2000). Bernardi et al. (2003) had used a PCA-based classification to select "passive" galaxies based on the spectra of SDSS galaxies. Although, this classification scheme was fast, and easy to implement, it probably excluded early-type galaxies that had signatures of ongoing star formation. It was therefore realized that to find such objects would require a sample of early type galaxies based solely on their morphological visual appearance, without the use of spectral or color information, that is that could include normal, passive "red" early types, as well as the possibility of "bluer" star-forming early types. Kevin Schawinski, as part of his PhD thesis work at Oxford University, under the supervision of Daniel Thomas, took on the task to build such a complete sample of early type galaxies based solely on their visual appearance and started by inspecting 50,000 SDSS galaxies to create the MOrphologically Selected Ellipticals in SDSS (MOSES) sample; an order of magnitude more than any visually inspected sample created to that point. The MOSES sample has resulted in a number of interesting results (e.g., Schawinski et al., 2007a,b, 2009b; Thomas et al., 2010) and in particular shows that there is a significant fraction of early-type galaxies that show recent star-formation activity.

The experience with MOSES proved the need for independent morphological classifications for galaxies, while also demonstrating that scaling the MOSES methodology to all SDSS galaxies was unfeasible for a small number of researchers to manage. At this point, Kevin Schawinski and Chris Lintott (a researcher at Oxford also involved in MOSES) became motivated to find a way to visually classify all SDSS galaxies in a reasonable amount of time, thus creating the initial "Galaxy Zoo" concept. They concluded that the only reasonable way to approach this problem was to "outsource" the visual inspection task and put it on the Internet inviting volunteers to participate. At the time, the Stardust@Home Project[†] was using the

---

[*] http://www.sdss.org/
[†] http://stardustathome.ssl.berkeley.edu/

internet to recruit volunteers to identify tracks made by interstellar dust in samples that were flown on NASA's Stardust sample-return mission to Comet Wild-2 (Westphal et al., 2006). Stardust@Home had ~20,000 volunteers, and by extrapolation, Lintott and Schawinski figured that if even one quarter of 20,000 volunteers did one galaxy classification per day, the full SDSS Main Galaxy Sample (approximately a million galaxies) could have secure galaxy classifications in 3 years (assuming each galaxy was visually inspected five times each).

At the same time, another researcher at Oxford, Kate Land was planning a similar interface to classify, and charactersize the sense of rotation of spiral galaxies. She was interested in an article that suggested that there was a correlation between the "handedness" of spiral arms in the SDSS disk galaxies and their position on the sky, that is that the direction of the rotation of disk galaxies did not appear to be random (Longo, 2007). Land had planned to build an interface on a laptop computer and then place it in the canteen of the Oxford Physics Department, hoping to enlist the help of her fellow scientists. However, after a fortuitous meeting of the two groups, it became clear that the projects could be merged into a single interface addressing both questions.

Phil Murray and Dan Andreescu of Fingerprint Digital Media were recruited to design the Galaxy Zoo website and the initial success of Galaxy Zoo can probably be credited to the visual appeal and ease of use of the interface design, combined with a relatively easy classification scheme. The user was asked if the galaxy image they saw was "Spiral" or "Elliptical," followed by the classification of the apparent spin direction of the spiral arms (clockwise or anticlockwise). Another key factor was that people could get started right away after a relatively short tutorial. Once a user had passed the tutorial, they were free to classify as many galaxies as they wished and could login and out of their account as they wished. The original Galaxy Zoo team along with Land, Lintott and Schawinski included experts from the SDSS (Alex Szalay, Bob Nichol, Steven Bamford, Anze Slozar) and the MOSES team (Daniel Thomas), as well as experts in astronomy outreach (Jordan Raddick) and data archives (Jan van den Berg).

Galaxy Zoo was launched on July 11, 2007 and introduced in a BBC online article that same day.[*] In the first 3 h after launch, classifications were coming in at such a high rate that the data servers located at Johns Hopkins University hosting the site and SDSS images were unable to meet the demand. Fortunately, additional capacity was brought online quickly and, within 12 h of the launch, the Galaxy Zoo site was receiving 20,000 classifications per hour. After 40 h, the classification rate had increased to 60,000 per hour. After 10 days, the public had submitted ~8 million classifications. By April 2008, when the Galaxy Zoo team submitted their first paper (Lintott et al., 2008), over 100,000 volunteers had classified each of the ~900,000 SDSS galaxy images an average of 38 times.

One of the unforeseen consequences of the Galaxy Zoo launch was the avalanche of email the team received from the public. Within 2 weeks, the original Galaxy Zoo team was swamped with requests for information and queries, and several additional people were recruited to help manage these requests. This need to communicate inspired the creation of

---

[*] *Scientists seek galaxy hunt help*, by Christine McGourty (http://news.bbc.co.uk/1/hi/sci/tech/6289474.stm)

a Galaxy Zoo Internet forum, which encouraged the Galaxy Zoo users to communicate with each other (overseen by the Galaxy Zoo team). This allowed many of the basic queries from the public to be answered by other members of the public more experienced with Galaxy Zoo, and also allowed the volunteers (who named themselves "Zooites") to share their thoughts and ideas with each other. Once the forum was established, several members of the public ("citizen scientists") quickly volunteered to moderate the forum and began to generate a variety of discussion threads which included basic help with understanding astronomy and Galaxy Zoo, and a repository for "weird and wonderful" images people found.

In addition to the forum, in December 2007 the team began to communicate with the volunteers through a series of blog messages about the progress of the project and science.[*] Also, thanks to collaboration with Polish partners, a translation of the site was made available. The strong response this received indicated the importance of internationalization in the future development of similar projects.

11.4 GALAXY ZOO 1

As described above, the first phase of Galaxy Zoo (now known as "Galaxy Zoo 1" or GZ1) asked volunteers to provide only basic morphological information on each galaxy. They were asked to identify if a galaxy was "spiral," "elliptical," "a merger," or "star/don't know" and additionally split the spiral category into "clockwise," "anticlockwise," and "edge-on/don't know." Galaxies for the GZ1 project were drawn from the Main Galaxy Sample of the sixth SDSS Data Release (Adelman-McCarthy et al., 2008; Strauss et al., 2002) and comprised all extended objects in the survey that were brighter than a Petrosian magnitude of $r < 17.77$ mag. All objects were included, whether or not they had an SDSS spectrum, giving a total of 893,212 images.

11.4.1 From Clicks to Classifications

The Galaxy Zoo project was extremely successful in recruiting volunteer classifiers thus providing each galaxy in the sample with multiple independent classifications; GZ1 has a mean of 38 classifications per galaxy, with at least 20 classifications for all galaxies. Most previous morphological classifications had been done by single experts (or small groups of experts) agreeing on a single answer, but in Galaxy Zoo the situation was more like a "vote" on the galaxy classification. Going from these votes, or "clicks," to classifications can be done in several ways.

The first step in processing the user-generated data was to "clean" them by removing the tiny fraction of potentially malicious users, and any chance multiple classifications of a single galaxy by a given classifier. Next, there was a decision about how much weight each vote should have. The simplest choice is to give all classifiers equal weight. This gives a distribution of classifications for a galaxy which encodes information about the most likely classification as well as some measure of how certain that is (in the spread of classifications).

---

[*] http://blogs.zooniverse.org/galaxyzoo/

In GZ1 a weighting scheme was also explored which weighted users based on how well they agreed with the majority (in practice this was applied iteratively). This was an attempt to give more weight to "better" classifiers, where "better" was defined as agreeing with the majority. These "weighted" classifications for the most part were similar to unweighted classifications.

11.4.1.1 Classification Biases

Several bias studies were run in the original GZ1 to test the effect of the interface and types of images shown to the volunteers on the classifications which were entered. The two main goals of the bias studies were to: (1) test the effect of using color images for the classifications, and (2) test if users could reliably identify the sense of the spiral arm winding. To achieve this, a small number of monochrome and mirrored images were added to the GZ1 sample and the clicks on these images were compared to the original, unperturbed images.

Interestingly, a change in the behavior of the volunteer classifiers was witnessed during these bias testing exercises, in the sense that users appeared to be more careful in their classifications during bias testing periods. Therefore, only clicks collected on the original images at the same time as the tests were being carried out could be used for the comparison between classifications. The results of the monochrome bias test showed that there were only small differences in the galaxy classifications between color and black and white images. Users were slightly more likely to classify objects as "elliptical" in monochrome images; 56% of the votes went to ellipticals in the monochrome images compared to 55% in the original color SDSS images.

The results of the mirror image bias testing are discussed extensively in Land et al. (2008). They showed a significant bias in favor of anti clockwise direction arms (in both the original and mirrored images). The interpretation of this bias could be due to psychological effects (possibly related to the preference for right handedness among the population), or possibly site design (it being easier to click the anti clockwise button for example). However, once this bias was corrected for, the data could still be used (see below).

Finally, another source of bias in the GZ classifications has to do with the distance to the observed galaxies. We expect that at some distance, features become harder to resolve, and more galaxies will be classified as ellipticals. This effect was indeed found in the GZ1 sample by Bamford et al. (2009) where a correction was derived as a function of redshift and galaxy luminosity. The conclusion was that the GZ1 classifications are reliable, and the bias correction is small, for redshifts below $z < 0.08$, but at higher redshifts there is a strong trend for galaxies to be classified preferentially as elliptical.

11.4.1.2 Comparison with Other Classifications

In Lintott et al. (2008), the GZ1 classifications were compared against three sets of independent galaxy classifications. These included early-type galaxies in the MOSES sample (Schawinski et al., 2007b), a set of 2275 SDSS galaxies of all galaxy types classified by Fukugita et al. (2007), and the sample of 2834 visually identified SDSS spiral galaxies from Longo (2007). In all cases, GZ1 classifications were found to agree remarkably well (better than 90% of the time in most cases), and the conclusion was that using data from volunteers did not substantially degrade the quality of classifications, while expanding the number of classified galaxies by a large factor and additionally reducing the scope for human error introducing erroneous classifications.

11.4.2 Science Results from Galaxy Zoo 1

Classifications from GZ1 have been used for a wide range of galaxy evolution studies. A full list of the peer-reviewed papers coming from within the Galaxy Zoo 1 team is provided in Table 11.1. We review some of these science results here and stress that the data from GZ1 is now publicly available (Lintott et al., 2011)[*] and being used by several scientists

TABLE 11.1   A Sample of Peer-Reviewed Papers Based on Classifications Collected in the First Phase of Galaxy Zoo (in Order of Publication)

Author (Year)	Title—Galaxy Zoo
Kate Land et al. (2008)	The large-scale spin statistics of spiral galaxies in the SDSS
Chris Lintott et al. (2008)	Morphologies derived from visual inspection of galaxies from the SDSS
Anze Slosar et al. (2009)	Chiral correlation function of galaxy spins
Steven Bamford et al. (2009)	The dependence of morphology and color on environment
Kevin Schawinski et al. (2009a)	A sample of blue early-type galaxies at low redshift
Chris Lintott et al. (2009)	"Hanny's Voorwerp," a quasar light echo?
Ramin Skibba et al. (2009)	Disentangling the environmental dependence of morphology and color
Carie Cardamone et al. (2009)	Green Peas: Discovery of a class of compact extremely star-forming galaxies
Daniel Darg et al. (2010a)	The fraction of merging galaxies in the SDSS and their morphologies
Daniel Darg et al. (2010b)	The properties of merging galaxies in the nearby Universe—local environments, colors, masses, star formation rates, and AGN activity
Kevin Schawinski et al. (2010a)	The Sudden death of the nearest quasar
Kevin Schawinski et al. (2010b)	The Fundamentally different co evolution of supermassive black holes and their early- and late-type host galaxies
Karen Masters et al. (2010a)	Dust in spiral galaxies
Raul Jimenez et al. (2010)	A correlation between the coherence of galaxy spin chirality and star formation efficiency
Karen Masters et al. (2010b)	Passive red spirals
Manda Banerji et al. (2010)	Reproducing galaxy morphologies via machine learning
Chris Lintott et al. (2011)	Data release of morphological classifications for nearly 900,000 galaxies
O. Ivy Wong et al. (2011)	Building the low-mass end of the red sequence with local post-starburst galaxies
Daniel Darg et al. (2011)	Multi mergers and the millennium simulation

[*] http://www.data.galaxyzoo.org

beyond the original GZ1 team. For example, Galaxy Zoo 1 was used to remove late-type contaminants from the study of Trujillo et al. (2011), and was compared against a new method for automated classification in Huertas-Company et al. (2011). Moreover, the Galaxy Zoo 1 classifications have now been included in the Eighth Data Release of the SDSS (see Aihara et al., 2011) and can be electronically accessed alongside other SDSS galaxy parameters in their Catalog Archive Server (CAS).[*]

11.4.2.1 Color and Morphology
The greatest legacy from GZ1 has been the decoupling of color and morphology with high statistical significance. We have demonstrated that 80% of galaxies follow the expected correlations between color and morphology, that is, either "red" early-type galaxies or "blue" spiral galaxies. Therefore, for a majority of galaxies, color can be used as a crude proxy for morphology. However, GZ1 also shows that there is a significant numbers of red (passive) spiral galaxies and blue early-type galaxies. These interesting sub populations of galaxies have been explored in a number of GZ papers (see Table 11.1).

This disentangling of morphology and color has been used to study the separate dependences of the properties on environment and provide evidence that the transformation of galaxies from "blue" to "red" proceeds faster than the transformation from spiral to early-type (see Bamford et al. (2009) and Skibba et al. (2009) which use different methods to quantify this effect). The properties of the "blue" early-type galaxies in Galaxy Zoo have been studied by Schawinski et al. (2009a) and "red" (passive) spirals has been explored further by Masters et al. (2010a,b).

11.4.2.2 Spiral Arm Directions
The clockwise/anti clockwise classifications of the spiral galaxies have been used to show that (as expected from the cosmological principle) there is no evidence for a preferred rotation direction in the universe, but that humans preferentially classify spiral galaxies as anti clockwise (Land et al., 2008); and hint at a local correlation of galaxy spins at distances less than ∼0.5 Mpc—the first experimental evidence for chiral correlation of spins (Slosar et al., 2009). Intriguingly, there are also hints of a correlation between star formation history and spin alignments (Jimenez et al., 2010).

11.4.2.3 Merging Galaxies
The sample of merging galaxies has been used to show that the local fraction of mergers is about 1–3% and to study the global properties of merging galaxies (Darg et al., 2010a,b). Multi mergers (where more than two galaxies are merging at once)—which are much rarer than binary mergers have also recently been studied (Darg et al., 2011).

11.4.2.4 Active Galaxies
The GZ1 classifications also revealed interesting correlations between galaxy morphology and black hole growth. By splitting both the normal galaxy population and the active galaxy population by morphology, two fundamentally different modes of black hole feeding and feedback in early- and late-type galaxies were found (Schawinski et al., 2010b). Early-type

---

[*] http://skyservice.pha.jhu.edu/casjobs/

Active Galactic Nucleus (AGN) host galaxies are systematically lower mass and bluer than the general early-type population. Black hole growth is concentrated strongly in the "green valley" between the blue cloud and the low-mass end of the red sequence. These early-type AGN host galaxies furthermore feature strong post starburst stellar populations (Schawinski et al., 2007b) and thus are migrating from the blue cloud to passive evolution at the low mass end of the red sequence—they are thus building up the red sequence today.

Late-type AGN host galaxies dominate by number (up to 90% if "indeterminate" are included) and reside predominantly in massive host galaxies with no indications of recent suppression of star formation. Black hole growth in these disk-dominated galaxies is likely stochastic and has no significant connection to the evolutionary trajectory of the host galaxy. Intriguingly, the Milky Way galaxy resides in the locus of mass and color where black hole growth is most likely, potentially making the Milky Way and Sagittarius A* a prototype for this "secular" mode of black hole feeding in late-type galaxies.

11.4.2.5 Rare and Unusual Objects

GZ1 has brought to light several rare classes of object. "Hanny's Voorwerp" is perhaps the most famous of such objects and many are familiar with the story of the Dutch school teacher Hanny, who first noted this object (she was not the first volunteer to see it, but the first to ask about it) which is now memorialized in a Comic Book.[*] The Voorwerp is an unusual emission line nebula neighboring the spiral galaxy IC 2497 and has been studied in several follow-up projects (e.g., Lintott et al., 2009; Rampadarath et al., 2010; Schawinski et al., 2010a), and also features in much of the education material from Galaxy Zoo.

Another unusual class of objects discovered by the Galaxy Zoo volunteers are the "Green Peas." The properties of these emission-line galaxies, which appear green in the SDSS composite *gri* color images because of their strong [OIII] emission, are studied in detail in Cardamone et al. (2009).

11.5 EVOLUTION OF GALAXY ZOO

11.5.1 Galaxy Zoo 2 and Hubble Zoo

As the original Galaxy Zoo was the first time such a project had been attempted, the Galaxy Zoo team was cautious with their classification scheme, only asking for simple information about the appearance of the galaxies. Thanks to the overwhelming response, and prompted by requests from the volunteers who wanted to provide more detailed classifications, the team realized they could harvest much more information from the SDSS images than in GZ1. Therefore, Galaxy Zoo 2 (GZ2) was designed around asking more detailed questions about the ~250,000 brightest SDSS galaxies from the original GZ1 sample of galaxies.[†] Once again, the response was tremendous and in the 14 months that the site was live, Galaxy Zoo 2 users provided over 60 million classifications. Along the way, deeper SDSS images were added for a subset of GZ2 galaxies, taken from a patch of the sky known as "Stripe 82" which allows fainter structures in these galaxies to be visible.

---

[*] See http://hannysvoorwerp.zooniverse.org/

[†] The Website (http://zoo2.galaxyzoo.org/) for this phase of Galaxy Zoo was designed by Phil Murray and implemented by Danny Locksmith and Arfon Smith.

The first science results from GZ2 classifications are now appearing. In Masters et al. (2011), we showed that the fraction of barred disk galaxies (as compared to unbarred galaxies) depends on other galaxy properties, especially the overall color of the galaxy and the size of the central bulge. As a satellite project, Ben Hoyle at Portsmouth University developed an additional web interface using Google Maps technologies to allow GZ2 volunteers to draw the shapes and sizes of bars on GZ2-selected disk galaxies (Hoyle et al., 2011). From September 2009 to January 2010, he received 16,551 bar drawings for 8180 galaxies, making it by far the largest sample of disk galaxies with known bar lengths; again demonstrating the attraction of Galaxy Zoo even for such a complex task. These studies combined show the strong connection between the bar of a disk galaxy and its overall color, that is disk galaxies with long bars also exhibit prominent bulges and have redder colors than galaxies with smaller bars.

After Galaxy Zoo 2, the team launched "Hubble Zoo." To really understand galaxy evolution, and to get a sense of how the color-morphology relation might change over time, it is important to be able to classify morphologies for galaxies that are much further away than those classified from the SDSS. The light from these galaxies has taken much longer to get to us and hence provide images of galaxies at a much earlier epoch in the history of the universe. Such a dataset will allow us to answer questions like: Are there more blue ellipticals compared to red ellipticals earlier on in the Universe? Does the number of irregularly shaped galaxies increase as we look back further in time? To compare the results from the GZ1 and GZ2 classifications of the SDSS galaxies to galaxies at an earlier epoch, the latest incarnation of Galaxy Zoo is using data from the Hubble Space Telescope (HST) which goes deeper than ever before, for example, HST (Cosmic Evolution Survey COSMOS) has over two million galaxies that cover 75% of the age of the universe (Scoville et al., 2007). Hubble Zoo is currently undergoing classification using HST data from GEMS (Rix et al., 2004), GOODS (Giavalisco et al., 2004), AEGIS (Davis et al., 2007), and COSMOS (Scoville et al., 2007). The decision tree is identical to that for GZ2 except that there is an additional branch that classifies the "clumpiness" and symmetry of each galaxy.

11.5.2 The Citizen Scientists: Motivation and Unexpected Outcomes

Within the first several days after launch, it was clear to the Galaxy Zoo team that they had hit a nerve with the public—classifying galaxies on Galaxy Zoo provided some sort of fulfillment for the volunteers. Many team members suspected that the popularity of the project relied on the beauty of the images, or that the project had benefited from particularly good and lucky publicity. Already, team members were thinking of other scholarly areas where applying the method of visually inspecting data could lead to publishable results beyond what could be accomplished by application of machine algorithms. But before any such steps could be taken, it was essential to understand the motivations for volunteers participating in Galaxy Zoo. A survey of the motivations of citizen scientists involved in Galaxy Zoo is presented in Raddick et al. (2010). The results show that by far, the most common motivation Galaxy Zoo volunteers cite for their involvement in the project is their desire to contribute to real scientific work.

Thus, it should not have come as a surprise that many Galaxy Zoo volunteers developed their own lines of inquiry off the main task page. The Galaxy Zoo forum acted as a clearing house for volunteers to describe and discuss objects that they felt were noteworthy. Several threads were devoted to collecting objects with specific characteristics, for example, triple mergers, or "overlapping" galaxies, or small, round, green galaxies dubbed "Peas." These three examples have all resulted in scientific papers (Cardamone et al., 2009; Darg et al., 2011; Keel et al., 2011, respectively).

But one of the critical aspects enabling the development of collections and further inquiry into an object's characteristics was the link from the main task page for each object to the SDSS SkyServer Object Explorer page.[*] This page aggregated information about the galaxy including the image and accompanying spectrum as well as information about its magnitude, redshift, cross-identifications in other wavelengths and a host of links to more information such as NASA's Extragalactic Database.[†] It is through the Object Explorer page that Galaxy Zoo volunteers began to notice that the "Peas" all had extraordinarily high fluxes in the [OIII] emission line. Eventually over 250 of these objects were found while volunteers taught each other through the forum what the characteristics of a "pea" were and began to trade literature searches on what [OIII] meant and possible interpretations of these galaxies. After several months collecting and interpreting on their own, a graduate student from Yale, Carie Cardamone was assigned to moderate the "Peas" forum, working with the volunteers while she developed the full analysis of these rare dwarf galaxies with extremely high star–formation rates (which was published as Cardamone et al., 2009).

The story of the Galaxy Zoo "Peas" inspired the team to ensure that future projects provide links to supporting information and analysis tools related to the objects shown in the primary task. This is to enable the users to conduct their own research and allows for users to learn the process of research aided by peer-mentoring.

The experience with the Galaxy Zoo forums and blogs shows that the citizen scientist volunteers wanted to do much more than classify objects. They built a community of the volunteers, by the volunteers and for the volunteers. Indeed, Galaxy Zoo belonged to the volunteers—it was their time just as much as it was the scientists time spent working on the project. The team understood this important fact and made it a point of principle to keep the volunteers informed about various aspects of the project from the technical to the social and scientific. Moreover, the team realized earlier on that they must respect the time and commitment of the volunteers, and should only harvest classifications for as long as they were scientifically useful.

In fact, the volunteers have set up several projects of their own using Galaxy Zoo infrastructure or methods. The largest example of such a project is probably the "Irregulars" project. Initiated by Galaxy Zoo volunteers Richard Proctor ("Waveney") and Julia Wilkinson ("Jules") on a forum thread,[‡] the aim of this project was initially to collect a sample of irregular galaxies, that is galaxies that did not fit in to the classification scheme at all. This project

---

[*] http://skyserver.sdss.org/dr8/en/tools/explore/obj.asp
[†] http://nedwww.ipac.caltech.edu/
[‡] http://www.galaxyzooforum.org/index.php?topic=273410.0

now uses a self-built web interface (similar in style to Galaxy Zoo)[*] to ask for classifications of the objects and has inspired several volunteer–led research papers. Richard Proctor has recently applied to do a part-time PhD at the Open University using the data collected in this project.

The Galaxy Zoo forum[†] has been a scientific gold mine on several occasions. Examples of science results coming directly from the forum include the discovery of the Voorwerp (Lintott et al., 2009), targeted and serendipitous searches for smaller versions of the Voorwerp ("voorwerpjes" Chojnowski and Keel, 2011; Gagne et al., 2011), the Peas (Cardamone et al. 2009), overlaps (Keel et al. 2011), ring galaxies, and so on. The depth of interest shown by some of the volunteers is extraordinary. Volunteer, Richard Proctor ("waveney") has set up web forms for several searches and sample evaluations (including the "Irregulars" project mentioned above). Massimo Mezzoprete ("Half65") was so interested in the overlapping-galaxy search (Keel et al. 2011) that he learned SQL and perl, creating a tool that he could point to a forum thread and have it parse for either kind of unique SDSS Object ID, then query the CAS and create a PDF with a page of finding chart, photometry, and positional data for each object (Mezzoprete is a co-author on the first overlapping-galaxy paper). These forum results clearly show that through the Galaxy Zoo project, citizen scientists have become research collaborators.

11.6 THE ZOONIVERSE

The extension of the Galaxy Zoo idea to other scientific domains is obvious in our data-rich world, especially given the desire of the public to be involved in scientific investigations of these data. Researchers across a diverse range of academic fields face the common problem of developing new strategies and modes of computational thinking needed to transform this data flood into knowledge. With the current moderate-sized databases (terabytes), citizen science methods like Galaxy Zoo can replace some aspects of machine algorithms. However, as the data deluge will only intensify in the next decades, machine algorithms must advance to meet the data processing demands, incorporating techniques based on developing areas such as computer vision. Instead of displacing the citizen science method, these new algorithms will need to be trained from, and tested by, human input (e.g., GZ1 classifications have been used for machine learning in Banerji et al. (2010)). Thus, the visual processing methods of Galaxy Zoo will become essential to fully extract information from the data.

It is tempting to think of Galaxy Zoo purely as an Education and Outreach endeavor with all its successes in garnering publicity and focus on a community of non expert volunteers. And with that temptation, one might imagine applying the Galaxy Zoo method to an indiscriminate array of projects with the idea that the public would be engaged in the process and so it does not matter if the scientific outputs were "real" or whether the data processing could have been better accomplished through standard computational methods. What must be made clear is that Galaxy Zoo turned citizen science into a data processing *method*—a data reduction tool for data-intensive science that when applied correctly provides the best

---

[*] http://www.wavwebs.com/GZ/Irregular/Hunt.cgi
[†] http://www.galaxyzooforum.org

possible data product from a set of "raw" data. The genius in this method lies in the fact that the public actually prefer to participate in a meaningful set of tasks where they know their work is useful. Galaxy Zoo established this coupling between high-priority science output and the public engagement in science. Once it became clear that the appetite of the volunteer classifiers could crunch significantly more data, the question became one of how this new citizen science method could be made available across different disciplines and data products. And how to begin the process of developing the machine algorithms trained by the human classifiers.

The first step in providing access to more, and varied, data-intensive projects was to aggregate individual citizen science projects onto a common web-based portal. Several objectives are met by establishing a centralized common entry point. First, a home base is provided for the volunteers so they can move with ease between projects. It encourages a sense of community as the same volunteers can share information about their work within, and across, projects. It builds confidence and "brand loyalty" allowing volunteers to become more willing to try new types of projects and progress in their learning. Second, aggregating projects allows for cyberinfrastructure that can take advantage of cross-project efficiencies while retaining the flexibility to provide individualized tools for specific projects. For example, shared software makes development of different projects possible on a reasonably short timescale thus reacting quicker to new opportunities. Among other advantages provided, these factors can then reduce the overhead on recruiting volunteers and allows for the possibility of deploying small and exploratory projects that would be prohibitive to create on their own. Building on the zoo aspect of the Galaxy Zoo brand, the "Zooniverse" became the answer to how to create a centralized portal to a universe of Zoo-like projects.

To turn the "Zooniverse" into reality, several new projects with data sets beyond the SDSS were developed. In order to help manage the Zooniverse and its expanding set of projects, in June 2009 the Citizen Science Alliance[*] was formed initially by Chris Lintott, Steven Bamford, Lucy Fortson and Arfon Smith. The Zooniverse Project[†] Website was launched in December 2009.

To shift from the original Galaxy Zoo to the Zooniverse, substantial technical changes were implemented in order to produce a robust and flexible system. The most important change was the shift from hosting on a single server to hosting in the "Cloud," that is making use of commercial services provided by Amazon Web Services. This technology allows new servers to be brought online in response to demand, and therefore allows the site to cope with spikes in Internet traffic due to the fluctuating media coverage. The new system is built in a "Ruby on Rails" framework with a restful API layer between a thin web layer and the database. Authentication of users is carried out by an implementation of the Central Authentication Service (CAS) single sign-on solution. This technology allows volunteers to use the same account for both the forum and the main Galaxy Zoo site, as well as between different projects. The use of an API allows the Zooniverse team to support not only the main website but also iPhone and Android applications, allowing mobile users to take part in

---

[*] See http://www.citizensciencealliance.org
[†] http://www.zooniverse.org

Galaxy Zoo. Early results suggest that this may be an effective way of increasing the number of classifications per user.

The Zooniverse codebase was designed with a flexible domain model and extensible reuse of code. These attributes allow features developed for new projects to be useable by all projects. The use of cloud computing services provides hosting scalability, while the virtual platform also handles content distribution and asynchronous classification processing. As of late 2011, the "Zooniverse" is running 10 Zoo projects and has handled many millions of classifications by more than 250,000 users. Several different task functions have been implemented through these projects including basic decision trees, drawing shapes on images ("MoonZoo," "MilkyWay" Project), real-time asset prioritization and alerts with the Galaxy Zoo Supernova project of Smith et al. (2011), manipulating simulated data parameters (Galaxy Zoo Mergers) and text transcription ("Old Weather").

To aid in the development of the Zooniverse as a community of citizen scientists, and to enable users to engage in inquiry related to the data for a given Zoo, a discussion tool was recently developed to replace the forum structure used in GZ. The new discussion tool (called "Talk") was launched with the Milky Way Project and encourages users to create collections of objects, share information and join in online discussions. Several social media features such as tagging, tag clouds, "trending," and "recent" toggles improve Talk over the older forum structure, while retaining the primary collaborative functions such as the discussion boards in Galaxy Zoo. The Zooniverse team has already seen a marked increase in traffic to Talk compared to the number of users navigating to the old forum structure.

The Zooniverse team also has developed numerous education resources and continues to conduct education research into the motivations of the volunteers to contribute to tasks, how usage patterns vary over different levels of engagement with the project and whether there is any gain in understanding the process of research—just to name a few of the topics. Further description of these efforts is outside the scope of this paper.

11.6.1 Tasks Suitable for the Zooniverse

One of the difficulties for the Zooniverse is understanding the types of tasks that are suitable for citizen scientists. The original Galaxy Zoo project primarily asked users to classify images. The interface was simple, with only a few buttons to click for every image. Some of the early success of this project might have been due to the simple requirements of this task.

The newer Galaxy Zoo 2, Hubble Zoo, and Galaxy Zoo Supernova project are also based on having volunteers do classifications on images. However, these projects use a context-based decision tree to ask more detailed morphological questions about the objects, rather than just using a single classification of an object. If the galaxy was a spiral, does it have a bar at the center? How many spiral arms are visible? Although each question only has a few possible answers, the data for each object has more detail than would be possible from a single question interface.

The GZ "Mergers" project operates in a fundamentally different way than the other Zoos. Users are not asked to classify images, but rather to match simulated images to data. In some cases, none of the simulations presented are similar to the target galaxy. In other cases, there are several selections possible within the main interface. After selecting an image, users then have the option of enhancing it using a Java applet. By using two-dimensional sliders, the users can generate new simulations to try to make their results match the target image more closely. The overall approach of this project has some similarities to online citizen science games like "Fold-It!."[*] The primary difference is the lack of an objective score for the goodness of fit. In the case of mergers, we do not have such an objective fitness function, as one of the goals of the project is to create a sufficiently large sample of galaxy mergers that such a function could be derived. The users therefore have to use their best judgment to determine the goodness of fit.

When the GZ "Mergers" project first started, there were a large number of images that were being viewed everyday. Of the images being viewed, ~5% were selected by the volunteers as possible matches to the target galaxy. Upon inspection, the science team found that a high fraction (up to 95%) of the simulations were not likely matches to the real galaxies, as it appeared that many simulations were inadvertently selected, or inexperienced users tried to select too many simulated galaxies as matches. To increase the fraction of good matches, the team created a second level interface called "Merger Wars." In this interface, volunteers were given the opportunity to select the best of the simulation images by allowing the full suite of simulated images to compete with each other in one-to-one competitions, for example, users were shown only two simulations at a time, and asked to pick the best one, and then iterate. Although some of this analysis is still underway, the science team believes that the selection rate for good matches has dramatically improved. A larger fraction of the originally selected simulations get zero votes in this second level competition.

In the Planet Hunters[†] site, users are not looking at images, but rather are presented with time series data on the light curves of nearby stars, and are tutored on how to recognize the signature of an extrasolar planet in that data. Despite the seemingly esoteric nature of light curve data, this Zooniverse project has been very successful, proving that citizen scientists are happy to deal with more complex types of data possibly because the scope for discovery is high.

In addition to classification and matching, citizen scientists are also being asked to do measurements on images. In the "Solar Storm Watch," "Moon Zoo," and "Milky Way" projects, volunteers use drawing tools to identify features. In the "Moon Zoo," for example, volunteers are asked to draw circles around the rims of craters. A similar process is used to identify bubbles in the interstellar medium in the "Milky Way" project.

In some ways, these last projects require more advanced skills and more patience than just clicking through a classification tree. However, with the right interface and the right users, very good results can be obtained on these types of projects.

A key observation from all of these Zooniverse projects, and from the forums and Talk interactions, is that some of the volunteers have very advanced abilities and interests. There is a great deal of effort being dedicated to develop a suite of tools that allow these users to carry out additional scientific investigations on their own and, as discussed above, some of the most interesting discoveries come from the users themselves.

---

[*] http://fold.it/portal/
[†] http://www.planethunters.org/

11.6.2 Data Mining the Zooniverse Results

One of the key features of the Zooniverse project is the application of machine learning (data mining) algorithms to the Zooniverse volunteer-contributed tags. These tag data themselves generate a significant volume of data (e.g., the many hundreds of millions of galaxy classifications from Galaxy Zoo). Finding correlations and trends among these user-contributed tags alongside automatically measured parameters of the same objects within the science database (e.g., the SDSS object catalog) will enable the development of improved classification and anomaly-detection algorithms for future sky surveys (such as the Large Synoptic Survey Telescope (LSST)), which will measure properties for at least 100 times more galaxies, 100 times more stars, and 100 thousand times more source observations.

For example, a preliminary study of the galaxy mergers found in the Galaxy Zoo I project was carried out (Baehr et al., 2010). It was found that certain science database parameters in the SDSS science database correlated strongly with how often Galaxy Zoo users identified an object as a merger. These database attributes included: (a) the log-likelihood that the galaxy's surface brightness profile was fit neither by an exponential disk (the `lnLExp_u` attribute in the `PhotoObjAll` table) that is typical of spiral/disk galaxies nor by a de Vaucouleurs profile (the `lnLDeV_u` attribute in the `PhotoObjAll` table) that is typical of elliptical galaxies; (b) a gradient in the position angle of the isophotal major axis of the galaxy (the `isoAGrad_u` attribute in the `PhotoObjAll` table up to Data Release 7); and (c) the galaxy's "texture" (the `texture_u` attribute in the `PhotoObjAll` table up to Data Release 7), which is essentially the root-mean-square variation of the galaxy's surface brightness profile relative to one of the standard galaxy profile-fitting functions. In hindsight, it could have been predicted that these parameters would be useful in distinguishing normal (undisturbed) galaxies from abnormal (merging, colliding, interacting, disturbed) galaxies. These results may now be applied to future sky surveys, to improve the automatic (machine-based) classification algorithms for colliding and merging galaxies. All of this was made possible by the fact that the galaxy classifications provided by Galaxy Zoo I participants led to the creation of the largest pure set of visually identified colliding and merging galaxies yet to be compiled for use by astronomers.

Another example of machine learning using Galaxy Zoo classifications is provided in Banerji et al. (2010) who trained a neural network on a subset of the GZ1 data, and (depending on the automatic measurements given to the algorithm) could reproduce the classifications to better than 90%. They concluded that Galaxy Zoo would provide an invaluable training set for future algorithms likely to be developed to classify the next generation of wide-field imaging surveys.

11.6.3 Future Citizen Science Projects

With all the recent activities in the Zooniverse, it is important to consider the implications that citizen science has for future astronomical projects [e.g., LOw Frequency ARray (LOFAR) (Falcke et al., 2007), and the Dark Energy Survey[*]]. For example, we briefly consider

---

[*] http://www.darkenergysurvey.org

here how volunteers might help with a project like the LSST (LSST Science Collaborations, 2009), when it comes online this decade.

During the first Galaxy Zoo project, volunteers examined images of approximately one million galaxies. The storage space needed for all these images was only a few Terabytes and therefore relatively easy to host and serve. In contrast, LSST will generate tens of Terabytes per day and over its approximate 10-year operational lifetime, it is estimated that it will generate tens of Petabytes.

Among the citizen science projects that may contribute to LSST science are those that explore the time series data from the survey. Since, LSST will do repeated imaging of the sky over the 10-year project duration, each of the roughly 50 billion objects observed by LSST will have ~1000 separate observations. These 50 trillion time-series data points will provide an enormous opportunity to discover all types of rare phenomena, rare objects, rare classes, and new objects, classes, and sub classes.

No group of volunteers could hope to view all such data being generated from LSST. At the same time, the science team on the project will have no hope to keep up with such a data flow from the system. Obviously, automatic algorithms need to be used to triage the data and do basic classification of events. However, even with automatic classification, it is anticipated that tens of thousands (or more) anomalous events will be detected everyday. Some of these might be astronomically significant (asteroids, supernovae, etc.). However, many will not fall into any particular category, and in many cases, they might be some kind of noise (an airplane flying in the field of view).

The contributions of human participants may include: detection of unusual light curves in rotating asteroids; human-assisted search for best-fit models of these asteroids (including shapes, spin periods, and varying surface reflection properties); discovery of unusual variations in known variable stars; discovery of interesting objects in the environments around variable objects; discovery of associations among multiple variable and/or moving objects in a field; and more. This is especially important for the nightly event stream—perhaps 100,000 new events will be detected each and every night for 10 years. There are not enough observing facilities or professional astronomers (or graduate students) in the world to follow-up on each of these events. Engaging a large cohort of willing participants to examine these events will contribute significantly to the scientific discovery efficiency and effectiveness of the LSST survey: citizen scientists may explore this massive event stream for novel and interesting features, thereby characterizing the behavior of each such object. The creation of a characterization database of time-varying objects (from which astronomers may query, search, and retrieve events based upon prescribed characteristic light curve behaviors) may prove to be one of the most significant contributions of citizen scientists to the LSST project that is the development of a major externally joined database component of the LSST science data collection.

Something like this approach was effectively used with the detection of the "Peas" described above (Section 11.4.2.4), that is once this class of objects was discovered, and determined to be interesting, a computer algorithm was developed to find them (Cardamone et al., 2009). By finding new classes of data, volunteers can make major contributions to science that would not be possible without their help. The Galaxy Zoo Supernova project also uses a similar methodology (Smith et al., 2011). During an observing run, the science team receives tens of thousands of possible supernova candidates, and automatic algorithms are used to reduce this to a few hundred events a day that are likely supernovae. With the help of citizen scientists, these candidate supernovae are visually checked and thus confirmed for follow-up observations in real time.

In summary, as the data rates increase, and we become further dependent on automatic classification algorithms, citizen scientists can play a crucial role in reviewing subsets of the data and identifying anomalies. The algorithms can then be adapted by the science teams to increase their success rate (based on the visual checks). The two methodologies will need to work in tandem and the process will likely be iterative.

11.7 GALAXY ZOO IN THE CONTEXT OF OTHER CITIZEN SCIENCE PROJECTS

Galaxy Zoo is certainly not the only citizen science project. As mentioned in the beginning of this chapter, one of the inspirations for Galaxy Zoo was the Stardust@Home project (West-phal et al., 2006). This was one of the few projects at the time where volunteers were asked to participate in the data *analysis* of a project rather than the data *collection* phase of a project. Many of the historically significant citizen science projects, such as the Audubon Society's Christmas Bird Count program (started in 1900) and the American Association of Variable Star Observers variable star observations project (starting in 1911), were based on data *collection*. With the advent of the internet as a distribution system, citizen science projects could move into work on data *analysis*. Here, we make a distinction between distributed analysis projects and distributed computation projects such as SETI@Home[*], which utilizes the computational power of over a million idle computers belonging to volunteers to process radio data looking for signals that could indicate extra-terrestrial intelligence. Distributed analysis projects require the brain—or the "wetware"—of the volunteer to be engaged, not just their computer. One of the earliest distributed data analysis projects (dating from 2001) was Clickworkers which asked the public to count the number of craters on maps of the Martian surface returned by the Mars Orbiter Camera (MOC). While the project was successful in recruiting sufficient volunteers to identify over 800,000 MOC craters (Gulick et al., 2010), there are few scientific results published as of yet on this body of work. Clickworkers has morphed into a project with a more game-like interface[†] where the public are currently asked to "tag" surface features on images from the Mars Rovers, Spirit, and Opportunity. Another project with a game-like interface is FoldIt![‡] which asks the public to help solve protein folding "puzzles." Both the new Clickworkers and the FoldIt! projects require the user to download an application. In the context of distributed data analysis projects, Galaxy Zoo (and its successor projects in the Zooniverse) is quite probably the largest, both in terms of number of registered volunteers world-wide as well as number of peer-reviewed papers published based on data processed by volunteers.

---

[*] http://setiathome.berkeley.edu/index.php
[†] http://beamartian.jpl.nasa.gov/welcome
[‡] http://fold.it/portal/

There are many excellent citizen science projects in ecology, animal studies, and other disciplines where the distributed nature of data *collection* is critical to the success of the project. For example, Cornell University's Lab of Ornithology FeederWatch Project[*] asks volunteers to enter counts of bird species into an online form to track winter bird populations, or the CoCoRaHS Network (Community Collaborative Rain, Hail and Snow Network)[†] has thousands of volunteers across all 50 of the United States who have installed sensors outside their homes and record amounts of rain, snow, and hail. Thus, one potential trend in citizen science will then be projects that link the distributed data collection and distributed data analysis aspects of their work.

11.8 CONCLUSIONS

In this chapter, we have presented a brief overview of the Galaxy Zoo and Zooniverse projects. We gave a short discussion on the history and motivation for the original Galaxy Zoo, as well as the motivations to extend it to "Galaxy Zoo 2," "Hubble Zoo," and an entire "Zooniverse" of citizen science projects. We have described the highlights of the many scientific results that have already come from Galaxy Zoo.

We move on to discuss what makes a good citizen science project, and why we think Galaxy Zoo was so successful. We describe the importance of having a central portal, the Zooniverse, as a gateway to citizen science projects across multiple disciplines. We then consider likely future applications of community-based science in the coming data-rich era. Finally, to provide a context for the importance of the Galaxy Zoo project, we present a short description of various other citizen science projects and modalities.

Galaxy Zoo and the many Zooniverse projects would not have been possible without the participation of now over 400,000 volunteers who have registered with the Zooniverse. The contributions of volunteers to Galaxy Zoo are individually acknowledged at http://www.galaxyzoo.org/Volunteers.aspx; and volunteers who classified in Galaxy Zoo 2 (and wished to be acknowledged) are listed at http://zoo2.galaxyzoo.org/authors.

REFERENCES

Adelman-McCarthy, J. K., Agüeros, M. A., Allam, S. S., Allende Prieto, C., Anderson, K. S. J., Anderson, S. F., Annis, J. et al. 2008. The sixth data release of the sloan digital sky survey, *ApJS*, 175, 297, doi: 10.1086/524984.

Aihara, H., Prieto, C. A., An, D., Anderson, S. F., Aubourg, E., Balbinot, E., Beers, T. C. et al. 2011. The eighth data release of the sloan digital sky survey: First data from SDSS-III, *ApJS*, 193(2), 29.

Baehr, S., Vedachalam, A., Borne, K., and Sponseller, D. 2010. Data mining the Galaxy Zoo Mergers. In *Proceedings of CIDU'2010*, https://c3.ndc.nasa.gov/dashlink/resources/223/, pp. 133–144.

Ball, N. M., Loveday, J., Fukugita, M., Nakamura, O., Okamura, S., Brinkmann, J., and Brunner, R. J. 2004. Galaxy types in the Sloan Digital Sky Survey using supervised artificial neural networks, *MNRAS*, 348, 1038.

Bamford, S. P., Nichol, R. C., Baldry, I. K., Land, K., Lintott, C. J., Schawinski, K., Slosar, A. et al. 2009. Galaxy Zoo: The dependence of morphology and colour on environment, *MNRAS*, 393, 1324.

---

[*] http://www.birds.cornell.edu/pfw/
[†] http://www.cocorahs.org/

Banerji, M., Lahav, O., Lintott, C. J., Abdalla, F. B., Schawinski, K., Bamford, S. P., Andreescu, D. et al. 2010. Galaxy Zoo: Reproducing galaxy morphologies via machine learning, *MNRAS*, 406, 342.

Bernardi, M., Sheth, R. K., Annis, J., Burles, S., Finkbeiner, D. P., Lupton, R. H., Schlegel, D. J. et al. 2003. Early-type galaxies in the sloan digital sky survey. IV. Colors and chemical evolution, *AJ*, 125, 1882.

Buta, R. J. 2011. Galaxy morphology. In *Planets, Stars, and Stellar Systems*, Vol. 6, Series Editor T. D. Oswalt, Vol. editor W. C. Keel, Springer Reference, 2011 (arXiv:1102.0550), to be published.

Cardamone, C., Schawinski, K., Sarzi, M., Bamford, S. P., Bennert, N., Urry, C. M., Lintott, C. et al. 2009. Galaxy Zoo green peas: Discovery of a class of compact extremely star-forming galaxies, *MNRAS*, 399, 1191.

Chojnowski, D. and Keel, W. C. 2011. Galaxy-scale clouds of ionized gas around agn—history and obscuration, *Bull. Am. Astron. Soc.*, 43, #142.07.

Conselice, C. J. 2006. The fundamental properties of galaxies and a new galaxy classification system, *MNRAS*, 373, 1389.

Darg, D. W., Kaviraj, S., Lintott, C. J., Schawinski, K., Sarzi, M., Bamford, S., Silk, J. et al. 2010a. Galaxy Zoo: The fraction of merging galaxies in the SDSS and their morphologies, *MNRAS*, 401, 1043.

Darg, D. W., Kaviraj, S., Lintott, C. J., Schawinski, K., Sarzi, M., Bamford, S., Silk, J. et al. 2010b. Galaxy Zoo: The properties of merging galaxies in the nearby universe—local environments, colours, masses, star formation rates and AGN activity, *MNRAS*, 401, 1552.

Darg, D. W., Kaviraj, S., Lintott, C. J., Schawinski, K., Silk, J., Lynn, S., Bamford, S., and Nichol, R. C. 2011. Galazy Zoo: Multi-mergers and the millennium simulation, *MNRAS*, 416, 1745.

Davis, M., Guhathakurta, P., Konidaris, N. P., Newman, J. A., Ashby, M. L. N., Biggs, A. D., Barmby, P. et al. 2007. The all-wavelength extended groth strip international survey (AEGIS) data sets, *ApJL*, 660, L1.

de Vaucouleurs, G., de Vaucouleurs, A., Corwin, H. G., Buta, R., Paturel, G., and Fouque, P. 1991, *Third Reference Catalogue of Bright Galaxies*, New York: Springer (RC3).

Falcke, H., van Haarlem, M. P., de Bruyn, A. G., Braun, R., Röttgering, H. J. A., Stappers, B., Boland, W. H. W. M. et al. 2007. A very brief description of LOFAR—the low frequency array, *HiA* 14, 386 [astro-ph/0610652].

Fukugita, M., Nakamura, O., Okamura, S., Yasuda, N., Barentine, J. C., Brinkmann, J., Gunn, J. E. et al. 2007. A catalog of morphologically classified galaxies from the sloan digital sky survey: North equatorial region, *AJ*, 134, 579.

Gagne, J., Crenshaw, D. M., Keel, W. C., and Fischer, T. C. 2011. Optical spectra of the teacup AGN, *Bull. Am. Astron. Soc.*, 43, #142.12.

Giavalisco, M., Ferguson, H. C., Koekemoer, A. M., Dickinson, M., Alexander, D. M., Bauer, F. E., Bergeron, J. et al. 2004. The great observatories origins deep survey: Initial results from optical and near-infrared imaging, *ApJL*, 600, L93.

Gulick, V. C., Deardorff, G., Kanefsky, B., Hirise Science Team. 2010. Online citizen science with clickworkers & MRO HiRISE E/PO. In *American Geophysical Union, Fall Meeting 2010*, abstract #ED13B-08 12/2010 2010AGUFMED13B..08G.

Hoyle, B., Masters, K. L., Nichol, R. C., Jimenez, R., and Bamford, S. P. 2011. The fraction of early-type galaxies in low redshift groups and clusters of galaxies, *MNRAS*, arXiv:1110.6320.

Hubble, E. P. 1926. Extragalactic nebulae, *ApJ*, 64, 321.

Hubble, E. P. 1927. The classification of spiral nebulae, *The Observatory*, 50, 276.

Huertas-Company, M., Aguerri, J. A. L., Bernardi, M., Mei, S., and Sánchez Almeida, J. 2011. Revisiting the Hubble sequence in the SDSS DR7 spectroscopic sample: A publicly available Bayesian automated classification, *A&A*, 525, A157.

Jimenez, R., Slosar, A., Verde, L., Bamford, S., Lintott, C., Schawinski, K., Nichol, R. et al. 2010. Galaxy Zoo: A correlation between the coherence of galaxy spin chirality and star formation efficiency, *MNRAS*, 404, 975.

Keel, W., Manning, A. M., Holwerda, B. W. et al. 2011. *MNRAS* (submitted).

Lahav, O., Naim, A., Buta, R. J., Corwin, H. G., de Vaucouleurs, G., Dressler, A., Huchra, J. P. et al. 1995. Galaxies, human eyes, and artificial neural networks, *Science*, 267, 859.

Land, K., Slosar, A., Lintott, C. J., Andreescu, D., Bamford, S., Murray, P., Nichol, R. et al. 2008. Galaxy Zoo: The large-scale spin statistics of spiral galaxies in the sloan digital sky survey, *MNRAS*, 388, 1686.

Lintott, C. J., Schawinski, K., Slosar, A., Land, K., Bamford, S., Thomas, D., Raddick, M. J. et al. 2008. Galaxy Zoo: Morphologies derived from visual inspection of galaxies from the sloan digital sky survey, *MNRAS*, 389, 1179.

Lintott, C. J., Schawinski, K., Keel, W., van Arkel, H., Bennert, N. Edmondson, E., Thomas, D. et al. 2009. Galaxy Zoo: 'Hanny's Voorwerp', a quasar light echo? *MNRAS*, 399, 129.

Lintott, C., Schawinski, K., Bamford, S., Slosar, A., Land, K., Thomas, D., Edmondson, E. et al. 2011. Galaxy Zoo 1: Data release of morphological classifications for nearly 900 000 galaxies, *MNRAS*, 410, 166.

Longo, M. J. 2007. Evidence for a preferred handedness of spiral galaxies, arXiv:0707.3793.

LSST Science Collaborations. 2009. LSST Science Book, Version 2.0, arXiv:0912.0201.

Masters, K. L., Nichol, R., Bamford, S., Mosleh, M., Lintott, C. J., Andreescu, D., Edmondson, E. M. et al. 2010a. Galaxy Zoo: Dust in spiral galaxies, *MNRAS*, 404, 792.

Masters, K. L., Mosleh, M., Romer, A. K., Nichol, R. C., Bamford, S. P., Schawinski, K., Lintott, C. J. et al. 2010b. Galaxy zoo: Passive red spirals, *MNRAS*, 405, 783.

Masters, K. L., Nichol, R. C., Hoyle, B., Lintott, C., Bamford, S. P., Edmondson, E. M., Fortson, L. et al. 2011. Galaxy Zoo: Bars in disc galaxies, *MNRAS*, 411, 2026.

Peng, C. Y., Ho, L. C., Impey, C. D., and Rix, H.-W. 2002. Detailed structural decomposition of galaxy images, *AJ*, 124, 266.

Raddick, M. J., Bracey, G., Gay, P. L., Lintott, C. J., Murray, P., Schawinski, K., Szalay, A. S., and Vandenberg, J. 2010. Galaxy Zoo: Exploring the motivations of citizen science volunteers, *Astron. Educ. Rev.*, 9, 010103.

Rampadarath, H., Garrett, M. A., Józsa, G. I. G., Muxlow, T., Oosterloo, T. A., Paragi, Z., Beswick, R., van Arkel, H., Keel, W. C., and Schawinski, K. 2010. Hanny's Voorwerp: Evidence of AGN activity and a nuclear starburst in the central regions of IC 2497, *A&A*, 517, L8.

Rix, H.-W., Barden, M., Beckwith, S. V. W., Bell, Eric F., Borch, A., Caldwell, J. A. R., Häussler, B. et al. 2004. GEMS: Galaxy evolution from morphologies and SEDs, *ApJS*, 152, 163.

Sandage, A. 1961, *The Hubble Atlas of Galaxies, Carnegie Institution of Washington Publication No. 618*.

Scarlata, C., Carollo, C. M., Lilly, S., Sargent, M. T., Feldmann, R., Kampczyk, P., Porciani, C. et al. 2007. COSMOS morphological classification with the Zurich estimator of structural types (ZEST) and the evolution since $z = 1$ of the luminosity function of early, disk, and irregular galaxies, *ApJS*, 172, 406.

Schawinski, K., Kaviraj, S., Khochfar, S., Yoon, S.-J., Yi, S. K., Deharveng, J.-M., Boselli, A. et al. 2007a. The effect of environment on the ultraviolet color-magnitude relation of early-type galaxies, *ApJS*, 173, 512.

Schawinski, K., Thomas, D., Sarzi, M., Maraston, C., Kaviraj, S., Joo, S.-J., Yi, S. K., and Silk, J. 2007b. Observational evidence for AGN feedback in early-type galaxies, *MNRAS*, 382, 1415.

Schawinski, K., Lintott, C. J., Thomas, D., Sarzi, M., Andreescu, D., Bamford, S. P., Kaviraj, S. et al. 2009a. Galaxy Zoo: A sample of blue early-type galaxies at low redshift, *MNRAS*, 396, 818.

Schawinski, K., Lintott, C. J., Thomas, D., Kaviraj, S., Viti, S., Silk, J., Maraston, C. et al. 2009b. Destruction of molecular gas reservoirs in early-type galaxies by active galactic nucleus feedback, *ApJ*, 690, 1672.

Schawinski, K., Evans, D. A., Virani, S., Urry, C. M., Keel, W. C., Natarajan, P., Lintott, C. J. et al. 2010a. The sudden death of the nearest quasar, *ApJL*, 724, L30.

Schawinski, K., Urry, M. C., Virani, S., Coppi, P., Bamford, S. P., Treister, E., Lintott, C. J. et al. 2010b. Galaxy Zoo: The fundamentally different co-evolution of supermassive black holes and their early- and late-type host galaxies, *ApJ*, 711, 284.

Scoville, N., Abraham, R. G., Aussel, H., Barnes, J. E., Benson, A., Blain, A. W., Calzetti, D. et al. 2007. COSMOS: Hubble space telescope observations, *ApJS*, 172, 38.

Simard, L., Willmer, C. N. A., Vogt, N. P., Sarajedini, V. L., Phillips, A. C., Weiner, B. J., Koo, D. C., Im, M., Illingworth, G. D., and Faber, S. M. 2002. The DEEP Groth strip survey. II. Hubble space telescope structural parameters of galaxies in the Groth strip, *ApJS*, 142, 1.

Skibba, R. A., Bamford, S. P., Nichol, R. C., Lintott, C. J., Andreescu, D., Edmondson, E. M., Murray, P. et al. 2009. Galaxy Zoo: Disentangling the environmental dependence of morphology and colour, *MNRAS*, 399, 966.

Slosar, A., Land, K., Bamford, S., Lintott, C., Andreescu, D., Murray, P., Nichol, R. et al. 2009. Galaxy Zoo: Chiral correlation function of galaxy spins, *MNRAS*, 392, 1225.

Smith, A. M., Lynn, S. P., Sullivan, M., Lintott, C. J., Nugent, P. E., Botyanszki, J., Kasliwal, M. et al. 2011. Galaxy Zoo supernovae, *MNRAS*, 150.

Strauss, M. A., Wienberg, D. H., Lupton, R. H., Narayanan, V. K., Annis, J., Bernardi, M., Blanton, M. et al. 2002. Spectroscopic target selection in the sloan digital sky survey: The main galaxy sample, *AJ*, 124, 1810.

Thomas, D., Maraston, C., Schawinski, K., Sarzi, M., and Silk, J. 2010. Environment and self-regulation in galaxy formation, *MNRAS*, 404, 1775.

Trujillo, I., Ferreras, I., and de La Rosa, I. G. 2011. Dissecting the size evolution of elliptical galaxies since $z \sim 1$: puffing up vs minor merging scenarios, *MNRAS*, 415, 3903.

Westphal, A. J., von Korff, J., Anderson, D. P., Alexander, A., Betts, B., Brownlee, D. E., Butterworth, A. L. et al. 2006. Stardust@home: Virtual microscope validation and first results. In *37th Annual Lunar and Planetary Science Conference*, March 13–17, 2006, League City, Texas, abstract no. 2225.

Wong, O. I., Schawinski, K., Kaviraj, S., Masters, K. L., Nichol, R. C., Lintott, C., Keel, W. C. et al. 2011. Galaxy Zoo: Building the low-mass end of the red sequence with local post-starburst galaxies, *MNRAS*, arXiv:1111.1785.

York, D. G., Adelman, J., Anderson, J. E. Jr., Anderson, S. F., Annis, J., Bahcall, N. A., Bakken, J. A. et al. 2000. The Sloan digital sky survey: Technical summary, *AJ*, 120, 1579.

CHAPTER 12

The Utilization of Classifications in High-Energy Astrophysics Experiments

Bill Atwood

University of California at Santa Cruz

CONTENTS

12.1   Introduction   237
12.2   A Classification Tree Primer   240
12.3   Measuring Performance   243
12.4   Relative Importance of the Independent Variables   245
12.5   Classification Trees in Practice   246
12.6   Training Bias   246
       12.6.1 Summary   250
12.7   Usage Examples   250
12.8   What can Go Wrong, Go Wrong, . . .   261
12.9   Summary for Fermi-LAT CT Usage   263
12.10 Usage of CTs in the Analysis of an IACT: MAGIC   263
12.11 Summary   264
Acknowledgments   265
References   265

12.1 INTRODUCTION

The history of high-energy gamma observations stretches back several decades. But it was with the launch of the Energetic Gamma Ray Experiment Telescope (EGRET) in 1991 onboard the Compton Gamma Ray Observatory (CGRO) [1], that the field entered a new era of discovery. At the high-energy end of the electromagnetic spectrum, incoming particles of light, photons, interact with matter mainly by producing electron–positron pairs and this process dominates above an energy of 10–30 MeV depending on the material. To a high degree the directionality of the incoming gamma ray is reflected in the e^+ and e^-, and hence the detection of the trajectories of the e^+e^- pair can be used to infer the direction of the originating photon. Measuring these high-energy charged particles is the domain of high-energy particle physics and so it should be of little surprise that particle physicists played a significant role in the design and construction of EGRET, as well as the design and implementation of analysis methods for the resulting data. Prior to EGRET, only a handful of sources in the sky were known as high-energy gamma-ray emitters. During EGRET's 9-years mission the final catalog included over 270 sources including new types such as Gamma Ray Bursts (GRBs). This set the stage for the next-generation mission, the Gamma-ray Large Area Space Telescope (GLAST) [2].

Very early in the EGRET mission, the realization that the high-energy gamma-ray sky was extremely interesting led to a competition to develop the next-generation instruments. The technology used in EGRET was frozen in the late 1970s and by 1992, enormous advances had been made in experimental particle physics. In particular the effort to develop solid state detectors, targeted for use at the Super Conducting Super Collider (SSC), had made the technology of silicon strip detectors (SSDs) commercially viable for use in large area arrays. Given the limitations imposed by the space environment (e.g., operate in a vacuum, scarce availability of electric power, etc.), this was the ideal technology for the next gamma-ray mission.

Consistent with contemporary practice in particle physics, a nearly complete and detailed computer model of GLAST was made to study performance and optimize the design. The jargon in the field refers to such models generically as "the Monte Carlo" (MC) and it included a complete suite of radiation transport codes modeling most of the known interactions that particles undergo upon passage through matter. The MC is also used to provide a randomized source of incoming particles which can be made to mimic celestial sources of gamma rays as well as background cosmic rays. The cosmic rays referred to here are comprised of two main components: trapped radiation in the earth's magnetosphere and a flux of primary charged particles originating from outside. Both fluxes contain a variety of particle types including protons, heavier nuclei, electrons, and positrons.

It cannot be emphasized too strongly the value that such a tool brings. Having the "Monte Carlo Truth" for each simulated event allows for the evaluation of what went right and what went wrong both at the detector level as well as at the data analysis level. The Monte Carlo simulations of GLAST are at the heart of its success today.

The simulations allowed for the development of the reconstruction analysis (RA) of the flight data prior to the existence of the instrument. The RA transforms the collections of sensor readouts in an event into tracks, energies, and other higher analysis entities. In the case of GLAST, there were many iterations of the RA, first to prove the merits of the design and then, post awarding of the flight instrument contract, the creation of the code to be used in the initial phases of the mission. Since then, the now renamed Fermi-LAT mission is engaged in the 8th such iteration.

Critical to the success of any experiment are the identification and quantification of the "signal." Mostly all experiments have backgrounds or artifacts which obscure a clear signal and in the area of high-energy gamma-ray astronomy the situation is extreme. By the very nature of the detection method for gamma rays, it leaves the apparatus vulnerable to interpret cosmic rays as "signal." In low earth orbit the incoming rate of cosmic rays can exceed the gamma-ray rate by over 10,000. To achieve residual background levels in the percent range requires an aggregate separation power of upwards to a million-to-one while at the same time preserving a high efficiency for capturing the signal. From the triggering, to onboard filtering and ultimately the analysis on the ground, the goal is the same: kill background—keep gamma rays.

The LAT, the hardware trigger, is a combination of sensor responses coincident at the microsecond timescale, causing all the sensors to be readout forming an "event." The LAT trigger was constructed to be highly efficient for gamma rays and as inefficient as possible for cosmic rays. The rejection power achieved by the trigger is about 5:1 while retaining over 98% efficiency for gamma rays.

This still leaves a data event stream hopelessly large to downlink to the ground. A bank of onboard computers does a preliminary event reconstruction to increase the rejection power. For the LAT, this resulted in another factor of ~5 while maintaining a gamma-ray efficiency of over 96%. Hence, what is downlinked to ground is an orbit averaged rate of about 400 cps of which only a few cps are gamma rays. All this complexity as to actual event composition (e.g., which cosmic ray events make it to the ground), is modeled in the simulations and provides a realistic facsimile to real data and this serves as the input to the ground analysis.

Part of the requirements for the mission was to demonstrate prior to launch that the science goals were achievable and this included background rejection at a specified level. During the several years of development working toward this goal it was realized that the science requirements as written were not achievable mainly due to a background which we termed as "irreducible." Positrons hitting the outer protective layers of the LAT (the thermal blanket and micrometeor shield) can annihilate with the atomic electrons producing gammas and cosmic ray protons interacting with the same material produce neutral pions which then promptly decay into two gamma rays. In some cases, the only particles entering the fiducial volume of the detector are gamma rays from these processes and they are indistinguishable from celestial gamma rays. The rejection was increased until the majority of the remainder was "irreducible" in origin along with a demonstration that this residual could be subtracted with appropriately small systematic errors.

The LAT instrument and reconstruction of the gamma-ray data from it fall into the realm of particle physics. Our first attempts at background rejection followed standard practice in particle physics: identify good discrimination variables and make cuts. What is meant by "cuts" is to accept (or reject) events for which a given variable (or variables) falls within (or outside) a certain range of values. This method, however, soon revealed itself to be inadequate especially when considering the resulting efficiency for retaining gamma rays. We then turned to data mining techniques which had seen considerable success in the life sciences and financial industries. In the end the classification tree technology was found to be inadequate by itself. A hybrid approach was developed in which first cuts are made using some of the discrimination variables to whittle down the problem and then the simplified problem is solved using a classification tree. In the next section we will see why classification trees offer a substantial increase in efficiency over the "cut and keep" paradigm. Other machine learning methods were tried, such as neural nets, but they were found to be inferior for this problem.

It should come as little surprise to find that other gamma-ray experiments also found the classification tree technology advantageous for similar reasons. In particular the ground-based Imaging Air Cherenkov Telescope (IACT), MAGIC [3], successfully developed a background rejection for its data based on classification trees.

There are several packages which provide classification tree (CT)-based technology [4]. Some of these are commercial while others are free. Please see the references for specifics. In the discussion which follows, the commercial product by TIBCO (Spotfire Miner) is used [5].

12.2 A CLASSIFICATION TREE PRIMER

The precursor to classification trees was published 1963 by Morgan and Sonquist: automated interaction detection (AID) [6]. In 1984 Breiman, Freidman, Olshen, and Stone published Classification and Regression Trees (CART) [7] which is still the foundational reference for the subject. This was one of the first "machine learning" approaches to data analysis. By "machine learning" what is meant is an algorithm which given **training data sets** could compute a network which can separate the input data sets into desired subsets. For the problem at hand, the training data sets are a mixture of background cosmic rays (BKG) and gamma rays (GAM). These techniques generalize to more than just two categories. For a more complete exposition of classification trees please see Chapter 23.

Input to the CT algorithm is a data set comprised of both BKG and GAM, with each event tagged accordingly. This categorical variable is considered to be the dependent variable, which may be separated according to type by a set of independent measures for each event derived from the instrument's responses.

Each independent variable in the data set is measured to determine its power in separating the various categories. This is commonly carried out by using either the **Gini** or **Entropy** metric.[*] The goal is to find the independent variable and the place to cut on that variable that minimizes the metric summed over the resultant nodes thereby maximizing the separation. These are simple and fast to compute:

$$\textbf{Gini} = \sum_{\substack{over\ all \\ categories\ i}} P_i(1 - P_i) = 1 - \sum_i P_i^2$$

and

$$\textbf{Entropy} = \sum_{\substack{over\ all \\ categories\ i}} P_i\ log(P_i)$$

where P_i is the fraction of events of category i in the node (for our case these sums have two terms). The *Gini* metric tends to favor splits that are somewhat imbalanced in terms of the number of events going to each of the resultant nodes, while *Entropy* favors more

---

[*] The term *Gini Index* is named after its creator, an Italian economist circa early 1900s, Corrado Gini. *Entropy* is a measure of order in a collection of objects and comes from *Statistical Mechanics and Thermodynamics in Physics*.

balanced splits. Which is best is dependent on the specific separation being attempted and so one should experiment. For the LAT analysis, *Entropy* is used. However, the differences in the resulting separation power with respect to using *Gini* were found to be small. The cut on the independent variables serves to divide the data set into two subsets. The initial data set is called the *root node* and these new resulting subsets of data are simply called *nodes*. The elegance of the method is that it is recursive: simply repeat the above procedure until the process results in nodes with too few events to continue or has reached some other termination criterion (e.g., has only one class left). These termination nodes are sometimes referred to as *leaves, leaf nodes*, or *termination nodes*. The overall structure is one of a series of connected nodes, each with an input and two outputs and is often called a *binary tree* given its resemblance to its biological namesake. An example of this structure is shown in Figure 12.1.

In this example the recursive splitting algorithm produced a total of 40 leaf nodes of which 24 had mostly events of category GAM. Only four leaves are shown in Figure 12.1.

At a qualitative level, it is now clear why classification trees are more efficient than a simple set of cuts which would describe only one branch in such a tree. Here the "discard" pile of events is subsequently searched again and again for other possible separations. Another attractive aspect of classification trees is the objectivity of the process. The human bias, so easily introduced into an analysis such as separating signal from background, is at least now

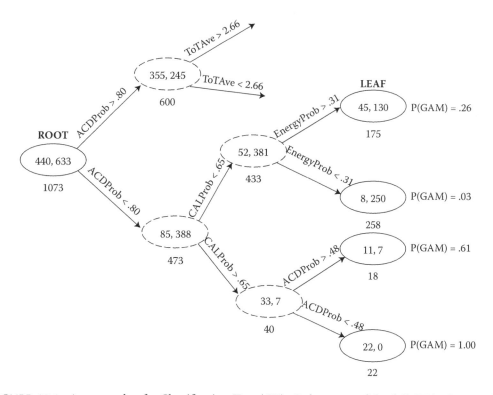

FIGURE 12.1 An example of a Classification Tree (CT). Only a part of the full CT is shown. The input is on the left in the node labeled **ROOT** and the outputs are the **LEAF**s. The resulting GAM probabilities are shown next to each LEAF. The cuts are given on the branches, the number of events at each node is given below the node, and the number of GAM, BKG are given inside the nodes.

"once removed." The results, however, specifically the leaf nodes, are statistical in nature. Repeating the process on an independent set of training data will result in a similar, but different, tree.

After a tree has been "grown" using training data, it can then be used to analyze data. Events starting at the root node will wind their way through the splits, ultimately ending up in a specific leaf node. The purity of each leaf node can be assessed from the training sample, allowing for a choice to be made as whether or not to keep events landing in it. This choice can take the form of a probability, derived from the training sample. In the tree shown in Figure 12.1, leaf nodes are typed GAM if they contain >50% gamma-ray events at the training stage. A more stringent requirement would be to accept only events which fall into leaf nodes with >75% probability for being of type GAM.

Of some concern in the creation of classification trees is that they are statistical and depend on the specific training sample. What we want is for the training process to reveal patterns in the data rather than the CTs essentially memorizing a specific training data set. If one allowed the branching to continue down to leafs containing one event, the fluctuations would be enormous and the resulting CT would in essence be just a memorized, event-by-event replay of the training sample. The performance on the training sample for such a tree would appear to achieve perfect separation. However, be prepared for disappointing performance on an independent data sample.

There are several ways of limiting this so-called training bias. Perhaps the simplest is to stop the branching process when the number of events falls below a preset limit when the tree is being grown. The next level of sophistication is to also limit branching when one of the new nodes would have too few events. Researchers in statistics have found that a ratio of 3:1 for these choices is somewhat optimal (e.g., stop branching at 30 events and do not produce nodes with fewer than 10 events). More sophisticated yet is to evaluate the output of the tree with and without including more nodes and continue the branching until the changes being made fall below a set limit. In practice, simply limiting the statistic at which to terminate branching has been found to be transparent and robust provided the counts are kept at a reasonably high level (e.g., in the LAT analysis we stop branching when nodes have fewer than 30 events or the split results in a node with fewer than 10 events).

An excellent way to minimize the training bias is to grow many independent trees, independent in the sense that they are derived from independent training samples. One can then average over the results. When such an ensemble of CTs are used to analyze data, each event is dropped into this "forest," it filters down to a set of leaf nodes in each of the trees (each with its own probability) and the output is averaged over these. Such sets of CTs are often referred to as *ensembles* or *forests*. For the types of problems being addressed here after you have more than a dozen trees, the gains become small. When using CTs in an ensemble, deeper trees can be used: you can allow the branching in individual trees to continue past the point that is prudent in the case of a single CT. Averaging over the ensemble will wash out some of the statistical fluctuation associated with low counts in the leaf nodes.

If one has a single CT, then the probabilities (derived from the training sample) will be quite granular (you get one probability per leaf node). However, in an ensemble, the averaging over the set of CTs washes this less then desirable feature out, resulting in a

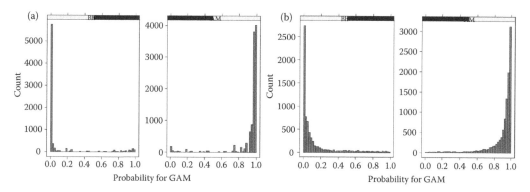

FIGURE 12.2 On the left are the results for a single CT. On the right are the results for an ensemble of 17 CTs. The training sample and testing sample were the same for both. The upper banner indicates BKG (a) and GAM (b) and so the probabilities for background and signal are shown separately.

smooth distribution (albeit sometimes with sharp kinks). The more CTs in the ensemble the smoother the resulting probability curves become. An example of these probabilities distributions for the case of separating GAM from BKG is shown in Figure 12.2.

The statistical nature of single trees is apparent in Figure 12.2. The ensemble shows a smoother probability distribution compared to the granularity of the single CT.

An advancement in CT technology was made in the 1990s which is often referred to as *boosting*. There are two separate aspects to this. The first was to improve a CT's performance by iterating the training. In the iterations, events which were misclassified by the previous iteration, were given increased weight so as to force the training process to better reject them. The second was based on the realization that not all trees in the ensemble were of equal quality. Post training, one can measure each tree's performance (discussed next) and assign to each a weight used during the "average over the ensemble" process when applied to data. These have been found to give superior results to nonboosted ensembles of CTs.

12.3 MEASURING PERFORMANCE

Independent of whether you are dealing with a single CT or an ensemble or whether or not the trees are boosted, measuring the resulting performance (separation power) is essential. It is important that the data set used in this evaluation be independent of the training data sets so as to remove the possibility of training bias in the assessment. If an evaluation event winds up in a leaf node which had at least 50% signal during training, it will be declared signal (a GAM event). Of course one can require a higher or lower percentage.

Perhaps the simplest evaluation of a classifiers performance is to tabulate the number of times "it is right" versus "it is wrong." With just two categories there are four possible outcomes:

	Classifier prediction (positive and negative)	
	False negative (fn)	False positive (fp)
True classification (true and false)	True negative (tn)	True positive (tp)

The diagonal elements show when the classifier produces the correct result, while the off-diagonal elements quantify the error rate. An example of how this might look in an analysis separating background (BKG) from the gamma-ray signal (GAM) is

		Classifier		Totals
		BKG	GAM	
MC truth	BKG	13719	892	14611
	GAM	663	14037	14700
Totals		14382	14929	29311

There are at least two measures of practical importance: *recall* (efficiency) and *precision* (purity). *Recall* is simple: it is the fraction of events which are correctly classified (and presumably kept). CTs quite often achieve efficiencies of 80–90%. In the aforementioned example, GAM has a *recall* of 95.5% and BKG has a *recall* of 93.9%. *Precision* is the residual background contamination in the data classified as signal. For our example we see that in the Classifier GAM column there are 892 BKG events and hence this CT ensemble has a precision of $14037/14929 = 94\%$ for identifying the GAM signal.

To score the overall performance of a classifier, the so-called *F-Measure* is often used [8]. The F-Meas. is the harmonic mean (as distinguished from the arithmetic mean and the geometric mean) of the *recall* and *precision*:

$$\frac{1}{F} = \frac{1}{2}\left(\frac{1}{Recall} + \frac{1}{Precision}\right)$$

A perfect classifier would have $Recall = 1$ and $Precision = 1$ hence the *F-Meas.* $= 1$.

This can be generalized to emphasize either the *recall* or *precision* by including a weighting parameter β:

$$\frac{1}{F_\beta} = \frac{1}{1+\beta^2}\left(\frac{\beta^2}{Recall} + \frac{1}{Precision}\right)$$

For example for $\beta = 2$ *recall* is favored while for $\beta = 0.5$ *precision* is emphasized. For our example, $F = F_1 = (0.5(1/.955 + 1/.94))^{-1} = 0.947$, $F_{0.5} = 0.943$ (emphasizing *precision*) and $F_2 = 0.952$ (emphasizing *recall*). In this example since the recall and the precision are both very high, there is not a lot of difference in these various performance measures. This will not be the case with a classifier with worse performance. The aforementioned F-Meas. relationships can easily be recast in terms of the true positives (tp), false positives (fp), and so on:

$$F_\beta = \frac{(1-\beta^2)tp}{(1+\beta^2)tp + \beta^2 fn + fp}$$

Earlier it was stated that CTs seemed to be the best classifier for the gamma-ray/cosmic ray separation problem. We can now quantify this in terms of the F-Meas. The data used here are the same GAM/BKG data set that used in our previous examples. Below the F-Meas.

is shown for a Neural Net Classifier and a Logistic Regression Classifier in addition to the CT Ensemble Classifier. The F-Meas. is F_1 (i.e., $\beta = 1$).

Neural net		
Recall	Precision	F-Measure
89.3%	87.5%	88.4%

Logistics regression		
Recall	Precision	F-Measure
91.5%	89.4%	90.5%

Classification trees (ensemble)		
Recall	Precision	F-Measure
95.5%	94.0%	94.8%

12.4 RELATIVE IMPORTANCE OF THE INDEPENDENT VARIABLES

A very useful aspect of classifiers, and in particular CTs, is that during the training process, the relative importance of each of the independent variables used in the problem can be assessed. Such an assessment for the example we have been following is shown in Figure 12.3.

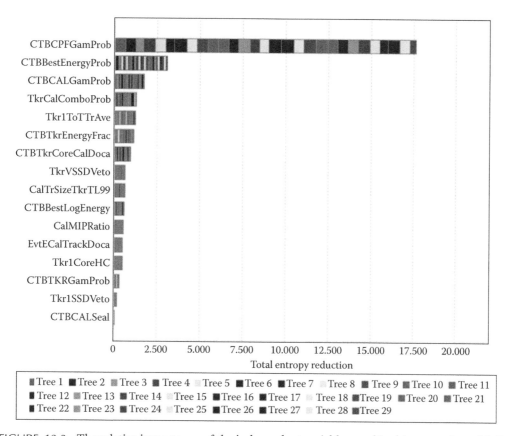

FIGURE 12.3 The relative importance of the independent variables used in this separation of GAM from BKG is shown. For the LAT analysis, *Entropy* (as opposed to *Gini*) was used. The colors in the bar graph indicate the contributions to individual *Trees* within the ensemble of 29 *Trees*.

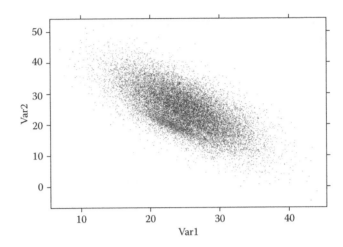

FIGURE 12.4 (**See color insert.**) A toy data set with two correlated variables and two event types: RED and BLUE.

The variable labeled CTBCPFGamProb plays the dominant role in the separation yielding by far the largest *entropy* reduction. This variable represents as a probability the chance that the incoming particle to the detector was charged: in short it is playing the role of a veto signal. And, yes it is also the output from an ensemble of CTs through which the data was processed prior to the training of this CT ensemble. So it is not surprising that this variable is key to separating the signal from the background.

12.5 CLASSIFICATION TREES IN PRACTICE

Before engaging in the complexity of background rejection in an instrument such as LAT, it is worth exploring the classification technology with a toy analysis. Our toy will consist of two types of events: RED and BLUE. Each event will have two independent variables associated with it and these variables will be correlated. The generated ratio of RED:BLUE is 1:10. This is shown in Figure 12.4. With this toy data set we can explore many aspects of CTs and the influence of training techniques on the resulting performance.

Our toy data set was generated with 2000 RED and 20,000 BLUE events. This was then split 50–50 into a training sample and a testing sample.

12.6 TRAINING BIAS

First we compare a single CT with a branching cut-off at 30 events in a node and no fewer than 10 events in a split. The independent variables are Var1 and Var2. From the tabulations given here, considerable training bias is revealed by the F-Meas.: it is 0.75 on the training data but falls to 0.67 on the test data.

Training sample		Predicted BLUE	Predicted RED	Totals
Observed	BLUE	9768	228	9996
Observed	RED	260	744	1004
Totals		10028	972	11000

	Observed BLUE	Observed RED	Overall
% Agree	97.7%	74.1%	95.6%

Positive category-RED		
Recall	Precision	F-Measure
74.1%	76.5%	75.3%

Testing sample		Predicted BLUE	Predicted RED	Totals
Observed	BLUE	9705	299	10004
Observed	RED	342	654	996
Totals		10047	953	11000

	Observed BLUE	Observed RED	Overall
% Agree	97.0%	65.7%	94.2%

Positive category-RED		
Recall	Precision	F-Measure
65.7%	68.6%	67.1%

Now, let us compare the results when using an ensemble of five CTs (see tables provided). The change in the F-Meas. is now ~0.03 as opposed to ~0.08 when a single tree was used. The training bias is not entirely eliminated, but is substantially reduced. As the number of trees in the ensemble is increased the bias decreases.

Training sample		Predicted BLUE	Predicted RED	Totals
Observed	BLUE	9715	281	9996
Observed	RED	304	700	1004
Totals		10019	981	11000

	Observed BLUE	Observed RED	Overall
% Agree	97.2%	69.7%	94.7%

Positive category-RED		
Recall	Precision	F-Measure
69.7%	71.4%	70.5%

Testing sample		Predicted BLUE	Predicted RED	Totals
Observed	BLUE	9697	307	10004
Observed	RED	327	669	996
Totals		10024	976	11000

	Observed BLUE	Observed RED	Overall
% Agree	96.9%	67.2%	94.2%

Positive category-RED		
Recall	Precision	F-Measure
67.2%	68.5%	67.8%

Next we examine the independent variables. Var1 and Var2 were purposely created with a high degree of correlation. One would guess the relative importance of these variables would be comparable and it is. The question of interest is how much better might a CT do if we remove this correlation. Plotted in Figure 12.5 is the sum (Var2 + Var1) versus Var1.

If we include VarSum (= Var2 + Var1) as an independent variable, then the resulting ensemble CT achieves a bit better separation power (larger F-Meas.) with fewer leaf nodes and the VarSum variable is by far the most important. This somewhat over-simplified example illustrates an important technique. Specifically, use the CTs to find the important variables and then explore the correlations among these attempting to undo or at least reduce them. There are automated ways of finding correlations (e.g., principle components analysis) and these can be of use prior to the application of a classifier. However, these programs only consider linear correlation and often the correlations are more complex such as the energy dependence on the errors associated with track extrapolations.

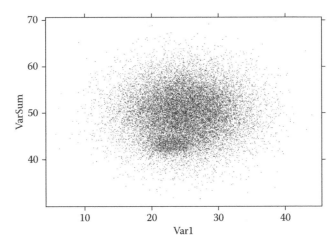

FIGURE 12.5 (**See color insert.**) The toy data set with the correlation between Var1 and Var2 undone by making a new variable VarSum = Var1 + Var2.

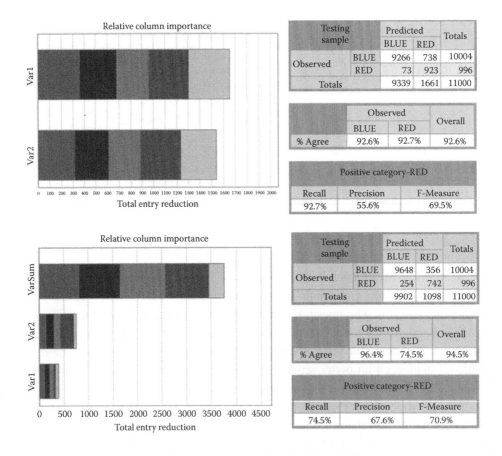

Next we explore what the influence on the ratio of RED:BLUE in the training sample has on the resulting performance. We can easily take just a subset of the training sample BLUE events, thereby making the ratio of RED:BLUE whatever we would like. Here, the results of using a 1:2 mix for training is given. The F-Meas. for the training sample has soared; however, when this ensemble of CTs is applied to the test sample, the F-Meas. is essentially unchanged from that obtained with the 1:10 ratio of the original sample. However, if you look at the results more closely, you will see that this F-Meas. is achieved largely at the expense of precision which has dropped by over 20%. By changing the relative proportion of RED:BLUE in the training sample we are skewing the resulting probabilities in the leaves of the resulting CTs.

Training sample 2:1		Predicted		Totals
		BLUE	RED	
Observed	BLUE	1839	161	2000
	RED	58	946	1004
Totals		1897	1107	3004

Testing sample		Predicted		Totals
		BLUE	RED	
Observed	BLUE	9266	738	10004
	RED	73	923	996
Totals		9339	1661	11000

	Observed		Overall
	BLUE	RED	
% Agree	92.0%	94.2%	92.7%

	Observed		Overall
	BLUE	RED	
% Agree	92.6%	92.7%	92.6%

Positive category-RED		
Recall	Precision	F-Measure
94.2%	85.5%	89.6%

Positive category-RED		
Recall	Precision	F-Measure
92.7%	55.6%	69.5%

The last item we explore with our toy data set is the F-Meas. In all of the calculations of the F-Meas. so far, an event being classified as RED will fall into leaves with an average probability >50% of being RED based on the training sample. The F-Meas. depends on this.

Figure 12.6 shows the dependence as well as what F_β looks like for $\beta = 0.5$ and $\beta = 2$.

In the left-hand plot F_2 peaks around 0.2 in the probability cut which shows its bias toward emphasizing recall whereas $F_{0.5}$ peaks around 0.65 showing the bias toward favoring precision.

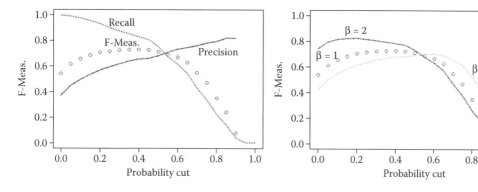

FIGURE 12.6 The F-Meas. dependence on the CT-generated probability that an event is RED. As the cut is increased, RED-recall decreases while its precision increases as shown on the left. The plot on the right shows how F_2 and $F_{0.5}$ behave.

12.6.1 Summary

1. To minimize training bias, keep reasonable statistics at leaf nodes (e.g., 30/10).

2. Using ensembles (forests) of CTs helps to limit training bias and to make the leaf node probabilities more continuous.

3. Use relative variable importance as a guide to designing better independent variables for the CTs. Taking out correlations *a priori* improves performance.

4. Relative balance of event classes in the training sample will skew the resultant performance to favor either recall or precision.

12.7 USAGE EXAMPLES

Any analysis is only as good as the input (yes, GIGO* applies everywhere). The selection of input variables to a classification analysis is perhaps the hardest part of the process. Ideally you want a set of variables that are as independent from each other as possible, each having a fair degree of separation power. In the rejection of backgrounds in gamma-ray astrophysics, a start is to divide the problem up according to the various separate detector components. For a conventional pair conversion telescope, the usual subsystems are an anticoincident detector (ACD) used to flag incoming charge particles (cosmic rays), a gamma-ray converter/tracker (TKR) where the incoming photon pair converts and the subsequent e^+e^- pair are tracked, followed by a calorimeter (CAL) to measure the energy in the electromagnetic shower. A sketch of the LAT instrument showing the various components is shown in Figure 12.7.

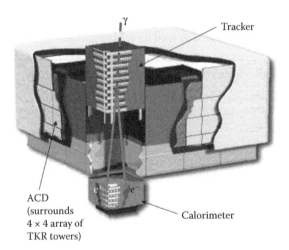

FIGURE 12.7 A sketch of the LAT instrument showing the three principle subsystems: ACD, TKR (Tracker), and CAL (Calorimeter). The instrument was built as a 4 × 4 array of identical tower modules each comprised of a TKR and CAL unit. Covering the entire field of view is an array of scintillation tiles used to flag events caused by incident charged cosmic rays.

---

* Garbage In—Garbage out!

The amount of information recorded for each event can have many hundreds, sometimes thousands, of readouts from the sensors in these three subsystems. This information is processed on the ground through an involved "reconstruction" program (Recon), which knits all the bits and pieces of data together. The hits on each charged trajectory, recorded by the TKR, are associated together to form "tracks." The Recon subsequently tries to pair tracks together to find the classic inverted "V" signature of a pair conversion (this stage is called "Vertexing"). These trajectories are projected back into the anti coincident system to see how close they come to ACD veto counters registering "hits" as well as forward into the CAL for association with energy deposits. The intent here is not to fully describe the reconstruction, but from this thumbnail it should be apparent that dozens of quantities can be selected with which to address the issue of background. In the case of LAT the list is close to 100 variables. It is also far from reality to think that all of these variables were conceived of *a priori*: instead there was a long evolution and learning process to understand exactly where the discrimination power of the detector lay and how best to use it.

In the case of space-based gamma-ray astronomy, the incoming noise-to-signal can exceed 10^4:1 (dependent mainly on orbit location). Also the separation of the signal must be extremely efficient so as to provide the largest effective collecting area for the very faint high-energy gamma-ray signals. This sets the requirement that the background rejection must exceed 10^5:1 rejection while retaining most of the gamma rays which convert in the TKR.

Another issue concerns the energy range over which the background rejection needs to work. For LAT the lowest energy is $\sim$20 MeV while the high end is $\sim$1 TeV. This is almost five decades and over this range the characteristics of the events changes a lot. At low energies there are few hits in the tracker, the calorimeter has only sporadic energy deposits mostly in the first layers, and the ACD system is quiet. At high energies, there are many hits in the tracker of which many are associated with back-splash x-rays from the shower in the calorimeter. The ACD also suffers many such hits as well and this was a problem for EGRET where the event trigger had the veto hardwired in. The calorimeter has a well-defined energy profile with lots of energy depositions, so it is tempting to divide the analysis up into energy bands. However, this risks creating artifacts at the boundaries and is particularly annoying when plotting spectra for celestial sources. Examples of a low-energy event and a high-energy event from the simulations are shown in Figure 12.8.

Since the rejection analysis will be developed using simulated events, the fidelity of the simulated performance is a key to success. In the case of LAT, an extensive set of "beam tests" were performed at accelerators to provide verification. These tests used a "calibration unit" made from flight spares and the engineering tower. During these tests, beams of various particle types (photons, protons, and electrons) and energies (from <1 GeV to >100 GeV) were directed at the calibration unit and the resulting sensor responses were recorded. This provided a direct point of comparison for the simulations and thereby served to validate them.

The point cannot be over emphasized that in complex instruments such as LAT, the simulations (MC) represents your best understanding of the instrument overall. It will determine the instrumental response functions such as effective area as a function of energy, angles, and conversion point. It will calibrate the residual backgrounds underlying the physics

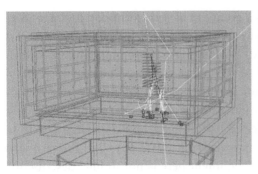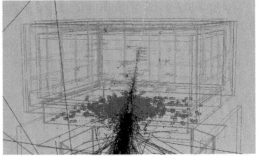

FIGURE 12.8 **(See color insert.)** On the left is the simulation of a low-energy (∼100 MeV) gamma-ray event in LAT. The black lines show the paths of charged particles and the white lines show those of neutral particles (photons). Blue lines indicate the tracks found by the reconstruction analysis and the yellow line shows the reconstructed gamma-ray direction. On the right is a high-energy event (∼100 GeV). Drawing of the photon trajectories is suppressed, otherwise they would obscure the other details. In this wireframe event display, the outlines of the various pieces of the instrument are shown along with markers as to where hits were registered by the detectors.

measurements. And it will be the guide to explore systematic errors. In the case of LAT, the instrument model has over 50,000 volumes. Even details such as the screw holes in the plastic ACD scintillators which are used to attach to tiles to their support structure are present and the carbon fiber lattice supporting the CsI crystals are all represented in the simulation. This level of detail might seem excessive, but for example, the screw holes in the ACD scintillators can be seen in the data(!) and the design of the ACD system was to achieve better than $1:10^4$ rejection of entering cosmic rays.

The LAT event analysis begins post Recon. Events without found tracks are not considered further here and events with tracks are first analyzed for the accuracy of the energy reconstruction and pointing resolution. Both of these analyses use an ensemble of CTs and the training is done using simulated gamma-ray events.

For the energy analysis, the reconstructed energy is compared to the Monte Carlo truth. For training purposes, if the event's energy was within 1 σ_{Energy} of the generated energy, the event was classified as "Well Meas." otherwise it was tagged "Poorly Meas." The model for the energy resolution was an analytic function which mimics the performance of the calorimeter and is shown in Figure 12.9. This dependence results at low energies from the large fraction of energy left in the TKR and at high energies from the increasing unaccounted for leakage fluctuations of the showers escaping out the back of the CAL. It was necessary to have this energy dependence of the resolution "built-in" for the training samples in order to minimize making a "knob" with large energy dependence. The term "knob" is used to label variables which can be used in high level science analysis to explore dependencies of the results on global properties of the instrument/analysis, in this case the overall energy resolution of a data sample.

The independent variables were drawn from both the tracker and calorimeter subsystems and were mainly quantifying the shower shape. The resulting ensemble of CTs was then used to produce a probability indicating "good energy reconstruction." More precisely this is the

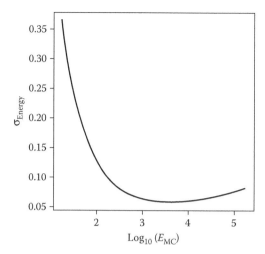

FIGURE 12.9 A plot of the analytic model for the energy resolution used in the CT analysis is shown. This is a heuristic model and reflects the actual performance of the instruments.

probability, given the details of the event that the resulting energy falls within 1 σ_{Energy} of the true energy. We envisioned that providing end users with an "energy resolution knob" could be of benefit particularly when spectral features were present (i.e., lines, cutoffs, etc.). The effectiveness of this good-energy probability is shown in Figure 12.10 for simulated gamma-ray events with $18\,\text{MeV} < E_{MC} < 562\,\text{GeV}$ with a generated spectrum of $1/E$.

For simplicity, McEnergySigma is defined as

$$McEnergySigma = \frac{ReconEnergy - McEnergy}{.1 * McEnergy}$$

McEnergy is the Monte Carlo truth energy while *ReconEnergy* is the energy estimated from the sensor responses by Recon.

Each plot is for a quartile of GoodEnergyProb (i.e., first quartile is GoodEnergyProb < 0.25 etc.). The percentages shown in each plot indicates the relative number of the events falling into that quartile. The effect of increasing the cut on GoodEnergyProb is to reduce both the high-side and low-side tails and significantly narrow the distribution albeit at the expense of efficiency. There remains some energy dependence and this is shown in Figure 12.11. The columns are decades in energy starting at $18\,\text{MeV}$ and the rows are the quartiles in GoodEnergyProb.

A similar strategy was applied to determining the accuracy of the direction reconstruction. This attribute of an imaging instrument is referred to as its Point Spread Function (PSF). Training classes were defined according to whether or not each event's direction was reconstructed within a modeled 68% containment angle (PSF_{68}):

$$PSF_{68}(E_\gamma) = \frac{3.5^\circ}{\left(\frac{E_\gamma}{100\,MeV}\right)^{.72}}$$

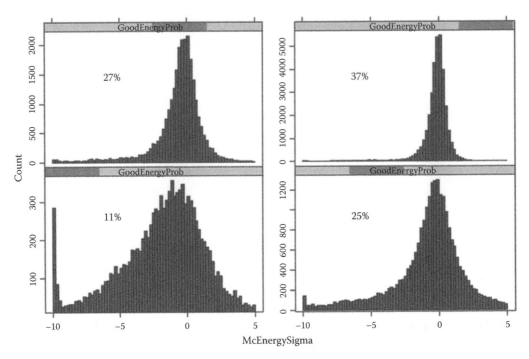

FIGURE 12.10 The energy resolution for four different ranges of the "energy resolution knob" and are shown symbolically in the banner at the top of each plot. The percentages shown in each plot indicate the fraction the events falling into that range of the resolution knob.

There was no need to simulate a point source since in the Monte Carlo the true direction of each photon is known *a priori*. The events were generated over the full field-of-view (FoV). No compensation was made for how far off the instrument axis the events were since the variation of the PSF is small over the FoV. The independent variables used to discriminate *Good* from *Poor* events were based on the details of the recorded hit structure in the TKR, particularly near the start of the tracks and hence the presumed conversion point and if there was a vertex found, its characteristics. In addition, the alignment of the TKR event with the CAL response were included and play a significant role at energies >1 GeV.

Shown in Figure 12.12 are four PSF plots with increasing cuts provided by the Image Resolution Knob (i.e., the CT generated probability that the event direction was reconstructed accurately). Figure 12.13 shows the resulting trends with increasing cuts on the Image Resolution Prob. Tightening the cut on the IR Prob improves the overall PSF not only by sharpening the core of the PSF distribution, but also by greatly reducing the non-Gaussian tails of the PSF as demonstrated by the 95/68 ratio quoted on the plots.

Now we come to the background rejection. This was divided into three parts with a fourth section which established the event analysis classes. The three main sections are divided up along the instrument's hardware subsystems: ACD, TKR, and CAL. In each, the same strategy was followed: eliminate as much of the background with simple cuts as possible which have little impact on the recall (efficiency) and follow this with a CT analysis. The first step was accomplished with a series of cuts dubbed "prefilters."

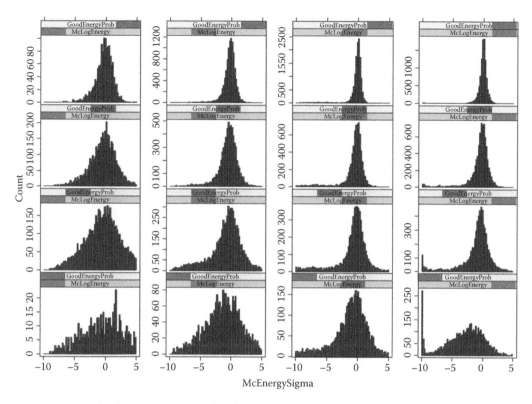

FIGURE 12.11 Similar to Figure 12.10, but here each column is approx. one decade in energy, lowest energies on the left, highest on the right.

For the ACD subsystem, there were several prefilters. This veto shield guarding against the entry of charged particles ideally should be hermetic. Reality makes achieving this all but impossible especially when coupled with the constraints imposed by make the system space-flight qualified. In the LAT ACD this required compromises allowing for a gap between scintillation tiles in one dimension. These were subsequently "covered" by ribbons of scintillating fibers. However, this technology does not result in 100% particle detection efficiency: that is, the ribbons "leak" a little. In addition, along each of the four vertical corners of the LAT, it was not possible to provide a true overlap of the scintillation tiles leaving gaps. The prefilters dealing with these problems use projections of the found tracks back to the ACD tile layer and if the track passed close to a corner the event is eliminated. If the track passed close to a ribbon, tight hard cuts are imposed in the ribbon's response. These are two of six prefilter cuts in the ACD section of the background rejection. Combined they sealed the ACD system over the FoV and provided a rejection for charged particles entering the gamma-ray conversion volume of $1:10^4$.

This progression can be summarized by the following sequence of images shown in Figure 12.14. What is shown in this figure are the entry points for charged particles at various stages of the prefiltering process. The first image shows what arrives on the ground. One clearly sees the image of the ribbons (the obvious stripes on the top and sides) as they are not used in either the Hardware Trigger or onboard filtering. The ACD tiles are

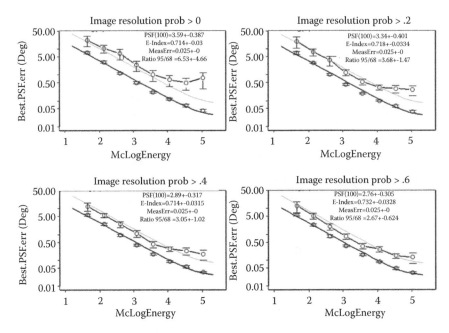

FIGURE 12.12 The effect of cuts on the Image Resolution determined for the CT analysis is shown for events that are within 20° of the instrument axis. In each plot the lower line is the 68% containment and the upper line is the 95% containment (the light gray line is 3 times the 68% containment and is shown for reference). The PSF is fit to a power-law with a prefactor ($PSF_{68}(E = 100\,\text{MeV})$ labeled PSF(100) on the plot) and energy power law (E-Index). Also the ratio of the 95–68% containment radius is shown.

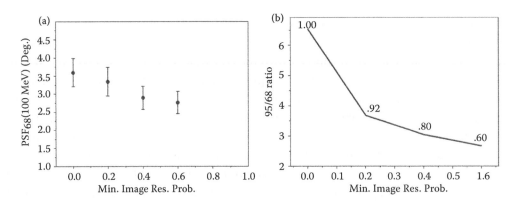

FIGURE 12.13 The plot (a) shows how the prefactor ($PSF_{68}(E = 100\,\text{MeV})$) improves with the minimum allowed value of the Image Resolution Prob. The plot (b) shows the more dramatic improvement in minimizing the tails of the PSF (the 95/68 ratio). In addition, the fraction of remaining events is shown on the ratio plot.

included in the Hardware Trigger as well as the onboard filter. The corner gaps and ribbons are "plugged" by prefilter cuts and the next image now appears without noticeable traces of these features. Much of the remaining background is eliminated by cutting events which have reconstructed tracks that point to ACD tiles which registered a significant response.

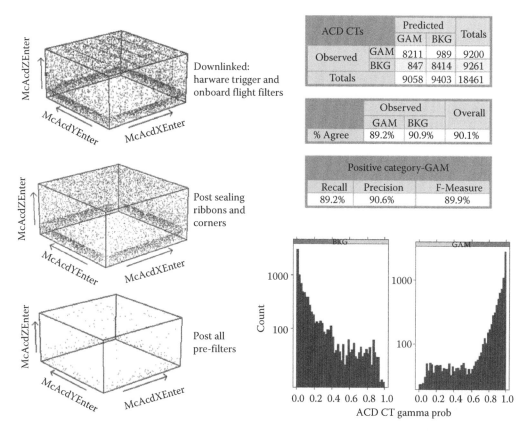

ACD CTs		Predicted		Totals
		GAM	BKG	
Observed	GAM	8211	989	9200
	BKG	847	8414	9261
Totals		9058	9403	18461

	Observed		Overall
	GAM	BKG	
% Agree	89.2%	90.9%	90.1%

Positive category-GAM		
Recall	Precision	F-Measure
89.2%	90.6%	89.9%

Downlinked: harware trigger and onboard flight filters

Post sealing ribbons and corners

Post all pre-filters

FIGURE 12.14 Summary of the ACD background rejection analysis. In the left-hand column, the effect of the various prefilter stages is shown pictorially by plotting the entry point of BKG events into LAT. The right-hand column shows the overall performance of the final CT ensemble and the probability distributions for both GAM and BKG events.

BKG and GAM events surviving these cuts are used to train the CTs for the final piece of the ACD subsystem background rejection.

On the right-hand side of Figure 12.14 the results of CT training are shown. Limitation of computer time to simulate backgrounds resulted in only enough events to train 4 CTs for this ensemble. This is less than optimal as one needs >10 to exhaust the advantages of using forests/ensembles. At the top of the right-hand column is the overall assessment showing the breakdown for the evaluation sample to correctly or incorrectly being classified as a gamma ray. Below that, is a plot for the BKG and GAM samples showing the CT-generated probability. In this instance, both the training and evaluation samples had equal numbers of GAM and BKG events.

The main reason for a failure of the ACD subsystem to reject an incoming charged cosmic ray is that it either interacts early on in its passage through the tracker and/or the track reconstruction fails to find correctly the trajectory corresponding to the entering particle. This mistracking should correlate with the Image Resolution (IR) Knob. Plotting the IR Knob versus ACD GamProb is shown in Figure 12.15.

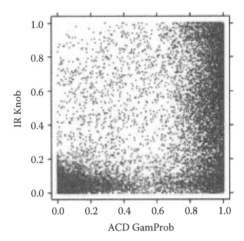

FIGURE 12.15 BKG events are shown in blue and GAM events are shown in red. The background is seen to predominantly populate the lower left-hand corner: low IR Knob (poorly reconstructed track) and low ACD GamProb (likely background events).

In training the ACD GamProb CTs, the IR Knob was not used as an independent variable and there is no correlation for the GAM sample between these two CT-generated probabilities in Figure 12.15. More examples of cross-correlations between CT-generated probabilities are given subsequently.

The TKR and CAL subsystems are treated similar to that mentioned earlier. However, the training samples are required to pass the ACD prefilter cuts: there is little sense in eliminating obvious BKG events multiple times and their presence clouds the tasks at hand! We first describe the TKR analysis.

The TKR is at the heart of the LAT instrument. The charged particle trajectories are key to understanding and linking the response recorded in the other subsystems. Perhaps the most valuable piece of information from the track reconstruction is the topological characterization of the overall event. The gamma-ray pair conversion process has a classic signature of an inverted "V." Background events rarely result in this topology. However, as the energy increases the e^+ and e^-, the angle between them becomes smaller until the granularity of the tracking detectors becomes insufficient to resolve them as separate tracks. Also at low energies, one of the tracks may be so low in energy that it fails to cross and register hits in at least three tracking planes (the minimum need to find a track).

For the LAT analysis it was found that using the event topology was very useful. Events with a reconstructed vertex (an inverted "V"—category called "Vtx") were analyzed separately from events with just one found track (category "1Tkr"). Events with more than one found track, but no reconstructed "V," formed a third topology class (category "nTkr"). The level of background contamination in the Vtx category has almost an order of magnitude lower background rate, while the category with only one found track has the highest background rates. Another issue with these different topological types is that the set of independent variables is not the same. For example, if the event is of type Vtx, then quantities from the pairing of the two tracks are present (e.g., the opening angle between the tracks, etc.). For

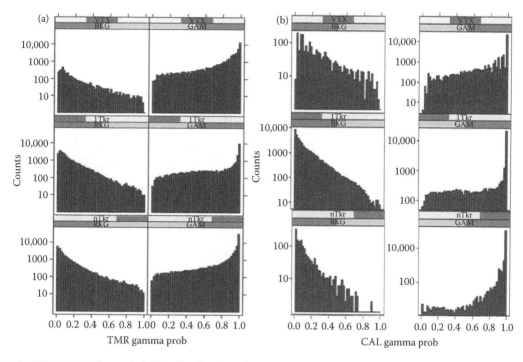

FIGURE 12.16 The probability distributions for events categories BKG and GAM are plotted for the TKR (a) and CAL (b) subsystems. Each row shows the results for a different event topology: Vtx at the top, then 1Tkr and nTkr at the bottom.

these reasons each of these three topologies is analyzed separately; however, the paradigm was similar to that used in the already-discussed ACD analysis.

Shown in Figure 12.16 are the CT-generated probability distributions divided up according to topology type. On the left are results of TKR analysis while on the right the results are given for the CAL analysis. The events used in training the ensembles of CT used here all passed the relevant prefilters for that subsystem and topology class.

With the three subsystem analyses in hand, a final level of analysis is used to optimize the event classes. A class of events with loose cuts was given the name "Transient Class" reflecting the intended usage for the analysis of GRBs. Here the emphasis was on preserving effective area (emphasize *recall*) particularly at low energies. High levels of background contamination could be tolerated since events of interest come from a small area in the sky as well as being highly localized in time. The next class was intended for studying sources (named "Source Class"). This class has lower levels of background but still strives to attain the largest possible effective area. The final class maximized background rejection (emphasize *precision*) and was designed for studies on extended objects and diffuse radiation (named "Diffuse Class").

Inputs to the formulation of these classes were the prefilters and probabilities for each subsystem. In addition, a handful of other useful independent variables were selected mainly to provide cross-subsystem correlations. It was determined that after the ACD subsystem had provided rejection of the vast majority of charge particles triggering the instrument,

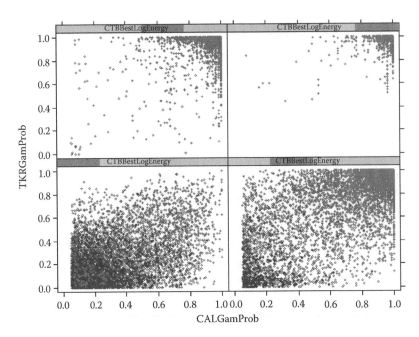

FIGURE 12.17 Plotted are the two main CT-generated probabilities used to define the various event classes. Each scatter plot covers a decade in energy starting at 20 MeV (lower left-hand plot). The BKG events preferentially populate low values of each probability while GAM events populate high values.

its further utility was marginal. On the other hand, the TKR and CAL subsystems provided somewhat independent information on the event type. In Figure 12.17 the good-gamma probabilities is shown for these variables for both BKG events (blue) and GAM events (red) over four decades in energy (one decade per plot) ranging from 20 MeV to 200 GeV.

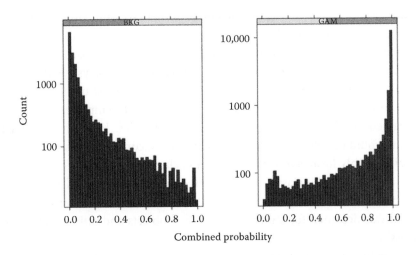

FIGURE 12.18 The final event selection probability distributions for BKG and GAM events. The main independent variables selected for this classification separation were the GAM probs. from the TKR and CAL analysis along with a few others designed to capture cross-subsystem correlations.

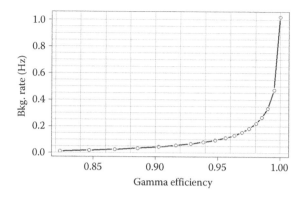

FIGURE 12.19 The contour generated using the CT-generated Combined Probability as a parametric variable. No cuts other than the various prefilters is applied for the right-most point (Gamma Efficiency = 1) and is increased in increments of 0.05. The background rate (Bkg. Rate) does not include the irreducible component mentioned earlier.

Various ways of combining the information from these two subsystems was explored, including simply making a correlated variable by averaging the two probabilities. Ultimately, these probabilities along with a handful of other independent variables were used as input to a global CT analysis. The resulting final background rejection CT probabilities are shown in Figure 12.18.

This background-rejection knob provided a convenient way to adjust background levels using a single variable. The trade-off between gamma-ray detection efficiency and background rate can be visualized by tracing out the contour generated by placing increasing cuts on this combined probability as shown in Figure 12.19.

In the plot here, a certain amount of the improvement in background rejection as the cut on the Combine Probability knob is increased, which is simply the elimination of the effective area at low energy. That is the area most polluted with backgrounds as was already apparent in Figure 12.17. A more complete exposition would expand this plot to include the event energy.

12.8 WHAT CAN GO WRONG, GO WRONG, …

The LAT analysis described so far was in place prior to launch and is still in use. The success is due to the fidelity of the Monte Carlo simulations. However, there were notable shortcomings. The biggest omission was not including the fact that the instrument's sensors have a finite persistence after the passage of a charged particle and for the TKR this is ~10 μs. In an environment with sometimes in excess of 10^4 particles/s passing through the detector, it inevitably leads to an event overlap probability of ~10%. Hence, in otherwise good gamma events occasionally there would be an out-of-time track as well. These were called "ghosts" and their presence decreased the efficiency beyond that anticipated from the Monte Carlo. The reason was simple: the demands placed on events to pass the background rejection were such that events with "extra stuff" were quite likely to fail. The decreased efficiency lead to overestimations of the effective area and the overestimation was background rate dependent. Algorithms to correct for this were constructed and are currently in use to compensate for this.

This is brought up here as an example of what can go wrong with the Monte Carlo. So, how can one correct or compensate for the inevitable discrepancy between the machine model of an instrument and the "real McCoy"? For the LAT analysis the first step was to use real data to provide the background on top of which the simulated events were placed. Among the various LAT triggers, there is a Periodic Trigger (PT) which is generated by the onboard computers without regard to the state of the sensors or other triggers. This takes events at the rate of 2 Hz and provides a nonsensor-triggered sample of what the "quiescent" instrument looks like. Simulated events were then overlaid on top of these PT events. In this way, the real detector noise along with the ghost track phenomenon was incorporated. The background analysis CTs were retrained with the PT overlaid simulations for both TKR and GAM samples, and this eliminated the problem.

Implicit in the LAT analysis is a reliance on the Monte Carlo to a part in 10^5 or more and this is risky. An alternative is to try to use real data in place of the simulations. For example, one could select a sample of data for which the known gamma-ray sources on the sky have been masked out. This would leave a data set completely dominated by background albeit with a tiny fraction of gamma rays. This data set could be labeled as BKG and used in conjunction with simulated gamma-ray events to train the CTs. Hence the question becomes what kind of corruption does the small mixture of signal with the background data set introduce and this can be directly explored using the simulations.

The investigation of contamination of signal (GAM) into the BKG set used the CT ensemble from the CAL nTkr analysis. Here one expects a difficult separation (although not as challenging as for the 1Tkr topology class). A $1/E^2$ sample of simulated gammas was introduced into the BKG training sample. The CT ensemble was regenerated and then tested. The contamination was flagged through a categorical variable classifying what type of background the event actually was. The contamination level was varied from zero to the equivalent of a few Hz. In the table that follows, the results on the testing sample is shown for a contamination level of 0.5 Hz (the BKG rate was ~300 Hz). Circled in red is the entry for this Gamma Contamination.

Quite noticeable is that the CT ensemble is for the most part separating out the contamination gamma rays and correctly calling them signal(!): 39 of the Gamma Contamination events are classified as BKG while 251 of them are correlated indentified as gamma rays. This is quite remarkable in that it suggests if you know what you are looking for (i.e., can accurately describe it in terms of a set of variables) you can then apply the CT methodology and find it. It is a little like saying you know what ball point pens look like and then proceed to find lots of them in the city dump without having to know what everything else in the dump is.

The following plot shows the effect of contamination on the F-Meas. Here one must take care in light of the above observation. The F-Meas. can be calculated against the characterization of the events presented at training (i.e., contamination GAMs are labeled as BKG) or more correctly, one can take the true event types from the MC truth. The points in red show the latter while those in black show the former. The results are that the apparent F-Meas. immediately starts to deteriorate within the presence of contamination while the "true" F-Meas. remains almost unchanged for contaminations levels of up to 1 Hz in this example. This bodes well for directly using measured background samples for training CTs.

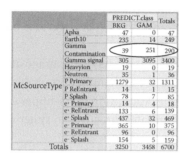

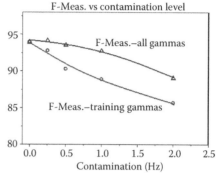

12.9 SUMMARY FOR FERMI-LAT CT USAGE

The usage of CTs in the Fermi-LAT event level analysis ranges from providing the end user with an energy resolution knob and image sharpening knob to adjustments of background levels at the expense of detection efficiency. The CTs by themselves were found insufficient in the case of background rejection to achieve the desired results without the use of preemptive cuts to eliminate the "easy events." In all cases, ensembles (forests) of CTs were used to provide quasi-continuous probability variables as well as to minimize the effects of training bias.

12.10 USAGE OF CTS IN THE ANALYSIS OF AN IACT: MAGIC

The MAGIC collaboration uses CTs in the analysis of their air Imaging Air Chernkov Telescope (IACT) data [9] for a similar purpose as in Fermi-LAT: background rejection. IACTs detect very high energy gamma rays by observing the Chernkov light produced when these particles interact in the upper atmosphere. The electromagnetic shower resulting from an incoming high-energy gamma ray produces a cone of light about the shower axis. This light pulse is quite short (~2 ns) and quite compact and these characteristics allow for both triggering focal plane detector arrays as well as image analysis to distinguish these events from the much more prevalent hadronic cosmic ray showers. Again, detailed Monte Carlos are used to optimize hardware designs, triggers, and ultimately set the stage for the background rejection. The source of light in these gamma-ray showers is quite distant from the detectors and hence the signal is faint. Ultimately, this sets the lower limit in energy for the incoming gamma rays to be around 100 GeV. Similar to the space-based telescope the ratio of incident gamma rays to cosmic ray backgrounds is very small: typically 1:100 or worse even for bright sources.

The MAGIC analysis uses the Random Forest (RF) package with some modifications and is detailed in Ref. [9]. This software uses the *gini* metric as opposed to the *entropy* metric used in the LAT example. This analysis finds similar results as to those found in the LAT. In particular,

1. The improvement in the RF decreases with number of CTs past ~10.

2. The introduction of signal in the background data set does not affect the overall performance for contamination levels less than ~1%.

3. The RF are robust and require little special "tuning" and hence "operator bias."

The two aspects of the MAGIC analysis are now highlighted. First is the usage of data from the telescopes directly in the training process. For IACTs the simulation problem has two significantly more difficult problems. First, the accurate description of the upper atmosphere where the interactions occur has uncertainty. This translates directly into systematic errors. Using the real data negates this.

Second, in general as the energy of the incoming particles increases, so does the necessary computing resources, most notably the CPU time needed. This results in a limitation on the number of events that can be simulated. Using real data allows for much large training samples and better accuracy in the resulting CT generated.

While real data were used for the background samples in the MAGIC analysis, the gamma-ray sample came from the simulations. There are not really, really bright >100 GeV source in the sky! They found that they could even use real data where the telescope had a moderately bright gamma-ray source within the FoV and this is consistent with finding in the LAT analysis that small levels of contamination in the BKG are tolerable.

The second aspect of the MAGIC analysis that is quite innovative is the use of the CT-based analysis to determine the energies of the events. They divided up the energy range into ~10 bins. These bins were used as classes in the training sample (recall that in a CT analysis the number of categories (classes) is not limited to 2). The subsequent averaging over the forest of CTs when analyzing data thus provided a continuous measure of the energy.

12.11 SUMMARY

The usage of classification techniques in high-energy astrophysics has been found to be of significant value. In the case of LAT, this approach allowed for a highly efficient and robust technique for dealing with the extraction of the gamma-ray signal from the cosmic ray background. In addition, it provided a way to characterize both the quality of the image information as well as the accuracy of the energy determination. Similarly, for the MAGIC analysis, CTs were applied to both the problem of background identification and rejection as well as energy determination.

Both of these applications were at the underlying analysis level. Efforts are underway to utilize CTs in the higher-level science evaluation and analysis. For example, many sources observed can wind up as being "unidentified." Given training samples of known source types (e.g., pulsars, AGN, etc.), one can identify the most probable source category of these unidentified sources using a CT-based analysis.

I, for one, am sure that usage of techniques which can cross-correlate across many parameters will find many more uses in the future.

ACKNOWLEDGMENTS

The author gratefully acknowledges his daughter, Liz Atwood's detailed read through and comments. Also many thanks for the careful reading provided by the referee who caught several errors as well as flagged portions where jargon obscured the text. Finally the bulk of the knowledge presented here was gained as a result of my work within the Fermi-LAT collaboration. The feedback, suggestions, and better-ways-to-do-it over the years played an enormous role in shaping the outcome and I gratefully acknowledge this.

REFERENCES

1. Thompson, D. J. et al. 1993, *ApJS*, 86, 629.
2. Atwood, W. B. et al. 2009, *ApJ*, 697, 1071.
3. Lorenz, E. 2004, *New Astron. Rev.*, 48, 339.
4. There are numerous packages available, some free. TMVA—a Root Based Package with many bells and whistles: http://tmva.sourceforge.net/ Random Forest™ by Leo Breiman and Adele Cutler: http://www.stat.berkeley.edu/~breiman/RandomForests/cc_home.htm RPART—T.Therneau and B. Atkinson, R port by B. Ripley: http://cran.r-project.org/web/packages/rpart/index.html
5. TIBCO Spotfire Miner—a complete data mining package with supporting graphics and easy to use graphical interface: http://spotfire.tibco.com
6. Morgan, J.N. and Sonquist, J.A. 1963, *J. Am. Stat. Assoc.*, 58, 415.
7. Breiman, L., Freidman, J.H., Olshen, R.A., and Stone, C.J. 1984, Classification and Regression Trees (CART).
8. van Rijsbergen, C. J. 1979, *Information Retrieval*, 2nd Edition, Butterworth, Newton, MA, USA.
9. Albert, J. et al. 2008, *Nucl. Inst. Meth. A*, 588, 424–432 (this is an excellent paper).

CHAPTER 13

Database-Driven Analyses of Astronomical Spectra

Jan Cami

University of Western Ontario
SETI Institute

CONTENTS

13.1	Introduction	267
13.2	Methods: Fitting and Modeling	270
	13.2.1 Nonnegative Least-Squares Spectral Fits	270
	13.2.2 Molecular Modeling	271
13.3	PAHS and the Unidentified Infrared Bands	272
	13.3.1 PAHs in Space	273
	13.3.2 The PAH Database and Model	274
	13.3.3 Results	275
13.4	Circumstellar Envelopes of AGB Stars	277
	13.4.1 Problem Setting	277
	13.4.2 Spectroscopic Databases	279
	13.4.3 Methods	279
	13.4.4 Results	282
13.5	Outlook: Toward Automated Spectral Analyses	283
References		283

13.1 INTRODUCTION

Spectroscopy is one of the most powerful tools to study the physical properties and chemical composition of very diverse astrophysical environments. In principle, each nuclide has a unique set of spectral features; thus, establishing the presence of a specific material at astronomical distances requires no more than finding a laboratory spectrum of the right material that perfectly matches the astronomical observations. Once the presence of a substance is established, a careful analysis of the observational characteristics (wavelengths or frequencies, intensities, and line profiles) allows one to determine many physical parameters of the environment in which the substance resides, such as temperature, density, velocity, and so on. Because of this great diagnostic potential, ground-based and space-borne astronomical observatories often include instruments to carry out spectroscopic analyses of various celestial objects and events.

Of particular interest is molecular spectroscopy at infrared wavelengths. From the spectroscopic point of view, molecules differ from atoms in their ability to vibrate and rotate, and quantum physics inevitably causes those motions to be quantized. The energies required to excite vibrations or rotations are such that vibrational transitions generally occur at infrared wavelengths, whereas pure rotational transitions typically occur at sub-mm wavelengths. Molecular vibration and rotation are coupled though, and thus at infrared wavelengths, one commonly observes a multitude of ro-vibrational transitions (see Figure 13.1). At lower spectral resolution, all transitions blend into one broad ro-vibrational molecular band. The precise shape of molecular bands depends primarily on the temperature of the gas, and infrared observations of molecular bands can thus be used to probe gas at temperatures as low as 10 K and up to a few thousand K. Similarly, molecular band shapes contain some information about the gas densities. The resulting ro-vibrational spectrum is truly unique for each molecular species, just as is the case for atoms. As a telling example, Figure 13.2 illustrates how we can even discriminate between different *isotopologues*—two species with the same chemical formula and structure, but where one (or more) of the atoms is a different

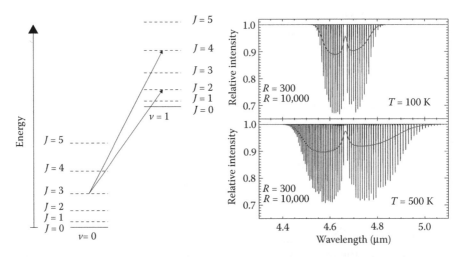

FIGURE 13.1 An illustration of infrared ro-vibrational transitions in the carbon monoxide (CO) molecule. *(Left)* Each vibrational energy level (characterized by the vibrational quantum number v) contains many rotational energy levels (characterized by the quantum number J). Quantum physics generally only allows ro-vibrational transitions if the rotational quantum number between the upper and lower states differs by $\Delta J = \pm 1$; for some other species, a third type of transition is sometimes possible where $\Delta J = 0$. *(Right)* Transitions arising from all rotational levels within a vibrational level then result in a multitude of closely spaced spectral lines (black), divided in two branches that blend into broad bands when observed at low resolving powers (illustrated by the smooth curves). Note how the width and the shape of the band clearly change with temperature—noticeable even at low resolving powers.

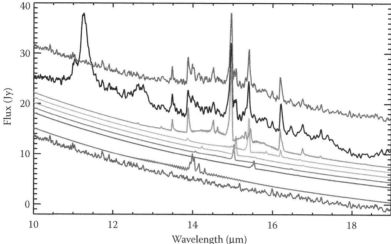

FIGURE 13.2 (**See color insert.**) Shown here (in black) is part of the infrared spectrum of the peculiar star HR 4049 (an old and evolved star in a binary system), obtained with the Spitzer Space Telescope. In this wavelength range, most emission peaks are due to carbon dioxide (CO_2). Remarkably, spectral features can be distinguished of all possible varieties of CO_2 containing different oxygen and carbon isotopes [see also 6]: $^{12}CO_2$ (blue); $^{13}CO_2$ (green); $^{18}O^{12}C^{16}O$ (orange); $^{17}O^{12}C^{16}O$ (magenta); and even $^{18}O^{13}C^{16}O$ and $^{17}O^{13}C^{16}O$ (dark blue). The spectrum furthermore contains hydrogen cyanide (HCN, purple) and water vapor (H_2O, maroon, bottom). The full model containing all gases is shown in red above the observations. The broad and strong features at 11.2 and 12.7 μm are PAH bands and were not included in this analysis.

isotope. Molecular spectroscopy thus allows us to see a difference of one neutron in an atomic nucleus that is located at astronomical distances!

Since the detection of the first interstellar molecules (the CH [21] and CN [14] radicals), more than 150 species have been detected in space, ranging in size from diatomic species to the fullerene species C_{60} and C_{70} [4]. Given the large number and variety of molecules detected in space, molecular infrared spectroscopy can be used to study pretty much any astrophysical environment that is not too energetic to dissociate the molecules. At the lowest energies, it is interesting to note that molecules such as CN have been used to measure the temperature of the Cosmic Microwave Background (see e.g., Ref. 15).

The great diagnostic potential of infrared molecular spectroscopy comes at a price though. Extracting the physical parameters from the observations requires expertise in knowing how various physical processes and instrumental characteristics play together in producing the observed spectra. In addition to the astronomical aspects, this often includes interpreting and understanding the limitations of laboratory data and quantum–chemical calculations; the study of the interaction of matter with radiation at microscopic scales (called radiative transfer, akin to ray tracing) and the effects of observing (e.g., smoothing and resampling) on the resulting spectra and possible instrumental effects (e.g., fringes). All this is not trivial. To make matters worse, observational spectra often contain many components, and might include spectral contributions stemming from very different physical conditions. Fully analyzing such observations is thus a time-consuming task that requires mastery of several techniques. And with ever-increasing rates of observational data acquisition, it seems clear that in the near future, some form of automation is required to handle the data stream.

It is thus appealing to consider what part of such analyses could be done without too much human intervention. Two different aspects can be separated: the first step involves simply identifying the molecular species present in the observations. Once the molecular inventory is known, we can try to extract the physical parameters from the observed spectral properties. For both steps, good databases of molecular spectroscopic information is vital; the second step furthermore requires a good understanding of how these spectral properties change for each of the physical parameters of interest.

Here, we will first present a general strategy using a convenient and powerful computational tool. We will then show two different, specific examples of analyses where some fundamental spectroscopic information is available in (a) database(s), and where that information has been successfully used to analyze observations. Whereas these examples do not directly address the question of automated analyses, they offer some insight into the possibilities and difficulties one can expect to encounter on this quest.

In Section 13.2, we first describe the general formalism and method that we will be using, and we offer a few general considerations about what is involved in modeling. In Section 13.3, we illustrate our method for the case of polycyclic aromatic hydrocarbons (PAHs). In Section 13.4, we show how the same approach can be slightly modified to extract physical parameters from infrared observations of red giant stars. We then come back to the question of automated spectral analyses in Section 13.5.

13.2 METHODS: FITTING AND MODELING

13.2.1 Nonnegative Least-Squares Spectral Fits

The basic principles for fitting models to observational data are well known and quite simple. For a well-constructed model spectrum (see below), one compares the observed flux value y_d to the calculated model value y_m point by point at the same wavelengths, and calculates the χ^2 statistic, that is,

$$\chi^2 = \sum_i \frac{(y_d - y_m)^2}{\sigma^2} \tag{13.1}$$

where the sum runs over each data point, and σ represents the uncertainties on the measurements. A standard approach is then to calculate many different models, and calculate the χ^2 value for each of those models. The best model is the one that minimizes the χ^2 value; the absolute value of the χ^2 then offers some insight into the overall quality of the fit.

Of particular interest for what follows is the case where the overall model spectrum y_m is a linear combination of n contributing spectral components:

$$y_m = \sum_n f_j y_j \tag{13.2}$$

If the individual spectral components y_j are known, finding the best fit in this case simplifies to finding the scale factors f_j that minimize the χ^2 value. These can all be calculated at once by simply substituting Equation 13.2 into Equation 13.1 and setting $d\chi^2/df_k = 0$ for each spectral component k, yielding a set of n equations with n unknown scale factors:

$$\sum_i \left[-\frac{2y_k}{\sigma^2} \left(y_d - \sum_n f_j y_j \right) \right] = 0 \qquad (13.3)$$

If we further define

$$M_{jk} = \sum_i \frac{y_j y_k}{\sigma^2} \quad \text{and} \quad D_j = \sum_i \frac{y_d y_j}{\sigma^2} \qquad (13.4)$$

the system becomes

$$\begin{pmatrix} M_{11} & M_{12} & \cdots & M_{1n} \\ M_{21} & M_{22} & \cdots & M_{2n} \\ \vdots & & & \vdots \\ M_{n1} & M_{n2} & \cdots & M_{nn} \end{pmatrix} \begin{pmatrix} f_1 \\ f_2 \\ \vdots \\ f_n \end{pmatrix} = \begin{pmatrix} D_1 \\ D_2 \\ \vdots \\ D_n \end{pmatrix} \qquad (13.5)$$

On first sight, the scale factors f_j can then be obtained by a simple-matrix inversion. However, for our spectroscopic analyses, the scale factors represent a flux scaling, and thus cannot be negative. Fortunately, a beautiful geometric algorithm exists that solves the set of equations represented by Equation 13.5, subject to the constraint that all the scale factors are positive (or zero). This is called the *nonnegative least-squares* (NNLS) problem [13]. The system is solved iteratively, and always converges and terminates. The algorithm has long been used in imaging applications that require a positive kernel (e.g., deconvolution), but has only been adopted in astronomical spectral analyses for a few years.

We should point out though that the NNLS algorithm is certainly not the only available method, and often not even the best one to determine these scale factors. For instance, the precise characteristics of the noise on the data (entering through the σ values) can greatly complicate the problem and result in multiple local minima in the χ^2 hypersurface. In such cases, the NNLS algorithm might terminate in one of the local minima, and thus not necessarily arrive at the best solution. This algorithm might thus not be the best choice for a full, detailed analysis of a specific data set. The more appealing (and generally more reliable) alternatives in such cases are Bayesian methods, and several examples are available in the literature of such methods in the context of spectral analyses (see e.g., Refs. 8,12,16).

However, in the context of automated first analyses of archival data, the emphasis is not so much on the most accurate analysis, but on the most efficient way to obtain a first (order of magnitude) overview of the information contained in the data. That makes the quick NNLS algorithm the method of choice here; in most cases that we have studied, we find that the NNLS alogrithm gives indeed reasonable results in a fairly short time.

13.2.2 Molecular Modeling

One might be tempted to directly compare astronomical observations to the huge amount of molecular information contained in spectroscopic database such as the ones described below. However, a crucial implicit assumption whose importance is often underestimated in χ^2 calculations is that the model is well constructed.

In the first place, this means that any model spectrum must in essence be a realistic simulated observation. Thus, the wavelengths and the intensities of the individual transitions must be determined by *physical* parameters that describe how to transform the molecular properties from their values in the database to the model values, and not by *ad-hoc* methods. Moreover, these physical parameters must correspond to a description that is appropriate for the observed object. Such a description can include microscopic (e.g., detailed population over the molecular energy levels) as well as macroscopic (e.g., velocity) effects. In the most general terms, the physics that determines the intensities can be broadly divided in two different categories: local thermodynamic equilibrium (LTE) or non-LTE distributions. LTE conditions describe gases where collisions between the gas particles are frequent enough to thermalize the gas. Under such conditions, the populations over the molecular levels follow a simple law (the Boltzmann distribution) that only depends on the local temperature. A fairly simple formalism then allows one to calculate the correct intensities for each temperature starting from reference values given in databases. For non-LTE situations, there is no such simple formalism, and a numerical approach is required for which additional data is needed (e.g., collisional excitation rates) that is not generally available. For most astrophysical environments, the densities are too low to assume that gases will be in LTE. However, even in such situations, an LTE approximation can often yield useful order-of-magnitude estimates.

When the proper physical model is calculated and radiative transfer has been carried out, the final step in creating a model is to mimic the observations: convolution with the instrumental spectral response function, and rebinning the result to a wavelength grid that is precisely the same as that of the observations. Doing this properly is often somewhat overlooked—but certainly important in the case of molecular spectra where, for example, regular, recurring line patterns can result in systematically incorrect models if the line width or resolving power is not properly chosen or treated.

Even for the simplest possible parametrized models, the entire process to go from molecular properties to final simulated observations is already quite involved, and requires more expertise in molecular spectroscopy than can be expected from most astronomers. To facilitate such analyses, we thus set up the SpectraFactory database [5]. That database does not contain any new spectroscopic information in itself, but presents available molecular data from other databases in the form of parametrized simulated model spectra. The database currently contains more than half-a-million model spectra, corresponding to a wide array of instruments at various astronomical observatories, for many molecular species and a wide range of physical parameters.

13.3 PAHs AND THE UNIDENTIFIED INFRARED BANDS

One of the most direct applications of the NNLS algorithm is the analysis of the so-called unidentified infrared (UIR) bands using the National Aeronautics and Space Administration (NASA) Ames database of infrared spectroscopic properties of PAH molecules.

13.3.1 PAHs in Space

Since they were first seen in the early 1970s [9], infrared instruments have revealed in a wide variety of sources a number of strong emission features at 3.3, 6.2, 7.7, 8.6, and 11.2 µm (see Figure 13.3). At first, it was unclear as to what was causing these emission bands, and thus they were dubbed the UIR bands. Most astronomers would now agree that these bands are due to so-called fluorescent emission of PAH molecules [see e.g., Refs. 1,22, for reviews].

PAHs are large, aromatic molecules containing several fused benzene rings in a generally planar configuration with hydrogen attached to the carbon atoms at the edges (see Figure 13.4). PAHs are formed in combustion processes of many carbonaceous substances, and are thus found on Earth in auto exhaust, candle soot and tobacco smoke—and are generally infamous for their known carcinogenic properties. The smallest PAH is naphthalene ($C_{10}H_8$), with just two benzene rings; but the largest PAHs can contain up to hundreds of carbon atoms. As a class, PAHs are among the most abundant molecules in space, and their infrared spectral signatures are observed in pretty much any astronomical line of sight.

Just like any other molecule, PAHs exhibit vibrational transitions at infrared wavelengths. As it turns out, a molecule with N atoms has no less than $3N-6$ vibrational modes. But since all PAHs share some of the same chemical structure, they also share some vibrational properties. Thus, the strong IR emission features are the superposition of the same vibrational mode in a large collection of different molecular species. For example, all PAHs have vibrational modes in which a H atom oscillates in a direction perpendicular to the plane of the PAH molecule—the so-called out-of-plane bending mode. For all PAHs, the energies required to excite this mode is similar, and corresponds to a wavelength of about 11.2 µm.

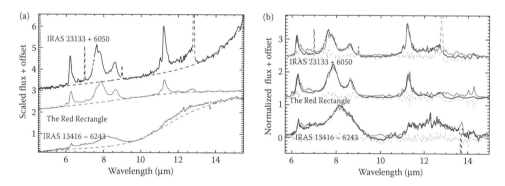

FIGURE 13.3 (**See color insert.**) (a) Three examples of some of the PAH bands in astronomical observations. All three spectra were recorded with the Short-Wavelength Spectrometer [SWS, 7] on board the Infrared Space Observatory [ISO, 11]. The spectrum of the HII region infrared astronomical satellite (IRAS) 23133+6050 (top) is a class A PAH source; the spectrum of the evolved object known as the Red Rectangle (middle) is a class B PAH source and the post-asymptotic giant branch (AGB) star IRAS 13416−6243 (bottom) represents a class C PAH profile. The dashed lines indicate the dust continuum that was subtracted from the observations to obtain the pure PAH emission. (b) The continuum-subtracted observations (black) and the best fit PAH models using the NASA Ames Astrochemistry database (red). Residuals are shown in green. (From Cami, J., *EAS Publications Series*, 46:117–122, 2011. With permission.)

$C_{20}H_{10}$ $C_{120}H_{36}$ $C_{23}H_{12}N$

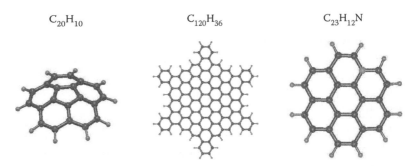

FIGURE 13.4 A few examples of specific PAH molecules. The building block for all species are benzene rings—carbon hexagons. Note how in the case of corannulene ($C_{20}H_{10}$, left), the center is made up of a carbon pentagon which gives the molecule a distinct curvature. The species on the right is a polycyclic aromatic nitrogen heterocycle—a PAH species where one of the carbon atoms is replaced by a nitrogen atom. The infrared signatures of PAHs are due to vibrations in which either the bond length ("stretching modes") or the bond angle ("bending modes") between the atoms change. The species shown here are all fairly symmetric; that, however, is not a general property of PAHs. (From Ehrenfreund, P. and Cami, J., *Cold Spring Harb. Perspect. Biol.*, 2(12):a002097, 2010. With permission.)

Thus, the observed 11.2 μm band in a given astronomical spectrum is the sum of the C–H out-of-plane bending modes of all PAHs that exist in that astrophysical environment. Similarly, other common vibrational modes are responsible for the other features.

Figure 13.3 shows three examples of astronomical PAH spectra and illustrates some of the (sometimes subtle) variations in terms of peak positions of the individual features and intensity ratios. Based on those observed variations, it was shown that astronomical PAH spectra can be classified in just three classes dubbed A, B, and C [see e.g., Refs. 18,23]. The three objects shown in Figure 13.3 are in fact representatives of each of those classes.

13.3.2 The PAH Database and Model

In an effort that has taken many years, the NASA Ames Astrochemistry group has combined expertise in PAH infrared laboratory measurements and quantum–chemical calculations to produce what is now the largest database of PAH spectroscopic properties [3]. Although the database contains the molecular and spectroscopic properties of well over 550 species, the database is highly biased toward neutral and especially fairly small PAHs: half of the species in the database contain fewer than 21 carbon atoms; about 20% of the species have >50 carbon atoms, and the largest species currently in the database has 130 carbon atoms. This is not quite representative of the PAH population expected in space, but obtaining these data for such large species through quantum–chemical calculations is time–consuming and resource-intensive process.

For each of the species in the database, frequencies and intrinsic intensities are given of the pure vibrational, fundamental transitions. For the fluorescent emission at mid-IR wavelengths treated here, it turns out that the rotational aspects are not important. What is also lacking in the database is information on overtones and hot bands—vibrational transitions that do not involve the ground state. As we will see, this is a major issue for a proper treatment of the PAH emission.

The excitation and de-excitation mechanisms that lead to the observed infrared PAH emission is a clear example of a non-LTE situation. PAHs are excited by absorption of a single UV photon, which brings the PAH molecule to an excited electronic state. A rapid conversion of the absorbed energy into vibrational energy leaves the molecule in the electronic ground state, but in highly excited vibrational states. The molecules then cool by emitting the infrared photons that we observe in astronomical observations. These processes are often referred to as stochastic heating and fluorescent cooling.

As a first step, we thus need a model that properly describes these physical processes, and results in the correct intensities for the vibrational modes of individual PAH molecules based on only one parameter: the average energy of the absorbed UV photon. Such calculations are somewhat tedious but straightforward, and we have indeed carried out a full treatment of stochastic heating and fluorescent cooling for each PAH species individually. However, there are two related difficulties. As explained above, the infrared emission starts with PAH molecules in highly excited vibrational states. But the PAH database only contains the fundamental transitions—that is, the transitions involving the ground state. We are thus missing part of the molecular information that we need for proper modeling. However, transitions from the excited states occur at slightly different wavelengths from the fundamental modes, and have similar intensities. We can thus approximate transitions from the excited states by using the properties of the fundamental modes. As a consequence of these slight wavelength shifts (and also more generally because of anharmonicities in the potential energy of the PAH molecule), the observed band profiles are asymmetric. Since there is no known formalism for obtaining the frequencies of the excited states, and since such information is only available experimentally for a handful of species, we have no information on these wavelength shifts, and thus we cannot easily incorporate such asymmetries. In our current models, we therefore necessarily adopted a symmetric Lorentzian profile for the individual PAH bands. This is one of the major shortcomings of our models.

13.3.3 Results

The PAH-UIR problem and the availability of the NASA Ames PAH database offer the ideal conditions to test the NNLS principles outlined above, and illustrate some of the pitfalls.

The PAH model calculations (including smoothing and rebinning to instrumental parameters corresponding to the observations) result in a single model spectrum for each species in the database. If the PAH database contains PAH species that are representative for the astrophysical PAH population, we should be able to obtain overall good fits to the observations. Thus, for the y_j components from Equation 13.2, we use the individual PAH model spectra. For a given observation, we can then determine the scale factors and the best fit from inverting Equation 13.5. As a testimony to the discriminatory power of spectroscopy, the matrix $\mathbf{M}$ is in fact invertible, which indicates that no two PAH molecules have the same spectrum—even though some PAH species in the database only differ by the position of a single H atom.

The best-fit PAH models that we obtained for each of the three observational PAH classes in this way is shown in the right panel of Figure 13.3. Clearly, the first conclusion is that a collection of PAH molecules can indeed reproduce the observed PAH profiles and their spectral variations quite well. The diagnostic power of this approach though lies not in the fit itself, but in the decomposition of the fit into different PAH classes. Two such examples are shown in Figure 13.5. Even though the PAH features are due to vibrational modes that are shared between all PAHs, there are apparently enough constraints for the NNLS algorithm to pick out very specific subclasses of molecules to include in the fit. For instance, even though the PAH database is highly biased toward smaller species, the PAH bands at 8.6 and 11.2 μm are reproduced by the much smaller number of large PAHs in the database. Clearly, that must indicate that the properties of those features cannot adequately be reproduced—within constraints—by any combination of the smaller species.

That kind of conclusion is important when considering the uniqueness of the fit. Mathematically, only one single solution can be the best fit given the database and an observation. But if the emission is really due to a collection of PAH molecules, then a realistic model cannot be highly dependent on the precise contents of a database. In other words, if we were to slightly alter the database and for instance remove some of the contributing species,

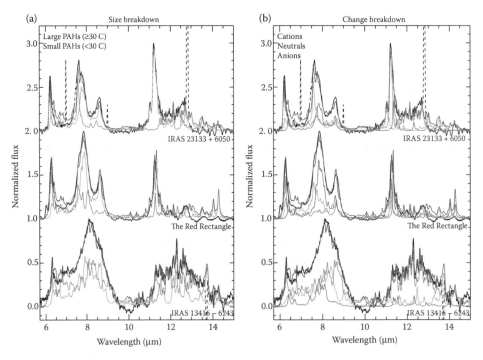

FIGURE 13.5 (**See color insert.**) These figures show the breakdown of the best fit into various constituent classes of PAHs. (a) In spite of the strong bias toward small PAHs in the NASA Ames database, the best fits show that the 11.2 μm feature is solely due to large molecules. (b) Similarly, the 11.2 μm band is primarily due to neutral PAHs, whereas PAH cations and anions contribute significantly to the 7.7 μm PAH complex. (From Cami, J., *EAS Publications Series*, 46:117–122, 2011. With permission.)

would a radically different model be the result, or would a new fit keep most of the PAH molecules in the mixture?

Such questions are easy to address, and we have carried out several such experiments where we have removed even the most contributing species from the database, and then redone the fit. Although the precise mixture that comes out is then obviously different, there is not much difference in the breakdown plots such as presented in Figure 13.5, provided that a large–enough spectral range is used for the fitting. Indeed, from the spectroscopic point of view, that does make sense: an individual vibrational transition might not be unique to a species, but the entire *set* of transitions is.

The necessity of using large spectral ranges is also clear when looking at the overall properties of transitions in the database. About 60% of all transitions in the PAH database fall in the 6–14 μm range shown in Figures 13.3 and 13.5, and thus fitting this entire range ensures that the fit is fairly constrained. Now suppose on the other hand that we were to restrict our fitting to a very small wavelength range around a single feature such as the 11.2 μm band (e.g., take a fitting range from 11.0 μm–11.5 μm). Since this corresponds to a C–H mode, all PAHs will have a transition in this range. Indeed, the database contains >1700 individual transitions in this small wavelength range, belonging to 506 different species. When fitting observational data in such a small wavelength range, it would thus be fairly easy for the algorithm to find a very good fit indeed, since there are transitions at pretty much any wavelength point and the routine is unrestricted in its choice of spectral components. The PAH mixture that would yield the best fit in this case would most likely exhibit strong emission bands *outside* the fit range where no observed features are located. Thus, for analyses such as the one presented here, it is crucial to include as large a wavelength range as possible.

We also have done other tests to assess the uniqueness of the method and to validate our approach. Suppose that the astrophysical PAH mixture that is responsible for the observed emission would be made up of individual species that are in fact part of the database at this point. Would we be able to find that exact mixture back with our approach? To test this, we created several artificial observations. These were created by picking out a different number of PAH species from the database at random, and assigning random abundances to them. The total PAH spectrum was calculated, and we added random noise to the observations. Each of the resulting spectra we thus obtained, we fed into our algorithm, and even at signal-to-noise ratios as low as 10, we did manage to find exactly the input mixture back, with the proper scale factors—provided again we used the entire 5–14 μm wavelength range.

13.4 CIRCUMSTELLAR ENVELOPES OF AGB STARS

13.4.1 Problem Setting

A more complex situation is encountered when studying infrared observations of stars on the AGB. Having exhausted hydrogen in their cores, such stars are now in the last stages of their evolution. Most of their mass is contained in a small and dense core, but they have very extended, cool ($T_{eff} \sim 3000$ K) outer envelopes that have swollen to giant proportions—their radii are comparable to the orbit of the Earth around the Sun. Their large size makes their outer envelopes only loosely gravitationally bound, and stellar pulsations initiate a process that leads to severe mass loss: the outer envelope is lifted up and thus cools down, leading to the formation of molecules and dust above the stellar surface. Once dust is formed, radiation pressure blows the dust away, and the gas is dragged along. Since molecules as well as dust have their main spectral characteristics at infrared wavelengths, studies of the mass loss and dust formation process have traditionally focused on infrared studies.

Infrared spectra of stars on the AGB show indeed copious amounts of molecular absorption and emission superposed on a stellar continuum that resembles a blackbody curve, and dust features start dominating the spectra at wavelengths longer than typically $10\,\mu\text{m}$. However, both the amounts of molecular gas and dust and its composition can be wildly different between different AGB stars.

The problem is now quite different from the PAH fitting problem. Some of the molecular bands originate from layers fairly close (or in some cases, even in) the stellar atmosphere (see Figure 13.6). From our point of view, these are located in front of the stellar emission which they will thus (partly) absorb. That absorption is not only dependent on the properties of the molecular gas anymore, but is also—at each wavelength—proportional to the intensity of the stellar flux. That makes it a nonlinear problem, and thus it is impossible to separate out the spectral components for each molecular species as we did for the PAH problem.

However, there is another complication to consider. One of the components that is often present in large quantities in these circumstellar environments is water vapor. Water vapor is spectroscopically quite different from most other molecules in that its spectral features are not limited to a narrow spectral range, but water vapor has a large amount of strong transitions over a very large wavelength range. If a thick water layer is present around an

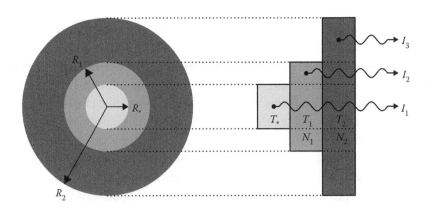

FIGURE 13.6 This sketch illustrates the geometry and approximation used for the circumstellar AGB star envelopes. The star itself as well as the layers surrounding it are spherically symmetric. Here, we show a configuration with just two molecular layers, but many more could be included. The layers can differ in temperature (T), density (N), and radius (R), and can have a different chemical composition. We approximate that geometry by a plane-parallel geometry using isothermal slabs as shown on the right. We can then identify three different rays: I_1 represents the stellar flux that is absorbed by the material in layers 1 and 2; I_2 is the ray that originates from molecular emission in layer 1, but that may be absorbed by layer 2; and I_3 represents pure molecular emission from layer 2. Depending on the radii of the molecular layers, these rays will be scaled with a different scale factor.

AGB star, it will have two different effects. First, it will absorb stellar radiation over pretty much the entire infrared part of the spectrum, and in the process change the apparent stellar continuum. Indeed, in some cases the infrared continuum is entirely dominated by the spectral properties of water rather than by the stellar flux. In addition to that, the part of the layer that—from our vantage point—extends beyond the star, will also cause **emission** over a large spectral range. If other molecules are present further out, they then will partly absorb this emitted water flux.

13.4.2 Spectroscopic Databases

A crucial first ingredient in the development of automated spectral analysis methods is detailed knowledge of the spectroscopic properties of all astrophysically relevant molecular species. Generally, such information is presented in the form of "line lists," that detail for each individual transition, the transition frequency, some measure of the intrinsic line strength (e.g., Einstein A coefficient or oscillator strength) and some other quantities required to calculate model spectra (e.g., statistical weight and energies of the levels involved in the transition). It is surprisingly hard and labor intensive to create such lists—even for fairly small molecules like water—since they necessarily require extremely complex quantum–chemical calculations and need to be gauged by accurate laboratory experiments and their analyses. How daunting this task is, may be illustrated by the sheer size of the recently computed line list for the water molecule [2]. That line list contains over 500 million individual transitions from the UV range to millimeter wavelengths. Consequently, fairly complete line lists are only available for a handful of molecules, even though we have the technological means to obtain them for many more astrophysically relevant species.

Some good sources of line information do exist though, and are often the result of interdisciplinary collaborations. For instance, they can contain information that is suitable for studies of the Earth's atmosphere, and thus should also be useful for astronomical observations up to temperatures of about 300 K, provided that the density is not too high. The most commonly used databases in this context are the HITRAN database [20] and the GEISA database [10]. Additionally, we have also used the Cologne Database for Molecular Spectroscopy [17] and a catalog (for sub-mm wavelengths and beyond) at the Jet Propulsion Laboratory [19].

13.4.3 Methods

Here, we assume that the molecular gas is in LTE, and thus we can always calculate the line intensities from the properties in the line lists and the given temperature.

We can then still use a similar approach as before, by simplifying the geometry of the problem, and carrying out ray-tracing after considering which rays contribute to the observed spectrum. A sketch of the situation is presented in Figure 13.6, and in the case of two molecular layers, can be characterized by ray-tracing only three different rays. The central ray (I_1) starts with the stellar surface emitting the continuum flux. The first molecular layer absorbs (part of) this stellar emission depending on the stellar intensity and the properties of the molecular layer. The resulting intensity can then be subject to further absorption by more molecular layers further out. Right next to the stellar surface, the second ray (I_2)

originates from the first molecular layer. Since for that ray, there is no stellar emission to absorb, the first layer can only cause molecular emission, which can be absorbed by the molecules further out (in layer 2). Finally, the second layer can produce emission in its own right (ray I_3).

For a given set of parameters (that define the properties of the stellar flux as well as the properties of the molecular layer), each of the rays is thus well determined, and we can use each of the rays now as the spectral components y_j in Equation 13.2, and use the NNLS algorithm to determine the scale factors for each of the three components that provides the best fit to the observations. The scale factors we obtain after the fitting are then easily interpreted in terms of the relative radii of each of the molecular layers. For instance, from Figure 13.6, it is clear that ray I_1 needs to be scaled by $\pi R_\star^2$ to reproduce the full stellar flux; and ray I_2 needs to be scaled by $\pi R_1^2 - \pi R_\star^2$. Thus, after the fitting, we can determine the relative extent of the molecular layer R_1 from the scale factors as follows:

$$\frac{f_2}{f_1} = \frac{\pi R_1^2 - \pi R_\star^2}{\pi R_\star^2} = \left(\frac{R_1}{R_\star}\right)^2 - 1, \tag{13.6}$$

where f_1 and f_2 are the scale factors corresponding to, respectively, I_1 and I_2. Clearly, if the NNLS algorithm finds that $f_2 = 0$, it implies that layer 1 is exactly the same size as the stellar background. The nonnegativity boundary condition thus means that each successive layer must be at least the same size as the previous layer.

A full analysis of a single observational spectrum then still requires a grid search over the relevant part of the parameter space. That in itself can already be quite computationally expensive. The temperatures of the molecular layers can range from a few 100 K in the outermost layers to about 2000 K for the layers very close to the star, and a difference of 100 K can be very noticeable in the spectrum. Similarly, column densities[*] are in the range 10^{16}–10^{22} cm^{-2} (from barely detectable to almost fully opaque), and typically, noticeable spectral changes correspond to change of the order of $\log N \sim 0.1$. Finally, stellar temperatures can range from about 2500 K to about 4000 K. If we were to evaluate models at each possible combination of parameters, a model for just a single molecular layer with six molecular components as in Figures 13.7 through 13.9 would already correspond to $\sim 10^{13}$ individual grid points in an eight-dimensional parameter space; a two-layer model to $\sim 10^{25}$ points in a 15-dimensional parameter space. Reasonable additional constraints are required to reduce the parameter space—for instance, the further away a molecular layer is from the star, the cooler it should be. If we furthermore assume that all parameters are independent and well behaved (which certainly might be questionable in some cases), then we can use an adaptive mesh approach to find the minimum in the χ^2 over the relevant parameter space.

---

[*] The column density measures the number of absorbers per cross-sectional area, and is therefore the total density along the length of a ray. Thus, a short ray with a high density can have the same column density as a long ray with a low density—and their resulting interaction with light would be exactly the same in that case.

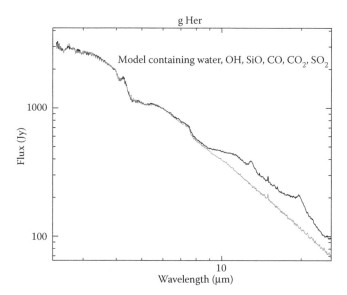

FIGURE 13.7 The infrared spectrum of the AGB star g Her (black) as observed by the SWS on board the ISO and a model fit following the procedures outlined in the text (gray). In this case, the molecular layers are not very dense, and thus the spectrum shows primarily the blackbody-like spectrum of the star, with some molecular absorption superposed. Note that longward of about 10 μm, dust emission starts dominating the spectrum which we have not included in our model.

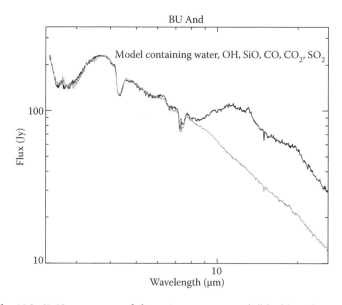

FIGURE 13.8 The ISO-SWS spectrum of the AGB star BU And (black) and a model fit following the procedures outlined in the text (gray). Although the spectrum is still dominated by the stellar contribution (up to 10 μm), the molecular layers are now much more dense, resulting in much deeper absorption bands. Note that longward of about 10 μm, dust emission starts dominating the spectrum which we have not included in our model.

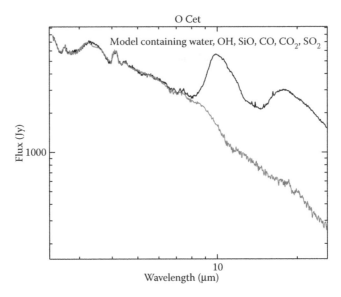

FIGURE 13.9 The ISO-SWS spectrum of the AGB star O Cet (aka Mira, black) and a model fit following the procedures outlined in the text (gray). Here, the water layers surrounding this object are very dense, and in fact do not allow any of the stellar radiation to pass through in this wavelength range. The continuum radiation is now due to emission of a hot and dense layer of water vapor (layer 1), against which other molecules can still absorb.

13.4.4 Results

Figures 13.7 through 13.9 show a few examples of the fits obtained in such a way. All three models were calculated with the same basic setup in terms of layers and molecular components; the only difference between the models is the exact value of temperatures, column densities, and radii. Despite the approximations, the method does a remarkably good job at reproducing the observations even though the spectral appearance can be very different. The case of o Cet (Mira) is especially interesting (see Figure 13.9). In this object, there is so much water present in the first layer, that all stellar light is absorbed and thus cannot penetrate any further. Most of the continuum emission is thus due to water! It is this emission which is then partly absorbed by the second, cooler layer. This explains why, in this region, the continuum seems to correspond to blackbody with a temperature of about 1500 K, which is too cold to be of a stellar nature.

Note that all three observations show that in addition to molecular gas, there is also emission from circumstellar dust at wavelengths longer than about 10 μm. Here, we have not taken dust into account; however, it is perfectly possible. In many cases, the dust emission of these AGB stars is optically thin, meaning that it is very transparent. In such cases, we can treat the dust components in a very similar way as we treated the PAHs (indeed, the dust emission would then simply be additive to the flux from the stellar and molecular layers). If the dust layers are more opaque, we could consider including dust optical properties in our ray-tracing calculations. However, given the large extent of the dusty envelopes around AGB stars, a good representation would require a fairly large number of dust layers of quite different temperatures. In such cases, geometrical effects do become much more important, and it is questionable as to whether our approximations would still be valid.

13.5 OUTLOOK: TOWARD AUTOMATED SPECTRAL ANALYSES

What we have presented here might seem to be very specific cases (PAHs and molecular layers surrounding AGB stars) for which no more general applications are possible. However, the methods described here can be generalized and could lead to a powerful aid in analyzing any astronomical observations. We have in fact used similar approaches successfully to analyze observations of objects as diverse as protostars, supernova remnants, and Ultraluminous Infrared Galaxies. Clearly, this indicates that the method we outlined here has a much larger applicability.

The PAH example illustrates the case where a spectrum is a linear combination of a potentially large number of individual spectral components. The AGB star example on the other hand illustrates different cases where ray-tracing implies nonlinear behavior. Any astronomical spectrum in essence contains a bit of both, but we may not know *a priori* what the shape of the continuum is against which absorption occurs, nor what molecular species may be present at what temperatures and densities. An iterative procedure following some ingredients illustrated above could certainly prove helpful in the analysis.

To identify what molecular components are present in the spectrum, we would first need to approximate the continuum. That can be achieved by simply fitting a low-order polynomial (or a cubic spline) through the observations that results in the least absolute deviation. Once the continuum is estimated, we can try a first attempt at fitting the molecular contributions in the observations. Although the precise shape of molecular bands is very dependent on the temperature and the column density, we could conceivably determine the *presence* of a species even we do not use a model at precisely the right temperature or column density. A good temperature to use for such purposes would be the temperature at which a blackbody peaks in the wavelength range of interest, and column densities that are just large enough to cause a clearly detectable feature. In such a case, a first χ^2 minimization would just determine what species are likely present in the spectrum.

Once the molecular inventory is determined, we can use the methods outlined above to find the proper parameters that best reproduce the observations. Since this requires many approximations and assumptions, it can only be expected to be a first step. However, it could be an automated step leading to a good selection of observations for which a complete scientific analysis is warranted with a specific geometry in which radiative transfer is included that is appropriate for the source.

REFERENCES

1. L. J. Allamandola, G. G. M. Tielens, and J. R. Barker. Interstellar polycyclic aromatic hydrocarbons—The infrared emission bands, the excitation/emission mechanism, and the astrophysical implications. *ApJS*, 71:733–775, 1989.
2. R. J. Barber, J. Tennyson, G. J. Harris, and R. N. Tolchenov. A high-accuracy computed water line list. *MNRAS*, 368:1087–1094, 2006.

3. C. W. Bauschlicher, C. Boersma, A. Ricca, A. L. Mattioda, J. Cami, E. Peeters, F. Sánchez de Armas, G. Puerta Saborido, D. M. Hudgins, and L. J. Allamandola. The NASA Ames polycyclic aromatic hydrocarbon infrared spectroscopic database: The computed spectra. *ApJS*, 189:341–351, 2010.

4. J. Cami, J. Bernard-Salas, E. Peeters, and S. E. Malek. Detection of C_{60} and C_{70} in a young planetary nebula. *Science*, 329:1180–ff., 2010.

5. J. Cami, R. van Malderen, and A. J. Markwick. SpectraFactory.net: A database of molecular model spectra. *ApJS*, 187:409–415, 2010.

6. J. Cami and I. Yamamura. Discovery of anomalous oxygen isotopic ratios in HR 4049. *A&A Letters*, 367:L1–L4, 2001.

7. T. de Graauw, L. N. Haser, D. A. Beintema, P. R. Roelfsema, H. van Agthoven, L. Barl, O. H. Bauer et al., Observing with the iso short-wavelength spectrometer. *A&A*, 315:L49–L54, 1996.

8. D. Gençağa, D. F. Carbon, and K. H. Knuth. Characterization of interstellar organic molecules. In M. D. S. Lauretto, C. A. D. B. Pereira, and J. M. Stern, editors, *American Institute of Physics Conference Series*, American Institute of Physics Conference Series, Vol. 1073, pp. 286–293, 2008.

9. F. C. Gillett, W. J. Forrest, and K. M. Merrill. 8–13-Micron spectra of NGC 7027, BD +30 3639, and NGC 6572. *ApJ*, 183:87–93, 1973.

10. N. Jacquinet-Husson, N. A. Scott, A. Chédin, L. Crépeau, R. Armante, V. Capelle, J. Orphal et al., The GEISA spectroscopic database: Current and future archive for Earth and planetary atmosphere studies. *J. Quant. Spectrosc. Radiat. Transfer*, 109:1043–1059, 2008.

11. M. F. Kessler, J. A. Steinz, M. E. Anderegg, J. Clavel, G. Drechsel, P. Estaria, J. Faelker et al., The infrared space observatory (iso) mission. *A&A*, 315:L27–L31, 1996.

12. K. H. Knuth, M. Kit Tse, J. Choinsky, H. A. Maunu, and D. F. Carbon. Bayesian source separation applied to identifying complex organic molecules in space. In *Proceedings of the 2007 IEEE/SP 14th Workshop on Statistical Signal Processing*, pp. 346–350, Washington, DC, USA, 2007. IEEE Computer Society.

13. C. L. Lawson and R. J. Hanson. *Solving Least Squares Problems*. Prentice-Hall Series in Automatic Computation, Englewood Cliffs: Prentice-Hall, 1974.

14. A. McKellar. Evidence for the molecular origin of some hitherto unidentified interstellar lines. *PASP*, 52:187–ff., 1940.

15. D. M. Meyer, K. C. Roth, and I. Hawkins. A precise CN measurement of the cosmic microwave background temperature at 1.32 millimeters. *ApJL*, 343:L1–L4, 1989.

16. S. Moussaoui, D. Brie, A. Mohammad Djafari, and C. Carteret. Separation of non-negative mixture of non-negative sources using a Bayesian approach and MCMC sampling. *IEEE Trans. Signal Process.*, 54:4133–4145, 2006.

17. H. S. P. Müller, F. Schlöder, J. Stutzki, and G. Winnewisser. The Cologne database for molecular spectroscopy, CDMS: A useful tool for astronomers and spectroscopists. *J. Mol. Struct.*, 742:215–227, 2005.

18. E. Peeters, S. Hony, C. Van Kerckhoven, A. G. G. M. Tielens, L. J. Allamandola, D. M. Hudgins, and C. W. Bauschlicher. The rich 6 to 9 μm PAH spectrum. *A&A*, 390:1089, 2002.

19. H. M. Pickett, I. R. L. Poynter, E. A. Cohen, M. L. Delitsky, J. C. Pearson, and H. S. P. Muller. Submillimeter, millimeter and microwave spectral line catalog. *J. Quant. Spectrosc. Radiat. Transfer*, 60:883–890, 1998.

20. L. S. Rothman, I. E. Gordon, A. Barbe, D. Chris Benner, P. F. Bernath, M. Birk, V. Boudon et al., The HITRAN 2008 molecular spectroscopic database. *J. Quant. Spectrosc. Radiat. Transfer*, 110:533–572, 2009.

21. P. Swings and L. Rosenfeld. Considerations regarding interstellar molecules. *ApJ*, 86:483–486, 1937.

22. A. G. G. M. Tielens. Interstellar polycyclic aromatic hydrocarbon molecules. *ARA&A*, 46:289–337, 2008.

23. B. van Diedenhoven, E. Peeters, C. Van Kerckhoven, S. Hony, L. J. Allamandola, D. M. Hudgins, and A. G. G. M. Tielens. The profiles of the 3.3 to 12 μm PAH features. *ApJ*, 611:928–939, 2004.

24. J. Cami. Analyzing astronomical observations with the NASA Ames PAH database. *EAS Publications Series*, 46:117–122, 2011.

25. P. Ehrenfreund and J. Cami. Cosmic carbon chemistry: From the interstellar medium to the early Earth. *Cold Spring Harb. Perspect. Biol.*, 2(12):a002097, 2010.

CHAPTER 14

Weak Gravitational Lensing

Sandrine Pires, Jean-Luc Starck, Adrienne Leonard,
and Alexandre Réfrégier

CEA/DSM–CNRS–Université Paris Diderot

CONTENTS

14.1	Introduction	288
14.2	Weak Lensing Theory	289
	14.2.1 The Deflection Angle	290
	14.2.2 The Lens Equation	291
	14.2.3 The Distortion Matrix A	292
	14.2.4 The Gravitational Shear γ	293
	14.2.5 The Convergence κ	293
14.3	Shear Estimation	294
	14.3.1 Basics	294
	14.3.2 Biases	295
	14.3.2.1 Instrumental and Atmospheric Bias	295
	14.3.2.2 Intrinsic Alignments and Correlations	297
	14.3.3 Challenges	298
14.4	Two-Dimensional (2D) Mapping of the Dark Matter	299
	14.4.1 Inversion Problem	299
	14.4.1.1 Global Inversion	299
	14.4.1.2 Local Inversion	300
	14.4.2 E and B Modes Decomposition	300
	14.4.3 Missing Data	301
	14.4.4 Filtering	303
	14.4.4.1 Linear Filters	303
	14.4.4.2 Bayesian Methods	303
14.5	3D Mapping of the Dark Matter	306
	14.5.1 Formalism	307
	14.5.2 Filtering	308
	14.5.2.1 Linear Filters	308
	14.5.2.2 Nonlinear Filters	309
14.6	Cosmological Model Constraints	310

        14.6.1 Methodology   310
        14.6.2 Second-Order Statistics   311
        14.6.3 Non-Gaussian Statistics   314
14.7 Conclusion   316
Acknowledgment   317
References   317

14.1 INTRODUCTION

This chapter reviews the data mining methods recently developed to solve standard data problems in weak gravitational lensing. We detail the different steps of the weak lensing data analysis along with the different techniques dedicated to these applications. An overview of the different techniques currently used will be given along with future prospects.

Until about 30 years ago, astronomers thought that the Universe was composed almost entirely of ordinary matter: protons, neutrons, electrons, and atoms. The field of weak lensing has been motivated by the observations made in the last decades showing that visible matter represents only about 4–5% of the Universe (see Figure 14.1). Currently, the majority of the Universe is thought to be dark, that is, does not emit electromagnetic radiation. The Universe is thought to be mostly composed of an invisible, pressureless matter—potentially relic from higher energy theories—called "dark matter" (20–21%) and by an even more mysterious term, described in Einstein equations as a vacuum energy density, called "dark energy" (70%). This "dark" Universe is not well described or even understood; its presence is inferred indirectly from its gravitational effects, both on the motions of astronomical objects and on light propagation. So this point could be the next breakthrough in cosmology.

Today's cosmology is based on a cosmological model that contains various parameters that need to be determined precisely, such as the matter density parameter Ω_m or the dark energy density parameter Ω_Λ. Weak gravitational lensing is believed to be the most promising tool to understand the nature of dark matter and to constrain the cosmological parameters used to describe the Universe because it provides a method to directly map the distribution of dark matter (see [1,6,60,63,70]). From this dark matter distribution, the nature of dark matter can be better understood and better constraints can be placed on dark energy, which

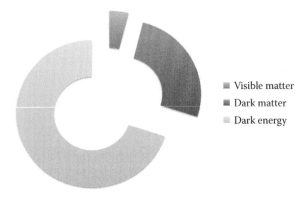

  ■ Visible matter
  ■ Dark matter
  ■ Dark energy

FIGURE 14.1 Universe content.

FIGURE 14.2 Strong gravitational lensing effect observed in the Abell 2218 cluster (W. Couch et al., 1975—HST). Photo used with permission from NASA, ESA and the ERO team.

affects the evolution of structures. Gravitational lensing is the process by which light from distant galaxies is bent by the gravity of intervening mass in the Universe as it travels toward us. This bending causes the images of background galaxies to appear slightly distorted, and can be used to extract important cosmological information.

In the beginning of the twentieth century, A. Einstein predicted that massive bodies could be seen as gravitational lenses that bend the path of light rays by creating a local curvature in space time. One of the first confirmations of Einstein's new theory was the observation during the 1919 solar eclipse of the deflection of light from distant stars by the sun. Since then, a wide range of lensing phenomena have been detected. The gravitational deflection of light by mass concentrations along light paths produces magnification, multiplication, and distortion of images. These lensing effects are illustrated by Figure 14.2, which shows one of the strongest lenses observed: Abell 2218, a very massive and distant cluster of galaxies in the constellation Draco. The observed gravitational arcs are actually the magnified and strongly distorted images of galaxies that are about 10 times more distant than the cluster itself.

These strong gravitational lensing effects are very impressive but they are very rare. Far more prevalent are weak gravitational lensing effects, which we consider in this chapter, and in which the induced distortion in galaxy images is much weaker. These gravitational lensing effects are now widely used, but the amplitude of the weak lensing signal is so weak that its detection relies on the accuracy of the techniques used to analyze the data. Future weak lensing surveys are already planned in order to cover a large fraction of the sky with high accuracy, such as Euclid [68]. However, improving accuracy also places greater demands on the methods used to extract the available information.

14.2 WEAK LENSING THEORY

A gravitational lens is formed when the light from a very distant galaxy is deflected around a massive object between the source and the observer. The properties of the gravitational lensing effect depend on the projected mass density integrated along the line of sight and on the cosmological angular distances between the observer, the lens and the source (see Figure 14.3). The bending of light rays around massive objects makes the images of

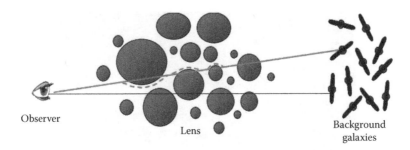

FIGURE 14.3 Illustration of the gravitational lensing effect by large-scale structures: the light coming from distant galaxies (on the right) traveling toward the observer (on the left) is bent by the structures (in the middle). This bending causes the image of background galaxies to appear slightly distorted. The structures causing the deformations are called gravitational lenses by analogy with classical optics.

distant galaxies appear deformed. The gravitational lensing causes a tangential alignment of the source. From the measurement of these distortions, the distribution of the intervening mass can be inferred.

14.2.1 The Deflection Angle

Light rays propagate along null geodesics, which correspond to the shortest path between two points in a curved spacetime. Therefore, to exactly compute the deflection of light rays around a massive object, it is necessary to determine the geometry of the Universe around that object by solving General-Relativistic equations. However, the problem can be simplified by making use of Fermat's principle, which states that light rays propagate along the path that minimizes the light travel time between to points. Even though, according to the formalism of General Relativity, the path followed by light rays is an intrinsic property of space time, while the travel time is an observer-dependent notion, the motion of light rays in a curved space time can still be described by this principle. A possible interpretation is to consider that light slows down in a gravitational field. The refractive index n in a gravitational field Φ is given by

$$n = 1 + \frac{2}{c^2}|\Phi| \tag{14.1}$$

where c is the speed of light and Φ the 3-dimensional (3D) Newtonian potential, supposed to be weak ($\Phi \ll c^2$).

Although the light speed in a vacuum is a constant c in General Relativity, we assume that the speed of light in this disturbed region becomes

$$c' = \frac{c}{n} = \frac{c}{1 + 2/c^2|\Phi|} \tag{14.2}$$

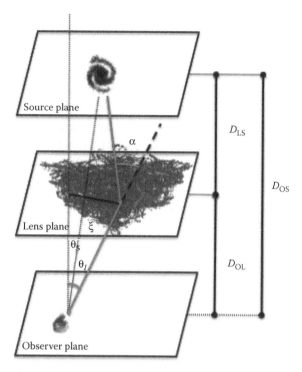

FIGURE 14.4   The thin lens approximation.

Then, the light ray is bent and the deflection angle α can be obtained by integrating the (perpendicular component of the) refractive index along the light path:

$$\vec{\alpha}(\vec{\xi}) = -\int \nabla_\perp n(\vec{r})\, dz \tag{14.3}$$

where $\vec{\xi}$ is the impact parameter in the lens plane (see Figure 14.4) and $\nabla_\perp$ is the perpendicular component of the gradient operator.

14.2.2 The Lens Equation

In theory, light rays are bent by all the matter encountered along the light path between the source and the observer. Given that mass concentrations tend to be localized within the Universe, we may use the so-called thin lens approximation to simplify the problem. In this approximation, the lensing effect is supposed to come from a single matter inhomogeneity located on a plane between the source and the observer. This approximation is valid as long as the physical extent of the mass concentration is small compared to the lens–source, lens–observer, and source–observer distances. The system is then divided into three planes: the source plane, the lens plane, and the observer plane. The light ray is assumed to travel without deflection between these planes with just a slight deflection α while crossing the lens plane (see Figure 14.4). In the limit of a thin lens, all the physics of the gravitational lensing effect is contained in the lens equation, which relates the true position of the source θ_S to its observed position(s) on the sky θ_I:

$$\vec{\theta}_S = \vec{\theta}_I - \frac{D_{LS}}{D_{OS}}\vec{\alpha}(\vec{\xi}) \tag{14.4}$$

where $\vec{\xi} = D_{OL}\vec{\theta}_I$ and D_{OL}, D_{LS} and D_{OS} are the distance from the observer to the lens, the lens to the source, and the observer to the source, respectively. From Equation 14.3, the deflection angle α is related to the projected gravitational potential ϕ obtained by the integration of the 3D Newtonian potential $\Phi(\vec{r})$ along the line of sight:

$$\vec{\alpha}(\vec{\xi}) = \frac{2}{c^2}\int \nabla_\perp \Phi(\vec{r})\,dz = \nabla_\perp \underbrace{\left(\frac{2}{c^2}\int \Phi(\vec{r})\,dz,\right)}_{\phi} \tag{14.5}$$

We can distinguish two regimes of gravitational lensing. In most cases, the bending of light is small and the background galaxies are just slightly distorted. This corresponds to the weak lensing effect. Occasionally (as seen previously), the bending of light is so extreme that the light travels along several different paths to the observer, and multiple images of one single source appear on the sky. Such effects are collectively termed as strong lensing, and are typically seen where the angular position of the source is closely aligned with that of the center of the mass concentration. In this chapter, we address only the weak gravitational lensing regime, in which sources are singly imaged and weakly distorted.

14.2.3 The Distortion Matrix A

The weak gravitational lensing effect results in both an isotropic dilation (the convergence, κ) and an anisotropic distortion (the shear, γ) of the source. To quantify this effect, the lens equation has to be solved. Assuming θ_I is small, we may approximate the lens equation by a first-order Taylor-series expansion:

$$\theta_{S,i} = A_{ij}\theta_{I,j} \tag{14.6}$$

where

$$A_{i,j} = \frac{\partial \theta_{S,i}}{\partial \theta_{I,j}} = \delta_{i,j} - \frac{\partial \alpha_i(\theta_{I,i})}{\partial \theta_{I,j}} = \delta_{i,j} - \frac{\partial^2 \phi(\theta_{I,i})}{\partial \theta_{I,i}\partial \theta_{I,j}} \tag{14.7}$$

$A_{i,j}$ are the elements of the matrix A and $\delta_{i,j}$ is the Kronecker delta. The first-order lensing effects (the convergence κ and the shear γ) can be described by the Jacobian matrix A, called the distortion matrix:

$$A = (1-\kappa)\begin{pmatrix} 1 & 0 \\ 0 & 1 \end{pmatrix} - \gamma\begin{pmatrix} \cos2\varphi & \sin2\varphi \\ \sin2\varphi & -\cos2\varphi \end{pmatrix} \tag{14.8}$$

where $\gamma_1 = \gamma\cos2\varphi$ and $\gamma_2 = \gamma\sin2\varphi$ are the two components of the gravitational shear γ.

The convergence term κ magnifies the images of background objects, and the shear term γ stretches them tangentially around the foreground mass.

14.2.4 The Gravitational Shear γ

The gravitational shear γ describes the anisotropic distortions of background galaxy images. It corresponds to a two-component field, γ_1 and γ_2, that can be derived from the shape of observed galaxies: γ_1 describes the shear in the x and y directions and γ_2 describes the shear in the $x = y$ and $x = -y$ directions. Using the lens equation, the two shear components γ_1 and γ_2 can be related to the gravitational potential ϕ by

$$\gamma_1 = \frac{1}{2}\left[\frac{\partial^2 \phi(\vec{\theta}_I)}{\partial \theta_{I,1}^2} - \frac{\partial^2 \phi(\vec{\theta}_I)}{\partial \theta_{I,2}^2}\right]$$

$$\gamma_2 = \frac{\partial^2 \phi(\vec{\theta}_I)}{\partial \theta_{I,1} \partial \theta_{I,2}} \tag{14.9}$$

If a galaxy is initially circular with a diameter equal to 1, the gravitational shear will transform the galaxy image to an ellipsoid with a major axis $a = 1/1 - \kappa - |\gamma|$ and a minor axis $b = 1/1 - \kappa + |\gamma|$. The eigenvalues of the amplification matrix (corresponding to the inverse of the distortion matrix A) provide the elongation and the orientation produced on the images of lensed sources [60]. The shear γ is frequently represented by a line segment representing the amplitude and the direction of the distortion (see Figure 14.5).

14.2.5 The Convergence κ

The convergence κ, corresponding to the isotropic distortion of background galaxy images, is related to the trace of the distortion matrix A by

$$\text{tr}(A) = \delta_{1,1} + \delta_{2,2} - \frac{\partial^2 \phi(\vec{\theta}_I)}{\partial \theta_{I,1}^2} - \frac{\partial^2 \phi(\vec{\theta}_I)}{\partial \theta_{I,2}^2}$$

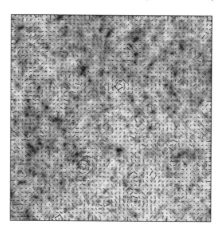

FIGURE 14.5 (**See color insert.**) Simulated convergence map by [92] covering a 2° × 2° field with 1024 × 1024 pixels. The shear map is superimposed to the convergence map. The size and the direction of the line segments represent the amplitude and the direction of the deformation locally.

$$\text{tr}(A) = 2 - \Delta\phi(\vec{\theta}_I) = 2(1 - \kappa) \tag{14.10}$$

$$\kappa = \frac{1}{2}\left(\frac{\partial^2\phi(\vec{\theta}_I)}{\partial\theta_{I,1}^2} + \frac{\partial^2\phi(\vec{\theta}_I)}{\partial\theta_{I,2}^2}\right) \tag{14.11}$$

The convergence κ is defined as half the Laplacian of the projected gravitational potential $\Delta\phi$, and is directly proportional to the projected matter density of the lens (see Figure 14.5). For this reason, κ is often referred to as the mass distribution of the lens.

The dilation and distortion of images of distant galaxies are directly related to the distribution of the (dark) matter and thus to the geometry and the dynamics of the Universe. As a consequence, weak gravitational lensing offers unique possibilities for probing the statistical properties of dark matter and dark energy in the Universe. The weak lensing effect is typically very small, and therefore the constraints that can be obtained on cosmology from the weak lensing effect rely strongly on the quality of the techniques used to analyze the data. In the following, an overview of the different techniques currently used to process weak lensing data will be presented.

14.3 SHEAR ESTIMATION

As described previously, the weak gravitational lensing effect distorts the images of background galaxies. The deformations can be split into two terms, the shear γ and the convergence κ. The shear term stretches the background galaxies tangentially around the foreground mass and the convergence term magnifies (*or demagnifies*) them by increasing (*or decreasing*) their size.

In this section, we describe the methods used to measure the shear field γ. In the next section, we will explain how the convergence field κ is derived from the shear field. Either the shear field or the convergence field can be used to constrain the cosmological model (see Section 14.6), but whatever the method used, the constraints on the cosmological model will depend on the shear measurement accuracy.

14.3.1 Basics

To first approximation, the gravitational shear γ can be traced from the ellipticity of the galaxies that can be expressed as a function of the quadrupole moments of the light distribution of the galaxy image $M_{i,j}$:

$$\epsilon_1 = \frac{M_{1,1} - M_{2,2}}{M_{1,1} + M_{2,2}} \tag{14.12}$$

$$\epsilon_2 = \frac{2M_{1,2}}{M_{1,1} + M_{2,2}} \tag{14.13}$$

where $M_{i,j}$ are defined by

$$M_{i,j} = \frac{\int d^2\theta W(\theta)I(\theta)\theta_i\theta_j}{\int d^2\theta W(\theta)I(\theta)} \tag{14.14}$$

where W is a Gaussian function of scale length r estimated from the object size, I is the surface brightness of the object, and θ is the angular distance from the object center.

The induced shear is small, typically around an order of magnitude smaller than the RMS ellipticity seen in background galaxies, and therefore the weak lensing effect cannot be measured on a single galaxy. To measure the shear field, it is necessary to measure the ellipticities of many background galaxies and construct a statistical estimate of their systematic alignment. The fundamental problem is that galaxies are not intrinsically circular, and so the measured ellipticity is a combination of their intrinsic ellipticity and the gravitational shear. By assuming that the orientations of the intrinsic ellipticities of galaxies is random, any systematic alignment between multiple galaxies is assumed to be caused by gravitational lensing.

The amplitude of the cosmic shear can be quantified statistically by computing the two-point correlation functions of the shear:

$$\xi_{i,j}(\theta) = \langle \gamma_i(\vec{\theta'}) \gamma_j(\vec{\theta'} + \vec{\theta}) \rangle \tag{14.15}$$

where $i, j = 1, 2$ correspond, respectively, to the tangential and radial component of the shear and the averaging is done over pairs of galaxies separated by angle $\theta = |\vec{\theta}|$. By isotropy $\xi_{1,1}$ and $\xi_{2,2}$ are functions only of θ and $\xi_{1,2} = \xi_{2,1} = 0$ is due to the scalar origin of the gravitational lensing effect and to the fact that galaxy ellipticity components are uncorrelated. This measurement demonstrates that the component of the galaxy ellipticities of well-separated galaxies are uncorrelated, and it is in some sense a strong indication that our signal at small scales is of cosmological origin.

However, the weak lensing effect is so small that it requires the control of any systematic error that can mimic the lensing signal.

14.3.2 Biases

14.3.2.1 Instrumental and Atmospheric Bias

A major source of systematic errors in surveys of weak gravitational lensing comes from the point spread function (PSF) due to instrumental and atmospheric effects, which causes the observed images to be smeared. Each background galaxy image is convolved by the PSF of the imagery system H to produce the image that is seen by the instrument. The PSF describes the response of an imaging system to a point source. For ground-based observations, atmospheric turbulence dominates the contribution to the PSF. For space-based observations, the PSF is essentially dependent on the quality of the imaging system. An ideal PSF for weak lensing observations should be small and isotropic. A large PSF tends to make small objects appear more isotropic, destroying some of the information about their true ellipticity. An anisotropic PSF adds a small level of ellipticity to observed background galaxies that can mimic a true lensing signal. Even for the most modern telescopes, the anisotropy of the PSF is usually at least of the same order of magnitude as the weak gravitational shear, and is often much larger.

PSF debiasing requires to have an estimation of the PSF at the location of each galaxy in the field. Stars present in the field (which correspond to point sources) provide a direct measurement of the PSF, and they can be used to model the variation of the PSF across the field, as long as a sufficient number of stellar images are present in the field.

Broadly, there are four groups of methods to correct for instrumental and atmospheric distortions [55] distinguished by their solution to the two most important tasks in shear estimation. The first task is how to correct for the PSF. Some methods subtract the ellipticity of the PSF from the ellipticity of each galaxy, while other methods attempt to deconvolve each galaxy from the PSF before measuring the ellipticity. The second task is how to measure the shear. Some methods do a direct measurement of the shear while others shear a model until it closely resembles the observed galaxy.

The most widely adopted method belongs to the first category, and is the result of a series of successive improvements of the original Kaiser-Squires-Broadhurst (KSB) method proposed by Kaiser, Squires, and Broadhurst [41]. The core of the method is based on the measurement of the weighted ellipticity of the background galaxies and stars from Equation 14.13. The PSF correction is obtained by subtracting the star-weighted ellipticity ϵ_i^* from the observed galaxy-weighted ellipticity ϵ_i^{obs}. The corrected galaxy ellipticity ϵ_i is given by

$$\epsilon_i = \epsilon_i^{obs} - P^{sm}(P^{sm*})^{-1}\epsilon_i^* \tag{14.16}$$

where $i = 1, 2$ and P^{sm} and P^{sm*} are the smear susceptibility tensors for the galaxy and star given by Kaiser [41], which can be derived from higher-order moments of the images [31,41, 49]. This method has been used by many authors, although different interpretations of the method have introduced differences between the various implementations. One drawback of the KSB method is that for non-Gaussian PSFs, the PSF correction is poorly defined, mathematically. In [43], the authors propose a method to better account for realistic PSF by convolving images with an additional kernel to eliminate the anisotropic component of the PSF. Nevertheless, KSB method is thought to have reached its limits and it has been superseded by new competitive methods.

On the recent methods, some methods attempt a direct deconvolution of each galaxy from the PSF. The deconvolution requires a matrix inversion, and becomes an ill-posed inverse problem if the matrix H describing the PSF is singular (i.e., cannot be inverted). The [56] shear-measurement method that belongs to a second class of methods is attempting a full deconvolution by decomposing each galaxy in shapelet basis functions convolved by the PSF. Some other methods belonging to a third class, such as [11,48], have been developed to correct for the PSF without a direct deconvolution. These methods try to reproduce the observed galaxies by modeling each unconvolved background galaxy (the background galaxy as it would be seen without a PSF). The model galaxy is then convolved by the PSF estimated from the stars present in the field and the galaxy model tuned in such way that the convolved model reproduces the observed galaxy. Following a similar approach, the lensfit method [46,61] measures the shear by fitting realistic galaxy profiles in a fully Bayesian way. There is a last class of methods [12] where the PSF correction is obtained by subtraction after the images have been convolved with an additional kernel to eliminate the anisotropic component of the PSF. The shear is obtained by decomposing each galaxy in a distorted shapelet basis function.

Further improvements of these methods will be required to ensure high accuracy and reliability in a the future weak lensing studies.

14.3.2.2 Intrinsic Alignments and Correlations

The intrinsic alignment of galaxies constitutes the major astrophysical source of systematic errors in surveys of weak gravitational lensing. Usually, the measurement of the gravitational shear is obtained by averaging the ellipticity of several nearby galaxies by assuming that the mean intrinsic ellipticity tend to zero. But in practice, the assumption of randomly distributed galaxy shapes is unrealistic. Nearby galaxies are expected to have experienced the same tidal gravitational forces, which is likely to cause a radial alignment of their intrinsic ellipticities, and therefore correlations in their observed shapes and orientations.

There are two types of intrinsic alignments. The first type of alignment appears adjacent background galaxies that form in the same large-scale gravitational potential, and which therefore share a preferred intrinsic ellipticity orientation. This is the so-called Intrinsic–Intrinsic (II) correlation (see Figure 14.6).

The second type corresponds to the case where a matter structure causes a radial alignment of a foreground galaxy and contributes at the same time to the lensing signal of a background galaxy causing a tangential alignment of this background galaxy. In this case, there is an anticorrelation between the foreground intrinsic ellipticity and the background galaxy shear, known as the gravitational–intrinsic (GI) alignment (see Figure 14.7).

Intrinsic alignments are believed to introduce a significant bias in high-precision weak lensing surveys that leads to a bias on cosmological parameters. For example, the shear power spectrum $P_\gamma(l)$, which is the most common method for constraining cosmological parameters, is biased by the previously described intrinsic alignments:

$$P_\gamma(l) = P_\gamma^{GG}(l) + P_\gamma^{II}(l) + P_\gamma^{GI}(l) \qquad (14.17)$$

where the first term is the usual gravitational lensing contribution, the second term arises from the intrinsic alignment of physically close galaxies and the last term arises from GI alignments.

FIGURE 14.6 II alignment: When a large cloud of dark matter, gas, and dust is collapsing, the surrounding matter falls in to form a disc. The resulting galaxy becomes aligned in the direction of the tidal field. Consequently, the intrinsic ellipticity of nearby galaxies will be aligned with the local tidal field like the two upper-most galaxies on the right with vertical lines through them.

FIGURE 14.7 GI alignment: When the tidal field generated by a (dark) matter structure aligns a foreground galaxy (the vertical "oval" shaped structure circled above the cloud) and at the same time generates a gravitational shear on a background galaxy (circled right-most spherical object), there is an anticorrelation between the intrinsic ellipticity of a foreground galaxy (vertical "oval" shaped structure in the screen) and the gravitational shear of a background galaxy (horizontal "oval" shaped structure in the screen).

The shear power spectrum $P_\gamma(l)$ is the Fourier transform of the shear two-point correlation $\xi_{i,j}(\theta)$ defined in the previous section.

The use of galaxy distance information, obtained from either photometric or spectroscopic redshifts,[*] can be used to reduce these systematic errors. In studies [55,75], the signal is evaluated as a function of the galaxy redshift, thanks to the photometric redshift information. But the cosmological constraints in these recent papers have been calculated assuming that there are no intrinsic alignments, although the bias is nonnegligible. To increase the precision of future surveys, it is essential to eliminate intrinsic alignments. It has been shown [28,44,45,87] that the II correlations can be almost completely removed by identifying physically close pairs of galaxies and applying a weighting scheme based on photometric redshifts. The GI effect is more problematic as it affects pairs of galaxies which are not physically close. There are two main approaches to deal with GI alignments. The first approach is model independent; it is a purely geometrical method that removes the GI effect by using a particular linear combination of tomographic shear power spectra, assuming the redshifts of the galaxies are known [39,40]. The other approach tries to model the expected intrinsic alignment contribution for a particular survey. The parameters of the model can thus be varied and marginalized over to constrain cosmological parameters [10,13,38].

14.3.3 Challenges

All methods involve estimating an ellipticity ϵ_i for each galaxy, whose definition can vary between the different methods. The accuracy of the shear measurement method depends on the technique used to estimate the ellipticity of the galaxies in the presence of observational and cosmological effects. In the KSB method [41], the ellipticity is derived from quadrupole moments weighted by a Gaussian function. This method has been used by many authors

---

[*] Photometric and spectroscopic redshifts are estimates for the distance of an astronomical object such as a galaxy. The photometric redshift measurement uses the brightness of the object viewed through various standard filters while the spectroscopic redshift measurement is obtained from the spectroscopic observation of an astronomical object.

but it is not sufficiently accurate for future surveys. The extension of KSB to higher-order moments has been done to allow more complex galaxy and PSF shapes [11,14,48]. The shapelets (see Ref. [57]) can be seen as an extension of KSB to higher order; the first few shapelet basis functions are precisely the weighted functions used in KSB. However, shapelet basis functions are constructed from Hermite polynomials weighted by a Gaussian function and are not optimal at representing galaxy shapes that are closer to exponential functions. By consequence, in the presence of noise, shear measurement methods based on a shapelet decomposition are not optimal.

Many other methods have been developed to address the global problem of shear estimation. To prepare for the next generation of wide-field surveys, a wide range of shear estimation methods have been compared blindly in the Shear Testing Program ([29,55]). Several methods have achieved an accuracy of a few percent. However, the accuracy required for future surveys is of the order of 0.1%. Another challenge, called GREAT08 [15], has been set outside the weak lensing community as an effort to spur further developments. The primary goal was to stimulate new ideas by presenting the problem to researchers outside the shear measurement community. A number of fresh ideas have emerged, especially to reduce the dependence on realistic galaxy modelling. The most successful astronomical algorithm has been found to be the lensfit method [46,61]. In the upcoming GREAT10 challenge, the image simulations are more sophisticated and complex, in order to further improve the accuracy of the methods.

14.4 TWO-DIMENSIONAL (2D) MAPPING OF THE DARK MATTER

The problem of mass reconstruction has become a central topic in weak lensing since the very first maps have demonstrated that this method can be used to visualize the dark side of the Universe. Indeed, weak gravitational lensing provides a unique way to map directly the distribution of dark matter in the Universe. This is done by estimating the convergence κ, which is directly proportional to the projected matter distribution along the line of sight.

14.4.1 Inversion Problem

The convergence, which corresponds to the isotropic distortion of background galaxies images, is not easy to estimate directly because the initial size of galaxies is not known. However, the convergence κ can be derived by inversion of the shear field. In order to do this, a shear field averaged on a regular grid is required. The grid needs to be sufficiently coarse that several galaxies fall in each cell of the grid, otherwise the shear field is only defined on an irregular grid defined by the galaxy positions, yet not so coarse that the shear changes sufficiently across the grid cell.

14.4.1.1 Global Inversion

A global relation between κ and γ can be derived from the relations (14.9) and (14.10). Indeed, it has been shown by Kaiser and Squires [42] that the least-squares estimator $\hat{\tilde{\kappa}}_n$ of the convergence $\hat{\kappa}$ in the Fourier domain is

$$\hat{\tilde{\kappa}}_n(k_1, k_2) = \hat{P}_1(k_1, k_2)\hat{\gamma}_1^{\text{obs}}(k_1, k_2) + \hat{P}_2(k_1, k_2)\hat{\gamma}_2^{\text{obs}}(k_1, k_2) \tag{14.18}$$

where γ_1^{obs} and γ_2^{obs} represent the noisy shear components. The hat symbols denote the Fourier transform and

$$\hat{P}_1(k_1, k_2) = \frac{k_1^2 - k_2^2}{k_1^2 + k_2^2}$$

$$\hat{P}_2(k_1, k_2) = \frac{2k_1 k_2}{k_1^2 + k_2^2} \tag{14.19}$$

with $\hat{P}_1(k_1, k_2) \equiv 0$ when $k_1^2 = k_2^2$, and $\hat{P}_2(k_1, k_2) \equiv 0$ when $k_1 = 0$ or $k_2 = 0$. The most important drawback of this method is that it requires a convolution of shears to be performed over the entire sky. As a result, if the observed shear field has a finite size or a complex geometry, the method can produce artifacts on the reconstructed convergence distribution near the boundaries of the observed field. Masking out the bright stars in the field is common practice in weak lensing analysis, and therefore the global inversion requires a proper treatment of the gaps in the shear map.

14.4.1.2 Local Inversion
There exist local inversions, which have the advantage to address the problems encountered by global relations: the missing data problem and the finite size of the field. A relation between the gradient of $K = \log(1 - \kappa)$ and combinations of first derivatives of $g = \gamma/(1 - \kappa)$ has been derived by Kaiser [41]

$$\nabla K \equiv u$$

$$\frac{-1}{1 - |g|^2} \begin{pmatrix} 1 - g_1 & -g_2 \\ -g_2 & 1 + g_1 \end{pmatrix} \begin{pmatrix} g_{1,1} + g_{2,2} \\ g_{2,1} - g_{1,2} \end{pmatrix} \equiv u \tag{14.20}$$

This equation can be solved by line integration and there exist an infinite number of local inverse formulae which are exact for ideal data on a finite-size field, but which differ in their sensitivity to observational effects such as noise. The reason why different schemes yield different results can be understood by noting that the vector field u (the right-hand side of Equation 14.20) contains a rotational component due to noise because it comes from observational estimates. In Ref. [78], the authors have split the vector field u into a gradient part and a rotational part and they derive the best formula that minimizes the sensitivity to observational effects by convolving the gradient part of the vector field u with a given kernel.

The local inversions reduce the unwanted boundary effects but whatever formula is used, the reconstructed field will be more noisy than that obtained with a global inversion. Note also that the reconstructed dark matter mass map still has a complex geometry that will complicate the later analysis.

14.4.2 E and B Modes Decomposition
Just as a vector field can be decomposed into a gradient or electric (E) component, and a curl or magnetic (B) component, the shear field $\gamma_i(\theta)$ can be decomposed into two components

which for convenience we also call (E) and (B). The decomposition of the shear field into each of these components can be easily performed by noticing that a pure E mode can be transformed into a pure B mode by a rotation of the shear by $45°$: $\gamma_1 \to -\gamma_2, \gamma_2 \to \gamma_1$.

Because the weak lensing arises from a scalar potential (the Newtonian potential Φ), it can be shown that weak lensing only produces E modes. As a result, the least square estimator of the E mode convergence field is simply

$$\tilde{\kappa}^{(E)} = \frac{\partial_1^2 - \partial_2^2}{\partial_1^2 + \partial_2^2}\gamma_1 + \frac{2\partial_1\partial_2}{\partial_1^2 + \partial_2^2}\gamma_2 \tag{14.21}$$

On the other hand, residual systematics arising from imperfect correction of the instrumental PSF, telescope aberrations or complex geometry, generally generates both E and B modes. The presence of B modes can thus be used to test for the presence of residual systematic effects in current weak lensing surveys.

14.4.3 Missing Data

As mentioned previously, analyzing an image for weak lensing inevitably involves the masking out of regions to remove bright stars from the field. The measured shear field is then incomplete. Although the masking out of the stars is common practice, depending on the tools used to analyze this incomplete field, the gaps present in the field will require proper handling.

At present, the majority of lensing analyses use the two-point statistics of the cosmic shear field introduced Section 14.3.1 because this method is not biased by missing data. However, this method is computationally intensive and could not be used for future ultrawide lensing surveys. Measuring the power spectrum is significantly less demanding computationally, but is strongly affected by missing data. Higher-order statistical measures of the cosmic shear field, such as three- or four-point correlation functions have been studied and have shown to provide additional constraints on cosmological parameters but could not be reasonably estimated in future survey.

A solution that has been proposed by Pires et al. [66] to deal with missing data consists in judiciously filling in the masked regions by using an "inpainting" method simultaneously with a global inversion. Inpainting techniques are methods by which an extrapolation of the missing information is carried out using some priors on the solution. This new method uses a prior of sparsity in the solution introduced by Elad et al. [22]. It assumes that there exists a dictionary (i.e., a representation) $\mathcal{D}$ in which the complete data are sparse while the incomplete data are less sparse. This means that we seek a dictionary $\alpha = \Phi^T X$ of the signal X in the representation Φ where most coefficients α_i are close to zero, while only a few have a significant absolute value. If the signal is a sinusoid, its sparsest representation will be the Fourier representation because the signal can be represented by a unique coefficient in the Fourier domain. In many applications—such as compression, denoising, source separation and, of course, inpainting—a sparse representation of the signal is necessary to improve the quality of the processing. Over the past decade, traditional signal representations have been replaced by a large number of new multiresolution representations. Instead of representing signals as a superposition of sinusoids using classical Fourier representation, we now have many available alternative representations such as wavelets [52], ridgelets [18], or curvelets [17,83], most, of which are overcomplete. This means that some elements of the dictionary can be described in terms of other ones, and therefore a signal decomposition in such a dictionary is not unique. Although this can increase the complexity of the signal analysis, it gives us the possibility to select among many possible representations the one which gives the sparsest representation of our data.

The weak lensing inpainting method consists in recovering a complete convergence map κ from the incomplete measured shear field γ_i^{obs}. The solution is obtained by minimizing

$$\min_\kappa \|\mathcal{D}^T \kappa\|_0 \text{ subject to } \sum_i \| \gamma_i^{obs} - M(P_i * \kappa) \|^2 \leq \sigma \qquad (14.22)$$

where P_i is defined by the relation 14.20, $\mathcal{D}^T$ is the Digital Cosine Transform, σ stands for the noise standard deviation, and M is the binary mask (i.e., $M_i = 1$ if we have information at pixel i, $M_i = 0$ otherwise). We denote by $\|z\|_0$ the l_0 pseudo-norm, that is the number of nonzero entries in z, and by $\|z\|$ the classical l_2 norm (i.e., $\|z\|^2 = \sum_k (z_k)^2$).

If $\mathcal{D}^T \kappa$ is sparse enough, the l_0 pseudo-norm can be replaced by the convex l_1 norm (i.e., $\|z\|_1 = \sum_k |z_k|$) [21]. The solution of such an optimization task can be obtained through an iterative thresholding algorithm called Morphological Component Analysis (MCA) [22]:

$$\kappa^{n+1} = \Delta_{\mathcal{D},\lambda_n} \left(\kappa^n + M[P_1 * (\gamma_1^{obs} - P_1 * \kappa^n) + P_2(\gamma_2^{obs} - P_2 * \kappa^n)] \right) \qquad (14.23)$$

where the nonlinear operator $\Delta_{\mathcal{D},\lambda}(Z)$ consists in

- Decomposing the signal Z on the dictionary $\mathcal{D}$ to derive the coefficients $\alpha = \mathcal{D}^T Z$

- Thresholding the coefficients with hard-thresholding ($\tilde{\alpha} = \alpha_i$ if $|\alpha_i| > \lambda$ and 0 otherwise). The threshold parameter λ_n decreases with the iteration number

- Reconstructing $\tilde{Z}$ from the thresholded coefficients $\tilde{\alpha}$

The MCA algorithm has been originally designed for the separation of linearly combined texture and cartoon layers in a given image. By incorporating a binary mask in the model, it leads to an inpainting method. MCA relies on an iterative thresholding algorithm, using a threshold which decreases linearly toward zero along the iterations. This algorithm requires to perform at each iteration a forward transform, a thresholding of the coefficients and an inverse transform.

This method enables one to reconstruct a complete convergence map κ that can be used to perform further analyses of the field. The method can also be used to reconstruct the dark matter distribution and draw comparisons with other probes. These comparisons are usually done after a filtering of the dark matter map, whose quality is improved by the absence of missing data.

14.4.4 Filtering

The convergence map obtained by inversion of the shear field is very noisy (infinite variance) [42] even with a global inversion. The noise comes from the shear measurement errors and the residual intrinsic ellipticities present in the shear maps that propagate during the weak lensing inversion. An efficient filtering is required to map the dark matter and to compare its distribution with other probes.

14.4.4.1 Linear Filters

Linear filters are often used to eliminate or attenuate unwanted frequencies.

- Gaussian filter:
  The standard method [42] consists in convolving the noisy convergence map κ with a Gaussian window G with standard deviation σ_G:

$$\kappa_G = G * \kappa_n = G * P_1 * \gamma_1^{obs} + G * P_2 * \gamma_2^{obs} \tag{14.24}$$

  The Gaussian filter is used to suppress the high frequencies of the signal. However, a major drawback of this method is that the quality of the result depends strongly on the value of the width σ_G of the Gaussian filter, which controls the level of smoothing.

- Wiener filter:
  An alternative to Gaussian filter is the Wiener filter [3,91] obtained by assigning the following weight to each k-mode:

$$w(k_1, k_2) = \frac{|\hat{\kappa}(k_1, k_2)|^2}{|\hat{\kappa}(k_1, k_2)|^2 + |\hat{N}(k_1, k_2)|^2} \tag{14.25}$$

In theory, if the noise and the signal follow a Gaussian distribution, the Wiener filter provides the minimum variance estimator. However, the signal is not Gaussian. In fact, on small scales, the convergence map deviates significantly from Gaussianity. However, the Wiener filter leads to reasonable results, with a better resolution than a simple Gaussian filter.

Linear filters are very common and easy to design, but improved results can be obtained using nonlinear filtering.

14.4.4.2 Bayesian Methods
- Bayesian filters
  Some recent filters are based on Bayesian theory which considers that some prior information can be used to improve the solution. Bayesian filters search for a solution that maximizes the *a posteriori* probability using Bayes' theorem:

$$P(\kappa|\kappa_n) = \frac{P(\kappa_n|\kappa)P(\kappa)}{P(\kappa_n)} \tag{14.26}$$

where

- $P(\kappa_n|\kappa)$ is the likelihood of obtaining the data κ_n given a particular convergence distribution κ.

- $P(\kappa_n)$ is the *a priori* probability of the data κ_n. This term, called the evidence, is simply a constant that ensures that the *a posteriori* probability is correctly normalized.

- $P(\kappa)$ is the *a priori* probability of the estimated convergence map κ. This term codifies our expectations about the convergence distribution before acquisition of the data κ_n.

- $P(\kappa|\kappa_n)$ is called the *a posteriori* probability.

Searching for a solution that maximizes $P(\kappa|\kappa_n)$ is the same that searching for a solution that minimizes the quantity Q:

$$
\begin{aligned}
Q &= -\log(P(\kappa|\kappa_n)) \\
Q &= -\log(P(\kappa_n|\kappa)) - \log(P(\kappa))
\end{aligned}
\tag{14.27}
$$

If the noise is uncorrelated and follows a Gaussian distribution, the likelihood term $P(\kappa_n|\kappa)$ can be written as

$$
P(\kappa_n|\kappa) \propto \exp\left(-\frac{1}{2}\chi^2\right)
\tag{14.28}
$$

with

$$
\chi^2 = \sum_{x,y} \frac{(\kappa_n(x,y) - \kappa(x,y))^2}{\sigma_{\kappa_n}^2}
\tag{14.29}
$$

Equation (14.28) can then be expressed as follows:

$$
Q = \frac{1}{2}\chi^2 - \log(P(\kappa)) = \frac{1}{2}\chi^2 - \beta H
\tag{14.30}
$$

where β is a constant that can be seen as a parameter of regularization and H represents the prior that is added to the solution.

If we have no expectations about the convergence distribution, the *a priori* probability $P(\kappa)$ is uniform and the maximum *a posteriori* is equivalent to the well-known maximum likelihood. This maximum likelihood method has been used by Bartelmann et al. and Seljak [5,80] to reconstruct the weak lensing field, but the solution needs to be regularized in some way to prevent overfitting the data. This has been done via the *a priori* probability of the convergence distribution. The choice of this prior is one of the most critical aspects of the Bayesian analysis. An entropic prior is frequently used but there exists many definitions of the entropy [25]. One that is currently used is the maximum entropy method (MEM) [16,54].

Some authors [5,79] have also suggested reconstructing the gravitational potential ϕ instead of the convergence distribution κ, still using a Bayesian approach, but it is clearly better to reconstruct the mass distribution κ directly because it allows a more straightforward evaluation of the uncertainties in the reconstruction.

- Multiscale Bayesian filters
A multiscale maximum entropy prior has been proposed by Maoli et al. [54] which uses the intrinsic correlation functions (ICF) with varying width. The multichannel MEM-ICF method consists in assuming that the visible-space image I is formed by a weighted sum of the visible-space image channels I_j, $I = \sum_{j=1}^{N_c} p_j I_j$ where N_c is the number of channels and I_j is the result of the convolution between a hidden image h_j with a low-pass filter C_j, called the ICF (i.e., $I_j = C_j * o_j$). In practice, the ICF is a Gaussian. The MEM-ICF constraint is

$$
H_{ICF} = \sum_{j=1}^{N_c} |o_j| - m_j - |o_j| \log\left(\frac{|o_j|}{m_j}\right)
\tag{14.31}
$$

Another approach, based on a sparse representation of the data, has been used by Pantin and Starck [64] that consists of replacing the standard entropy prior by a wavelet-based prior. Sparse representations of signals have received considerable interest in recent years. The problem solved by the sparse representation is to search for the most compact representation of a signal in terms of linear combination of atoms in an overcomplete dictionary.

The entropy is now defined as

$$
H(I) = \sum_{j=1}^{J-1} \sum_{k,l} h(w_{j,k,l})
\tag{14.32}
$$

In this approach, the information content of an image I is viewed as the sum of the information at different scales w_j of a wavelet transform. The function h defines the amount of information relative to a given wavelet coefficient. Several functions have been proposed for h.

In Ref. [84], the most appropriate entropy for the weak lensing reconstruction problem has been found to be the NOISE-multiscale entropy (MSE) entropy:

$$
h(w_{j,k,l}) = \frac{1}{\sigma_j^2} \int_0^{|w_j|} u \operatorname{erfc}\left(\frac{|w_{j,k,l}| - u}{\sqrt{2}\sigma_j}\right) du
\tag{14.33}
$$

where σ_j is the noise standard deviation at scale j. The NOISE-MSE exhibits a quadratic behavior for small coefficients and is very close to the l_1 norm (i.e., the absolute value of the wavelet coefficient) when the coefficient value is large, which is known to produce good results for the analysis of piecewise smooth images.

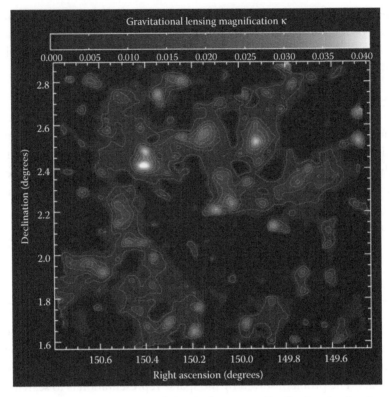

FIGURE 14.8 (**See color insert.**) Map of the dark matter distribution in the 2-square degrees COSMOS field by [58]: the linear blue scale shows the convergence field κ, which is proportional to the projected mass along the line of sight. Contours begin at 0.004 and are spaced by 0.005 in κ. (From R. Massey et al. *Nature*, 445:286–290, 2007. With permission.)

A Multiscale Bayesian filter, called Multiresolution for weak Lensing (MRLens) [84], based on the above method, has shown to outperform other techniques (Gaussian, Wiener, MEM, MEM-ICF) in the reconstruction of dark matter. It has been used to reconstruct the largest weak lensing survey ever undertaken with the Hubble Space Telescope [58]. The result is shown in Figure 14.8, this map is the most precise and detailed dark matter mass map, covering a large-enough area to see extended filamentary structures.

In the paper [91], a study using a simulated full-sky weak lensing map has shown that the Wiener filter is optimal to filter the large-scale structures but not the non-Gaussianity of the field such as the clusters. On the contrary, the MRLens filter is very efficient to reconstruct the clusters but not the large-scale structures. From these results, the authors indicate that an optimal method could be obtained by a smart combination of these two filters.

14.5 3D MAPPING OF THE DARK MATTER

In this section, we will discuss methods to reconstruct the 3D dark matter field from shear measurements.

14.5.1 Formalism

Following the flat-sky approach, the 3D convergence field can be reconstructed by sorting the galaxies in several redshift bins. A convergence map can be computed from the shear at each redshift bin using the relation 14.18. But, this reconstruction is unsatisfactory because there are correlations between the convergence maps reconstructed at different redshift bins. The reconstruction of a 3D dark matter map without correlation requires to estimate the 3D density contrast δ or the 3D gravitational potential Φ, which is related to the density contrast δ by Poisson's equation:

$$\nabla^2\Phi = 4\pi G\rho_m \delta a^2 = \frac{3}{2}\lambda_H^{-2}\Omega_m a^{-1}\delta \tag{14.34}$$

where a is the cosmological scale factor, $\delta = (\rho - \bar{\rho})/\bar{\rho}$ is the density contrast, $\lambda_H = 1/H_0$ is the Hubble length, and Ω_m is the present-day mass-density parameter.

The 3D gravitational potential Φ can be reconstructed using the weak lensing measurements together with the redshifts for all galaxies of the field [3,27,88–90]. Indeed, assuming the Born approximation (i.e., assuming the light path is unperturbed), the lensing potential ϕ is given by

$$\phi(r) = 2\int_0^r dr' \left(\frac{f_K(r) - f_K(r')}{f_K(r)f_K(r')}\right)\Phi(r') \tag{14.35}$$

where $f_K(r)$ is the angular diameter distance, which is a function of the comoving radial distance r and the curvature K. This relation may be inverted to yield [89]:

$$\Phi(r) = \frac{1}{2}\partial_r r^2 \partial_r \phi(r) \tag{14.36}$$

But the lensing potential ϕ is not an observable. The observable is the reduced shear $g_i(\theta) = (\gamma_i(\theta)/1 - \kappa(\theta))$. In the weak lensing regime, κ can be neglected and the observable becomes the shear γ_i corresponding to the distortion of the lensed images of background galaxies.

An estimate of the lensing potential $\tilde{\phi}$ from the shear field is given by the following relation [42]:

$$\tilde{\phi}(\mathbf{r}) = 2\partial^{-4}\partial_i\partial_j\gamma_{ij}(\mathbf{r}) \tag{14.37}$$

with $\gamma_{ij} = \begin{pmatrix} \gamma_1 & \gamma_2 \\ \gamma_2 & -\gamma_1 \end{pmatrix} = (\partial_i\partial_j - \frac{1}{2}\delta_{ij}\partial^2)\phi$ is the shear matrix.

The lensing potential ϕ is related to the convergence field by the following relation:

$$\nabla^2\phi = 2\kappa(\theta) \tag{14.38}$$

The density contrast δ is related to the convergence κ by a line-of-sight integral over the lensing efficiency function $\bar{W}$:

$$[\kappa^i]_k = \frac{3H_0^2}{2c^2}\Omega_m \int_0^\infty dr \frac{\bar{W}^i(r)f_K(r)}{a(r)}\delta(f_K(r)\theta_k, r)$$

$$\bar{W}^i(r) = \int_0^\infty dr' \frac{f_K(r'-r)}{f_K(r')}\left[p_z^{(i)}(z)\frac{dz}{dr}\right]_{z=z(r')} \tag{14.39}$$

where p_z^i represents the ith bin of the probability distribution of sources as a function of redshift.

With these sets of equations the 3D lensing potential, the 3D gravitational potential, the 3D convergence and the 3D matter density fields can all be generated from combined shear and redshift information.

Some authors have attempted to reconstruct the 3D lensing potential or the 3D gravitational potential [3,58] in order to reconstruct the 3D matter density field. But in the majority of papers relating to 3D lens mapping [35,82,94], the reconstruction of the 3D matter density field is derived from the convergence tomography. It consists of two linear steps: a 3D convergence field is first derived from the shear measurements sorted in several redshift bins using relation 14.18; then, the 3D density contrast is obtained by inversion of relation 14.39. The resulting reconstruction is very noisy and requires a regularization of the solution.

14.5.2 Filtering
Linear map-making can be expressed in terms of the general inverse problem:

$$d_b = P_{ba}s_a + n_b \tag{14.40}$$

where we seek an estimate $\hat{s}_a$ from a data vector d_b that is a linear projection P_{ba} of the signal with measurement noise n_b. s_a can be the 3D density contrast or the 3D gravitational potential, and d_b can be the 3D shear field or the 3D convergence field.

14.5.2.1 Linear Filters
A first linear approach is the maximum likelihood that searches for an estimate of the signal $\hat{s}_a$ that minimizes χ^2:

$$\chi^2 = (d_b - P_{ba}\hat{s}_a)^t N_{bb}^{-1}(d_b - P_{ba}\hat{s}_a) \tag{14.41}$$

where N_{bb} is the noise covariance. Minimizing χ^2 returns the linear estimator:

$$\hat{s}_a = R_{ab}d_b \tag{14.42}$$

where

$$R_{ab} = [P_{ba}^t N_{bb}^{-1} P_{ba}]^{-1} P_{ba}^t N_{bb}^{-1} \tag{14.43}$$

If P_{ba} is invertible, then $R_{ab} = P_{ba}^{-1}$ and the estimator becomes independent of both the signal and the noise. Note that minimizing χ^2 is not the same as minimizing the reconstruction noise $N_{aa} = \langle (\hat{s}_a - s_a)(\hat{s}_a - s_a)^t \rangle$. It minimizes N_{aa} subject to the constraint $R_{ab}P_{ba} = I$.

A Bayesian approach can be used to set a penalty function H on the solution using prior knowledge of the statistical properties of the signal and the noise. An estimate of the signal can then be sought by minimizing $\chi^2 + H$. Defining the penalty function $H = \hat{s}_w^t S_{aa}^{-1} \hat{s}_w$, the minimization of $\chi^2 + H$ returns the Wiener filtered estimate of the signal $\hat{s}_w = R_{wb}d_b$, where:

$$R_{wb} = [S_{aa}^{-1} + P_{ba}^t N_{bb}^{-1} P_{ba}]^{-1} P_{ba}^t N_{bb}^{-1}$$
$$R_{wb} = S_{aa}P_{ba}^t [P_{ba}S_{aa}P_{ba}^t + N_{bb}]^{-1} \tag{14.44}$$

The classical Wiener filter definition (see Equation 14.25) can be recovered by assuming the data may be expressed as $d_b = Is_a + n_b$. Equation 14.44 then becomes

$$R_{wb} = S_{aa}[S_{aa} + N_{bb}]^{-1} \tag{14.45}$$

where $S_{aa} = \langle s_a s_a^t \rangle$ and $N_{bb} = \langle n_b n_b^t \rangle$. The Wiener filter reduces the reconstruction noise N_{aa} by using the expected noise properties as a prior on the solution [35].

In the literature, no 3D Wiener filter has yet been proposed to process the full 3D matter density field. In Ref. [35], the authors apply a Wiener filter along each individual line of sight, ignoring the correlation between different lines of sight (radial Wiener filter). In Ref. [82], the authors use two types of pseudo-3D Wiener filter: a "radial Wiener filter" and a "transverse Wiener filter." Both filters are well adapted to filter the large-scale regime because the signal is Gaussian, but small scales require a more efficient filtering. Moreover, both filters show a systematic shift and stretch of the structures in the radial direction, and fail to account for correlations either along the line of sight (transverse filter) or in the transverse direction (radial filter).

14.5.2.2 Nonlinear Filters

Several nonlinear filters have been proposed to filter the 3D matter density field. The MEM filter that has been presented in Section 14.4.4 has been introduced by Hu and Keeton [35] to reconstruct dark matter halos in 3D matter distribution. The MEM filter that is based on the Bayesian theory has already been applied in 2D matter distribution (see [16,54]). In this filter, the penalty function is an entropy that is estimated from the full 3D weak lensing field. A fundamental problem with the MEM filter is the difficulty in assessing the errors in the reconstruction.

Another nonlinear approach that is a full-3D filter has been proposed by Hu and Keeton, VanderPlas et al. [35,94], and consists in expressing the data in a new set of uncorrelated orthogonal basis elements that are ranked by their signal-to-noise ratio. This method is known as the Karhunen–Loeve transform or singular-value decomposition, assuming the modes are the eigenvectors of the covariance matrix. A direct noise reduction which does not require knowledge of the statical properties of the signal is obtained by eliminating the lower signal-to-noise singular values.

The full reconstruction of the 3D dark matter distribution from weak lensing has only been considered recently [35,88] because the finite size of the earlier surveys did not allow for 3D mapping. Recently, the COSMOS survey [58,59] has been able to reconstruct the first 3D weak lensing map. Although the results were limited by the finite size of the survey, they provide a framework for future large weak lensing surveys like the future Euclid mission. The next step, is to develop a full 3D Wiener filter, to extend the MRLens method from 2D to 3D, and possibly to develop a combined filtering method.

14.6 COSMOLOGICAL MODEL CONSTRAINTS

Measurement of the distortions in images of background galaxies caused by large-scale structures provides a direct way to study the statistical properties of the evolution of structures in the Universe. Weak gravitational lensing measures the mass and can thus be directly compared to theoretical models of structure formation. But because we have only one realization of our Universe, a statistical analysis is required to do the comparison. The estimation of the cosmological parameters from weak lensing data can be seen as an inverse problem. The direct problem consisting of deriving weak lensing data from cosmological parameters can be solved using numerical simulations, but the inverse problem cannot be solved so easily because the N-body equations used by the numerical simulations cannot be inverted.

14.6.1 Methodology

A solution is to use analytical predictions for statistics of cosmic shear and to compare with the values estimated from the data. A statistical analysis of the weak lensing field is then required to constrain the cosmological parameters.

The method that is usually used to constrain cosmological parameters from statistical estimation is the maximum likelihood (see Section 14.4.4). Let us estimate the statistic $\eta(x, y)$ in the weak lensing field to constrain the cosmological parameters p_1 and p_2. By assuming the noise follows an uncorrelated and Gaussian distribution, the likelihood function $\mathcal{L}$ is defined as follows:

$$\mathcal{L}(p_1, p_2, p_3) \equiv P(\eta^{obs} | \eta^{mod}) \propto \exp\left(-\frac{1}{2} \chi^2(p_1, p_2, p_3)\right) \quad (14.46)$$

with

$$\chi^2(p_1, p_2, p_3) = \sum_{x,y} \frac{(\eta^{obs}(x, y) - \eta^{mod}(x, y; p_1, p_2, p_3))^2}{\sigma^2(x, y)} \quad (14.47)$$

where η^{mod} is the analytic prediction depending on the two cosmological parameters (p_1, p_2) and a nuisance parameter (p_3). η^{obs} is the statistic estimation from the data. χ^2 is then a function of three parameters. A 2D probability distribution can be obtained by marginalizing the 3D probability distribution function over p_3:

$$\mathcal{L}(p_1, p_2) = \int dp_3 P(p_3) \exp\left(-\frac{1}{2}\chi^2(p_1, p_2, p_3)\right) \qquad (14.48)$$

Here $P(p_3)$ is the prior distribution function for the parameter p_3, which is assumed to be a known Gaussian distribution. Using $\mathcal{L}(p_1, p_2)$, we can define the 1, 2, and 3σ contours on the 2D (p_1, p_2) parameter space. The best-fit values for p_1 and p_2 can also be easily determined by maximizing the likelihood function. Recent cosmic shear results obtained from maximum likelihood maximization are given in [24,75].

The different cosmological parameters can be quantified using a variety of statistics estimated either in the shear field or in the convergence field. Most lensing studies do the statistical analysis in the shear field to avoid the inversion. But most of the following statistics can also be estimated in the convergence field if the missing data are carefully accounted for. A description of the different statistics that can be used to constrain cosmological parameters is provided below.

14.6.2 Second-Order Statistics

The most common method for constraining cosmological parameters uses second-order statistics of the shear field calculated either in real or Fourier space (or Spherical harmonic space). Whatever the second-order statistic that is considered, it can be easily related to the theoretical 3D matter power spectrum $P(k, \chi)$ by means of the 2D convergence power spectrum $P_\kappa(l)$.

- The convergence and shear power spectra $P_\kappa(l)$ and $P_\gamma(l)$:

  The 2D convergence power spectrum $P_\kappa(l)$ only depends on $l = |\vec{l}|$, and is defined by

$$\langle \hat{\kappa}(\vec{l})\hat{\kappa}(\vec{l}')\rangle = (2\pi)^2 \delta(\vec{l} - \vec{l}') P_\kappa(l) \qquad (14.49)$$

where $\hat{\kappa}$ is the Fourier transform of the 2D convergence κ.

The 2D convergence power spectrum $P_\kappa(l)$ can be expressed as a function of the 3D matter power spectrum $P(k, \chi)$, the mass fluctuations $\delta\rho/\rho$ and cosmological parameters [67]:

$$P_\kappa(l) = \frac{9}{16}\left(\frac{H_0}{c}\right)^4 \Omega_{\rm m}^2 \int d\chi \left[\frac{g(\chi)}{ar(\chi)}\right]^2 P\left(\frac{l}{r}, \chi\right) \qquad (14.50)$$

where a is the cosmological scale factor, H_0 is the Hubble constant, $\Omega_{\rm m}$ is the matter density parameter, χ the comoving distance, $r = a^{-1}D_A$ with D_A being the angular diameter distance, and g the lensing efficiency function.

Furthermore, in the case of weak lensing, the power spectra for the shear $P_\gamma(l)$ and the convergence $P_\kappa(l)$ are the same.

In general, there are advantages for cosmological parameter estimation in using Fourier (or Spherical harmonic) statistics, because the Fast Fourier transform (FFT) algorithm can be used to estimate these power spectra rapidly, and the correlation properties are more convenient to express in Fourier space. However, for surveys with complicated geometry due to the removal of bright stars, the spatial stationarity is not satisfied and the missing data need proper handling. The problem of power spectrum estimation from weak lensing data with missing data has been studied by Hikage et al. [30]. But real space statistics are largely used because they are easier to estimate, although statistical error bars are harder to estimate.

- Shear variance $\langle\gamma^2(\theta)\rangle$:

An example of real space second-order statistic is the shear variance $\langle\gamma^2(\theta)\rangle$, defined as the variance of the average shear $\bar{\gamma}$ evaluated in circular patches of varying radius θ_s. The shear variance $\langle\gamma^2(\theta)\rangle$ can be related to the underlying 3D matter power spectrum via the 2D convergence power spectrum P_κ by the following relation:

$$\langle\gamma^2(\theta)\rangle = \int \frac{dl}{2\pi} l P_\kappa(l) \frac{J_1^2(l\theta_s)}{(l\theta_s)^2} \tag{14.51}$$

where J_n is the Bessel function of order n. The shear variance has been frequently used in weak lensing analysis to constrain cosmological parameters [24,32,53].

- Shear two-point correlation function $\xi_{i,j}(\theta)$:

Another real space statistic is the shear two-point correlation function $\xi_{i,j}(\theta)$ defined in Section 14.3, which is widely used because it is easy to implement and can be estimated even for complex geometry. The shear two-point correlation function can also be related to the underlying 3D matter power spectrum via the 2D convergence power spectrum P_κ. The shear two-point correlation function is the Fourier transform of the convergence power spectrum P_κ, which becomes a Hankel transform (also called Fourier–Bessel transform) considering the isotropy of the Universe:

$$\xi_+(\theta) = \xi_{1,1}(\theta) + \xi_{2,2}(\theta) = \int_0^\infty \frac{dl}{2\pi} l P_\kappa(l) J_0(l\theta) \tag{14.52}$$

where J_0 corresponds to the Bessel function at zero order.

The two-point correlation function is the most popular statistical tool used in weak lensing analysis. It has been used in many recent weak lensing analyses to constrain cosmological parameters (see, e.g. [7,24,32]).

The simplest way to compute a two-point correlation function consists in counting the number of background galaxy pairs separated by a distance d. But this brute force approach has a complexity of $O(N^2)$, which becomes an important computational load for large surveys. Recently, some progress has been made in this field, and new methods have been developed to speed up the calculation of the two-point correlation function by reducing the complexity to $O(N \log N)$. These algorithms are based on ordered binary trees that are an interesting data structure to scan the galaxy pairs [62,95]. But to reach this complexity some approximations are required, such as neglecting large scales or binning the two-point correlation function [37,95].

- Shear Tomography $\xi_{i,j}^{k,l}(\theta)$:

The constraints on cosmological parameters can be significantly improved if the shape measurements are combined with photometric redshifts. The 2D shear correlation formalism can be extended to 3D shear by splitting the galaxy sample into redshift bins from which the auto- and cross-correlations can be calculated. It is the so-called cosmic shear tomography [2,34]. The shear cross-correlation functions $\xi_{i,j}^{k,l}(\theta)$ between the bins k and l are defined as follows:

$$\xi_{i,j}^{k,l}(\theta) = \langle \gamma_i^k(\vec{\theta}')\gamma_j^l(\vec{\theta}' + \vec{\theta}) \rangle \tag{14.53}$$

The shear cross-correlation functions $\xi_{i,j}^{k,l}(\theta)$ can be related to the underlying 3D matter power spectrum via the 2D convergence cross-power spectrum $P_\kappa^{k,l}$ by the following relation:

$$\xi_+^{k,l}(\theta) = \int_0^\infty \frac{dl}{2\pi} l P_\kappa^{k,l}(l) J_0(l\theta) \tag{14.54}$$

with

$$P_\kappa^{k,l}(l) = \frac{9}{16}\left(\frac{H_0}{c}\right)^4 \Omega_m^2 \int d\chi \frac{g^k(\chi)g^l(\chi)}{(ar(\chi))^2} P\left(\frac{l}{r}, \chi\right) \tag{14.55}$$

Shear tomography has been applied to real data (e.g., [4,75]) to improve the constraints on cosmological parameters. The separation of source galaxies into tomographic bins improves significantly the constraints on cosmological parameters, and particularly those of dark energy that drives cosmic expansion.

- Second-order statistics to separate E and B modes:

If the shear, estimated from the image shapes of distant galaxies, was only due to gravitational lensing, then it should consist only of a pure E-mode shear (see Section 14.4.2). But the estimated shear may contains systematic errors. Therefore, the splitting of the observed shear field into its E- and B-modes is of great importance to isolate the gravitational shear from the shear components most likely not due to lensing. The standard technique for this separation is the variance of the aperture mass $\langle M_{ap}^2 \rangle$ [74], which corresponds to an average shear two-point correlation. This statistic is the result of the convolution of the shear two-point correlation with a compensated filter. Several forms of filters have been suggested which trade locality in real space with locality in Fourier space. By considering the filter defined by Schneider et al. [69] with a scale of θ_s, the variance of the aperture mass can be expressed as a function of the 2D convergence power spectrum as follows:

$$\langle M_{ap}^2(\theta_s) \rangle = \int \frac{dl}{2\pi} l P_\kappa(l) \frac{576 J_4^2(l\theta_s)}{(l\theta_s)^4} \tag{14.56}$$

This method has been used in many weak lensing analyses to constrain cosmological parameters (see, e.g., [24,32,81,93]). However, the lack of knowledge of the shear correlation function on very small scales introduces a systematic bias on the variance aperture method. More recently, new second-order statistic, that is the ring statistic [72] and Complete Orthogonal Sets of E-/B-mode Integrals [71] have been proposed to overcome the practical problem encountered by the aperture mass method.

Second-order statistics measure the Gaussian properties of the field, which limits the amount of information extracted, since it is known that the low redshift Universe is highly non-Gaussian on small scales. Indeed, gravitational clustering is a nonlinear process and, in particular, at small scales the mass distribution is highly non-Gaussian. Consequently, if only second-order statistics are used to place constraints on the cosmological model, degenerate constraints are obtained between some important cosmological parameters (see, e.g., [24,33,75]).

14.6.3 Non-Gaussian Statistics

If the weak lensing signal was Gaussian, it would be fully described by its angular power spectrum. However, in the standard model of structure formation, fluctuations that are initially Gaussian are amplified by gravitational collapse to produce a highly non-Gaussian matter distribution. Thus, except at a large scale, the convergence field is highly non-Gaussian. On small scales, we can observe structures like galaxies and clusters of galaxies, and on intermediate scales, we observe some filamentary structures. The characterization of this non-Gaussianity can be used to constrain the cosmological parameters. A possible solution is to consider higher-order statistics of the shear or convergence field. The statistics presented below are estimated in the convergence field because the non-Gaussianity is clearly visible in the field.

Third-order statistics are the lowest-order statistics that can be used to detect non-Gaussianity. Many authors have already addressed the problem of three-point statistics and semianalytical predictions for the three-point correlation function and the bispectrum have already been derived (e.g., [19,50,51,77]). However, the correction for missing data in the full higher-order Fourier statistics remains an outstanding issue. An alternative solution (see Section 14.4.3) has been proposed by Pires et al. [66] to derive second-order and third-order statistics and possibly higher-order statistics from an incomplete shear map.

- The convergence bispectrum $B_\kappa(|\vec{l}_1|, |\vec{l}_2|, |\vec{l}_3|)$:

  By analogy with second-order statistics, whatever the third-order statistic considered, it can be easily related to the convergence bispectrum $B_\kappa(|\vec{l}_1|, |\vec{l}_2|, |\vec{l}_3|)$, which is the

Fourier analog of the three-point correlation function:

$$B_\kappa(|\vec{l_1}|, |\vec{l_2}|, |\vec{l_3}|) \propto \langle \hat{\kappa}(|\vec{l_1}|) \hat{\kappa}(|\vec{l_2}|) \hat{\kappa}^*(|\vec{l_3}|) \rangle \qquad (14.57)$$

where $\kappa*$ is the complex conjugate of κ. The bispectrum only depends on distances $|\vec{l_1}|$, $|\vec{l_2}|$, and $|\vec{l_3}|$. Scoccimarro and Colombi [76] proposed an algorithm to compute the bispectrum from numerical simulations using a FFT but without considering the case of incomplete data. This method is used by Fosalba [23] to estimate the bispectrum from numerical simulations in order to compare it with the semi-predictions of the analytic halo model.

- Convergence three-point correlation function $\xi_{i,j,k}(\theta)$:

The three-point correlation function $\xi_{i,j,k}$ is easy to estimate and can be estimated even for complex geometry. It is defined as follows:

$$\xi_{i,j,k}(\theta) = \langle \kappa(\vec{\theta_1}) \kappa(\vec{\theta_2}) \kappa(\vec{\theta_3}) \rangle \qquad (14.58)$$

The same relation can be derived for the shear. It has been shown that tighter constraints can be obtained with the three-point correlation function [86]. Estimating three-point correlation function from data has already be done [9] but cannot be considered in future large data sets because it is computationally too intensive. In the conclusion of [85], the authors briefly suggested to use the p-point correlation functions with implementations that are at best $O(N(\log N)^{p-1})$. However, it was not clear as to whether this suggestion is valid for the missing data case.

- Third-order moment of the convergence S_κ:

A simpler quantity than the three-point correlation function is provided by measuring the third-order moment of the convergence κ, also named the skewness, that measures the asymmetry of the distribution [65]:

$$S_\kappa = \sum_{i=1}^{N} \frac{(\kappa_i - \bar{\kappa})^3}{(N-1)\sigma^3} \qquad (14.59)$$

The convergence skewness is primarily due to rare and massive dark matter halos. The distribution will be more or less skewed positively depending on the abundance of rare and massive halos.

- Fourth-order moment of the convergence K_κ:

We can also estimate the fourth-order moment of the convergence, also known as kurtosis, that measures the peakiness of a distribution [65]: The kurtosis of the convergence κ is defined as follows:

$$K_\kappa = \sum_{i=1}^{N} \frac{(\kappa_i - \bar{\kappa})^4}{(N-1)\sigma^4} - 3 \qquad (14.60)$$

which is known as excess kurtosis. A high kurtosis distribution has a sharper "peak" and flatter "tails," while a low kurtosis distribution has a more rounded peak.

- The skewness and the kurtosis of the aperture mass $\langle M_{ap}^n \rangle$:

Similar to the aperture mass variance described in the previous section, the skewness and the kurtosis of the aperture mass have also been introduced to weak lensing analyses to constrain cosmological parameters (see e.g., [37,73]).

- Peak counting (PC)

As said previously, the convergence field is highly non-Gaussian. Another approach to look for non-Gaussianity is to perform a statistical analysis directly on the non-Gaussian structures present in the convergence field. For example, galaxy clusters, which are the largest virialized cosmological structures in the Universe, can provide a unique way to focus on non-Gaussianity present at small scales. One interesting method is the PC, which searches the number of peaks detected on the field that differs from the cluster abundance because of the projection of the large–scale structures. The PC has been used to measure the number of peaks detected on the convergence field by [26,36,47]. Peaks can be also counted on the shear field using a filtered version of the shear [20].

It has been proposed by Pires et al. [65] to drew a comparison between several statistics and several representations. The comparison shows that the wavelet transform makes statistics more sensitive to the non-Gaussianity present in the convergence field. In the same paper, several non-Gaussian statistics have been compared and the PC estimated in a wavelet representation, called Wavelet PC, has been found to be the best non-Gaussian statistic to constrain cosmological parameters. In this chapter, the comparison with bispectrum was restricted to the equilateral configuration. In Ref. [8], the authors shows that bispectrum using all triangle configurations outperforms PC as a function of clusters' mass and redshift.

14.7 CONCLUSION

The weak gravitational lensing effect, which is directly sensitive to the gravitational potential, provides a unique method to map the 3D dark matter distribution and thus understand the process of structure formation. This can be used to set tighter constraints on cosmological models and to better understand the nature of dark matter and dark energy. But the constraints derived from this weak lensing effect depend on the techniques used to analyze the weak lensing signal, which is very weak.

The field of weak gravitational lensing has recently seen great success in mapping the distribution of dark matter (Figure 14.8). The next breakthrough will certainly happen in the next decade, thanks to the future full-sky missions designed to weak gravitational lensing: ground-based missions like Large Synoptic Survey Telescope or space-based missions like Euclid or Wide Field Infrared Survey Telescope. The primary goal of these missions is precisely to map very accurately the geometry of the dark universe and its evolution by measuring shapes and redshifts of galaxies over the entire sky. It will provide a 3D full-sky map of the dark and visible matter in the Universe and will permit one to set tighter constraints on Dark Energy and other cosmological parameters. For this to happen, new methods are now necessary to reach the accuracy required by this survey, and ongoing efforts are needed to improve the standard analyses. This chapter attempts to give an overview of the techniques that are currently used to analyze the weak lensing signal along with future directions. It shows that weak lensing is a dynamic research area in constant progress.

In this chapter, we have detailed the different steps of the weak lensing data analysis, thus presenting various aspects of signal processing. For each problem, we have systematically presented a range of methods currently used, from earliest to up-to-date methods. This chapter highlights the introduction of Bayesian ideas that have provided a way to incorporate prior knowledge in data analysis as a major milestone in weak lensing analysis. The next major step might possibly be the introduction of sparsity. Indeed, we have presented new methods based on sparse representations of the data that have already had some success.

ACKNOWLEDGMENT

This work has been supported by the European Research Council grant SparseAstro (ERC-228261).

REFERENCES

1. A. Albrecht, G. Bernstein, R. Cahn, W. L. Freedman, J. Hewitt, W. Hu, J. Huth et al. Report of the Dark Energy Task Force. *ArXiv Astrophysics e-prints : astro-ph/0609591*, September 2006.
2. A. Amara and A. Réfrégier. Optimal surveys for weak-lensing tomography. *MNRAS*, 381:1018–1026, 2007.
3. D. J. Bacon and A. N. Taylor. Mapping the 3D dark matter potential with weak shear. *MNRAS*, 344:1307–1326, 2003.
4. D. J. Bacon, A. N. Taylor, M. L. Brown, M. E. Gray, C. Wolf, K. Meisenheimer, S. Dye, L. Wisotzki, A. Borch, and M. Kleinheinrich. Evolution of the dark matter distribution with three-dimensional weak lensing. *MNRAS*, 363:723–733, 2005.
5. M. Bartelmann, R. Narayan, S. Seitz, and P. Schneider. Maximum-likelihood cluster reconstruction. *ApJ*, 464:L115–ff., 1996.
6. M. Bartelmann and P. Schneider. Weak gravitational lensing. *Phys. Rep.*, 340:291–472, 2001.
7. J. Benjamin, C. Heymans, E. Semboloni, L. van Waerbeke, H. Hoekstra, T. Erben, M. D. Gladders, M. Hetterscheidt, Y. Mellier, and H. K. C. Yee. Cosmological constraints from the 100 deg 2 weak-lensing survey. *MNRAS*, 381:702–712, 2007.
8. J. Bergé, A. Amara, and A. Réfrégier. Optimal capture of non-Gaussianity in weak-lensing surveys: Power spectrum, bispectrum, and Halo counts. *ApJ*, 712:992–1002, 2010.
9. F. Bernardeau, Y. Mellier, and L. van Waerbeke. Detection of non-Gaussian signatures in the VIRMOS-DESCART lensing survey. *A&A*, 389:L28–L32, 2002.
10. G. M. Bernstein. Comprehensive two-point analyses of weak gravitational lensing surveys. *ApJ*, 695:652–665, 2009.
11. G. M. Bernstein and M. Jarvis. Shapes and shears, stars and smears: Optimal measurements for weak lensing. *AJ*, 123:583–618, 2002.
12. G. M. Bernstein and M. Jarvis. Shapes and shears, stars and smears: Optimal measurements for weak lensing. *AJ*, 123:583–618, 2002.
13. S. Bridle and L. King. Dark energy constraints from cosmic shear power spectra: Impact of intrinsic alignments on photometric redshift requirements. *New J. Phys.*, 9:444–ff., 2007.

14. S. Bridle, J.-P. Kneib, S. Bardeau, and S. Gull. Bayesian galaxy shape estimation. In P. Natarajan (ed.), *The Shapes of Galaxies and their Dark Halos*, World Scientific, New Haven, CT, USA, pp. 38–ff., 2002.

15. S. Bridle, J. Shawe-Taylor, A. Amara, D. Applegate, S. T. Balan, G. Bernstein, J. Berge et al. Handbook for the GREAT08 challenge: An image analysis competition for cosmological lensing. *Ann. Appl. Stat.*, 3:6–37, 2009.

16. S. L. Bridle, M. P. Hobson, A. N. Lasenby, and R. Saunders. A maximum-entropy method for reconstructing the projected mass distribution of gravitational lenses. *MNRAS*, 299:895–903, 1998.

17. E. Candès, L. Demanet, D. Donoho, and Y. Lexing. Fast discrete curvelet transforms. *SIAM. Multiscale Model. Simul.*, 5:861–899, 2006.

18. E. Candès and D. Donoho. Ridgelets: A key to higher dimensional intermittency. *Philos. Trans. R. Soc. A*, 357:2495–2509, 1999.

19. A. Cooray and W. Hu. Weak gravitational lensing bispectrum. *ApJ*, 548:7–18, 2001.

20. J. P. Dietrich and J. Hartlap. Cosmology with the shear-peak statistics. *MNRAS*, 402:1049–1058, 2010.

21. D. L. Donoho and X. Huo. Uncertainty principles and ideal atomic decomposition. *Trans. Inf. Theory*, 47:2845–2862, 2001.

22. M. Elad, J.-L. Starck, P. Querre, and D. L. Donoho. Simultaneous cartoon and texture image inpainting using morphological component analysis (MCA). *J. Appl. Comput. Harmon. Anal.*, 19(3):340–358, 2005.

23. P. Fosalba, J. Pan, and I. Szapudi. Cosmological three-point function: Testing the Halo model against simulations. *ApJ*, 632:29–48, 2005.

24. L. Fu, E. Semboloni, H. Hoekstra, M. Kilbinger, L. van Waerbeke, I. Tereno, Y. Mellier et al. Very weak lensing in the CFHTLS wide: Cosmology from cosmic shear in the linear regime. *A&A*, 479:9–25, 2008.

25. S.-F. Gull. Developments in maximum entropy data analysis. In J. Skilling (ed.), *Maximum Entropy and Bayesian Methods*, Springer-Verlag, Cambridge, UK, 1989.

26. T. Hamana, M. Takada, and N. Yoshida. Searching for massive clusters in weak lensing surveys. *MNRAS*, 350:893–913, 2004.

27. A. Heavens. 3D weak lensing. *MNRAS*, 343:1327–1334, 2003.

28. C. Heymans and A. Heavens. Weak gravitational lensing: Reducing the contamination by intrinsic alignments. *MNRAS*, 339:711–720, 2003.

29. C. Heymans, L. Van Waerbeke, D. Bacon, J. Berge, G. Bernstein, E. Bertin, S. Bridle et al. The shear testing programme—I. Weak lensing analysis of simulated ground-based observations. *MNRAS*, 368:1323–1339, 2006.

30. C. Hikage, M. Takada, T. Hamana, and D. Spergel. Shear power spectrum reconstruction using the pseudo-spectrum method. *MNRAS*, 412:65–74, 2011.

31. H. Hoekstra, M. Franx, K. Kuijken, and G. Squires. Weak lensing analysis of CL 1358+62 using Hubble space telescope observations. *ApJ*, 504:636–ff., 1998.

32. H. Hoekstra, Y. Mellier, L. van Waerbeke, E. Semboloni, L. Fu, M. J. Hudson, L. C. Parker, I. Tereno, and K. Benabed. First cosmic shear results from the Canada-France-Hawaii telescope wide synoptic legacy survey. *ApJ*, 647:116–127, 2006.

33. H. Hoekstra, H. K. C. Yee, and M. D. Gladders. Constraints on Ω_m and σ_8 from weak lensing in red-sequence cluster survey fields. *ApJ*, 577:595–603, 2002.

34. W. Hu. Power spectrum tomography with weak lensing. *ApJ*, 522:L21–L24, 1999.

35. W. Hu and C. R. Keeton. Three-dimensional mapping of dark matter. *Phys. Rev. D*, 66(6):063506–ff., 2002.

36. B. Jain and L. Van Waerbeke. Statistics of dark matter halos from gravitational lensing. *ApJ*, 530:L1–L4, 2000.

37. M. Jarvis, G. Bernstein, and B. Jain. The skewness of the aperture mass statistic. *MNRAS*, 352:338–352, 2004.

38. B. Joachimi and S. L. Bridle. Simultaneous measurement of cosmology and intrinsic alignments using joint cosmic shear and galaxy number density correlations. *A&A*, 523:A1–ff., 2010.

39. B. Joachimi and P. Schneider. The removal of shear-ellipticity correlations from the cosmic shear signal via nulling techniques. *A&A*, 488:829–843, 2008.

40. B. Joachimi and P. Schneider. The removal of shear-ellipticity correlations from the cosmic shear signal. Influence of photometric redshift errors on the nulling technique. *A&A*, 507:105–129, 2009.

41. N. Kaiser. Nonlinear cluster lens reconstruction. *ApJ*, 439:L1–L3, 1995.

42. N. Kaiser and G. Squires. Mapping the dark matter with weak gravitational lensing. *ApJ*, 404:441–450, 1993.

43. N. Kaiser, G. Wilson, and G. A. Luppino. Large-scale cosmic shear measurements. *ArXiv e-prints*, arXiv:astro-ph/0003338, 2000.

44. L. King and P. Schneider. Suppressing the contribution of intrinsic galaxy alignments to the shear two-point correlation function. *A&A*, 396:411–418, 2002.

45. L. J. King and P. Schneider. Separating cosmic shear from intrinsic galaxy alignments: Correlation function tomography. *A&A*, 398:23–30, 2003.

46. T. D. Kitching, L. Miller, C. E. Heymans, L. van Waerbeke, and A. F. Heavens. Bayesian galaxy shape measurement for weak lensing surveys—II. Application to simulations. *MNRAS*, 390:149–167, 2008.

47. J. M. Kratochvil, Z. Haiman, and M. May. Probing cosmology with weak lensing peak counts. *Phys. Rev. D*, 81(4):043519–ff., 2010.

48. K. Kuijken. Shears from shapelets. *A&A*, 456:827–838, 2006.

49. G. A. Luppino and N. Kaiser. Detection of weak lensing by a cluster of galaxies at $Z = 0.83$. *ApJ*, 475:20–ff., 1997.

50. C.-P. Ma and J. N. Fry. Deriving the nonlinear cosmological power spectrum and bispectrum from analytic dark matter halo profiles and mass functions. *ApJ*, 543:503–513, 2000.

51. C.-P. Ma and J. N. Fry. What does it take to stabilize gravitational clustering? *ApJ*, 538:L107–L111, 2000.

52. S. Mallat. A theory for multiresolution signal decomposition the wavelet representation. *IPAMI*, 11:674–693, 1989.

53. R. Maoli, L. Van Waerbeke, Y. Mellier, P. Schneider, B. Jain, F. Bernardeau, T. Erben, and B. Fort. Cosmic shear analysis in 50 uncorrelated VLT fields. Implications for Omega0, sigma8. *A&A*, 368:766–775, 2001.

54. P. J. Marshall, M. P. Hobson, S. F. Gull, and S. L. Bridle. Maximum-entropy weak lens reconstruction: Improved methods and application to data. *MNRAS*, 335:1037–1048, 2002.

55. R. Massey, C. Heymans, J. Bergé, G. Bernstein, S. Bridle, D. Clowe, H. Dahle et al. The shear testing programme 2: Factors affecting high-precision weak-lensing analyses. *MNRAS*, 376:13–38, 2007.

56. R. Massey and A. Refregier. Polar shapelets. *MNRAS*, 363:197–210, 2005.

57. R. Massey, A. Refregier, D. J. Bacon, R. Ellis, and M. L. Brown. An enlarged cosmic shear survey with the William Herschel Telescope. *MNRAS*, 359:1277–1286, 2005.

58. R. Massey, J. Rhodes, R. Ellis, N. Scoville, A. Leauthaud, A. Finoguenov, P. Capak et al. Dark matter maps reveal cosmic scaffolding. *Nature*, 445:286–290, 2007.

59. R. Massey, J. Rhodes, A. Leauthaud, P. Capak, R. Ellis, A. Koekemoer, A. Réfrégier et al. COSMOS: Three-dimensional weak lensing and the growth of structure. *ApJS*, 172:239–253, 2007.

60. Y. Mellier. Probing the universe with weak lensing. *Annu. Rev. of Astron. Astrophys.*, 37:127–189, 1999.

61. L. Miller, T. D. Kitching, C. Heymans, A. F. Heavens, and L. van Waerbeke. Bayesian galaxy shape measurement for weak lensing surveys—I. methodology and a fast-fitting algorithm. *MNRAS*, 382:315–324, 2007.

62. A. W. Moore, A. J. Connolly, C. Genovese, A. Gray, L. Grone, N. I. Kanidoris, R. C. Nichol et al. Fast algorithms and efficient statistics: N-Point correlation functions. In A. J. Banday, S. Zaroubi, and M. Bartelmann, editors, *Mining the Sky*, Springer, Garching, Germany, pp. 71–ff., 2001.

63. D. Munshi, P. Valageas, L. van Waerbeke, and A. Heavens. Cosmology with weak lensing surveys. *Phys. Rep.*, 462:67–121, 2008.

64. E. Pantin and J.-L. Starck. Deconvolution of astronomical images using the multiscale maximum entropy method. *A&AS*, 118:575–585, 1996.

65. S. Pires, J.-L. Starck, A. Amara, A. Réfrégier, and R. Teyssier. Cosmological model discrimination with weak lensing. *A&A*, 505:969–979, 2009.

66. S. Pires, J.-L. Starck, A. Amara, R. Teyssier, A. Réfrégier, and J. Fadili. FAst STatistics for weak Lensing (FASTLens): Fast method for weak lensing statistics and map making. *MNRAS*, 395:1265–1279, 2009.

67. A. Refregier. Weak gravitational lensing by large-scale structure. *ARA&A*, 41:645–668, 2003.

68. A. Refregier, A. Amara, T. D. Kitching, A. Rassat, R. Scaramella, J. Weller, and for the Euclid Imaging Consortium. Euclid imaging consortium science book. *ArXiv e-prints*, arXiv:1001.0061, 2010.

69. P. Schneider. Detection of (dark) matter concentrations via weak gravitational lensing. *MNRAS*, 283:837–853, 1996.

70. P. Schneider. Weak gravitational lensing. In G. Meylan, P. Jetzer, and P. North, editors, *Gravitational Lensing: Strong, Weak & Micro. Lecture Notes of the 33rd Saas-Fee Advanced Course*. Springer-Verlag, Berlin, 2003.

71. P. Schneider, T. Eifler, and E. Krause. COSEBIs: Extracting the full E-/B-mode information from cosmic shear correlation functions. *A&A*, 520:A116–ff., 2010.

72. P. Schneider and M. Kilbinger. The ring statistics—How to separate E- and B-modes of cosmic shear correlation functions on a finite interval. *A&A*, 462:841–849, 2007.

73. P. Schneider, M. Kilbinger, and M. Lombardi. The three-point correlation function of cosmic shear. II. Relation to the bispectrum of the projected mass density and generalized third-order aperture measures. *A&A*, 431:9–25, 2005.

74. P. Schneider, L. van Waerbeke, B. Jain, and G. Kruse. A new measure for cosmic shear. *MNRAS*, 296:873–892, 1998.

75. T. Schrabback, J. Hartlap, B. Joachimi, M. Kilbinger, P. Simon, K. Benabed, M. Bradač et al. Evidence of the accelerated expansion of the Universe from weak lensing tomography with COSMOS. *A&A*, 516:A63–ff., 2010.

76. R. Scoccimarro, S. Colombi, J. N. Fry, J. A. Frieman, E. Hivon, and A. Melott. Nonlinear evolution of the bispectrum of cosmological perturbations. *ApJ*, 496:586–ff., 1998.

77. R. Scoccimarro and H. M. P. Couchman. A fitting formula for the non-linear evolution of the bispectrum. *MNRAS*, 325:1312–1316, 2001.

78. S. Seitz and P. Schneider. Cluster lens reconstruction using only observed local data: An improved finite-field inversion technique. *A&A*, 305:383–ff., 1996.

79. S. Seitz, P. Schneider, and M. Bartelmann. Entropy-regularized maximum-likelihood cluster mass reconstruction. *A&A*, 337:325–337, 1998.

80. U. Seljak. Weak lensing reconstruction and power spectrum estimation: Minimum variance methods. *ApJ*, 506:64–79, 1998.

81. E. Semboloni, Y. Mellier, L. van Waerbeke, H. Hoekstra, I. Tereno, K. Benabed, S. D. J. Gwyn et al. Cosmic shear analysis with CFHTLS deep data. *A&A*, 452:51–61, 2006.

82. P. Simon, A. N. Taylor, and J. Hartlap. Unfolding the matter distribution using three-dimensional weak gravitational lensing. *MNRAS*, 399:48–68, 2009.

83. J. L. Starck, D. L. Donoho, and E. J. Candès. Astronomical image representation by the curvelet transform. *A&A*, 398:785–800, 2003.

84. J.-L. Starck, S. Pires, and A. Réfrégier. Weak lensing mass reconstruction using wavelets. *A&A*, 451:1139–1150, 2006.

85. I. Szapudi, S. Prunet, D. Pogosyan, A. S. Szalay, and J. R. Bond. Fast cosmic microwave background analyses via correlation functions. *ApJ*, 548:L115–L118, 2001.

86. M. Takada and B. Jain. Three-point correlations in weak lensing surveys: Model predictions and applications. *MNRAS*, 344:857–886, 2003.

87. M. Takada and M. White. Tomography of lensing cross-power spectra. *ApJ*, 601:L1–L4, 2004.

88. A. Taylor. New dimensions in cosmic lensing. In *The Davis Meeting On Cosmic Inflation*, 2003.

89. A. N. Taylor. Imaging the 3-D cosmological mass distribution with weak gravitational lensing. *ArXiv Astrophysics e-prints: astro-ph/0111605*.

90. A. N. Taylor, D. J. Bacon, M. E. Gray, C. Wolf, K. Meisenheimer, S. Dye, A. Borch, M. Kleinhein-rich, Z. Kovacs, and L. Wisotzki. Mapping the 3D dark matter with weak lensing in COMBO-17. *MNRAS*, 353:1176–1196, 2004.

91. R. Teyssier, S. Pires, S. Prunet, D. Aubert, C. Pichon, A. Amara, K. Benabed, S. Colombi, A. Refregier, and J.-L. Starck. Full-sky weak-lensing simulation with 70 billion particles. *A&A*, 497:335–341, 2009.

92. C. Vale and M. White. Simulating weak lensing by large-scale structure. *ApJ*, 592:699–709, 2003.

93. L. Van Waerbeke, Y. Mellier, R. Pelló, U.-L. Pen, H. J. McCracken, and B. Jain. Likelihood analysis of cosmic shear on simulated and VIRMOS-DESCART data. *A&A*, 393:369–379, 2002.

94. J. T. VanderPlas, A. J. Connolly, B. Jain, and M. Jarvis. Three-dimensional reconstruction of the density field: An SVD approach to weak-lensing tomography. *ApJ*, 727:118–ff., 2011.

95. L. L. Zhang and U.-L. Pen. Fast n-point correlation functions and three-point lensing application. *Nature*, 10:569–590, 2005.

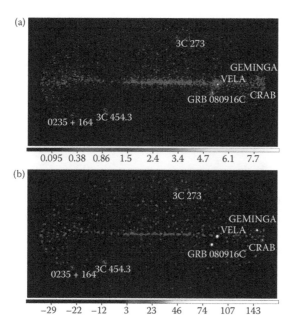

FIGURE 4.6 (a) All-sky counts map of the data for the interval 00:00:00–06:00:00, September 16, 2008. During this interval, GRB 080916C was detected. This map is binned in an Aitoff-Hammer projection with 0.5° × 0.5° pixels at the image center. The map has been smoothed with a 1° width Gaussian kernel. The bar along the bottom edge shows the mapping between counts per pixel and color. (b) The wavelet-transformed map produced by the pgwave program. The sources found by ASP for this interval are indicated in both maps. The bright pulsars, Vela, Geminga, and Crab are easily seen, as is GRB 080916C. 0235 + 164, 3C 454.3, and 3C 273 are blazars. The three remaining sources cannot be sufficiently well localized with these data to provide a firm identification.

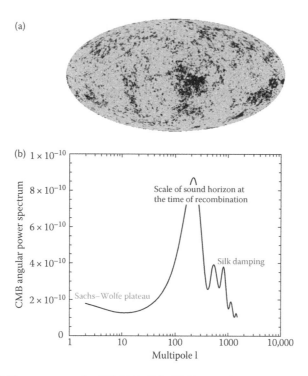

FIGURE 5.1 (a) CMB map seen by WMAP. (b) CMB angular power spectrum, showing the important scales discussed in the text.

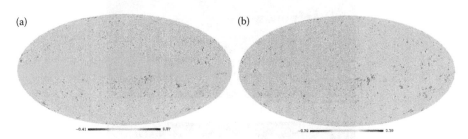

FIGURE 5.3 (a) CMB data map provided by the WMAP team. Areas of significant foreground contamination in the galactic region and at the locations of strong radio point sources have been masked out. (b) Map obtained by applying the MCA-inpainting algorithm on the sphere to the former incomplete WMAP CMB data map.

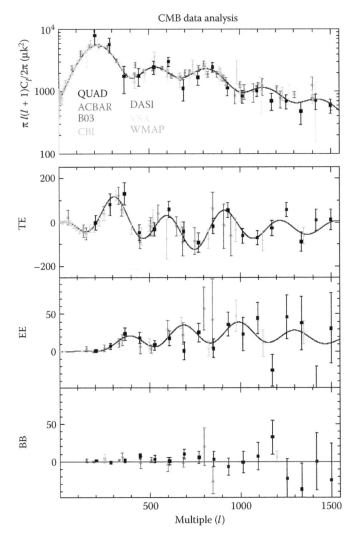

FIGURE 5.4 CMB temperature and polarization power spectra measured by different CMB experiments, each covering different range of scales [4]. (From P. Ade et al. *APJ*, 674: 22–28, 2008. With permission.)

FIGURE 6.2 "The Astronomer," by Johannes Vermeer, c. 1668. In many ways, the epoch of the armchair astronomer is returning to primacy.

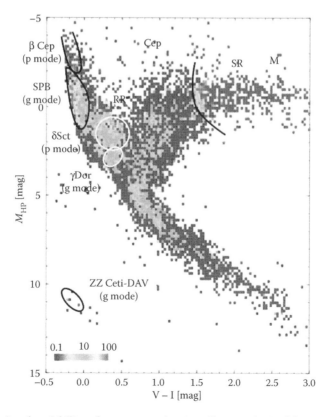

FIGURE 6.4 Fractional variability of stars across the H–R diagram derived from Hipparcos data. Red indicates significant variability and blue low-amplitude variability (10% peak-to-peak). Identification of colors coupled with distances provide a rather clean path to classification. (From L. Eyer and N. Mowlavi. *J. Phys. Conf. Ser.*, 118(1):012010, 2008. With permission.)

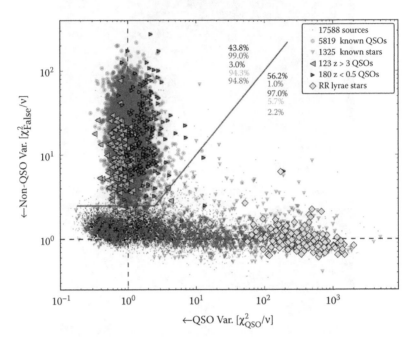

FIGURE 6.6 Variability selection of quasars. Using a Bayesian framework to connect light curves of point sources to damped random walk behavior, statistics that account for uncertainty and covariance can be developed to find QSO-like behavior. This selection (green line) is highly efficient at finding known QSOs (∼99%) and impure at the 3% level. (From N. R. Butler and J. S. Bloom. *AJ*, 141:93, 2011. With permission.)

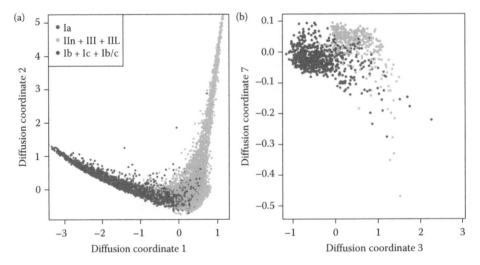

FIGURE 6.8 Light-curve distance measures can be used in conjunction with spectral methods, such as diffusion map, to compute informative features. In this example, a spline-based distance between supernova light curves, designed to capture both shape and color differences, was used. In the first two diffusion map coordinates (a), Type Ia and II SNe are distinguished, whereas higher features (b) reveal some separation between Ia and Ib/c supernovae in the subset of spectroscopic supernovae. (From J. W. Richards et al., *ArXiv e-prints*, 1103.6034, 2011. With permission.)

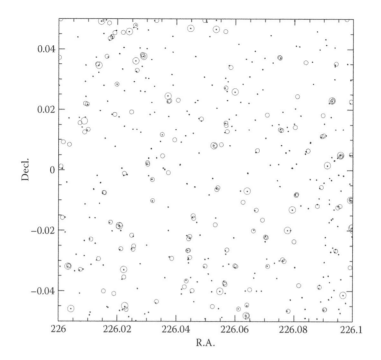

FIGURE 7.1 Select SDSS, GALEX, and 2MASS sources are shown on a random patch of the sky in black, blue, and red, respectively. The size of the radii of the circles correspond to the astrometric uncertainties of $\sigma = 0.1$, 0.5, and 0.8 arc seconds in the three surveys. Here we plot their 5σ contours for better visibility. Cross-identification is about finding the detections in multiple datasets that belong to the same astronomical object.

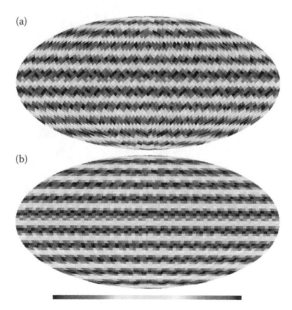

FIGURE 8.7 Schematic representation of two types of pixelization on sphere: HEALPix (a) and GLESP (b). Various colors of pixels are used to show their shape.

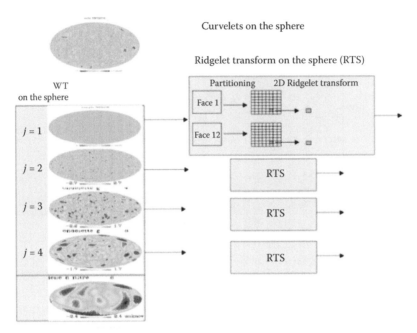

Curvelets on the sphere

Ridgelet transform on the sphere (RTS)

FIGURE 10.3 Principle of curvelets transform on the sphere.

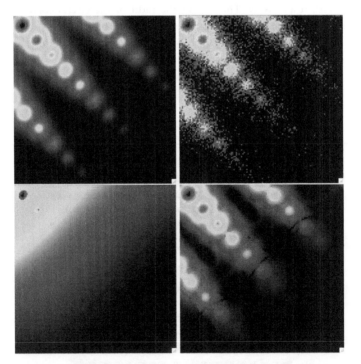

FIGURE 10.5 Comparison of MS-VSTS with Anscombe + wavelet shrinkage on a single face of the first scale of the HEALPix pixellization (angular extent: $\pi/3\,\mathrm{sr}$). *Top left*: Sources of varying intensity. *Top right*: Sources of varying intensity with Poisson noise. *Bottom left*: Poisson sources of varying intensity reconstructed with Anscombe + wavelet shrinkage. *Bottom right*: Poisson sources of varying intensity reconstructed with MS-VSTS.

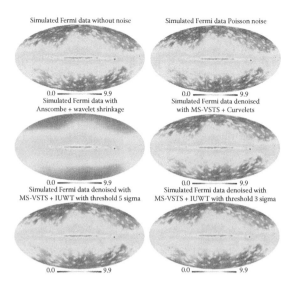

FIGURE 10.6 *Top left*: Fermi simulated map without noise. *Top right*: Fermi simulated map with Poisson noise. *Middle left*: Fermi simulated map denoised with Anscombe VST + wavelet shrinkage. *Middle right*: Fermi simulated map denoised with MS-VSTS + curvelets (Algorithm 10.3). *Bottom left*: Fermi simulated map denoised with MS-VSTS + IUWT (Algorithm 10.1) with threshold $5\sigma_j$. *Bottom right*: Fermi simulated map denoised with MS-VSTS + IUWT (Algorithm 10.1) with threshold $3\sigma_j$. Pictures are in logarithmic scale.

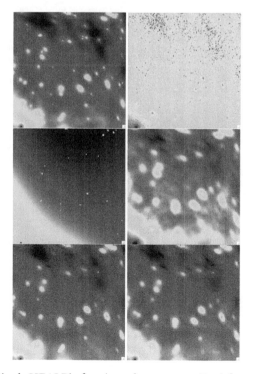

FIGURE 10.7 View of a single HEALPix face (angular extent: $\pi/3$ sr) from the results of Figure 10.6. *Top left*: Fermi simulated map without noise. *Top right*: Fermi simulated map with Poisson noise. *Middle left*: Fermi simulated map denoised with Anscombe VST + wavelet shrinkage. *Middle right*: Fermi simulated map denoised with MS-VSTS + curvelets (Algorithm 10.3). *Bottom left*: Fermi simulated map denoised with MS-VSTS + IUWT (Algorithm 10.1) with threshold $5\sigma_j$. *Bottom right*: Fermi simulated map denoised with MS-VSTS + IUWT (Algorithm 10.1) with threshold $3\sigma_j$. Pictures are in logarithmic scale.

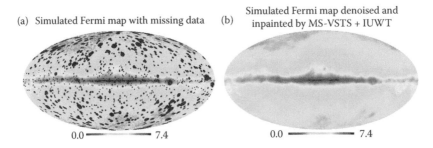

(a) Simulated Fermi map with missing data   (b)   Simulated Fermi map denoised and inpainted by MS-VSTS + IUWT

0.0 ━━━ 7.4   0.0 ━━━ 7.4

FIGURE 10.8   MS-VSTS-Inpainting. (a) Fermi simulated map with Poisson noise and the most luminous sources masked. (b) Fermi simulated map denoised and inpainted with wavelets (Algorithm 10.4). Pictures are in logarithmic scale.

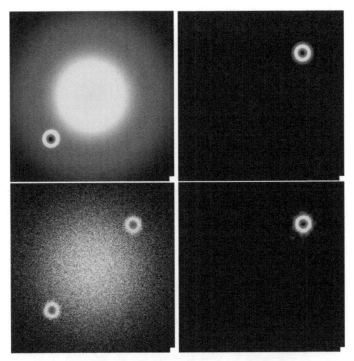

FIGURE 10.9   Theoretical testing for MS-VSTS + IUWT denoising + background removal algorithm (Algorithm 10.5). View on a single HEALPix face. *Top left*: Simulated background : sum of two Gaussians of standard deviation equal to 0.1 and 0.01, respectively. *Top right*: Simulated source: Gaussian of standard deviation equal to 0.01. *Bottom left*: Simulated poisson data. *Bottom right*: Image denoised with MS-VSTS + IUWT and background removal.

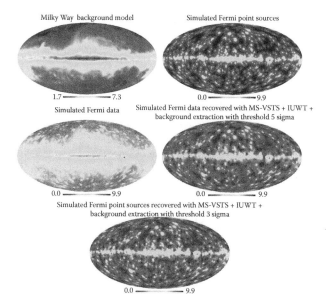

Milky Way background model

1.7 — 7.3

Simulated Fermi point sources

0.0 — 9.9

Simulated Fermi data

0.0 — 9.9

Simulated Fermi data recovered with MS-VSTS + IUWT + background extraction with threshold 5 sigma

0.0 — 9.9

Simulated Fermi point sources recovered with MS-VSTS + IUWT + background extraction with threshold 3 sigma

0.0 — 9.9

FIGURE 10.10 *Top left*: Simulated background model. *Top right*: Simulated γ-ray sources. *Middle left*: Simulated Fermi data with Poisson noise. *Middle right*: Reconstructed γ-ray sources with MS-VSTS + IUWT + background removal (Algorithm 10.5) with threshold $5\sigma_j$. *Bottom*: Reconstructed γ-ray sources with MS-VSTS + IUWT + background removal (Algorithm 10.5) with threshold $3\sigma_j$. Pictures are in logarithmic scale.

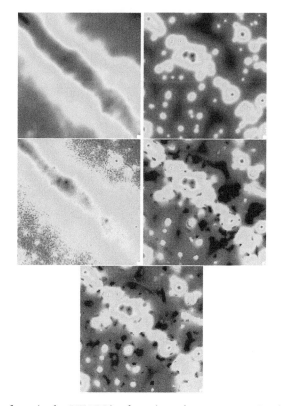

FIGURE 10.11 View of a single HEALPix face (angular extent: $\pi/3\,\mathrm{sr}$) from the results of Figure 10.10. *Top left*: Simulated background model. *Top right*: Simulated γ-ray sources. *Middle left*: Simulated Fermi data with Poisson noise. *Middle right*: Reconstructed γ-ray sources with MS-VSTS + IUWT + background removal (Algorithm 10.5) with threshold $5\sigma_j$. *Bottom*: Reconstructed γ-ray sources with MS-VSTS + IUWT + background removal (Algorithm 10.5) with threshold $3\sigma_j$. Pictures are in logarithmic scale.

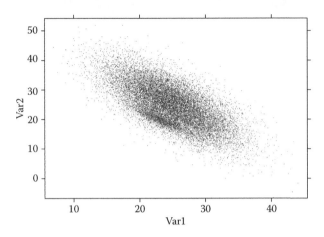

FIGURE 12.4 A toy data set with two correlated variables and two event types: RED and BLUE.

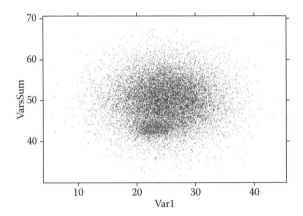

FIGURE 12.5 The toy data set with the correlation between Var1 and Var2 undone by making a new variable VarSum = Var1 + Var2.

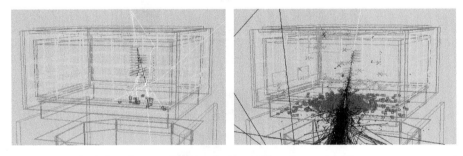

FIGURE 12.8 On the left is the simulation of a low-energy (∼100 MeV) gamma-ray event in LAT. The black lines show the paths of charged particles and the white lines show those of neutral particles (photons). Blue lines indicate the tracks found by the reconstruction analysis and the yellow line shows the reconstructed gamma-ray direction. On the right is a high-energy event (∼100 GeV). Drawing of the photon trajectories is suppressed, otherwise they would obscure the other details. In this wireframe event display, the outlines of the various pieces of the instrument are shown along with markers as to where hits were registered by the detectors.

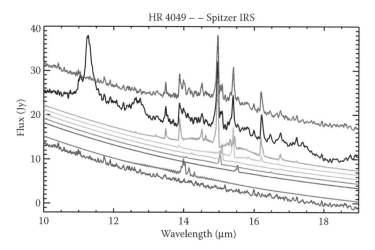

FIGURE 13.2 Shown here (in black) is part of the infrared spectrum of the peculiar star HR 4049 (an old and evolved star in a binary system), obtained with the Spitzer Space Telescope. In this wavelength range, most emission peaks are due to carbon dioxide (CO_2). Remarkably, spectral features can be distinguished of all possible varieties of CO_2 containing different oxygen and carbon isotopes [see also 6]: $^{12}CO_2$ (blue); $^{13}CO_2$ (green); $^{18}O^{12}C^{16}O$ (orange); $^{17}O^{12}C^{16}O$ (magenta); and even $^{18}O^{13}C^{16}O$ and $^{17}O^{13}C^{16}O$ (dark blue). The spectrum furthermore contains hydrogen cyanide (HCN, purple) and water vapor (H_2O, maroon, bottom). The full model containing all gases is shown in red above the observations. The broad and strong features at 11.2 and 12.7 μm are PAH bands and were not included in this analysis.

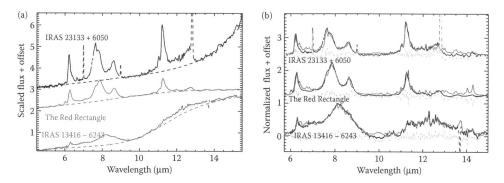

FIGURE 13.3 (a) Three examples of some of the PAH bands in astronomical observations. All three spectra were recorded with the Short-Wavelength Spectrometer [SWS, 7] on board the Infrared Space Observatory [ISO, 11]. The spectrum of the HII region infrared astronomical satellite (IRAS) 23133 + 6050 (top) is a class A PAH source; the spectrum of the evolved object known as the Red Rectangle (middle) is a class B PAH source and the post-asymptotic gaint branch (AGB) star IRAS 13416 − 6243 (bottom) represents a class C PAH profile. The dashed lines indicate the dust continuum that was subtracted from the observations to obtain the pure PAH emission. (b) The continuum-subtracted observations (black) and the best fit PAH models using the NASA Ames Astrochemistry database (red). Residuals are shown in green. (From Cami, J., *EAS Publications Series*, 46:117–122, 2011. With permission.)

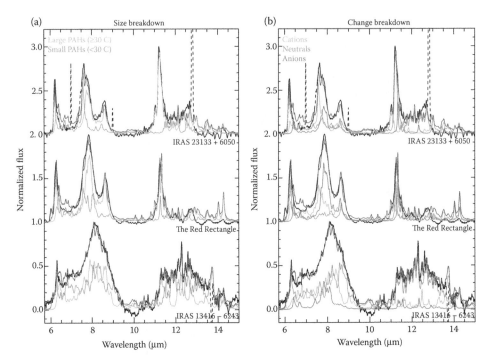

FIGURE 13.5 These figures show the breakdown of the best fit into various constituent classes of PAHs. (a) In spite of the strong bias toward small PAHs in the NASA Ames database, the best fits show that the 11.2 μm feature is solely due to large molecules. (b) Similarly, the 11.2 μm band is primarily due to neutral PAHs, whereas PAH cations and anions contribute significantly to the 7.7 μm PAH complex. (From Cami, J., *EAS Publications Series*, 46:117–122, 2011. With permission.)

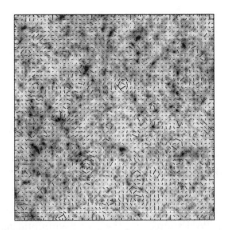

FIGURE 14.5 Simulated convergence map by [92] covering a 2° × 2° field with 1024 × 1024 pixels. The shear map is superimposed to the convergence map. The size and the direction of the line segments represent the amplitude and the direction of the deformation locally.

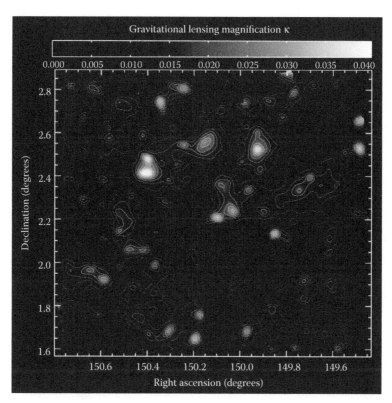

FIGURE 14.8 Map of the dark matter distribution in the 2-square degrees COSMOS field by [58]: the linear blue scale shows the convergence field κ, which is proportional to the projected mass along the line of sight. Contours begin at 0.004 and are spaced by 0.005 in κ. (From R. Massey et al. *Nature*, 445:286–290, 2007. With permission.)

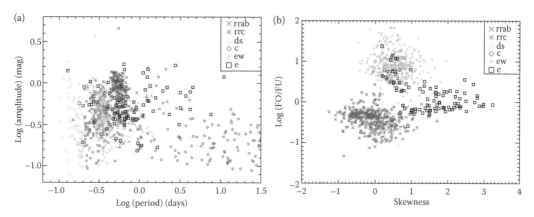

FIGURE 18.1 Projections of the feature space for classification of periodic variable stars in the ROTSE sample. Only six classes are shown for clarity. The symbols in the legend correspond to variability types from top to bottom: RR Lyrae AB, RR Lyr C, Delta Scuti, Cepheids, W UMa eclipsing binaries, and detached eclipsing binaries. Panel (a) shows the period-amplitude locus of the training data set. The ratio of the first overtone to fundamental mode (FO/FU) and the skewness of the magnitude distribution are plotted in (b). Table 18.1 contains performance comparisons for selected machine learning classifiers applied to these data.

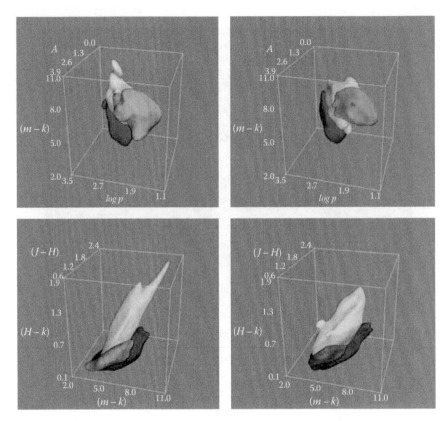

FIGURE 18.4 Feature space for classification of red variables in the NSVS. Training data (left panels) are compared to the results of classification on the entire catalog (right panels). Two 3D projections are shown: the log period-amplitude-color relation (top), and the space of three independent colors (bottom). Density contours are plotted at 10% of the peak density within each of the three classes: regular Miras M (red), carbon Miras C (yellow), and other red variables SR+L (green). The corresponding confusion matrix is shown in Figure 18.3. Table 18.2 summarizes the details of this classification experiment.

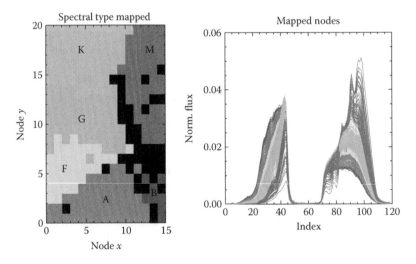

FIGURE 18.12 Mapping of the spectral type onto the SOM from Figure 18.9. The map is mostly sensitive to the effective temperature and clearly capable of distinguishing major spectral types and subtypes.

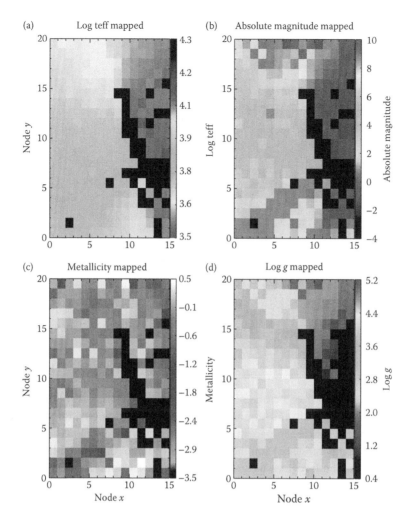

FIGURE 18.13 Mapping of the stellar parameters onto the SOM from Figure 18.9. Maps are color coded with: logarithm of the effective temperature $\log T_{\text{eff}}$ (a), absolute magnitude (b), metallicity (c), and surface gravity g (d). Augmented maps of this kind can be used to infer astrophysical properties of newly discovered objects from other inputs—in this case of low-resolution spectra.

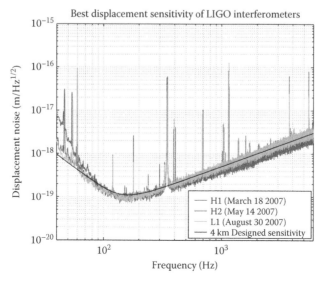

FIGURE 19.4 Best (one-sided) noise PSD measured during LIGO's S5 data taking.

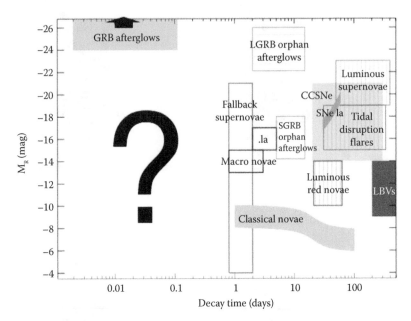

FIGURE 28.1 Discovery space of astronomical variability. Numerous classes of flux variability have been identified through analysis of astronomical time-series observations. But there is still a vast unexplored region of this parameter space that the LSST survey aims to explore. The axes on this plot are time scale of variability (abscissa) and the negative logarithm of the flux (ordinate), measured in magnitudes, where magnitudes decrease as an object gets brighter. (From LSST Science Collaborations and LSST Project, *LSST Science Book*, version 2.0, arXiv:0912.0201, http://www.lsst.org/lsst/scibook, 2009 and adapted from Rau, A. et al. *Exploring the Optical Transient Sky with the Palomar Transient Factory*. Astronomical Society of the Pacific, 886, 1334–1351, 2009.)

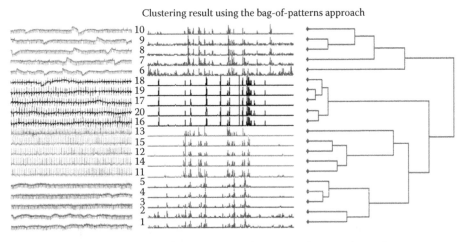

FIGURE 28.9 Clustering result on the same ECG datasets, using our bag-of-patterns approach. All data are clustered correctly. This figure is best viewed in color.

CHAPTER 15

Photometric Redshifts

50 Years After

Tamás Budavári

The Johns Hopkins University

CONTENTS

15.1 Introduction   323
15.2 Pragmatism of the Pioneers   324
    15.2.1 Empirical Approaches   324
    15.2.2 Template Fitting   325
15.3 A Unified View from First Principles   326
    15.3.1 Physical Properties versus Observables 326
    15.3.2 Mapping Measurements with Errors   327
    15.3.3 Traditional Methods as Corner Cases 328
    15.3.4 Which Training Set?   329
15.4 Towards Advanced Methods   330
    15.4.1 Photometric Errors   330
    15.4.2 Estimating the Relation   331
    15.4.3 Priors for the Models   331
    15.4.4 Optimal Filter Design   332
15.5 Concluding Remarks   332
References   333

15.1 INTRODUCTION

Almost half a century has passed since Baum [1] first applied a novel method to a handful of galaxies. Using the mean spectral energy distribution (SED) of six bright ellipticals in the Virgo cluster, he could accurately estimate the redshifts of other clusters in a comparison that we today call SED fitting or, more generally, photometric redshifts.

Owing to the expansion of the Universe, galaxies farther away appear to be redder. Their observed colors are a combination of this redshift and their intrinsic properties. Thanks to the latest detector technology, today we can undertake deep, multicolor surveys to probe statistically meaningful volumes. The yield of photometry in terms of the number of sources is over two orders of magnitude higher than what is achievable by (the more accurate) spectroscopic follow-ups. To exploit the information in the photometric data themselves, several new methods have been developed over the years. One particular successful example is the estimation of photometric redshifts. Baum's motivation for using photometric measurements instead of spectroscopy was the same back then as ours is now: to push the analyses to uncharted territories. His original idea has grown into a research area of its own, which is more important now than ever before. In this chapter, we look at some of the recent advancements of the field. In Section 15.2 we briefly highlight some of the original ideas and the current state of the art in estimating the photometric redshifts. Section 15.3 introduces a Bayesian framework for discussing the traditional methods within a unified context; it explicitly enumerates and identifies their (missing) ingredients. Section 15.4 aims to plant seeds for new ideas for future directions, and Section 15.5 offers some concluding remarks.

15.2 PRAGMATISM OF THE PIONEERS

The usual way of determining the redshift of an extragalactic object is to collect its photons into narrow wavelength bins in a spectrograph, and compare its atomic lines to the wavelengths measured in the laboratory. Assuming one sees the spectral lines and can accurately identify them, the method's precision is set by the wavelength resolution. Photometric passbands are typically a few thousand Ångstrom wide, which makes the detection of individual lines practically impossible. Thus, the photometric redshift estimators mostly leverage the broader spectral features, such as the overall shape, the Ly-α or the Balmer breaks of red galaxies. Figure 15.1 illustrates two sets of photometric filters: the *UBVI* passbands on top (deployed on the Hubble Space Telescope's WFPC2 camera) and the *ugriz* below (used by the Sloan Digital Sky Survey). These passbands cover similar wavelength ranges, and hence capture similar information but are clearly not identical. Typical blue (top) and red (bottom) galaxies are plotted at rest and redshifted to $z = 0.2$ to show the effect of the shift on the integrated fluxes. The forward process is simple: multiply the spectrum with the filter curve and integrate. We would like to solve the inverse problem.

In a review article [2], Koo offers the following working definition: "photometric redshifts are those derived from *only* images or photometry with spectral resolution $\lambda/\Delta\lambda \lesssim 20$ (p. 4). This choice of 20 is intended to exclude redshifts derived from slit and slitless spectra, narrow band images, ramped-filter imager, Fabry-Perot images, Fourier transform spectrometers, etc." The definition leaves room for a wide variety of approaches that are actively being explored by members of the community. The estimators traditionally fall into one of two broad categories: empirical and template fitting.

15.2.1 Empirical Approaches

One class of techniques relies on training sets of objects with known photometry and spectroscopic redshifts. The usual assumption is that galaxies with the same colors are at identical redshifts. The redshift of a new object is derived based on its vicinity in observable space to the calibrators of the training set.

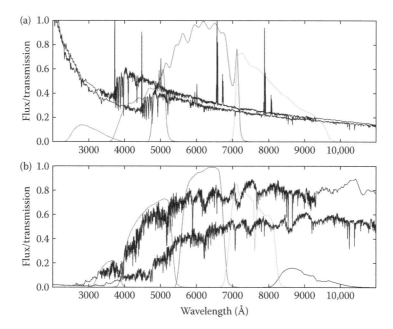

FIGURE 15.1 The top panel illustrates in thick, solid lines the spectral energy distribution of a late-type, blue galaxy locally and at a higher redshift. The photometric filters, shown in thin lines, sample different part of the restframe spectrum and provide leverage to measure the change. The bottom panel shows a typical red elliptical galaxy along with a different set of filters on the same wavelength range.

The earliest studies were colorful proofs of concepts, since then most machine learning techniques have been applied to the problem. Connolly et al. [3] illustrated iso-z layers in the multidimensional color space and successfully used polynomial fitting functions up to 3rd order to calibrate the redshifts. Piecewise fitting was introduced by Brunner et al. [4] and showed promising accuracies with locally linear approximations. Csabai et al. [5] produced a photometric redshift catalog for the Early Data Release of the sloan digital sky survey (SDSS) using linear fits to k nearest neighbors, and compared them to other methods. Several other machine learning techniques were adopted, such as neural networks [6], support vector machines [7], and more recently Gaussian processes [8], diffusion maps [9,10], and random forests [11].

The advantage of these methods lies in their empirical nature: all the effects of dust and spectral evolution that are present in the training set are implicitly accounted for. The drawback is that it can only reliably predict redshifts for objects that are similar to those in the training set. Also the applicability is limited to query sets that have identical photometric measurements to the calibrators. Many of them rely on ad hoc distance metrics in magnitude space, and they almost never utilize the photometric uncertainties. Further challenges would often present themselves when new input parameters are introduced.

15.2.2 Template Fitting
The other class of methods use high-resolution template spectra along with a detailed model of the telescopes photometric system. By minimizing the difference between the observed

and synthetic colors, one can find the most likely type and redshift of a galaxy. The method is simple to implement, and does not rely on additional observations or any spectroscopic measurements. The implementations differ mainly in the templates and the way object type is parameterized. Theoretical models, for example, [12], have physical parameters to adjust, and empirical spectra, for example, [13], are mostly just enumerated and sometimes interpolated. Generalizations have been proposed to use linear combinations of a few components [14–16]. Orthogonal eigenspectra can span a variety of types and still only use a few parameters that are easy to solve.

The main limitation of these methods is that the underlying SEDs of galaxies within the sample must be well known. Comparisons of the colors of galaxies with spectral synthesis models have shown that the modeling of the ultra-violet spectral interval for galaxies is uncertain. Consequently, photometric-redshift estimates are most accurate when we apply empirical SEDs derived from observations of local galaxies. These empirical templates are, however, constructed from a small number of nearby galaxies, and there is no guarantee that they represent the full distribution of galaxy types. Early attempts to improve on the templates [15,17–19] adjusted the spectra to better match the photometric measurements of the calibrators at known redshifts. These repaired spectra can, in turn, be used for redshift estimation and provide better quality.

15.3 A UNIFIED VIEW FROM FIRST PRINCIPLES

Having glanced at the plethora of photometric redshift methods, one might be inclined to conclude that the possibilities for invention have been exhausted. The truth is that the techniques are essentially built on the same assumptions and share more features than what meets the eye. The approach discussed in this section sheds light on these relations from a new angle.

Recently, the focus of the development has shifted from providing redshift estimates to deriving the full probability density functions. Taking it one step further, the mapping can also be done consistently for all parameters of interest and not only for redshift. It is especially important considering that these properties are often known to be correlated, for example, redshift and galaxy type. Simply put, a distant blue galaxy might have colors similar to that of a nearby elliptical. Now we turn to discuss this more general photometric inversion. Formulating the problem as Bayesian inference, the physical properties are constrained from first principles. The resulting framework not only explains the connection between the traditional methods but also points out the directions of possible further improvement.

15.3.1 Physical Properties versus Observables

Broadband photometry cannot possibly contain all the information about the objects. In general, it is too naive to assume that there is only a single possible redshift for a given set of measurements. The regression model, where the redshift is a function of the magnitudes, is simply too restrictive. Instead, one should consider their general relation. If x represents the photometric measurements, the conditional probability density $p(z|x)$ describes the possible redshift values (z) that one might see for objects with given x. Depending on what colors are measured and how accurately, this distribution might be wider or narrower. When it is sufficiently narrow, function fitting $z = \hat{z}(x)$ may be a good approximation of the relation; formally $p(z|x) = \delta(z - \hat{z}(x))$.

For a training set of objects with both redshifts z and photometric measurements x, we can not only derive fitting functions, but also map out their full relation. For example, the densities of the ratio

$$p(z|x) = \frac{p(z, x)}{p(x)} \tag{15.1}$$

can be measured by kernel density estimation or Voronoi tessellation. Although conceptually straightforward, conditional density estimation is a hard problem in general. Sharp discontinuities are typically difficult to manage. Practical implementations of more sophisticated direct methods are needed for robust results. With the appropriate statistical tools in place, this automatically works for more than just redshift. In other words, z can really represent the entire set of physical properties.

In practice, the training sets are limited and the observed relation is not necessarily identical to the underlying distribution. Let $P(T|x, z)$ represent the selection function of the calibrators. This is the probability that an object with x and z will appear in the training set T. Bayes' rule provides insight into the dependence of the observed and true relations,

$$p(z|x, T) = \frac{p(z|x)\, P(T|x, z)}{P(T|x)} \tag{15.2}$$

We see that when the training set is selected only on the photometry and no other parameters that could introduce an implicit dependence on the redshift, $P(T|x, z) = P(T|x)$, the underlying relation is identical to that seen on the training set. The implications are surprisingly strong. To measure the unbiased relation, one cannot just rely on the colors, if the training set has a magnitude threshold as the selection function depends on the neglected quantity. Similarly, one cannot just use the magnitudes, if the calibrators are limited in surface brightness or include morphological cuts.

15.3.2 Mapping Measurements with Errors

The other component of the our framework is even more straightforward, and yet it holds the key to generalization and various improvements as well as to establishing a unified view of the traditional methods. It is the proper treatment of photometric uncertainties.

Empirical estimators can traditionally only be applied to query sets with the same type of observations as the calibrators, that is, when the filters are truly identical. For instance, if the calibrator galaxies have SDSS *ugriz* measurements then the method cannot be applied to a new galaxy with any other observables even if they cover a similar wavelength range, like in the two panels of Figure 15.1. Converting from a photometric system y to another x requires a spectral model. Depending on the passbands, the model could simply consist of fitted formulas with interpolated color terms [20], (which is what astronomers often resort to), more elaborate SED modeling or even galaxy simulations. In general, any such model M has a set of parameters, say θ, for which it can predict the photometric measurements in two separate systems, that is, x and y. The former is for the training set, $p(x|\theta, M)$, and the latter for the query set via a similar $p(y|\theta, M)$. A given set of measurements y_q of a query point can be mapped onto the training set using the model

$$p(x|y_q, M) = \int d\theta\, p(x|\theta, M)\, p(\theta|y_q, M) \qquad (15.3)$$

where $p(\theta|y_q, M) \propto p(\theta|M)p(y_q|\theta, M)$ and $p(\theta|M)$ is the parameter's prior. The likelihood function for a given observation is given by the photometric uncertainty. For example, the usual Gaussian assumption with C_q covariances yields $p(y_q|\theta, M) = N(y_q|\bar{y}(\theta), C_q)$, where $\bar{y}(\theta)$ is the simulated photometry.

The final results is the conditional probability density of the redshift (and other physical parameters) z given the y_q observations. Between the model and the training set, we have all the ingredients,

$$p(z|y_q, M) = \int dx\, p(z|x)\, p(x|y_q, M) \qquad (15.4)$$

In practice, the integral can be evaluated, for example, on a grid of z_r locations as a sum over (the nearby) training points [21],

$$p(z_r|y_q, M) \approx \sum_{t \in T} p(z_r|x_t)p(x_t|y_q, M)\Delta x_t \qquad (15.5)$$

The $(t|q)$ matrix is calculated from the photometric uncertainties, the relation is measured on the training set, and the volume element is inversely proportional to the density of the training set. Several optimizations are possible starting from an adaptive redshift grid to use multidimensional indices to find the neighbors with nonzero contributions. The pdf can be characterized in different ways. If unimodal, a Gaussian might be a reasonable approximation. Its mean

$$\bar{z}_q = \int dz\, z\, p(z|y_q, M) \approx \sum_r z_r P_{rq} \qquad (15.6)$$

could also serve as an estimate. Alternatively, one could use another family of curves or some mixture models. In 1D the pdf could even be conceivably tabulated, if necessary.

15.3.3 Traditional Methods as Corner Cases

The aforementioned approach is *not* a method. It is a framework that prescribes the ingredients for any photometric inversion technique. Naturally, the traditional empirical and template fitting are also included. They are special cases.

Template fitting has no empirical training set; however, we can generate one from the spectra. The set (or grid) of parameters $\{\theta_t\}$ used in the fitting is technically a calibrator set: $\{x_t, z_t\} = \{\bar{x}(\theta_t), \bar{z}(\theta_t)\}$, where $\bar{z}(\theta_t)$ is often simply a subset of θ_t, for example, the redshift is just one of the parameters in the models of SEDs. If we assume that the artificial training set has no errors associated with the reference points, and that $\bar{x}(\theta)$ is invertible, we can formulate this with Dirac's δ symbol as

$$p(x|\theta, M) = \delta(x - \bar{x}(\theta)) \tag{15.7}$$

$$p(z|x_t, M) = \delta(z - z_t) \tag{15.8}$$

An analytical calculation yields the intuitive final result where the grid points are weighted by their likelihood multiplied by the corresponding prior

$$p(z|y_q, M) \propto \sum_{t \in T} \delta(z - z_t)\, p(\theta_t|M) N(y_q|\bar{y}(\theta_t), C_q) \tag{15.9}$$

The usual SED fitting applications apply the same formula to find the best redshift on the grid or perform interpolations. Usually, brute force maximum likelihood estimation is used on the Gaussian likelihood function $N(y_q|\bar{y}(\theta_t), C_q)$ with the photometric covariances. With a proper $p(\theta_t|M)$ prior on the redshift and type, this same formula has been referred to as "Bayesian Photo-Z" [22,23].

Empirical methods typically stop at establishing the regression, which is formally a fitting function $z_{\text{phot}} = \hat{z}(x)$. Even in situations when that is a good approximation, they often neglect to fold in the photometric errors of the query points. In a minimalist case with identical x and y photometric systems and a flat prior, the mapping is as simple as the convolution of the Gaussian photometric errors [21]

$$p(x_t|y_q, M) = \int d\theta\, N(x_t|\theta, C_t) N(\theta|y_q, C_q) \tag{15.10}$$

In general, however, we could even connect different types of photometric systems, hence extending the applicability of the training set to other datasets. If the regression model in Equation 15.8 is a reasonable approximation, the redshift estimate is a weighted average over the training set

$$\bar{z}_q \approx \sum_{t \in T} z_t\, p(x_t|y_q, M)\, \Delta x_t = \sum_{t \in T} z_t P_{tq} \tag{15.11}$$

A number of successful empirical methods use this weighting scheme, and differ only in the way the weights are derived. The randomized trees of a random forest [11] are piecewise constant fits, whose average is essentially a weighted average of the calibrator redshifts. Other examples include local linear fits [4,5], kernel regression [24] or the so–called "weighting method" [25]. We emphasize the importance of the Δx_t term, without which a method becomes (more) sensitive to the sampling of the color space, and would formally require a representative training set.

15.3.4 Which Training Set?

Using better spectral models allows us to utilize a variety of training sets. Their coverage of the color space, however, would be different and may not probe the regime of a given query point. The Bayesian approach also has an answer for that. If a training set is selected to have sources $r < 17.7$ magnitudes, for example, the SDSS spectroscopic sample [26], clearly there is no guarantee that objects much fainter than that threshold will obey the measured color–redshift relation. The *window function* of the training set $P(W|x)$ can be used to pinpoint when these problems occur. In our previous example, the threshold is $r = 17.7$, so $P(W|x) = 1$ for the brighter sources and 0 otherwise. Using Equation 15.3 we can test a query point against the training set by calculating

$$P(W|y_q, M) = \int dx\, P(W|x)p(x|y_q, M) \tag{15.12}$$

When this value is around 1, we can rely on the constraints from the training set but if it is close to 0, the results cannot be trusted. A large probability of coverage, however, does not mean high accuracy. The constraining power depends on the passbands and photometric uncertainties. For example, a deep training set with only two ultraviolet bands (e.g., GALEX [27]) will not provide precision redshifts but may reliably constrain dropouts to be beyond a certain distance. These are called the dropout techniques, for example, [28]. When no empirical dataset can provide good coverage, artificial training sets generated from reference templates or other spectral models can also be used interchangeably. While formally they do not necessarily have a window function, their limitation is in the goodness of the models that may be tested on real observations beyond which the model's validity is unknown.

15.4 TOWARDS ADVANCED METHODS

Bayesian inference provides a simple, coherent view of the traditional methods. Thinking in these terms, one can discover the hidden relations of the techniques and gain insight into the similarities (or differences) of their performance. By explicitly enumerating the ingredients of a method, we can not only pinpoint missing elements of the various approaches but can also improve upon them. We already discussed some of the most straightforward ways but there are several other general areas in machine learning and statistics with rooms for improvements.

15.4.1 Photometric Errors

One of the most fundamental issues is the photometric uncertainty. The more accurate the photometry, the better our estimation. This only works if our understanding of the errors is correct. Today astronomy catalogs typically list an error value for each flux or magnitude measurement, which is not capturing all the details as the measurements in separate bands are not necessarily independent. Multicolor measurements with a common morphological model or aperture provide the best colors but their magnitudes inevitably become correlated. Any mistake in the aperture would directly affect all the fluxes. On the other hand, the colors could remain unchanged. For example, if a galaxy were cut in to half, its fluxes would be a factor of 2 lower but the colors would be the same.

Such scenarios can still be described by multidimensional Gaussians but we need the full covariance matrix. The analysis of repeat observations can help to empirically solve for the off-diagonal elements of the covariance matrix. For the SDSS, a fitting formula was developed to fill in the missing elements based on the quoted magnitudes and errors [29]. Although the upcoming surveys will all have time-domain data that will accommodate such studies, it should be possible to extract that information directly as part of the photometric pipeline.

15.4.2 Estimating the Relation

Using training sets to estimate the relation between photometric parameters and physical properties is a difficult problem, especially in higher dimensions. The generic relation is formally a conditional density $p(z|x)$ that is to be determined in the presence of sharp boundaries and sparse marginal densities. The introduction of new parameters becomes cumbersome if fine-tuning is required each time a promising candidate is considered. Experimenting with different parameters, for example, surface brightness or inclination, in addition to magnitudes could become an automated procedure with generic tools for estimating conditional densities.

15.4.3 Priors for the Models

Not all objects are equally frequent in the Universe. The SED models account for these differences using a prior on their parameters, that is, the aforementioned $p(\theta|M)$ function. In a first approximation, the prior is often negligible when the measurements provide strong constraints. On a closer look, we see that its gradients would enter the formulas.

In principle, we can learn about the model priors by studying the ensemble of objects. A good prior of a reasonable model should reproduce the density of sources in the observable space. For example, if we measure the density on the query set Q, that is, $p(y|Q)$, it should be the same as the model's prediction $p(y|M)$, which can be calculated from the prior

$$p(y|Q) = p(y|M) = \int d\theta \, p(y|\theta, M)p(\theta|M) \tag{15.13}$$

If the likelihood function is known, the prior is formally obtained by deconvolution. In the presense of degeneracies, scientific judgment, a prior is required for the inversion. In lower dimensions the problem is tractable, for example, using the trick by Richardson [30] and Lucy [31], but becomes problematic in many dimensions. Many SED models used for photometric redshifts have just a few parameters, and can be computationally approachable from this aspect. If one is only interested in the local gradient of the prior, a parametric solution might also work.

The template repair methods developed for the Hubble Space Telescope and Sloan Digital Sky Survey observations [15,17,18] and more recently as part of other photo-z toolkits, for example, [19], can be considered a naive attempt to fine tune the SED prior for the observations. The goal of these projects was to incrementally alter the high-resolution template spectra to better match observations with known redshift. The more accurate photometrically calibrated templates can be used for analyzing the larger photometric datasets. Photometric redshift estimation is one such application and the improvement in precision is a natural side-effect of the better modeling.

15.4.4 Optimal Filter Design

The accuracy of the redshift (and other parameter) estimates is primarily determined by the signal-to-noise ratio (SNR) of the observations and the shape of the photometric passbands.* The former can be improved by increasing the exposure time but the latter is usually given. Photometric surveys are designed with a number of science goals in mind. When the photometric inversion is an integral part of the studies, its accuracy is tested with different filter sets to pick the "best" one. Single-pass broadband filters are usually the choice for keeping the efficiency high. The narrower the wavelength range, the longer the integration time to reach the same SNR. In addition to the wavelength coverage, the shape of the filter curve is also an important factor. For accurate estimates of the physical properties, the photometric observables need to vary significantly with those quantities. In other words, the photo-z precision will be lower in the regimes, where the colors only evolve slowly. For example, the red elliptical galaxies have a sharp step-like feature at 4000 Å (what we referred to as the Balmer break in Section 15.2) that helps the estimation for most filter sets. However, if two consecutive passbands had a long gap between them, the redshifts will be proportionally more uncertain. For example, when the 4000 Å break is transitioning between the filters, as none of the fluxes changes much in comparison with the photometric errors.

For specific redshift ranges and object types, it is possible, in principle, to derive optimal filter shapes. A generic filter profile may not be possible to manufacture but the technology allows for more complicated constructions than a single band. An early study to use photometric redshift accuracy as an objective for filter set design [32] looked at adding a special multihump or rugate filter (with up to three disjoint wavelength ranges) to the collection of the Hubble Space Telescope's Advanced Camera for Surveys. The result was that one extra exposure could cut down the mean square error on photometric redshifts by 34% over the $z < 1.3$ redshift range. Recently other studies, for example, [33], started to look at designing an entire filter set for photometric redshifts. While none of these approaches has been followed up with actual observations, there is definitely room for improvement along these lines. With the strong emphasis on the photometric techniques today, such optimizations might result in a welcome breakthrough.

15.5 CONCLUDING REMARKS

The early success of the photometric techniques was followed by a number of new science results and reassuring confirmations. The statistical power of the large photometric datasets overcomes the method's relative imprecision. Yet, a significant residual scatter remains in all methods. On the SDSS datasets, the accuracy of the better empirical methods is approximately the same; the root mean square (RMS) is $\sigma_z \approx 0.02$. This is likely to be due to the limitations in the photometry. The magnitudes themselves do not provide all the information needed for a precise inversion. New observational parameters are considered to break the degeneracies and reduce the scatter. Spectral lines can contribute a significant fraction of the flux in even broad passbands. The relation of the lines and the continuum features

---

* Reviewer's comment: assuming that the model misspecification is small.

of galaxy spectra can be modeled [34] and could be used as part of the photometric inversion. The spatial orientation also plays an important role for spiral galaxies, whose colors are modulated by the dust in their disc as a function of inclination angle [35]. Surface brightness can provide a constraint on the redshift through Tolman's test [36]. As well, radial profile information has been reported to help with blue galaxies [37].

One of the major limitations of today's methods lies in the available spectroscopic training sets. Sampling of the galaxy populations has not yet been really optimized for photometric science. In addition, there is also a natural limit to what can be achieved with spectroscopy, and its degree of completeness affects the inversion. Often we can explore the incompleteness and can try to approximate the systematics by constraints on the observables but such tweaks of the window function (see Section 15.3.4) cannot fully resolve the issue. Ideally one could perform a joint analysis of the relevant effects including but not limited to the aforementioned list of physical properties. The clustering of galaxies is clearly an important aspect to consider. The redshift distribution of spatially overlapping photometric and spectroscopic samples can be consistently calibrated [38] but it is not yet clear as to how such methods could provide realistic constraints on individual sources. Ultimately, we would like to connect photometric redshift analyses to more advanced clustering studies to understand the astronomical objects in the context of the cosmic web.

REFERENCES

1. W. A. Baum. Photoelectric magnitudes and red-shifts. In G. C. McVittie, editor, *Problems of Extra-Galactic Research, IAU Symposium*, Vol. 15, pp. 390–ff., 1962.
2. D. C. Koo. Overview—Photometric redshifts: A perspective from an old-timer[!] on their past, present, and potential. In R. Weymann, L. Storrie-Lombardi, M. Sawicki, and R. Brunner, editors, *Photometric Redshifts and the Detection of High Redshift Galaxies*, Astronomical Society of the Pacific Conference Series, Vol. 191 pp. 3–ff., 1999.
3. A. J. Connolly, I. Csabai, A. S. Szalay, D. C. Koo, R. G. Kron, and J. A. Munn. Slicing through multicolor space: Galaxy redshifts from broadband photometry. *AJ*, 110:2655–ff., 1995.
4. R. J. Brunner, A. J. Connolly, A. S. Szalay, and M. A. Bershady. Toward more precise photometric redshifts: Calibration via CCD photometry. *ApJ*, 482:L21–ff., 1997.
5. I. Csabai, T. Budavári, A. J. Connolly, A. S. Szalay, Z. Győry, N. Benítez, J. Annis et al. The application of photometric redshifts to the SDSS early data release. *AJ*, 125:580–592, 2003.
6. A. A. Collister and O. Lahav. ANNz: Estimating photometric redshifts using artificial neural networks. *PASP*, 116:345–351, 2004.
7. Y. Wadadekar. Estimating photometric redshifts using support vector machines. *PASP*, 117:79–85, 2005.
8. M. J. Way, L. V. Foster, P. R. Gazis, and A. N. Srivastava. New approaches to photometric redshift prediction via Gaussian process regression in the Sloan digital sky survey. *ApJ*, 706:623–636, 2009.
9. J. W. Richards, P. E. Freeman, A. B. Lee, and C. M. Schafer. Exploiting low-dimensional structure in astronomical spectra. *ApJ*, 691:32–42, 2009.
10. P. E. Freeman, J. A. Newman, A. B. Lee, J. W. Richards, and C. M. Schafer. Photometric redshift estimation using spectral connectivity analysis. *MNRAS*, 398:2012–2021, 2009.
11. S. Carliles, T. Budavári, S. Heinis, C. Priebe, and A. S. Szalay. Random forests for photometric redshifts. *ApJ*, 712:511–515, 2010.
12. G. Bruzual and S. Charlot. Stellar population synthesis at the resolution of 2003. *MNRAS*, 344:1000–1028, 2003.

13. G. D. Coleman, C.-C. Wu, and D. W. Weedman. Colors and magnitudes predicted for high redshift galaxies. *ApJS*, 43:393–416, 1980.

14. A. J. Connolly, A. S. Szalay, M. A. Bershady, A. L. Kinney, and D. Calzetti. Spectral classification of galaxies: An orthogonal approach. *AJ*, 110:1071–ff., 1995.

15. T. Budavári, A. S. Szalay, A. J. Connolly, I. Csabai, and M. Dickinson. Creating spectral templates from multicolor redshift surveys. *AJ*, 120:1588–1598, 2000.

16. G. B. Brammer, P. G. van Dokkum, and P. Coppi. EAZY: A fast, public photometric redshift code. *ApJ*, 686:1503–1513, 2008.

17. I. Csabai, A. J. Connolly, A. S. Szalay, and T. Budavári. Reconstructing galaxy spectral energy distributions from broadband photometry. *AJ*, 119:69–78, 2000.

18. T. Budavári, I. Csabai, A. S. Szalay, A. J. Connolly, G. P. Szokoly, D. E. Vanden Berk, G. T. Richards et al. Photometric redshifts from reconstructed quasar templates. *AJ*, 122:1163–1171, 2001.

19. R. Feldmann, C. M. Carollo, C. Porciani, S. J. Lilly, P. Capak, Y. Taniguchi, O. Le Fèvre et al. The Zurich extragalactic Bayesian redshift analyzer and its first application: COSMOS. *MNRAS*, 372:565–577, 2006.

20. K. Jordi, E. K. Grebel, and K. Ammon. Empirical color transformations between SDSS photometry and other photometric systems. *A&A*, 460:339–347, 2006.

21. T. Budavári. A unified framework for photometric redshifts. *ApJ*, 695:747–754, 2009.

22. N. Benítez. Bayesian photometric redshift estimation. *ApJ*, 536:571–583, 2000.

23. T. Dahlen, B. Mobasher, M. Dickinson, H. C. Ferguson, M. Giavalisco, N. A. Grogin, Y. Guo et al. A detailed study of photometric redshifts for GOODS-south galaxies. *ApJ*, 724:425–447, 2010.

24. D. Wang, Y. X. Zhang, C. Liu, and Y. H. Zhao. Kernel regression for determining photometric redshifts from Sloan broad-band photometry. *MNRAS*, 382:1601–1606, 2007.

25. C. E. Cunha, M. Lima, H. Oyaizu, J. Frieman, and H. Lin. Estimating the redshift distribution of photometric galaxy samples—II. Applications and tests of a new method. *MNRAS*, 396:2379–2398, 2009.

26. M. A. Strauss, D. H. Weinberg, R. H. Lupton, V. K. Narayanan, J. Annis, M. Bernardi, M. Blanton et al. Spectroscopic target selection in the sloan digital sky survey: The main galaxy sample. *AJ*, 124:1810–1824, 2002.

27. D. C. Martin, J. Fanson, D. Schiminovich, P. Morrissey, P. G. Friedman, T. A. Barlow, T. Conrow et al. The galaxy evolution explorer: A space ultraviolet survey mission. *ApJ*, 619:L1–L6, 2005.

28. C. C. Steidel, M. Giavalisco, M. Pettini, M. Dickinson, and K. L. Adelberger. Spectroscopic confirmation of a population of normal star-forming galaxies at redshifts $Z < 3$. *ApJ*, 462:L17–ff., 1996.

29. R. Scranton, A. J. Connolly, A. S. Szalay, R. H. Lupton, D. Johnston, T. Budavari, J. Brinkman, and M. Fukugita. Photometric covariance in multi-band surveys: Understanding the photometric error in the SDSS. *ArXiv Astrophysics e-prints*, 2005.

30. W. H. Richardson. Bayesian-based iterative method of image restoration. *J. Opt. Soc. Am. (1917–1983)*, 62:55–ff., 1972.

31. L. B. Lucy. An iterative technique for the rectification of observed distributions. *AJ*, 79:745–ff., 1974.

32. T. Budavári, A. S. Szalay, I. Csabai, A. J. Connolly, and Z. Tsvetanov. An optimal multihump filter for photometric redshifts. *AJ*, 121:3266–3269, 2001.

33. N. Benítez, M. Moles, J. A. L. Aguerri, E. Alfaro, T. Broadhurst, J. Cabrera-Caño, F. J. Castander et al. Optimal filter systems for photometric redshift estimation. *ApJ*, 692:L5–L8, 2009.

34. Z. Győry, A. S. Szalay, T. Budavári, I. Csabai, and S. Charlot. Correlations between nebular emission and the continuum spectral shape in SDSS galaxies. *AJ*, 141:133–ff., 2011.

35. C.-W. Yip, A. S. Szalay, S. Carliles, and T. Budavári. Effect of inclination of galaxies on photometric redshift. *ApJ*, 730:54–ff., 2011.

36. M. J. Kurtz, M. J. Geller, D. G. Fabricant, W. F. Wyatt, and I. P. Dell'Antonio. μ-PhotoZ: Photometric redshifts by inverting the Tolman surface brightness test. *AJ*, 134:1360–1367, 2007.

37. J. J. Wray and J. E. Gunn. A new technique for galaxy photometric redshifts in the Sloan digital sky survey. *ApJ*, 678:144–153, 2008.

38. J. A. Newman. Calibrating redshift distributions beyond spectroscopic limits with cross-correlations. *ApJ*, 684:88–101, 2008.

CHAPTER 16

Galaxy Clusters

Christopher J. Miller

University of Michigan

CONTENTS

16.1	Introduction	337
	16.1.1 The Halo Concept	338
	16.1.2 The Dark Matter Particle Halo	340
	16.1.3 Galaxy Clusters versus Simulated Halos	340
16.2	Halo Finding Techniques	340
	16.2.1 Friends-of-Friends	341
	16.2.2 Spherical Overdensity	341
	16.2.3 Identifying the Subhalos	341
16.3	Galaxy Cluster Finding Techniques	343
	16.3.1 Matched Filter	344
	16.3.2 Color Clustering	344
	16.3.3 Geometric Clustering	347
16.4	Discussion	348
	References	348

16.1 INTRODUCTION

There are many examples of clustering in astronomy. Stars in our own galaxy are often seen as being gravitationally bound into tight globular or open clusters. The Solar System's Trojan asteroids cluster at the gravitational Langrangian in front of Jupiter's orbit. On the largest of scales, we find gravitationally bound clusters of galaxies, the Virgo cluster (in the constellation of Virgo at a distance of ~50 million light years) being a prime nearby example. The Virgo cluster subtends an angle of nearly 8° on the sky and is known to contain over a thousand member galaxies.

Galaxy clusters play an important role in our understanding of the Universe. Clusters exist at peaks in the three-dimensional large-scale matter density field. Their sky (2D) locations are easy to detect in astronomical imaging data and their mean galaxy redshifts (redshift is related to the third spatial dimension: distance) are often better (spectroscopically) and cheaper (photometrically) when compared with the entire galaxy population in large sky surveys. Photometric[*] redshift (z) determinations of galaxies within clusters are accurate to better than $\delta z = 0.05$ [7] and when studied as a cluster population, the central galaxies form a line in color–magnitude space (called the the E/S0 ridgeline and visible in Figure 16.3) that contains galaxies with similar stellar populations [15]. The shape of this E/S0 ridgeline enables astronomers to measure the cluster redshift to within $\delta z = 0.01$ [23]. The most accurate cluster redshift determinations come from spectroscopy of the member galaxies, where only a fraction of the members need to be spectroscopically observed [25,42] to get an accurate redshift to the whole system.

If light traces mass in the Universe, then the locations of galaxy clusters will be at locations of the peaks in the true underlying (mostly) dark matter density field. Kaiser (1984) [19] called this the high-peak model, which we demonstrate in Figure 16.1. We show a two-dimensional representation of a density field created by summing plane-waves with a predetermined power and with random wave-vector directions. In the left panel, we plot only the largest modes, where we see the density peaks (black) and valleys (white) in the combined field. In the right panel, we allow for smaller modes. You can see that the highest density peaks in the left panel contain smaller-scale, but still high-density peaks. These are the locations of future galaxy clusters. The bottom panel shows just these cluster-scale peaks. As you can see, the peaks themselves are clustered, and instead of just one large high-density peak in the original density field (see the left panel), the smaller modes show that six peaks are "born" within the broader, underlying large-scale density modes. This exemplifies the "bias" or amplified structure that is traced by galaxy clusters [19].

Clusters are rare, easy to find, and their member galaxies provide good distance estimates. In combination with their amplified clustering signal described above, galaxy clusters are considered an efficient and precise tracer of the large-scale matter density field in the Universe.

Galaxy clusters can also be used to measure the baryon content of the Universe [43]. They can be used to identify gravitational lenses [38] and map the distribution of matter in clusters. The number and spatial distribution of galaxy clusters can be used to constrain cosmological parameters, like the fraction of the energy density in the Universe due to matter (Ω_{matter}) or the variation in the density field on fixed physical scales (σ_8) [26,33]. The individual clusters act as "Island Universes" and as such are laboratories where we can study the evolution of the properties of the cluster, like the hot, gaseous intra-cluster medium or shapes, colors, and star-formation histories of the member galaxies [17].

16.1.1 The Halo Concept

In a famous paper written by William Press and Paul Schechter in 1974 [44, p. 425], the concept of the "halo" was introduced. The abstract states:

> We consider an expanding Friedmann cosmology containing a "gas" of self-gravitating masses. The masses condense into aggregates which (when sufficiently bound) we identify as single particles of a larger mass [a *halo*]. We propose that after this process

---

[*] Photometric techniques use the broad band filter magnitudes of a galaxy to estimate the redshift. Spectroscopic techniques use the galaxy spectra and emission/absorption line features to measure the redshift.

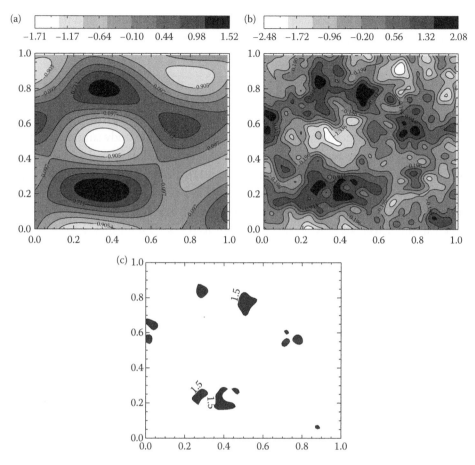

FIGURE 16.1 This figure is a toy model which graphically represents the amplification of clustering for predefined high-amplitude peaks in an underlying density field. In the panel (a), we plot only the largest modes, where we see density peaks (black) and valleys (white). The panel (b) includes smaller-scale modes, where the highest peaks (black) sit on top of the larger-scale modes (from panel (a)). These are the locations of future galaxy clusters. The panel (c) shows just those rare and high-density peaks (black) from the panel (b). Note that there are more of these high-density small-scale peaks than there are of the high-density large-scale modes from the top panel. Also note that the density peaks in the right panel are strongly clustered (i.e., the clustering is amplified).

has proceeded through several scales, the mass spectrum of condensations becomes "self-similar" and independent of the spectrum initially assumed. Some details of the self-similar distribution, and its evolution in time, can be calculated with linear perturbation theory.

The importance of this discovery is that, if observed galaxy clusters map directly to the Press and Schechter matter halos in the Universe, they can be used to tell us about the initial conditions and evolution of structure (i.e., cosmology). Perhaps more importantly, the halo concept allows us to analytically understand the growth and evolution of structure in the Universe. Over the last 35 years, the Press–Schechter formalism (and its modern-day variants) has been ubiquitous in the study of the Universe using galaxy clusters. For a review of the halo concept, see Ref. [8].

16.1.2 The Dark Matter Particle Halo

Press and Schechter [44] conducted a small computer simulation to prove the merits of their thesis. Over the years, researchers have grown the size of these N-body simulations from the 1000 particles used in Press and Schechter to more than 10 billion particles [35]. Within the distribution of these 10 billion dark matter particles are the Press–Schechter mass halos that would be the locations of galaxy clusters in the real Universe. It is this final mapping from the dark matter halo to the real galaxy cluster (in a statistical sense) that is a major topic of modern cosmological cluster research. The trick is to properly identify the halos in the dark matter particle simulations.

16.1.3 Galaxy Clusters versus Simulated Halos

N-body simulations use point-mass particles to trace the evolving gravitational field. Traditionally, the number of particles in simulations has always been larger than the number of cataloged real galaxies. And there are many important differences between the simulations and the real Universe.

- Particles in the simulations trace the matter density field, which is dominated by dark matter. Whereas galaxies are observed as baryonic matter (even though they mostly contain dark matter).

- Particles in the N-body simulations are not galaxies. There are no galaxies in the N-body simulations. However, if the resolution of the simulation is high (i.e., the number of particles is large), the smallest halos could represent galaxies.

- Particles in the simulations have exact physical locations and velocities. In most cases, that exact position and velocity of galaxies are unknown. Galaxy redshifts (which act as a proxy for the third spatial dimension) are a combination of the recessional velocity due to the expansion of the Universe and the local peculiar velocity of the galaxy.

These differences between simulations and the real Universe imply that there will also be differences in how clustering is accomplished in each.

16.2 HALO FINDING TECHNIQUES

In the simulations, we carefully use the term "halo finding" as opposed to clustering. Arguably, the halos themselves are clustered collections of dark matter particles. But the context is with respect to the Press & Schechter formalism described above. In other words, halos define the theoretical entities which are the foundation of the galaxy clusters. Halos are not galaxy clusters, since they contain no galaxies.

Definition 16.1 *A halo is defined to be a region that is a peak in the matter density field. It has a radius that defines a density contrast today such that inside the collapsed region, linear perturbation theory can no longer explain the dynamics.*

Like the galaxies in the real Universe, the dark matter particles in simulations have positions and velocities. Besides the particle residing within the radius of a halo, one could also require that the velocity of the particle be small enough so that it is gravitationally bound to the halo. Typically, a radius for the halo is defined such that the region within that radius is some multiple of the matter density of the Universe which ensures that the particles within that radius are in fact gravitationally bound to the halo. In the following subsections, we describe two of the most common techniques used to find halos in the particle simulations of the Universe. Knebe et al. [22] present a nice comparison between many different implementations of these techniques.

16.2.1 Friends-of-Friends

This algorithm links all particles that fall within a prespecified linkage length, b. It is a simple algorithm which was given the name "friend-of-friend" in the astronomy literature by Press & Davis (1982). The choice of b is very important and small changes in b can have large effects on the final catalog of halos.

Over the years, the friends-of-friends algorithm has been improved in various ways. DenMax [5] applies a fixed kernel smoothing to the data and then the friends-of-friends chains are split at saddle-points in the smoothed data. SKID [41] adds an adaptive kernel and calculates the Lagrangian density to account for the motion (as well as the position) of the particles. HOP [13] adds in the ability of particles to "hop" into neighboring local density maxima to better separate overlapping halos.

16.2.2 Spherical Overdensity

Earlier on in the development of halo theory, spherical collapse via gravity was the preferred heuristic to describe the growth of the peaks in the density field [18]. The spherical overdensity algorithm assumes that all halos are spherical and that the centers of halos have the highest local densities. This algorithm requires a mapping of the density field, an assumption of halo shape, and a density threshold [24]. The densities around each particle are calculated and sorted. Particles that lie within a sphere and which are in regions above some prespecified density threshold define the halo.

16.2.3 Identifying the Subhalos

Cold Dark Matter (CDM) paradigm [6] relies on nonrelativistic noninteracting dark matter to follow the standard laws of gravity. In this paradigm, small regions of high density will collapse first and merge into larger collapsed regions. Most N-body simulations are based on the CDM paradigm.

The N-body simulations have shown us that small halos often reside within the radii of larger halos. This is a result of hierarchical structure formation: subhalos accrete onto larger halos. While the subhalos are undoubtedly altered during this accretion, they are seen to survive in the simulations. Thus, a large and recently formed dark matter halo can contain smaller subhalos. Observationally, the definition of a subhalo is not well determined, although real-world candidates have been found [31]. However, in the simulations, the clustering usually includes a second clustering step to identify the subhalos within the main halos [10]. For example, the algorithm *SUBFIND* presented in [36], separates a group of N-body dark matter particles into a set of disjoint and self-bound substructures. If certain criteria are met, these substructures become the subhalos. They are locally over-dense regions within the density field of a larger background halo.

In Figure 16.2a, we plot the dark matter particles from a small portion of the Millenium Simulations [35]. In Figure 16.2b, we plot those particles identified as part of a single friends-of-friends main halo in this region. Note that the friends-of-friends particles map

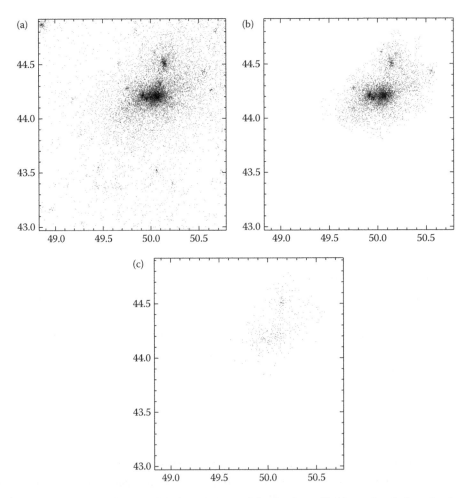

FIGURE 16.2 These figures are of dark matter particles in the Millenium simulation. In the panel (a), we show all particles in a region around a halo in the simulation. In the panel (b), we show the particles belonging to the largest friends-of-friends halo. In the panel (c), we show those particles in the halo that are also identified as being part of a subhalo. Nowadays, a typical simulation will have billions of these particles and millions of halos and subhalos.

onto multiple subhalos for this single halo. Finally, we have plotted (Figure 16.2c) just those dark matter particles which belong to a single smaller subhalo within this region.

16.3 GALAXY CLUSTER FINDING TECHNIQUES

Observationally, galaxy clusters contain both galaxies and hot gas. The galaxies are visible in the optical and infrared (among other) wavelength regimes. The hot gas emits in the x-ray and millimeter wavelength regimes. We do not yet understand how either the galaxies or the gas became the visible components. The detection of clusters via their dark matter (which is actually the dominant contribution of mass in a cluster) is in its infancy, and will no doubt become a vital technique when the data allow. For this chapter, we focus on the detection of galaxy clusters via their member galaxies which act as point-like tracers of the density field. For a review of recent cosmological results covering all aspects of observed galaxy clusters, see Ref. [2].

Definition 16.2 A galaxy cluster is defined to be gravitationally bound collection of galaxies, gas, and dark matter.

While the N-body simulators find halos through exact and known particle positions, galaxy clustering algorithms work differently. First, while the sky-positions of galaxies can be precisely measured, their distances are always *unknown* to some degree. The redshift of a galaxy includes both its recessional velocity due to the expansion of the Universe, and also its local (or peculiar) velocity due to gravitational effects (e.g., infall into the nearest cluster). It is difficult to separate these two components and isolate the recessional velocity, which in turn can be related to a proper distance given an underlying cosmology. Because of this difficulty, galaxy clustering is often done in two dimensions (or the third dimension is somehow weighted differently). For instance, a friend-of-friend cluster finder might use a linkage length that is different in the direction perpendicular to the line of sight versus along the line of sight.

Second, unlike dark matter which we cannot directly see, galaxies have intrinsic brightnesses. In fact, the luminosity function (number of galaxies as a function of brightness) is a well-measured quantity in the local Universe (obviously, it is a harder measurement to make at large distances). This means that galaxy clustering is not a simple point process like it is in N-body simulations, where the particles have fixed identical "masses." In real galaxy data, the "points" (i.e., the galaxies) have intrinsically variable scalar brightnesses. There are more dim galaxies than there are bright galaxies. This becomes especially important since an observed galaxy dataset has an inherent flux limit.

Finally, galaxies have colors and shapes and it is known that these correlate well with the locally measured density field [11,16]. Therefore, galaxy cluster algorithms can be designed to take advantage of properties of the galaxies themselves. This is not possible in the N-body simulations. In the following subsections, we discuss three different forms of clustering algorithms that are common in the literature. We note that before the modern era of observational galaxy clustering algorithms (which started in the mid-1990s), most scientific studies of galaxy clusters were based on catalogs generated "by eye" (e.g., the famous Abell catalog [1,3]).

16.3.1 Matched Filter

A matched filter method simply uses a convolution of the data with a filter that is based on the assumed underlying shape of the signal (i.e., the galaxy overdensities which are the clusters). Mathematically, one convolves the observed signal with a filter in order to maximize the signal-to-noise of the underlying signal. In the case of galaxy clusters, the signal is the two-dimensional overdensity of galaxies in a background of noncluster foreground or background galaxies. Postman et al. [32] first applied a matched filter technique via flux and radial (2D shapes) filters. In their method, the filter is further tuned adaptively to the redshift of each signal (or cluster).

The original matched filter formalism of [32] required enough counts in the signal+background to utilize Gaussian statistics and maximize a likelihood function given by the one-dimensional radial profile of galaxies in clusters, the global number count of galaxies per unit magnitude, and the sky density of galaxies (all as a function of galaxy redshift and brightness). The counts are always relative to a background count, which can be modeled or measured. This likelihood is measured over the whole dataset, and it is tuned at each step to maximize the count relative to the background. There are many variants on this technique [14,20].

16.3.2 Color Clustering

Using the observed fact that the dominant form of galaxies which reside within the radii of massive clusters are known to be of a homogeneous color (red) and shape (elliptical), [15] created an algorithm which identifies peaks in color as well as positional space. The original "red sequence" algorithm of [15] specifically modeled the expected colors of galaxies in clusters based on simulated galaxy colors (i.e., spectral synthesis models).

The concept of color clustering can be defined between two extremes: one where the model is based on the precise expected color and luminosity of a single central cluster galaxy (e.g., the brightest cluster galaxy (maxBCG) catalog [23]) and another where no models for the galaxy colors or luminosities are assumed at all, but simply the fact that galaxies cluster in color as well as position (e.g., the sloan digital sky survey (SDSS)-C4 Cluster Catalog [28]).

The maxBCG algorithm [23] determines a likelihood for each galaxy to be a BCG. This likelihood is constructed from a model of the distribution of intrinsic brightnesses and colors of brightest cluster galaxies, as well as the number of neighbors around the galaxy that have the "right" colors (i.e., those associated with the E/S0 ridgeline in models and previously observed clusters). The C4 algorithm [28] places a four-dimensional color "probability box" around every galaxy and compares the count of galaxies within that color box to a random selection of galaxies drawn from the larger sample. Galaxies are identified as "cluster-like" if they are not drawn from the random distribution of colors (i.e., the null). Both of these algorithms require galaxy datasets that have well-measured multifilter brightnesses. And while both require the clusters to also be spatially clustered, the emphasis is on the galaxy colors (and brightnesses, in the case of maxBCG).

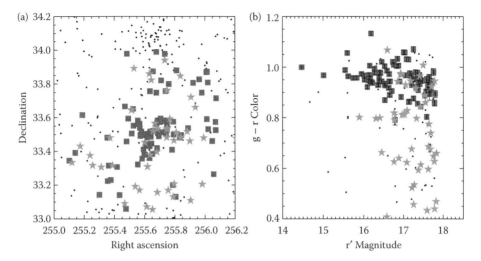

FIGURE 16.3 This is a galaxy cluster in the Sloan digital sky survey spectroscopic redshift catalog. The panel (a) shows the sky positions of the galaxies. Squares are cluster members (identified using their observed redshifts) which have little active star formation. Stars are also cluster members, but have ongoing star formation. The small dots are back ground/foreground galaxies. In the panel (b), we show the color versus magnitude relation. Note how the E/S0 ridgeline galaxies (the squares) are clustered in both position and color.

In Figure 16.3, we show the positions and the color versus magnitude diagram of a galaxy cluster in real data [37]. The squares represent typical red and elliptical cluster galaxies that have little active star formation. The stars are galaxies that have a fair amount of actively forming stars. The small circles are foreground/background galaxies not associated with the cluster. The E/S0 ridgeline is clearly seen in the right panel of Figure 16.3. This is the foundation for most color-based clustering schemes.

An example of the power of color clustering can be seen in Figure 16.4. This is Abell 1882, originally identified as a large, massive, x-ray bright, and rich (in terms of the number of galaxies) cluster. We applied the C4 color clustering algorithm to these data (right panel) and one can immediately see the sub-structure in this complex system. It is not a simple halo-like sphere. In fact, the density field of the "cluster-like" galaxies identified by the C4 algorithm lie in three separate density peaks (the contours in Figure 16.4 bottom). These galaxy density peaks align with the locations of the hot x-ray gas (shaded ellipses). The ability to "defragment" clusters into their constituent components is a problem similar to that of identifying subhalos in the N-body simulations described earlier.

In Figure 16.4, we use large circles to note members identified via their spectroscopic redshifts, and boxes indicate galaxies with active nuclei [e.g., active galactic nuclei (AGN) or galaxies with strong line emission created in the region surrounding the black-hole engines at the centers of galaxies]. Star-symbols represent galaxies that appear to have ongoing star formation (see also stars in Figure 16.3). Note that the star-forming galaxies avoid the cluster cores, while the AGN and other galaxies reside in all regions (see e.g., [27,40]). This highlights another important avenue in the determination of the spatial locations of galaxy clusters:

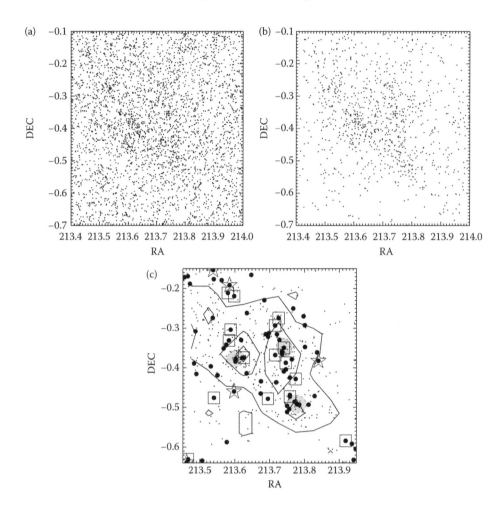

FIGURE 16.4 This is the Abell 1882 system. It was originally identified as a single large cluster, but the C4 algorithm and follow-up x-ray observations of the hot intracluster gas showed that it is actually three smaller clusters. The panel (a) shows all galaxies in the region, while the panel (b) shows only those that are "cluster-like" according to their colors via the C4 clustering algorithm (Adapted from C.J. Miller et al. *Astronomical Journal*, 130:968–1001, 2005.). The panel (c) shows these same galaxies with isodensity contours which overlap with the clumps of x-ray gas (the shaded ellipses).

the astrophysical properties of galaxies are related to the environment in which they live. Besides cosmology, an important goal in the study of galaxy clusters is explaining the origin and evolution of the E/S0 ridgeline and how the environment might have evolved galaxies over time.

It is worth noting that the true population of galaxy morphologies and colors within galaxy clusters is not precisely known. In addition, certain types of smaller galaxy groups contain mostly blue spiral galaxies [29]. So while the use of galaxy colors in cluster-finding is an important component, it might be the case that strong restrictions on the allowable colors are producing biases samples of clusters.

16.3.3 Geometric Clustering

The Voronoi Tessellation and its variants have been applied to galaxy cluster catalogs and is an example of geometric clustering. Briefly, a set of points in a plane can be used to define Voronoi sites, around which cells are identified which contain all points closer to the site than any other point in the plane. These Voronoi cells are convex polygons. Alternatively, one can define a segment in a Voronoi diagram that is comprised of all the points that are equidistant to their nearest Voronoi sites. Okabe et al. [30] provide a nice review of Voronoi Tessellations, including their history, theory, and a wide range of applications.

In the distribution of galaxies, a picture is worth a thousand words. In Figure 16.5, we show the Voronoi Tessellation diagram of the Abell 1882 system described above. In the left panel, we tessellate all galaxies, while in the right panel we tessellate only those previously identified as "cluster-like" using the C4 algorithm. The size of the Voronoi cell is inversely proportional to the local point-wise density. Thus, this technique is really an adaptive-kernel like planar density estimation technique.

Weygaert et al. [39] first showed that the spatial distribution of the Voronoi nodes (which are points that are equidistant to three or more Voronoi cells) is similar to the spatial distribution of Abell clusters [1]. Since then, others have used the Voronoi technique to map the galaxy density field and identify the galaxy clusters [21,34].

There are numerous other techniques that have been used to find galaxy clusters in observed data. For instance, multiscale wavelets, percolation, friend-of-friend techniques have all been utilized at some point to create galaxy cluster catalogs. However, the matched-filter (and its variants), color-clustering, and geometric techniques like the Voronoi Tessellation are the most common techniques applied to rich galaxy datasets such as the Sloan digital sky survey [37].

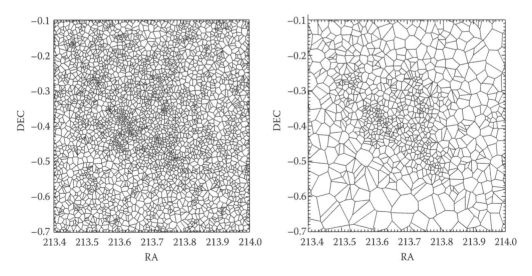

FIGURE 16.5 This is the Voronoi Tessellation of the Abell 1882 system. The Voronoi cells are inversely proportional to the point-wise density. The left panel shows the tessellation on all galaxies in the region. The right panel shows only those that have been identified as "cluster-like" using the C4 algorithm.

16.4 DISCUSSION

In this chapter, we presented the concept of clustering with respect to two different but related disciplines: halos in cosmological N-body simulations and galaxy clusters in real data. The algorithms that each discipline relies on do not necessarily overlap. This is a natural consequence of the different types of point-wise data each discipline has to work with.

From a machine-learning perspective, one might classify both halo finding and galaxy cluster finding as important and challenging research topics. In most cases, speed is not the main issue. Instead, the issue is the ability to find the clusters and halos among a sea of noisy backgrounds. When does a subhalo become part of the background? When do large single entities get split into smaller constituents?

I expect that current research will lead us toward the development of metrics which can compare not only the qualities of the different algorithms, but also the ability to better integrate the concept of the halo and the reality of the galaxy cluster. As such, the ability to determine a truth table is all the more important. However, even in the simulations, the truth is hard to define. Simulations are not able to make catalogs of galaxies from first principles. At the moment, researchers utilize various techniques to paste galaxies onto the locations of dark matter particles. Croton et al. [9] use what astronomers call "semianalytic" techniques to model the properties of galaxies in the Universe. These techniques are physically well motivated and, for example, are able to turn star formation off once a galaxy enters a dense region (as is seen in the real Universe). An alternate technique is to paste galaxies onto dark matter particles in such a way that simultaneously constrains multiple parameters as observed in the real Universe (see e.g., the description in Ref. [28]). Regardless, while no one is yet able to simulate a fully self-consistent galaxy/halo catalog, these advances in mock galaxy catalogs are enabling researchers to test their independent clustering techniques. In other words, clustering algorithms based on the halos and the observed galaxy clusters can be compared in a closed loop.

On another front, researchers will soon focus on understanding their clustering algorithms to a degree which specifically enables the science one wants to conduct. If the completeness and false-detection rates are well known, they can fold that knowledge back into the scientific analyses, even if the clusters themselves were not found via an "optimal" clustering algorithm.

There is no shortage of point-wise datasets in astronomy that require clustering algorithms. Similarly, there seems to be no shortage on ideas on how to best cluster these data. What might be lacking is a more rigorous approach to clustering which utilizes recent advances in machine-learning clustering techniques. The challenge is to mold those techniques to work with the "real" data, or to determine what is needed in the data to best apply new techniques.

REFERENCES

1. G. O. Abell. The distribution of rich clusters of galaxies. *Astrophysical Journal Supplement*, 3:211–ff., 1958.
2. S. W. Allen, A. E. Evrard, and A. B. Mantz. Cosmological parameters from observations of galaxy clusters. *Annual Review of Astronomy and Astrophysics*, 49(1):409–470, 2011.

3. N. A. Bahcall. Abell's catalog of clusters of galaxies. *Astrophysical Journal*, 525:C873–ff., 1999.

4. W. A. Barkhouse, P. J. Grean, A. Vikhlinin, D.-W. Kim, D. Perley, R. Cameron, J. Silverman et al. ChaMP Serendipitous galaxy cluster survey. *Astrophysical Journal*, 645:955–976, 2006.

5. E. Bertschinger and J. M. Gelb. Cosmological N-body simulations. *Computers in Physics*, 5:164–175, 1991.

6. G. R. Blumenthal, S. M. Faber, J. R. Primack, and M. J. Rees. Formation of galaxies and large-scale structure with cold dark matter. *Nature*, 311:517–525, 1984.

7. A. Collister, O. Lahav, C. Blake, R. Cannon, S. Croom, M. Drinkwater, A. Edge et al. MegaZ-LRG: A photometric redshift catalogue of one million SDSS luminous red galaxies. *Monthy Notices of the Royal Astronomical Society*, 375:68–76, 2007.

8. A. Cooray and R. Sheth. Halo models of large scale structure. *Physics Reports*, 372:1–129, 2002.

9. D. J. Croton, V. Springel, S. D. M. White, G. De Lucia, C. S. Frenk, L. Gao, A. Jenkins, G. Kauffmann, J. F. Navarro, and N. Yoshida. The many lives of active galactic nuclei: Cooling flows, black holes and the luminosities and colours of galaxies. *Monthly Notices of the Royal Astronomical Society*, 365:11–28, 2006.

10. G. De Lucia, G. Kauffmann, V. Springel, S. D. M. White, B. Lanzoni, F. Stoehr, G. Tormen, and N. Yoshida. Substructures in cold dark matter haloes. *Monthly Notices of the Royal Astronomical Society*, 348:333–344, 2004.

11. A. Dressler. Galaxy morphology in rich clusters—Implications for the formation and evolution of galaxies. *Astrophysical Journal*, 236:351–365, 1980.

12. P. R. M. Eisenhardt, M. Brodwin, A. H. Ganzales, S. A. Stanford, D. Stern, P. Barmby, M. J. I. Brown et al. Clusters of galaxies in the first half of the universe from the IRAC Shallow survey. *Astrophysical Journal*, 684:905–932, 2008.

13. D. J. Eisenstein and P. Hut. HOP: A new group-finding algorithm for N-body simulations. *Astrophysical Journal*, 498:137–ff., 1998.

14. R. R. Gal, R. R. de Carvalho, S. C. Odewahn, S. G. Djorgovski, and V. E. Margoniner. The Northern sky optical cluster survey. I. Detection of galaxy clusters in DPOSS. *Astronomical Journal*, 119:12–20, 2000.

15. M. D. Gladders and H. K. C. Yee. A new method for galaxy cluster detection. I. The algorithm. *Astronomical Journal*, 120:2148–2162, 2000.

16. P. L. Gómez, R. C. Nichol, C. J. Miller, M. L. Balogh, T. Goto, A. I. Zabludoff, A. K. Romer et al. Galaxy star formation as a function of environment in the early data release of the Sloan digital sky survey. *Astrophysical Journal*, 584:210–227, 2003.

17. M. E. Gray, C. Wolf, K. Meisenheimer, A. Taylor, S. Dye, A. Borch, and M. Kleinheinrich. Linking star formation and environment in the A901/902 supercluster. *Monthy Notices of the Royal Astronomical Society*, 347:L73–L77, 2004.

18. J. E. Gunn and J. R. Gott, III. On the infall of matter into clusters of galaxies and some effects on their evolution. *Astrophysical Journal*, 176:1–ff., August 1972.

19. N. Kaiser. On the spatial correlations of Abell clusters. *Astrophysical Journal*, 284:L9–L12, 1984.

20. J. Kepner, X. Fan, N. Bahcall, J. Gunn, R. Lupton, and G. Xu. An automated cluster finder: The adaptive matched filter. *Astrophysical Journal*, 517:78–91, 1999.

21. R. S. J. Kim, J. V. Kepner, M. Postman, M. A. Strauss, N. A. Bahcall, J. E. Gunn, R. H. Lupton et al. Detecting clusters of galaxies in the sloan digital sky survey. I. Monte Carlo comparison of cluster detection algorithms. *Astronomical Journal*, 123:20–36, 2002.

22. A. Knebe, S. R. Knollmann, S. I. Muldrew, F. R. Pearce, M. A. Aragon-Calvo, Y. Ascasibar, P. S. Behroozi et al. Haloes gone MAD: The halo-finder comparison project. *Monthly Notices of the Royal Astronomical Society*, 415(3): 2293–2318, 2011.

23. B. P. Koester, T. A. McKay, J. Annis, R. H. Wechsler, A. Evrard, L. Bleem, M. Becker et al. A MaxBCG catalog of 13,823 galaxy clusters from the Sloan digital sky survey. *Astrophysical Journal*, 660:239–255, 2007.

24. C. Lacey and S. Cole. Merger rates in hierarchical models of galaxy formation—Part two—Comparison with N-body simulations. *Monthy Notices of thr Royal Astronomical Society*, 271:676–ff., 1994.

25. C. J. Miller, D. J. Batuski, K. A. Slinglend, and J. M. Hill. Projection, spatial correlations, and anisotropies in a large and complete sample of Abell clusters. *Astrophysical Journal*, 523:492–505, 1999.

26. C. J. Miller, R. C. Nichol, and D. J. Batuski. Possible detection of baryonic fluctuations in the large-scale structure power spectrum. *Astrophysical Journal*, 555:68–73, 2001.

27. C. J. Miller, R. C. Nichol, P. L. Gómez, A. M. Hopkins, and M. Bernardi. The environment of active galactic nuclei in the Sloan digital sky survey. *Astrophysical Journal*, 597:142–156, 2003.

28. C. J. Miller, R. C. Nichol, D. Reichart, R. H. Wechsler, A. E. Evrard, J. Annis, T. A. McKay et al. The C4 clustering algorithm: Clusters of galaxies in the Sloan digital sky survey. *Astronomical Journal*, 130:968–1001, 2005.

29. A. Oemler, Jr. The systematic properties of clusters of galaxies. Photometry of 15 clusters. *Astrophysical Journal*, 194:1–20, 1974.

30. A. Okabe, B. Boots, and K. Sugihara. *Spatial Tessellations. Concepts and Applications of Voronoi Diagrams.* John Wiley & Sons Ltd., Chichester, New York, Brisbane, Toronto, and Singapore, 1992.

31. N. Okabe, Y. Okura, and T. Futamase. Weak-lensing mass measurements of substructures in coma cluster with Subaru/Suprime-cam. *Astrophysical Journal*, 713:291–303, 2010.

32. M. Postman, L. M. Lubin, J. E. Gunn, J. B. Oke, J. G. Hoessel, D. P. Schneider, and J. A. Christensen. The Palomar distant clusters survey. I. The cluster catalog. *Astronomical Journal*, 111:615–ff., 1996.

33. E. Rozo, R. H. Wechsler, E. S. Rykoff, J. T. Annis, M. R. Becker, A. E. Evrard, J. A. Frieman et al. Cosmological constraints from the Sloan digital sky survey maxBCG cluster catalog. *Astrophysical Journal*, 708:645–660, 2010.

34. M. Soares-Santos, R. R. de Carvalho, J. Annis, R. R. Gal, F. La Barbera, P. A. A. Lopes, R. H. Wechsler, M. T. Busha, and B. F. Gerke. The Voronoi Tessellation cluster finder in 2+1 dimensions. *Astrophysical Journal*, 727:45–ff., 2011.

35. V. Springel, S. D. M. White, A. Jenkins, C. S. Frenk, N. Yoshida, L. Gao, J. Navarro et al. Simulations of the formation, evolution and clustering of galaxies and quasars. *Nature*, 435:629–636, 2005.

36. V. Springel, S. D. M. White, G. Tormen, and G. Kauffmann. Populating a cluster of galaxies—I. Results at [formmu2]z = 0. *Monthly Notices of the Royal Astronomical Society*, 328:726–750, 2001.

37. C. Stoughton, R. H. Lupton, M. Bernardi, M. R. Blanton, S. Burles, F. J. Castander, A. J. Connolly et al. Sloan digital sky survey: Early data release. *Astronomical Journal*, 123:485–548, 2002.

38. J. A. Tyson, F. Valdes, J. F. Jarvis, and A. P. Mills, Jr. Galaxy mass distribution from gravitational light deflection. *Astrophysical Journal*, 281:L59–L62, 1984.

39. R. van de Weygaert and V. Icke. Fragmenting the universe. II—Voronoi vertices as Abell clusters. *Astronomy and Astrophysics*, 213:1–9, 1989.

40. M. J. Way, R. A. Flores, and H. Quintana. Statistics of active galactic nuclei in rich clusters revisited. *Astrophysical Journal*, 502:134–ff., 1998.

41. D. H. Weinberg, L. Hernquist, and N. Katz. Photoionization, numerical resolution, and galaxy formation. *Astrophysical Journal*, 477:8–ff., 1997.

42. S. D. M. White, D. I. Clowe, L. Simard, G. Rudnick, G. De Lucia, A. Aragón-Salamanca, R. Bender et al. EDisCS—the ESO distant cluster survey. Sample definition and optical photometry. *Astronomy and Astrophysics*, 444:365–379, 2005.

43. S. D. M. White, J. F. Navarro, A. E. Evrard, and C. S. Frenk. The baryon content of galaxy clusters: A challenge to cosmological orthodoxy. *Nature*, 366:429–433, 1993.

44. W. H. Press and P. Schechter. Formation of galaxies and clusters of galaxies by self-similar gravitational condensation. *Astrophysical Journal*, 187:425–438, 1974.

3

Signal Processing (Time-Series) Analysis

CHAPTER 17

Planet Detection

The Kepler Mission

Jon M. Jenkins, Jeffrey C. Smith, Peter Tenenbaum,
Joseph D. Twicken, and Jeffrey Van Cleve

SETI Institute/NASA Ames Research Center

CONTENTS

17.1 Introduction   355
17.2 An Overview of the Kepler SOC Pipeline   358
    17.2.1 Pixel-Level CAL   359
    17.2.2 Photometric Analysis   360
    17.2.3 Systematic Error Corrections   361
    17.2.4 Transiting Planet Search   362
        17.2.4.1 A Wavelet-Based Matched Filter   362
    17.2.5 Data Validation   366
17.3 A Bayesian Approach to Correcting Systematic Errors   368
    17.3.1 The Problem with PDC   368
    17.3.2 Why Should We Hope to Do Better?   370
    17.3.3 The MAP Approach   371
    17.3.4 An Application of MAP to Kepler Data   373
    17.3.5 Empirical MAP Implementation Considerations   377
17.4 Conclusions   377
Acknowledgments   378
References   379

17.1 INTRODUCTION

The search for exoplanets is one of the hottest topics in astronomy and astrophysics in the twenty-first century, capturing the public's attention as well as that of the astronomical community. This nascent field was conceived in 1989 with the discovery of a candidate planetary companion to HD114762 [35] and was born in 1995 with the discovery of the first extrasolar planet 51 Peg-b [37] orbiting a main sequence star.

As of March, 2011, over 500 exoplanets have been discovered[*] and 106 are known to transit or cross their host star, as viewed from Earth. Of these transiting planets, 15 have been announced by the Kepler Mission, which was launched into an Earth-trailing, heliocentric orbit in March, 2009 [1,4,6,15,18,20,22,31,32,34,36,43]. In addition, over 1200 candidate transiting planets have already been detected by Kepler [5], and vigorous follow-up observations are being conducted to vet these candidates. As the false-positive rate for Kepler is expected to be quite low [39], Kepler has effectively tripled the number of known exoplanets. Moreover, Kepler will provide an unprecedented data set in terms of photometric precision, duration, contiguity, and number of stars.

Kepler's primary science objective is to determine the frequency of Earth-size planets transiting their Sun-like host stars in the habitable zone, that range of orbital distances for which liquid water would pool on the surface of a terrestrial planet such as Earth, Mars, or Venus. This daunting task demands an instrument capable of measuring the light output from each of over 100,000 stars simultaneously with an unprecedented photometric precision of 20 parts per million (ppm) at 6.5-h intervals. The large number of stars is required because the probability of the geometrical alignment of planetary orbits that permit observation of transits is the ratio of the size of the star to the size of the planetary orbit. For Earth-like planets in 1-astronomical unit (AU) orbits[†] about sun-like stars, only ~0.5% will exhibit transits. By observing such a large number of stars, Kepler is guaranteed to produce a robust null result in the unhappy event that no Earth-size planets are detected in or near the habitable zone. Such a result would indicate that worlds like ours are extremely rare in the Milky Way galaxy and perhaps the cosmos, and that we might be solitary sojourners in the quest to answer the age-old question: "Are we alone?"

Kepler is an audacious mission that places rigorous demands on the science pipeline used to process the ever-accumulating, large amount of data and to identify and characterize the minute planetary signatures hiding in the data haystack. Kepler observes over 160,000 stars simultaneously over a field of view (FOV) of 115 square degrees with a focal plane consisting of 42 charge-coupled devices[‡] (CCDs), each of which images 2.75 square degrees of sky onto 2200×1024 pixels.[§] The photometer, which contains the CCD array, reads out each CCD every 6.54 s [10,11] and co-adds the images for 29.4 min, called a long cadence (LC) interval. Due to storage and bandwidth constraints, only the pixels of interest, those that contain images of target stars, are saved onboard the solid–state recorder (SSR), which can store 66+ days of data. An average of 32 pixels per star is allowed for up to 170,000 stellar target definitions. In addition, a total of 512 targets are sampled at 58.85-s short cadence (SC) intervals, permitting further characterization of the planet-star systems

---

[*] Almost all of these have mass estimates provided by spectroscopic observations, and ~144 of these have had their orbital inclinations determined.

[†] The Earth orbits the Sun at an average distance of 149,597,870.7 km, called an AU.

[‡] CCDs are the light-sensitive devices used as "electronic film" in common consumer electronics such as digital cameras, camcorders, and cell phones.

[§] An electronics failure incurred the loss of the two CCDs on module 3 in January 2010, reducing the active FOV to 110 square degrees.

for the brighter stars with a Kepler magnitude,[*] Kp, brighter than 12 ($Kp < 12$) stars via asteroseismology [17], and more precise transit timing. In addition to the stellar images, collateral data used for calibration (CAL) are also collected and stored on the SSR. For each of the 84 CCD readout channels these data include up to 4500 background sky pixels used to estimate and remove diffuse stellar background and zodiacal light; 1100 pixels containing masked smear measurements and another 1100 pixels containing virtual smear measurements used to remove artifacts caused by the lack of a shutter and a finite 0.51-s readout time; and 1070 trailing black measurements that monitor the bias voltage presented at the input of the analog-to-digital converter so that the zero point can be restored to the digitized data during processing [24].

There are a total of up to 6,092,680 pixels containing the stellar and collateral data collected for each LC, with 48 LCs/day. While only 512 SC targets are defined at any given time, there are 30 SC intervals for each LC interval, and an average of 85 pixels are allocated for each SC target star. Smear and black-level measurements are collected for each SC target, but only for the rows and columns occupied by SC stellar target pixels. Approximately 21% of the pixel data returned by Kepler are SC data. The total data rate for both LC and SC data is 1.3 GB/day when the data are expanded to 4 bytes/pixel from the compressed bit stream.

Raw pixel data are downlinked at monthly intervals through National Aeronautics and Space Administration's (NASA's) Deep Space Network[†] (DSN) and routed through the ground system to the Kepler Science Operations Center (SOC) at NASA Ames Research Center. The SOC performs a number of critical functions for the mission, including management of the target definitions which specify the pixels needed for each stellar target and the compression tables that allow a ~5:1 compression of the science data onboard the SSR (from 23 bits/pixel to 4.6 bits/pixel), but its two major tasks are to:

1. Process raw pixel data to produce archival science data products, including calibrated pixels, measurements of the location or centroid of each star in each frame, flux time series representing the brightness of each star in each data frame, and systematic error-corrected flux time series that have instrumental artifacts removed.

2. Search each target-star light curve to identify transit-like features and to perform a suite of diagnostic tests on each such event to make or break confidence in each transit-like signature by eliminating eclipsing binaries and other false positives.

This chapter focuses on two of the most important subtasks, as they represent the most challenging ones from the perspective of the major themes of this book: machine learning and data mining. First, Section 17.2 gives a brief overview of the SOC science processing pipeline. This includes a special subsection detailing the adaptive, wavelet-based transit detector in the transiting planet search (TPS) pipeline component that performs the automated search through each of the hundreds of thousands of light curves for transit

---

[*] Kp, or Kepler magnitude, is a measure of the intensity of the observed source. A $Kp = 12$ star generates 4.5×10^9 photoelectrons in 6.5 h on Kepler's detectors.
[†] The DSN is operated by the Jet Propulsion Laboratory for NASA.

signatures of Earth-size planets. Following the overview, Section 17.3 describes an approach under development to improve the science pipeline's ability to identify and remove instrumental signatures from the light curves while minimizing distortion of astrophysical signals in the data and preventing the introduction of additional noise that may mask small transit features. The chapter concludes with some thoughts about the future of large transit surveys in the context of the Kepler experience.

17.2 AN OVERVIEW OF THE KEPLER SOC PIPELINE

Kepler's observations are organized into three-month intervals called quarters, which are defined by the roll maneuvers the spacecraft executes about its boresight to keep the solar arrays pointed toward the sun and the radiator used to cool the focal plane pointed to deep space [18]. In addition to the science data, the SOC receives a selected set of ancillary engineering data containing any parameters likely to have a bearing on the quality of the science data, such as temperature measurements of the focal plane and readout electronics.

The processing of the data is controlled by the Kepler pipeline framework [30], a software platform designed for building applications that run as automated, distributed pipelines. The framework is designed to be reusable, and therefore contains no application-specific code. Application-specific components, such as Pipeline modules, parameters used by the application, and the unit-of-work specification for each module, integrate with the framework via interfaces and abstract classes [29]. The SOC uses this framework to build the pipelines that process the 23 GiB of raw pixel data downlinked from the Kepler spacecraft every month on four computing clusters containing a total of 64 nodes, 584 CPU cores, 2.9 TiB of RAM, and 198 TiB of raw disk storage [19].[*]

The Science Pipeline is divided into several components in order to allow for efficient management and parallel processing of data, as shown in Figure 17.1. Raw pixel data downlinked

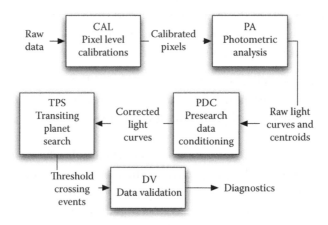

FIGURE 17.1 Data flow diagram for the Kepler science pipeline. (From J. M. Jenkins et al., *ApJL* 713(2), L87–L91, 2010.)

---

[*] 1 GiB = $(2^{10})^3$ bytes, and 1 TiB = $(2^{10})^4$ bytes, according to NIST standards (http://physics.nist.gov/cuu/Units/binary.html).

from the Kepler photometer are calibrated by the CAL module to produce calibrated target and background pixels and their associated uncertainties [40]. The calibrated pixels are processed by the photometric analysis module (PA) to fit and remove sky background and extract simple aperture photometry from the background-corrected, calibrated target pixels[*] [45]. PA also measures the centroid locations of each star in each frame. Finally, systematic error-corrected light curves are produced in the presearch data conditioning module (PDC), where signatures in the light curves correlated with systematic error sources from the telescope and spacecraft, such as pointing drift, focus changes, and thermal transients are removed [44]. Output data products include raw and calibrated pixels, raw and systematic error-corrected flux time series, and centroids and associated uncertainties for each target star, which are archived to the data management center (DMC)[†] and made available to the public through the Multimission Archive at Space Telescope Science Institute (STScI)[‡] [38].

In the TPS module [26], a wavelet-based, adaptive matched filter is applied to identify transit-like features with durations in the range of 1–16 h (see Section 17.2.4). Light curves with transit-like features whose combined signal-to-noise ratio (SNR) exceeds 7.1σ for a specified trial period and epoch are designated as threshold crossing events (TCEs) and subjected to further scrutiny by the data validation module (DV). This threshold ensures that no more than one false positive will occur due to random fluctuations over the course of the mission, assuming nonwhite, nonstationary Gaussian observation noise [21,23]. DV performs a suite of statistical tests to evaluate the confidence in the transit detection, to reject false positives by eclipsing binaries, and to extract physical parameters of each system (along with associated uncertainties and covariance matrices) for each planet candidate [42,47]. After the planetary signature is fitted, DV removes it from the light curve and searches over the residual time series for additional transiting planets. This process repeats until no further TCEs are identified. The DV results and diagnostics are then furnished to the science team to facilitate disposition by the Follow-up Observing Program [16].

17.2.1 Pixel-Level CAL

The pipeline module CAL corrects the raw Kepler photometric data at the pixel level prior to the extraction of photometry and astrometry [40]. Several of the processing steps given in Figure 17.2 are familiar to ground-based photometrists. However, a few are peculiar to Kepler due to its lack of a shutter and unique features in its analog electronics chains. Details of these instrument characteristics and how they were determined and updated in flight are discussed in Ref. [10] and are comprehensively documented in the Kepler Instrument Handbook[§] [46].

The sequence of processing steps in CAL that produce calibrated pixels and associated uncertainties is: (a) the 2-D black level (CCD bias voltage) structure (fixed pattern noise) is removed, followed by fitting and removing a dynamic estimate of the black level; (b) gain

---

[*] In simple aperture photometry, the brightness of a star in a given frame is measured by summing up the pixel values containing the image of the star.

[†] The DMC is located at the STScI in Baltimore, MD.

[‡] http://stdatu.stsci.edu/kepler/

[§] Available via http://keplergo.arc.nasa.gov/calibration/KSCI-19033-001.pdf.

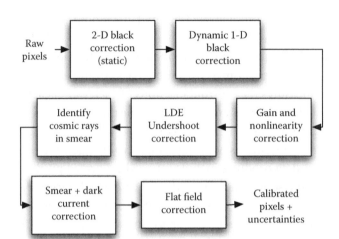

FIGURE 17.2 Data flow diagram for the CAL pipeline module. (From J. M. Jenkins et al., *ApJL* 713(2), L87–L91, 2010.)

and nonlinearity corrections are applied; (c) the analog electronics chain exhibits memory, called local detector electronics (LDE) undershoot, necessitating the application of a digital filter to remove this effect; (d) cosmic ray events in the black and smear measurements are removed prior to subsequent corrections; (e) the smear signal caused by operating in the absence of a shutter and the dark current for each CCD readout channel are estimated from the masked and virtual smear collateral data measurements; (f) a flat field correction is applied.

17.2.2 Photometric Analysis

The PA module extracts brightness and location measurements for each star for each time step, producing flux time series and centroid time series [45] along with associated uncertainties [12].

Before photometry and astrometry can be extracted from the calibrated pixel time series, the pipeline detects "Argabrightening" events in the background pixel data. These mysterious transient increases in the background flux were identified early in mission commissioning.[*] Argabrightenings that affect 10 or more CCD readout channels occur ∼15 times per month, but the rate is dropping over time. The most egregious of these events cannot be perfectly corrected by the current background correction. We discard the data when the excess background flux exceeds the 100 median absolute deviation (MAD) level.

PA then robustly fits a 2-D surface to ∼4500 background pixels on each channel to estimate the sky background, which is evaluated at each target-star pixel location and subtracted from the calibrated pixel values. Cosmic ray events are also identified in the calibrated background pixel time series and flagged, but since the sky background uses a robust fit, the resulting correction is not sensitive to the effectiveness of the cosmic ray identification.

---

[*] The current hypothesis is that these transient events are due to small dust particles from Kepler achieving escape velocity after micrometeorite hits, which then, in turn reflect sunlight into the barrel of the telescope for a few seconds as they drift across the telescope's FOV.

Each target pixel time series is scanned for cosmic rays by first detrending the time series with a moving median filter with a width of five cadences (time steps) and examining the residuals for outliers as compared to the MAD of the residuals for each pixel. Care is taken not to remove clusters of outliers that might be due to astrophysical signatures such as flares or transits that are intrinsic to the target star.

The photocenters of the 200 brightest unsaturated stars on each channel are fitted using the pixel response functions (PRFs)[*] and then used to define the ensemble motion of the stars over the observations. The aggregate star motion is used along with the PRFs to define the optimal aperture as the collection of pixels that maximizes the mean SNR of the flux measurement for each star [8]. The background and cosmic ray-corrected pixels are then summed over the optimal aperture to define a flux estimate for each cadence frame.

17.2.3 Systematic Error Corrections

The PDC module's task is to remove systematic errors from the raw flux time series [44]. These include telescopic and spacecraft effects such as pointing errors, focus changes, and thermal effects on instrument properties. PDC cotrends each flux time series against ancillary engineering data, such as temperatures of the focal plane and electronics, reconstructed pointing, and focus variations, to remove signatures correlated with these proxy systematic error source measurements. A singular-value decomposition (SVD) is applied to the design matrix containing the ancillary data to identify the most significant independent components and to stabilize the matrix inversion inherent in the fit to the data. Additionally, PDC identifies residual isolated outliers and fills intra-quarter gaps so that the data for each quarterly segment are contiguous when presented to TPS. Finally, PDC adjusts the light curves to account for excess flux in the optimal apertures due to starfield crowding and the fraction of the target star flux in the aperture, to make apparent transit depths uniform from quarter to quarter as the stars move from detector to detector with each roll maneuver.

The systematic errors observed in flight exhibit a range of different timescales, from a few hours to several days to many days and weeks. Such phenomena include the intermittent modulation of the focus by $\sim 1\,\mu$m every 3.2 h by a heater on one of the reaction wheel assemblies during the first quarter (Q1). One of the Fine Guidance Sensor's guide stars through the first quarter of observations was an eclipsing binary whose 30% eclipses induced a 1-millipixel pointing excursion lasting ~ 8 h every 1.7 days [18]. By far the strongest systematic effects in the data so far have occurred after safe mode events [18] during which the photometer was shut off, the telescope cooled, and the focus changed by $\sim 2.2\,\mu$m/°C. One of these occurred at the end of Q1 and the second ~ 2 weeks into Q2. Thermal effects can be observed in the science data for ~ 5 days after each safe mode recovery. The fact that most systematics such as these affect all the science data simultaneously, and that there is a rich amount of ancillary engineering data and science diagnostics available, provides significant leverage in dealing with these effects.

---

[*] Each PRF is a super-resolution representation of the brightness distribution of a point source convolved with the spatial responsiveness of a single pixel, and were developed from test data acquired during the science commissioning phase prior to the start of science operations [9].

Some systematic phenomena are specific to individual stars and cannot be corrected by cotrending against ancillary data. The first issue is the occasional abrupt drop in pixel sensitivity that introduces a step discontinuity in an affected star's light curve (and associated centroids). This is often preceded immediately by a cosmic ray event, and is sometimes followed by an exponential recovery over the course of a few hours, but usually not to the same flux level as before. The typical drop in sensitivity is 1%, which is unmistakable in the flux time series. Such step discontinuities are identified separately from those due to operational activities, such as safe modes and pointing tweaks, and are mended by raising the light curve after the discontinuity for the remainder of the quarter. These events do not mimic transits, since they do not recover to the same pre-event flux level and few transits, if any, are affected by this correction. The second issue is that many stars exhibit coherent or slowly evolving variations that interfere with systematic error removal. The approach taken is to identify and remove strong coherent components in the frequency domain prior to cotrending, and then to restore these components to the residuals after cotrending.

Figure 17.3 shows the results of running two flux time series obtained during Quarter 2 through PDC, demonstrating PDC's effectiveness for some target stars. Unfortunately, PDC does not perform well in all cases, especially for stars that exhibit strong, noncoherent oscillations or variability. Furthermore, PDC can treat the same star differently in each quarter depending on the details of the star's variability and the positioning of the star on the CCD pixels, as discussed in Section 17.3.1. Learning to deal with the various systematic errors will probably consume a great deal of effort over the lifetime of the mission as the detection limit is pushed to smaller and smaller planets.

17.2.4 Transiting Planet Search

TPS searches for transiting planets by "stitching" the quarterly segments of data together to remove gaps and edge effects and then applies the wavelet-based, adaptive matched filter of Jenkins [21]. This approach is a time-adaptive approach that estimates the power spectrum of the observational noise as a function of time. This approach was developed specifically for solar-like stars with colored, broad-band power spectra. Some modifications to the original approach have been developed to accommodate target stars that exhibit coherent structure in the frequency domain. Similar to the approach adopted in PDC, we fit and remove strong harmonics that are inconsistent with transit signatures prior to applying the wavelet-based filter. This significantly increases the sensitivity of the transit search for such stars and also provides photometric precision estimates (as by-products of the search) that are more realistic for such targets. If the transit-like signature of a given target star exceeds 7.1σ, then a TCE is recorded and sent to DV for additional scrutiny. The heart of TPS is the adaptive wavelet-based detector discussed in the following subsection.

17.2.4.1 A Wavelet-Based Matched Filter

The optimal detector for a deterministic signal in colored Gaussian noise is a prewhitening filter followed by a simple matched filter [27]. In TPS, we implement the wavelet-based matched filter as per Jenkins [21] using Debauchies' 12-tap wavelets [14]. An octave-band

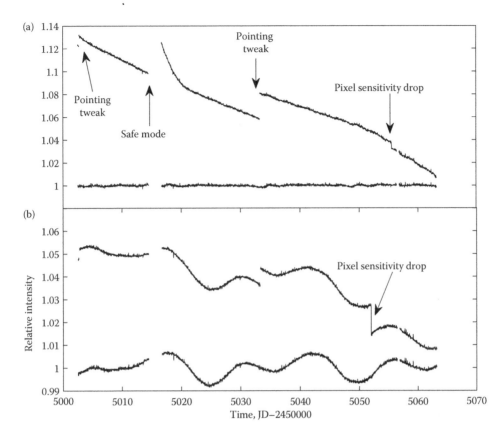

FIGURE 17.3 Raw and systematic error-corrected light curves for two different stars. The raw light curves exhibit step discontinuities due to pointing offsets, thermal transients due to safe modes and pixel sensitivity changes, as well as pointing errors, as indicated. The long–term drop in brightness over the 89-day period is due to a change in focus due to thermal effects. Panel (a) shows the raw flux time series (top curve, offset) and the PDC flux time series (bottom curve) of a $Kp = 15.6$ dwarf star. Panel (b) shows the raw and PDC flux time series for a $Kp = 14.9$ dwarf star that displays less sensitivity to the thermal transients. (From J. M. Jenkins et al., *ApJL* 713(2), L87–L91, 2010.)

filter bank separates the input flux time series into different band passes to estimate the power spectral density (PSD) of the background noise process as a function of time. This scheme is analogous to a graphic equalizer for an audio system. TPS constantly measures the "loudness" of the signal in each bandpass and then dials the gain for that channel so that the resulting noise power is flat across the entire spectrum. Flattening the power spectrum transforms the detection problem for colored noise into a simple one for white Gaussian noise (WGN), but also distorts transit waveforms in the flux time series. TPS correlates the trial transit pulse with the input flux time series in the whitened domain, accounting for the distortion resulting from the prewhitening process. This is analogous to visiting a funhouse "hall of mirrors" with a friend of yours and seeking to identify your friend's face by looking in the mirrors. By examining the way that your own face is distorted in each mirror, you can predict what your friend's face will look like in each particular mirror, given that you know what your friend's face looks like without distortion. The wavelet-based matched filter is developed as follows.

Let $x(n)$ be a flux time series. The over-complete wavelet transform of $x(n)$ is defined as

$$\mathbb{W}\{x(n)\} = \{x_1(n), x_2(n), \ldots, x_M(n)\}, \tag{17.1}$$

where

$$x_i(n) = h_i(n) * x(n), \quad i = 1, 2, \ldots, M, \tag{17.2}$$

"$*$" denotes convolution, and $h_i(n)$ for $i = 1, \ldots, M$ are the impulse responses of the filters in the filter bank implementation of the wavelet expansion with corresponding frequency responses $H_i(\omega)$ for $i = 1, \ldots, M$.

Figure 17.4 is a signal flow graph illustrating the process. The first filter, H_1, is a high-pass filter that passes frequency content from half the Nyquist frequency, $f_{Nyquist}$, to the Nyquist frequency ($[f_{Nyquist}/2, f_{Nyquist}]$). The next filter, H_2, passes frequency content in the interval ($[f_{Nyquist}/4, f_{Nyquist}/2]$), as illustrated in Figure 17.5. Each successive filter passes frequency content in a lower bandpass until the final filter, H_M, the lowest bandpass, which passes DC content as well. The number of filters is dictated by the number of observations and the length of the mother wavelet filter chosen to implement the filterbank. In this wavelet filter bank, there is no decimation of the outputs so that there are M times as many points in the wavelet expansion of a flux time series, $\{x_i(n)\}, i = 1, \ldots, M$, as there were in the original flux time series $x(n)$. This representation has the advantage of being shift invariant, so that the wavelet expansion of a trial transit pulse, $s(n)$, need only be calculated once. The noise in each channel of the filter bank is assumed to be white and Gaussian and its power is

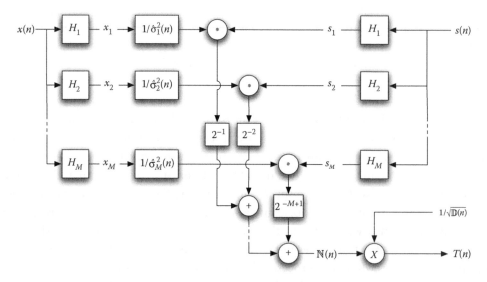

FIGURE 17.4 Signal flow diagram for TPS. (From J. M. Jenkins et al. *Society of Photo-Optical Instrumentation Engineers (SPIE) Conference Series*, Vol. 7740, pp. 77400D-1–77400D-11, 2010.)

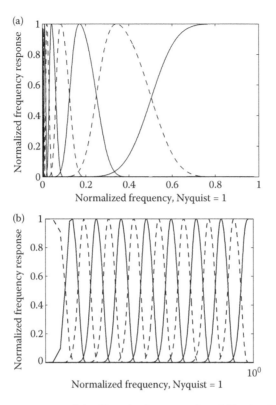

FIGURE 17.5 Frequency responses of the filters in the octave-band filterbank for a wavelet expansion corresponding to the signal flow graph in Figure 17.4 using Debauchies' 12-tap filter. (a) Frequency responses on a linear frequency scale. (b) Frequency response on a logarithmic frequency scale, illustrating the "constant-Q" property of an octave-band wavelet analysis. (From J.M. Jenkins. *ApJ* 575:493–505, 2002.)

estimated as a function of time by a moving variance estimator (essentially a moving average of the squares of the data points) with an analysis window chosen to be significantly longer than the duration of the trial transit pulse.

The detection statistic is computed by multiplying the whitened wavelet coefficients of the data by the whitened wavelet coefficients of the transit pulse:

$$T = \frac{\tilde{x} \cdot \tilde{s}}{\sqrt{\tilde{s} \cdot \tilde{s}}} = \frac{\sum_{i=1}^{M} 2^{-\min(i,M-1)} \sum_{n=1}^{N} [x_i(n)/\hat{\sigma}_i(n)] \, [s_i(n)/\hat{\sigma}_i(n)]}{\sqrt{\sum_{i=1}^{M} 2^{-\min(i,M-1)} \sum_{n=1}^{N} s_i^2(n)/\hat{\sigma}_i^2(n)}}, \tag{17.3}$$

where the time-varying channel variance estimates are given by

$$\hat{\sigma}_i^2(n) = \frac{1}{2^i K + 1} \sum_{k=n-2^{i-1}K}^{n+2^{i-1}K} x_i^2(k), \quad i = 1, \dots, M, \tag{17.4}$$

where each component $x_i(n)$ is periodically extended in the usual fashion and $2K+1$ is the length of the variance estimation window for the shortest time scale. In TPS, K is a parameter set to typically 50 times the trial transit duration.

To compute the detection statistic, $T(n)$, for a given transit pulse centered at all possible time steps, we simply "doubly whiten" $\mathbb{W}\{x(n)\}$ (i.e., divide $x_i(n)$ point-wise by $\hat{\sigma}_i^2(n)$, for $i = 1, \ldots, M$), correlate the results with $\mathbb{W}\{s(n)\}$, and apply the dot product relation, performing the analogous operations for the denominator, noting that $\hat{\sigma}_i^{-2}(n)$ is itself a time series:

$$T(n) = \frac{\mathbb{N}(n)}{\sqrt{\mathbb{D}(n)}} = \frac{\sum_{i=1}^{M} 2^{-\min(i,M-1)} [x_i(n)/\hat{\sigma}_i^2(n)] * s_i(-n)}{\sqrt{\sum_{i=1}^{M} 2^{-\min(i,M-1)} \hat{\sigma}_i^{-2}(n) * s_i^2(-n)}}. \tag{17.5}$$

Note that the "$-$" in $s_i(-n)$ indicates time reversal. The numerator term, $\mathbb{N}(n)$, is essentially the correlation of the reference transit pulse with the data. If the data were WGN, then the result could be obtained by simply convolving the transit pulse with the flux time series. The expected value of Equation 17.5 under that alternative hypothesis for which $x_i(n) = s_i(n)$ is $\sqrt{\sum_{i=1}^{M} 2^{-\min(i,M-1)} \hat{\sigma}_i^{-2}(n) * s_i^2(-n)}$. Thus, $\sqrt{\mathbb{D}(n)}$ is the expected SNR of the reference transit in the data as a function of time.

We define the combination of all noise sources as the Combined Differential Photometric Precision (CDPP), consisting basically of stellar noise, photon counting shot noise, and instrumental noise, which is the strength of the combined noise sources as seen by a transit pulse, and is obtained as

$$CDPP(n) = 1 \times 10^6 / \sqrt{\mathbb{D}(n)} \text{ ppm.} \tag{17.6}$$

Note that in the calculation of the in-band noise $\hat{\sigma}_i(n)$ in Equation 17.4, both the stellar background variation and the variation due to transits are included. For transits which are comparable in amplitude to the high-frequency stellar noise, the presence of the transits is a perturbation and the calculation still produces appropriate results. For transits which are extremely deep, the noise estimated in Equation 17.4 will be too large, and the detection statistic computed in Equation 17.5 will be too small. This infelicity is mitigated by replacing the root mean square (RMS) calculation in Equation 17.4 with the median absolute deviation (MAD) multiplied by 1.4826, which is the ratio of the RMS to the MAD for a Gaussian distribution. The use of MAD allows a more robust estimate of the stellar noise which is more nearly independent of the depth of transits, eclipses, outliers, and other low duty-cycle phenomena. Figure 17.6 illustrates the process of estimating CDPP for a star exhibiting strong transit-like features. Once the time-varying power spectral analysis has been performed, TPS searches for periodic transit pulses. Figure 17.7 shows the calculated 3-h CDPP for all targets for Module 7, Output 3 for one quarter.

17.2.5 Data Validation

The DV module performs a suite of tests to establish or break confidence in each TCE flagged by TPS, as well as to fit physical parameters to each transit-like signature [42,47].

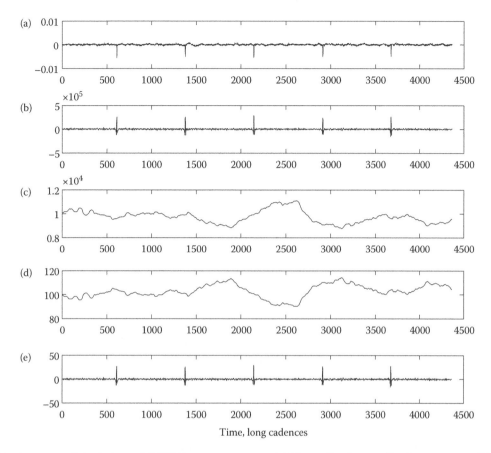

FIGURE 17.6 Calculation of CDPP for one target star. (a) Normalized target flux in parts per million (ppm). (b) Correlation time series $\mathbb{N}(n)$ from Equation 17.5. (c) Normalization time series $\mathbb{D}(n)$ from Equation 17.5. (d) 3-h CDPP time series. (e) Single-event statistic time series, $T(n)$. In all cases, the trial transit pulse, $s(n)$, is a square pulse of unit depth and 3-h duration. (From J. M. Jenkins et al. *Society of Photo-Optical Instrumentation Engineers (SPIE) Conference Series*, Vol. 7740, pp. 77400D-1–77400D-11, 2010.)

The statistical confidence in the TCE is examined by performing a bootstrap test [21,23] to take into account non-Gaussian statistics of the individual light curves. A transiting planet model is fitted to the transit signature as a joint noise characterization/parameter estimation problem. That is, the observation noise is *not* assumed to be white and its characteristics are estimated using the wavelet-based approach employed in TPS, but as an estimator, rather than as a detector. This process yields a set of physical parameters and an associated covariance matrix.

To eliminate false positives due to eclipsing binaries, the planet model fit is performed again to only the even transits, and then only to the odd transits, and the resulting odd/even depths and epochs are compared in order to see if the results indicate the presence of secondary eclipses. After the multi-TPS is complete, the periods of all candidate planets are compared to detect eclipsing binaries with significant eccentricity causing TPS to detect two transit pulse trains at essentially the same period, but at a phase other than 0.5.

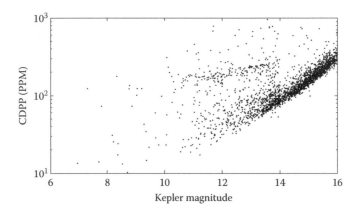

FIGURE 17.7 3-h Photometric precision as a function of Kepler magnitude for 2286 stars on Module 7, Output 3, for one representative quarter. There is an apparent drop in the density of the stars for $Kp > 14$ due to the manner in which target stars were selected, which penalized targets dimmer than 14th mag [2]. (From J. M. Jenkins et al. *Society of Photo-Optical Instrumentation Engineers (SPIE) Conference Series*, Vol. 7740, pp. 77400D-1–77400D-11, 2010.)

To guard against background eclipsing binaries, a centroid motion test is performed to determine whether the centroids shifted during the transit sequence. If so, the source position (right ascension and declination) can be estimated by the measured in- versus out-of-transit centroid shift normalized by the fractional change in brightness of the system (i.e., the transit "depth") [3,22].

17.3 A BAYESIAN APPROACH TO CORRECTING SYSTEMATIC ERRORS

This section details a Bayesian approach to identifying and removing instrumental signatures from the Kepler light curves that is under development now and promises to ameliorate many of the shortcomings of the current approach implemented in the SOC pipeline.

17.3.1 The Problem with PDC

Kepler is opening up a new vista in astronomy and astrophysics and is operating in a new regime where the instrumental signatures compete with the miniscule signatures of terrestrial planets transiting their host stars. The dynamic range of the intrinsic stellar variability observed in the Kepler light curves is breathtaking: RR Lyrae stars explosively oscillate with periods of ∼0.5 days, doubling their brightness over a few hours. Some flare stars double their brightness on much shorter timescales at unpredictable intervals. At the same time, some stars exhibit quasi-coherent oscillations with amplitudes of 50 ppm that can be seen by eye in the raw flux time series [25]. The richness of Kepler's data lies in the huge dynamic range for the variations in intensity (10^4) and the large dynamic range of timescales probed by the data, from a few minutes for SC data to weeks, months, and ultimately, to years. Given that Kepler was designed to be capable of resolving small 100-ppm changes in brightness over several hours, it is remarkably rewarding that it is revealing so much more. The challenge is that an instrument so sensitive to the amount of light from a star striking a small collection of pixels is also very sensitive to small changes in its environment.

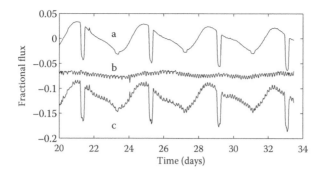

FIGURE 17.8 The light curve (a) for an eclipsing binary exhibiting spot variations on channel 2.1, along with (b) an LS fit to the original light curve using 10 systematic error components identified from the ensemble of stars, and (c) an LS-corrected light curve obtained by removing the model fit from the original light curve. Curves (b) and (c) have been offset from 0 for clarity.

PDC was conceived and tested against synthetic Kepler data that contained solar-like variability modeled on variability of solar irradiance and gravity oscillations (VIRGO)/Solar and Heliospheric Observatory (SOHO) data,[*] planetary transits, eclipsing binaries, and pointing errors. We did not expect focus variations on timescales relevant for transit detection to occur. In the face of the diversity of the photometric behavior of the stars and the complex signatures due to thermal changes in the instrument, we can only conclude that PDC is fundamentally flawed in its present incarnation.

The approach to correcting systematic errors in the pipeline suffers from the fact that the implicit model fitted to the data for each star is incomplete. Cotrending projects the data vector onto the selected basis vectors and removes the components that are parallel to any linear combination of the basis vectors. This process is guaranteed to reduce the bulk RMS residuals, but may do so at the cost of injecting additional noise or distortion into the flux time series. Indeed, this occurs frequently for stars with high intrinsic variability, such as RR Lyrae stars, eclipsing binaries, and classical pulsators, as demonstrated in Figure 17.8. The situation is analogous to opening a jigsaw puzzle box and finding only 30% of the pieces present. PDC gamely tries to put the jigsaw pieces together in order to match the picture on the box cover by stretching, rotating, and translating the pieces that were present in the box. The result is a set of pieces that roughly overlap the picture on the box cover, but one where the details do not necessarily match up well, even though individual pieces may obviously fit. In order to improve the performance of PDC, we need to provide it with constraints on the magnitudes and signs of the fit coefficients.

The problem with PDC is a common problem with systematic error removal that use a least squares approach. Methods such as SysRem [41] and trend filtering algorithm (TFA) [33] fall into this category and the resulting light curves tend to be over-fitted for Kepler data. For example, if one of the model terms is strongly related to focus variations and the long-term trend is for the width of the stellar point spread function to broaden over

[*] VIRGO is an investigation on the solar and heliospheric observatory SOHO, a mission of international cooperation between European space agency (ESA) and NASA.

the observation interval, then the flux for all stars should decrease over time. A least squares fit, however, may invert the focus-related model term for a star whose flux increases over the observation interval, thereby removing the signature of intrinsic stellar variability from this light curve because there is a coincidental correlation between the observed change in flux and the observed change in focus. Given that the star would be expected to dim slightly over time, due to the focus change, PDC should be correcting the star so that it brightens slightly more than the original flux time series indicates, rather than "flattening" the light curve.

17.3.2 Why Should We Hope to Do Better?

While PDC is currently not providing uniformly good results for all stellar targets, there is good reason to believe that it can be made significantly better. The instrument behavior that imputes artifacts into the light curves does so across the board. That is, all the stars' flux measurements are responding in some correlated way to these systematic errors. Improving the quality of the results requires an underlying physical model describing how a change in pointing or the shape of the telescope will induce a change in an individual light curve. Unfortunately, given that the thermal state of the telescope cannot be described by a model with a small or even finite number of parameters, and the fact that the temperature measurements on the spacecraft are few and coarsely quantized, where they do exist, we do not have access to such a model. However we can look to the ensemble behavior of the stars to develop an empirical model of the underlying physics. For example, the photometric change that can be induced by a pointing change of 0.1 arcsec must be bounded, and this bound can be estimated by looking at how the collection of stars behaves for a pointing change of this magnitude.

Channel 2.1 is the most thermally sensitive CCD channel in Kepler's focal plane. Nearly all stars on this channel exhibit obvious focus- and pointing-related instrumental signatures in their pixel time series and flux time series. Figure 17.9 shows the light curve for a typical star on channel 2.1 for Q1. Note the long-term decrease in flux over the 34-day interval. This is due to seasonal changes in the shape of the telescope and therefore its focus as the Sun's incidence angle rotates about the barrel of the telescope while the spacecraft orbits the Sun and maintains its attitude fixed on the FOV. The short-term oscillations evident in the time series are due to focus changes driven by a heater cycling on and off to condition the temperature of the box containing reaction wheels 3 and 4 on the spacecraft

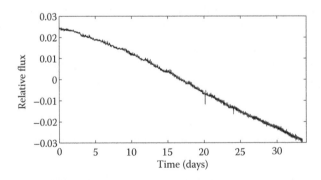

FIGURE 17.9 The light curve for a typical star on channel 2.1.

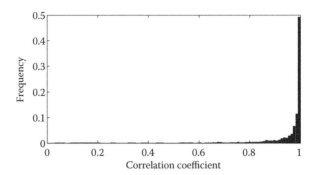

FIGURE 17.10 Histogram of the absolute value of the correlation coefficient matrix for 2000 light curves on channel 2.1.

bus. This component was receiving more and more shade throughout this time interval and the thermostat actuated more frequently over time.

How can we separate intrinsic stellar variability from instrumental signatures? We do not expect intrinsic stellar variability to be correlated from star to star, except for rare coincidences, and even then one would not expect a high degree of correlation for all timescales. However, we *do* expect instrumental signatures to be highly correlated from star to star and can exploit this observation to provide constraints on the cotrending that PDC performs.

Figure 17.10 shows a histogram of the absolute value of the correlation coefficient matrix for 2000 stars on channel 2.1. The stars' light curves are highly correlated as evidenced by the pile-up near an absolute correlation coefficient of 1. Examination of individual light curves indicates that these light curves are contaminated to a large degree by instrumental signatures, as evidenced in Figure 17.9. Not all the stars are dominated by systematic errors. The stars that are poorly correlated with other stars tend to be dominated by large amplitude, intrinsic stellar variability. The trick is to come up with a method that can distinguish between intrinsic stellar variability and chance correlations with linear combinations of the instrumental diagnostic time series used to cotrend out systematic errors.

17.3.3 The MAP Approach

A Bayesian approach called the Maximum A Posteriori (MAP) method should allow us to provide PDC with constraints on the fitted coefficients to help prevent over-fitting and distortion of intrinsic stellar variability. In this exposition we follow the notation of Kay [28].

The MAP technique examines the behavior of the fit coefficients across an ensemble of stars on each CCD readout channel in order to develop a description for the "typical" value for each model term. This description is a probability density function (PDF) that can be used to constrain the coefficients fitted in a second pass. To develop this approach, we build on the current maximum likelihood approach of PDC.

The current PDC models each light curve, $\mathbf{y}$, as a linear combination of instrumental systematic vectors arranged as the columns of a design matrix, $\mathbf{H}$, plus zero-mean, Gaussian observation noise, $\mathbf{w}$:

$$\hat{\mathbf{y}} = \mathbf{H}\boldsymbol{\theta} + \mathbf{w}. \tag{17.7}$$

The Maximum Likelihood Estimator (MLE) seeks to find the solution, $\hat{\boldsymbol{\theta}}_{\text{MLE}}$, that maximizes the likelihood function, $p(\mathbf{y}; \boldsymbol{\theta})$, given by

$$p(\mathbf{y}; \boldsymbol{\theta}) = \frac{1}{(2\pi)^{N/2} |\mathbf{C_w}|^{1/2}} \exp\left[-\frac{1}{2}(\mathbf{y} - \mathbf{H}\boldsymbol{\theta})^{\text{T}} \mathbf{C_w^{-1}}(\mathbf{y} - \mathbf{H}\boldsymbol{\theta})\right], \qquad (17.8)$$

where $\mathbf{C_w}$ is the covariance of $\mathbf{w}$. Taking the gradient of the log of Equation 17.8, setting it equal to zero, and solving for $\boldsymbol{\theta}$ yields the familiar least squares solution,

$$\hat{\boldsymbol{\theta}}_{\text{MLE}} = \left(\mathbf{H}^{\text{T}}\mathbf{C_w^{-1}}\mathbf{H}\right)^{-1} \mathbf{H}^{\text{T}}\mathbf{C_w^{-1}}\mathbf{y}. \qquad (17.9)$$

Adopting the Bayesian approach allows us to incorporate side information, such as knowledge of prior constraints on the model, in a natural way. Bayesianists view the underlying model as being drawn from a distribution and the data as being one realization of this process. In this case we wish to find the MAP estimator of the model coefficients given the observations (data):

$$\hat{\boldsymbol{\theta}}_{\text{MAP}} = \arg\max_{\boldsymbol{\theta}} p(\boldsymbol{\theta}|\mathbf{y}) = \arg\max_{\boldsymbol{\theta}} p(\mathbf{y}|\boldsymbol{\theta}) p(\boldsymbol{\theta}), \qquad (17.10)$$

where we have applied Bayes' rule to simplify the expression. In this equation, $p(\boldsymbol{\theta})$ is the prior PDF of the model coefficients. The mathematical form for $p(\mathbf{y}|\boldsymbol{\theta})$ is the same as for the non-Bayesian likelihood function $p(\mathbf{y}; \boldsymbol{\theta})$ in Equation 17.8. For illustration purposes, if we adopt a Gaussian PDF for $\boldsymbol{\theta}$,

$$p(\boldsymbol{\theta}) = \frac{1}{(2\pi)^{M/2} |\mathbf{C_\theta}|^{1/2}} \exp\left[-\frac{1}{2}(\boldsymbol{\theta} - \boldsymbol{\mu_\theta})^{\text{T}} \mathbf{C_\theta^{-1}}(\boldsymbol{\theta} - \boldsymbol{\mu_\theta})\right], \qquad (17.11)$$

where $\mathbf{C_\theta}$ and $\boldsymbol{\mu_\theta}$ are the covariance and mean of $\boldsymbol{\theta}$, respectively, we can then maximize Equation 17.10 by maximizing its log likelihood to obtain

$$\hat{\boldsymbol{\theta}}_{\text{MAP}} = \left(\mathbf{H}^{\text{T}}\mathbf{C_w^{-1}}\mathbf{H} + \mathbf{C_\theta^{-1}}\right)^{-1} \left(\mathbf{H}^{\text{T}}\mathbf{C_w^{-1}}\mathbf{y} + \mathbf{C_\theta^{-1}}\boldsymbol{\mu_\theta}\right). \qquad (17.12)$$

Note that if the observation noise, $\mathbf{w}$, is zero-mean, WGN with variance σ^2, then Equation 17.12 can be rewritten as

$$\hat{\boldsymbol{\theta}}_{\text{MAP}} = \left(\mathbf{H}^{\text{T}}\mathbf{H} + \sigma^2\mathbf{C_\theta^{-1}}\right)^{-1} \left(\mathbf{H}^{\text{T}}\mathbf{y} + \sigma^2\mathbf{C_\theta^{-1}}\boldsymbol{\mu_\theta}\right). \qquad (17.13)$$

If the uncertainties in the data are large compared to the "spread" allowed by the prior PDF for the model, then the MAP estimator gives more weight to the prior so that $\hat{\boldsymbol{\theta}}_{\text{MAP}} \rightarrow \boldsymbol{\mu_\theta}$ as $\sigma^2 \rightarrow \infty$. This case would correspond, for example, to targets with large stellar variability such as with the target given in Figure 17.8. The MAP weighting would constrain the fitter from distorting the light curve and introducing noise on a short timescale. Conversely, if the uncertainties in the data are small compared to the degree to which the prior PDF confines the model, the MAP estimator "trusts" the data over the prior knowledge and $\hat{\theta}_{MAP} \rightarrow \hat{\theta}_{MLE}$ as $\sigma^2 \rightarrow 0$. This case would correspond to targets with small stellar variability where there is little risk of over-fitting and distortion of the light curves and it is a "safe" bet to use the conditional, least squares fit.

In order to apply MAP to Kepler data we need to develop knowledge of the prior PDF of the model terms representing instrumental signatures that are to be fitted to the data. The following section details a prototype that demonstrates how MAP can help PDC.

17.3.4 An Application of MAP to Kepler Data

This section describes a prototype for applying MAP to Kepler data for channel 2.1 from Q1, the first 34 days of science operations of Kepler. Rather than using ancillary engineering data or other instrumental diagnostics, we extract model vectors representing instrumental systematic effects from the light curves themselves using SVD. We also construct a Gaussian prior for the model that is a function of Kepler magnitude. While this is not the most accurate prior PDF possible, it allows us to determine whether the MAP approach will be helpful in reducing the over-fitting and distortion of intrinsic stellar variability.

We first select a subset of light curves from this channel upon which to perform SVD by choosing light curves that are highly correlated with other light curves. Specifically, we take the 10th percentile of the absolute value of the correlation coefficients for each star relative to each other star in the set as the measure of the degree to which each star is correlated with the other stars as a group. The 1400 light curves with the highest "typical" absolute correlation coefficients (>0.92) form the basis for the characterization of systematic errors.

An SVD decomposition of these 1400 highly correlated light curves was performed to extract the vectors most responsible for the correlated features across these stars' light curves. Figure 17.11 shows the first three SVD components responsible for most of the variability

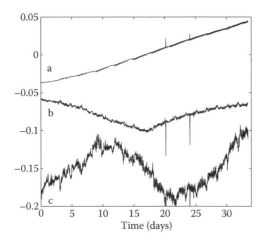

FIGURE 17.11 First three singular components for highly correlated light curves on channel 2.1, offset for clarity. These singular components, obtained by performing SVD on the 1400 most highly correlated light curves for stars on this channel are plotted. (a) first component, (b) second component, and the (c) third component.

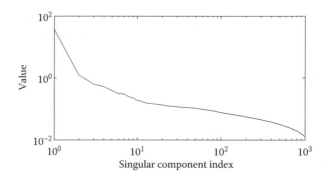

FIGURE 17.12 The singular values for highly correlated light curves on channel 2.1. The strength of the singular values for the light curves for channel 2.1 drops quickly, indicating that the signatures correlated across the light curves can be represented by a small number of components.

across the flux time series. These components exhibit features that are associated with thermal and pointing changes during Q1 observations, and include impulses corresponding to times when Argabrightenings occurred. As Figure 17.12 shows, the correlated signatures across the flux time series can be explained by a small number (~10) of components. Since the columns of the resulting design matrix are orthogonal, the prior PDF has a diagonal covariance matrix, simplifying the construction of the prior.

Figure 17.13 shows the values of the first coefficient corresponding to the first singular vector in Figure 17.11 as a function of stellar magnitude. The 600 least-correlated stars are plotted with cross symbols and many of them fall outside the core of points rendered for the 1400 most-correlated stars. We fit a 4th-order robust polynomial[*] to the points corresponding to the 1400 most-correlated stars, constraining the fitted polynomial to be zero and have zero slope at $K_p = 6$. Similar fits were performed for each of the 10 singular vectors included in the design matrix, A. The resulting polynomials describe the mean value for each model term as a function of Kepler magnitude, $\mu_\theta (K_p)$. The covariance matrix diagonal entries, C_θ, were defined as the MAD of the residuals of the polynomial fits to the 1400 data points for each term. While this model for the prior PDF is simplistic, it allows a demonstration of the capability of MAP to prevent overfitting.

To complete the description of the statistical quantities, we estimate the observation noise variance, σ_w^2, as the standard deviation of each uncorrected light curve. A MAP fit was performed for each star on channel 2.1 using these descriptions for the prior PDF and for the noise variance of each light curve using Equation 17.13.

The results of this experiment demonstrate the potential of the MAP approach to solving PDC's major weakness. Figure 17.14 shows the light curves of two stars that are dominated by intrinsic stellar variations. One star is an RR Lyrae star that oscillates every ~0.5 days, doubling its brightness over the span of a few hours with each cycle. The other star is an eclipsing binary that exhibits spot variations that change slightly with each orbit of the stars.

---

[*] This polynomial is the product of a quadratic polynomial and the term $(K_p - 6)^2$.

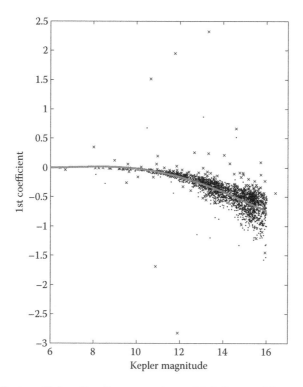

FIGURE 17.13 The first coefficient for all stars on channel 2.1. Some of the more egregious outliers have been clipped by the y-axis limits. This first term in the model (see Figure 17.11) is dependent on the star's brightness. The crosses indicate stars that are not well correlated with other stars, while the dots indicate stars that are highly correlated with other stars' light curves. The curve is a polynomial fit to the data for highly correlated stars.

In both cases an MLE estimator for the systematic errors over-fits the data, distorting the signatures of the intrinsic stellar astrophysics as well as introducing additional noise on short timescales. Such distortions destroy the usefulness of the PDC-corrected light curves for astrophysical investigations. In contrast, the MAP approach recognizes that the fitted coefficients for an MLE estimator are not typical and the prior PDF "pulls" the MAP coefficients in toward the core distribution to provide a more reasonable fit given the behavior of the ensemble of stars. The MAP fits do not over-fit and distort the corrected light curves, while still correcting for the long-term focus drift and compensating for the heater cycle-induced focus variations on timescales of hours.

The light curves of "quiet" stars that are strongly influenced by systematic errors are treated benignly by the MAP approach as well, as evidenced by the light curve in Figure 17.15. Here there is little difference between the MLE estimate and the MAP estimate. Both do a good job of removing the long-term trends due to the drift in focus and the shorter timescale variations due to the heater cycle. The MLE estimator does a better job of fitting the Argabrightening events near days 20 and 24, and provides slightly less noisy light curves than the MAP estimates for the highly correlated stars. For this experiment the tuning of

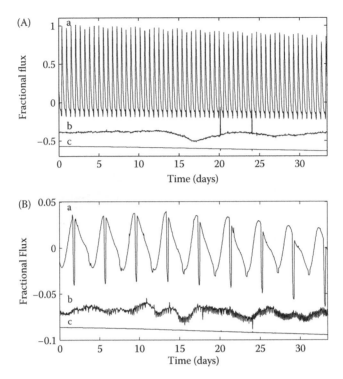

FIGURE 17.14 The light curves (a) for two stars on channel 2.1, along with (b) an LS fit to instrumental components extracted from the light curves and (c) a Bayesian fit to the same instrumental components. Curves (b) and (c) have been offset from 0 for clarity. Panel A shows the results for an RR Lyrae star while panel B shows them for an eclipsing binary. Both light curves are dominated by intrinsic stellar variability rather than by instrumental signatures. The RR Lyrae doubles its brightness every 0.5 days, while the eclipsing binary exhibits spot variations that change slowly over time.

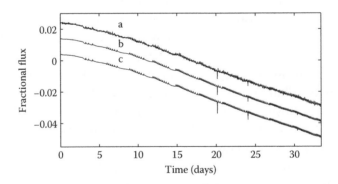

FIGURE 17.15 The light curve (a) for a quiet star on channel 2.1, along with (b) an LS fit to instrumental components extracted from the light curves and (c) a Bayesian fit to the same instrumental components. Curves (b) and (c) have been offset from 0 for clarity. This is one of the 1400 stars whose light curves are most highly correlated with each other, indicating that they are strongly influenced by systematic errors. In this case, there is not much noticeable difference between the MAP approach and a robust, least–squares approach.

the prior PDF was driven by the desire to demonstrate the MAP approach's ability to ignore strong intrinsic stellar variations unlikely to be caused by instrumental signatures.

17.3.5 Empirical MAP Implementation Considerations

The softest aspect of the prototype is the use of a Gaussian prior PDF. While the distributions of the coefficients for all 10 model terms in this experiment were unimodal, there are other CCD channels for which this is not the case. There may be other independent variables, such as location on the sky (right ascension and declination) that would allow us to tease apart the multiple peaks apparent in the distributions when viewed simply as functions of stellar magnitude. However, even if this proves to be the case, there is no reason to believe that the prior PDFs should have Gaussian profiles. The most direct way to estimate the prior PDFs is to use an empirical method, such as that of Bowman and Azzalini [7]. One significant disadvantage of the use of an empirical prior is computational speed. Priors that are "bumpy" or multimodal complicate the process of finding the global minimum of the joint PDF of Equation 17.10. The solution must be found numerically by iterative methods and care must be taken not to get "stuck" in a local minimum. The beauty of the simple Gaussian prior is that the resulting log-likelihood function is quadratic in the unknown parameters and admits a closed-form solution. An approach that makes use of empirical priors must yield a solution that is practical given that it must be applied to the light curves of ∼160,000 stars on a monthly basis, and then separately on a quarterly basis when new data arrive in the Kepler SOC. It is also evident from Figure 17.13 that the width of the distribution of the fit coefficients increases toward dimmer stars. The broadening of the distribution of the coefficients cannot be explained by the greater uncertainties in the coefficients themselves, even though there is more noise in the dimmer stars' light curves and one would expect more scatter. We are also investigating which of the ∼40 ancillary data vectors PDC is using are most helpful in modeling the systematic errors as part of the overfitting problem that may be due to having an insufficiently parsimonious model to begin with. The prototype described in the last section points the way to an approach that promises to significantly improve PDC's handling of stars with significant stellar variability and the treatment of systematic noise in general, but significant development will be required to realize this promise.

17.4 CONCLUSIONS

The Kepler mission is unveiling an astonishing array of astrophysical mysteries with its unprecedented photometric precision and near-continuous observations. The SOC Science Pipeline is a crucial component of the system driving the discovery process. It is required to process the pixel data for ∼200,000 stars observed during the course of the mission to date at a rate of 6×10^{12} pixels/year to derive light curves and centroid time series for each star, accumulating 18,000 samples/year. This system must deal with idiosyncratic behavior on the part of the stars of so much interest to the astrophysics community, as well as the complex behavior of the photometer that imputes signatures into the photometric measurements that compete for attention with those of the stars themselves, and more importantly, with the planetary signatures so eagerly sought by the mission. The systematic error correction afforded by the current algorithm is providing a rich and large assortment of candidate planets to validate. However, it can clearly be improved by the incorporation of information about the behavior of the ensemble of targets on each CCD camera in order to develop an implicit model of the effects of the various and important sources of instrumental systematic errors. An effort is currently under way to develop an implementation of a Bayesian approach called the MAP estimator for the Kepler pipeline in order to realize this goal. Furthermore, we bring a wavelet-based technique for adaptively characterizing the observation noise and stellar variability to perform transit detection on the systematic error-corrected flux time series.

The pipeline must process the data with no intervention from humans in a parallel manner on a cluster of worker machines controlled by a Pipeline infrastructure, and it must do so in a timely fashion in order to keep up with the ever-accumulating data and to support the follow-up observing season for the Kepler FOV, which can be observed from the ground between April and November each year. The current Pipeline is doing an admirable job of keeping up with the data and providing the science team and the community with extraordinary light curves and astrophysical discoveries. The complex behavior of the instrument and the stars poses an interesting and important set of puzzles to solve as the mission continues.

The Kepler mission will not be the last spaceborne astrophysics mission to scan the heavens for planetary abodes. ESA is considering the PLATO mission [13], an ambitious transit survey to discover Earth-size planets around the closest and brightest main sequence stars. It would launch in 2017 or 2018 into an orbit at L2 to survey several large >800 deg^2 FOV with an instrument composed of ~ 40 small telescopes. Massachusetts Institute of Technology (MIT) has proposed the Transiting Exoplanet Survey Satellite (TESS)[*] mission to NASA's Explorer program, an all-sky transit survey satellite to be launched into orbit in 2016 for a two-year mission with an instrument composed of multiple telescopes with disjointed FOV. TESS will observe as many as 2.5 million stars over its mission lifetime. Clearly, these future missions can benefit from the lessons learned from the Kepler mission and will face many of the same challenges that may be more difficult to solve given the significantly larger volume of data to be collected on a far greater number of stars than Kepler has had to deal with. Given the intense interest in exoplanets by the public and by the astronomical community, the future for exoplanet science appears to just be dawning with the initial success of the Kepler mission.

ACKNOWLEDGMENTS

Funding for this discovery mission is provided by NASA's Science Mission Directorate. We thank the thousands of people whose efforts made Kepler's grand voyage of discovery possible. We especially want to thank the Kepler Science Operations Center staff who helped design, build, and operate the Kepler Science Pipeline for putting their hearts into this endeavor.

---

[*] See http://space.mit.edu/TESS/Mission_presentation/Welcome.html.

REFERENCES

1. N. M. Batalha, W. J. Borucki, S. T. Bryson, L. A. Buchhave, D. A. Caldwell, J. Christensen-Dalsgaard, D. Ciardi et al., Kepler's first rocky planet: Kepler-10b. *ApJ*, 729:27–47, 2011.

2. N. M. Batalha, W. J. Borucki, D. G. Koch, S. T. Bryson, M. R. Haas, T. M. Brown, D. A. Caldwell et al., Selection, prioritization, and characteristics of Kepler target stars. *ApJL*, 713:L109–L114, 2010.

3. N. M. Batalha, J. F. Rowe, R. L. Gilliland, J. J. Jenkins, D. Caldwell, W. J. Borucki, D. G. Koch et al., Pre-spectroscopic false-positive elimination of Kepler planet candidates. *ApJL*, 713:L103–L108, 2010.

4. W. J. Borucki, D. Koch, G. Basri, N. Batalha, T. Brown, D. Caldwell, J. Caldwell et al., Kepler planet-detection mission: Introduction and first results. *Science*, 327:977–980, 2010.

5. W. J. Borucki, D. G. Koch, G. Basri, N. Batalha, T. M. Brown, S. T. Bryson, D. Caldwell et al., Characteristics of planetary candidates observed by Kepler, II: Analysis of the first four months of data. *ApJ*, 736:19–40, 2011.

6. W. J. Borucki, D. G. Koch, T. M. Brown, G. Basri, N. M. Batalha, D. A. Caldwell, W. D. Cochran et al., Kepler-4b: A hot Neptune-like planet of a G0 star near main-sequence turnoff. *ApJL*, 713:L126–L130, 2010.

7. A. W. Bowman and A. Azzalini., *Applied Smoothing Techniques for Data Analysis*. New York: Oxford University Press, 1987.

8. S. T. Bryson, J. M. Jenkins, T. C. Klaus, M. T. Cote, E. V. Quintana, J. R. Hall, K. Ibrahim et al., Selecting pixels for Kepler downlink. In N. M. Radziwill and A. Bridger, editors, *Society of Photo-Optical Instrumentation Engineers (SPIE) Conference Series*, Vol. 7740, pp. 77401D-1–77401D-12, 2010.

9. S. T. Bryson, P. Tenenbaum, J. M. Jenkins, H. Chandrasekaran, T. Klaus, D. A. Caldwell, R. L. Gilliland et al., The Kepler pixel response function. *ApJL*, 713:L97–L102, 2010.

10. D. A. Caldwell, J. J. Kolodziejczak, J. E. Van Cleve, J. M. Jenkins, P. R. Gazis, V. S. Argabright, E. E. Bachtell et al., Instrument performance in Kepler's first months. *ApJL*, 713:L92–L96, 2010.

11. D. A. Caldwell, J. E. van Cleve, J. M. Jenkins, V. S. Argabright, J. J. Kolodziejczak, E. W. Dunham, J. C. Geary et al., Kepler instrument performance: An in-flight update. In J. M. Oschmann Jr., M. C. Clampin, and H. A. MacEwen, editors, *Society of Photo-Optical Instrumentation Engineers (SPIE) Conference Series*, Vol. 7731, pp. 773117-1–773117-11, 2010.

12. B. D. Clarke, C. Allen, S. T. Bryson, D. A. Caldwell, H. Chandrasekaran, M. T. Cote, F. Girouard et al., A framework for propagation of uncertainties in the Kepler data analysis pipeline. In N. M. Radziwill and A. Bridger, editors, *Society of Photo-Optical Instrumentation Engineers (SPIE) Conference Series*, Vol. 7740, pp. 774020-1–774020-12, 2010.

13. R. Claudi., A new opportunity from space: PLATO mission. *Ap&SS*, 328:319–323, 2010.

14. I. Debauchies. Orthonormal bases of compactly supported wavelets. *Comm. Pure Appl. Math.*, 41:909–996, 1988.

15. E. W. Dunham, W. J. Borucki, D. G. Koch, N. M. Batalha, L. A. Buchhave, T. M. Brown, D. A. Caldwell et al., Kepler-6b: A transiting hot Jupiter orbiting a metal-rich star. *ApJL*, 713:L136–L139, 2010.

16. T. N. Gautier, III, N. M. Batalha, W. J. Borucki, W. D. Cochran, E. W. Dunham, S. B. Howell, D. G. Koch et al., The Kepler follow-up observation program. *ArXiv e-prints*, 2010 (arXiv:1001.0352).

17. R. L. Gilliland, T. M. Brown, J. Christensen-Dalsgaard, H. Kjeldsen, C. Aerts, T. Appourchaux, S. Basu et al., Kepler Asteroseismology program: Introduction and first results. *PASP*, 122:131–143, 2010.

18. M. R. Haas, N. M. Batalha, S. T. Bryson, D. A. Caldwell, J. L. Dotson, J. Hall, J. M. Jenkins et al., Kepler science operations. *ApJL*, 713:L115–L119, 2010.

19. J. R. Hall, K. Ibrahim, T. C. Klaus, M. T. Cote, C. Middour, M. R. Haas, J. L. Dotson et al., Kepler science operations processes, procedures, and tools. In D. R. Silva, A. B. Peck, and B. T. Soifer, editors, *Society of Photo-Optical Instrumentation Engineers (SPIE) Conference Series*, Vol. 7737, pp. 77370H-1–77370H-12, 2010.

20. M. J. Holman, D. C. Fabrycky, D. Ragozzine, E. B. Ford, J. H. Steffen, W. F. Welsh, J. J. Lissauer et al., Kepler-9: A system of multiple planets transiting a sun-like star, confirmed by timing variations. *Science*, 330:51–54, 2010.

21. J. M. Jenkins., The impact of solar-like variability on the detectability of transiting terrestrial planets. *ApJ*, 575:493–505, 2002.

22. J. M. Jenkins, W. J. Borucki, D. G. Koch, G. W. Marcy, W. D. Cochran, W. F. Welsh, G. Basri et al., Discovery and Rossiter-McLaughlin effect of exoplanet Kepler-8b. *ApJ*, 724:1108–1119, 2010.

23. J. M. Jenkins, D. A. Caldwell, and W. J. Borucki., Some tests to establish confidence in planets discovered by transit photometry. *ApJ*, 564:495–507, 2002.

24. J. M. Jenkins, D. A. Caldwell, H. Chandrasekaran, J. D. Twicken, S. T. Bryson, E. V. Quintana, B. D. Clarke et al., Overview of the Kepler science processing pipeline. *ApJL*, 713:L87–L91, 2010.

25. J. M. Jenkins, D. A. Caldwell, H. Chandrasekaran, J. D. Twicken, S. T. Bryson, E. V. Quintana, B. D. Clarke et al., Initial characteristics of Kepler long cadence data for detecting transiting planets. *ApJL*, 713:L120–L125, 2010.

26. J. M. Jenkins, H. Chandrasekaran, S. D. McCauliff, D. A. Caldwell, P. Tenenbaum, J. Li, T. C. Klaus, M. T. Cote, and C. Middour., Transiting planet search in the Kepler pipeline. In N. M. Radziwill and A. Bridger, editors, *Society of Photo-Optical Instrumentation Engineers (SPIE) Conference Series*, Vol. 7740, pp. 77400D-1–77400D-11, 2010.

27. S. Kay., Adaptive detection for unknown noise power spectral densities. *IEEE Trans. Sig. Proc.*, 47(1):10–21, 1999.

28. S. M. Kay., *Fundamentals of Statistical Signal Processing: Estimation Theory*. New Jersey: Prentice-Hall PTR, 1993.

29. T. C. Klaus, M. T. Cote, S. McCauliff, F. R. Girouard, B. Wohler, C. Allen, H. Chandrasekaran et al., The Kepler science operations center pipeline framework extensions. In N. M. Radziwill and A. Bridger, editors, *Society of Photo-Optical Instrumentation Engineers (SPIE) Conference Series*, Vol. 7740, pp. 774018-1–774018-11, 2010.

30. T. C. Klaus, S. McCauliff, M. T. Cote, F. R. Girouard, B. Wohler, C. Allen, C. Middour, D. A. Caldwell, and J. M. Jenkins., Kepler science operations center pipeline framework. In N. M. Radziwill and A. Bridger, editors, *Society of Photo-Optical Instrumentation Engineers (SPIE) Conference Series*, Vol. 7740, pp. 77401B-1–77401B-12, 2010.

31. D. G. Koch, W. J. Borucki, G. Basri, N. M. Batalha, T. M. Brown, D. Caldwell, J. Christensen-Dalsgaard et al., Kepler mission design, realized photometric performance, and early science. *ApJL*, 713:L79–L86, 2010.

32. D. G. Koch, W. J. Borucki, J. F. Rowe, N. M. Batalha, T. M. Brown, D. A. Caldwell, J. Caldwell et al., Discovery of the transiting planet Kepler-5b. *ApJL*, 713:L131–L135, 2010.

33. G. Kovács, G. Bakos, and R. W. Noyes. A trend filtering algorithm for wide-field variability surveys. *MNRAS*, 356:557–567, 2005.

34. D. W. Latham, W. J. Borucki, D. G. Koch, T. M. Brown, L. A. Buchhave, G. Basri, N. M. Batalha et al., Kepler-7b: A transiting planet with unusually low density. *ApJL*, 713:L140–L144, 2010.

35. D. W. Latham, R. P. Stefanik, T. Mazeh, M. Mayor, and G. Burki., The unseen companion of HD114762—A probable brown dwarf. *Nature*, 339:38–40, 1989.

36. J. J. Lissauer, D. C. Fabrycky, E. B. Ford, W. J. Borucki, F. Fressin, G. W. Marcy, J. A. Orosz et al., A closely packed system of low-mass, low-density planets transiting Kepler-11. *Nature*, 470:53–58, 2011.

37. M. Mayor and D. Queloz., A Jupiter-mass companion to a solar-type star. *Nature*, 378:355–359, 1995.

38. S. McCauliff, M. T. Cote, F. R. Girouard, C. Middour, T. C. Klaus, and B. Wohler., The Kepler DB: A database management system for arrays, sparse arrays, and binary data. In N. M. Radziwill and A. Bridger, editors, *Society of Photo-Optical Instrumentation Engineers (SPIE) Conference Series*, Vol. 7740, pp. 77400M-1–77400M-12, 2010.

39. T. D. Morton and J. A. Johnson., On the low false positive probabilities of Kepler planet candidates. *ApJ*, 738:170–181, 2011.

40. E. V. Quintana, J. M. Jenkins, B. D. Clarke, H. Chandrasekaran, J. D. Twicken, S. D. McCauliff, M. T. Cote et al., Pixel-level calibration in the Kepler science operations center pipeline. In N. M. Radziwill and A. Bridger, editors, *Society of Photo-Optical Instrumentation Engineers (SPIE) Conference Series*, Vol. 7740, pp. 77401X-1–77401X-12, 2010.

41. O. Tamuz, T. Mazeh, and S. Zucker., Correcting systematic effects in a large set of photometric light curves. *MNRAS*, 356:1466–1470, 2005.

42. P. Tenenbaum, S. T. Bryson, H. Chandrasekaran, J. Li, E. Quintana, J. D. Twicken, and J. M. Jenkins., An algorithm for the fitting of planet models to Kepler light curves. In N. M. Radziwill and A. Bridger, editors, *Society of Photo-Optical Instrumentation Engineers (SPIE) Conference Series*, Vol. 7740, pp. 77400J-1–77400J-12, 2010.

43. G. Torres, F. Fressin, N. M. Batalha, W. J. Borucki, T. M. Brown, S. T. Bryson, L. A. Buchhave et al., Modeling Kepler transit light curves as false positives: Rejection of blend scenarios for Kepler-9, and validation of Kepler-9d, a super-Earth-size planet in a multiple system. *ApJ*, 727:24–41, 2010.

44. J. D. Twicken, H. Chandrasekaran, J. M. Jenkins, J. P. Gunter, F. Girouard, and T. C. Klaus., Presearch data conditioning in the Kepler science operations center pipeline. In N. M. Radziwill and A. Bridger, editors, *Society of Photo-Optical Instrumentation Engineers (SPIE) Conference Series*, Vol. 7740, pp. 77401U-1–77401U-12, 2010.

45. J. D. Twicken, B. D. Clarke, S. T. Bryson, P. Tenenbaum, H. Wu, J. M. Jenkins, F. Girouard, and T. C. Klaus., Photometric analysis in the Kepler science operations center pipeline. In N. M. Radziwill and A. Bridger, editors, *Society of Photo-Optical Instrumentation Engineers (SPIE) Conference Series*, Vol. 7740, pp. 774023-1–774023-12, 2010.

46. J. Van Cleve and D. A. Caldwell. *Kepler Instrument Handbook, KSCI 19033-001*. Moffett Field, CA: NASA Ames Research Center, 2009. http://archive.stsci.edu/kepler/manuals/KSCI-19033-001.pdf

47. H. Wu, J. D. Twicken, P. Tenenbaum, B. D. Clarke, J. Li, E. V. Quintana, C. Allen et al., Data validation in the Kepler science operations center pipeline. In N. M. Radziwill and A. Bridger, editors, *Society of Photo-Optical Instrumentation Engineers (SPIE) Conference Series*, p. 42W, Vol. 7740, pp. 774019-1–774019-12, 2010.

Classification of Variable Objects in Massive Sky Monitoring Surveys

Przemek Woźniak

Los Alamos National Laboratory

Łukasz Wyrzykowski and Vasily Belokurov

Cambridge University

CONTENTS

18.1 Introduction   383
18.2 Machine Learning Algorithms for Classification of Periodic Variability   384
18.3 Identifying Miras with SVM   385
  18.3.1 Red Variables in the Northern Sky Variability Survey   385
  18.3.2 Support Vector Machines   388
  18.3.3 Initial Classification   388
  18.3.4 Identifying Carbon Stars   389
  18.3.5 Validation of Results and Possible Misclassifications   391
18.4 Microlensing Event Selection Using Scan Statistics   391
18.5 Neural Networks for Classification of Transients   393
18.6 Self-Organizing Maps Applied to OGLE Microlensing and Gaia Alerts 397
  18.6.1 Self-Organizing Maps   397
  18.6.2 Classification and Novelty Detection   398
  18.6.3 Spectral Classification of Flux-Based Gaia Science Alerts   399
18.7 Conclusions   402
References   404

18.1 INTRODUCTION

The era of great sky surveys is upon us. Over the past decade we have seen rapid progress toward a continuous photometric record of the optical sky. Numerous sky surveys are discovering and monitoring variable objects by hundreds of thousands. Advances in detector, computing, and networking technology are driving applications of all shapes and sizes ranging from small all sky monitors, through networks of robotic telescopes of modest size, to big glass facilities equipped with giga-pixel CCD mosaics. The Large Synoptic Survey Telescope will be the first peta-scale astronomical survey [18]. It will expand the volume of the parameter space available to us by three orders of magnitude and explore the mutable heavens down to an unprecedented level of sensitivity.

Proliferation of large, multidimensional astronomical data sets is stimulating the work on new methods and tools to handle the identification and classification challenge [3]. Given exponentially growing data rates, automated classification of variability types is quickly becoming a necessity. Taking humans out of the loop not only eliminates the subjective nature of visual classification, but is also an enabling factor for time-critical applications. Full automation is especially important for studies of explosive phenomena such as γ-ray bursts that require rapid follow-up observations before the event is over. While there is a general consensus that machine learning will provide a viable solution, the available algorithmic toolbox remains underutilized in astronomy by comparison with other fields such as genomics or market research. Part of the problem is the nature of astronomical data sets that tend to be dominated by a variety of irregularities. Not all algorithms can handle gracefully uneven time sampling, missing features, or sparsely populated high-dimensional spaces. More sophisticated algorithms and better tools available in standard software packages are required to facilitate the adoption of machine learning in astronomy.

The goal of this chapter is to show a number of successful applications of state-of-the-art machine learning methodology to time-resolved astronomical data, illustrate what is possible today, and help identify areas for further research and development. After a brief comparison of the utility of various machine learning classifiers, the discussion focuses on support vector machines (SVM), neural nets, and self-organizing maps. Traditionally, to detect and classify transient variability astronomers used ad hoc scan statistics. These methods will remain important as feature extractors for input into generic machine learning algorithms. Experience shows that the performance of machine learning tools on astronomical data critically depends on the definition and quality of the input features, and that a considerable amount of preprocessing is required before standard algorithms can be applied. However, with continued investments of effort by a growing number of astro-informatics savvy computer scientists and astronomers the much-needed expertise and infrastructure are growing faster than ever.

18.2 MACHINE LEARNING ALGORITHMS FOR CLASSIFICATION OF PERIODIC VARIABILITY

One of the early demonstrations of the potential of machine learning algorithms for classifying astronomical objects was in the domain of periodic variability [29]. Standard machine learning tools have been tested, for example, on variable stars discovered by the ROTSE-I all sky survey [1]. The accuracy of the best algorithms proved to be comparable to that of human experts. In a set of nine 16×16 deg fields, 1781 periodic variable stars were identified with mean magnitudes between $m_v = 10.0$ and $m_v = 15.5$. This pilot study analyzed only 5.6% of the total ROTSE-I sky coverage. Stars were divided into nine classes: 186 RR Lyr AB (rrab), 113 RR Lyr C (rrc), 91 Delta Scuti stars (ds), 201 intermediate period pulsators (c), 382 close eclipsing binaries (ew), 109 other eclipsing binaries (e), 146 Mira variables (m), 534 long-period variables (LPV), and 19 other objects (o). Classifications were based on periods, amplitudes, three ratios formed using amplitudes of the first three Fourier components of the light curve, and the sign of the largest deviation from the mean. The final classifications were corrected visually and supplemented with information from the *General Catalog of the Variable Stars* (GCVS; [14]).

Several supervised and unsupervised algorithms were then applied to the same data set taking visual classifications as the ground truth. Machine learning classifications were based on a light curve in a single photometric band only (period, amplitude, the ratio of first overtone to fundamental frequencies and the skewness of the magnitude distribution). The location of the training data in this feature space is shown in Figure 18.1. Table 18.1 summarizes the results for SVM [27], a decision tree builder (J4.8[*]; [25]), a five nearest-neighbor classifier (5-NN; [10]), a k-means clusterer [17] and a Bayesian modeling system called Autoclass [9]. In addition to classification accuracy on training data, the table also lists results of cross validation testing based on a five-fold scheme, where randomly selected 4/5 of the sample is used for training and then the accuracy is evaluated on the remaining 1/5. A state-of-the-art SVM method delivers 90% accuracy for the full problem (nine classes). The corresponding confusion matrix is shown in Figure 18.2. It was estimated that two human analysts would agree at a similar level. In fact, even the same person making the same classification twice, may arrive at a different conclusion each time. This problem is eliminated when working with machine classifiers. The performance can be as good as 95% or even 98% when larger, more general classes are considered, or objects of one class are detected against the background of "everything else." Notably, decision trees are among the best-performing algorithms. This type of algorithm is particularly attractive because it gives the full insight into how the features were used to compute the classification, and the machine can then convey the knowledge back to a human. The trees constructed on the ROTSE data correctly reproduce many of the features found in the human-made algorithm, despite the fact that the adopted feature space was somewhat different from the one used in the benchmark visual study. The final classification tree must be trimmed of the least populated branches and leaves to control overfitting.

18.3 IDENTIFYING MIRAS WITH SVM

18.3.1 Red Variables in the Northern Sky Variability Survey

Classification of red variables in the Northern Sky Variability Survey (NSVS) is an example of a successful application of SVMs to astrophysical problems involving time-series and multicolor photometric data [32]. The NSVS catalog of red variables contains 8678 slowly varying stars with near-infrared colors corresponding to the evolved asymptotic giant branch (AGB) population. NSVS is based on the ROTSE-I data and covers the entire sky above declination −38° in a single unfiltered photometric band corresponding to the V-band magnitude range of 8–15.5 mag [31]. After quality cuts, the number of measurements for

---

[*] Implementation of the C4.5 algorithm in the WEKA package.

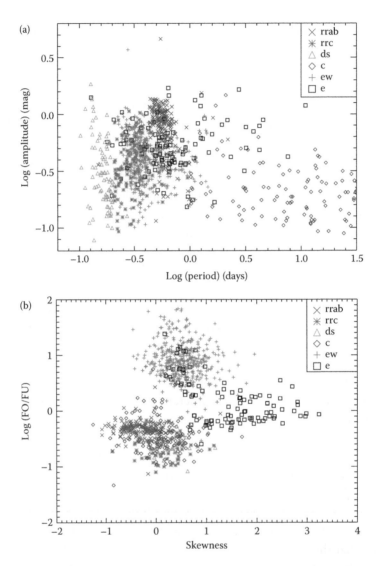

FIGURE 18.1 (**See color insert.**) Projections of the feature space for classification of periodic variable stars in the ROTSE sample. Only six classes are shown for clarity. The symbols in the legend correspond to variability types from top to bottom: RR Lyrae AB, RR Lyr C, Delta Scuti, Cepheids, W UMa eclipsing binaries, and detached eclipsing binaries. The top panel (a) shows the period-amplitude locus of the training data set. The ratio of the first overtone to fundamental mode (FO/FU) and the skewness of the magnitude distribution are plotted in the bottom (b). Table 18.1 contains performance comparisons for selected machine learning classifiers applied to these data.

a typical star was up to 1000 for high declination stars with the mean around 150. The study clearly demonstrated that a modern machine-learning algorithm such as SVM can reliably distinguish Mira variables from other types of red variables, namely, semiregular and irregular. The machine also identified the region of the parameter space dominated by carbon stars. In fact, the inspection of the confusion matrix for the initial classification problem suggested that the class of carbon Miras should be separated from regular Miras.

TABLE 18.1 Comparison of Machine Learning Classifiers on Variable Stars from the ROTSE-I Sample Columns[a]

Algorithm	Training Accuracy (%)	Cross-Validation Accuracy (%)
Supervised		
SVM	95	90
J4.8	92	90
5-NN	86	81
Unsupervised		
Autoclass	80	80
k-Means		70

[a] Classifier (1), classification accuracy on the entire training data set (2), classification accuracy from a five-fold cross-validation experiment (3).

		Machine								
		rrab	rrc	ds	c	ew	e	m	lpv	o
	rrab	176	1	1	1	0	1	0	0	6
	rrc	4	88	15	1	0	3	0	0	2
H	ds	0	3	87	0	0	0	0	0	1
u	c	4	0	0	194	1	1	0	1	0
m	ew	2	0	0	0	370	10	0	0	0
a	e	0	2	4	2	40	61	0	0	0
n	m	0	0	0	0	0	0	123	23	0
	lpv	0	0	0	3	0	0	34	497	0
	o	7	5	6	0	0	0	0	0	1

FIGURE 18.2 Confusion matrix for one of the best SVM runs compared to the results of visual inspection using the same set of ROTSE-I variables. The overall accuracy of SVM relative to the human expert is 90%. Most of the mixing occurs between subclasses of the major classes, for example, between various types of pulsators or eclipsing binaries. Variability types: RR Lyrae AB and C (rrab, rrc), delta scuti (ds), Cepheids (c), W Uma eclipsing binaries (ew), detached eclipsing binaries (e), Miras (m), semiregular and irregular LPV, and other (o).

This particular classification was based on period, amplitude, and three independent colors possible with the photometry from NSVS and the Two Micron All Sky Survey (2MASS). The overall classification accuracy was ~90% despite the relatively short survey baseline of 1 year and a limited set of features.

Variable stars in the red and luminous part of the H–R diagram are often collectively referred to as red variables. They are mostly AGB stars of K and M spectral types and are traditionally classified into Mira (M), semiregular (SR), and slow irregular (L) variables. Pulsation instability is the most likely explanation of variability in those objects. Much of their importance comes from the fact that Mira variables and Mira-like semiregulars obey period–luminosity relations and therefore can be used as distance indicators and tracers of the intermediate-age-to-old populations [12]. AGB stars are also intrinsically bright. A relatively shallow survey to the flux limit of 16 mag can detect them throughout the Galaxy.

The primary means to find and initially classify red variables are time-resolved photometry and light-curve morphology. A considerable amount of preprocessing is typically required in order to preselect objects of interest from a very large survey. Red variables in

NSVS are first identified by their slow variability pattern, well resolved by daily observations. The use of the near-infrared colors greatly improves the fidelity of the classification. Single epochs near IR colors are still useful because of the much lower level of variability of long period pulsators toward the red part of the spectrum. Using combined NSVS and 2MASS data we can form three independent colors. After preprocessing there are five features to use in the classification: $P, A, (J - H), (H - K_s)$, and $(m_R - K_s)$, that is, period, amplitude, and three colors. The last color is a broad band optical to IR color formed as a difference between the 2MASS K_s magnitude and the median NSVS baseline magnitude m_R.

18.3.2 Support Vector Machines

SVMs are a state-of-the-art method of supervised learning that requires a training set of data with known class membership. For a thorough introduction to SVMs we refer the interested reader to specialized texts [11,27]. With a clever use of the kernel functions to transform the data into a high-dimensional feature space, SVMs are capable of finding highly nonlinear class boundaries using only hyperplanes. SVMs are inherently resistant to overfitting. The objective of the SVM is to maximize the so-called margin, a generalized orthogonal distance between the class boundary and data vectors closest to the boundary on both sides. For a given set of input parameters, SVMs guarantee that the final result represents the global minimum of the objective function rather than one of the local minima. This is one of the reasons behind excellent generalization properties of SVMs on previously unseen data. Red variables in NSVS were classified using LIBSVM [8], a publicly available implementation of SVMs that reduces N-class problems to N binary classifications by maximizing the margin of each class against all others. The package offers soft-margin SVMs capable of working with data that are not fully separable into a hyperplane in the n-dimensional space of transformed features. This capability is very important in the presence of noise, where the simpler maximal-margin machine usually breaks down. For a Gaussian kernel function the training algorithm employs two parameters, the width of the Gaussian kernel and the amplitude of the penalty term for misclassification, referred to as γ and C, respectively. They are somewhat correlated and measure the level of coupling between the data vectors sensed by the algorithm (γ) and how hard the SVM algorithm should try to avoid misclassification by making the boundary more flexible (C).

18.3.3 Initial Classification

A subset of red variables in common with the GCVS and previously classified as M, SR, or L type was used for the purpose of training the machine and obtaining the classifier. The sample contains 2095 such stars, all uniquely identified within 14.4 arcsec (1 ROTSE-I pixel). Because SVM utilizes the distance in a multidimensional feature space to derive the classifier, rescaling of the input features usually improves results. Periods were replaced with their logarithm and feature distributions were renormalized to zero mean and unit variance.

The first classification problem to consider is a three-class problem with types M, SR, and L. In this case, the results suffered from a significant confusion between SR and L types at the level of 30%. Classification was sensitive primarily to the gap in the bimodal amplitude distribution of the training set, as previously discussed by Mattei et al. [20]. There is a slight progression of colors toward the red when moving from type L through SR and to the M stars of the training set. Nevertheless, the overlap between classes remains very large. Unfortunately, statistical dereddening of colors does not change this situation. The available photometric data did not permit a reliable separation of variables of types SR and L.

A two-class problem to classify Mira variables (M) versus "other" variables (SR+L) was a much better match for the SVM algorithm. The confusion matrices M_{ij} for this sample, which measures the fraction of training instances in class i assigned to class j, are given in Figure 18.4. The final classification accuracy for this problem was about 87%. Both the confusion matrix and the accuracy were estimated using a 10-fold cross-validation scheme, where the SVM is repeatedly trained using 90% of the data, and its accuracy is estimated using the remaining 10% of the sample. This estimate is a better predictor of performance on new data than the accuracy on the training set. The best results were obtained by resampling the training set to achieve equal weighting of all classes. This was particularly important in the next attempt to identify carbon stars, which are rare in the GCVS and the training set.

18.3.4 Identifying Carbon Stars

In various color–color projections of the NSVS/GCVS training data one notices a distinct tail of objects extending toward the reddest part of the (H, K) plane—the locus of carbon stars [6,28]. There are only 647 stars with spectral information among M, SR, and L types from the GCVS that were also selected in NSVS. Of those, 53 have carbon spectra (regardless of variability type). The subsample known to have carbon spectra almost exclusively falls within the red tail, while objects with other spectral types rarely overlap with the tail. An SVM classifier applied to a three-class problem with types M, C, and SR+L for Mira, "carbon," and "other" readily recognizes those features. Only stars with known spectra were used in this training run. A reweighting of the samples further improves results by reducing systematics due to large differences in the number of available objects in each class. With the introduction of the third class of carbon stars, the final accuracy increased to about 90% ($\gamma = 0.1$ and $C = 1.0$). The confusion matrix for this classification is shown in Figure 18.3. Figure 18.4 shows the locus of the training data and the full classified catalog in two 3-D projections of the feature space. Table 18.2 lists the number of objects available for training in various

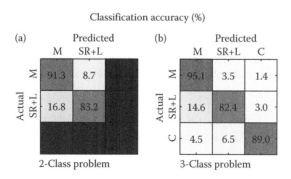

FIGURE 18.3 Confusion matrix for classification of red variables in the NSVS using SVM. The two-class problem (a) attempts to divide the data set between Miras (M) and other LPVs (SR+L). The preferred classifier (b) separates stars with carbon spectra into a distinct class (C).

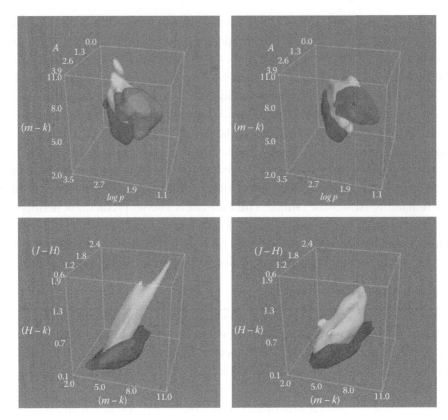

FIGURE 18.4 (**See color insert.**) Feature space for classification of red variables in the NSVS. Training data (left panels) are compared to the results of classification on the entire catalog (right panels). Two 3D projections are shown in each case: the log period-amplitude-color relation (top), and the space of three independent colors (bottom). Density contours are plotted at 10% of the peak density within each of the three classes: regular Miras M (red), carbon Miras C (yellow), and other red variables SR+L (green). The corresponding confusion matrix is shown in Figure 18.3. Table 18.2 summarizes the details of this classification experiment.

classes, the input parameters, and the resulting classifications for both two-class and three-class runs. As expected, the amplitudes of carbon stars are lower than those of Mira-like variables at the same period [20,21,28]. These objects tend to populate the amplitude gap between types M and SR, and confuse the separation of Miras from other stars. Separating them in color dimensions makes for a better–defined gap in the period–amplitude diagram

TABLE 18.2   SVM Classification of Red Variables in NSVS Columns[a]

Classes	γ	C	Training Set M/SR+L/C	Target Set M/SR+L/C	Accuracy (%)
2	0.04	1.0	1221/874/	2565/6113/	87
3	0.10	1.0	417/177/53	2276/5719/683	90

[a]   Number of classes (1), width of the Gaussian kernel (2), amplitude of the penalty term for misclassification (3), number of objects of each class in the training set (4) and the full catalog (5) given as Mira/Semiregular-irregular/Carbon, classification accuracy (6).

and allows better discrimination between the two remaining classes. Statistically, the group of stars classified as C by the SVM is dominated by objects with carbon spectra. It must be stressed, however, that each individual case requires a spectrum for actual confirmation.

18.3.5 Validation of Results and Possible Misclassifications

Confirmation of known correlations is a good way to ensure that the algorithm selects objects of a certain type. In case of Mira variables one expects an approximately linear relationship between $\log P$ and the near-IR colors. The agreement between existing samples of Mira variables and the one selected using SVMs is very good, confirming the utility of the machine learning classifier. The M sequence in the period–color relations closely follows the published fits [28].

Although the level of contamination in the considered sample of M/SR/L variables is low, it is interesting to see what kinds of contaminants we may expect when an SVM classifier is applied to a large catalog of red variables. There were 78 stars in the NSVS catalog with GCVS classifications of types other than M, SR, and L. About half of the corresponding GCVS entries evidently required reclassification. Severe blending is the most likely explanation for a majority of these misclassifications. The single largest contaminating group was composed of IS-type rapid irregulars (16) and IN Ori variables (10) related to young systems of the T Tauri type and distinguished based on hard-to-obtain information such as the presence of the nebula.

18.4 MICROLENSING EVENT SELECTION USING SCAN STATISTICS

Transient variability phenomena present additional challenges for detection and classification. While periodic variability is best captured by the shape parameters of the period-folded light curve, for example, Fourier components, detection of transient behavior is better accomplished using scan statistics evaluated in running windows. A classical example is the microlensing light curve with hundreds of observations in the baseline spread over many years and typically 1–2 dozen data points covering an event lasting several weeks (cf. Figure 18.6). Global statistics such as the goodness of fit over the entire light curve tend to dilute localized signals resulting in loss of sensitivity. The Optical Gravitational Lensing Experiment (OGLE) microlensing search provides an example.

Each of the 49 OGLE-II [26] galactic bulge fields was observed ~200 times between 1996 and 1999 observing seasons. The Difference Image Analysis pipeline returned 220,000 candidate variable objects. Selection of microlensing events begins with a set of 4424 light curves showing episodic variability on top of a constant baseline [30]. In addition to a well-defined baseline, the most important property of the observed microlensing light curve is the presence of several consecutive high S/N points which are well fitted by the microlensing model. Therefore, the main parameter distinguishing single microlensing events from other types of variability was chosen as

$$
R = \max_k \left[\frac{\sum_{i=k}^{k+11} (f_i - F_0)^2}{\sum_{i=k}^{k+11} (f_i - M_i)^2} \right]^{1/2},
\tag{18.1}
$$

where f_i is the flux of the ith observation, F_0 is the fitted baseline flux, M_i is the best-fit microlensing model, the sums are over a 12-point window, and R is the maximum value of the ratio over the entire light curve. In case of a good fit R is roughly proportional to the S/N ratio, while for a bad fit R is much lower, even in cases of high S/N variability. The test runs on cyclic light curves with the beginning and the end joined together.

As emphasized before, in microlensing light curve one expects a baseline which is long compared to the event duration, making statistics such as the total χ^2 or χ_ν^2 per degree of freedom inefficient. It is important to have a handle on the goodness of fit separately near the event and far from the event, in the baseline. The event is taken as the region of the light curve, where the flux from the best-fit model is at least one median error $d_{1/2}$ above F_0. Two approximately χ^2 distributed sums are calculated:

$$s_1 = \frac{1}{(m_1 - 4)} \sum_{M_i - F_0 > d_{1/2}} \frac{(f_i - M_i)^2}{\sigma_i^2}, \tag{18.2}$$

for m_1 points in the event region and

$$s_2 = \frac{1}{(m_2 - 1)} \sum_{M_i - F_0 < d_{1/2}} \frac{(f_i - M_i)^2}{\sigma_i^2} \tag{18.3}$$

for m_2 points of the baseline region, both of which should stay near one for normal cases of microlensing and properly normalized errors. The total duration of the event T is then the time between the beginning and the end of the event region, as given by the best-fit model. Long-lasting events are ambiguous, because their baselines are short relative to the duration of the experiment. Setting a higher S/N threshold for longer events compensates for this effect.

The left panel of Figure 18.5 shows the distribution in the (s_1, R) plane of candidate light curves with $s_2 < 2.0$, and $0 < T < 500$ days, and minimum eight high points in the event region. A high point is simply the one detected at more than 3σ above the baseline. Single microlensing events are relatively well separated by R alone, with all remaining types of variability and artifacts located at low R and spanning a large range in s_1. Above $R = 15$ there are no light curves of unwanted types, and there are 134 events with reasonably good fits, that is $s_1 < 2.0$. It is possible to decrease the limit on signal to noise to $R = 10$ for shorter events with $T < 200$ days, as shown in the right panel of Figure 18.5. This relaxed cut selects 188 events.

The above criteria strongly reject events with any departures from a single-point mass microlensing model. In Figure 18.5, some events from the visual sample are located in the region of $R < 10$ and have poor fits with $s_1 > 3.0$. These are mostly cases of binary lensing and possibly binary sources. Other departures from a perfect fit to the model are expected too. The cumulative error distribution revealed a weak non-Gaussian tail at the level of $\sim 1\%$. Outlying points can also be produced by parallax effects due to a changing viewing angle from the Earth traveling around the Sun, as well as weakly binary events with very low mass companions such as Jovian planets.

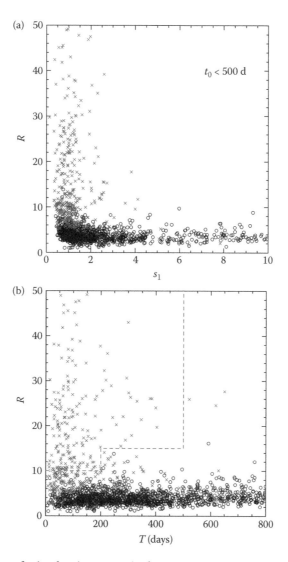

FIGURE 18.5 Selection of microlensing events in the OGLE survey using scan statistics. The bumpiness of the light curve R is plotted against the goodness of fit in the microlensing bump s_1 (a) and event duration T (b). Microlensing events (crosses) are well separated from other light curves consistent with transient behavior (circles). All quantities must be computed in running windows that are much shorter than the survey baseline to avoid diluting the signal.

18.5 NEURAL NETWORKS FOR CLASSIFICATION OF TRANSIENTS

The utility of neural networks for automatic classification of transient variability in astronomical objects has been demonstrated by Belokurov et al. [4,5]. A good introduction to neural network computation can be found in Ref. [13]. The NETLAB package [22] offers a convenient implementation for experiments and development. Here, both simulated data and actual observed light curves were used to train various networks to recognize microlensing events against the background of stellar variability, as well as instrumental artifacts

(Figure 18.6). Several types of eruptive behaviors were considered in addition to periodic variability.

The inputs to the networks include parameters describing the shape and the size of the light curve. The distribution of the training and test data sets in the feature space is shown in Figure 18.7. Each input pattern consists of five numbers $(x_1, \ldots, x_5)$ defined as: (1) the maximum of the autocorrelation function; (2) the ratio of the mean fluxes for measurements above and below the median baseline; (3) the peak value of the cross-correlation between the light curve and its time-reversed version; (4) the mean frequency weighted by the power spectrum; and (5) the standard deviation of the auto-correlation function. These features are chosen to capture, correspondingly, the presence of variability signal, positive bumpiness, the symmetry of the transient event, the nonzero mean frequency for periodic variables, and the time scale of the variability event. This preprocessing of light curves is applied primarily to reduce the amount of data that must be examined. The range of features x_1 and x_4 was compressed using the logistic function, followed by the usual normalization of all dimensions to zero mean and unit variance. A careful choice of inputs is the key to obtaining good performance. Feature normalization promotes better stability in the training stage. With the appropriate choice of the activation functions for neurons (logistic function), the network computes the posterior probability of microlensing given the inputs and the weights. The single-output neuron and the input layer of five neurons are fully connected with the hidden layer of 5–7 neurons. Additional neurons must only be added if their presence significantly improves the classification error. Network architecture should be optimized

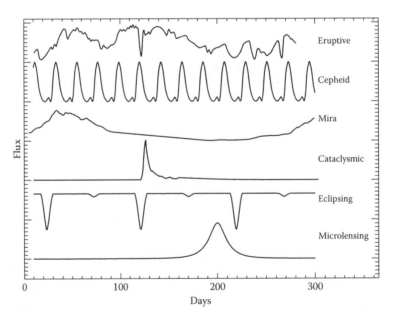

FIGURE 18.6 Template light curves for class prototypes in neural network classification of variability in the MACHO survey. The main objective is to separate microlensing events from other types of transients and periodic sources. (From V. Belokurov, N. W. Evans, and Y. L. Du. *MNRAS*, 341:1373–1384, 2003. With permission.)

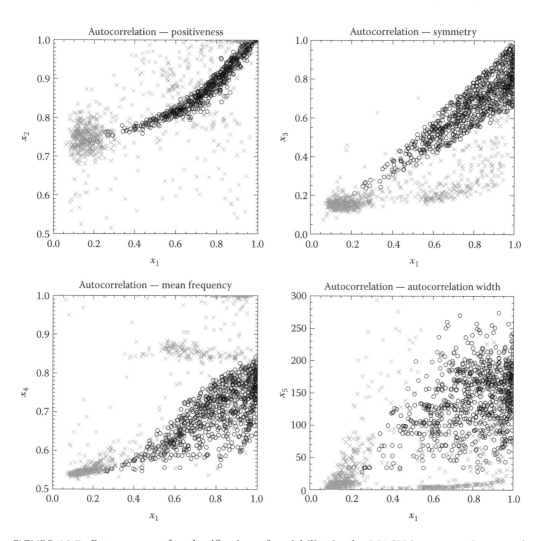

FIGURE 18.7 Feature space for classification of variability in the MACHO survey using neural networks. Shown are four projections onto principal planes of the five-dimensional space of inputs. Microlensing events (black circles) are selected against variable stars and noise (gray crosses) in the training and validation data sets. Features: (1) the maximum of the autocorrelation function; (2) the ratio of the mean fluxes for measurements above and below the median baseline; (3) the peak value of the cross-correlation between the light curve and its time-reversed version; (4) the mean frequency weighted by the power spectrum; and (5) the standard deviation of the auto-correlation function. (From V. Belokurov, N. W. Evans, and Y. L. Du. *MNRAS*, 341:1373–1384, 2003. With permission.)

toward modeling the decision boundaries between the patterns rather than modeling the patterns themselves. Large neural networks have a tendency to overfit the data, that is, "memorize" the patterns, and therefore may perform poorly on previously unseen data. This can be mitigated with the appropriate choice of stopping criteria. Here, the training process was monitored with the use of an independent validation data set, and terminated right before the classification error began to increase. Training was performed using a slight modification of the standard back-propagation algorithm [7]. The error function to be minimized consists of the standard cross-entropy term and the weight decay term. Adjusting the coefficient in front of the weight decay term enables one to control the magnitude of weights and provides an alternative way to control overfitting without the need for a validation data set. This can be done automatically during training. Further reduction of the variance in network predictions can be achieved by using a committee of networks. A very inexpensive, but efficient, way of introducing the committee involves simply taking the output of the committee to be the average of the outputs of the individual networks. The members of the committee are competing solutions of the classification problem, which occurred as a result of starting the search in the parameter space from different (random) initial weights. It is also beneficial to combine networks with different numbers of neurons in the hidden layer.

The classifier is then applied to the test set in order to estimate the rate of false negatives (microlensing events misclassified as not microlensing) and false positives (nonmicrolensing events misclassified as microlensing). The standard cross-entropy error evaluated against the number of neurons in the hidden layer using the test set begins to flatten around six or seven neurons. Therefore, equal numbers of networks with six and seven neurons in the hidden layer were used to form a committee of 50 networks. Figure 18.8 summarizes the final results.

Once the new input pattern has been transformed into the posterior probability, it is important to estimate the error of the output. The error arises through variance and through undersampling in the parameter space during training. The variance part of the output error is easiest to deal with. It can be approximated by taking the standard deviation of the output

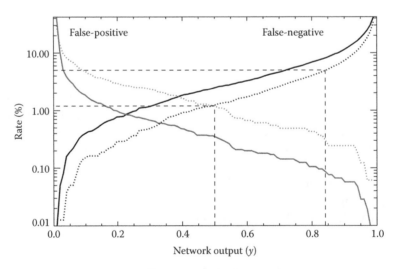

FIGURE 18.8 False-positive and false-negative rates for a committee of neural networks applied to MACHO light curves of stars in the Large Magellanic Cloud. The horizontal axis is the network output. The solid lines apply to the raw data without any cleaning and the dotted lines correspond to light curves with at least five data points with signal-to-noise ratio above 5 in the microlensed region. (From V. Belokurov, N. W. Evans, and Y. L. Du. *MNRAS*, 352:233–242, 2004. With permission.)

of a committee of neural networks. The second part of the output error is more awkward, but can be approximated [19]. There will always be regions in input space with low training data density. Typically the network will give overconfident predictions in such regions. A more representative output is obtained after averaging over the distribution of network weights. This marginalization step always drives the output closer to the formal decision boundary. When the network is applied to real data after training, it is confronted with more complex light curves which inevitably extend beyond the data domain encountered during training. Neural networks sometimes classify these in an unpredictable manner, as this amounts to an extrapolation of the decision boundaries. The use of marginalized or moderated output guards against such artifacts, as unexpected or unpredicted patterns are then driven back toward the formal decision boundary [5,19].

Neural networks can be arranged sequentially in a cascade to perform complicated pattern recognition tasks. For example, the light-curve data are examined first with neural networks which eliminate the contaminating variable stars. Then, light curves successfully passing this first stage are analyzed anew with neural networks which eliminate contaminating super-novae using additional information such as the color of the object and color evolution during the event. Excellent microlensing candidates must pass both stages. The neural net-work classifiers described above have been used to process light curves of 22,000 objects in the direction of the LMC collected by the MACHO collaboration [2]. The original analysis of these data selected 13 events for which the microlensing interpretation was believed to be secure and four additional detections labeled as possible microlensing. Neural networks confirmed the microlensing nature of only seven of those 17 events. The difference is highly significant and very consequential since the reduced event sample no longer requires a dark matter contribution to the microlensing optical depth toward LMC, in line with the results obtained by the EROS and OGLE microlensing groups. Although the relative frequency of microlensing in the training data is many orders of magnitude larger than in nature, a simple correction can be devised for a direct calculation of the microlensing rate from the output of the neural networks. In principle, this allows a measurement of the microlensing optical depth without the need for cumbersome efficiency calculations using Monte Carlo simulations.

18.6 SELF-ORGANIZING MAPS APPLIED TO OGLE MICROLENSING AND GAIA ALERTS

18.6.1 Self-Organizing Maps

Self-organizing maps (SOMs) are two-dimensional lattices of weight vectors represent-ing patterns in the parameter space under consideration. They were discovered by Teuvo Kohonnen, a Finnish scientist who studied self-organization phenomena, such as associative memory [15]. SOMs can be viewed as neural networks with just the input and output layers, and their essential use is in unsupervised learning. They are simple, fast, deliver intuitive visualizations, and are therefore great tools for classification work on very large data sets.

The number of nodes in the map should be chosen to be roughly equal or somewhat larger than the number of distinct classes and subclasses resolved by the training data. SOM construction starts from initialization of all nodes $\mathbf{w}_0$, typically with random values. Then, in iteration t each training pattern $\mathbf{x}_t$ is assigned to the closest node. Minimization of the Euclidean distance $\min\|\mathbf{x}_t - \mathbf{w}_t\|$ is commonly employed to find the best matching node, but other metrics can be used as necessary. Finally, the weights in the neighborhood $O(\mathbf{w}, t)$ of the winning node are adjusted according to the formula:

$$\mathbf{w}_{t+1} = \mathbf{w}_t + \Theta_t(r)\alpha(t)(\mathbf{x}_t - \mathbf{w}_t).$$

This requires choosing the proximity function $\Theta_t(r)$ of the distance r in the (x, y) space of the map, and the learning rate $\alpha(t)$. The learning function α is monotonically decreasing, so the fast learning of crude features is followed by fine tunning of details that distinguish the subclasses. After all patterns have been presented to the SOM, the learning process is typically repeated many times, perhaps with different parameters, to reinforce the clustering of patterns. The nodes of the SOM encode a topologically correct mapping of the data space onto a low-dimensional visualization space (2-D maps are by far the most popular because they are easy to plot). The convergence of this process depends on the choice of parameters for the algorithm. In order to get a continuous map that does not twist and fold on itself, the training process should start with a large region of influence $O(\mathbf{w}, t)$ that contains most of the nodes and $\Theta_t(r)$ that is not strongly peaked toward the center. Perhaps the easiest way to get more insight into what information is captured by the SOM is to study in detail the process of training on a 2-D or 3-D input data space.[*]

Once the training is completed, we are ready to perform all sorts of "cartography" experiments, that is, derive inverse mappings of various parameters for objects classified on the SOM. The data set is assigned to the map nodes one more time (without adjusting the node weights), and all parameters of interest are averaged within each node. The resulting new map can be used to predict the values of parameters for input vectors that were not used in training. This includes parameters that are not part of the feature vector used in the original clustering. Of particular interest is the class membership mapping that gives an SOM classifier. Another possibility is using calibrated SOMs to interpolate parameters over areas not covered by the training set, or recovery of missing features. When analyzing the examples that follow, one must keep in mind that the arrangement of nodes in an SOM is topological and in principle unrelated to the distance in the feature space. In fact, a good way to quickly find the most important clusters is to plot the distance to the nearest node on an SOM, and find groups of nodes with low values. A trained SOM contains the full information on the hierarchical clustering structure of the data set.

18.6.2 Classification and Novelty Detection

SOMs are well suited for incremental and iterative classification problems. In a typical time-domain survey, light curves can be presented to an SOM classifier as soon as the first few measurements are available. With each incoming measurement, objects advance along tracks on the map and move closer to their final classification. This approach has been applied to classification of temporal variability [33,34] with results indicating good convergence

---

[*] http://www.jjguy.com/som/

properties. It is also relatively straightforward to modify and adapt the basic SOM training algorithm for specific needs. A good example is preprocessing on the input node to take out phases of periodic signals with the use of a cross-correlation step.

The distance between a given pattern and the nearest node, referred to as quantization error, carries important information on clustering. Data vectors "well known" to the map are part of some cluster and therefore a short distance away from the closest node. Additionally, the distribution of node distances in the vicinity of cluster members shows a steep rise as we move further out. Anomalous feature vectors, on the other hand, are distant from their best matching node and tend to have a flat distribution of neighbor distances. These properties are the basis for the use of SOMs as alert systems. In order to find rare objects against some background, the SOM should be trained on background objects and then used to classify all incoming data. Unusual and interesting objects will stand out as patterns with a large quantization error and a flat distribution of neighbor distances. A prototype system for Gaia science alerts based on this method is described in Ref. [33]. Trial applications to actual OGLE light curves show very promising results. The feature space spanned 65 dimensions with most features describing the power spectrum of variability, and supplemented with additional information such as color and the presence of explosive/dipping behavior. The main conclusion is that SOMs are capable of not only reliably classifying periodic variable stars of various types, but also selecting rare objects such as Nova explosions for detailed follow-up studies.

18.6.3 Spectral Classification of Flux-Based Gaia Science Alerts

The science alert system of the Gaia space mission is based on a collection of SOM classifiers. An overview of the Gaia project can be found in Ref. [23]. While the main objective of Gaia is to provide micro-arcsecond astrometry for milions of stars in the Milky Way, the instrument will also deliver time-resolved photometry and low-resolution spectra. The Gaia science alert system operates mainly on the basis of the G-band fluxes. Any flux anomaly with respect to previous flux history or appearance of a new bright object becomes a potential alert. Spectral classification is the key to understanding the physics of the object before a reliable request for follow-up can be generated. For example, spectroscopic confirmation is necessary in order to classify a new object detected photometrically as a supernova explosion. On the other hand, spectral information is used to rule out some false alarms, for example, by recognizing the source of the flux variation or appearance as a long period variable of the Mira type. The following discussion of the Gaia spectral classifier illustrates the main concepts and methods in SOM classification as applied to astronomy.

As the mission progresses, clustering maps will be gradually constructed based on the incoming spectra of bright stars. Prior to the mission, the input to the Gaia spectral classifier consists of simulated spectra combined with additional information such as published supernova observations. In one of the development studies a set of 2700 synthetic stellar spectra from the Gaia Object Generator is presented 100,000 times, in a random order, to an SOM with 15×20 nodes. Each spectrum sampled at 120 wavelengths incorporates realistic noise and instrumental signatures. The original data are from the Basel library of synthetic spectra and colors for Gaia [16]. The SOM was trained using a Gaussian distance function with $FWHM = 2 \times r_0$, a learning parameter $\alpha = 0.3$, and a neighborhood radius $r_0 = 7$.

The resulting map and the input spectra can be compared in Figure 18.9. The magnitude limit of this sample is 15 mag. The input space in this case is spanned by 120 dimensions corresponding to the number of sampled wavelengths, but the topological ordering of the data can still be captured in a 2-D structure of the SOM. Figure 18.10 shows the correspnding U-matrix and helps in the interpretation of the clustering. The U-matrix value of a node

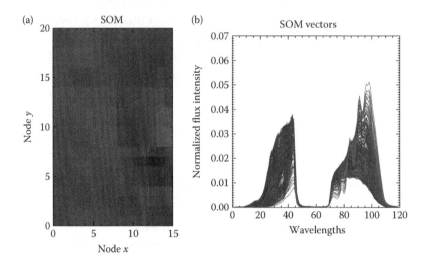

FIGURE 18.9 Self-organizing map trained on simulated Gaia spectra. Color-coded node membership (a) is traced back to the shape of the feature vectors (b). The shape and size of the map are chosen arbitrarily to facilitate visualization of particular data sets (here 15 × 20).

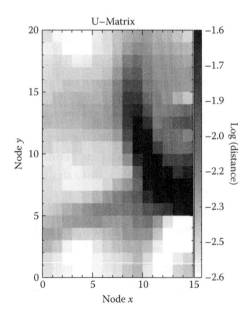

FIGURE 18.10 U-matrix corresponding to a trained SOM in Figure 18.9. The intensity level corresponds to the logarithm of the mean distance between each node and its neighbors. Plots of this type reveal the location of major clusters.

is the average distance in the feature space of that node to the adjacent nodes on the map. There is a continuous gradient of spectra starting with red at [5,14] and looping upwards, then to the left and down. The main boundary passes through the center of the map and is also visible in the U-matrix.

Figure 18.11 shows the H–R diagram of all simulated stars (absolute V magnitude vs. log effective temperature). Individual spectral types are marked with different colors and luminosity classes are marked with different symbols. This data set is then used to map the spectral type on the SOM and turn it into a classifier. Figure 18.12 shows the result of this transformation. The nodes of the map are now assigned to class labels and color coded according to the scheme from the H–R diagram. Clearly, the SOM correctly grouped stars by spectral type and can be used to reliably classify new spectra. Equal representation of the spectral types in the training sample did not have much impact on the outcome, which is an important result showing the level of convenience afforded by SOMs. Rare examples of spectra in the training sample are recognized by the map without additional effort.

Further insight into the classification can be gained from the mapping of astrophysical parameters in Figure 18.13. The single feature with the maximum influence on the SOM and the node class is the effective temperature. Absolute magnitude is also well correlated with the node membership and provides additional information. The SOM does a good job of estimating the luminosity class as well as the surface gravity $\log g$. The upper-left green to light blue part of the map corresponds mainly to G and F-type stars, which spread over a range in M_V between 6 and −2 mag. Spectral types A and B are concentrated at similar magnitudes. The M stars are well represented at both bright and faint magnitudes due to

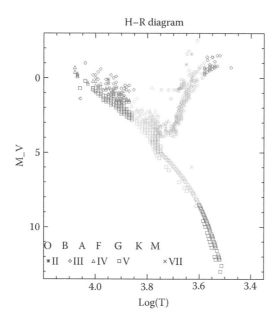

FIGURE 18.11 The locus of simulated spectra in the H–R diagram for classification of Gaia science alerts using self-organizing maps. Spectral types and luminosity classes are coded correspondingly by color and symbol shape.

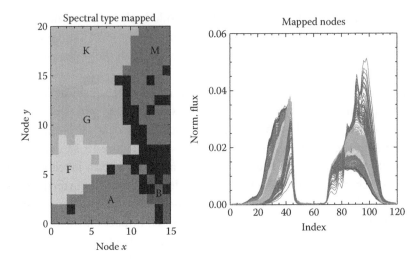

FIGURE 18.12 (**See color insert.**) Mapping of the spectral type onto the SOM from Figure 18.9. The map is mostly sensitive to the effective temperature and clearly capable of distinguishing major spectral types and subtypes.

the depth of the sample. In case of metallicity, the mapping is quite noisy, but is still useful in statistical applications.

In another exploratory study an attempt was made to include supernova spectra in the classification process. The Gaia DPAC software, and particularly the XpSim package by A.G.A. Brown, were used extensively in this work. A set of patterns representing the spectrum of SN2002dj at various stages of evolution [24] was appended to the sample of stellar spectra from the Basel library. The spectral type of the supernova was set to 10+epoch/1000 and the age covered the range between −10 and +274 days after the peak. Figure 18.14 shows the results of this classification. All objects brighter than 18 mag were classified within one class of their actual spectral type, for example, an F5 star at magnitude 18 could be classified as F7 (type 5.5 vs. 5.7). The SOM was capable of distinguishing young supernovae from the old ones using low-resolution spectra and could even predict the age of the brighter supernovae. These results demonstrate a great potential of self-organizing maps as classification and anomaly detection engines for a variety of astrophysical problems.

18.7 CONCLUSIONS

Machine learning algorithms have already proven their utility in astronomy and bear the promise of eliminating human judgement from classification and anomaly detection tasks. This chapter presented a selection of machine learning techniques relevant to classification of variable sources in massive sky monitoring surveys and examples of their use in real-world applications. The main conclusion from this work is that given a properly constructed feature space the classification accuracy of existing machine learning tools on this type of data can match that of a human classifier. While supervised algorithms typically deliver the best overall classification accuracy on samples composed of known classes, unsupervised learning methods are better suited for anomaly detection tasks. For a typical sampling and S/N

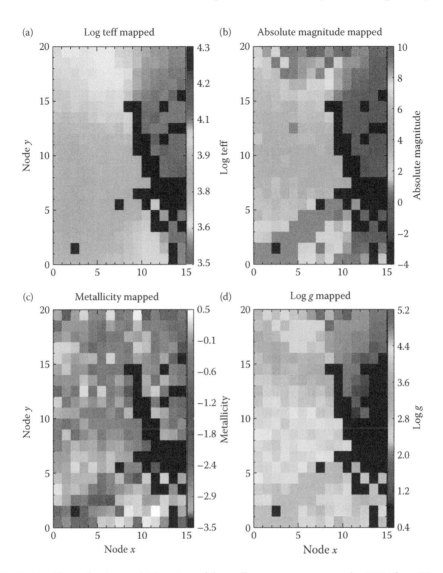

FIGURE 18.13 (**See color insert.**) Mapping of the stellar parameters onto the SOM from Figure 18.9. Maps are color coded with: logarithm of the effective temperature log T_{eff} (a), absolute magnitude (b), metallicity (c), and surface gravity g (d). Augmented maps of this kind can be used to infer astrophysical properties of newly discovered objects from other inputs—in this case, low-resolution spectra.

ratio offered by contemporary surveys good results can usually be obtained by combining photometric time-series data with additional information such as color or low-to-medium-resolution spectra. The existing machine learning toolbox can be trusted to perform well on a relatively routine task such as providing class labels for a catalog of periodic variable stars from a completed survey. Automated classification of non-periodic variability such as microlensing events and explosive transients is currently a very active area of research. This chapter discussed the prospect of solutions based on neural computation. Self-organizing

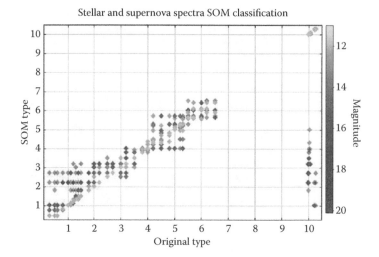

FIGURE 18.14 Classification accuracy for an SOM trained on stellar and supernova spectra. The numerical spectral type ranges from 0 to 6 for stars of type O to M and is around 10 for supernova depending on its age.

maps are a natural choice for unsupervised learning on incrementally updated data streams. They are particularly promising as anomaly detection engines for very large collections of objects with hundreds of features monitored over time. Scientific classification schemes tend to take the form of a tree. In the future, we are likely to see more focus on the development of hierarchical classifiers that reflect domain knowledge and utilize output from a number of lower-level algorithms. As the current trend toward wider, faster, and deeper surveys interacting with real-time follow-up is expected to accelerate, the importance of machine learning tools for time-domain astronomy will continue to grow in the future.

REFERENCES

1. C. Akerlof, S. Amrose, R. Balsano, J. Bloch, D. Casperson, S. Fletcher, G. Gisler et al. ROTSE All-sky surveys for variable stars. I. Test fields. *Astron. J.*, 119:1901–1913, 2000.
2. C. Alcock, R. A. Allsman, D. R. Alves, T. S. Axelrod, A. C. Becker, D. P. Bennett, K. H. Cook et al. The MACHO project: Microlensing results from 5.7 years of large magellanic cloud observations. *Astrophys. J.*, 542:281–307, 2000.
3. N. M. Ball and R. J. Brunner. Data mining and machine learning in astronomy. *Int. J. Mod. Phys. D*, 19:1049–1106, 2010.
4. V. Belokurov, N. W. Evans, and Y. L. Du. Light-curve classification in massive variability surveys—I. Microlensing. *MNRAS*, 341:1373–1384, 2003.
5. V. Belokurov, N. W. Evans, and Y. Le Du. Light-curve classification in massive variability surveys—II. Transients towards the large magellanic cloud. *MNRAS*, 352:233–242, 2004.
6. M. S. Bessell and J. M. Brett. JHKLM photometry—Standard systems, passbands, and intrinsic colors. *PASP*, 100:1134–1151, 1988.
7. C. Bishop. *Neural Networks for Pattern Recognition*. Oxford University Press, New York, 1996.
8. Chang, C. C. and Lin, C. J. LIBSVM: A library for support vector machines: http://www.csie.ntu.edu.tw/~cjlin/libsvm, 2001.

9. P. Cheeseman and J. Stutz. Bayesian classification (AutoClass): Theory and results. In U. M. Fayyad, G. Piatetsky-Shapiro, P. Smyth, and R. Uthurusamy, editors, *Advances in Knowledge Discovery and Data Mining*, pp. 61–83, 1996.

10. T. M. Cover and P. E. Hart. Nearest neighbor pattern classification. *IEEE Trans. Inf. Theory*, 13:21–27, 1967.

11. N. Cristianini and J. Shawe-Taylor. *An Introduction to Support Vector Machines and Other Kernel-Based Learning Methods*. Cambridge University Press, 2000.

12. M. Feast. AGB variables as distance indicators. In D. W. Kurtz and K. R. Pollard, editors, *IAU Colloq. 193: Variable Stars in the Local Group*, Astronomical Society of the Pacific Conference Series, (Vol. 310), pp. 304–ff., May 2004.

13. S. Haykin. *Neural Networks: A Comprehensive Foundation*. Prentice Hall, Englewood Cliffs, NJ, 1998.

14. P. N. Kholopov, N. N. Samus, M. S. Frolov, V. P. Goranskij, N. A. Gorynya, E. A. Karitskaya, E. V. Kazarovets et al. *General Catalogue of Variable Stars*. 4th ed., Nauka, Moscow, 1998.

15. T. Kohonen. Self-organized formation of topologically correct feature maps. *Biological Cybernetics*, 43:59, 1982.

16. E. Lastennet, T. Lejeune, E. Oblak, P. Westera, and R. Buser. BaSeL: A library of synthetic spectra and colours for GAIA. *A&A Suppl.*, 280:83–87, 2002.

17. S. P. Lloyd. Least squares quantization in PCM. *IEEE Trans. Inf. Theory*, 28:129–137, 1982.

18. LSST Science Collaborations, P. A. Abell, J. Allison, S. F. Anderson, J. R. Andrew, J. R. P. Angel, L. Armus, D. Arnett et al. *LSST Science Book*, Version 2.0. *ArXiv e-prints*, December 2009.

19. D. J. C. MacKay. The evidence framework applied to classification networks. *Neural Comput.*, 4:720, 1992.

20. J. A. Mattei, G. Foster, L. A. Hurwitz, K. H. Malatesta, L. A. Willson, and M. O. Mennessier. Classification of red variables. In R. M. Bonnet, E. Høg, P. L. Bernacca, L. Emiliani, A. Blaauw, C. Turon, J. Kovalevsky et al., editors, *Hipparcos—Venice '97*, ESA Special Publication, (Vol. 402), pp. 269–274, August 1997.

21. M. O. Mennessier, H. Boughaleb, and J. A. Mattei. Mean light curves of long-period variables and discrimination between carbon- and oxygen-rich stars. *A&A Suppl.*, 124:143–151, 1997.

22. I. T. Nabney. *Algorithms for Pattern Recognition*. Springer–Verlag, New York, 2002.

23. M. A. C. Perryman. Overview of the Gaia mission. In C. Turon, K. S. O'Flaherty, and M. A. C. Perryman, editors, *The Three-Dimensional Universe with Gaia*, ESA Special Publication, (Vol. 576), pp. 15–ff., January 2005.

24. G. Pignata, S. Benetti, P. A. Mazzali, R. Kotak, F. Patat, P. Meikle, M. Stehle et al. Optical and infrared observations of SN 2002dj: Some possible common properties of fast-expanding type Ia supernovae. *MNRAS*, 388:971–990, 2008.

25. J. R. Quinlan. *C4.5: Programs for Machine Learning*. Morgan Kaufmann, New York, 1993.

26. A. Udalski, M. Kubiak, and M. Szymanski. Optical gravitational lensing experiment. OGLE-2—the second phase of the OGLE project. *Acta Astron.*, 47:319–344, 1997.

27. V. Vapnik. *Statistical Learning Theory*. Wiley, New York, 1998.

28. P. Whitelock, F. Marang, and M. Feast. Infrared colours for Mira-like long-period variables found in the Hipparcos Catalogue. *MNRAS*, 319:728–758, 2000.

29. P. R. Wozniak, C. Akerlof, S. Amrose, S. Brumby, D. Casperson, G. Gisler, R. Kehoe et al., Classification of ROTSE variable stars using machine learning. In *Bulletin of the American Astronomical Society*, (Vol. 33), pp. 1495–ff., 2001.

30. P. R. Wozniak, A. Udalski, M. Szymanski, M. Kubiak, G. Pietrzynski, I. Soszynski, and K. Zebrun. Difference image analysis of the OGLE-II bulge data. II. Microlensing events. *Acta Astron.*, 51:175–219, 2001.

31. P. R. Woźniak, W. T. Vestrand, C. W. Akerlof, R. Balsano, J. Bloch, D. Casperson, S. Fletcher et al., Northern sky variability survey: Public data release. *Astron. J.*, 127:2436–2449, 2004.

32. P. R. Woźniak, S. J. Williams, W. T. Vestrand, and V. Gupta. Identifying red variables in the northern sky variability survey. *Astron. J.*, 128:2965–2976, 2004.

33. N. Wyn Evans and V. Belokurov. A prototype for science alerts. In C. Turon, K. S. O'Flaherty, and M. A. C. Perryman, editors, *The Three-Dimensional Universe with Gaia*, ESA Special Publication, (Vol. 576), pp. 385–ff., January 2005.

34. Ł. Wyrzykowski and V. Belokurov. Self-organizing maps in application to the OGLE data and Gaia science alerts. In C. A. L. Bailer-Jones, editor, *Classification and Discovery in Large Astronomical Surveys*, American Institute of Physics Conference Series, (Vol. 1082), pp. 201–206, December 2008.

CHAPTER 19

Gravitational Wave Astronomy

Lee Samuel Finn
Pennsylvania State University

CONTENTS

19.1 Introduction   408
19.2 Gravitational Waves and Their Detection   408
    19.2.1 What Are Gravitational Waves?   409
    19.2.2 How Are Gravitational Waves Connected with Their Source?   411
    19.2.3 How Does Propagation Affect Gravitational Waves?   413
    19.2.4 Detecting Gravitational Waves   414
        19.2.4.1 10 Hz–1 kHz: Laser Interferometry   414
        19.2.4.2 0.01 mHz–1 Hz: Spacecraft Doppler Tracking   419
        19.2.4.3 1 nHz–1 μHz: Pulsar Timing Arrays   421
19.3 Data Analysis for Gravitational Wave Astronomy   424
    19.3.1 Overall Characterization of the Analysis Problem   425
    19.3.2 Noise Characterization   426
    19.3.3 Detecting a Stochastic Gravitational Wave Background   426
    19.3.4 Detecting Periodic Gravitational Waves   427
    19.3.5 Detecting Gravitational Waves from Inspiraling Compact Binaries   427
    19.3.6 Detecting Gravitational Wave Bursts   428
19.4 Machine Learning for Gravitational Wave Astronomy   428
    19.4.1 Noise Characterization   428
        19.4.1.1 Noise Nonstationarity   428
        19.4.1.2 Glitch Identification and Classification   429
        19.4.1.3 Directions for Future Work   429
    19.4.2 Response Function Estimation: LISA Recovery from Antenna Pointing   429
    19.4.3 Source Science   430
        19.4.3.1 Inspiraling Massive Black-Hole Binaries   430
        19.4.3.2 Emerging Source Population Statistics   430
19.5 Conclusions   431

Acknowledgments   431
References   432

19.1 INTRODUCTION

If two black holes collide in a vacuum, can they be observed?

Until recently, the answer would have to be "no." After all, how would we observe them? Black holes are "naked" mass: pure mass, simple mass, mass devoid of any matter whose interactions might lead to the emission of photons or neutrinos, or any electromagnetic fields that might accelerate cosmic rays or leave some other signature that we could observe in our most sensitive astronomical instruments.

Still, black holes do have mass. As such, they interact—like all mass—gravitationally. And the influence of gravity, like all influences, propagates no faster than that universal speed we first came to know as the speed of light. The effort to detect that propagating influence, which we term as *gravitational radiation* or *gravitational waves*, was initiated just over 50 years ago with the pioneering work of Joe Weber [1] and has been the object of increasingly intense experimental effort ever since.

Have we, as yet, detected gravitational waves? The answer is still "no." Nevertheless, the accumulation of the experimental efforts begun fifty years ago has brought us to the point where we can confidently say that gravitational waves will soon be detected and, with that first detection, the era of *gravitational wave astronomy*—the observational use of gravitational waves, emitted by heavenly bodies—will begin.

Data analysis for gravitational wave astronomy is, today, in its infancy and its practitioners have much to learn from allied fields, including machine learning. Machine learning tools and techniques have not yet been applied in any extensive or substantial way to the study or analysis of gravitational wave data. It is fair to say that this owes principally to the fields relative youth and not to any intrinsic unsuitability of machine learning tools to the analysis problems the field faces. Indeed, the nature of many of the analysis problems faced by the field today cry-out for the application of machine learning techniques. My principal goal in this chapter is to (i) describe the gravitational wave astronomy problem domain and associated analysis challenges, and (ii) identify some specific problem areas where the application of machine learning techniques may be employed to particular advantage. In Section 19.2, I describe what gravitational waves are, how they are generated and propagated, and the several different observational technologies through which we expect, over the next decade or so, gravitational wave astronomy will exploit. I have written this section for the nonastronomer; however, I think that even the gravitational wave astronomer may find the viewpoint taken here to be of interest. In Section 19.3, I deconstruct the work involved in the analysis of gravitational wave data and describe (briefly!) the techniques currently used for data analysis. The focus of Section 19.4 is on the application of machine learning tools and techniques in gravitational wave data analysis. I conclude with some closing remarks in Section 19.5.

19.2 GRAVITATIONAL WAVES AND THEIR DETECTION

The value to the astronomer of observing gravitational waves, electromagnetic waves, neutrinos, cosmic rays, or any other field or particle arises from the connection between the particles or waves and one or more of (i) its source and the source environment, (ii) the medium the radiation has propagated through, or (in the case of so-called "fundamental physics" investigations) (iii) the character of the radiation itself (e.g., propagation speed or number of polarization states). Observations or analyses that enable the astronomer to infer more effectively or efficiently the properties of the source and its environment, the intervening medium, or the character of the radiation itself, will compel his or her immediate interest and attention. In this section I have two goals: first, to describe the character of the information carried by gravitational waves about the source, the intervening medium, or the character of physical law; second, to describe how the radiation is detected and, thus, the character of the observational data available to achieve those goals.

19.2.1 What Are Gravitational Waves?

Among Einstein's greatest insights was the intimate connection between space–time structure and what we call gravity. In Einstein's conceptualization, three-dimensional space and one-dimensional time are cross-sections through an intrinsically four-dimensional space–time and what we call gravity is a manifestation of space–time geometry.

A flatland analogy may be helpful in gaining the gist of Einstein's vision. A sheet of paper and the surface of a sphere are both two-dimensional surfaces. The paper sheet's intrinsic geometry is flat while the corresponding geometry of the sphere's surface is curved.[*] Using only measurements made on the surface—either the sheet's or the sphere's—how can we tell whether the surface is curved? Trace pairs of straight lines on our surface that begin at nearby points and start out in parallel directions. On the sheet of paper we find that the lines always remain parallel, while on the sphere's surface they begin to converge upon each other, crossing after traveling a distance $\pi R/2$, where R is the sphere's radius.

Completing our analogy, straight lines on the sphere, or the sheet of paper, are the paths of freely falling bodies in space–time. Turning Newton's First Law[†] inside out, Einstein said that a straight line in space–time is the trajectory of an object at rest or in uniform motion upon which no external forces act: that is, it is the trajectory of an object in "free fall." Release an object and watch it fall: it is in free fall, traveling in a straight line in space-time with no forces acting on it until it strikes the ground. Gravity is *not* responsible for the falling. Drop two objects, side by side, and watch them fall. Note that they start out on parallel space–time trajectories. Note, too, that they accelerate toward each other as each falls separately toward the Earth's center. *Gravity is responsible for the relative acceleration in the separation between nearby objects in free fall.* A cannonball, following a parabolic trajectory, is actually travelling in a straight line through a curved space–time; Earth, in orbit about Sun, is also traveling along a straight line trajectory in a curved, four-dimensional space–time.

---

[*] In this example it is important to distinguish between the *intrinsic* and *extrinsic* geometry of the surfaces. Intrinsic geometry refers only to the surface; extrinsic geometry refers to the relationship between the surface and the higher-dimensional space in which it is embedded. The sheet of paper may, for example, be wrapped about a cylinder, in which case it acquires an *extrinsic* curvature; however, its *intrinsic* geometry remains flat. The sheet's extrinsic curvature is apparent only in reference to the three-dimensional space in which it is embedded.

[†] "An object will remain at rest or in uniform motion in a straight line unless acted upon by an external force."

If **n** is the vector separation between two (infinitesimally) separated particles, both in free fall with paths initially parallel, then we may write the acceleration of the separation as

$$\frac{d^2 n^j}{dt^2} = -\Phi^j_k n^k \tag{19.1}$$

where we are introducing Φ to represent a particular projection of the space–time curvature tensor (the *Riemann*) and there is an implied summation over repeated indices (e.g., $\sum_{k=1}^3$, corresponding to the x, y, and z spatial directions).* In Newton's theory of gravity the same equation holds when we make the substitution

$$\Phi^j_k = -\frac{\partial \phi}{\partial x^j \partial x^k} \tag{19.2}$$

where ϕ is the Newtonian gravitational potential. Where Newton sees a position-dependent force, Einstein sees space–time curvature. These differences in outlook lead to startlingly different theories, whose differences become apparent when we make predictions about gravitational waves (which do not exist in Newton's theory), or regions of very strong gravity (e.g., near neutron stars or black holes, where the theories make different predictions and Einstein's can be seen to be correct), or when we talk about the Universe at large (e.g., its expansion, a concept that cannot even be formulated in Newton's theory).

Return now to our two-dimensional planar surface. Imagine two long, straight, parallel "roads" on this plane, along which you and a colleague each travel at constant velocity. Suppose that a mountain interrupts the otherwise flat plane, but only between the two roads, as in Figure 19.1. The mountain introduces curvature into the geometry of the sheet, which is otherwise flat. Each of the roads traverse the flat part of the surface, are straight, and, setting aside the region where the plane is interrupted by the mountain, parallel. Focus attention on the left-hand-side traveller. Suppose that every time interval $\delta\tau$ that traveler sends a runner, who travels at constant velocity, ahead to intercept the traveler on the right-hand road with a message. When the runner's path must traverse the mountain the path is longer than when it remains entirely along the plane. Correspondingly, even though the

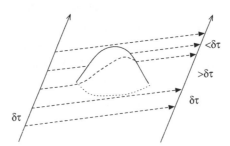

FIGURE 19.1 Curvature, here depicted in the form of a mountain, affects the length of the path between travelers who pass on either side of it.

---

* Do not look for much rigor in this discussion: I am introducing many conceptual, notational and other conveniences in this overview of gravitational wave physics in order to get the important ideas across.

messages are sent at constant intervals of time, they are received at intervals that vary owing to the curvature of the path along which the runner must travel. So it is in space–time: the length of the path between two points in space–time depends upon the space–time curvature along the path. Given a collection of paths and the time it takes to traverse them, we can measure the curvature. For weak gravitational waves it is helpful to separate the tensor Φ that appears in Equation 19.1 into two pieces: the average curvature Φ_0 and the short-wavelength gravitational wave contribution $\delta\Phi$:

$$\Phi \sim \Phi_0 + \delta\Phi \qquad (19.3)$$

19.2.2 How Are Gravitational Waves Connected with Their Source?

To this point we have emphasized a holistic view of space–time. Let us now adopt a different perspective and think about space–time as a sequence of nonintersecting spatial cross-sections, each indexed by time.[*] From this perspective, a particle's trajectory through space–time can be described by its location on successive spatial cross-sections: that is, as a function of time. In a similar fashion, some of the change in electric and magnetic fields from cross-section to cross-section can be associated with electromagnetic waves and, most importantly for us, some aspects of the change in curvature from cross-section to cross-section can be thought of as gravitational waves.

It is through this perspective that we say that, for example, electric and magnetic fields are "generated" by a source, and propagate away from their source. Acceleration of charges or currents lead to changes in the electric and magnetic fields that move through space with time: that is, they lead to a propagating disturbance in the electromagnetic field. Furthermore, this disturbance propagates in a way that, once having left its source, is and always remains independent of the source. These "free-standing," propagating fields are what we call electromagnetic waves and they can be primordial in origin, or sourced by the dynamics of a charge or current distribution. Even when sourced, however, once generated they propagate freely, following their own laws of motion and without any further reference to their source.

In the same way, space–time curvature is generated mass-energy density, flux, and stresses in any form: for example, the matter under pressure that makes up the sun is inextricably linked with the space–time curvature that leads to Earth's orbital trajectory. Space–time curvature can also exist without any stress-energy to support its existence. One example, with no electromagnetic analog, is the black hole: a localized region of space–time curvature without any accompanying matter, or energy density or flux, or stresses. Another example— the one that concerns us here—is gravitational radiation, which appears as propagating waves of space–time curvature. Like electromagnetic waves, gravitational waves can be primordial in origin, or sourced by the dynamics of mass-energy. Even when sourced, however, once generated they propagate freely through space–time following their own laws of motion and without any further reference to the source kinematics of dynamics.

---

[*] It does not matter how we choose these cross-sections as long as every point in space–time is in one, and only one, cross-section; and that the paths traveled by imaginary light rays intersect any cross-section at no more than one point.

Carrying our electromagnetic analogy further we gain insight into what gravitational waves can tell us about their source. The generation and propagation of electromagnetic and (weak) gravitational waves are both governed by the wave equation. In the electromagnetic case the source is the electric charge and current density; in the gravitational case the source is the stress-energy of matter and fields. The solution to either wave equation can be written as an expansion in powers of the source's internal velocities. From this expansion we find that the energy associated with the propagating fields is a sum over the squared magnitude of the time derivatives of the source's spatial multipole moments: for example, the first derivative of the source's monopole moment, the second time derivative of its dipole moment.

The monopole moment of a charge distribution is just the total charge. In the electromagnetic case we know that the total electric charge is conserved; so, there is no monopole electromagnetic radiation. Only source motions that lead to accelerating electric (or electric current) dipole moments or higher can lead to propagating electromagnetic waves. Conversely, observation of propagating electromagnetic waves can tell us about, for example, the corresponding time derivatives of source charge or current multipoles.

In the gravitational case the monopole moment of the "gravitational charge" distribution is the total mass, which is also (for weak waves) conserved: that is, there is no monopole gravitational radiation. The dipole moment we can write as the total mass times the location of the source center-of-mass. The first derivative of the mass dipole is just thus the total source momentum, which is also conserved. There is, thus, no gravitational wave dipole radiation. *The leading-order contribution to the gravitational waves radiated by a source arises from the time-varying acceleration of the source's quadropole moment.* Conversely, observation of gravitational waves can tell us about the internal dynamics of the source's mass-energy distribution. Since these, in turn, are governed by the forces acting within the source, gravitational waves provide a truly unique insight into the dynamical source structure and the totality of forces and their distribution, which governs its internal evolution. Through gravitational waves we can thus "observe" two black holes orbiting or colliding with each other, the collapse of a stellar core as the precursor to a γ-ray burst, the swirling of currents superdense matter in a rapidly rotating neutron star, kinks and cusps on cosmic strings, and properties of space–time that cannot be studied in any other way.

Dotting the i's and crossing the t's we can write the leading-order contribution to the "gravitational wave" tensor $\delta\boldsymbol{\Phi}$ introduced in Equation 19.3 as

$$\delta\Phi^{j}_{k}(t,\mathbf{x}) = -\frac{2G_N}{c^6 r}\frac{\mathrm{d}^4}{\mathrm{d}t^4}\left[I_{jk}(t - cr)\right]^{\mathrm{TT}} \tag{19.4a}$$

where r is the distance from the source to $\mathbf{x}$, G_N and c are Newton's constant of Universal Gravitation and the speed of light, the source's quadrupole moment is

$$I_{jk} = \int \mathrm{d}^3 x \rho(x)\left(x^j x^k - \frac{r^2}{3}\delta_{jk}\right) \tag{19.4b}$$

with ρ the source mass density, δ_{jk} the Kronecker delta function, and $[]^{\mathrm{TT}}$ denotes the taking of the *transvere-traceless* part of I_{jk}: that is, if we are talking about waves propagating in the (unit) direction $\hat{n}^j$

$$\left[I_{jk}\right]^{\text{TT}} = \left(\delta_j^l - \frac{1}{2}\hat{n}_j\hat{n}^l\right) I_{lm} \left(\delta_k^m - \frac{1}{2}\hat{n}^m\hat{n}_k\right) \qquad (19.4c)$$

Once again, there is an implied summation over the repeated indices l and m. Equation 19.4 is our representation of what gravitational wave scientists call the *quadrupole formula*. Owing to the weakness of the gravitational "force" it plays a more central role in gravitational wave astronomy than the corresponding electric dipole formula plays in electromagnetic astronomy.

19.2.3 How Does Propagation Affect Gravitational Waves?

To understand how gravitational radiation is affected by its propagation between their source and Earth it is helpful to again consider the electromagnetic case first. The interpretation of electromagnetic observations of distant sources must account for a number of different effects associated with the propagation of the radiation between the source and Earth. These may be separated into two categories. The first category we take to be the effects associated with space–time curvature itself. These include the expansion of the universe and the associated cosmological redshift, the so-called Sachs–Wolfe effect, and gravitational lensing. The second category are effects associated with the interaction between the propagating radiation and intervening matter and its environment: for example, gas, dust, and the magnetic fields of the Interstellar Medium (ISM). These include the overall absorption of energy by the ISM (what astronomer call *extinction*), the differential absorption of higher-frequency light compared to lower-frequency light (the same phenomenon that makes the sky blue, and what astronomers call *reddening*), the frequency-dependent slowing of light in propagating through the ISM (what astronomers call *dispersion*), and the rotation of the wave polarization plane owing to magnetic fields lacing the ISM (the *Faraday effect*).

Gravitational waves are affected by everything in the first category in the same way as electromagnetic waves: that is, they are affected by the cosmological redshift, the Sachs–Wolfe effect, and gravitational lensing just as are electromagnetic waves.[*] When it comes to the effects in the second category, however, significant differences emerge.

Compared to electromagnetic forces the coupling between matter and gravity is *very* weak: the electric repulsion between two protons is 10^{40} times greater than their gravitational attraction. Owing to this great disparity of scales gravitational waves propagate from their source to our detectors effectively undisturbed by any intervening matter. This is true even for waves generated at the very earliest moments of creation, when the universe was no more than 10^{-43} s old!

Owing to the strong interaction between matter and electromagnetic fields we can often infer properties of the intervening matter when we observe the light from a distant source. These inferences include the matter density, temperature, contained average magnetic energy,

---

[*] There are some subtle differences associated with the difference between the vector-character of the electromagnetic field and the second-rank tensor character of the gravitational wave contribution to the overall space–time curvature.

dust particle size, distance, composition, and so on. Gravitational wave observations will not add to our knowledge of any of these things.

There are, however, offsetting advantages to the weakness of the gravitational wave interaction with intervening matter. For a vivid example, consider a core-collapse supernova explosion. In this especially energetic event the core of a massive star—the innermost ~1.4 solar masses of an ~10 solar mass star—undergoes a sudden collapse when its nuclear fire is no longer able to support its mass against its own gravity. During this collapse the stellar core shrinks from a radius of thousands of kilometers to kilometers in under a second. The subsequent "bounce" of the now superdense core starts a shock wave that travels outward through the rest of the star, emerging ~10 h later and spewing tremendous amounts of energy and matter throughout the universe. It is the light from the emergence of the shock that we see electromagnetically: What has gone on in the interior is completely obscured from our direct view. The gravitational waves from the core collapse and bounce, however, will pass through the rest of the matter of the star without being disturbed or reprocessed in any way, reaching our detector hours before the electromagnetic signature of the collapse appears, and providing us with an unobstructed view of the core's collapse, bounce, and (perhaps) recollapse to form a central black hole.

19.2.4 Detecting Gravitational Waves

As I write this, the effort to detect gravitational waves is entering its 51st year. Over that period, different techniques and technologies have been introduced, vigorously developed, and discarded in favor of newer and more promising techniques and technologies.[*] Direct detection of gravitational waves is currently being pursued by three different technologies in three different wavebands. The character of the data and the way in which gravitational waves make their presence known is different for each method. In the remainder of this section I describe, for each waveband, the anticipated sources; the techniques that are being used or proposed to detect them; and the character of the observational data.

19.2.4.1 10 Hz–1 kHz: Laser Interferometry

About the band. The highest-frequency band currently being explored for gravitational waves extends from the 10s of Hz to several kHz. In this band the principal anticipated sources are orbiting neutron star or black hole binary systems in the tens of seconds prior to and including their coalescence. Other sources that may be observed include circulating "ocean currents"[†] or cm-sized "mountains" on rapidly rotating neutron stars, or the burst of radiation from the collapse of a dense stellar core (See Ref. [5] for a more complete list and discussion of potential sources).

Black hole/neutron star or neutron star/neutron star binary systems are believed to be the progenitors of short γ-ray bursts while hypernovae—associated with the collapse of

---

[*] The history of how we have gotten to where we are is a fascinating story. The effort to detect gravitational waves has a certain "War and Peace" quality to it—dramatic, epic in scale, with false starts, grand and petty competitions within laboratories and spanning states, countries, and continents; and rife with the tragedies, human frailties and greatness of character that are the ingredients of great novels. For 35 of its years it has been the subject of a study in the sociology of science whose research has been published in learned journals and three (very approachable) books [2–4].

[†] The so-called *r*-modes.

massive, rapidly rotating stellar cores—are believed to be the progenitors of long γ-ray bursts. Correspondingly, the gravitational waves associated with binary coalescence or stellar core collapse may be shortly followed by an electromagnetic burst [6–26]. Joint gravitational wave and electromagnetic observations will allow us to better characterize γ-ray burst progenitors. The nuclear density equation of state directly relates to the size of the mountain that a neutron star crust can maintain against gravity and the character of the radiation associated with any circulating currents; correspondingly, observation of the gravitational radiation from rapidly rotating neutron stars will provide a critical diagnostic of the property of matter at extremes of density unreachable in any terrestrial laboratory. Gravitational wave observations of binary coalescence can reveal the mass and absolute distance to the observed systems [27–29]. The distribution of these masses and distances can be used to infer the black hole and neutron star birth rate and provide clues to the paths by which they form and evolve. With an electromagnetic counterpart whose redshift can be measured the absolute distance measurement to the source will provide a calibration of the cosmic distance ladder on distance scales orders of magnitude greater than those now possible.

Detector technology and status. Laser interferometry is the principal technology employed in the effort to detect gravitational waves in this band. "First generation" observatory-scale detectors have been constructed in the United States (the Laser Interferometer Gravitational-Wave Observatory LIGO project [30]) and in Europe by a French/Italian Consortium (the Virgo Project [31]). Prototype and development instruments have been pursued by a German/British Consortium (the GEO600 project [32]), Japan (TAMA300 [33]), and Australia (Australian Consortium for Interferometric Gravitational Astronomy).

The LIGO Laboratory operates observatory-scale detectors: two located at the same site on in Hanford, Washington State and one located in Livingston, Louisiana. Virgo operates a single detector in Cascina, Italy. LIGO reached its design sensitivity in late 2005 and Virgo reached its peak sensitivity in 2010. In 2007, LIGO and Virgo entered into an agreement to analyze their data and publish their results jointly.

LIGO and Virgo have separately and jointly published the results of analyses aimed at bounding the gravitational wave power in a stochastic wave background [34–38] or from pulsars and other periodic sources [39–50], the rate of neutron star or black hole binary coalescences [51–60], gravitational waves associated with γ-ray bursts [22,24–26,61,62] and other astrophysical events identified via their electromagnetic signature [26,54,63], exotic physics phenomena [64], and gravitational wave bursts of unknown origin [65–73].

In late 2010 both LIGO and Virgo—the only observatory-scale gravitational wave detectors operating in the "high-frequency" band—were taken "off the air" for instrumentation upgrades that are expected to increase their sensitivity to gravitational wave burst events by a factor of several thousand. As currently planned, construction of the first of the Advanced LIGO (advLIGO) detectors will end in early 2015. Experience with the commissioning of first-generation LIGO detectors suggests that advLIGO science operations will begin some years after that, with design sensitivity requiring as many as 5 years of commissioning. Advanced Virgo development can be expected to follow a similar schedule.

When complete, advLIGO will include three detectors. One detector will be located at the Livingston site and at least one detector at the Hanford site. The LIGO Laboratory, with the concurrence of the United States National Science Foundation (NSF), has proposed that the third advLIGO detector be located in Gingin, Australia. The Australian government is currently studying this proposal. If an agreement is not reached, then two advLIGO detectors will be located at the Hanford site.

In the later part of the current decade a new observatory-scale detector will join advLIGO and advanced Virgo. In June 2010 the Japanese Government funded the construction of the LCGT [74]. LCGT construction is currently expected to end in early 2017. Like advLIGO and Virgo, LCGT is likely to require several years to reach its design sensitivity.

AdvLIGO, advanced Virgo, and LCGT are all expected to have equivalent sensitivity. Working together at their design sensitivity these detectors should be capable of observing neutron-star or stellar-mass black hole binary inspiral and coalescence events with an estimated rate ranging from several per year to as great as once per day [75].

How the detector works. Figure 19.2 shows, in outline, how a laser interferometric gravitational wave detector works. In an archetypical laser interferometric gravitational wave detector light from an ultrastable laser is directed toward a beamsplitter. The beamsplitter directs the incident light equally along the interferometer's two arms, which are typically oriented at right angles to each other. At the end of each arm (in current detectors 3–4 km from the beamsplitter) the light is reflected back toward the beamsplitter. The returning beams are recombined at the beamsplitter and the recombined light observed. The intensity of the recombined light is proportional to the cosine of the difference in the round-trip light travel time along the two arms, measured in units of the laser frequency. Passing gravitational waves disturb the space–time paths traveled by the beams along the two arms differently, leading to differential changes in the round-trip light travel time and corresponding changes in the intensity of the recombined light. Light travel time disturbances amounting to as little as 10^{-27} s (a *milli-yocto second!*) are measurable today using gravitational wave detectors based on this technology.

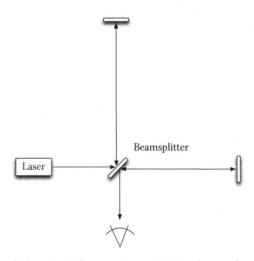

FIGURE 19.2 Schematic of a laser interferometric gravitational wave detector.

A detector network. Individual interferometric detectors are not capable of identifying the propagation direction of detected gravitational waves (correspondingly, the direction to the radiation source). Figure 19.3 shows the antenna pattern (or beam) of a single interferometric detector. The radius–vector distance of the surface from the origin is proportional to the detector's sensitivity to sources in that direction; the lines show the interferometer arms. A useful measure of a detector's ability to resolve the location of a source is the area of the sky within which 99.73% (i.e., 3-σ) of the detectors integrated sensitivity lies. For a single interferometric detector this is 95% of the sky!

The ability to resolve sources on the sky improves greatly when we work with the data from a geographically distributed *network* of detectors. Gravitational waves propagating past Earth interact with any given detector at a time determined by the wave propagation direction and the location of the detector on the globe. Similarly, the amplitude response of any given detector depends on the location of the source relative to the detector's plane. When the radiation from a single source is observed in a network of detectors the relative response amplitude and time delay in the different detectors will constrain the source's location on the sky.

Data character. The data stream in a detector network is a collection of time series. The "science data" consist of a single time series from each detector in the network (in the advanced detector era the three advLIGO detectors and the single advanced Virgo and LCGT detectors). The signal content of each time-series varies linearly with the incident gravitational wave ($\delta\Phi$). Each science data time-series is uniformly sampled at 16 kHz with a 16-bit resolution. In addition to the "science data," time-series data from a large number of instrument diagnostics and detector pem (e.g., seismometers, anemometers, microphones, magnetometers) are recorded and used to truncate or veto data epochs when a detector may be misbehaving or its physical environment may be exceptionally noisy.

Each detector in a network responds to the same gravitational wave signal; however, since the radiation interacts with each detector at times that vary with the wave propagation direction and the detector location on Earth the time series are effectively *nonsimultaneously sampled* with an offset that varies with wave propagation direction.

FIGURE 19.3 The antenna pattern of a laser interferometric detector.

Regrettably little can be said about the data quality from any of the interferometric gravitational wave detectors (see Box 19.1). Each detector's science data noise is characterized by its PSD. In the current instrumentation noise nonstationarity and non-Gaussian tails are significant; however, quantitative information on these aspects of the data is embargoed by the projects. Neither the detector response nor the detector noise PSD are separately available: only figures showing the ratio of the noise PSD to the amplitude-squared response have been released. Figure 19.4 shows an authorized PSD of the best performance from the LIGO S5 data taking (Nov 2005–Nov 2007) [76].[*]

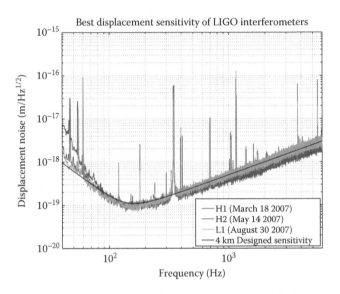

FIGURE 19.4 (**See color insert.**) Best (one-sided) noise PSD measured during LIGO's S5 data taking.

BOX 19.1 A PLEA FOR OPEN DATA, OPENLY ARRIVED AT

Remarkably, while laser interferometry is the most developed of all detection technologies and detectors built on this technology have gathered the most data and published the most "results" papers, it is also the technology about whose data the least can be said publicly! In 1996 the NSF, which funds this effort in the United States, commissioned a Blue Ribbon Panel to recommend how LIGO should be operated as a national facility.

---

[*] Passing gravitational waves disturb the light travel time in the interferometers arms. This difference in light travel time can be expressed as an equivalent change in the detector arm length by the simple device of multiplying by the speed of light. The sensitivity of interferometric gravitational wave detectors is most often expressed in terms of their ability to measure a differential change in detector arm length: hence, *displacement sensitivity*. An alternative convention, used equally often, is to divide the displacement sensitivity by the arm length for a dimensionless *strain sensitivity*. In any event, the actual effect of the passing gravitational waves is to disturb the light travel time in the detector.

That report [77] recommended that neither the data nor any data products be made available for study or investigation outside the LIGO Laboratory and a scientific collaboration (which became the LIGO Scientific Collaboration (LSC)) to be formed for that purpose. The NSF adopted that recommendation. (Full disclosure: I am a member of the LSC with unencumbered data access). Laboratory and Collaboration rules forbid any discussion of actual science data or data quality beyond carefully guarded remarks in select publications, and in 10 years of operations no release of scientific data from the detectors has ever been made.

The reason given for withholding data access is that, as first instruments of their kind, the data from these detectors, at least initially, cannot be interpreted except by those who have actually participated in detector construction and operation. As it has evolved, however, the practice of data acquisition and analysis within LIGO and the LSC does not reflect the assumption made by the NSF's advisory committee in reaching their recommendation. Even if current practices were similar to those supposed by the panel, the large number of publications emanating from the LSC claiming scientifically meaningful upper limits on the number of gravitational wave sources of different kinds, or the amplitude of gravitational waves from specific sources and the corresponding implications these have for example, neutron star structure, suggest strongly that the stated reasons for withholding access to the data no longer hold.

Restricting access to data is inimical to quality science. It prevents the independent replication or checking of analysis results; more insidiously, it shields disagreements and alternative interpretations of observations from the broader scientific community. It breeds insularity in the privileged group, impeding the introduction of new insights, ideas and concepts into the analysis. The absence of competition between truly independent groups leads to weaker analyses that take longer to complete. In short, restricting access to data severely limits the quality of the analysis, the impact of the observations, and the return-on-investment provided the taxpayers who fund the work.

Beginning in 2011 the NSF is requiring all proposals include a data management plan, which will be considered as part of the peer-review process. This new requirement is part of a growing recognition within the NSF that data from publicly funded research needs to be made public. How the new policies will develop and affect access to LIGO data remains to be seen. At present, and into the forseeable future, open scientific investigations involving realistic detector data, including investigations focused on the introduction of new analysis techniques, are not possible.

19.2.4.2   0.01 mHz–1 Hz: Spacecraft Doppler Tracking
About the band. The 0.01 mHz–1 Hz band is truly a "target rich" environment for gravitational wave detection. Upwards of tens of millions of galactic white dwarf binary systems have periods in this band and radiate gravitationally with a strength observable by the proposed Laser Interferometer Space Antenna (LISA). Of these, tens of thousands will

be individually resolvable, with the remainder contributing to a confusion-limit background [78–83]. Burst of radiation from the close passage of compact objects (neutron stars or stellar mass black holes) by our Galaxy's central black hole will deliver bursts of radiation approximately once per month, and several bursts per year can be expected to arise from a similar process in the Large Magellanic Cloud, a nearby satellite galaxy of our own [84,85]. Moving further outward, nearly all galaxies have massive ($10^{4.5}\,M_\odot \lesssim M \lesssim 10^{7.5}\,M_\odot$) or supermassive ($10^{7.5}\,M_\odot \lesssim M$) black holes in their central nuclei.[*] The capture of stellar mass compact objects by massive black holes in nearby galaxies, referred to as emri events, will lead to quasi-periodic gravitational wave signals may be capable of observing in this band [86–89]. Nearly all galaxies have undergone in their past at least one merger with a similarly sized galaxy. In these mergers the central black holes quickly "sink" to the center of the merged galaxies and form a binary that later coalesces. The radiation from the black hole coalescence events involving massive (but not supermassive) black holes in galactic nuclei will be visible by LISA throughout the universe wherever they occur [90,91].

White dwarf stars are electromagnetically faint and, consequently, difficult to observe and characterize. The number of white dwarf stars that gravitational wave observations will make possible surpass by orders of magnitude the sample size we have from electromagnetic observations. The spatial and frequency distribution of the confusion background from the large number of unresolvable systems is related to the spatial and period distribution of the binary systems that contribute to the background, provide an important diagnostic of our galaxies shape and stellar population. The frequency and character of the gravitational wave bursts associated with close stellar "fly-bys" of our Galaxy's central black hole will provide clues to the structure of our galactic nucleus [84,85]. In cosmology especially distance is time: events observed at greater distances occurred further in the universe's past. Tracking the evolution of black hole mass with distance thus traces the formation and evolution of galaxies [92–94]. As with ground-based observations in the high-frequency band (see Section 19.2.4.1) distances to events with an electromagnetic counterpart whose redshift can be measured will provide a measurement of the cosmic distance ladder on unprecedented scales [95,96].

Detector technology and status. The gravitational "attraction" between, for example, a tumbleweed rolling past the end-mirror of a ground-based laser interferometer will cause the end-mirror to move and lead to a signal in the detector that is, in that detector, indistinguishable from the effect of a passing gravitational wave [97]. Similarly, for moving fluctuations in the air density, Earth tidal and thermal distortions, and so on. The amplitude of these noises, which cannot be shielded against, climbs rapidly with decreasing frequency and render terrestrial gravitational wave detectors useless at frequencies much below 10 Hz [98]. To detect gravitational waves in this band we must move away from these disturbances: that is, move to a space-based detector system.

The LISA is a proposed space-based gravitational wave observatory [99,100]. The proposal for a space-based gravitational wave detector is over 30 years old [101]. In its current form, LISA has been the subject of intense study and technology development by European Space Agency (ESA) and National Aeronautics and Space Administration (NASA) for the last 20

---

[*] The massive versus supermassive nomenclature is not well defined; we adopt the definitions provided in the text.

years [102–112]. A flight technology demonstration mission—LISA Pathfinder [113–116]—is nearly complete and expected to launch in 2014. At this writing the ESA–NASA partnership has been suspended, with ESA examining a scaled-down, ESA-only mission (not yet named) that would be do-able within the current decade, and nasa looking at a scaled-down mission that might be possible on a longer timescale.

How the detector works. The LISA mission concept is straightforward. LISA will consist of three identical spacecraft, each in a solar orbit with an approximately 1 au radius. The initial location of the spacecraft on their orbits, and the choice of orbital eccentricity and relative inclination, are carefully chosen so that the spacecraft mark the vertices of an equilateral triangle with 5×10^6 km arms whose plane is inclined 60 deg with respect to the Earth's orbital plane [117–119]. In this configuration the triangular configuration is relatively stable and appears to rotate once around its center in the course of a year's revolution of the three spacecraft about the sun [120]. Once on orbit, the LISA spacecraft track their separation and relative velocities using laser transponders on each science craft. Passing gravitational waves affect the light travel time between the different science craft in the LISA constellation, leading to disturbances in their apparent relative velocities, from which the incident gravitational wave signals can be determined.

A detector network. It is sometimes helpful to think of LISA as a set of three different laser interferometric detectors of the kind discussed in Section 19.2.4.1. In this picture each spacecraft plays the role of beamsplitter and the other two spacecraft act as the mirrors at the end of the interferometer's arms. LISA can thus be viewed as a network of three detectors (albeit three detectors whose noise is correlated). Furthermore, as LISA orbits the sun, its orientation with respect to any given source will change with time, leading to modulations in the amplitude and arrival time of gravitational wave signals, which depends on each source's location on the sky. In this way, LISA observations will be capable of localizing on the sky sources whose signal duration is a substantial fraction of a year.

Data character. LISA "science data" will consist of six time-series data streams. Each time series corresponds to a Doppler measurement made on the transponded signal from one of its sister spacecraft. There are six time series because, for example, the round trip from spacecraft *a* to *b* and back traverses space–time differently than the round trip from *b* to *a* and back. By combining these Doppler velocity measurements in different ways the gravitational wave contribution can be enhanced and different system noises suppressed, and vice versa. The relative advantage of different combinations may change with solar activity, aging of the spacecraft lasers, source location on the LISA constellation's sky, and the distribution of the gravitational wave signal power with frequency. Choosing, at any given time and for any given purpose, the appropriate combination of Doppler links that will lead to the best characterization of the system noise and the most sensitive gravitational wave detector is an opportunity for the use of machine learning technology.

19.2.4.3 *1 nHz–1 μHz: Pulsar Timing Arrays*

Gravitational wave detection is an exercise in precision measurement. The standard against which measurements are compared must be stable in the measurement band. In the case of ground-based detectors the measurement standard is the laser phase, which must be stable in the 10 Hz–10 kHz band. In the case of LISA the measurement standard is again the laser phase, which now must be stable in the 10 μHz–1 Hz band. Verifying—let alone achieving—laser stability on time scales of days is an almost impossible task and LISA depends upon different combinations of the Doppler measurements to separate the correlated gravitational wave signal from the laser frequency noise in the measurement band [121–145]. Constructing, verifying and operating a precision measurement standard with sufficient stability on timescales of tens of years is simply beyond our ability.

Nature labors under no such limitations. In the guise of Millisecond Pulsars (MSPs) she provides us with metronomes whose beat is stable on nHz timescales over tens of years. By careful monitoring and correlation of the arrival times at Earth of the beats from a Pulsar Timing Array (PTA)—a network of MSPs—the signature of gravitational waves with periods ranging from days to years can be observed.

About the band. Detectable gravitational waves in the nHz–μHz band may arise from the inflationary epoch in the early universe [146–150]; cosmic strings or other exotic physics phenomena [64,151–158]; supermassive (i.e., $M \gtrsim 10^{7.5} M_{\odot}$) black hole binary or triplet systems [152–154]; and the confusion limit of a large number of weaker sources like supermassive black hole binaries and cosmic strings. As with LISA, the spatial and frequency distribution of the confusion background from these large number of unresolvable systems is related to the spatial and period distribution of sources that contribute to the background, providing an important source diagnostic. The detection of supermassive binary black hole systems complements lisa ability to do the same for massive black holes, extending our ability to track the evolution of galaxy structure to more recent times (corresponding to larger nuclear black hole masses).

Detector technology and status. Three major efforts to detect gravitational waves via pulsar timing are now underway. These are the European PTA [155,156], North American Nanohertz Observatory for Gravitational Waves (NANOGrav) [157,158], the Parkes PTA [159,160]. These three largely "continental" efforts work cooperatively—sharing expertise, personnel and data—under the guise of the International PTA [161].

The ability to accurately time MSPs depends on radio telescope signal-to-noise ratio (SNR), which (all other things being equal) is directly proportional to telescope collecting area; modeling of the solar system ephemeris (in order to take into account time delays owing to Earth's location in its orbit relative to the pulsar's location); propagation effects (most especially dispersion associated with the ISM); modeling of the pulsar ephemeris (many MSP are located in binary systems and their motion as a member of the binary must be accounted for); and timing noise intrinsic to the pulsar and its emission mechanism. With ground-based detectors like LIGO, Virgo or the LCGT, or the space-based LISA detector system, active measures can be taken to mitigate the principal sources of noise. In the case of PTA-based observations, however, only the telescope is under our active control and, once telescope noise is no longer dominant, only observing strategies are available to mitigate against the important noise sources that we cannot control.

The principal telescopes used for pulsar timing observations are the 100 m Green Bank Telescope [162], the 300 m Arecibo Telescope [163], the 100 m Effelsberg Telescope [164], the 76 m Lovell Telescope at Jodrell Bank [165], the 94 m Nançy Radio Telescope [166], the 64 m Sardinia Telescope [167], and the Westerbork Synthesis Radio Telescope [168], and the Parkes Observatory Telescope [169]. With these radio telescopes and associated timing instrumentation the arrival time of the beats from the "best" MSPs are predictable to better than 50 ns over the course of years. With improved instrumentation—principally the larger radio telescope collecting areas associated with the 500 m Aperture Spherical Radio Telescope, expected to begin operations in China in 2014, and the Square Kilometer Array (SKA) [170] pathfinders Australian Square Kilometer Array (ASKAP) [171,172] and Karoo Array Telescope (MeerKat) [173–175]—and better use of more extensive multifreqency observations—it is expected that this can be improved to 15 ns or better for many MSP [158,176]. In addition, dedicated searches with telescopes of larger collecting area are expected to increase the number of MSPs suitable for use as part of a PTA by as much as an order of magnitude.

Timing of individual pulsars has been used to bound the gravitational wave power associated with stochastic gravitational waves of cosmological origin [177–179]; timing residual correlations among pulsar pairs have been used to bound the power in an isotropic stochastic gravitational wave background [180]; the absence of evidence for gravitational waves in the timing of a single pulsar has been used to rule out the presence of a proposed supermassive black hole binary in 3C 66B [181]; observations from multiple pulsars have been used to place limits on periodic gravitational waves from the galactic center [182]; and timing analyses of individual pulsars have been combined to place a sky-averaged constraint on the merger rate of nearby black-hole binaries in the early phases of coalescence [183].

How the detector works. Electromagnetic emission from a pulsar is tightly beamed from a region of the pulsar magnetosphere that is effectively tied to the underlying neutron star. The neutron star's rotation carries the beam with it, which thus sweeps through space much like a lighthouse beam sweeps along the horizon. When the beam is aligned with the pulsar Earth line of sight an electromagnetic pulse is launched toward Earth.

At the pulsar the interval between pulses launched toward Earth is tied to the rotation of the pulsar, which acts as a huge flywheel and keeps the pulse rate regular. Successive pulses travel along different space–time paths (i.e., they leave the pulsar at different times). In the presence of gravitational waves the pulse propagation time along these different paths is different. Correspondingly, even though successive pulses leave the pulsar at regular intervals they arrive at Earth at intervals whose regularity is disturbed by the gravitational wave distortions of the space–time along their propagation path. While pulse arrival time disturbances—referred to as *timing residuals*—may arise from many causes, few are correlated between different pulsars whereas timing residuals owing to gravitational waves are correlated in a manner that depends on the wave propagation direction and the pulsar Earth line of sight.[*] Gravitational wave astronomy practiced via pulsar timing involves finding this correlation and inferring from it the radiation and source properties.

A detector network. The initial proposals for using pulsar timing to detect gravitational waves focused on single pulsars [184–186] and either periodic sources or limits on the

---

[*] There is an additional contribution that depends on the distance to the pulsars, which can be exploited if the Earth-pulsar distances is known to sufficient accuracy.

strength of a stochastic gravitational wave background. Later analyses looked for periodic gravitational waves from periodic sources [181,182]. Detection of gravitational waves using a single pulsar cannot identify the direction to the gravitational wave source: the antenna pattern corresponding to a single pulsar baseline is proportional to $(1 - \hat{k} \cdot \hat{n})^2$, where $\hat{k}$ is the wave propagation direction and $\hat{n}$ is the direction to the pulsar. In 1990, Foster and Backer [187] proposed timing an array of pulsars as a means of increasing sensitivity; however, it has only been in the last few years that analyses making full use of all the information in the multipulsar correlations have been described [188–190]. No analysis searching for gravitational wave point sources that makes use of the full power of a timing array has been carried out as yet.

Data character. PTA data are nonuniformly and nonsimultaneously sampled. The time between observation of any particular pulsar is, at present, measured in weeks to months and depends on the availability of radio telescope time and visibility of the pulsar in the sky (which is seasonal). Different pulsars are also timed at different times on different days with different telescopes. Significant gaps in the timing records exist owing to telescope upgrades. Corresponding to these upgrades are significant changes in the noise amplitude associated with timing observations before and after the upgrades. Pulsar timing records extend, for some pulsars, more than 40 years (the first pulsar was discovered in 1967 [191]); however, timing records for the class of MSPs, whose stability is exceptional among all pulsars, extends back only to the discovery of the first MSP in 1982 [192].

19.3 DATA ANALYSIS FOR GRAVITATIONAL WAVE ASTRONOMY

Irrespective of the source, signal, detector, or science, the analysis of observational or experimental data has a common set of requirements. Of central importance is a quantitative characterization of the detector system: in particular, how it responds to incident signals and the character of the associated system noise. If a description of an anticipated signal is available that description can be used to develop an analysis targeted toward detection of the signal. Finally, if the signal is tied to a source or to some source science, an analysis that speaks directly to the source science, without the intermediate step of characterizing the signal, can be deployed. In this section I briefly describe the analysis methods currently in use by the gravitational wave data analysis community for noise characterization, and signal detection and characterization.

Detectors exist and data have been collected and analyzed in two of the wavebands, and using two of the technologies, described in the previous section. Data from ground-based laser interferometers is, regrettably, not available for study outside the LIGO and Virgo collaborations. Data from PTA are more readily available, though—at present—not centrally distributed or well documented (though this is rapidly improving). While LISA and LISA data are in the future, the community of physicists and astronomers interested in the opportunities LISA presents have been the most aggressive in exploring a wide range of different techniques for addressing the analysis challenges that can be anticipated [193–209]. There has been surprisingly little transfer of ideas between researchers addressing analysis problems in the three different wavebands.

Gravitational wave astronomy is in its infancy—better said, waiting to be born. If, as you review the survey that follows, you find yourself wondering why it is that some analysis methodology is in use and another is not, keep in mind that the field is young, that sequestration of the data from the premier experiments has impeded the exploitation of knowledge and expertise from more mature, experienced and well-established fields, and that—for all these reasons, and more—not all choices are as well informed as you might hope or expect.

19.3.1 Overall Characterization of the Analysis Problem

Gravitational wave data analysts divide the space of signals into four categories: a diffuse *stochastic* gravitational wave background, assumed to be stationary; (nearly) *periodic* signals from point sources; the *chirp* signals [210,211] characteristic of coalescing compact objects (e.g., relativistic stars or black holes); and general gravitational wave *bursts*, defined as unmodeled signals of finite duration on order several to several thousand times times the detector's inverse bandwidth. In each band there are sources that we expect, based on other observational modalities, to observe; in the remainder of this subsection I describe the character of the analysis problem in the each of the three wavebands and the analysis techniques currently being exploited to detect sources of these different types. Nevertheless, *caelum*[*] *icognita* is the by-word for gravitational wave astronomy. The possibility of detecting sources not anticipated is, for many like myself, the prospect most exciting in this emerging observational discipline.

For ground-based detectors and the high-frequency band the combination of detector sensitivity and anticipated sources conspire to make the analysis problem one of identifying and characterizing rare, and isolated signals at low SNR. (In the high-frequency band we do not anticipate finding evidence for any gravitational wave stochastic background.)[†] Gravitational wave bursts in this band are expected to have durations ranging from tens of milliseconds to seconds. For the advanced detectors now being developed, coalescing compact binary systems whose signal amplitude is great enough to be detected are anticipated with rates that may be as large as several per day to as small as a few per year. Periodic sources, if detectable, are expected to be sufficiently weak that integration times of months to years will be required to accumulate sufficient signal-to-noise to identify them; correspondingly, observable sources will be well separated in the frequency domain.

For the low-frequency band associated with the proposed LISA project the combination of detector sensitivity and anticipated sources conspire to make the analysis problem one of identifying and characterizing large numbers of signals with (amplitude) SNR ranging into many thousands. Of the tens of millions of binary Galactic white dwarf systems, some tens of thousands will contribute gravitational wave signals with amplitudes great enough to be individually identified in frequency and on the sky, with the remainder contributing to an overall diffuse, stationary stochastic background that traces the shape of our Galaxy.

---

[*] Latin for "sky."

[†] This expectation is based both on theory and on experiment. In the case of experiment, particularly, it owes to (i) the impossibility of shielding a gravitational wave detector and, so, distinguishing between a stochastic background and instrumental noises *in a single detector*, and (ii) the wide separation between the different ground-based detectors, which strongly suppresses the in-band correlation between the signal in the different detectors in a network [212–214].

The chirping signal from individual extragalactic massive black hole binary systems will be detectable anywhere from several months to several years before coalescence. Finally, gravitational waves from of order tens of EMRI—stellar mass compact objects in orbit about much more massive black holes—may appear as either bursts or "periodic" signals with a spectrum whose lines are evolving rapidly in frequency and amplitude.

In the ultralow-frequency band probed by PTA observations the analysis problem suggested by the combination of array sensitivity and anticipated sources is one of identifying and characterizing moderate numbers of signals with varying amplitude. The principal anticipated source of gravitational waves in this band are supermassive black-hole binary systems with periods on order years. Several of these sources will have signal strengths great enough to be individually resolved in frequency and on the sky; however, the larger fraction will contribute to an unresolved, diffuse, and isotropic gravitational wave background [152,154,215,216].

19.3.2 Noise Characterization

At present, detector noise for all analyses and in all wavebands, whether operative or proposed, is quantitatively characterized as normally distributed and stationary.

In the case of LISA the assumption of normally distributed and stationary measurement noise is a reasonable assumption until such time as actual data is available.

For the ground-based detectors, whose data are uniformly sampled, the noise PSD associated with each detector's data is estimated using a modification of Welch's method over several minutes of data* and the cross-spectral densities (i.e., the correlations between the noise associated with detectors at different sites) are assumed to vanish. Detector noise is nonstationary on timescales of tens of minutes, with the overall amplitude of the noise PSD varying by factors of ∼30%. Day-to-day variations in the noise amplitude may amount to as much as a factor of two. On shorter durations (≲1 s) large-amplitude noise transients—*glitches*—arise relatively frequently in the current ground-based detectors. These transients are identified through a number of ad hoc tools [217]. Hand-sorting of transients so identified leads to a list of "vetoes": that is, intervals of time that are either excluded from analysis or suspect. The total duration excluded from analysis in this way—the *dead-time*—is ∼1% of each detector's operating time.

Pulsar timing data are irregularly sampled and the timing noise can vary significantly from pulsar to pulsar. As a general rule the timing noise is white on timescales of a few years, and red on longer timescales [218,219]. The timing noise PSD is estimated using a variety of ad hoc methods, all developed with an eye toward addressing the red noise component [220].

19.3.3 Detecting a Stochastic Gravitational Wave Background

The principal challenge in the detection of a stochastic background is distinguishing the signal from detector noise. In the case of gravitational waves this is complicated by the inability of taking the detector "off-source" (i.e., shielding it from the background that one is trying to detect). Correspondingly, a minimum of two independent detectors are required so that the

---

* In this modification the final power spectrum estimate is the median, and not the mean, of individual estimates.

stochastic background can be made evident in the correlation of their data [213,221,222]. In the case of the ground-based detectors this is further complicated by the large separation of the detectors relative to their bandwidth. The closest pair of ground-based detectors are the two LIGO detectors, which are separated by 3000km.[*] For gravitational waves with wavelength shorter than 3000km, corresponding to frequencies greater than 10Hz, the detectors no longer respond coherently to the incident radiation, strongly suppressing the correlated signal [213,214]. Similar considerations also hold for stochastic background detection using LISA and PTAs.

19.3.4 Detecting Periodic Gravitational Waves

Periodic signals may arise from rapidly rotating neutron stars, including those known via electromagnetic observations as pulsars. In either case the periodic signal is phase modulated owing to Earth's orbital motion about the sun. In searching for periodic gravitational waves two basic approaches are taken. For sources that can be observed electromagnetically the source's precisely known location on the sky and time-dependent phase evolution are used to search for a coherent signal in the observed data. For "blind" searches—that is, searches for sources not otherwise known—the observational data set is divided into a series of short segments whose upon which a maximum-likelihood statistic is calculated for an exactly periodic signal propagating from a source in particular direction on the sky. The log-likelihoods are summed to form an overall Frequentist detection statistic. For searches aimed at discovering gravitational wave emission from a particular source—that is, a particular pulsar—the known pulsar ephemeris and timing model are used to demodulate the observations and recover any signal power. In this latter case the statistical analysis is typically Bayesian.

19.3.5 Detecting Gravitational Waves from Inspiraling Compact Binaries

The gravitational wave signature of an inspiraling compact binary system is nearly monochromatic, with an instantaneous frequency and amplitude that increases as the time to coalescence decreases. In the case of the ground-based detectors and solar mass compact object binaries the signal crosses the detector band on timescales of seconds; in the case of LISA and massive black hole binaries the signal crosses the detector band on timescales of years; and in the case of supermassive black hole binaries and PTA the signal crosses the detector band on timescales of tens of millions of years. Thus, for ground-based detectors the signal has the character of a short burst of radiation, for LISA a long-lived quasi-periodic signal, and for PTAs almost all the signals will be effectively monochromatic over the timescale of the observations.

The gravitational wave signal for inspiraling binaries is well modeled by a 15-parameter function [223]. Analyses using ground-based detectors, and those proposed for LISA data, seek to make use of the well-modeled character of the signal via matched filtering.[†] In the

---

[*] In fact, there are LIGO detectors at the Hanford site; however, these detectors utilize the same facilities, making it impossible to distinguish correlated noises from a stochastic gravitational wave signal.

[†] The gravitational wave data analysis literature generally does not distinguish between the matched filter and the Wiener matched filter, often using the terms interchangeably. To be clear, the analyses currently used and proposed employ the matched filter—which is constructed to maximize the SNR for a given signal—and not the Wiener filter, which is constructed to minimize the mean-square difference between the observed and expected system response.

analysis of data from the ground-based detectors the data from each detector are analyzed separately and then compared for coincidence in the parameters of the matched filters that give produce the greatest amplitude response. An overall detection statistic is formed from the corresponding SNR and compared against a threshold computed to insure an acceptable false alarm rate.

19.3.6 Detecting Gravitational Wave Bursts

A separate effort in the collaborations with access to data from the ground-based detectors is focused on identifying gravitational wave bursts where no parameterized signal model exists and, hence, matched filter cannot be applied.[*] The principal analysis in this band is the so-called "WaveBurst Algorithm" [224–226]. WaveBurst forms a wavelet decomposition of the input data stream from each detector, evaluates the power in the wavelet coefficients, and identifies intervals of time when the power in the wavelet coefficients exceeds its expectation by some threshold. Other techniques for analysis have been proposed and applied to the analysis of data, in varying degrees, including changepoint [227] techniques, analyses that address the joint response of all detectors in the network [228,229] and the use of Bayesian analysis techniques that explore (using Markov Chain Monte Carlo techniques) the posterior probability density over the source parameters [190,230–236].

19.4 MACHINE LEARNING FOR GRAVITATIONAL WAVE ASTRONOMY

Machine learning tools and techniques have not yet been applied in any extensive or substantial way to the study or analysis of gravitational wave data. It is fair to say that this owes principally to the fields relative youth and not to any intrinsic unsuitability of machine learning tools to the analysis problems the field faces. Indeed, the nature of many of the analysis problems faced by the field today cry-out for the application of machine learning techniques. In this section I point toward some of those problems and review the few cases where machine learning tools have been applied.

19.4.1 Noise Characterization

Machine learning techniques have made only the narrowest of inroads into gravitational wave data analysis. These inroads have taken place in the area of detector noise characterization.

19.4.1.1 Noise Nonstationarity

Perhaps the earliest investigation application of machine learning techniques to problems in gravitational wave data analysis focused on the tracking the evolving character of certain spectral line features in the data stream of the large, ground-based laser interferometric detectors [237,238]. The mirrors in these detectors are suspended on fine wires, which are constantly vibrating owing to thermal and other excitations. These "violin mode" vibrations lead to similar vibrations in the location of the mirrors and, thus, to noise in the data stream.

---

[*] It should be noted, however, that the leadership of the LSC is split on the question of whether signals for which matched filtering cannot be applied, or for which a concurrent observation in the electromagnetic spectrum has also been made, can every be said to have been "detected" [87, p. 87].

In this application of machine learning technology a Kalman filter was designed to follow and predict the amplitude and phase of the so-called suspension "violin-modes" with the goal of regressing that noise from the data stream.

19.4.1.2 Glitch Identification and Classification
The system noise in the large, ground-based gravitational wave detectors is nonstationary and plagued by short-duration noise transients—*glitches*—associated with the physical environment (e.g., seismic noise from passing trains or acoustic noise from overflying planes) or instrumental misfunction. The nonstationarity is manifest as an evolution in the noise amplitude and, to a lesser extent, noise spectrum on scales of tens of minutes. The noise transients often show characteristic spectral and/or temporal features, or appear in a correlated way in several different detector control or monitoring channel time series.

More recently, a multidimensional classification analysis has been applied to the study of glitches in the ground-based interferometers [239,240]. In this application the glitches are resolved via a wavelet decomposition and the multirate channels are then classified using an analysis designed for multivariate time series. More recently, Longest Common Subsequence [241] is being used for unsupervised classification of glitch waveforms [242].

19.4.1.3 Directions for Future Work
The LIGO and Virgo detectors record many hundreds of data channels, corresponding to everything from state monitors to control loop error signals to PEMs (e.g., seismometers, acoustic microphones, magnetometers, etc.). These monitors provide, in principle, a valuable diagnostic tool for understanding glitches and their cause. Owing principally to their shear number and the overwhelming human effort that would be involved, only a fraction of these detector monitors are reviewed on a regular basis and even fewer are correlated against each other. Automated review and supervised or unsupervised learning techniques (especially principal component analysis or independent component analysis) applied to these channels could provide a *causal* diagnosis of glitches arising in the detector data stream, reveal more subtle detector misfunctions that are not readily apparent as data glitches, and identify a subset or combination of channels that would be especially valuable for human monitoring.

19.4.2 Response Function Estimation: LISA Recovery from Antenna Pointing
LISA's gravitational wave response involves a time-dependent combination of the several Doppler data streams recorded at each science-craft. The details of the combination involve the relative velocities and accelerations of the separate science-craft, which are themselves determined from the Doppler data. These relative velocities and accelerations are subtly affected by the periodic repointing of the science-craft antenna, which is required so that the detector system maintains contact with Earth. Following an antenna repointing our knowledge of LISA's response is lost and must be recomputed [243].

As the time from the last antenna repointing increases the data available to reevaluate LISA's response—and, correspondingly, our knowledge of the response—increases. The determination of LISA's response following an antenna repointing can thus be posed and approached as a problem in machine learning, with the continuously increasing data set allowing for the identification of an increasingly accurate response and, correspondingly, an increasingly sensitive detector.

In fact, LISA's sensitivity following a repointing event is not time-dependent: only our knowledge of its response depends on time. If the corresponding latency is acceptable each full postrepointing data set could be used to a posteriori to analyze the postrepointing data. Often, however, accepting this latency would compromise LISA's science potential: e.g., it would delay the identification of gravitational wave bursts which, if more quickly identified, might be used to point more conventional telescopes in a search for an electromagnetic counterpart to the gravitational wave burst. Since it is our knowledge of LISA's response that is evolving, and not the response itself, a separate problem, itself ideal for machine learning, is to make to provide the best current or up-to-date assessment of the gravitational wave content of the LISA data stream. Viewed in this way it is also likely that the signal from known or previously characterized gravitational wave signals that span antenna repointing events can be used to speed response function reestimation following sciencecraft disturbances.

19.4.3 Source Science

Beyond detection of gravitational radiation is the characterization of gravitational wave sources and the use of observations to improve our understanding of, for example, source astrophysics, populations and their evolution, and so on. For long-lived sources the constantly increasing data set allows for a constantly improving characterization of the source characteristics. Observation of a *population* of long-lived sources introduces a new set of challenges: as data accumulate sources that may be indistinguishable initially will become resolvable, characterizations will improve and, in some cases, may change discontinuously. Machine learning techniques are particularly well-suited for analyses whose aim is to map an incremental increase in data to an incremental improvement in understanding.

19.4.3.1 Inspiraling Massive Black-Hole Binaries

LISA will detect binary systems of black holes, with comparable masses in the range $10^{4.5}\,\mathrm{M_\odot} \lesssim \mathrm{M} \lesssim 10^{7.5}$, in the months to years before their rapid loss orbital energy to gravitational waves leads to their coalescence and the formation of a single, larger black hole. As data accumulate for each binary in the months before its coalescence, the binary system's component masses, distance and location on the sky, expected time to coalescence, and other characteristic will gradually emerge and become increasingly well determined. In particular, as the binary system's three-dimensional location and expected coalescence become increasingly well-determined electromagnetic telescopes can be pointed toward the events location in the hope of observing any associated electromagnetic events associated with the merger and its aftermath. The ability to have, with minimum latency, a continuously updated assessment of a binary system's location and its expected coalescence time is of critical importance to the success of the LISA science mission.

19.4.3.2 Emerging Source Population Statistics

As described in Section 19.2.4.2, LISA will observe the gravitational waves from tens of millions of galactic white dwarf binary systems. Of these millions of binary systems, the radiation from some tens of thousands will be resolvable as individual sources, with the radiation of the remainder characterizable as a diffuse gravitational wave background whose specific intensity maps the spatial and period evolution of these binary systems throughout the galaxy. The number of individually resolved sources and the structure of the residual diffuse background will increase as the data increase with time. Some sources will be identifiable as strong spectral lines in the first few minutes of LISA observations while others will require a year or more of data. Sources that initially are resolvable only in frequency will become resolved on the sky after several months of observations; others will require years of observation. The residual background will initially be best characterized as spatially isotropic and spectrally smooth; over time, however, it will resolve itself into a lumpy planar component and nuclear component overlapping with the galactic plane and nucleus. The time-dependent number of resolved sources, their period and their sky location, together with the frequency-dependent spatial distribution of the residual unresolved background radiation, maps the three-dimensional structure of our galaxy. Machine learning technologies are well suited to the problem of evolving understanding with continuously accumulated data.

19.5 CONCLUSIONS

In 2008, the United States National Academies were commissioned to survey the field of space- and ground-based astronomy and astrophysics, recommending priorities or the most important scientific and technical activities for the decade just now beginning. In their August 2010 report, gravitational wave detection, which they identified as a bold new frontier of astronomical discovery, figured prominently in their recommended priorities space- and ground-based activities [244]. Such a high profile given to such a new subdiscipline of astronomy is all the more remarkable because no positive detection of gravitational waves has as yet been made!

The high profile given gravitational wave astronomy in the National Academies report owes everything to the field's extraordinary relevance to the most pressing questions and highest priorities of astronomy for the next decade. Owing to the field's youth its challenges have not yet been fully surveyed and its practitioners have much to learn from allied fields: including, especially, modern time series analysis and machine learning. In describing gravitational wave astronomy's "what" and "how" this chapter carries with it my hope that it draws the attention of the machine learning community to the problems of gravitational wave astronomy, and that their expertise will speed the day when gravitational wave observations achieve their promise and take their place among the tools of the Twenty-first-century astronomer.

ACKNOWLEDGMENTS

I thank the editors of this volume—Michael Way, Jeff Scargle, Kamal Ali, and Ashok Srivastava—for the opportunity to share this overview of the gravitational wave astronomy problem domain and some of the analysis problems it raises that are ripe for the application of machine learning methods. This material is based upon work supported by the National Science Foundation under Grant Numbers PHY-0653462, CBET-0940924, and PHY-0969857. Any opinions, findings, and conclusions or recommendations expressed in this material are those of the author(s) and do not necessarily reflect the views of the National Science Foundation.

REFERENCES

1. J Weber. Detection and generation of gravitational waves. *Phys. Rev.* 117, 1960, 306–313. doi: 10.1103/PhysRev.117.306.
2. H Collins. *Gravity's Shadow*. Chicago, IL: University of Chicago Press, 2004.
3. D Kennefick. *Traveling at the Speed of Thought*. Princeton, NJ: Princeton University Press, 2007.
4. H Collins. *Gravity's Ghost*. Chicago, IL: University of Chicago Press, 2010.
5. C Cutler and K Thorne. An overview of gravitational-wave sources. In: *General Relativity and Gravitation*. N. T. Bishop and S. D. Maharaj (eds). Singapore: World Scientific Press, 2002, pp. 72–111, doi: 10.1142/9789812776556_0004.
6. M Ruffert and HT Janka. Colliding neutron stars. Gravitational waves, neutrino emission, and gamma-ray bursts. *Astron. Astrophys.* 338, 1998, 535–555.
7. LS Finn, SD Mohanty, and JD Romano. Detecting an association between gamma ray and gravitational wave bursts. *Phys. Rev. D* 60(12), 1999, 121101–ff.
8. F Barone, B Bartoli, E Calloni, S Catalanotti, S Cavaliere, B D'Ettorre Piazzoli, T di Girolamo et al. Gravitational waves and gamma-ray bursts. *Mem. Soc. Astron. Ital.* 71, 2000, 1081–1084.
9. M Ruffert and HT Janka. Gamma-ray bursts and gravitational waves. In: *Astrophysical Sources for Ground-Based Gravitational Wave Detectors*. Ed. by JM Centrella. Vol. 575. American Institute of Physics Conference Series. Melville, NY: American Institute of Physics, 2001, 143–ff.
10. P Tricarico, A Ortolan, A Solaroli, G Vedovato, L Baggio, M Cerdonio, L Taffarello et al. Correlation between gamma-ray bursts and gravitational waves. *Phys. Rev. D* 63(8), 2001, 082002–ff.
11. G Modestino and A Moleti. Cross-correlation between gravitational wave detectors for detecting association with gamma ray bursts. *Phys. Rev. D* 65(2), 2002, 022005–ff.
12. S Kobayashi and P Mészáros. Polarized gravitational waves from gamma-ray bursts. *Astrophys. J. Lett.* 585, 2003, L89–L92, doi: 10.1086/374307.
13. S Kobayashi and P Mészáros. Polarized gravitational waves from gamma-ray bursts. In: The Astrophysics of Gravitational Wave Sources. Vol. 686. *AIP Conference Proceedings*. Melville, NY: American Institute of Physics Press, 2003, pp. 84–87.
14. S Kobayashi and P Mészáros. Gravitational radiation from gamma-ray burst progenitors. *Astrophys. J.* 589, 2003, 861–870, doi: 10.1086/374733.
15. P Sutton, LS Finn, and B Krishnan. Swift pointing and gravitationalwave bursts from gamma-ray burst events. *Class. Quantum Grav.* 20(17, Sp. Iss. SI), 2003, S815–S820. Issn: 0264-9381.
16. LS Finn, B Krishnan, and P Sutton. Swift pointing and the association between gamma-ray bursts and gravitational wave bursts. *Astrophys. J.* 607(1, Part 1), May 2004, 384–390. Issn: 0004-637X.
17. S Kobayashi and P Miszaros. Gravitational radiation from gamma-ray burst progenitors. In: *AIP Conference Proceedings. 727: Gamma-Ray Bursts: 30 Years of Discovery*. 2004, pp. 136–140.
18. P Mészáros, S Kobayashi, S Razzaque, and B Zhang. High energy neutrinos and gravitational waves from gamma-ray bursts. *Balt. Astron.* 13, 2004, 317–323.
19. P Mészáros, S Kobayashi, S Razzaque, and B Zhang. Ultra-high energy gamma-rays, neutrinos, and gravitational waves from GRBs. In: *AIP Conference Proceedings 727: Gamma-Ray Bursts: 30 Years of Discovery*. 2004, pp. 125–130.
20. N Sago, K Ioka, T Nakamura, and R Yamazaki. Gravitational wave memory of gamma-ray burst jets. *Phys. Rev. D* 70(10), 2004, 104012–ff, doi: 10.1103/PhysRevD.70.104012.

21. MH van Putten, A Levinson, HK Lee, T Regimbau, M Punturo, and GM Harry. Gravitational radiation from gamma-ray burst-supernovae as observational opportunities for LIGO and Virgo. *Phys. Rev. D* 69(4), 2004, 044007.

22. B Abbott, R Abbott, R Adhikari, A Ageev, B Allen, R Amin, SB Anderson et al. Search for gravitational waves associated with the gamma ray burst GRB030329 using the LIGO detectors. *Phys. Rev. D* 72(4), 2005, 042002, doi: 10.1103/PhysRevD.72.042002.

23. T Hiramatsu, K Kotake, H Kudoh, and A Taruya. Gravitational wave background from neutrino-driven gamma-ray bursts. *Mon. Not. R. Astron. Soc.* 364, 2005, 1063–1068, doi: 10.1111/j.1365-2966.2005.09643.x. eprint: arXiv:astro-ph/0509787.

24. B Abbott, R Abbott, R Adhikari, J Agresti, P Ajith, B Allen, R Amin et al. Implications for the origin of GRB 070201 from LIGO observations. *Astrophys. J.* 681(2), 2008, 1419–1430. Issn: 0004-637X.

25. B Abbott, R Abbott, R Adhikari, J Agresti, P Ajith, B Allen, R Amin et al. Search for gravitational waves associated with 39 gamma-ray bursts using data from the second, third, and fourth LIGO runs. *Phys. Rev. D* 77(6), 2008. Issn: 1550-7998. doi: 10.1103/PhysRevD.77.062004.

26. J Abadie, B Abbott, R Abbott, M Abernathy, T Accadia, F Acernese, C Adams et al. Search for gravitational waves from compact binary coalescence in LIGO and Virgo data from S5 and VSR1. *Phys. Rev. D* 82(10), 2010, 102001–ff, doi: 10.1103/PhysRevD.82.102001.

27. BF Schutz. Determining the Hubble constant from gravitational-wave observations. *Nature* 323, 1986, 310–311.

28. DF Chernoff and LS Finn. Gravitational radiation, inspiraling binaries, and cosmology. *Astrophys. J. Lett.* 411, 1993, L5–L8, doi: 10.1086/186898.

29. LS Finn. Binary inspiral, gravitational radiation, and cosmology. *Phys. Rev. D* 53(6), 1996, 2878–2894. Issn: 0556-2821.

30. URL: http://ligo.caltech.edu.

31. URL: http://w.virgo.infn.it.

32. URL: http://www.geo600.org.

33. URL: http://tamago.mtk.nao.ac.jp.

34. BP Abbott, R Abbott, F Acernese, R Adhikari, P Ajith, B Allen, G Allen et al. An upper limit on the stochastic gravitational-wave background of cosmological origin. *Nature* 460(7258), 2009, 990–994. Issn: 0028-0836, doi: 10.1038/nature08278.

35. B Abbott, R Abbott, R Adhikari, J Agresti, P Ajith, B Allen, R Amin et al. Upper limit map of a background of gravitational waves. *Phys. Rev. D* 76(8), 2007. Issn: 1550-7998. doi: 10.1103/PhysRevD.76.082003.

36. B Abbott, R Abbott, R Adhikari, J Agresti, P Ajith, B Allen, R Amin et al. Searching for a stochastic background of gravitational waves with the laser interferometer gravitational-wave observator. *Astrophys. J.* 659(2, Part 1), 2007, 918–930. Issn: 0004-637X.

37. B Abbott, R Abbott, R Adhikari, J Agresti, P Ajith, B Allen, J Allen et al. Upper limits on a stochastic background of gravitational waves. *Phys. Rev. Lett.* 95(22), 2005. Issn: 0031-9007, doi: 10.1103/PhysRevLett.95.221101.

38. B Abbott, R Abbott, R Adhikari, A Ageev, B Allen, R Amin, SB Anderson et al. Analysis of first LIGO science data for stochastic gravitational waves. *Phys. Rev. D* 69(12), 2004, 122004, doi: 10.1103/PhysRevD.69.122004.

39. J Abadie, B Abbott, R Abbott, M Abernathy, C Adams, R Adhikari, P Ajith et al. First search for gravitational waves from the youngest known neutron star. *Astrophys. J.* 722, 2010, 1504–1513, doi: 10.1088/0004-637X/722/2/1504. arXiv:1006.2535 [gr-qc].

40. BP Abbott, R Abbott, F Acernese, R Adhikari, P Ajith, B Allen, G Allen et al. Searches for gravitational waves from known pulsars with science run 5 LIGO data. *Astrophys. J.* 713(1), 2010, 671–685. Issn: 0004-637X. doi: 10.1088/0004-637X/713/1/671.

41. B Abbott, R Abbott, R Adhikari, P Ajith, B Allen, G Allen, R Amin et al. Beating the spin-down limit on gravitational wave emission from the crab pulsar (vol 683, pg L45, 2008). *Astrophys. J. Lett.* 706(1), 2009, L203–L204, doi: 10.1088/0004-637X/706/1/L203.

42. BP Abbott, R Abbott, R Adhikari, P Ajith, B Allen, G Allen, RS Amin et al. Einstein@Home search for periodic gravitational waves in early S5 LIGO data. *Phys. Rev. D* 80(4), 2009. Issn: 1550-7998, doi: 10.1103/ PhysRevD.80.042003.

43. BP Abbott, R Abbott, R Adhikari, P Ajith, B Allen, G Allen, RS Amin et al. All-sky LIGO search for periodic gravitational waves in the early fifth-science-run data. *Phys. Rev. Lett.* 102(11), 2009. Issn: 0031-9007, doi: 10.1103/PhysRevLett.102.111102.

44. B Abbott, R Abbott, R Adhikari, P Ajith, B Allen, G Allen, R Amin et al. Einstein@Home search for periodic gravitational waves in LIGO S4 data. *Phys. Rev. D* 79(2), 2009. Issn: 1550-7998, doi: 10.1103/PhysRevD.79.022001.

45. B Abbott, R Abbott, R Adhikari, J Agresti, P Ajith, B Allen, R Amin et al. All-sky search for periodic gravitational waves in LIGO S4 data. *Phys. Rev. D* 77(2), 2008. issn: 1550-7998, doi: 10.1103/PhysRevD.77.022001.

46. B Abbott, R Abbott, R Adhikari, J Agresti, P Ajith, B Allen, R Amin et al. Searches for periodic gravitational waves from unknown isolated sources and Scorpius X-1: Results from the second LIGO science run. *Phys. Rev. D* 76(8), 2007. Issn: 1550-7998, doi: 10.1103/PhysRevD.76.082001.

47. B Abbott, R Abbott, R Adhikari, J Agresti, P Ajith, B Allen, R Amin et al. Upper limits on gravitational wave emission from 78 radio pulsars. *Phys. Rev. D* 76(4), 2007. Issn: 1550-7998. doi: 10.1103/PhysRevD.76.042001.

48. B Abbott, R Abbott, R Adhikari, A Ageev, J Agresti, B Allen, J Allen. First all-sky upper limits from LIGO on the strength of periodic gravitational waves using the Hough transform. *Phys. Rev. D* 72(10), 2005. Issn: 1550-7998, doi: 10.1103/PhysRevD.72.102004.

49. B Abbott, R Abbott, R Adhikari, A Ageev, B Allen, R Amin, S Anderson et al. Limits on gravitational-wave emission from selected pulsars using LIGO data. *Phys. Rev. Lett.* 94(18), 2005. Issn: 0031-9007, doi: 10.1103/PhysRevLett.94.181103.

50. B Abbott, R Abbott, R Adhikari, A Ageev, B Allen, R Amin, SB Anderson et al. Setting upper limits on the strength of periodic gravitational waves from PSR J1939+2134 using the first science data from the GEO 600 and LIGO detectors. *Phys. Rev. D* 69(8), 2004, 082004–ff, doi: 10.1103/PhysRevD.69.082004.

51. J Abadie, BP Abbott, R Abbott, M Abernathy, T Accadia, F Acerneseac, C Adams et al. Predictions for the rates of compact binary coalescences observable by ground-based gravitational-wave detectors. *Class. Quantum Grav.* 27(17), 2010, 173001. Issn: 0264-9381, doi: 10.1088/0264- 9381/27/17/173001. URL: http://stacks.iop.org/0264-9381/27/ i=17/a=173001.

52. BP Abbott, R Abbott, R Adhikari, P Ajith, B Allen, G Allen, RS Amin et al. Search for gravitational wave ringdowns from perturbed black holes in LIGO S4 data. *Phys. Rev. D* 80(6), 2009. Issn: 1550-7998, doi: 10.1103/PhysRevD.80.062001.

53. BP Abbott, R Abbott, R Adhikari, P Ajith, B Allen, G Allen, RS Amin et al. Search for gravitational waves from low mass compact binary coalescence in 186 days of LIGO's fifth science run. *Phys. Rev. D* 80(4), 2009. Issn: 1550-7998, doi: 10.1103/PhysRevD.80. 047101.

54. BP Abbott, R Abbott, R Adhikari, P Ajith, B Allen, G Allen, RS Amin et al. Search for gravitational waves from low mass binary coalescences in the first year of LIGO's S5 data. *Phys. Rev. D* 79(12), 2009. Issn: 1550-7998, doi: 10.1103/PhysRevD.79.122001.

55. B Abbott, R Abbott, R Adhikari, J Agresti, P Ajith, B Allen, R Amin et al. Search for gravitational waves from binary inspirals in S3 and S4 LIGO data. *Phys. Rev. D* 77(6), 2008. Issn: 1550-7998, doi: 10.1103/PhysRevD.77.062002.

56. B Abbott, R Abbott, R Adhikari, A Ageev, J Agresti, P Ajith, B Allen et al. Joint LIGO and TAMA300 search for gravitational waves from inspiralling neutron star binaries. *Phys. Rev. D* 73(10), 2006. Issn: 1550-7998, doi: 10.1103/PhysRevD.73.102002.

57. B Abbott, R Abbott, R Adhikari, A Ageev, J Agresti, P Ajith, B Allen et al. Search for gravitational waves from binary black hole inspirals in LIGO data. *Phys. Rev. D* 73(6), 2006. Issn: 1550-7998, doi: 10.1103/PhysRevD.73.062001.

58. B Abbott, R Abbott, R Adhikari, A Ageev, B Allen, R Amin, SB Anderson et al. Search for gravitational waves from galactic and extra-galactic binary neutron stars. *Phys. Rev. D* 72(8), 2005, 082001, doi: 10.1103/PhysRevD.72.082001.

59. B Abbott, R Abbott, R Adhikari, A Ageev, B Allen, R Amin, S Anderson et al. Search for gravitational waves from primordial black hole binary coalescences in the galactic halo. *Phys. Rev. D* 72(8), 2005. Issn: 1550-7998, doi: 10.1103/PhysRevD.72.082002.

60. B Abbott, R Abbott, R Adhikari, A Ageev, B Allen, R Amin, S Anderson et al. Analysis of LIGO data for gravitational waves from binary neutron stars. *Phys. Rev. D* 69(12), 2004. Issn: 0556-2821, doi: 10.1103/PhysRevD.69.122001.

61. J Abadie, BP Abbott, R Abbott, T Accadia, F Acernese, R Adhikari, P Ajith et al. Search for gravitational-wave bursts associated with gamma-ray bursts using data from LIGO Science Run 5 and Virgo Science Run 1. *Astrophys. J.* 715, 2010, 1438–1452, doi: 10.1088/0004-637X/715/2/1438. arXiv:0908.3824 astro-ph.HE].

62. B Abbott, R Abbott, R Adhikari, J Agresti, P Ajith, B Allen, R Amin et al. Search for gravitational wave radiation associated with the pulsating tail of the SGR 1806 20 hyperflare of 27 December 2004 using LIGO. *Phys. Rev. D* 76(6), 2007. Issn: 1550-7998. doi: 10.1103/PhysRevD.76.062003.

63. BP Abbott, R Abbott, R Adhikari, P Ajith, B Allen, G Allen, RS Amin et al. Stacked search for gravitational waves from the 2006 SGR 1900+14 storm. *Astrophys. J. Lett.* 701(2), 2009, L68–L74, doi: 10.1088/0004-637X/701/2/L68.

64. BP Abbott, R Abbott, R Adhikari, P Ajith, B Allen, G Allen, RS Amin et al. . First LIGO search for gravitational wave bursts from cosmic (super)strings. *Phys. Rev. D* 80(6), 2009. Issn: 1550-7998, doi: 10.1103/PhysRevD.80.062002.

65. J Abadie, BP Abbott, R Abbott, T Accadia, F Acernese, R Adhikari, P Ajith et al. All-sky search for gravitational-wave bursts in the first joint LIGO-GEO-Virgo run. *Phys. Rev. D* 81(10), 2010. Issn: 1550-7998, doi: 10.1103/PhysRevD.81.102001.

66. BP Abbott, R Abbott, R Adhikari, P Ajith, B Allen, G Allen, RS Amin et al. Search for gravitational-wave bursts in the first year of the fifth LIGO science run. *Phys. Rev. D* 80(10), 2009, 102001, doi: 10.1103/PhysRevD.80.102001.

67. BP Abbott, R Abbott, R Adhikari, P Ajith, B Allen, G Allen, RS Amin et al. Search for high frequency gravitational-wave bursts in the first calendar year of LIGO's fifth science run. *Phys. Rev. D* 80(10), 2009. Issn: 1550-7998, doi: 10.1103/PhysRevD.80.102002.

68. B Abbott, R Abbott, R Adhikari, P Ajith, B Allen, G Allen, R Amin et al. First joint search for gravitational-wave bursts in LIGO and GEO 600 data. *Class. Quantum Grav.* 25(24), 2008. Issn: 0264-9381, doi: 10.1088/0264-9381/25/24/245008.

69. L Baggio, M Bignotto, M Bonaldi, M Cerdonio, M De Rosa, P Falferi, S Fattori et al. A joint search for gravitational wave bursts with AURIGA and LIGO. *Class. Quantum Grav.* 25(9), 2008. Issn: 0264-9381, doi: 10.1088/0264-9381/25/9/095004.

70. B Abbott, R Abbott, R Adhikari, J Agresti, P Ajith, B Allen, R Amin et al. Search for gravitational-wave bursts in LIGO data from the fourth science run. *Class. Quantum Grav.* 24(22), 2007, 5343–5369. Issn: 0264-9381, doi: 10.1088/0264-9381/24/22/002.

71. B Abbott, R Abbott, R Adhikari, J Agresti, P Ajith, B Allen, J Allen et al. Search for gravitational-wave bursts in LIGO's third science run. *Class. Quantum Grav.* 23(8, Sp. Iss. SI), 2006, S29–S39. Issn: 0264-9381, doi: 10.1088/0264-9381/23/8/S05.

72. B Abbott, R Abbott, R Adhikari, A Ageev, J Agresti, P Ajith, B Allen et al. Upper limits from the LIGO and TAMA detectors on the rate of gravitational-wave bursts. *Phys. Rev. D* 72(12), 2005. Issn: 1550-7998, doi: 10.1103/PhysRevD.72.122004.

73. B Abbott, R Abbott, R Adhikari, A Ageev, B Allen, R Amin, SB Anderson et al. First upper limits from LIGO on gravitational wave bursts. *Phys. Rev. D* 69(10), 2004, 102001–ff, doi: 10.1103/PhysRevD.69.102001.

74. URL: http://gw.icrr.u-tokyo.ac.jp/lcgt.

75. J Abadie, BP Abbott, R Abbott, M Abernathy, T Accadia, F Acernese, C Adams etal. Topical Review: Predictions for the rates of compact binary coalescences observable by ground-based gravitational-wave detectors. *Class. Quantum Grav.* 27(17), 2010, 173001–ff, doi: 10.1088/0264-9381/27/17/173001, arXiv:1003.2480 [astro-ph.HE].

76. J Abadie, B Abbott, R Abbott, M Abernathy, C Adams, R Adhikari, P Ajith et al. Calibration of the LIGO gravitational wave detectors in the fifth science run. *Nucl. Instrum. Methods Phys. Res. A* 624, 2010, 223–240, doi: 10.1016/j.nima.2010.07.089. arXiv:1007.3973 [gr-qc].

77. W Frazer, EL Godwasser, RA Hulse, BD McDaniel, P Oddone, PR Saulson, and SC Wolff. Report of the panel on the use of the laser interferometer gravitational-wave observatory (LIGO). Tech. rep. 1996.

78. PL Bender and D Hils. Confusion noise level due to galactic and extragalactic binaries. *Class. Quantum Grav.* 14, 1997, 1439–1444.

79. G Nelemans, LR Yungelson, and SF Portegies Zwart. The gravitational wave signal from the Galactic disk population of binaries containing two compact objects. *Astron. Astrophys.* 375, 2001, 890–898, doi: 10.1051/0004–6361:20010683, eprint: astroph/0105221.

80. F Verbunt and G Nelemans. Binaries for LISA. *Class. Quantum Grav.* 18, 2001, 4005–4011. Issn: 0264–9381.

81. G Nelemans. Gravitational waves from double white dwarfs and AM CVn binaries. *Class. Quantum Grav.* 20, 2003, 81–ff, doi: 10.1088/0264–9381/20/10/310.

82. A Stroeer, A Vecchio, and G Nelemans. LISA astronomy of double white dwarf binary systems. *Astrophys. J.* 633, 2005, L33–L36. Issn: 0004–637X.

83. M Benacquista and K Holley-Bockelmann. Consequences of disk scale height on LISA confusion noise from close white dwarf binaries. *Astrophys. J.* 645, 2006, 589–596. Issn: 0004–637X, eprint: astro-ph/0504135.

84. LJ Rubbo, K Holley-Bockelmann, and LS Finn. Event rate for extreme mass ratio burst signals in the Laser Interferometer Space Antenna band. *Astrophys. J.* 649(1, Part 2), 2006, L25–L28. Issn: 0004-637X.

85. LJ Rubbo, K Holley-Bockelmann, and LS Finn. Event rate for extreme mass ratio burst signals in the LISA band. In: *Laser Interferometer Space Antenna.* S Merkowitz and J Livas (eds). Vol. 873. *AIP Conference Proceedings. 6th International Laser Interferometer Space Antenna,* Greenbelt, MD, Jun 19–23, 2006, pp. 284–288. Isbn: 978-0-7354-0372-7.

86. LS Finn and K Thorne. Gravitational waves from a compact star in a circular, inspiral orbit, in the equatorial plane of a massive, spinning black hole, as observed by LISA. *Phys. Rev. D* 62(12), 2000. Issn: 0556-2821, doi: 10.1103/PhysRevD.62.124021, http://link.aps.org/doi/10.1103/PhysRevD.62.124021.

87. SA Hughes. Evolution of circular, nonequatorial orbits of Kerr black holes due to gravitational-wave emission. *Phys. Rev. D* 61(8), 2000, 084004–ff.

88. SA Hughes. Gravitational waves from inspiral into massive black holes. In: *Gravitational Waves.* AIP Conference Proceedings, Vol. 523. Melville NY: American Institute of Physics Press, 2000, 76–ff.

89. L Barack and C Cutler. Confusion noise from LISA capture sources. *Phys. Rev. D* 70, 2004, 122002–ff. Issn: 0556-2821, doi: 10.1103/PhysRevD.70.122002, eprint: gr-qc/0409010.

90. A Sesana, F Haardt, P Madau, and M Volonteri. The gravitational wave signal from massive black hole binaries and its contribution to the LISA data stream. *Astrophys. J.* 623, 2005, 23–30. Issn: 0004–637X, eprint: astro-ph/0409255.

91. A Sesana, F Haardt, P Madau, and M Volonteri. Low-frequency gravitational radiation from coalescing massive black holes. *Class. Quantum Grav.* 22, 2005, 363–ff, doi: 10.1088/0264-9381/22/10/030, eprint: astro-ph/0502462.

92. SA Hughes. Untangling the merger history of massive black holes with LISA. *Mon. Not. R. Astron. Soc.* 331, 2002, 805–816. Issn: 0035–8711.

93. SA Hughes and RD Blandford. Black hole mass and spin coevolution by mergers. *Astrophys. J. Lett.* 585, 2003, L101–L104, doi: 10.1086/375495.

94. P Madau, T Abel, PL Bender, T Di Matteo, Z Haiman, SA Hughes, A Loeb et al. Massive black holes across cosmic time. *Astronomy 2010*, 2009, 189–ff, arXiv:0903.0097.

95. SA Hughes and DE Holz. Cosmology with coalescing massive black holes. *Class. Quantum Grav.* 20, 2003, 65–ff. doi: 10. 1088/0264-9381/20/10/308, eprint: astro-ph/ 0212218.

96. C Hogan, B Schutz, C Cutler, S Hughes, and D Holz. Precision cosmology with gravitational waves. *The Astron. Astrophys. Decadal Surv.* 2010, 2009, 130–ff. ArXiv Astrophysics e-prints.

97. T Creighton. Tumbleweeds and airborne gravitational noise sources for LIGO. *Class. Quantum Grav.* 25(12), 2008, 125011–ff, doi: 10.1088/0264-9381/25/12/ 125011, eprint: arXiv: gr-qc/0007050.

98. PR Saulson. Terrestial gravitational noise on a gravitational wave antenna. *Phys. Rev. D* 30(4), 1984, 732–736.

99. URL: http://www.rssd.esa.int/index.php?project=LISA.

100. URL: http://lisa.nasa.gov.

101. R Decher, J Randall, P Bender, and J Faller. Design aspects of a laser gravitational wave detector in space. In: *Society of Photo-Optical Instrumentation Engineers (SPIE) Conference Series*. WJ Cuneo Jr. (ed.) Vol. 228. Presented at the Society of Photo-Optical Instrumentation Engineers (SPIE) Conference. Jan. 1980, pp. 149–153.

102. J Faller, P Bender, Y Chan, J Hall, D Hills And, and J Hough. Laser-gravitational wave experiment in space. In: *General Relativity and Gravitation*. B. Bertotti, F. de Felice, and A. Pascolini (eds). Volume 1. Dordrecht, Holland: Reidel, 1983, 960–ff.

103. J Faller, P Bender, J Hall, D Hils, and M Vincent. Space antenna for gravitational wave astronomy. In: *Kilometric Optical Arrays in Space*. N. Longdon and O. Melita (eds). Vol. 226. Cargèse: ESA Special Publication. Apr. 1985, pp. 157–163.

104. JE Faller, PL Bender, JL Hall, D Hils, and RT Stebbins. An antenna for laser gravitational-wave observations in space. *Adv. Space Res.* 9, 1989, 107–111, doi: 10.1016/0273-1177(89) 90014-8.

105. J Faller and P Bender. A possible laser gravitational wave experiment in space. In: *Precision Measurement and Fundamental Constants II*. B. N. Taylor and W. D. Phillips (eds). Gaitheational Bureau of Standards, 1984, pp. 689–690.

106. K Danzmann, A Rüdiger, R Schilling, W Winkler, J Hough, GP Newton, D Roberson et al. LISA: Proposal for a laser-interferometric gravitational wave detector in space. *Tech. rep. MPQ177*. Garching: Max Planck Institute für Quantenoptik, 1993.

107. P Bender, I Ciufolini, K Danzmann,W Folkner, J Hough, D Robertson, A Rüdiger et al. LISA: Laser Interferometer Space Antenna for the detection and observation of gravitational waves. Tech. rep. MPQ 208. Garching: Max-Planck-Institut für Quantenoptik, Feb. 1996.

108. P Bender, A Brillet, I Ciufolini, AM Cruise, C Cutler, K Danzmann, F Fidecaro et al. LISA: Pre-Phase A Report. Tech. rep. MPQ233. Garching: Max-Planck-Institut für Quantenoptic, 1998.

109. WM Folkner, PL Bender, and RT Stebbins. LISA mission concept study, laser interferometer space antenna for the detection and observation of gravitational waves. In: *NASA STI/Recon Technical Report N*, 1998, 55623–ff.

110. R Stebbins. LISA mission tutorial. In: *Laser Interferometer Space Antenna: 6th International LISA Symposium*. SM Merkowitz and JC Livas (eds). Vol. 873. AIP Conference Proceedings. Melville NY: American Institute of Physics, 2006, pp. 3–12.

111. P Bender, P Binetruy, S Buchman, J Centrella, M Cerdonio, N Cornish, M Cruise, C Cutler et al. *LISA Science Requirements Document*. Tech. rep. LISA-ScRD-004. NASA GSFC, Greenbelt, MD 20771: LISA Project Office, 2007.

112. RT Stebbins. Rightsizing LISA. *Class. Quantum Grav.* 26(9), 2009, 094014, doi: 10.1088/0264-9381/26/9/094014.

113. URL: http://www.rssd.esa.int/index.php?project=LISAPATHFINDER.

114. J Hough, S Phinney, T Prince, D Richstone, D Robertson, M Rodrigues, A Rudiger et al. LISA and its in-flight test precursor SMART-2. In: *Nuclear Physics B Proceedings Supplements* 110, 2002, pp. 209–216.

115. S Vitale. *The LTP Experiment on the LISA Pathfinder Mission.* 2005, eprint: arXiv:gr-qc/0504062.

116. M Landgraf, M Hechler, and S Kemble. Mission design for LISA pathfinder. *Class. Quantum Grav.* 22, 2005, S487–S492. Issn: 0264–9381, eprint: gr-qc/0411071.

117. WM Folkner, F Hechler, TH Sweetser, MA Vincent, and PL Bender. LISA orbit selection and stability. *Class. Quantum Grav.* 14, 1997, 1405–1410. Issn: 0264–9381.

118. WM Folkner. LISA orbit selection and stability. *Class. Quantum Grav.* 18, 2001, 4053–4057. Issn: 0264–9381.

119. S Hughes. Preliminary optimal orbit design for the laser interferometer space antenna(LISA). *Adv. Astronaut. Sci.* 111, 2002, 61–78.

120. SV Dhurandhar, KR Nayak, S Koshti, and JY Vinet. Fundamentals of the LISA stable flight formation. *Class. Quantum Grav.* 22, 2005, 481–487. Issn: 0264–9381. eprint: gr-qc/0410093.

121. JW Armstrong, FB Estabrook, and M Tinto. Time-delay interferometry for space-based gravitational wave searches. *Astrophys. J.* 527, 1999, 814–826, doi: 10.1086/308110.

122. FB Estabrook, M Tinto, and JW Armstrong. Time-delay analysis of LISA gravitational wave data: Elimination of spacecraft motion effects. *Phys. Rev. D* 62, 2000, art. no.–042002. Issn: 0556–2821.

123. M Tinto, JW Armstrong, and FB Estabrook. Discriminating a gravitational-wave background from instrumental noise using timedelay interferometry. *Class. Quantum Grav.* 18, 2001, 4081–4086.

124. SV Dhurandhar, KR Nayak, and JY Vinet. Algebraic approach to time-delay data analysis for LISA. *Phys. Rev. D* 65, 2002, 024015–ff. Issn: 0556–2821, eprint: gr-qc/0112059.

125. SL Larson, RW Hellings, and WA Hiscock. Unequal arm space-borne gravitational wave detectors. *Phys. Rev. D* 66(6), 2002, 062001–ff, eprint: gr-qc/0206081.

126. M Tinto, FB Estabrook, and JW Armstrong. Time-delay interferometry for LISA. *Phys. Rev. D* 65, 2002, 082003–ff. Issn: 0556–2821.

127. JW Armstrong, FB Estabrook, and M Tinto. Time delay interferometry. *Class. Quantum Grav.* 20, 2003, 283–ff, doi: 10.1088/0264-9381/20/10/331.

128. AC Kuhnert, R Spero, AR Abramovici, BL Schumaker, and DA Shaddock. LISA laser noise cancellation test using time-delayed interferometry. In: *Gravitational-Wave Detection*. Cruise, M., Saulson, P. (eds). Proceedings of the SPIE, Vol. 4856, pp. 74–77, 2003, doi: 10.1117/12.461501.

129. KR Nayak, A Pai, SV Dhurandhar, and JY Vinet. Improving the sensitivity of LISA. *Class. Quantum Grav.* 20, 2003, 1217–1231. Issn: 0264–9381.

130. DA Shaddock, M Tinto, FB Estabrook, and JW Armstrong. Data combinations accounting for LISA spacecraft motion. *Phys. Rev. D* 68, 2003, 061303–ff. Issn: 0556–2821, eprint: gr-qc/0307080.

131. M Tinto, DA Shaddock, J Sylvestre, and JW Armstrong. Implementation of time-delay interferometry for LISA. *Phys. Rev. D* 67, 2003, 122003–ff. Issn: 0556–2821.

132. KR Nayak and JY Vinet. Algebraic approach to time-delay data analysis for orbiting LISA. *Phys. Rev. D* 70, 2004, 102003–ff. Issn: 0556–2821, doi: 10.1103/PhysRevD.70.102003.

133. DA Shaddock. Operating LISA as a Sagnac interferometer. *Phys. Rev. D* 69, 2004, 022001–ff. Issn: 0556–2821, eprint: grqc/0306125.

134. J Sylvestre. Simulations of laser locking to a LISA arm. *Phys. Rev. D* 70, 2004, . 102002–ff. Issn: 0556–2821, doi: 10.1103/PhysRevD.70.102002.

135. M Tinto and SL Larson. LISA time-delay interferometry zero-signal solution: Geometrical properties. *Phys. Rev. D* 70, 2004, 062002–ff. Issn: 0556–2821, doi: 10.1103/PhysRevD.70.062002.

136. M Tinto, FB Estabrook, and JW Armstrong. Time delay interferometry with moving spacecraft arrays. *Phys. Rev. D* 69(8), 2004, 082001–ff, doi: 10.1103/PhysRevD.69.082001.

137. M Vallisneri. Synthetic LISA: Simulating time delay interferometry in a model LISA. gr-qc/0407102; submitted to *The Astrophysical Journal*; cf. Synthetic LISA homepage at: http://www.vallis.org/syntheticlisa/. 2004.

138. SV Dhurandhar and M Tinto. Time-delay interferometry. *Living Rev. Relativ.* 8, 2005, 4–ff.

139. KR Nayak and JY Vinet. Algebraic approach to time-delay data analysis: Orbiting case. *Class. Quantum Grav.* 22, 2005, 437–ff, doi: 10.1088/0264-9381/22/10/040.

140. M Tinto and S Larson. The LISA zero-signal solution. *Class. Quantum Grav.* 22(10, Sp. Iss. SI), 2005. 5th International LISA Symposium/38th ESLAB Symposium, Noordwijk, Netherlands, Jul 12- 15, 2004, S531–S535. Issn: 0264-9381. doi: 10.1088/0264-9381/22/10/054.

141. M Tinto, M Vallisneri, and JW Armstrong. Time-delay interferometric ranging for space-borne gravitational-wave detectors. *Phys. Rev. D,* 71(4), 2005, 041101–ff, doi: 10.1103/PhysRevD.71.041101. eprint: gr-qc/0410122.

142. M Vallisneri. Synthetic LISA: Simulating time delay interferometry in a model LISA. *Phys. Rev. D* 71, 2005, 022001–ff. Issn: 0556–2821, doi: 10.1103/PhysRevD.71.022001.

143. M Vallisneri. Geometric time delay interferometry. *Phys. Rev. D* 72(4), 2005, 042003–ff, doi: 10.1103/PhysRevD.72.042003, eprint: gr-qc/0504145.

144. JD Romano and GWoan. Principal component analysis for LISA: The time delay interferometry connection. *Phys. Rev. D* 73, 2006, 102001–ff, Issn: 1550–7998, doi: 10.1103/PhysRevD.73.102001.

145. SV Dhurandhar. Time-delay interferometry and the relativistic treatment of LISA optical links. *J. Phys. Conf. Ser.* 154, 2009, 012047, doi: 10.1088/17426596/154/1/012047.

146. B Allen. Stochastic gravity-wave background in inflationary universe models. *Phys. Rev. D* 37, 1988, 2078.

147. MS Turner. Detectability of inflation-produced gravitational waves. *Phys. Rev. D* 55, 1997, 435–ff, eprint: astro-ph/9607066.

148. C Ungarelli and A Vecchio. High energy physics and the very early universe with LISA. *Phys. Rev. D* 63, 2001, 064030–ff. Issn: 0556–2821.

149. C Ungarelli, P Corasaniti, RA Mercer, and A Vecchio. Gravitational waves, inflation and the cosmic microwave background: Towards testing the slow-roll paradigm. *Class. Quantum Grav.* 22, 2005, 955–ff, doi: 10.1088/0264- 9381/22/18/S09, eprint: astroph/0504294.

150. LP Grishchuk. Relic gravitational waves and cosmology. *Phys.-Usp.,* 48(12), 2005, 1235–1247, doi: 10. 1070/PU2005v048n12ABEH005795.

151. T Damour and A Vilenkin. Gravitational radiation from cosmic (super) strings: Bursts, stochastic background, and observational windows. *Phys. Rev. D* 71(6), 2005, 063510–ff, doi: 10. 1103/PhysRevD.71.063510, eprint: hep-th/0410222.

152. V Berezinsky, B Hnatyk, and A Vilenkin. Gamma ray bursts from superconducting cosmic strings. *Phys. Rev. D* 64(4), 2001, 043004–ff, eprint: astro-ph/0102366.

153. CJ Hogan. Gravitational waves from light cosmic strings: Backgrounds and bursts with large loops. *Phys. Rev. D* 74(4), 2006, 043526–ff, doi: 10.1103/PhysRevD.74.043526, eprint: astroph/0605567.

154. JS Key and NJ Cornish. Characterizing the gravitational wave signature from cosmic string cusps. *Phys. Rev. D* 79(4), 2009, 043014–ff, doi: 10.1103/PhysRevD.79.043014, arXiv:0812.1590.

155. L Leblond, B Shlaer, and X Siemens. Gravitational waves from broken cosmic strings: The bursts and the beads. *Phys. Rev. D* 79, 2009, 123519.

156. F Accetta and L Krauss. Can millisecond pulsar timing measurements rule out cosmic strings? *Phys. Lett. B* 233, 1989, 93–98, doi: 10.1016/03702693(89)90622-9.

157. FA Jenet, GB Hobbs, W van Straten, RN Manchester, M Bailes, JPW Verbiest, RT Edwards, AWHotan, JM Sarkissian, and SM Ord. Upper bounds on the low-frequency stochastic gravitational wave background from pulsar timing observations: current limits and future prospects. *Astrophys. J.* 653, 2006, 1571–1576. doi: 10.1086/508702, eprint: arXiv:astro-ph/0609013.

158. X Siemens, V Mandic, and J Creighton. Gravitational-wave stochastic background from cosmic strings. *Phys. Rev. Lett.* 98(11), 2007, 111101, doi: 10.1103/PhysRevLett.98.111101, eprint: astro-ph/0610920.

159. AH Jaffe and DC Backer. Gravitational waves probe the coalescence rate of massive black hole binaries. *Astrophys. J.* 583, 2003, 616.

160. P Amaro-Seoane, A Sesana, L Hoffman, M Benacquista, C Eichhorn, J Makino, and R Spurzem. Triplets of supermassive black holes: Astrophysics, gravitational waves and detection. *Mon. Not. R. Astron. Soc.* 402, 2010, 2308–2320. doi: 10.1111/j.1365-2966.2009.16104.x. ArXiv:0910.1587.

161. A Sesana and A Vecchio. Gravitational waves and pulsar timing: Stochastic background, individual sources and parameter estimation. *Class. Quantum Grav.* 27(8), 2010, 084016–ff. Doi: 10.1088/0264-9381/27/8/084016, arXiv:1001.3161.

162. URL: http://www.epta.eu.org/.

163. R Ferdman, R van Haasteren, C Bassa, M Burgay, I Cognard, A Corongiu, N D'Amico et al. The European pulsar timing array: Current efforts and a LEAP toward the future. *Class. Quantum Grav.* 27(8), 2010, 084014–ff, doi: 10.1088/0264-9381/27/8/084014, arXiv:1003.3405 [astro-ph.HE].

164. URL: http://www.nanograv.org.

165. F Jenet, LS Finn, J Lazio, A Lommen, M McLaughlin, I Stairs, D Stinebring et al. *The North American Nanohertz Observatory for Gravitational Waves*. 2009, arXiv:0909.1058.

166. URL: http://www.atnf.csiro.au/research/pulsar/ppta/.

167. J Verbiest, M Bailes, N Bhat, S Burke-Spolaor, D Champion, W Coles, G Hobbs. Status update of the Parkes pulsar timing array. *Class. Quantum Grav.* 27(8), 2010, 084015–ff, doi: 10.1088/0264- 9381/27/8/084015, arXiv:0912.2692 [astro-ph.GA].

168. G Hobbs, A Archibald, Z Arzoumanian, D Backer, M Bailes, NDR Bhat, M Burgay et al. The international pulsar timing array project: Using pulsars as a gravitational wave detector. *Class. Quantum Grav.* 27(8), 2010, 084013–ff, doi: 10.1088/0264-9381/27/8/084013, arXiv:0911.5206 [astro-ph.SR].

169. URL: http://www.gb.nrao.edu/gbt/.

170. URL: http://www.naic.edu/.

171. URL: http://www.mpifr.de/english/radiotelescope/index.html.

172. URL: http://www.jb.man.ac.uk/aboutus/lovell/.

173. URL: http://www.obs-nancay.fr/.

174. URL: http://www.srt.inaf.it/.

175. URL: http://www.astron.nl/radio-observatory/public/public.

176. URL: http://www.parkes.atnf.csiro.au/.

177. URL: http://www.skatelescope.org.

178. URL: http://www.atnf.csiro.au/SKA.

179. I Stairs, M Keith, Z Arzoumanian, WBecker, A Berndsen, A Bouchard, N Bhat et al. Pulsars with the Australian Square Kilometre Array Pathfinder. *ArXiv e-prints*, 2010, arXiv:1012.3507 [astro-ph.IM].

180. URL: http://www.ska.ac.za/meerkat.

181. R Booth, W de Blok, J Jonas, and B Fanaroff. MeerKAT key project science, specifications, and proposals. *ArXiv e-prints*, 2009, arXiv:0910.2935 [astro-ph.IM].

182. J Jonas. The MeerKAT SKA precursor telescope. In: *Panoramic Radio Astronomy: Wide-field 1-2 GHz Research on Galaxy Evolution*. Published on-line: SISSA Proceedings of Science, 2009. URL: http: //pos.sissa.it/cgi-bin/reader/conf.cgi?confid=89.

183. JM Cordes, M Kramer, TJW Lazio, BW Stappers, DC Backer, and S Johnston. Pulsars as tools for fundamental physics and astrophysics. In: *New Astron. Rev.* 2004, 1413, doi: 10.1016/j.newar.2004.09.040.

184. R Romani and J Taylor. An upper limit on the stochastic background of ultralow-frequency gravitational waves. *Astrophys. J. Lett.* 265, 1983, L35–L37, doi: 10.1086/183953.

185. VM Kaspi, JH Taylor, and M Ryba. High-precision timing of millisecond pulsars. III. Long-Term Monitoring of PSRs B1855+09 and B1937+21. *Astrophys. J.* 428, 1994, 713–728.

186. AN Lommen. New limits on gravitational radiation using pulsars. In: *Neutron Stars, Pulsars, and Supernova Remnants*. W Becker, H Lesch, and J Trümper (eds). Garching bei München: Max-Planck-Institut für Extraterrestrische Physik, 2002, pp. 114–ff.

187. FA Jenet, A Lommen, SL Larson, and L Wen. Constraining the properties of supermassive black hole systems using pulsar timing: Appliation to 3C 66B. *Astrophys. J.* 606, 2004, 799–803.

188. AN Lommen and DC Backer. Using pulsars to detect massive black hole binaries via gravitational radiation: Sagittarius A* and nearby galaxies. *Astrophys. J.* 562, 2001, 297–302, doi: 10.1086/323491, eprint: arXiv:astro-ph/0107470.

189. DRB Yardley, GB Hobbs, FA Jenet, JPW Verbiest, ZL Wen, RN Manchester, WA Coles et al. The sensitivity of the Parkes pulsar timing array to individual sources of gravitational waves. *ArXiv e-prints* 2010, arXiv:1005.1667.

190. MV Sazhin. Opportunities for detecting ultralong gravitational waves. *Sov. Astron.* 22, 1978, 36–38.

191. S Detweiler. Pulsar timing measurements and the search for gravitational waves. *Astrophys. J.* 234, 1979, 1100–1104, doi: 10.1086/157593.

192. DC Backer and RW Hellings. Pulsar timing and general relativity. *Annu. Rev. Astron. Astrophys.* 24, 1986, 537–575, doi: 10.1146/annurev.aa.24.090186.002541.

193. RS Foster and DC Backer. Constructing a pulsar timing array. *Astrophys. J.* 361, 1990, 300–308.

194. R van Haasteren and Y Levin. Gravitational-wave memory and pulsar timing arrays. *ArXiv e-prints* 2009, arXiv:0909.0954.

195. MS Pshirkov, D Baskaran, and KA Postnov. Observing gravitational wave bursts in pulsar timing measurements. *Mon. Not. R. Astron. Soc.* 402, 2010, 417–423, doi: 10.1111/j.1365-2966.2009.15887.x, arXiv:0909.0742.

196. LS Finn and AN Lommen. Detection, localization, and characterization of gravitational wave bursts in a pulsar timing array. *Astrophys. J.* 718(2), 2010, 1400–1415. Issn: 0004-637X, doi: 10.1088/0004-637X/718/2/1400.

197. A Hewish, S Bell, J Pilkington, P Scott, and R Collins. Observation of a rapidly pulsating radio source. *Nature* 217, 1968, 709–713, doi: 10.1038/217709a0.

198. DC Backer, SR Kulkarni, C Heiles, MM Davis, and WM Goss. A millisecond pulsar. *Nature* 300, 1982, 615–618.

199. KA Arnaud, S Babak, JG Baker, MJ Benacquista, NJ Cornish, C Cutler, SL Larson et al. An overview of the Mock LISA data challenges. In: *Laser Interferometer Space Antenna: 6th International LISA Symposium*. Vol. 873. American Institute of Physics Conference Series, Nov. 2006, pp. 619–624, doi: 10.1063/1.2405108.

200. KA Arnaud, G Auger, S Babak, JG Baker, MJ Benacquista, E Bloomer, DA Brown et al. Report on the first round of the mock LISA data challenges. *Class. Quantum Grav.* 24(19, Sp. Iss. SI), 2007, S529–S539. Issn: 0264-9381, doi: 10.1088/0264-9381/24/19/S16.

201. KA Arnaud, S Babak, JG Baker, MJ Benacquista, NJ Cornish, C Cutler, LS Finn et al. An overview of the second round of the mock LISA data challenges. In: *Class. Quantum Grav.* 24(19, Sp. Iss. SI), 2007, S551–S564. Issn: 0264-9381, doi: 10.1088/0264-9381/24/19/S18.

202. NJ Cornish and EK Porter. Searching for massive black hole binaries in the first Mock LISA data challenge. *Class. Quantum Grav.* 24, 2007, 501–ff, doi: 10.1088/0264-9381/24/19/S13, eprint: arXiv:gr-qc/0701167.

203. J Crowder and NJ Cornish. Extracting galactic binary signals from the first round of Mock LISA data challenges. *Class. Quantum Grav.* 24, 2007, 575–ff, doi: 10.1088/0264-9381/24/19/S20, arXiv:0704.2917.

204. R Prix and JT Whelan. F-statistic search for white-dwarf binaries in the first Mock LISA Data challenge. *Class. Quantum Grav.* 24, 2007, 565–ff, doi: 10.1088/0264-9381/24/19/S19, arXiv:0707.0128.

205. C Rover, A Stroeer, E Bloomer, N Christensen, J Clark, M Hendry, C Messenger et al. Inference on inspiral signals using LISA MLDC data. *Class. Quantum Grav.* 24(19), 2007, S521–S527. URL: http://stacks.iop.org/0264-9381/24/S521.

206. A Stroeer, J Veitch, C Rover, E Bloomer, J Clark, N Christensen, M Hendry et al. Inference on white dwarf binary systems using the first round Mock LISA Data Challenges data sets. *Class. Quantum Grav.* 24(19), 2007, S541–S549. URL: http://stacks.iop.org/0264-9381/24/S541.

207. KR Nayak, SD Mohanty, and K Hayama. The tomographic method for LISA binaries: Application to MLDC data. *Class. Quantum Grav.* 24, 2007, 587–ff, doi: 10.1088/0264-9381/24/19/S21.

208. S Babak, JG Baker, MJ Benacquista, NJ Cornish, J Crowder, SL Larson, E Plagnol et al. The Mock LISA data challenges: From challenge 1B to challenge 3. *Class. Quantum Grav.* 25(18), 2008, 184026–ff, doi: 10.1088/0264-9381/25/18/184026, arXiv:0806.2110.

209. JR Gair, E Porter, S Babak, and L Barack. A constrained metropolis hastings search for EMRIs in the mock LISA data challenge 1B. *Class. Quantum Grav.* 25(18), 2008, 184030–ff, doi: 10.1088/0264-9381/25/18/184030. arXiv:0804.3322.

210. EL Robinson, JD Romano, and A Vecchio. Search for a stochastic gravitational-wave signal in the second round of the mock LISA data challenges. *Class. Quantum Grav.* 25(18), 2008, 184019–ff, doi: 10.1088/0264-9381/25/18/184019, arXiv:0804.4144.

211. M Trias, A Vecchio, and J Veitch. Markov chain Monte Carlo searches for galactic binaries in mock LISA data challenge 1B data sets. *Class. Quantum Grav.* 25(18), 2008, 184028–ff, doi: 10.1088/0264-9381/25/18/184028, arXiv:0804.4029.

212. J Whelan, R Prix, and D Khurana. Improved search for galactic white-dwarf binaries in mock LISA data challenge 1B using an F-statistic template bank. *Class. Quantum Grav.* 25(18), 2008, 184029–ff, doi: 10.1088/0264-9381/25/18/184029, arXiv:0805.1972 [gr-qc].

213. A Blaut, A Królak, and S Babak. Detecting white dwarf binaries in mock LISA data challenge 3. *Class. Quantum Grav.* 26(20), 2009, 204023–ff, doi: 10.1088/0264-9381/26/20/204023.

214. S Babak, J Baker, M Benacquista, N Cornish, S Larson, I Mandel, S McWilliams et al. The mock LISA data challenges: From challenge 3 to challenge 4. *Class. Quantum Grav.* 27(8), 2010, 084009–ff, doi: 10.1088/0264-9381/27/8/084009, arXiv:0912.0548 [gr-qc].

215. A Blaut, S Babak, and A Królak. Mock LISA data challenge for the galactic white dwarf binaries. *Phys. Rev. D* 81(6), 2010, 063008–ff, doi:10.1103/PhysRevD.81.063008, arXiv:0911.3020 [gr-qc].

216. LS Finn and D Chernoff. Observing binary inspiral in gravitational radiation—One interferometer. *Phys. Rev. D* 47(6), 1993, 2198–2219. Issn: 0556-2821.

217. C Cutler, LS Finn, E Poisson, and GJ Sussman. Gravitational radiation from a particle in circular orbit around a black hole. II. Numerical results for the nonrotating case. *Phys. Rev. D* 47, 1993, 1511–1518.

218. RW Hellings and GS Downs. Upper limits on the isotropic gravitational radiation background from pulsar timing analysis. *Astrophys. J. Lett.* 265, 1983, L39–L42, doi: 10.1086/183954.

219. ÉÉ Flanagan. Sensitivity of the laser interferometer gravitational wave observatory to a stochastic background, and its dependence on the detector orientations. *Phys. Rev. D* 48(6), 1993, 2389–2407.

220. LS Finn, SL Larson, and JD Romano. Detecting a stochastic gravitational-wave background: The overlap reduction function. *Phys. Rev. D* 79(6), 2009. Issn: 1550-7998, doi: 10.1103/PhysRevD.79.062003.

221. A Sesana, A Vecchio, and CN Colacino. The stochastic gravitationalwave background from massive black hole binary systems: Implications for observations with Pulsar Timing Arrays. *Mon. Not. R. Astron. Soc.* 390(1), 2008, 192–209. Issn: 0035-8711, doi: 10.1111/j.1365-2966.2008.13682.x.

222. A Sesana, A Vecchio, and M Volonteri. Gravitational waves from resolvable massive black hole binary systems and observations with pulsar timing arrays. *Mon. Not. R. Astron. Soc.* 394(4), 2009, 2255–2265. Issn: 0035-8711, doi: 10.1111/j.1365-2966.2009.14499.x, eprint: 0809.3412.

223. L Blackburn, L Cadonati, S Caride, S Caudill, S Chatterji, N Christensen, J Dalrymple et al. The LSC glitch group: Monitoring noise transients during the fifth LIGO science run. *Class. Quantum Grav.* 25(18), 2008. Issn: 0264-9381, doi: 10.1088/0264-9381/25/18/184004.

224. F D'Alessandro, P McCulloch, P Hamilton, and A Deshpande. The timing noise of 45 southern pulsars. *Mon. Not. R. Astron. Soc.* 277, 1995, 1033–1046.

225. F D'Alessandro, AA Deshpande, and PM McCulloch. Power spectrum analysis of the timing noise in 18 southern pulsars. *J. Astrophys. Astron.* 18, 1997, 5, doi: 10.1007/BF02714849. URL: http://adsabs.harvard.edu/cgi-bin/nph-data_query?bibcode=1997JApA...18....5D&link_type=ABSTRACT.

226. A Deshpande, F Alessandro, and P McCulloch. Application of CLEAN in the power spectral analysis of non-uniformly sampled pulsar timing data. *J. Astrophys. Astron.* 17, 1996, 7–16, doi: 10.1007/BF02709341.

227. PF Michelson. On detecting stochastic background gravitational radiation with terrestrial detectors. In: *Mon. Not. R. Astron. Soc.* 227, 1987, 933–941.

228. SW Ballmer. A radiometer for stochastic gravitational wave. *Class. Quantum Grav.* 23, 2006, S179–S185, doi: 10.1088/0264-9381/23/8/S23.

229. L Blanchet. Gravitational radiation from post-Newtonian sources and inspiralling compact binaries. *Living Rev. Rel.* 9, 2006, and references therein, p. 4, eprint: gr-qc/0202016.

230. S Klimenko and G Mitselmakher. A wavelet method for detection of gravitational wave bursts. *Class. Quantum Grav.* 21, 2004, 1819–ff, doi: 10.1088/0264-9381/21/20/025.

231. S Klimenko, I Yakushin, M Rakhmanov, and G Mitselmakher. Performance of the WaveBurst algorithm on LIGO data. *Class. Quantum Grav.* 21, 2004, 1685–ff, doi: 10.1088/0264-9381/21/20/011.

232. S Klimenko, I Yakushin, A Mercer, and G Mitselmakher. A coherent method for detection of gravitational wave bursts. *Class. Quantum Grav.* 25, 2008, 114209, doi: 10.1088/0264-9381/25/11/114029.

233. J McNabb, M Ashley, LS Finn, E Rotthoff, A Stuver, T Summerscales, P Sutton, M Tibbits, K Thorne, and K Zaleski. Overview of the BlockNormal event trigger generator. *Class. Quantum Grav.* 21(20, Sp. Iss. SI), 2004, S1705–S1710. Issn: 0264-9381.

234. LS Finn. Aperture synthesis for gravitational-wave data analysis: Deterministic sources. *Phys. Rev. D* 63(10), 2001, 102001–ff.

235. TZ Summerscales, A Burrows, LS Finn, and CD Ott. Maximum entropy for gravitational wave data analysis: Inferring the physical parameters of core-collapse supernovae. *Astrophys. J.* 678(2), 2008, 1142–1157. Issn: 0004-637X.

236. WG Anderson, PR Brady, JDE Creighton, and ÉÉ Flanagan. *An excess power statistic for detection of burst sources of gravitational radiation.* gr-qc/0008066. Aug. 2000.

237. WG Anderson, PR Brady, JD Creighton, and É Flanagan. Excess power statistic for detection of burst sources of gravitational radiation. *Phys. Rev. D* 63(4), 2001, 042003–ff.

238. N Christensen and R Meyer. Using Markov chain Monte Carlo methods for estimating parameters with gravitational radiation data. *Phys. Rev. D* 64, 2001, 22001, doi: 10.1103/PhysRevD.64.022001.

239. N Christensen, R Dupuis, G Woan, and R Meyer. Metropolis-Hastings algorithm for extracting periodic gravitational wave signals from laser interferometric detector data. *Phys. Rev. D* 70(2), 2004, 022001–ff, doi: 10.1103/PhysRevD.70.022001, eprint: arXiv: gr-qc/0402038.

240. N Christensen, R Meyer, and A Libson. A Metropolis Hastings routine for estimating parameters from compact binary inspiral events with laser interferometric gravitational radiation data. *Class. Quantum Grav.* 21, 2004, 317–330, doi: 10.1088/0264-9381/21/1/023.

241. R Umstaetter and M Tinto. Bayesian comparison of post-Newtonian approximations of gravitational wave chirp signals. *Phys. Rev. D* 77, 2008, 82002, doi: 10.1103/PhysRevD.77.082002.

242. M van der Sluys, V Raymond, I Mandel, C Röver, N Christensen, V Kalogera, R Meyer, and A Vecchio. Parameter estimation of spinning binary inspirals using Markov chain Monte Carlo. *Class. Quantum Grav.* 25(18), 2008, 184011–ff, doi: 10.1088/0264-9381/25/18/184011, arXiv:0805.1689.

243. S Mukherjee and LS Finn. Removing instrumental artifacts: Suspension violin modes. In: *Gravitational Waves.* S Meshkov (ed). AIP Conference Proceedings 523. Proceedings of the Third Edoardo Amaldi Conference. Melville, New York: American Institute of Physics, 2000, 362–ff.

244. LS Finn and S Mukherjee. Data conditioning for gravitational wave detectors: A Kalman filter for regressing suspension violin modes. *Phys. Rev. D* 63(6), 2001, 062004–ff.

245. S Mukherjee and The LIGO Scientific Collaboration. Multidimensional classification of kleine Welle triggers from LIGO science run. *Class. Quantum Grav.* 23(19), 2006, S661–S672.

246. S Mukherjee and the LIGO Scientific Collaboration. Preliminary results from the hierarchical glitch pipeline. *Class. Quantum Grav.* 24, 2007, 701–ff, doi: 10.1088/0264-9381/24/19/S32.

247. S Mallat. *A Wavelet Tour of Signal Processing*, Second Edition. New York: Academic Press, 1999.

248. S Mukherjee, R Obaid, and B Matkarimov. Classification of glitch waveforms in gravitational wave detector characterization. *J. Phys.Conf. Ser* 243(1), 2010, 012006–ff, doi: 10.1088/1742-6596/243/1/012006.

249. SE Pollack. LISA science results in the presence of data disturbances. *Class. Quantum Grav.* 21, 2004, 3419–3432. Issn: 0264–9381.

250. Committee for a Decadal Survey of Astronomy and Astrophysics and National Research Council. Washington, DC: National Academies Press, 2010. Isbn: 9780309157995. URL: http://www.nap.edu/openbook.php?record_id=12951.

4

The Largest Data Sets

CHAPTER 20

Virtual Observatory and Distributed Data Mining

Kirk D. Borne

George Mason University

CONTENTS

20.1 Introduction	447
20.2 Astrophysics at the Crossroads of Science and Technology	448
20.2.1 Virtual Observatories: Federated Data Collections from Multiple Sky Surveys	450
20.2.2 Mining of Distributed Data	452
20.2.3 VO-Enabled Data Mining Use Cases	454
20.2.4 Distributed Mining of Data	456
20.3 Concluding Remarks	459
Acknowledgments	459
References	459

20.1 INTRODUCTION

New modes of discovery are enabled by the growth of data and computational resources (i.e., cyberinfrastructure) in the sciences. This cyberinfrastructure includes structured databases, virtual observatories (distributed data, as described in Section 20.2.1 of this chapter), high-performance computing (petascale machines), distributed computing (e.g., the Grid, the Cloud, and peer-to-peer networks), intelligent search and discovery tools, and innovative visualization environments. Data streams from experiments, sensors, and simulations are increasingly complex and growing in volume. This is true in most sciences, including astronomy, climate simulations, Earth observing systems, remote sensing data collections, and sensor networks. At the same time, we see an emerging confluence of new technologies and approaches to science, most clearly visible in the growing synergism of the four modes of scientific discovery: sensors–modeling–computing–data (Eastman et al. 2005). This has been driven by numerous developments, including the information explosion, development of large-array sensors, acceleration in high-performance computing (HPC) power, advances in algorithms, and efficient modeling techniques. Among these, the most extreme is the growth in new data. Specifically, the acquisition of data in all scientific disciplines is rapidly accelerating and causing a data glut (Bell et al. 2007).

It has been estimated that data volumes double every year—for example, the NCSA (National Center for Supercomputing Applications) reported that their users cumulatively generated one petabyte of data over the first 19 years of NCSA operation, but they then generated their next one petabyte in the next year alone, and the data production has been growing by almost 100% each year after that (Butler 2008). The NCSA example is just one of many demonstrations of the exponential (annual data-doubling) growth in scientific data collections. In general, this putative data-doubling is an inevitable result of several compounding factors: the proliferation of data-generating devices, sensors, projects, and enterprises; the 18-month doubling of the digital capacity of these microprocessor-based sensors and devices (commonly referred to as "Moore's law"); the move to digital for nearly all forms of information; the increase in human-generated data (both unstructured information on the web and structured data from experiments, models, and simulation); and the ever-expanding capability of higher density media to hold greater volumes of data (i.e., data production expands to fill the available storage space). These factors are consequently producing an exponential data growth rate, which will soon (if not already) become an insurmountable technical challenge even with the great advances in computation and algorithms. This technical challenge is compounded by the ever-increasing geographic dispersion of important data sources—the data collections are not stored uniformly at a single location, or with a single data model, or in uniform formats and modalities (e.g., images, databases, structured and unstructured files, and XML data sets)—the data are in fact large, distributed, heterogeneous, and complex. The greatest scientific research challenge with these massive distributed data collections is consequently extracting all of the rich information and knowledge content contained therein, thus requiring new approaches to scientific research. This emerging data-intensive and data-oriented approach to scientific research is sometimes called *discovery informatics* or *X-informatics* (where X can be any science, such as bio, geo, astro, chem, eco, or anything; Agresti 2003; Gray 2003; Borne 2010). This data-oriented approach to science is now recognized by some (e.g., Mahootian and Eastman 2009; Hey et al. 2009) as the fourth paradigm of research, following (historically) experiment/observation, modeling/analysis, and computational science.

20.2 ASTROPHYSICS AT THE CROSSROADS OF SCIENCE AND TECHNOLOGY

In astronomy in particular, rapid advances in three technology areas (observatory facilities, detectors, and computational resources) have continued unabated (e.g., Gray and Szalay 2004; with continuing evidence for this trend in the very large SPIE Astronomical Telescopes and Instrumentation conferences that take place each year[*]), all of which are leading to more data (Becla et al. 2006). The scale of data-capturing capabilities grows at least as fast as the underlying microprocessor-based measurement system (Gray et al. 2005). For example, the fast growth in CCD detector size and sensitivity has seen the average data set size of a

---

[*] See a partial list of the 2350 related search results on the SPIE website summarized here: http://bit.ly/f3WN21

typical large astronomy sky survey project grow from hundreds of gigabytes in the 1990s (e.g., the MACHO survey), to tens of terabytes in the 2000s (e.g., 2-Micron All-Sky Survey [2MASS] and Sloan Digital Sky Survey [SDSS]; Brunner et al. 2001; Gray and Szalay 2004), up to a projected size of tens of petabytes in the next decade (e.g., Pan-STARRS and LSST [Large Synoptic Survey Telescope]; Becla et al. 2006; Bell et al. 2007). As the giant among astronomical sky surveys, the LSST will produce one 56K × 56K (3-gigapixel) image of the sky every 20 s, generating nearly 15 TB of data every day for 10 years. The progression in the growth of sky survey sizes over the past 40 years, and then projected forward to the LSST, is depicted in Figure 9.1 of Tyson and Borne (this volume, Chapter 9)—this chart clearly demonstrates the exponential growth in capacities and outputs of these surveys.

Increasingly sophisticated computational and data science approaches will be required to discover the wealth of new scientific knowledge hidden within these new massive, distributed, and complex scientific data collections (Gray et al. 2002; Szalay et al. 2002; Borne 2006, 2009; Graham et al. 2007; Kegelmeyer et al. 2008). As illustrated schematically in Figure 20.1, not only are the scientific data collections multimodal and heterogeneous, but the types of information extracted from the data are also quite diverse, as are the types of machine

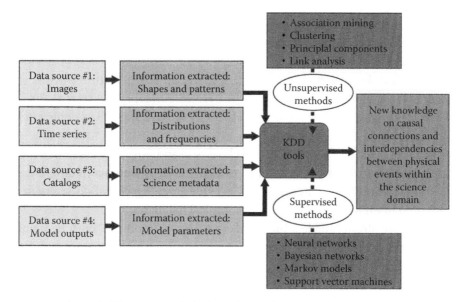

FIGURE 20.1 Schematic illustration of the knowledge discovery in databases (KDD) process flow for scientific discovery. Numerous astronomical databases, generated from large sky surveys, provide a wealth of data and a variety of data types for mining. Data mining proceeds first with data discovery and access from multiple distributed data repositories, then continues with the information extraction step, and then is enriched through the application of machine learning algorithms and data mining methods for scientific knowledge discovery. Finally, the scientist evaluates, synthesizes, and interprets the newly discovered knowledge in order to arrive at a greater scientific understanding of the Universe and its constituent parts, processes, and phenomena. *Supervised methods* are those based on training data—these are primarily classification methods. *Unsupervised methods* are those that do not rely on training data or preknown classes—these include traditional clustering methods.

learning algorithms applied to the data collections in order to achieve scientific knowledge discovery from the data (KDD), particularly from distributed databases.

20.2.1 Virtual Observatories: Federated Data Collections from Multiple Sky Surveys

Astronomers now systematically study the sky with large surveys. These sky surveys make use of uniform calibrations and well-engineered pipelines for the production of a comprehensive set of quality-controlled data products. Surveys are used to measure and collect data from all objects that are visible within large regions of the sky, in a systematic and repeatable fashion. These statistically robust procedures thereby generate very large unbiased samples of many classes of astronomical objects.

A common feature of modern astronomical sky surveys is that they are producing massive databases. Surveys produce hundreds of terabytes (TB) up to 100 (or more) petabytes (PB) both in the image data archive and in the object catalogs (databases). These include SDSS (York et al. 2000), the 2MASS (Skrutskie et al. 2006), and numerous others. Historically, most of these have been single-epoch (or *static*) surveys (i.e., they do not explicitly address the ongoing long-term time variability of individual sources in the sky). Notable exceptions include the following surveys (many of which cover only a tiny fraction of the sky): the SDSS Stripe 82 subsurvey (Bramich et al. 2008), numerous proper motion surveys[*] (e.g., Hipparchos; Perryman 1991; see also Munn et al. 2004), variable star surveys (e.g., CoRot; Debosscher et al. 2009), supernova surveys (e.g., SDSS-II; Frieman et al. 2008), the gravitational lens surveys (e.g., MACHO, Alcock et al. 2001; OGLE, Paczynski 1996; and others[†]), and extrasolar planet searches (e.g., Koch et al. 2010). Future (or just starting) large sky surveys—such as Pan-STARRS (Jewitt 2003), SkyMapper (Keller et al. 2007), Dark Energy Survey,[‡] GAIA (Perryman 2002), and LSST (Strauss et al. 2009)—deliberately focus on the time-varying sky, thereby enormously expanding the exciting field of time-domain astronomy (Borne 2008; Ivezic et al. 2008; see also Tyson and Borne, this volume, Chapter 9). In any case, wide-area or limited-area, large or small, the sheer number of different sky surveys (as the above abridged list indicates) can provide a rich harvest of discovery from data and information in multiple distributed astronomical databases.

Sky surveys have already made a huge impact on research output, both by members of the respective survey teams, as well as by the astronomical research community at large. For example, the SDSS project reported (as of late 2008) that SDSS data had contributed to over 2000 refereed science papers, with more than 70,000 citations,[§] a remarkable publication rate that appears to be continuing unabated (e.g., nearly 1200 additional refereed SDSS-related papers appeared in 2009–2010, already with over 10,000 citations). This scientific research productivity has been made possible because these sky survey projects have developed effective web-based mechanisms for the world scientific community to access these knowledge-rich sky survey databases and data products.

---

[*] http://spider.seds.org/spider/Misc/star_cats.html#survey
[†] http://timemachine.iic.harvard.edu/surveys/
[‡] http://www.darkenergysurvey.org/
[§] http://www.sdss.org/signature.html

One of the most significant technological developments for astronomy in the past decade has been the e-Science cyberinfrastructure for data discovery and access from all geographically distributed astronomical data collections. This framework is called the Virtual Observatory (VO)—in the United States, the implementation is referred to as the Virtual Astronomical Observatory[*] (VAO), which was developed initially as the National Virtual Observatory (NVO), an NSF-funded ITR (information technology research) project (Graham et al. 2007). The VO spans multiple international projects that have invested significant resources into creating and maintaining a rich e-Science infrastructure.[†] The various national VO projects have shared the development of a few dozen standards and protocols for metadata, data modeling, data description, data discovery, data access, data sharing, and integration of distributed data into versatile astronomical research tools and interfaces.[‡] One of these VO protocols is *VO-Event*,[§] which is used as a messaging standard and information payload for notification of temporal astronomical events. The VO-Event messaging protocol is already an important enabling technology for time-based (temporal) sky surveys, and the future promises even greater strides forward in these areas as time-domain astronomy takes center stage in the coming years (Bloom et al. 2008; Williams et al. 2010).

Another significant accomplishment of the VO community has been the development of registries for data discovery, access, and integration. Various data portals take advantage of these registries in order to provide "one-stop shopping" for all of your data needs. One particularly useful data portal for data mining specialists is the OpenSkyQuery.net site.[¶] This site enables multidatabase queries, even when (i.e., specifically when) the databases are geographically distributed around the world and when the databases have a variety of schema and access mechanisms. This facility enables distributed data mining of joined catalog data from numerous astrophysically interesting databases. This facility does not mine astronomical images (i.e., the pixel data, stored in image archives), nor does it mine all VO-accessible databases, but it does query and join on many of the most significant astronomical object catalogs. In this context, we consider "distributed data mining" to mean the "mining of distributed data (MDD)" (Section 20.2.2), as opposed to the algorithmically interesting and very challenging context of "distributed mining of data (DMD)" (Section 20.2.4). A set of VO-inspired astronomical data mining contexts and use cases are presented that are applicable to either type of distributed data mining (Section 20.2.3).

While the VO has been remarkably successful and productive, its initial focus historically was primarily on infrastructure, upon which a variety of application frameworks are being built.[‖] Time-domain astronomy is the next great application domain for which a VO-enabled framework is now required (Becker 2008; Djorgovski et al. 2008; Drake et al. 2009). As already described, another important application domain for VO is scientific data

---

[*] http://www.us-vo.org/
[†] http://www.ivoa.net/
[‡] http://www.ivoa.net/Documents/
[§] http://www.ivoa.net/cgi-bin/twiki/bin/view/IVOA/IvoaVOEvent
[¶] http://www.openskyquery.net/
[‖] http://www.ivoa.net/cgi-bin/twiki/bin/view/IVOA/IvoaApplications?rev=1.67

mining, specifically distributed data mining across multiple geographically dispersed data collections.[*]

The VO makes available to any researcher anywhere a vast selection of data for browsing, exploration, mining, and discovery. The VO protocols enable a variety of user interactions: query distributed databases through a single interface (e.g., VizieR,[†] or the next-generation OpenSkyQuery.net [A. Thakar, private communication]); search distributed image archives with a single request [e.g., DataScope[‡]]; discover data collections and catalogs by searching their descriptions within a comprehensive registry [e.g., the IVOA searchable registries[§]]; and integrate and visualize data search results within a single GUI or across a set of linked windows [e.g., VIM (visual integration and mining),[¶] Aladin,[‖] or SAMP (simple application messaging protocol)[**]]. The scientific knowledge discovery potential of the worldwide VO-accessible astronomical data and information resources is almost limitless—the sky is the limit! The VO has consequently been described as an "efficiency amplifier" for data-intensive astronomical research (A. Szalay, private communication), due to the fact that the efficiency of data discovery, access, and integration has been greatly amplified by the VO protocols, standards, tools, registries, and resources. This has been a great leap forward in the ability of astronomers to do multiarchive distributed data research. However, while the data discovery process has been made much more efficient by the VO, the KDD process can also be made much more efficient (measured by "time-to-solution") and more effective (measured by depth and breadth of solution). Improved KDD from VO-accessible data can be achieved particularly through the application of sophisticated data mining and machine learning algorithms,[††] whether operating on the distributed data *in situ* (DMD), or operating on data subsets retrieved to a central location from distributed sources (MDD). Numerous examples and detailed reviews of these methods, applications, and science results can be found in the review papers by Borne (2009) and by Ball and Brunner (2010), as well as in the other chapters of this book (which will not be repeated here).

20.2.2 Mining of Distributed Data

Anecdotal evidence suggests that most astronomers think of distributed data mining as analytic operations on data subsets that have been retrieved to a central location from distributed sources, as a plethora of research papers would indicate: a query of astronomy research papers at the NASA ADS (Astrophysics Data System) that included the words "data mining" in the title returned 128 instances,[‡‡] many of which mined multiple data sets, but not in a distributed manner. An early example of such MDD research was carried out

---

[*] http://www.ivoa.net/cgi-bin/twiki/bin/view/IVOA/IvoaKDD

[†] http://vizier.u-strasbg.fr/cgi-bin/VizieR

[‡] http://heasarc.gsfc.nasa.gov/vo/

[§] http://rofr.ivoa.net/

[¶] http://www.us-vo.org/vim/

[‖] http://aladin.u-strasbg.fr/aladin.gml

[**] http://www.ivoa.net/Documents/latest/SAMP.html

[††] http://www.ivoa.net/cgi-bin/twiki/bin/view/IVOA/IvoaKDDguide

[‡‡] See the results of the ADS "data mining" query here: http://bit.ly/gDPNGV

by Borne (2003)—the basic elements of that investigation are presented here. In this case study, some of the key aspects of MDD that relate to astrophysics research problems were presented. The case study focused on one particularly interesting rare class of galaxies that had not been well studied up to that time: the very luminous infrared galaxies (VLIRGs; with infrared luminosities between 10^{11} and 10^{12} solar luminosities). It was anticipated prior to the study that mining across a variety of multiproject multiwavelength multimodal (imaging, spectroscopic, catalog) databases would yield interesting new scientific properties and understanding of the VLIRG phenomenon.

VLIRGs represent an important phase in the formation and evolution of galaxies. On one hand, they offer the opportunity to study how the fundamental physical and structural properties of galaxies vary with IR (infrared) luminosity, providing a possible evolutionary link between the extremely chaotic and dynamic ultra-luminous IR galaxies (ULIRGs; with infrared luminosities between 10^{12} and 10^{13} solar luminosities) and the "boring" set of normal galaxies. On the other hand, VLIRGs are believed to be closely related to other significant cosmological populations in the sense that VLIRGs could be either the low-redshift analogs or the direct result of the evolution of those cosmologically interesting populations of galaxies that typically appear only at the earliest stages of cosmic time in our Universe. Several such cosmologically interesting populations have been identified by astronomers. These include (1) the galaxies that comprise the cosmic infrared background (CIB); (2) the high-redshift submillimeter sources at very large distance; (3) infrared-selected quasars; and (4) the "extremely red objects" (EROs) found in deep sky surveys. The primary questions pertaining to these cosmological sources are: What are they? Are they dusty quasars, or dusty starbursting galaxies, or massive old stellar systems (big, dead, and red galaxies)? Some of these classes of objects are undoubtedly members of the extremely rare class of ULIRGs, while the majority is probably related to the more numerous, though less IR-luminous, class of galaxies: the VLIRGs.

The Borne (2003) data mining research project was a proof of concept for a VO-enabled science search scenario, aimed at identifying potential candidate contributors to the CIB. This was significant since it was known that fully one half of all of the radiated energy in the entire Universe comes to us through the CIB! Our approach (Borne 2003) involved querying several distributed online databases and then examining the linkages between those databases, the image archives, and the published literature. As typical for most multimission multiwavelength astronomical research problems, our search scenario started by finding object cross-identifications across various distributed source lists and archival data logs. In a very limited sample of targets that we investigated in order to validate our VO-enabled distributed data mining (MDD) approach to the problem, we did find one object in common among three distributed databases. This object was a known hyperluminous infrared galaxy (HyLIRG; with infrared luminosity greater than 10^{13} solar luminosities) at moderate redshift, harboring a Quasar, which was specifically imaged by the Hubble Space Telescope because of its known HyLIRG characteristics. In this extremely limited test scenario, we did in fact find what we were searching for: a distant IR-luminous galaxy that is either a likely contributor to the CIB, or else it is an object similar in characteristics to the more distant objects that likely comprise the CIB.

While the aforementioned case study provides an example of distributed database querying, it is not strictly an example of data mining. It primarily illustrates the power of cross-identifications across multiple databases—that is, joins on object positions across various catalogs, subject to constraints on various observable properties (such as optical-to-IR flux ratios). After such query result lists are generated, any appropriate machine learning algorithm can then be applied to the data in order to find new patterns, new correlations, principal components, or outliers; or to learn classification rules or class boundaries from labeled (already classified) objects; or to classify new objects based upon previously learned rules or boundaries.

20.2.3 VO-Enabled Data Mining Use Cases

In general, astronomers will want to apply data mining algorithms on distributed multiwavelength (multimission) VO-accessible data within one of these four distinct astronomical research contexts (Borne 2003), where "events" refers to a class of astronomical objects or phenomena (e.g., supernovae, or planetary transits, or microlensing):

1. *Known events/known classification rules*—use existing models (i.e., descriptive models, or classification algorithms) to locate known phenomena of interest within collections of large databases (frequently from multiple sky surveys).

2. *Known events/unknown rules*—apply machine learning algorithms to learn the classification rules and decision boundaries between different classes of events; or find new astrophysical properties of known events (e.g., known classes) by examining previously unexplored parameter spaces (e.g., wavelength or temporal domains) in distributed databases.

3. *Unknown events/known rules*—use expected relationships (from predictive algorithms, or physics-based models) among observational parameters of astrophysical phenomena to predict the presence of previously unseen events within various databases or sky surveys.

4. *Unknown events/unknown rules*—use pattern detection, correlation analysis, or clustering properties of data to discover new observational (in our case, astrophysical) relationships (i.e., unknown connections) among astrophysical phenomena within multiple databases; or use thresholds or outlier detection techniques to identify new transient or otherwise unique ("one-of-a-kind") events, and thereby discover new astrophysical phenomena (the "unknown unknowns")—we prefer to call this "surprise detection."

This list represents four distinct contexts for astronomical knowledge discovery from large distributed databases. Another view of the same problem space is to examine the basis set of data mining functions (use cases) that astronomers apply to their data—these are the functional primitives, not the actual algorithms, which might be quite numerous for any given data mining function (Borne 2001a,b). These basic VO-driven data mining use cases include:

1. *Object cross-identification* (or *cross-matching*)—It is a well-known, notorious "problem" in astronomy that individual objects are often called by many names. This is an even more serious problem now with the advent of all-sky surveys in many wavelength bands—the same object will appear in a multitude of VO-accessible catalogs, observation logs, and databases; and the object will almost certainly be cataloged uniquely in each case according to a survey-specific naming convention. Identifying the same object across these distributed data collections is not only crucial to understanding its astrophysical properties and its physical nature, but this use case is also one of the most basic inquiries that an astronomer will make of these all-sky surveys. The problem of isolating these object cross-IDs across multiple surveys reduces in its simplest form to finding high-likelihood *spatial associations* of given objects among a collection of catalogs. Some databases, such as the SDSS, provide cross-matching services,[*] as does OpenSkyQuery.net (Budavari et al. 2009; Budavari, this volume, Chapter 7).

2. *Object cross-correlation*—After the cross-database identifications of the same object have been established, there are a wealth of astrophysical "What if?" queries that can be applied to the observational parameters of objects that are present in multiple databases. In the most general sense, these correspond to the application of *classification*, *clustering*, and *regression* (cross-correlation or anticorrelation) algorithms among the various database parameters. These correlations need not be confined to two parameters; they may be highly dimensional (e.g., the fundamental plane of elliptical galaxies; Djorgovski and Davis 1987; Dressler et al. 1987). Correlating the observational properties of a class of objects in one astronomical database with a different set of properties for the same class of objects in another database is the essence of VO-style distributed data mining. The results of these inquiries correspond to *pattern associations*.

3. *Nearest-neighbor identification*—In addition to identifying spatial associations (in real space) across multiple distributed astronomical data archives, it is also scientifically desirable to find "spatial" associations (*nearest neighbors*) in the highly dimensional space of complex astronomical observational parameters—to search for dense *clusterings* and *associations* among observational parameters in order to find new classes of objects or new properties of known classes. The subsets of objects that have such matching sets of parameters correspond to *coincidence associations*. They offer a vast potential for new scientific exploration and will provide a rich harvest of astronomical knowledge discovery from VO-accessible distributed databases.

4. *Systematic data exploration*—Within very large distributed databases, there are likely to be significant subsets of data (regions of observational parameter space) that have gone largely unexplored, even by the originating scientific research teams that produced the

---

[*] http://www.sdss.org/dr7/tutorials/crossid/index.html

data (Djorgovski et al. 2001). Archival data researchers accessing the databases will want to apply a wide variety of *parameter-based* and *correlation-based* constraints to the data. The exploration process involves numerous iterations on "What if?" queries. Some of these queries will not be well constrained or targeted, and will thus produce very large output sets. For most situations, this will be too large to be manageable by a typical user's desktop data analysis system. So, systematic data exploration includes browse and preview functions (to reveal qualitative properties of the results prior to retrieval) and allows iterations on preceding queries (either to modify or to tighten the search constraints).

5. *Surprise detection* (=outlier/anomaly/novelty discovery)—This corresponds to one of the more exciting and scientifically rewarding uses of large distributed databases: finding something totally new and unexpected. Outlier detection comes in many forms. One expression of outlyingness may be "interestingness," which is concept that represents how "interesting" (nontypical) are the parameters that describe an object within a collection of different databases. These may be outliers in individual parameter values, or they may have unusual correlations or patterns among the parameter values, or they have a significantly different distribution of parameter values from its peers (Borne and Vedachalam 2011).

20.2.4 Distributed Mining of Data

The DMD is a major research challenge area in the field of data mining. There is a large and growing research literature on this subject (Bhaduri et al. 2008). Despite this fact, there has been very little application of DMD algorithms and techniques to astronomical data collections, and even those few cases have not begun to tap the richness and knowledge discovery potential from VO-accessible data collections. We have applied DMD algorithms to distributed PCA (principal components analysis; Giannella et al. 2006, Das et al. 2009), distributed outlier detection (Dutta et al. 2007), distributed classification of tagged documents (Dutta et al. 2009), and systematic data exploration of astronomical databases using the PADMINI (peer-to-peer astronomy data mining) system (Mahule et al. 2010). In order to provide a specific detailed illustration of one of the astronomical research problems solved by these methods, the distributed PCA results will be presented here.

The class of elliptical galaxies has been known for more than 20 years to show dimension reduction among a subset of physical attributes, such that the three-dimensional distribution of three of those astrophysical parameters reduces to a two-dimensional plane (Djorgovski and Davis 1987; Dressler et al. 1987). The transformed coordinates within that plane represent the principal eigenvectors of the distribution, and it is found that the first two principal components capture significantly more than 90% of the variance among those three parameters.

By analyzing existing large astronomy databases (such as the SDSS and the 2MASS), we have generated a very large data set of elliptical galaxies with SDSS-provided spectroscopic redshifts. The data set consists of 102,600 cross-matched objects within our empirically determined volume-limited completeness cut-off: redshift $z < 0.23$. Each galaxy in this large data set was then assigned (labeled with) a new "local galaxy density" attribute, calculated through a volumetric Voronoi tessellation of the total galaxy distribution in space. The inverse of the Voronoi volume represents the local galaxy density (i.e., each galaxy occupies singly a well-defined volume that is calculated by measuring the distance to its nearest neighbors in all directions and then generating the three-dimensional convex polygon whose faces are the bisecting planes along the direction vectors pointing toward the nearest neighbors in each direction—the enclosed volume of the polygon is the Voronoi volume). Note that this is even possible only because every one of the galaxies in our sample has a measured redshift from the SDSS spectroscopic survey. Other groups who have tried to do this have used photometric redshifts (photo-zs; Rossi and Sheth 2008; Budavari, this volume, Chapter 15), which have (at best!) an uncertainty of $\Delta z = 0.01$ (especially when using a large number of smaller bandpasses; Wolf et al. 2004), corresponding to a distance uncertainty of $\sim 18\,h^{-1}\,\mathrm{Mpc}$, which is far above the characteristic scale of the large-scale structure ($\sim 5\,h^{-1}\,\mathrm{Mpc}$ comoving). Our work does not suffer from this deficiency since spectroscopic redshifts are at least one to two orders of magnitude better than photo-zs, and so our distance estimates and Voronoi volume calculations (hence, local galaxy density estimates) are well determined. It is also interesting to note that since the dynamical timescale (age) of a gravitating system is inversely proportional to the square root of the local galaxy density (e.g., Fellhauer and Heggie 2005), consequently the dynamical timescale is directly proportional to the square root of our calculated Voronoi volume. Therefore, studying the variation of galaxy parameters and relationships as a function of each galaxy's local Voronoi volume is akin to studying the time evolution of the ensemble population of galaxies.

For our initial correlation mining work, the entire galaxy data set was divided into 30 equal-sized partitions as a function of our independent variable: the local galaxy density. Consequently, each bin contains more than 3000 galaxies, thereby generating statistically robust estimators of the fundamental parameters in each bin. From the SDSS and 2MASS catalogs, we have extracted about four dozen measured parameters for each galaxy (out of a possible 800+ from the combined two catalogs): numerous colors (computed from different combinations of the 5-band SDSS ugriz photometry and the 3-band 2MASS jhk photometry), various size and radius measures, concentration indices, velocity dispersions (from SDSS), and various surface brightness measures. For those parameters that depend on distance, we have used the SDSS redshift to recalculate physical measures for those parameters (e.g., radii and surface brightness).

As a result of our data sampling criteria, we have been able to study eigenvector changes of the fundamental plane of elliptical galaxies as a function of density. This distributed data mining problem has also uncovered some new astrophysical results: we find that the variance captured in the first two principal components increases systematically from low-density regions to high-density regions (Figure 20.2), and we find that the direction of the principal eigenvector also drifts systematically in the three-dimensional parameter space from low-density regions to the highest-density regions (Figure 20.3).

Further PCA experiments with our cross-matched SDSS–2MASS data set have revealed some interesting trends of the variance captured by the first two principal components as a function of various color parameters: u–g, g–r, and j–k. Specifically, we again correlated

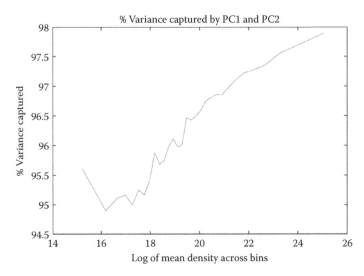

FIGURE 20.2 Variance captured by the first two principal components with respect to the log of the mean galaxy local density (for 30 different bins containing ~3000 galaxies each). The sample parameters used in this analysis are i-band Petrosian radius containing 50% of the galaxy flux (from SDSS), velocity dispersion (SDSS), and K-band mean surface brightness (2MASS). This plot clearly shows that the fundamental plane relation becomes tighter with increasing local galaxy density (inverse Voronoi volume, described in Section 20.2.4). (From Das, K. et al. 2009, *SIAM Conference on Data Mining SDM09*, pp. 247–258, http://www.siam. org/proceedings/datamining/2009/dm09.php. With permission.)

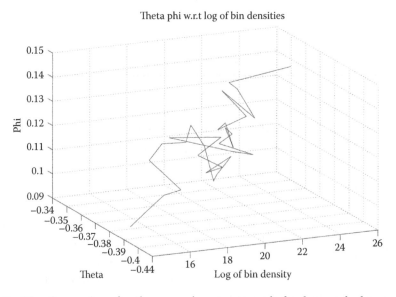

FIGURE 20.3 Direction cosines for the normal vector to each fundamental plane calculated as a function of local galaxy density (inverse Voronoi volume). Though there are some chaotic effects, in general there is a trend for the tilt of the fundamental plane to drift nearly systematically. (From Das, K. et al. 2009, *SIAM Conference on Data Mining SDM09*, pp. 247–258, http://www.siam.org/proceedings/datamining/2009/dm09.php. With permission.)

the standard fundamental plane galaxy parameters (radius, surface brightness, and central velocity dispersion), but this time as a function of the astronomical colors of the galaxies (in bins of ∼3000 galaxies each). The same trend that we see in Figure 20.2 is seen in the variance captured by the first two principal components as a function of increasing u–g color (i.e., as the galaxies become redder). The inverse trend is seen in the variance as a function of increasing j–k color. In contrast, there is no trend in the variance with changes in g–r color.

20.3 CONCLUDING REMARKS

The accessibility of large distributed repositories of heterogeneous data for millions (and soon, billions) of astronomical objects through Virtual Observatory protocols and portals will inspire a new generation of data-oriented scientists and data-oriented research in astronomy (astroinformatics; Borne 2010). The knowledge discovery potential from these vast data resources is equally vast. The hidden knowledge and information content stored within these data mountains can be exploited through standardized data formats, data models, and data access mechanisms. The astronomical Virtual Observatory is thus an effective tool for efficient astronomical data mining and scientific discovery.

ACKNOWLEDGMENTS

This research has been supported in part by NASA AISR grant number NNX07AV70G. The author is grateful to Dr. Hillol Kargupta and his research associates for many years of productive collaborations in the field of distributed data mining in Virtual Observatories.

REFERENCES

Agresti, W. W. 2003, Discovery informatics, *Communications of the ACM*, 46(8), 25.
Alcock, C. et al. 2001, The MACHO project microlensing detection efficiency, *ApJS*, 136, 439.
Ball, N. and Brunner, R. 2010, Data mining and machine learning in astronomy, *International Journal of Modern Physics D*, 19, 1049.
Becker, A. C. 2008, Transient detection and classification, *Astronomische Nachrichten*, 329, 280.
Becla, J., Hanushevsky, A., Nikolaev, S., Abdulla, G., Szalay, A., Nieto-Santisteban, M., Thakar, A., Gray, J. 2006, Designing a multi-petabyte database for LSST, SPIE Conference on Observatory Operations: Strategies, Processes, and Systems, 26–27 May 2006, Orlando, FL.
Bell, G., Gray, J. and Szalay, A. 2007, Petascale computational systems, *Computer*, 39, 110–112.
Bhaduri, K., Das, K., Liu, K., Kargupta, H., and Ryan, J. 2008. Distributed data mining bibliography (Release 1.8), http://www.cs.umbc.edu/~hillol/DDMBIB/.
Bloom, J. S. et al. 2008, Towards a real-time transient classification engine, *Astronomische Nachrichten*, 329, 284.
Borne, K. 2001a, Science user scenarios for a VO design reference mission: Science requirements for data mining, *Virtual Observatories of the Future*, ASP Conference Series 225, Brunner, R. J., Djorgovski, S.G., Szalay, A.S. (eds), Orem, UT: Astronomical Society of the Pacific, p. 333.
Borne, K. 2001b, Data mining in astronomical databases, *Mining the Sky*, Banday, A.J., Zaroubi, S., and Bartelmann, M. (eds), Berlin: Springer-Verlag, p. 671.
Borne, K. 2003, Distributed data mining in the National Virtual Observatory, *SPIE Data Mining and Knowledge Discovery V*, Vol. 5098, 211.
Borne, K. 2006, Data-driven discovery through e-science technologies, *IEEE International Conference on Space Mission Challenges for Information Technology*, pp. 251–256.

Borne, K. 2008, A machine learning classification broker for the LSST transient database, *Astronomische Nachrichten*, 329, 255.

Borne, K. 2009, Scientific data mining in astronomy, *Next Generation Data Mining*, Boca Raton, FL: CRC Press, p. 91

Borne, K. 2010, Astroinformatics: Data-Oriented astronomy research and education, *Journal of Earth Science Informatics*, 3, 5.

Borne, K. and Vedachalam, A. 2012, Surprise detection in multivariate astronomical data, in E. D. Feigelson and G. J. Babu (eds), *Statistical Challenges in Modern Astronomy V*, Springer, New York, pp. 275–290.

Bramich, D. et al. 2008, Light and motion in SDSS stripe 82: The catalogues, *MNRAS*, 386, 887.

Brunner, R., Djorgovski, S.G., Prince, T.A., and Szalay A.S. 2001, Massive datasets in astronomy, in Abello, J., Pardalos, P., Resende, M., (eds), *Handbook of Massive Datasets*, Boston: Kluwer Academic Publishers, 931.

Budavari, T. et al. 2009, GALEX: SDSS catalogs for statistical studies, *ApJ*, 694, 1281.

Butler, M. 2008, Hidden no more, NCSA news, http://gladiator.ncsa.illinois.edu/PDFs/access/fall08/hiddennomore.pdf.

Das, K., Bhaduri, K., Arora, S., Griffin, W., Borne, K., Giannella, C., and Kargupta, H. 2009, Scalable distributed change detection from astronomy data streams using local, asynchronous eigen monitoring algorithms, *SIAM Conference on Data Mining SDM09*, pp. 247–258, http://www.siam.org/proceedings/datamining/2009/dm09.php.

Debosscher, J. et al. 2009, Automated supervised classification of variable stars in the CoRoT programme, *A&A*, 506, 519.

Djorgovski, S. G. and Davis, M. 1987, Fundamental properties of elliptical galaxies, *ApJ*, 313, 59.

Djorgovski, S. G. et al. 2001, Exploration of parameter spaces in a virtual observatory, in Starck, J.-L. and Murtagh, F. D., (eds), *Astronomical Data Analysis*, Proc. SPIE, Vol. 4477, pp. 43–52.

Djorgovski, S. G. et al. 2008, The Palomar-Quest digital synoptic sky survey, *Astronomische Nachrichten*, 329, 263.

Drake, A. J. et al. 2009, First results from the Catalina Real-time Transient Survey, *ApJ*, 696, 870.

Dressler, A. et al. 1987, Spectroscopy and photometry of elliptical galaxies. I—A new distance estimator, *ApJ*, 313, 42.

Dutta, H. et al. 2007, Distributed top-k outlier detection from astronomy catalogs using the DEMAC system, *SIAM Conference on Data Mining SDM07*, pp. 473–476, http://www.siam.org/proceedings/datamining/2007/dm07.php.

Dutta, H., et al. 2009, TagLearner: A P2P classifier learning system from collaboratively tagged text documents, in *IEEE International Conference on Data Mining, Workshops*, pp. 495–500, http://ieeexplore.ieee.org/xpl/freeabs_all.jsp?arnumber=5360457.

Eastman, T. et al. 2005, eScience and archiving for space science, *Data Science Journal*, 4(1), 67.

Fellhauer, M. and Heggie, D. 2005, An exact equilibrium model of an unbound stellar system in a tidal field, *A&A*, 435, 875.

Frieman, J.A. et al. 2008, The SDSS-II supernova survey: Technical summary, *AJ*, 135, 338.

Giannella, C., Dutta, H., Borne, K., Wolff, R., and Kargupta, H. 2006, Distributed data mining for astronomy catalogs, *SIAM Conference on Data Mining SDM06, Workshop on Scientific Data Mining*, http://www.siam.org/meetings/sdm06/workproceed/Scientific%20Datasets/.

Graham, M., Fitzpatrick, M., and McGlynn, T. (eds) 2007, *The National Virtual Observatory: Tools and Techniques for Astronomical Research*, ASP Conference Series, Vol. 382.

Gray, J. et al. 2002, Data mining the SDSS SkyServer Database, arXiv:cs/0202014v1.

Gray, J. 2003, *Online Science*, http://research.microsoft.com/en-us/um/people/gray/talks/kdd_2003_OnlineScience.ppt.

Gray, J. and Szalay, A. 2004, Microsoft technical report MSR-TR-2004-110, http://research.microsoft.com/apps/pubs/default.aspx?id=64540.

Gray, J. et al. 2005, Scientific data management in the coming decade, Microsoft Technical Report MSR-TR-2005-10, arXiv:cs/0502008v1.

Hey, T., Tansley, S., and Tolle, K. (eds) 2009, *The Fourth Paradigm: Data-Intensive Scientific Discovery* Richmond, WA: Microsoft Research.

Ivezic, Z., Axelrod, T., Becker, A., Becla, J., Borne, K., et al. 2008, Parametrization and classification of 20 billion LSST objects: Lessons from SDSS, *Classification and Discovery in Large Astronomical Surveys*, AIP Conference Proceedings, Vol. 1082, p. 359.

Jewitt, D. 2003, Project Pan-STARRS and the outer Solar System, *Earth, Moon, and Planets*, 92, 465.

Keller, S. C. et al. 2007, The SkyMapper Telescope and the Southern Sky Survey, *Publications of the Astronomical Society of Australia*, 24, 1.

Kegelmeyer, P. et al. 2008, Mathematics for Analysis of Petascale Data: Report on a Department of Energy Workshop, Technical report 2007-2008-4349P, http://www.er.doe.gov/ascr/ProgramDocuments/Docs/PetascaleDataWorkshopReport.pdf.

Koch, D. G. et al. 2010, Kepler Mission design, realized photometric performance, and early science, *ApJ*, 713, L79.

Mahootian, F. and Eastman, T. 2009, Complementary frameworks of scientific inquiry: Hypothetico-deductive, hypothetico-inductive, and observational-inductive, *World Futures: The Journal of General Evolution*, 65, 61.

Mahule, T., Borne, K., Dey, S., Arora, S., and Kargupta, H. 2010, PADMINI: A peer-to-peer distributed astronomy data mining system and a case study, in *NASA Conference on Intelligent Data Understanding*, https://c3.ndc.nasa.gov/dashlink/resources/220/, pp. 243–257.

Munn, J. et al. 2004, An improved proper-motion catalog combining USNO-B and the Sloan Digital Sky Survey, *AJ*, 127, 3034–3042.

Paczynski, B. 1996, Gravitational microlensing in the local group, *Annual Reviews of Astronomy and Astrophysics*, 34, 419.

Perryman, M. A. C. 1991, Hipparcos: Revised mission overview, *Advances in Space Research*, 11, 15.

Perryman, M. A. C. 2002, GAIA: An astrometric and photometric survey of our galaxy, *Astrophysics and Space Science*, 280, 1.

Rossi, G. and Sheth, R. K. 2008, Reconstructing galaxy fundamental distributions and scaling relations from photometric redshift surveys. Applications to the SDSS early-type sample, *MNRAS*, 387, 735.

Skrutskie, M. F. et al. 2006, The Two Micron All Sky Survey (2MASS), *AJ*, 131, 1163.

Strauss, M., The LSST Science Collaborations, and LSST Project 2009, LSST Science Book, v2.0, arXiv:0912.0201 http://www.lsst.org/lsst/scibook.

Szalay, A., Gray, J., and vandenBerg, J. 2002, Petabyte scale data mining: Dream or reality? in Tyson, J. A., Wolff, S., (eds), *Survey and Other Telescope Technologies and Discoveries*, Proceedings of the SPIE, Vol. 4836, pp. 333–338, arXiv:cs/0208013v1.

Williams, R., Bunn, S., and Seaman, R. 2010, *Hot-Wiring the Transient Universe*, Available from http://hotwireduniverse.org/.

Wolf, C. et al. 2004, A catalogue of the Chandra Deep Field South with multi-colour classification and photometric redshifts from COMBO-17, *A&A*, 421, 913.

York, D. G. et al. 2000, The Sloan Digital Sky Survey: Technical summary, *AJ*, 120, 1579.

CHAPTER 21

Multitree Algorithms for Large-Scale Astrostatistics

William B. March, Arkadas Ozakin, Dongryeol Lee, Ryan Riegel, and Alexander G. Gray

Georgia Institute of Technology

CONTENTS

21.1 Introduction   463
21.2 All Nearest Neighbors   465
21.3 Euclidean Minimum Spanning Trees   468
21.4 N-Point Correlation Functions   471
21.5 Kernel Density Estimation   473
21.6 Kernel Regression   477
21.7 Kernel Discriminant Analysis   478
21.8 Discussion   480
References   482

21.1 INTRODUCTION

Common astrostatistical operations. A number of common "subroutines" occur over and over again in the statistical analysis of astronomical data. Some of the most powerful, and computationally expensive, of these additionally share the common trait that they involve distance comparisons between all pairs of data points—or in some cases, all triplets or worse. These include:

- *All Nearest Neighbors (AllNN)*: For each query point in a dataset, find the k-nearest-neighbors among the points in another dataset—naively $O(N^2)$ to compute, for $O(N)$ data points.

- *n-Point Correlation Functions*: The main spatial statistic used for comparing two datasets in various ways—naively $O(N^2)$ for the 2-point correlation, $O(N^3)$ for the 3-point correlation, etc.

- *Euclidean Minimum Spanning Tree (EMST)*: The basis for "single-linkage hierarchical clustering," the main procedure for generating a hierarchical grouping of the data points at all scales, aka "friends-of-friends"—naively $O(N^2)$.

- *Kernel Density Estimation (KDE)*: The main method for estimating the probability density function of the data, nonparametrically (i.e., with virtually no assumptions on the functional form of the pdf)—naively $O(N^2)$.

- *Kernel Regression*: A powerful nonparametric method for regression, or predicting a continuous target value—naively $O(N^2)$.

- *Kernel Discriminant Analysis (KDA)*: A powerful nonparametric method for classification, or predicting a discrete class label—naively $O(N^2)$.

(Note that the "two datasets" may in fact be the same dataset, as in two-point *auto*correlations, or the so-called *monochromatic* AllNN problem, or the leave-one-out cross-validation needed in kernel estimation.)

The need for fast algorithms for such analysis subroutines is particularly acute in the modern age of exploding dataset sizes in astronomy. The Sloan Digital Sky Survey yielded hundreds of millions of objects, and the next generation of instruments such as the Large Synoptic Survey Telescope will yield roughly this number every week, resulting in billions of objects. At such scales, even linear-time analysis operations present challenges, particularly since statistical analyses are inherently interactive processes, requiring that computations complete within some reasonable human attention span. The quadratic (or worse) runtimes of straightforward implementations become quickly unbearable.

Examples of applications. These analysis subroutines occur ubiquitously in astrostatistical work. We list just a few examples. The need to cross-match objects across different catalogs has led to various algorithms, which at some point perform an AllNN computation. 2-point and higher-order spatial correlations for the basis of spatial statistics, and are utilized in astronomy to compare the spatial structures of two datasets, such as an observed sample and a theoretical sample, for example, forming the basis for two-sample hypothesis testing. Friends-of-friends clustering is often used to identify halos in data from astrophysical simulations. Minimum spanning tree properties have also been proposed as statistics of large-scale structure. Comparison of the distributions of different kinds of objects requires accurate density estimation, for which KDE is the overall statistical method of choice. The prediction of redshifts from optical data requires accurate regression, for which kernel regression is a powerful method. The identification of objects of various types in astronomy, such as stars versus galaxies, requires accurate classification, for which KDA is a powerful method.

Overview. In this chapter, we will briefly sketch the main ideas behind recent fast algorithms which achieve, for example, *linear* runtimes for pairwise-distance problems, or similarly dramatic reductions in computational growth. In some cases, the runtime orders for these algorithms are mathematically provable statements, while in others we have only conjectures backed by experimental observations for the time being.

We briefly review the fastest algorithmic approaches to these problems to date. We begin with the AllNN computational problem, and introduce the idea of space-partitioning tree data structures, upon which all of the algorithms in this article are based. Here we also introduce the idea of traversing two trees simultaneously. Moving to n-point correlations, we see that the idea can be generalized to that of traversing n trees simultaneously, for problems that are (naively) not just $O(N^2)$, but $O(N^n)$. To treat the difficult problem in KDE of needing to approximate continuous functions of distance, rather than being able to simply cut off after certain distances as in spatial correlations, we briefly describe the idea of series expansions for sums of kernel functions. Similar approaches apply to the problem of kernel regression. In KDA, we see that we need to only compute the greater of two kernel summations, which leads to a different tree traversal strategy.

21.2 ALL NEAREST NEIGHBORS

Dual-tree algorithms represent an efficient class of divide-and-conquer algorithms introduced in [1]. We give a brief overview of these algorithms, illustrated using our simplest problem: the AllNN problem.

We are given two sets of points: a set of *queries* $\mathcal{Q}$ and a set of *references* $\mathcal{R}$, both embedded in a metric space with distance function $d(\cdot, \cdot)$. Our task is to compute the nearest neighbor in $\mathcal{R}$ of each point in $\mathcal{Q}$:

Definition 21.1 All Nearest-Neighbor Problem.

$$\forall q_i \in \mathcal{Q}, \text{ find } r^*(q_i) = \arg \min_{r_j \in \mathcal{R}} d(q_i, r_j)$$

Throughout our discussion, we will assume that $\mathcal{Q}$ and $\mathcal{R}$ are subsets of $\mathbb{R}^d$ and that the distance function is the usual Euclidean distance. However, this assumption is only made to simplify the descriptions of the algorithms. With a suitable choice of tree, dual-tree algorithms can be adapted to points in any metric space.

The naive approach. The first algorithmic solution for the AllNN problem that comes to mind is simply to examine every query in turn, compute the distance between it and every reference, and keep the closest reference. We refer to this as the "naive" or "exhaustive" algorithm. Since this method must compute all of the $\binom{N}{2}$ pairwise distances, it requires $O(N^2)$ time.

A key observation. Consider a point q_i and a group of points R in Figure 21.1. Assume that we have constructed a "bounding box" for the points R—that is a rectangular volume containing all the points. Using this box, we can quickly compute two distance bounds: $d^l(q_i, R)$—the closest any point in R can be to q_i, and $d^u(q_i, R)$—the farthest possible pairwise distance. Furthermore, consider the case where we have already computed the distance between q_i and at least one point in the reference set. The smallest distance we have seen so far provides an upper bound on the distance between q_i and the true nearest neighbor. Call this candidate nearest-neighbor distance $\hat{d}(q_i)$. If $\hat{d}(q_i) < d^l(q_i, R)$, then we know that none of the points in R can be the true nearest neighbor of q_i. We can therefore avoid computing all of the pairwise distances $d(q_i, R)$.

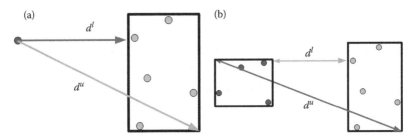

FIGURE 21.1 Pruning parts of the computation. (a) Comparing a point with a node. (b) Comparing two nodes.

In order to use this observation in a fast algorithm, we must consider possible query–reference pairs in a different order than in the naive method. Rather than iterate through the pairs one-by-one, we will employ a divide-and-conquer approach. Using a space-partitioning tree data structure (described in detail below), we can recursively split the computation into smaller, easier to handle subproblems. Using the bounding information in the tree, we can make use of the observation above. When the distance bounds indicate that no point in the reference subset being considered can be the nearest neighbor of the query, we can *prune* the subcomputation and avoid considering some of the pairs. If we are able to prune enough, we can improve substantially on the naive algorithm.

kd-trees. A *kd*-tree is a spatial data structure that is widely used for computational geometry problems [2]. We begin with the root node, which contains all of the data points and a bounding hyper-rectangle containing them. We obtain the next level of the tree by splitting the root's bounding box along its longest dimension. We now have two nodes, each of which contains (roughly) half the points. We shrink the bounding box for each of these nodes to be the smallest one that still contains all the node's points (see Figure 21.2). We repeat this procedure for all the nodes at a given level until each node contains some small number of points. At this point, we stop forming children and refer to the node as a *leaf*. The (one-time) cost of building the tree is $O(N \log N)$, and generally very fast in practice even for large datasets.

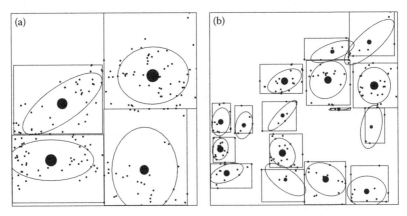

FIGURE 21.2 A *kd*-tree. (a) Level 2. (b) Level 4.

Other trees. Our examples focus on *kd*-trees. However, dual-tree algorithms can work with many kinds of space-partitioning trees. Ball trees [3] and quad- and oct-trees [4] are popular choices and are efficient for many problems. Cover trees [5] are a more recent data structure that combines efficiency in practice with theoretical performance guarantees. We use cover trees in the runtime theorems cited below.

Single-tree algorithm. We can now describe an improved divide-and-conquer algorithm using a *kd*-tree built on the set of references [2]. We consider each of the query points in turn along with the root node of our *kd*-tree. Assume that we are considering a point q_i and a node R, as in Figure 21.1. If the node is a leaf, we compute the distance between q_i and every point in R, and if one of these is the smallest distance seen so far, we update the candidate nearest neighbor. Otherwise, there are two possibilities. If the lower bound distance $d^l(q_i, R)$ is less than the candidate distance $\hat{d}(q_i)$, it is possible for R to contain the true nearest neighbor of q_i. Therefore, we recursively consider both children of R. If $d^l(q_i, R) > \hat{d}(q_i)$, then no point in R is the nearest neighbor of q_i. In this case, we can *prune* the rest of the tree under node R.

For a given query point q_i, this *single-tree algorithm* will find the nearest neighbor in $O(\log N)$ time. Since we must compute this algorithm for every query, the total running time is $O(N \log N)$.

An even better observation. We can make an even better use of the observation in Figure 21.1. Instead of considering the distance between a point and a bounding box, we can compute bounds on the distances between points in two bounding boxes—Figure 21.1b. We consider a query node Q and a reference node R, along with the largest candidate neighbor distance for the points in Q, $\hat{d}(Q) = \max_{q_i \in Q} \hat{d}(q_i)$. Now, if $d^l(Q, R)$ is greater than $\hat{d}(Q)$, no point in R can be the nearest neighbor of any point in Q, and we can prune both Q and R. When we pruned in the single tree algorithm, if the number of points in node R is $|R|$, we avoided computing $|R|$ distances. If we can make use of this new observation, we will save $|Q| \cdot |R|$ computations with each prune.

Dual-tree algorithm. We use this observation to improve on the single-tree algorithm. We construct two trees, one on references and one on queries (hence *dual-tree* algorithm). We consider pairs of nodes at a time and use the distance bounds shown in Figure 21.1b. If the bounds show that the nodes are too distant, then we can prune. Otherwise, we split one (or both) nodes, and recursively consider the two (or four) resulting pairs. We start by considering the root node twice.

Theoretical runtime guarantees. The cover tree data structure [5] is a space-partitioning tree that can provide rigorous bounds on runtimes for dual-tree algorithms. The *expansion constant* is a measure of the intrinsic dimensionality of a data set. The *degree of bichromaticity* bounds the difference in scale between the two sets. Both are defined in Ref. [6] along with the proof of the theorem.

Theorem 21.1 Given a reference set $\mathcal{R}$ of size N and expansion constant $c_{\mathcal{R}}$, a query set Q of size $\mathbf{O}(N)$ and expansion constant c_Q, and bounded degree of bichromaticity κ of the $(Q, \mathcal{R})$ pair, the **AllNN** algorithm (Alg. 21.1 computes the nearest neighbor in $\mathcal{R}$ of each point in Q in $\mathbf{O}(c_{\mathcal{R}}^{12} c_Q^{4\kappa} N)$ time.

Algorithm 21.1   AllNN (kd-tree Node Q, kd-tree Node R)

// $n(q_i)$ is the candidate nearest neighbor of point q_i

if Q and R are leaves **then**

3:      **for all** $q_i \in Q$ **do**

            for all $r_j \in R$ **do**

                if $d(q_i, r) < \hat{d}(q_i)$ **then**

6:                  $\hat{d}(q_i) = d(q_i, r_j); n(q_i) = r_j$

                end if

            end for

9:      **end for**

        $\hat{d}(Q) = \max_{q_i \in Q} \hat{d}(q_i)$

        else if $d^l(Q, R) > \hat{d}(Q)$ **then**

12:      // prune

        else

            AllNN($Q_{\text{left}}, R_{\text{left}}$)

15:      AllNN($Q_{\text{left}}, R_{\text{right}}$)

        AllNN($Q_{\text{right}}, R_{\text{left}}$)

        AllNN($Q_{\text{right}}, R_{\text{right}}$)

18:      $\hat{d}(Q) = \max\{\hat{d}(Q_{\text{left}}), \hat{d}(Q_{\text{right}})\}$

        end if

TABLE 21.1   Runtimes for Dual-Tree AllNN. The Data Are Generated from a Uniform Distribution in Two Dimensions

N	Dual-Tree (s)	Dual-Tree Speedup over Naive
100,000	1.038202	1.5×10^7
500,000	6.393331	9.3×10^7
1,000,000	14.078549	2.0×10^8
5,000,000	88.041680	1.3×10^9

Experimental performance. Practical speedups given by dual-tree methods over naive implementations are shown in Table 21.1, illustrating the significant gains that are typical of all the algorithms described in this chapter.

21.3 EUCLIDEAN MINIMUM SPANNING TREES

Given a set of points S in $\mathbb{R}^d$, the goal of the EMST problem is to find the lowest weight spanning tree in the complete graph on S with edge weights given by the Euclidean distances between points. A single-linkage hierarchical clustering (or "friends-of-friends" clustering) can be quickly obtained from the EMST by sorting the edges by increasing length, then deleting those edges longer than some threshold. The remaining components form the clusters.

Graph-based MST algorithms, such as the well-known Kruskal's [7] and Prim's [8] algorithms, have to consider the entire set of edges. In the Euclidean setting, there are $O(N^2)$ possible edges, rendering these algorithms too slow for large datasets.

Both algorithms maintain one or more components in spanning forest and use the cut between one component of the forest and the rest of the graph, adding the edges found in this way one at a time. In the Euclidean setting, there are $O(N^2)$ edges, rendering these algorithms too inefficient to be practical for large data. Thus, we need a way to efficiently identify edges (pairs of points) that belong in the MST.

Borůvka's algorithm. We focus on the earliest–known minimum spanning tree algorithm, Borůvka's algorithm, which dates from 1926. As we will show, this variation on the well-known Kruskal algorithm requires a computational step that is similar to the AllNN problem. Thus, we will be able to accelerate the computation with a dual-tree algorithm. See Ref. [9] for a translation and commentary on Borůvka's original papers.

As in Kruskal's algorithm, a minimum spanning forest is maintained throughout the algorithm. Kruskal's algorithm adds the minimum weight edge between any two components of the forest at each step, thus requiring $N - 1$ steps to complete. Borůvka's algorithm finds the minimum weight edge incident with each component, and adds all such edges. We define the *nearest-neighbor pair* of a component C as the pair of points $q_i \in C, r_j \notin C$ that minimizes $d(q_i, r_j)$. We denote the component containing a point q as $C^{(q)}$. Finding the nearest-neighbor pair for each component and adding the edges (q_i, r_j) to the forest is called a *Boruvka step*. Borůvka's algorithm then consists of forming an initial spanning forest with each point as a component and iteratively applying Boruvka steps until all components are joined. Since the number of components in the spanning forest is at least halved in each step, the algorithm requires at most $\log N$ iterations. Therefore, the total running time is $O(T(n) \log n)$, where $T(n)$ is the time for a Boruvka step. Therefore, we have reduced the EMST problem (and the associated clustering problem) to the Find Component Neighbors problem.

Definition 21.2 Find Component Neighbors Problem.

$$\text{For all components } C^{(q)}, \text{ find } (q^*, r^*) = \arg \min_{q_i \in C^{(q)}, r_j \notin C^{(q)}} d(q_i, r_j)$$

The Find Component Neighbors problem is clearly similar to the AllNN problem discussed previously. We update the AllNN algorithm (Alg. 21.1) so that it uses the same set for queries and references, only returns one nearest neighbor per component, and avoids returning pairs of points that belong to the same component. We can also add another opportunity to prune when all points in both Q and R belong to the same component. Our EMST algorithm, DUALTREEBORUVKA, uses a dual-tree method to find the nearest-neighbor pair for each component. Algorithm 21.2 gives the description of the outer loop. The subroutine UPDATETREE handles the propagation of any bounds up and down the tree and resets the upper bounds $d(C^{(q)})$ to infinity. We also make use of a disjoint set data structure [10] to store the connected components at each stage of the algorithm.

DUALTREEBORUVKA *on a **kd**-tree.* In the *kd*-tree version of DUALTREEBORUVKA, each node Q maintains an upper bound $\hat{d}(Q) = \max_{q_i \in Q} d(C^{(q_i)})$ for use in pruning. Each node can also store a flag to indicate whether all its points belong to the same component of the spanning

---

Algorithm 21.2   Dual-tree Borůvka (Tree root Q)

   $E = \emptyset$
   while $|E| < N - 1$ **do**
3:   **FindComponentNeighbors**(Q, Q)
     Add all new edges to E
     UpdateTree(Q)
6: **end while**

Algorithm 21.3   FindComponentNeighbors (kd-tree node Q, kd-tree node R)

---

   if $Q \bowtie R$ **then**
     return
3: **else if** $d(Q, R) > d(Q)$ **then**
     return
   else if Q and R are leaves **then**
6:   **for all** $q_i \in Q, r_j \in R, r_j \not\sim q_i$ **do**
       if $d(q_i, r_j) < d(C^{(q_i)})$ **then**
         $d(C^{(q_i)}) = d(q_i, r_j),\ e(C^{(q_i)}) = (q_i, r_j)$
9:     **end if**
     end for
     $d(Q) = \max_{q_i \in Q} d(C^{(q_i)})$
12: **else**
     FindComponentNeighbors$(Q_{\text{left}}, R_{\text{left}})$
     FindComponentNeighbors$(Q_{\text{right}}, R_{\text{left}})$
15:   **FindComponentNeighbors**$(Q_{\text{left}}, R_{\text{right}})$
     FindComponentNeighbors$(Q_{\text{right}}, R_{\text{right}})$
     $d(Q) = \max\{d(Q_{\text{left}}), d(Q_{\text{right}})\}$
18: **end if**

---

forest. We denote two points that belong to the same component by $q_i \sim r_j$. A node where all points belong to the same component is referred to as *fully connected*. If Q and R are fully connected and $q_i \sim r_j$ for $q_i \in Q, r_j \in R$, we write $Q \bowtie R$. With these records, we can prune when the either $d^l(Q, R) > \hat{d}(Q)$ or $Q \bowtie R$.

We can again prove a runtime theorem, using analysis similar to that for the AllNN problem. We require two additional parameters, the *cluster expansion* and *linkage expansion* constants, which describe the tightness of the clustering of the data in addition to the intrinsic dimension. The function $\alpha(N)$ is the inverse Ackermann function, which is formally unbounded, but extremely slow growing. It is much less than $\log N$. See Ref. [11] for the proof.

Theorem 21.2   For a set S of N points in a metric space with expansion constant c, cluster expansion constant c_p, and linkage expansion constant c_l, the DUALTREEBORUVKA algorithm using a cover tree requires

$$O(\max\{c^6, c_p^2 c_l^2\} \cdot c^{10} N \log N \alpha(N))$$

time.

21.4 N-POINT CORRELATION FUNCTIONS

We now consider the computation of n-**point Correlation Functions**, which for $n > 2$ will require a key generalization of our ideas so far, in order to efficiently deal not just with pairs of points, but arbitrary n-tuples.

Defining the n-point correlation function. Throughout, we assume that our data are generated by a homogeneous and isotropic spatial process—that is the distribution is the same at all points and in all directions. In this case, the probability of a point being located in some infinitesimal volume element depends only on the global mean density:

$$dP = \rho dV \tag{21.1}$$

Defining the 2-point correlation function. Consider two volume elements dV_1 and dV_2, separated by a distance r (Figure 21.3). The presence (or absence) of a point in dV_1 may be correlated, anticorrelated, or independent from a point in dV_2. This is captured more precisely by the *2-point correlation function*, $\xi(r)$:

$$dP_{12} = \rho^2 dV_1 dV_2 (1 + \xi(r)) \tag{21.2}$$

A positive value of ξ indicates correlation, negative anticorrelation, and zero independence.

Defining the 3-point correlation function. We can analogously define the 3-point correlation function, where r_{ij} denotes the distance between volume elements i and j (Figure 21.3).

$$dP_{123} = \rho^3 dV_1 dV_2 dV_3 (1 + \xi(r_{12}) + \xi(r_{23}) + \xi(r_{13}) + \zeta(r_{12}, r_{23}, r_{13})) \tag{21.3}$$

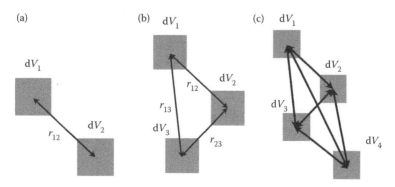

FIGURE 21.3 Definition of the n-point correlation functions. (a) 2-point; (b) 3-point; (c) 4-point.

The quantity in parentheses is sometimes called the *complete 3-point correlation function* and ζ is the *reduced 3-point correlation function*. We will often refer to ζ as simply the 3-point correlation function, since it will be the quantity of computational interest to us.

Defining the n-point correlation function. Higher-order correlation functions (such as the 4-point correlation in Figure 21.3c) are defined in the same fashion. The probability of finding n-points in a given configuration can be written as a summation over the n-point correlation functions. For example, in addition to the reduced 4-point correlation function, the complete 4-point correlation depends on the six 2-point terms (one for each pairwise distance), four 3-point terms (one for each triple of distances), and three products of two 2-point functions. These last terms arise from the possibility that of the four points, the positions of two pairs are correlated, but the pairs are not correlated with each other. The reduced 4-point function is in turn a function of the $\binom{4}{2} = 6$ pairwise distances.

Estimating the n-point correlation function. The n-point correlations are estimated using Monte Carlo sampling. For simplicity, we start with the 2-point correlation. We generate a random set of points from a Poisson distribution with the same (sample) density as our data. We choose a small bin size dr. Let $DD(r)$ denote the number of pairs of points (x_i, x_j) in our data whose pairwise distance $d(x_i, x_j)$ lies in the interval $(r - dr, r + dr)$. (DD denotes data-data pairs). Let $RR(r)$ be the number of points whose pairwise distances are in the same interval from the random sample (RR—random–random pairs). Then, a simple estimator for the 2-point correlation is

$$\hat{\xi}(r) = \frac{DD(r)}{RR(r)} - 1 \tag{21.4}$$

Many other estimators have been proposed with superior variance and error from edge effects. For example, the Landy–Szalay estimator has provably optimal variance [12]. It is given by

$$\hat{\xi}(r) = \frac{DD(r) - 2DR(r) + RR(r)}{RR(r)} \tag{21.5}$$

Here the notation $DR(r)$ denotes the number of pairs (x_i, y_j) within distance r where x_i is from the data and y_j is from the Poisson sample (DR—data-random pairs).

Estimators for the n-point case use counts of n-tuples of points of the form $D^{(i)}R^{(j)}(\cdot)$, where i and j indicate how many points come from the data and the random sample, respectively, and $i + j = n$. Rather than a single pairwise distance, we will need to specify $\binom{n}{2}$ distances. We refer to such a collection of distances as a *matcher* and write the lower- and upper-bound distances (analogous to $(r - dr, r + dr)$ from before) as $[l_{i,j}, u_{i,j}]$.

Therefore, our computational task is to count the number of points that "satisfy" the matcher.

Definition 21.3 *n-point correlation estimation.* Given Count the number of n-tuples of distinct points $(x_1, \ldots, x_n)$ with the property that there exists a permutation σ of $[1, \ldots, n]$ such that:

$$l_{\sigma(i),\sigma(j)} < d(x_i, x_j) < u_{\sigma(i),\sigma(j)} \tag{21.6}$$

for all distinct i and j.

Algorithm 21.4   NptEstimate (kd-tree Node $X^{(1)}, \ldots, kd$-tree Node $X^{(n)}$)

---

    if $X^{(1)}$ **and** $\cdots$ **and** $X^{(n)}$ are all leaves **then**
        Consider all n-tuples of points directly
3: **else if** CanPrune($X^{(1)}, \ldots, X^{(n)}$) **then**
        return 0
    else
6:    Let $R^{(k)}$ be the largest node
        $n_1 = $ NptEstimate($X^{(1)}, \ldots, R_{\text{left}}^{(k)}, \ldots, X^{(n)}$)
        $n_2 = $ NptEstimate($X^{(1)}, \ldots, R_{\text{right}}^{(k)}, \ldots, X^{(n)}$)
9:    **return** $n_1 + n_2$
    end if

Counting pairs of points. We first consider the task of estimating the 2-point correlation near a point r. Our computational task is to count pairs of points x_i, x_j such that $r - dr < d(x_i, x_j) < r + dr$. As in the AllNN case, we can construct a pruning rule. Given two tree nodes $X^{(1)}$ and $X^{(2)}$, we prune if either: (1) $d^u(X^{(1)}, X^{(2)}) < r - dr$ or 2) $d^l(X^{(1)}, X^{(2)}) > r + dr$. With this pruning rule, a dual-tree algorithm will quickly obtain the needed counts.

Efficiently estimating n-point correlation functions. We efficiently estimate the 2-point correlation by using two space-partitioning trees. Analogously, the key to a fast n-point estimation algorithm is to use n trees. Our algorithm for the n-point case uses the same basic observation: distance bounds obtained from bounding boxes can be used to quickly prune parts of the computation. Recall that for estimating the n-point correlation, we must count the number of n-tuples of (distinct) points that satisfy a given *matcher*—a set of $\binom{n}{2}$ pairwise distances that the points must satisfy in some order.

We consider n tree nodes at a time for our *multitree* algorithm. We compute an upper- and lower-bound distance for each pair of nodes using the bounding boxes in the same way as before. We use these $\binom{n}{2}$ pairwise upper and lower bounds obtained from the bounding boxes to check if the n-tuple of nodes can possibly satisfy the matcher. If not, we can immediately prune.

21.5 KERNEL DENSITY ESTIMATION

The problem of KDE will require extending our ideas in another direction. So far, we have presented dual-tree algorithms for computations with discrete pruning rules. When the distance bounds are large (or small) enough, we simply disregard any further computations between the pairs (or tuples). We now turn to a task where all query–reference pairs influence the final result. We will show how dual-tree methods can still be used to efficiently obtain approximate results with bounded error.

We are given a local *kernel function* $K_h(\cdot)$ centered upon each reference point in the reference set $\mathcal{R}$, and its scale parameter h (the "bandwidth"). The common choices for $K_h(\cdot)$ include the spherical, Gaussian, and Epanechnikov kernels. We are given the query dataset $\mathcal{Q}$ containing query points whose densities we want to predict. Our task is to compute the sum of the kernels at each of the query points.

Algorithm 21.5   CanPrune (kd-tree Node $X^{(1)}, \ldots, kd$-tree Node $X^{(n)}$)

---

for $i = 1 : n$ do
    for $j = i + 1 : n$ do
3:       for All permutations σ of $[1, \ldots, n]$ do
          if $d^u(X^{(i)}, X^{(j)}) < l_{\sigma(i),\sigma(j)}$ or $d^l(X^{(i)}, X^{(j)}) > u_{\sigma(i),\sigma(j)}$ then
            Mark σ as invalid
6:         end if
       end for
    end for
9: end for
    if All permutations invalid then
       Return true
12: else
       Return false
    end if

Definition 21.4   Kernel Density Estimation Problem. For all query points $q_i \in \mathcal{Q}$, compute:

$$\hat{\Phi}_h(q_i) = \frac{1}{|\mathcal{R}|} \sum_{r_j \in \mathcal{R}} \frac{1}{V_{Dh}} K_h \left(||q_i - r_j|| \right) \tag{21.7}$$

where $||q_i - r_j||$ denotes the Euclidean distance between the ith query point q_i and the jth reference point r_j, D the dimensionality of the data, $|\mathcal{R}|$ the size of the reference dataset, and $V_{Dh} = \int_{-\infty}^{\infty} K_h(z) - dz$, a normalizing constant depending on D and h.

If the kernel K_h has infinite support (such as the Gaussian kernel), then exactly evaluating the summation in Equation 21.7 will require us to compute every pairwise distance. This makes the quadratically scaling naive algorithm the fastest possible.

Computing approximate results. In order to achieve improved scaling, we will need to introduce *approximations* to dual-tree computations. We use the following two forms of error in our approximations:

Definition 21.5   (Bounding the absolute error) An approximation algorithm guarantees ϵ absolute error bound, if for each exact value $\Phi(q_i, \mathcal{R})$, it computes an approximation $\widetilde{\Phi}(q_i, \mathcal{R})$ such that $|\widetilde{\Phi}(q_i, \mathcal{R}) - \Phi(q_i, \mathcal{R})| \leq \epsilon$.

Definition 21.6   (Bounding the relative error) An approximation algorithm guarantees ϵ relative error bound, if for each exact value $\Phi(q_i, \mathcal{R})$, it computes an approximation $\widetilde{\Phi}(q_i, \mathcal{R})$ such that $|\widetilde{\Phi}(q_i, \mathcal{R}) - \Phi(q_i, \mathcal{R})| \leq \epsilon |\Phi(q_i, \mathcal{R})|$.

Guaranteeing error in dual-tree algorithms. We use the same distance bounds computed from the bounding boxes for nodes Q and R. We can generate an approximate value $\widetilde{\Phi}(Q, R)$

for the sum of kernels at points in R at each point in Q. We then use the distance bounds to bound the maximum error incurred by this approximation. By ensuring that the combined error from all prunes does not exceed the ϵ error allowed by our approximation scheme, we can efficiently find a valid approximation.

We conceptually allocate $\epsilon/|\mathcal{R}|$ error to each reference point. As long as the approximation for each reference at a query causes no more than this much error, the global error bound will be valid. We can therefore prune a pair of nodes Q and R if the error in our approximation is bounded by $|R|\epsilon/|\mathcal{R}|$.

Midpoint approximations. The simplest method for computing approximations is to use the midpoint. We assume that the kernel is monotonically decreasing. Then, we use the approximation:

$$\widetilde{K}(Q,R) = |R| \cdot \frac{1}{2}(K_h(d^{\mathrm{l}}(Q,R)) + K_h(d^{\mathrm{u}}(Q,R))) \tag{21.8}$$

In other words, we assume that all the reference points make the average possible contribution to the final result. This bounds the possible error at $0.5(K_h(d^{\mathrm{l}}(Q,R)) - K_h(d^{\mathrm{u}}(Q,R)))$.

Centroid approximation. We precompute the centroid of each node, denoted c_Q or c_R, by simply averaging the positions of the points. We compute the approximation by assuming that all points are located at the centroid:

$$\widetilde{K}(Q,R) = |R| \cdot K_h(d(c_Q, c_R)) \tag{21.9}$$

In Algorithm 21.6, the subroutine CANSUMMARIZE bounds the possible error. If the error is small enough to satisfy our global approximation scheme, we then compute the approximation (in SUMMARIZE), and apply it to the queries in Q.

Multipole approximation. We briefly sketch a much more sophisticated method of approximating kernel summations in the case of Gaussian kernels. We can view the centroid approximation above as the first term in a series expansion. By using more terms in this expansion, we can derive a method for kernel summations similar to the Fast Multipole Method [13] for electrostatic or gravitational interactions.

The basic idea is to express the kernel sum contribution of a reference node as a Taylor series of infinite terms and truncate it after some number of terms, given that the truncation error meets the desired absolute error tolerance. The following are two main types of Taylor-series representations for infinitely differentiable kernel functions $K_h(\cdot)$'s. The key difference between two representations is the location of the expansion center which is either in a reference region or a query region. The center of the expansion for both types of expansions is conveniently chosen to be the geometric center of the region.

Far-field expansion. A *far-field expansion* expresses the kernel sum contribution from the reference points in the reference node R for an arbitrary query point. It is expanded about $R.c$, a representative point of R (see Figure 21.4). Truncating to order p in D dimensions yields

---

Algorithm 21.6   KDE (*Q, R*)

---

if CANSUMMARIZE(Q, R, ϵ) **then**
  SUMMARIZE(Q, R)
else
  if Q is a leaf node **then**
    if R is a leaf node **then**
      Compute exhaustively
    else
      KDE(Q, R_{left}); **KDE**(Q, R_{right})
    end if
  else
    if R is a leaf node **then**
      KDE(Q_{left}, R); **KDE**(Q_{right}, R)
    else
      KDE($Q_{\text{left}}, R_{\text{left}}$); **KDE**($Q_{\text{left}}, R_{\text{right}}$)
      KDE($Q_{\text{right}}, R_{\text{left}}$); **KDE**($Q_{\text{right}}, R_{\text{right}}$)
    end if
  end if
end if

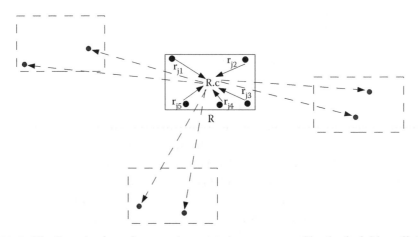

FIGURE 21.4 The Gaussian kernel sum series expansion represented by the farfield coefficients in R.

$$\widetilde{G}(q_i, R) = \sum_{r_j \in R} \prod_{d=1}^{D} \left(\sum_{\alpha[d] < p} \frac{1}{\alpha[d]!} \left(\frac{r_j[d] - R.c[d]}{\sqrt{2h^2}} \right)^{\alpha[d]} h_{\alpha[d]} \left(\frac{q_i[d] - R.c[d]}{\sqrt{2h^2}} \right) \right)$$

$$= \sum_{r_j \in R} \sum_{\alpha < p} \frac{1}{\alpha!} \left(\frac{r_j - R.c}{\sqrt{2h^2}} \right)^{\alpha} h_{\alpha} \left(\frac{q_i - R.c}{\sqrt{2h^2}} \right)$$

$$= \sum_{\alpha < p} \left[\sum_{r_j \in R} \frac{1}{\alpha!} \left(\frac{r_j - R.c}{\sqrt{2h^2}} \right)^{\alpha} \right] h_{\alpha} \left(\frac{q_i - R.c}{\sqrt{2h^2}} \right)$$

Local expansion. A *local expansion* is a Taylor expansion of the kernel sums about a representative point $Q.c$ in a query region Q. The kernel sum contribution of all reference points in a reference region R to a query point $q_i \in Q$ truncated at order p is given by

$$\widetilde{G}(q_i, R) = \sum_{r_j \in R} \prod_{d=1}^{D} \left(\sum_{n_d < p} \frac{(-1)^{n_d}}{n_d!} h_{n_d} \left(\frac{Q.c[d] - r_j[d]}{\sqrt{2h^2}} \right) \left(\frac{q_i[d] - Q.c[d]}{\sqrt{2h^2}} \right)^{\beta} \right)$$

$$= \sum_{r_j \in R} \sum_{\beta < p} \frac{(-1)^{\beta}}{\beta!} h_{\beta} \left(\frac{Q.c - r_j}{\sqrt{2h^2}} \right) \left(\frac{q_i - Q.c}{\sqrt{2h^2}} \right)^{\beta}$$

$$= \sum_{\beta < p} \left[\sum_{r_j \in R} \frac{(-1)^{\beta}}{\beta!} h_{\beta} \left(\frac{Q.c - r_j}{\sqrt{2h^2}} \right) \right] \left(\frac{q_i - Q.c}{\sqrt{2h^2}} \right)^{\beta}$$

We can then define far-to-local translation operators, analogous to those used in the FMM. We traverse the tree, building up farfield expansions for reference nodes. We then translate them to query nodes using the operator, and pass the resulting local expansion down the tree. Further details can be found in Refs. [14,15].

21.6 KERNEL REGRESSION

The simple, but powerful, idea of KDE can be extended to the problem of regression, leading to *Kernel regression* methods. One of the simplest nonparametric regression methods is *Nadaraya–Watson regression* (NWR). Roughly speaking, NWR is a generalization of moving averages. Suppose we are given a reference set that consists of pairs of real input/output values $\{(x_j^{(r)}, y_j^{(r)})\}, j = 1, \ldots, N_r$, where $y_j^{(r)} = g(x_j^{(r)}) + \epsilon_j$, g is an unknown function, and ϵ_j is noise, which we assume to be independent and identically distributed for each j. We would like to estimate the value of g at a query point $x_i^{(q)}$. Nadaraya–Watson estimate $\hat{y}_i^{(q)}$ of $g(x_i^{(q)})$ is a weighted average of the $y_j^{(r)}$s, with weights w_j being given in terms of the distances $|x_j^{(r)} - x_i^{(q)}|$.

Definition 21.7 *Nadaraya–Watson Regression.* Given reference inputs and outputs $(x_i^{(r)}, y_i^{(r)})$ and a set of query points $\{x_i^{(q)}\}$, compute an output for each query using

$$\hat{y}_i^{(q)} = \sum_j w_j y_j^{(r)} = \frac{\sum_j y_j^{(r)} K_h(\|x_i^{(q)} - x_j^{(r)}\|)}{\sum_j K_h(\|x_i^{(q)} - x_j^{(r)}\|)} \tag{21.10}$$

where $K_h(x)$ is a kernel function which we will assume to be positive and monotonically decreasing for $x \geq 0$.

Extensions. A straightforward generalization consists of allowing the bandwidth to depend on the reference point, which is useful when the data are not uniformly distributed. This approach is called *variable-bandwidth* NWR, and is given by

$$\hat{y}_i^{(q)} = \frac{\sum_j y_j^{(r)} K_{h_j}(|x_i^{(q)} - x_j^{(r)}|)}{\sum_j K_{h_j}(|x_i^{(q)} - x_j^{(r)}|)}$$

An alternative way of looking at NWR is through mean-squared error (MSE). Suppose we are trying to fit a function $\hat{g}^{(q)}(x_i^{(q)})$ to the reference data around a given query point $x_i^{(q)}$. For this purpose, consider a *weighted* mean-squared error (WMSE) over the reference set $\{(x_j^{(r)}, y_j^{(r)})\}$, given by $(1/N_r) \sum_j K_h(|x_j^{(r)} - x_i^{(q)}|)(y_j^{(r)} - \hat{g}^{(q)}(x_j^{(r)}))^2$, where K_h is once again a kernel function. If $\hat{g}_q$ is assumed to be a constant function, the choice that minimizes this WMSE turns out to be just the Nadaraya–Watson estimator at $x_i^{(q)}$, as can be seen by a differentiation. This suggests a generalization: the functions that we use for the local fit around a query point can be taken from a more general family, such as pth order polynomials.

This latter approach is called *local polynomial regression* (LPR). More explicitly, the LPR estimate of the underlying function at a given query point $x_i^{(q)}$ is done by first fitting a polynomial to the reference set around $x_i^{(q)}$ by finding the polynomial coefficients β_i that minimize the WMSE over the reference set,

$$\frac{1}{N_r} \sum_j K_h\left(|x_j^{(r)} - x_i^{(q)}|\right) \left[y_j^{(r)} - \left(\beta_0 + \beta_1 x_j^{(r)} + \beta_2 \left(x_j^{(r)}\right)^2 + \cdots + \beta_p \left(x_j^{(r)}\right)^p\right)\right]^2$$

and then plugging in $x_i^{(q)}$ as the argument of the local polynomial, that is

$$\hat{g}_q(x_i^{(q)}) = \left(\beta_0 + \beta_1 x_i^{(q)} + \beta_2 \left(x_i^{(q)}\right)^2 + \cdots + \beta_p \left(x_i^{(q)}\right)^p\right)$$

The case $p = 0$ is NWR, and the case $p = 1$, which is usually considered to be a sufficient improvement over NWR, is called local linear regression (LLR). The main advantage of LLR over NWR is related to the boundary behavior: NWR is particularly susceptible to boundary bias, which causes the estimates for query points near the boundary of the reference set to have low accuracy due to the nonsymmetrical distribution of the reference points near the boundary.

All of these methods ultimately depend on kernel summations. We can therefore adapt our dual-tree KDE algorithm for this problem as well. We omit details for space considerations, but the pruning rules can be updated in a similar fashion to our KDE algorithm. The full algorithm will be published elsewhere.

21.7 KERNEL DISCRIMINANT ANALYSIS

The idea of KDE can also be applied to the problem of classification, leading to the technique of KDA.

Bayes' rule in classification. The optimal classifier on M classes assigns observation $x \in \mathcal{X}$ to the class C_k, $1 \le k \le M$, which has the greatest posterior probability $P(C_k|x)$ [16]. Applying Bayes' rule,

$$P(A|B) = \frac{P(B|A)P(A)}{P(B)} \tag{21.11}$$

we assign x to C_k if, for all $l \ne k$, $1 \le l \le M$, we have

$$f(x|C_k)P(C_k) > f(x|C_l)P(C_l) \tag{21.12}$$

where $f(x|C)$ denotes the probability density of data sampled from C and $P(C)$ is the prior probability of C. (Note that Bayes' rule's denominator $f(x)$ cancels from either side.) It is typically given that $\sum_{k=1}^{M} P(C_k) = 1$, that is that there are no unexplained classes, and accordingly, that $f(x) = \sum_{k=1}^{M} f(x|C_k)P(C_k)$. In the case that $M = 2$, this implies that it is equivalent to classify x as C_1 when

$$P(C_1|x) = \frac{f(x|C_1)P(C_1)}{f(x|C_1)P(C_1) + f(x|C_2)P(C_2)} > 0.5 \tag{21.13}$$

In place of the unknown values of $f(x|C)$, KDA uses kernel density estimates of the form

$$\widehat{f_h}(x|C) = \frac{1}{N} \sum_{i=1}^{N} K_h(x, z_i) \tag{21.14}$$

Avoiding full summation. A key observation is that our computational problem can be solved more efficiently than by actually computing the full kernel summation needed for each class—to determine the class label for each query point, we need only to determine the greater of the two kernel summations.

In the $M = 2$ case, we can replace the 0.5 in Equation 21.13 with an alternate confidence threshold t. Rather than performing optimal classification, this has the effect of identifying points with very high (or low) probability of being in C_1, which can be a useful surrogate for fully computed class probabilities. Simple algebra and incorporation of the above gives us

$$(1 - t)\widehat{f_{h_1}}(x|C_1)P(C_1|y) > t\widehat{f_{h_2}}(x|C_2)P(C_2|y) \tag{21.15}$$

This gives us an additional opportunity to prune. If we can prove that Equation 21.15 holds for all the queries in a node, then we can prune all future computations on those queries, since their class is already known.

In order to facilitate this kind of pruning, we employ an alternative tree traversal pattern. Rather than the depth-first traversal in our previous examples, we use a hybrid bread and depth expansion. Query nodes are traversed in the same depth first expansion as before. References are expanded in a bread-first fashion. So, rather than a single query–reference pair, we consider a query along with a list of references. We consider

---

Algorithm 21.7   KDA (Query Q, References $\{R^{(i)}\}$)

---

for all $R^{(i)}$ **do**

   if CanSummarize$(Q, R^{(i)})$ **then**

      Summarize

   else if Q and $R^{(i)}$ are leaves **then**

      Compute interactions directly

   end if

end for

if $(1 - t)\Phi_1^l \Pi^l(X) > t\Phi_2^u(1 - \Pi^l(X))$ **then**

   label X as C_1; **return**

end if

if $t\Phi_2^l(1 - \Pi^u(X)) > (1 - t)\Phi_1^u \Pi^u(X)$ **then**

   label X as C_2; **return**

end if

KDA$(Q_{\text{left}}, \textbf{SplitAll}(\{R^{(i)}\}))$

KDA$(Q_{\text{right}}, \textbf{SplitAll}(\{R^{(i)}\}))$

---

the possible contribution to the kernel sums from all the references. If the upper and lower bounds are tight enough to allow pruning, we stop. Otherwise, we can either split the query and consider the children independently, or split *all* the references. Either case allows tighter bounds and eventual pruning of most computations. See Ref. [17] for further details.

21.8 DISCUSSION

History. The algorithms described for the AllNN and n-point correlation functions were first introduced in Ref. [1]. The n-point algorithm was described in the astronomy community in Ref. [18]. The dual-tree approach to KDE was explored in Ref. [19]. The multipole expansion-based approach for Gaussian summations was first proposed in Ref. [20] without a tree; the more efficient fully hierarchical (tree-based) approach was first introduced in Refs. [21,22]. The algorithm for KDA was introduced in Ref. [17]. The EMST algorithm was introduced in Ref. [11]. Some earlier approaches for computational physics problems [23] can be seen as kinds of dual-tree algorithms.

Status of formal results. In all these problems, the algorithms described are the fastest overall, in practice, for general-dimension problems. So far, runtimes have been formally established for the AllNN and KDE problems [24]. For KDE and n-point correlations, conjectures for favorable runtimes have been made, based on experimental behaviors, but remain to be established.

High-dimensional data. In high-dimensional data, the efficacy of the bounds used in such tree-based algorithms declines—however, this degradation occurs as a function of some notion of the *intrinsic* dimension of the data, not its explicit dimension. This concept is made formal in at least one sense, in terms of the expansion constant [5].

For such high-dimensional settings, adding a Monte Carlo idea yields orders of magnitude in speedup, at the cost of nondeterminism. By adding steps within the algorithms involving estimation of quantities using subsamples of the data, the algorithms can achieve large speedups even in very high dimensionalities, though the results are only guaranteed to hold with high (prespecified) probability. Such algorithms have been achieved for nearest-neighbor [25] and kernel summation problems [26].

Why not other statistical methods? Note that in some cases, the statistical methods listed here are not the most powerful which exist. Nonlinear support vector machines (SVMs), typically with a radial basis function (Gaussian) kernel, typically edge out even other nonparametric classification methods in accuracy. However, for the large-scale datasets of modern astronomy, the worst-case $O(N^3)$ runtime of SVM training is generally intractable. KDA, whose final functional form is similar to that of an SVM with radial basis function kernel, is often very close or comparable to it in accuracy, with the advantage that it becomes tractable using the fast algorithm described. Gaussian process regression represents a more powerful regression model in principle than the simplest kernel regression models, but is $O(N^3)$ as well. Higher-order and other extensions of kernel regression are likely close to or comparable in accuracy to Gaussian process regression, with the advantage that they can in principle be made fast using the kinds of tools described here. Methods superior in performance to KDE for nonparametric density estimation are more difficult to find—at least one has been shown recently [27], but its runtime is $O(N^4)$. In summary, it is difficult to significantly beat the nonparametric statistical methods shown here in statistical performance, but at the same time their relatively simple form makes them amenable to orders of magnitude in algorithmic speedup. We are also currently working on tree-based algorithms for the larger-than-random access memory (RAM) case (both out-of-core and parallel/distributed) which is to appear shortly.

Scientific results enabled. The acceleration of the computation of these basic statistical building blocks has opened up the possibility of much larger-scale astrostatistical analyses than have ever been possible before, on very wide fields or even the entire sky, which can in some cases represent a qualitative scientific leap. We give a few examples where the fast algorithms described here can enable first-of-a-kind scientific results. Large-scale exact 2-point correlations helped to validate evidence for dark energy using the Wilkinson microwave anisotropy probe (WMAP) cosmic microwave background (CMB) data based on the Integrated Sachs-Wolfe effect [28,29]. As cosmological parameter ranges tighten, to isolate increasingly subtle effects, 2-point correlations will no longer be enough, and 3-point or higher statistics will be needed. The largest 3-point correlation computation to date [30] was achieved using these fast algorithms. Very wide field studies of the density functions of galaxies of the kind done in Ref. [31] will lead to more greater understanding of these and other object types. The ability to perform large-scale nonparametric classification via KDA led to a significant leap in the size and accuracy of quasar catalogs [32,33]. Such powerful new quasar maps led in turn to the first wide-area validation of the cosmic magnification predicted by relativity [34,35]. We predict that the ability to perform large-scale nonparametric regression will significantly improve the quality and size of photometric redshift catalogs, which will lead in turn to numerous further possibilities in cosmology.

REFERENCES

1. A. Gray, A. W. Moore, 2001. N-body problems in statistical learning, in: T. K. Leen, T. G. Dietterich, V. Tresp (eds), *Advances in Neural Information Processing Systems* 13 (Dec 2000), MIT Press, Cambridge, MA.
2. J. H. Friedman, J. L. Bentley, R. A. Finkel, 1977. An algorithm for finding best matches in logarithmic expected time, *ACM Trans. Math. Softw.* 3(3), 209–226.
3. S. M. Omohundro, 1989. Five balltree construction algorithms, Technical Report TR-89-063, International Computer Science Institute.
4. H. Samet, 1984. The quadtree and related hierarchical data structures, *ACM Comput. Surv.* 16(2), 187–260.
5. A. Beygelzimer, S. Kakade, J. Langford, 2006. Cover trees for nearest neighbor, *23rd International Conference on Machine Learning*, Pittsburgh, PA, pp. 97–104.
6. P. Ram, D. Lee, W. B. March, A. G. Gray, 2009. Linear time algorithms for pairwise statistical problems, *Adv. Neural Inf. Process. Syst.* 23.
7. J. B. Kruskal, 1956. On the shortest spanning subtree of a graph and the traveling salesman problem, *Proc. Am. Math. Soc.* 7, 48–50.
8. R. C. Prim, 1957. Shortest connection networks and some generalizations, *Bell Syst. Tech. J.* 36, 1389–1401.
9. J. Nesetril, 2001. Otakar Boruvka on the minimum spanning tree problem, *Discrete Math.* 233, 3–36.
10. R. Tarjan, Data structures and network algorithms, 1988. *Soc. Ind. Appl. Math.* 44.
11. W. March, P. Ram, A. Gray, 2010. Fast Euclidean minimum spanning tree: Algorithm, analysis, and applications, in: *Proceedings of the 16th ACM SIGKDD International Conference on Knowledge Discovery and Data Mining, ACM*, Washington DC, pp. 603–612.
12. S. Landy, A. Szalay, 1993. Bias and variance of angular correlation functions, *Astrophys. J.* 412, 64–71.
13. L. Greengard, V. Rokhlin, 1987. A fast algorithm for particle simulations, *J. Computat. Phys.* 73, 325–248.
14. D. Lee, A. G. Gray, A. W. Moore, 2006. Dual-tree fast Gauss transforms, in: Y. Weiss, B. Schölkopf, J. Platt (eds), *Advances in Neural Information Processing Systems 18*, MIT Press, Cambridge, MA, pp. 747–754.
15. D. Lee, A. G. Gray, 2006. Faster Gaussian summation: Theory and experiment, in: *Proceedings of the Twenty-Second Conference on Uncertainty in Artificial Intelligence*, Cambridge, MA, p. 1.
16. C. Rao, 1973. *Linear Statistical Inference and Its Applications*, John Wiley & Sons.
17. R. Riegel, A. Gray, G. Richards, 2008. Massive-scale kernel discriminant analysis: Mining for quasars, in: *SIAM International Conference on Data Mining*, Atlanta, GA, pp. 208–218.
18. A. Moore, A. Connolly, C. Genovese, A. Gray, L. Grone, N. Kanidoris II, R. Nichol, J. et al. 2001. Fast algorithms and efficient statistics: N-point correlation functions, *Min. Sky* 71–82.
19. A. G. Gray, A. W. Moore, 2003. Nonparametric density estimation: Toward computational tractability, in: *SIAM International Conference on Data Mining* (SDM), San Francisco, CA, vol. 3.
20. L. Greengard, J. Strain, 1991. The fast Gauss transform, *SIAM J. Sci. Statist. Comput.* 12(1), 79–94.
21. D. Lee, A. Gray, A. Moore, Dual-tree fast Gauss transforms, in: Y. Weiss, B. Schölkopf, J. Platt (eds), 2006. *Advances in Neural Information Processing Systems 18*, MIT Press, Cambridge, MA, pp. 747–754.
22. D. Lee, A. Gray, 2006. Faster Gaussian summation: Theory and experiment, in: *Proceedings of the Twenty-Second Conference on Uncertainty in Artificial Intelligence*, Cambridge, MA.

23. A. Appel, 1985. An efficient program for many-body simulation, *SIAM J. Sci. Statist. Comput.* 6, 85.

24. P. Ram, D. Lee, W. March, A. G. Gray, 2010. Linear-time algorithms for pairwise statistical problems, in: *Advances in Neural Information Processing Systems (NIPS)* 22 (Dec 2009), MIT Press.

25. P. Ram, D. Lee, H. Ouyang, A. G. Gray, 2010. Rank-approximate nearest neighbor search: Retaining meaning and speed in high dimensions, in: *Advances in Neural Information Processing Systems (NIPS)* 22 (Dec 2009), MIT Press.

26. D. Lee, A. G. Gray, 2009. Fast high-dimensional kernel summations using the Monte Carlo multipole method, in: *Advances in Neural Information Processing Systems (NIPS)* 21 (Dec 2008), MIT Press.

27. R. Sastry, A. G. Gray, 2011. Convex adaptive Kernel estimation, in: *Proceedings of the Fourteenth International Conference on Artificial Intelligence and Statistics, 2011*, Fort Lauderdale, Florida, USA, vol. 15.

28. C. Seife, 2003. Breakthrough of the year: Illuminating the dark universe, *Science* 302 (5653), 2017–2172.

29. T. Giannantonio, R. Crittenden, R. Nichol, R. Scranton, G. Richards, A. Myers, R. Brunner, A. G. Gray, A. Connolly, D. Schneider, 2006. A high redshift detection of the integrated Sachs-Wolfe effect, *Phys. Rev. D*, 74, 063520 (pages).

30. G. Kulkarni, R. Nichol, R. Sheth, H. Seo, D. Eisenstein, A. G. Gray, 2007. The three-point correlation function of luminous red galaxies in the Sloan digital sky survey, *MNRAS* 378, 1196–1206.

31. M. Balogh, V. Eke, C. Miller, I. Lewis, R. Bower, W. Couch, R. Nichol, et al. 2004. Galaxy ecology: Groups and low-density environments in the SDSS and 2dFGRS, *MNRAS*, 348, 257–269.

32. G. Richards, R. Nichol, A. G. Gray, R. Brunner, R. Lupton, D. Vanden Berk, S. Chong et al. 2004. Efficient photometric selection of quasars from the Sloan Digital Sky Survey: 100,000 $z < 3$ quasars from data release one, *ApJS* 155, 257–269.

33. G. T. Richards, A. D. Myers, A. G. Gray, R. N. Riegel, R. C. Nichol, R. J. Brunner, A. S. Szalay, D. P. Schneider, S. F. Anderson, 2009. Efficient photometric selection of quasars from the Sloan Digital Sky Survey II. ∼1,000,000 quasars from data release six, *ApJS* 180, 67–83.

34. M. Peplow, 2005. Quasars reveal cosmic magnification, *Nature*, doi:10.1038/news050425-2.

35. R. Scranton, B. Menard, G. Richards, R. Nichol, A. Myers, J. Bhuvnesh, A. G. Gray, et al. 2005. Detection of cosmic magnification with the Sloan Digital Sky Survey, *ApJ* 633, 589–602.

III

Machine Learning Methods

CHAPTER 22

Time–Frequency Learning Machines for Nonstationarity Detection Using Surrogates

Pierre Borgnat and Patrick Flandrin
École Normale Supérieure de Lyon

Cédric Richard and André Ferrari
Université de Nice Sophia Antipolis

Hassan Amoud and Paul Honeine
Université de Technologie de Troyes

CONTENTS

22.1	Introduction	488
22.2	Revisiting Stationarity	488
	22.2.1 A Time–Frequency Perspective	488
	22.2.2 Stationarization via Surrogates	489
22.3	Time–Frequency Learning Machines	491
	22.3.1 Reproducing Kernels	492
	22.3.2 The Kernel Trick, the Representer Theorem	492
	22.3.3 Time–Frequency Learning Machines: General Principles	493
	22.3.4 Wigner Distribution versus Spectrogram	495
22.4	A Nonsupervised Classification Approach	495
	22.4.1 An Overview on One-Class Classification	496
	22.4.2 One-Class SVM for Testing Stationarity	496
	22.4.3 Spherical Multidimensional Scaling	498
22.5	Illustration	498
22.6	Conclusion	500
	References	501

22.1 INTRODUCTION

Time–frequency representations provide a powerful tool for nonstationary signal analysis and classification, supporting a wide range of applications [12]. As opposed to conventional Fourier analysis, these techniques reveal the evolution in time of the spectral content of signals. In Ref. [7,38], time–frequency analysis is used to test stationarity of any signal. The proposed method consists of a comparison between global and local time–frequency features. The originality is to make use of a family of stationary surrogate signals for defining the null hypothesis of stationarity and, based upon this information, to derive statistical tests. An open question remains, however, about how to choose relevant time–frequency features.

Over the last decade, a number of new pattern recognition methods based on reproducing kernels have been introduced. These learning machines have gained popularity due to their conceptual simplicity and their outstanding performance [30]. Initiated by Vapnik's support vector machines (SVM) [35], they offer now a wide class of supervised and unsupervised learning algorithms. In Ref. [17–19], the authors have shown how the most effective and innovative learning machines can be tuned to operate in the time–frequency domain. This chapter follows this line of research by taking advantage of learning machines to test and quantify stationarity. Based on one-class SVM, our approach uses the entire time–frequency representation and does not require arbitrary feature extraction. Applied to a set of surrogates, it provides the domain boundary that includes most of these stationarized signals. This allows us to test the stationarity of the signal under investigation.

This chapter is organized as follows. In Section 22.2, we introduce the surrogate data method to generate stationarized signals, namely, the null hypothesis of stationarity. The concept of time–frequency learning machines is presented in Section 22.3, and applied to one-class SVM in order to derive a stationarity test in Section 22.4. The relevance of the latter is illustrated by simulation results in Section 22.5.

22.2 REVISITING STATIONARITY

22.2.1 A Time–Frequency Perspective

Harmonizable processes define a general class of nonstationary processes whose spectral properties, which are potentially time dependent, can be revealed by suitably chosen time-varying spectra. This can be achieved, for example, with the Wigner–Ville spectrum (WVS) [12], defined as

$$W_x(t,f) := \int E\left\{x\left(t + \tau/2\right) x^*\left(t - \tau/2\right)\right\} e^{-j2\pi f\tau}\, d\tau \qquad (22.1)$$

where x stands for the analyzed process. Such a definition guarantees furthermore that second-order stationary processes, which are a special case of harmonizable processes, have a time-varying spectrum that simply reduces to the classical (stationary, time-independent) power spectrum density (PSD) at every time instant.

In practice, the WVS has to be estimated on the basis of a single observed realization, a standard procedure amounting to make use of spectrograms (or multitaper variations [4]) defined as

$$S_x(t,f;h) := \left| \int x(\tau)\, h^*(\tau - t)\, e^{-j2\pi f\tau}\, d\tau \right|^2 \qquad (22.2)$$

where h stands for some short-time observation window. In this case too, the concept of stationarity still implies time independence, the time-varying spectra identifying, at each time instant, to some frequency-smoothed version of the PSD. It follows, however, from this time–frequency (TF) interpretation that, from an operational point of view, stationarity cannot be an *absolute* property. A more meaningful approach is to switch to a notion of relative stationarity to be understood as follows: when considered over a given observed timescale, a process will be referred to as *stationary relative to this observation scale* if its time-varying spectrum undergoes no evolution or, in other words, if the local spectra $S_x(t_n, f; h)$ at all different time instants $\{t_n; n = 1, \dots, N\}$ are statistically similar to the global (average) spectrum

$$\bar{S}_x(t_n, f; h) := \frac{1}{N} \sum_{n=1}^{N} S_x(t_n, f; h) \qquad (22.3)$$

obtained by marginalization.

Based on this key point, one can imagine to design stationarity tests via some comparison between local and global features within a given observation scale [7] or, more generally, to decide whether an actual observation differs significantly from a stationary one within this time span. The question is therefore to have access to some stationary reference that would share with the observation the same global frequency behavior, while having a time-varying spectrum constant over time. As proposed in Ref. [7], an answer to this question can be given by the introduction of surrogate data.

22.2.2 Stationarization via Surrogates

The general idea is to build a reference of stationarity directly from the signal itself, by generating a family of stationarized signals which have the same density spectrum as the initial signal. Indeed, given a density spectrum, nonstationary signals differ from stationary ones by temporal structures encoded in the spectrum phase. The surrogate data technique [34] is an appropriate solution to generate a family of stationarized signals, by keeping unchanged the magnitude of the Fourier transform $X(f)$ of the initial signal $x(t)$, and replacing its phase by an i.i.d. one. Each surrogate signal $x_\ell(t)$ results from the inverse Fourier transform of the modified spectrum, namely,

$$X_\ell(f) = |X(f)|\, e^{j\phi_\ell(f)}$$

with $\phi_\ell(f)$ drawn from the uniform distribution over the interval $[-\pi, \pi[$. This leads to as many stationary surrogate signals, $x_1, \dots, x_n$, as phase randomizations $\phi_1(f), \dots, \phi_n(f)$ are operated. An illustration of the effectiveness of this approach in terms of its TF interpretation is given in Figure 22.1.

It has been first proved in Ref. [7] that surrogates are wide-sense stationary, that is, their first- and second-order moments are time-shift invariant. More recently, it has been

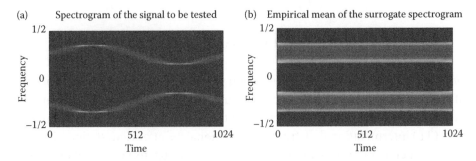

FIGURE 22.1 Spectrogram of an frequency-modulated (FM) signal (a) and empirical mean of the spectrograms of its surrogates (b).

established in Ref. [26] that surrogates are strict-sense stationary, the proof proceeding as follows. Let us derive the invariance with respect to time shifts of the $(L + 1)$-th order cumulant of the surrogate signal $x(t)$, where the subscript ℓ has been dropped for clarity

$$c(t; t_1, \ldots, t_L) = \text{cum}(x^{\epsilon_0}(t), x^{\epsilon_1}(t + t_1), \ldots, x^{\epsilon_L}(t + t_L))$$

where $\epsilon_i = \pm 1$ and $x^{\epsilon_i}(t) = x^*(t)$ when $\epsilon_i = -1$ (we suggest the reader to refer, for example, [1], for a detailed description of the tools related to high-order analysis of complex random processes). Let $\Phi(u) = E[e^{ju\phi}]$ be the characteristic function of the random phase ϕ. As it is uniformly distributed over $[-\pi, \pi[$, note that

$$\Phi(k) = 0, \forall k \in \mathcal{Z}^* \tag{22.4}$$

Using the multilinearity of the cumulants, we have

$$c(t; t_1, \ldots, t_L) = \int |X(f_0)| \cdots |X(f_L)| \, C(f_0, \ldots, f_L) \, e^{j2\pi t \sum_{i=0}^{L} \epsilon_i f_i} \, e^{j2\pi \sum_{i=1}^{L} \epsilon_i t_i f_i} \, df_0 \cdots df_L$$

where $C(f_0, \ldots, f_L) = \text{cum}(e^{j\epsilon_0 \phi(f_0)}, \ldots, e^{j\epsilon_L \phi(f_L)})$. If one variable f_i is different from the others, the corresponding random variable $e^{j\epsilon_0 \phi(f_i)}$ is independent from the others and $C(f_0, \ldots, f_L) = 0$. Consequently, the joint cumulant of the surrogate simplifies to

$$c(t; t_1, \ldots, t_L) = C_{L+1} \int |X(f)|^{L+1} e^{j2\pi ft \sum_{i=0}^{L} \epsilon_i} e^{j2\pi f \sum_{i=1}^{L} \epsilon_i t_i} \, df$$

where $C_{L+1} = \text{cum}(e^{j\epsilon_0 \phi}, \ldots, e^{j\epsilon_L \phi})$. Application of the Leonov–Shiryaev formula to this cumulant leads to

$$C_{L+1} = \sum_{\mathcal{P}} (|\mathcal{P}| - 1)! \, (-1)^{|\mathcal{P}|-1} \prod_{B \in \mathcal{P}} \Phi(\sum_{i \in B} \epsilon_i) \tag{22.5}$$

where $\mathcal{P}$ runs through the list of all the partitions of $\{0, \ldots, L\}$ and B runs through the list of all the blocks of the partition $\mathcal{P}$. This expression can be simplified using Equation 22.4

and noting that $\sum_{i \in B} \epsilon_i \in \mathcal{Z}$. Consequently, $\Phi\left(\sum_{i \in B} \epsilon_i\right)$ is nonzero, and necessarily equal to 1, if and only if $\sum_{i \in B} \epsilon_i = 0$.

- If L is even, whatever $\mathcal{P}$, at least one block B of $\mathcal{P}$ has an odd cardinal. For this block, we have $\sum_{i \in B} \epsilon_i \in \mathcal{Z}^*$ and, consequently, $C_{L+1} = 0$.

- If L is odd, the product in Equation 22.5 is nonzero, and thus equal to 1, if and only if $\sum_{i \in B} \epsilon_i = 0$ for all B of $\mathcal{P}$. Since $\sum_B \sum_{i \in B} \epsilon_i = \sum_{i=0}^{L} \epsilon_i$, this product is nonzero if, and only if, $\sum_{i=0}^{L} \epsilon_i = 0$.

As a conclusion, high-order cumulants of the surrogate signal $x(t)$ are nonzero only if $\sum_{i=0}^{L} \epsilon_i = 0$. This implies that $x(t)$ is a circular complex random signal. Moreover, substitution of this constraint in Section 22.2.2 leads to

$$c(t; t_1, \ldots, t_L) = C_{L+1} \int A(f)^{L+1} e^{j2\pi f \sum_{i=1}^{L} \epsilon_i t_i} \, df \qquad (22.6)$$

which proves that surrogates are strict-sense stationary.

Remark—Making use of strictly stationary surrogates proved effective for detecting nonstationarities in various scenarii, but the tests happen to be very sensitive. For instance, when applied to realizations of actual stationary processes, for example, auto-regressive (AR), the rejection rate of the null hypothesis turns out to be higher than the prescribed confidence level [7,37]. In a related way, one key point of the approach is to encompass in a common (time–frequency) framework stochastic and deterministic situations, stationarity referring to pure tones in the latter case. In this case too, surrogates cannot really reproduce the supposed stationarity of the observation. This is a natural outcome of the intrinsically stochastic generation of surrogates, but this makes again the test somehow pessimistic. The observation of such remaining limitations in the use of classical surrogates for testing stationarity prompts to think about related, possibly more versatile constructions. One possibility in this direction is, rather than *replacing* the spectrum phase by an i.i.d. sequence, to *modify* the original phase by adding some random phase noise to it. Depending on the nature and the level of this added phase noise, one can get this way a controlled transition from the original process (be it stationary or not) to its stationary counterpart [6].

Once a collection of stationarized surrogate signals has been synthesized, different possibilities are offered to test the initial signal stationarity [7,38]. A potential approach is to extract some features from the surrogate signals such as distance between local and global spectra, and to characterize the null hypothesis of stationarity by the statistical distribution of their variations in time. Another approach is based on statistical pattern recognition. It consists of considering surrogate signals as a learning set, and using it to estimate the support of the distribution of the stationarized signals. This will be detailed further in the next section.

22.3 TIME–FREQUENCY LEARNING MACHINES

Most pattern recognition algorithms can be expressed in terms of inner products only, involving pairs of input data. Replacing these inner products with a (reproducing) kernel provides an efficient way to implicitly map the data into a high-dimensional space, and apply the original algorithm in this space. Calculations are then carried out without making direct reference to the nonlinear mapping applied to input data. This so-called *kernel trick* is the main idea behind (kernel) learning machines. In this section, we show that learning machines can be tuned to operate in the time–frequency domain by a proper choice of kernel. Refer to Honeine et al. [18] for more details.

22.3.1 Reproducing Kernels

Let $\mathcal{X}$ be a subspace of $\mathcal{L}_2(\mathcal{C})$, the space of finite-energy complex signals, equipped with the usual inner product defined by $\langle x_i, x_j \rangle = \int_t x_i(t) x_j^*(t) \, dt$ and its corresponding norm. A kernel is a function $\kappa(x_i, x_j)$ from $\mathcal{X} \times \mathcal{X}$ to $\mathcal{C}$, with hermitian symmetry. It is said to be *positive definite* on $\mathcal{X}$ if [2]

$$\sum_{i=1}^{n} \sum_{j=1}^{n} a_i \, a_j^* \, \kappa(x_i, x_j) \geq 0 \tag{22.7}$$

for all $n \in \mathcal{N}, x_1, \ldots, x_n \in \mathcal{X}$ and $a_1, \ldots, a_n \in \mathcal{C}$. It can be shown that every positive-definite kernel κ is the reproducing kernel of a Hilbert space $\mathcal{H}$ of functions from $\mathcal{X}$ to $\mathcal{C}$, that is,

1. The function $\kappa_{x_j} : x_i \mapsto \kappa_{x_j}(x_i) = \kappa(x_i, x_j)$ belongs to $\mathcal{H}$, for all $x_j \in \mathcal{X}$

2. One has $\Theta(x_j) = \langle \Theta, \kappa_{x_j} \rangle_{\mathcal{H}}$ for all $x_j \in \mathcal{X}$ and $\Theta \in \mathcal{H}$

where $\langle \cdot, \cdot \rangle_{\mathcal{H}}$ denotes the inner product in $\mathcal{H}$. It suffices to consider the subspace $\mathcal{H}_0$ induced by the functions $\{\kappa_x\}_{x \in \mathcal{X}}$, and equip it with the following inner product

$$\langle \Theta_1, \Theta_2 \rangle_{\mathcal{H}_0} = \sum_{i=1}^{n} \sum_{j=1}^{m} a_{i,1} \, a_{i,2}^* \, \kappa(x_i, x_j) \tag{22.8}$$

where $\Theta_1 = \sum_{i=1}^{n} a_{i,1} \kappa_{x_i}$ and $\Theta_2 = \sum_{i=1}^{m} a_{i,2} \kappa_{x_i}$ are elements of $\mathcal{H}_0$. We fill this incomplete Hilbertian space according to Aronszajn [2], so that every Cauchy sequence converges in that space. Thus, we obtain the Hilbert space $\mathcal{H}$ induced by the reproducing kernel κ, called a *reproducing kernel Hilbert space* (RKHS). One can show that every reproducing kernel is positive definite [2]. An example of kernel is the Gaussian kernel defined by $\kappa(x_i, x_j) = \exp(-\|x_i - x_j\|^2 / 2\sigma^2)$, with σ the kernel bandwidth. Other examples of reproducing kernels, and rules for designing and combining them, can be found, for example, in Ref. [16,35].

22.3.2 The Kernel Trick, the Representer Theorem

Substituting Θ by κ_{x_i} in item 2 of the definition of RKHS in Section 22.3.1, we get the following fundamental property

$$\kappa(x_i, x_j) = \langle \kappa_{x_i}, \kappa_{x_j} \rangle_{\mathcal{H}} \tag{22.9}$$

for all $x_i, x_j \in \mathcal{X}$. Therefore, $\kappa(x_i, x_j)$ gives the inner product in $\mathcal{H}$, the so-called *feature space*, of the images κ_{x_i} and κ_{x_j} of any pair of input data x_i and x_j, without having to evaluate them explicitly. This principle is called the *kernel trick*. It can be used to transform any linear data processing technique into a nonlinear one, on the condition that the algorithm can be expressed in terms of inner products only, involving pairs of the input data. This is achieved by substituting each inner product $\langle x_i, x_j \rangle$ by a nonlinear kernel $\kappa(x_i, x_j)$, leaving the algorithm unchanged and incurring essentially the same computational cost. In conjunction with the kernel trick, the representer theorem is a solid foundation of kernel learning machines such as SVM [28]. This theorem states that any function Θ of $\mathcal{H}$ minimizing a regularized criterion of the form

$$J((x_1, y_1, \Theta(x_1)), \ldots, (x_n, y_n, \Theta(x_n))) + \rho(\|\Theta\|_{\mathcal{H}}^2) \tag{22.10}$$

with ρ a strictly monotonic increasing function on $\mathcal{R}_+$, can be written as a kernel expansion in terms of the available data, namely,

$$\Theta(\cdot) = \sum_{j=1}^{n} a_j \kappa(\cdot, x_j) \tag{22.11}$$

In order to prove this, note that any function Θ of the space $\mathcal{H}$ can be decomposed as $\Theta(\cdot) = \sum_{j=1}^{n} a_j \kappa(\cdot, x_j) + \Theta_\perp(\cdot)$, where $\langle \Theta_\perp(\cdot), \kappa(\cdot, x_j) \rangle_{\mathcal{H}} = 0$ for all $j = 1, \ldots, n$. Using this with Equation 22.9, we see that $\Theta_\perp$ does not affect the value of $\Theta(x_i)$, for all $i = 1, \ldots, n$. Moreover, we verify that (22.11) minimizes ρ since $\rho(\| \sum_{j=1}^{n} a_j \kappa(\cdot, x_j) \|_{\mathcal{H}}^2 + \|\Theta_\perp\|_{\mathcal{H}}^2) \geq \rho(\| \sum_{j=1}^{n} a_j \kappa(\cdot, x_j) \|_{\mathcal{H}}^2)$. This is the essence of the representer theorem.

22.3.3 Time–Frequency Learning Machines: General Principles

In this section, we investigate the use of kernel learning machines for pattern recognition in the time–frequency domain. To clarify the discussion, we shall first focus on the Wigner distribution. This will be followed by an extension to other time–frequency distributions, linear and quadratic. Below, $\mathcal{A}_n$ denotes a training set containing n instances $x_i \in \mathcal{X}$ and the desired outputs or labels $y_i \in \mathcal{Y}$.

Among the myriad of time–frequency representations that have been proposed, the Wigner distribution is considered fundamental in a number of ways. Its usefulness derives from the fact that it satisfies many desired mathematical properties such as the correct marginal conditions and the weak correct-support conditions. This distribution is also a suitable candidate for time–frequency-based detection since it is covariant to time shifts and frequency shifts and it satisfies the unitarity condition [12]. The Wigner distribution is given by

$$W_x(t, f) := \int x(t + \tau/2) \, x^*(t - \tau/2) \, e^{-2j\pi f \tau} \, d\tau \tag{22.12}$$

where x is the finite energy signal to be analyzed (one can remark that, under mild conditions, the WVS that has been previously considered, see Equation 22.1, is nothing but the ensemble average of the Wigner distribution 22.12). By applying conventional linear pattern recognition algorithms directly to time–frequency representations, we seek to determine a time–frequency pattern $\Phi(t,f)$ so that

$$\Theta(x) = \langle W_x, \Phi \rangle = \iint W_x(t,f)\,\Phi(t,f)\,dt\,df \qquad (22.13)$$

optimizes a given criterion J of the general form (Equation 22.10). The principal difficulty encountered in solving such problems is that they are typically very high dimensional, the size of the Wigner distributions calculated from the training set being quadratic in the length of signals. This makes pattern recognition based on time–frequency representations time consuming, if not impossible, even for reasonably sized signals. With the kernel trick and the representer theorem, kernel learning machines eliminate this computational burden. It suffices to consider the following kernel:

$$\kappa_W(x_i, x_j) = \langle W_{x_i}, W_{x_j} \rangle \qquad (22.14)$$

and note that W_{x_i} and W_{x_j} do not need to be computed since, by the unitarity of the Wigner distribution, we have

$$\kappa_W(x_i, x_j) = |\langle x_i, x_j \rangle|^2 \qquad (22.15)$$

We verify that κ_W is a positive-definite kernel by writing condition (Equation 22.7) as $\| \sum_j a_j W_{x_j} \|^2 \geq 0$, which is clearly verified. We are now in a position to construct the RKHS induced by this kernel, and denoted by $\mathcal{H}_W$. It is obtained by completing the space $\mathcal{H}_0$ defined below with the limit of every Cauchy sequence

$$\mathcal{H}_0 = \{\Theta : \mathcal{X} \to \mathcal{R} \mid \Theta(\cdot) = \sum_j a_j \, |\langle \cdot\,, x_j \rangle|^2, a_j \in \mathcal{R}, x_j \in \mathcal{X}\} \qquad (22.16)$$

Thus, we can use the kernel (Equation 22.15) with any kernel learning machine proposed in the literature to perform pattern recognition tasks in the time–frequency domain. Thanks to the representer theorem, solution (Equation 22.11) allows for a time–frequency distribution interpretation, $\Theta(x) = \langle W_x, \Phi_W \rangle$, with

$$\Phi_W = \sum_{j=1}^{n} a_j \, W_{x_j} \qquad (22.17)$$

This equation is directly obtained by combining (Equations 22.11 and 22.13). We should again emphasize that the coefficients a_j are estimated without calculating any Wigner distribution. The time–frequency pattern Φ_W can be subsequently evaluated with Equation 22.17, in an iterative manner, without suffering the drawback of storing and manipulating a large collection of Wigner distributions. The inherent sparsity of the coefficients a_j produced by most of the kernel learning machines, a typical example of which is the SVM algorithm, may speed up the calculation of Φ_W.

22.3.4 Wigner Distribution versus Spectrogram

Let $R_x(t,f)$ be a given time–frequency representation of a signal x. Proceeding as in the previous section with the Wigner distribution, we are led to optimization problems that only involve inner products between time–frequency representations of training signals:

$$\kappa_R(x_i, x_j) = \iint R_{x_i}(t,f)\, R_{x_j}(t,f)\, dt\, df = \langle R_{x_i}, R_{x_j} \rangle \tag{22.18}$$

This can offer significant computational advantages. A well-known time–frequency representation is the spectrogram (Equation 22.2), whose definition can be recast as

$$S_x(t,f;h) = \left| \langle x, h_{t,f} \rangle \right|^2$$

with $h_{t,f}(\tau) := h(\tau - t)\, e^{2j\pi f \tau}$. The inner product between two spectrograms, say S_{x_i} and S_{x_j}, is given by the kernel [18]

$$\kappa_S(x_i, x_j) = \iint \left| \langle x_i, h_{t,f} \rangle \langle x_j, h_{t,f} \rangle \right|^2 dt\, df$$

Computing this kernel for any pair of surrogate signals yields

$$\kappa_S(x_i, x_j) = \iint \left| \langle |X| e^{j\phi_i}, H_{t,f} \rangle\, \langle |X| e^{j\phi_j}, H_{t,f} \rangle \right|^2 dt\, df \tag{22.19}$$

where $H_{t,f}$ is the Fourier transform of $h_{t,f}$. This has to be contrasted with the Wigner distribution which, with its unitarity property, leads to some substantial computational reduction since

$$\kappa_W(x_i, x_j) = |\langle x_i, x_j \rangle|^2 = \left| \int |X(f)|^2\, e^{j(\phi_i(f) - \phi_j(f))}\, df \right|^2 \tag{22.20}$$

We emphasize here that there is no need to compute and manipulate the surrogates and their time–frequency representations. Given $|X(f)|$, only the random phases $\phi_i(f)$ and $\phi_j(f)$ are required to evaluate the kernel κ_W of Equation 22.20.

For the sake of simplicity, we illustrated this section with the spectrogram. However, kernel learning machines can use any time–frequency kernels to perform pattern recognition tasks in the time–frequency domain, as extensively studied in Ref. [17–19]. In the next section, we present the one-class SVM problem to test stationarity using surrogate signals.

22.4 A NONSUPERVISED CLASSIFICATION APPROACH

Adopting a viewpoint rooted in statistical learning theory by considering the collection of surrogate signals as a learning set, and using it to estimate the support of the distribution of stationarized data, avoids solving the difficult problem of density estimation that would be a prerequisite in parametric methods. Let us make this approach, which consists of estimating quantiles of multivariate distributions, more precise.

22.4.1 An Overview on One-Class Classification

In the context considered here, the classification task is fundamentally a one-class classification problem and differs from conventional two-class pattern recognition problems in the way how the classifier is trained. The latter uses only target data to perform outlier detection. This is often accomplished by estimating the probability density function of the target data, for example, using a Parzen density estimator [24]. Density estimation methods, however, require huge amounts of data, especially in high-dimensional spaces, which makes their use impractical. Boundary-based approaches attempt to estimate the quantile function defined by $Q(\alpha) := \inf\{\lambda(\mathcal{S}) : P(\mathcal{S}) := \int_{\omega \in \mathcal{S}} \mu(d\omega) \geq \alpha\}$ with $0 < \alpha \leq 1$, where $\mathcal{S}$ denotes a subset of the signal space $\mathcal{X}$ that is measurable with respect to the probability measure μ, and $\lambda(\mathcal{S})$ its volume. Estimators that reach this infimum, in the case where P is the empirical distribution, are called minimum volume estimators. The first boundary-based approach was probably introduced in Ref. [27], where the authors consider a class of closed convex boundaries in $\mathcal{R}^2$. More sophisticated methods were described in [21,22]. Nevertheless, they are based upon neural networks training and therefore suffer from the same drawbacks such as slow convergence and local minima. Inspired by SVM, the support vector data description algorithm proposed in Ref. [33] encloses data in a minimum volume hypersphere. More flexible boundaries can be obtained by using kernel functions, that map the data into a high-dimensional feature space. In the case of normalized kernel functions, such that $\kappa(x, x) = 1$ for all x, this approach is equivalent to the one-class SVM introduced in Ref. [29], which use a maximum margin hyperplane to separate data from the origin. The generalization performance of these algorithms was investigated in Ref. [29,30,36] via the derivation of bounds. In what follows, we shall focus on the support vector data description algorithm.

22.4.2 One-Class SVM for Testing Stationarity

Inspired by SVM for classification, the one-class SVM allows the description of the density distribution of a single class [32]. The main purpose is to enclose the training data into a minimum volume hypersphere, thus defining a domain boundary. Any data outside this volume may be considered as an outlier, and its distance to the center of the hypersphere allows a measure of its novelty. Here, we propose to use the set of surrogate signals to derive the hypersphere of stationarity, in the time–frequency domain defined by a reproducing kernel as given in Section 22.3.

Consider a set of n surrogate signals, $x_1, \ldots, x_n$, computed from a given signal x. Let $R_{x_1}, \ldots, R_{x_n}$ denote their time–frequency representations and κ the corresponding reproducing kernel. In this domain, we seek the hypersphere that contains most of these representations. Its center Ω and radius r are obtained by solving the optimization problem

$$\min_{\Omega, r, \xi} \; r^2 + \frac{1}{n\nu} \sum_{i=1}^{n} \xi_i$$

$$\text{s.t. } \|R_{x_i} - \Omega\|^2 \leq r^2 + \xi_i, \;\; \xi_i \geq 0, \;\; i = 1, \ldots, n$$

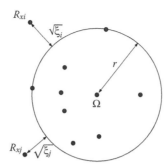

FIGURE 22.2 Support vector data description algorithm.

As illustrated in Figure 22.2, parameter $\nu \in [0,1]$ controls the trade-off between the radius r to be minimized, and the number of training data outside the hypersphere characterized by the slack variables $\xi_i = (\|R_{x_i} - \Omega\|^2 - r^2)_+$. Using Lagrangian principle, the optimization problem is reduced to

$$\max_{\alpha} \ \sum_{i=1}^{n} \alpha_i \kappa(x_i, x_i) - \sum_{i,j=1}^{n} \alpha_i \alpha_j \kappa(x_i, x_j)$$

$$\text{s.t.} \ \sum_{i=1}^{n} \alpha_i = 1, \ 0 \le \alpha_i \le \frac{1}{n\nu}, \ i = 1, \ldots, n$$

(22.21)

which can be solved with quadratic programming techniques. The resulting nonzero Lagrange multipliers α_i yield the center $\Omega = \sum_i \alpha_i R_{x_i}$, and the radius $r = \|R_{x_\ell} - \Omega\|$ with x_ℓ any data having $0 < \alpha_\ell < \frac{1}{n\nu}$.

For any signal x, the (squared) distance of its time–frequency representation to the center Ω can be written as

$$\|R_x - \Omega\|^2 = \kappa(x, x) - 2 \sum_{i=1}^{n} \alpha_i \kappa(x, x_i) + \sum_{i,j=1}^{n} \alpha_i \alpha_j \kappa(x_i, x_j)$$

As explained previously, we do not need to compute time–frequency representations to calculate this score, since only the values of the kernel are required. The coefficients α_i are obtained by solving (Equation 22.21), requiring only the evaluation of κ for training data. This *kernel trick* is also involved in the proposed decision function, defined by comparing the test statistics $\Theta(x) = \|R_x - \Omega\|^2 - r^2$ to a threshold γ:

$$\Theta(x) \underset{\text{stat.}}{\overset{\text{nonstat.}}{\gtrless}} \gamma$$

(22.22)

The signal x under study is considered as nonstationary if its time–frequency representation lies outside the hypersphere of squared radius $r^2 + \gamma$; otherwise, it is considered as stationary. The threshold γ has a direct influence upon the test performance [30]. For instance, with a probability $> 1 - \delta$, one can bound the probability of false positive by

$$\Delta = \frac{1}{\gamma n} \sum_{i=1}^{n} \xi_i + \frac{6\omega^2}{\gamma \sqrt{n}} + 3 \sqrt{\frac{\log(2/\delta)}{2n}} \qquad (22.23)$$

where ω is the radius of the ball centered at the origin containing the support of the probability density function of the class of surrogate signals. Here γ can be fixed arbitrarily in Equation 22.22, so as to set the required false-positive probability, for which Δ is an upper bound.

We shall now propose another use of Equation 22.23 as a measure of stationarity of the signal x under investigation. If x lies inside the hypersphere of the surrogate class, the score of stationarity is arbitrarily fixed to one. Else, one can set γ depending on the signal x to $\|R_x - \Omega\|^2 - r^2$, so that the signal would lie on the decision boundary. Then, Equation 22.23 gives a bound $\Delta(x)$ on the false-positive probability that should be assumed for the signal to be classified as stationary. The closer x is to the hypersphere boundary, the closer to one $\Delta(x)$ is; the closer $\Delta(x)$ is to zero, the greater the contrast between x and the surrogates. Hence, $\Delta(x)$ has the meaning of a stationarity score. See Ref. [7] for more details.

22.4.3 Spherical Multidimensional Scaling

Multidimensional scaling (MDS) is a classical tool in data analysis and visualization [9]. It aims at representing data in a d-dimensional space, where d is specified *a priori*, such that the resulting distances reflect in some sense the distances in the higher-dimensional space. The neighborhood between data is preserved, whereas dissimilar data tend to remain distant in the new space. MDS algorithm requires only the distances between data in order to embed them into the new space. Consider the set of time–frequency representations of surrogate signals $\{R_{x_1}, \ldots, R_{x_n}\}$, and the inner product between two representations defined as in Equation 22.18. We can apply classical MDS in order to visualize the data in a low-dimensional Euclidean space. On the condition that R_x satisfies the global energy distribution property $\iint R_x(t,f) \, dt \, df \propto \int |x(t)|^2 \, dt$, such as the spectrogram or the Wigner distribution, the time–frequency representations of surrogate signals lie on an hypersphere centered at the origin. This non-Euclidean geometry makes it desirable to place restrictions on the configuration obtained from the MDS analysis. This can be done by using a Spherical MDS technique, as proposed in Ref. [9], or more recently in Ref. [25], which forces points to lie on the the two-dimensional surface of a sphere.

In the experimental results section, we shall use spherical MDS analysis to visualize the time–frequency configuration of the signal x under study and the surrogate signals, and the decision boundary that discriminates between the null hypothesis of stationarity and its nonstationary alternative.

22.5 ILLUSTRATION

In order to test our method, we used the same two amplitude-modulated (AM) and FM signals as in Ref. [38]. While not covering all the situations of nonstationarity, these signals are believed to give meaningful examples. The AM signal is modeled as

$$x(t) = (1 + m \sin(2\pi t/T_0))\, e(t)$$

with $m \leq 1$, $e(t)$ a white Gaussian noise, and T_0 the period of the AM. In the FM case,

$$x(t) = \sin(2\pi(f_0 t + m \sin(2\pi t/T_0)))$$

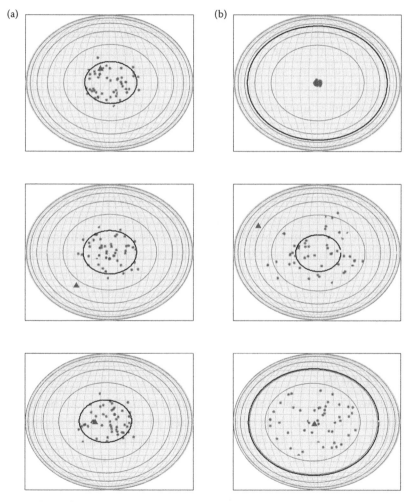

FIGURE 22.3 Spherical MDS representation of the surrogate signals (∗) and the test signal (▲), in AM (a) and FM (b) situations. From top to bottom, $T_0 = T/100$, $T = T_0$ and $T_0 = 100\,T$, with $T = 1024$. The minimum-volume domain of stationarity is represented by the thick curve. The tested signals are identified as nonstationary in the case $T = T_0$ (the triangle of the signal being outside the circle corresponding to the minimum-volume domain of stationarity), and can be considered as stationary in the cases $T_0 = T/100$ and $T_0 = 100\,T$ (the triangle of the signal being inside the circle). Other parameters are as follows – (AM): $m = 0.5$; (FM): $f_0 = 0.25$, $m = 0.02$, and SNR $= 10\,$dB.

with f_0 the central frequency. Based on the relative values of T_0 and the signal duration T, three cases can be distinguished for each type, AM and FM:

- $T \gg T_0$: The signal contains a large number of oscillations. This periodicity indicates a stationary regime.

- $T \approx T_0$: Only one oscillation is available. The signal is considered as nonstationary.

- $T \ll T_0$: With a small portion of a period, there is no change in the amplitude or the frequency. It is considered as a stationary signal.

For each experiment reported in Figure 22.3, 50 surrogate signals were generated from the AM or FM signal $x(t)$ to be tested. The results are displayed for $T_0 = T/100$, $T = T_0$, and $T_0 = 100\,T$, allowing one to consider stationarity relatively to the ratio between observation time T and modulation period T_0. The one-class SVM algorithm was run with the spectrogram kernel (Equation 22.19) and parameter $\nu = 0.15$. Then, spherical MDS analysis was applied for visualization purpose only. In each figure, the surrogate signals are shown with blue stars and the signal to be tested with a red triangle. The minimum-volume domain of stationarity is represented by the black curve. It should be noticed that this curve and the data are projections from the high-dimensional space of time–frequency distributions onto a sphere in $\mathcal{R}^3$ for visualization, meaning that the representation is inherently distorted. The tested signals are clearly identified as nonstationary in the case $T = T_0$ (the red triangle of the signal being outside the circle corresponding to the minimum-volume domain of stationarity), and can be considered as stationary in the cases $T_0 = T/100$ and $T_0 = 100\,T$ (the red triangle of the signal being inside the circle). These results are consistent with those obtained in previous works, using either the distance or the time–frequency feature extraction approach [7]. Here, the test is performed without suffering from the prior knowledge required to extract relevant features.

22.6 CONCLUSION

In this chapter, we showed how time–frequency kernel machines can be used to test stationarity. For any given signal, a set of stationarized surrogate signals is generated to train a one-class SVM, implicitly in the time–frequency domain by using an appropriate kernel. The originality here is the use of the whole time–frequency information, as opposed to feature extraction techniques where prior knowledge is required. This was proved effective for detecting nonstationarities with simulation results.

The resampling strategy actually used to generate surrogate signals is however quite strict in the sense that, after the phase replacement by some i.i.d. sequence, a possibly nonstationary signal is turned into a stationary one without any consideration about the fact that this property has to be understood in a relative sense that incorporates the observation scale. In this respect, strict stationarity may appear as a too strong commitment, and prompts us to think about more versatile constructions of stationary-relaxed surrogate signals. One perspective in this direction is to alter the original phase by embedding it into noise [6].

Depending on the nature and the level of these phase fluctuations, we could get this way a controlled transition from the original process to its stationary counterpart.

There has been preliminary attempts to use surrogates for testing the stationarity of some actual data: signals in biophysics [5], or in mechanics [8], and an adaptation to images is reported in Ref. [13].

Let us now turn to potential astronomical applications. Searching for evidence of nonstationarity in the temporal properties of astrophysical objects and phenomena are fundamental to their study, as in the case of galactic black holes switching from one spectral state to another. For instance, among the basic properties characterizing an active galactic nucleus, the x-ray variability is one of the most commonly used. This question has been explored by researchers in the case of NGC 4051 [15], Mrk 421 [10,11], 3C 390.3 [14], and PKS 2155-304 [39]. One of the most popular approaches is to measure the fluctuation of the PSD, by fitting a simple power law model [20], or by estimating the so-called excess variance [23]. As an alternative, analysis in the time domain using the structure function (SF) is also often considered [31]. Note that PSD with model fitting and SF should provide equivalent information as they are related to the auto-correlation function of the process at hand. It is unfortunate that, with this techniques, it is not possible to prove nonstationarity in a model-independent way. Nonlinear analysis using scaling index method, which measures both global and local properties of a phase-space portrait of time series, has also been considered in the literature [3].

Although its effective application to real astrophysical data has not yet been considered, it is believed that the approach proposed here could be a useful addition to such existing techniques by shedding a new light on stationarity versus nonstationarity issues.

REFERENCES

1. P. O. Amblard, M. Gaeta, and J. L. Lacoume. Statistics for complex variables and signals—Parts I and II. *Signal Processing*, 53(1):1–25, 1996.
2. N. Aronszajn. Theory of reproducing kernels. *Transactions of the American Mathematical Society*, 68:337–404, 1950.
3. H. Atmanspacher, H. Scheingraber, and G. Wiedenmann. Determination of f(.) for a limited random point set. *Physical Review A*, 40(7):3954–3963, 1989.
4. M. Bayram and R.G. Baraniuk. Multiple window time-varying spectrum estimation. In W.J. Fitzgerald, R.L. Smith, A.T. Walden, and P.C. Young, editors, *Nonlinear and Nonstationary Signal Processing*. Cambridge University Press, Cambridge, UK, 2000.
5. P. Borgnat and P. Flandrin. Stationarization via surrogates. *Journal of Statistical Mechanics: Theory and Experiment: Special Issue UPoN 2008*, P01001, 2009.
6. P. Borgnat, P. Flandrin, A. Ferrari, and C. Richard. Transitional surrogates. In *Proceedings of the IEEE International Conference on Acoustics, Speech and Signal Processing*, Prague, Czech Republic, pp. 3600–3603, 2011.
7. P. Borgnat, P. Flandrin, P. Honeine, C. Richard, and J. Xiao. Testing stationarity with surrogates: A time–frequency approach. *IEEE Transactions on Signal Processing*, 58(12):3459–3470, 2010.
8. M. Chabert, B. Trajin, J. Regnier, and J. Faucher. Surrogate-based diagnosis of mechanical faults in induction motor from stator current measurements (regular paper). In *Condition Monitoring, Stratford upon Avon, 22/06/2010-24/06/2010*, page (electronic medium). The British Institute of Non Destructive Testing, San Diego, CA, 2010.

9. T. F. Cox and M. A. A. Cox. Multidimensional scaling on a sphere. *Communications in Statistics: Theory and Methods*, 20:2943–2953, 1991.

10. D. Emmanoulopoulos, I. M. McHardy, and P. Uttley. Variability studies in blazar jets with SF analysis: Caveats and problems. In G. E. Romero, R. A. Sunyaev, and T. Belloni, editors, *IAU Symposium, IAU Symposium*, Vol. 275, pp. 140–144, 2011.

11. V. V. Fidelis and D. A. Iakubovskyi. The x-ray variability of Mrk 421. *Astronomy Reports*, 52:526–538, 2008.

12. P. Flandrin. *Time–Frequency/Time-Scale Analysis*. Academic Press Inc, San Diego, California, 1998.

13. P. Flandrin and P. Borgnat. Revisiting and testing stationarity. *Journal of Physics: Conference Series*, 139:012004, 2008.

14. M. Gliozzi, I. E. Papadakis, and C. Räth. Correlated spectral and temporal changes in 3C 390.3: A new link between AGN and Galactic black hole binaries? *Astronomy and Astrophysics*, 449:969–983, 2006.

15. A. R. Green, I. M. McHardy, and C. Done. The discovery of non-linear x-ray variability in ngc 4051. *Monthly Notices of the Royal Astronomical Society*, 305(2):309–318, 1999.

16. R. Herbrich. *Learning Kernel Classifiers. Theory and Algorithms*. The MIT Press, Cambridge, MA, 2002.

17. P. Honeine and C. Richard. Signal-dependent time–frequency representations for classification using a radially Gaussian kernel and the alignment criterion. In *Proceedings of the IEEE Statistical Signal Processing Workshop*, pp. 735–739, Madison, WI, 2007.

18. P. Honeine, C. Richard, and P. Flandrin. Time–frequency learning machines. *IEEE Transactions on Signal Processing*, 55:3930–3936, 2007.

19. P. Honeiné, C. Richard, P. Flandrin, and J.-B. Pothin. Optimal selection of time–frequency representations for signal classification: A Kernel-target alignment approach. In *Proceedings of the IEEE International Conference on Acoustics, Speech and Signal Processing*, pp. III.476–III.479, Toulouse, France, 2006.

20. A. Lawrence and I. Papadakis. X-ray variability of active galactic nuclei—A universal power spectrum with luminosity-dependent amplitude. *Astrophysical Journal*, 414:L85–L88, 1993.

21. M. Moya and D. Hush. Network contraints and multi-objective optimization for one-class classification. *Neural Networks*, 9(3):463–474, 1996.

22. M. Moya, M. Koch, and L. Hostetler. One-class classifier networks for target recognition applications. In *Proceedings of the World Congress on Neural Networks*, pp. 797–801, Portland, OR, 1993.

23. K. Nandra, I. M. George, R. F. Mushotzky, T. J. Turner, and T. Yaqoob. ASCA observations of Seyfert 1 galaxies. I. Data analysis, imaging, and timing. *The Astrophysical Journal*, 476(1):70, 1997.

24. E. Parzen. On estimation of a probability density function and mode. *Annals of Mathematical Statistics*, 33(3):1065–1076, 1962.

25. R. Pless and I. Simon. Embedding images in non-flat spaces. In H. R. Arabnia and Y. Mun, editors, *Proceedings of the International Conference on Imaging Science, Systems and Technology*, pp. 182–188, Las Vegas, CSREA Press, Athens, GA, 2002.

26. C. Richard, A. Ferrari, H. Amoud, P. Honeine, P. Flandrin, and P. Borgnat. Statistical hypothesis testing with time–frequency surrogates to check signal stationarity. In *Proceedings of the IEEE International Conference on Acoustics, Speech and Signal Processing*, pp. 3666–3669, Dallas, 2010.

27. T. W. Sager. An iterative method for estimating a multivariate mode and isopleth. *Journal of the American Statistical Association*, 74(366):329–339, 1979.

28. B. Schölkopf, R. Herbrich, and R. Williamson. A generalized representer theorem. Technical Report NC2-TR-2000-81, NeuroCOLT, Royal Holloway College, University of London, UK, 2000.

29. B. Schölkopf and A. J. Smola. *Learning with Kernels: Support Vector Machines, Regularization, Optimization and Beyond.* MIT Press, Cambridge, MA, 2001.

30. J. Shawe-Taylor and N. Cristianini. *Kernel Methods for Pattern Analysis.* Cambridge University Press, Cambridge, UK, 2004.

31. J. H. Simonetti, J. M. Cordes, and D. S. Heeschen. Flicker of extragalactic radio sources at two frequencies. *Astrophysical Journal*, 296:46–59, 1985.

32. D. M. J. Tax and R. P. W. Duin. Support vector domain description. *Pattern Recognition Letters*, 20(11–13):1191–1199, 1999.

33. D. M. J. Tax and R. P. W. Duin. Support vector data description. *Machine Learning*, 54(1):45–66, 2004.

34. J. Theiler, S. Eubank, A. Longtin, B. Galdrikian, and J. Doyne Farmer. Testing for nonlinearity in time series: The method of surrogate data. *Physica D: Nonlinear Phenomena*, 58(1–4):77–94, 1992.

35. V. Vapnik. *The Nature of Statistical Learning Theory.* Springer-Verlag, Berlin, Germany, 1995.

36. R. Vert. Theoretical insights on density level set estimation, application to anomaly detection. PhD thesis, Paris 11 - Paris Sud, 2006.

37. J. Xiao, P. Borgnat, and P. Flandrin. Sur un test temps-fréquence de stationnarité. *Traitement du Signal*, 25(4):357–366, 2008. (in French, with extended English summary).

38. J. Xiao, P. Borgnat, P. Flandrin, and C. Richard. Testing stationarity with surrogates—A one-class SVM approach. In *Proceedings of the IEEE Statistical Signal Processing*, pp. 720–724, Madison, WI, 2007.

39. Y. H. Zhang, A. Treves, A. Celotti, Y. P. Qin, and J. M. Bai. XMM-Newton view of PKS 2155–304: Characterizing the x-ray variability properties with EPIC PN. *The Astrophysical Journal*, 629(2):686, 2005.

CHAPTER 23

Classification

Nikunj Oza

NASA Ames Research Center

CONTENTS

23.1 Introduction   505
23.2 Features of Classification Model Learning Algorithms   506
23.3 Decision Trees and Decision Stumps   510
23.4 Naïve Bayes   513
23.5 Neural Networks   514
23.6 Support Vector Machines   517
23.7 Further Reading   521
References   521

23.1 INTRODUCTION

A supervised learning task involves constructing a mapping from input data (normally described by several features) to the appropriate outputs. A set of training examples—examples with known output values—is used by a learning algorithm to generate a model. This model is intended to approximate the mapping between the inputs and outputs. This model can be used to generate predicted outputs for inputs that have not been seen before. Within supervised learning, one type of task is a classification learning task, in which each output is one or more classes to which the input belongs. For example, we may have data consisting of observations of sunspots. In a classification learning task, our goal may be to learn to classify sunspots into one of several types. Each example may correspond to one candidate sunspot with various measurements or just an image. A learning algorithm would use the supplied examples to generate a model that approximates the mapping between each supplied set of measurements and the type of sunspot. This model can then be used to classify previously unseen sunspots based on the candidate's measurements. The generalization performance of a learned model (how closely the target outputs and the model's predicted outputs agree for patterns that have not been presented to the learning algorithm) would provide an indication of how well the model has learned the desired mapping.

More formally, a classification learning algorithm L takes a training set T as its input. The training set consists of $|T|$ *examples* or *instances*. It is assumed that there is a probability distribution $\mathcal{D}$ from which all training examples are drawn independently—that is, all the training examples are *independently and identically distributed* (i.i.d.). The ith training example is of the form $(\mathbf{x}_i, y_i)$, where $\mathbf{x}_i$ is a vector of values of several features and y_i represents the class to be predicted.* In the sunspot classification example given above, each training example would represent one sunspot's classification (y_i) and the corresponding set of measurements $(\mathbf{x}_i)$. The output of a supervised learning algorithm is a model h that approximates the unknown mapping from the inputs to the outputs. In our example, h would map from the sunspot measurements to the type of sunspot.

We may have a *test set* S—a set of examples not used in training that we use to test how well the model h predicts the outputs on new examples. Just as with the examples in T, the examples in S are assumed to be independent and identically distributed (i.i.d.) draws from the distribution $\mathcal{D}$. We measure the error of h on the test set as the proportion of test cases that h misclassifies:

$$\frac{1}{|S|} \sum_{(x,y) \in S} I(h(x) \neq y)$$

where $I(v)$ is the indicator function—it returns 1 if v is true and 0 otherwise.

In our sunspot classification example, we would identify additional examples of sunspots that were not used in generating the model, and use these to determine how accurate the model is—the fraction of the test samples that the model classifies correctly.

An example of a classification model is the decision tree shown in Figure 23.1. We will discuss the decision tree learning algorithm in more detail later—for now, we assume that, given a training set with examples of sunspots, this decision tree is derived. This can be used to classify previously unseen examples of sunpots. For example, if a new sunspot's inputs indicate that its "Group Length" is in the range 10–15, then the decision tree would classify the sunspot as being of type "E," whereas if the "Group Length" is "NULL," the "Magnetic Type" is "bipolar," and the "Penumbra" is "rudimentary," then it would be classified as type "C."

In this chapter, we will add to the above description of classification problems. We will discuss decision trees and several other classification models. In particular, we will discuss the learning algorithms that generate these classification models, how to use them to classify new examples, and the strengths and weaknesses of these models. We will end with pointers to further reading on classification methods applied to astronomy data.

23.2 FEATURES OF CLASSIFICATION MODEL LEARNING ALGORITHMS

As indicated above, classification learning aims to use a training set to generate a model that can be used to classify new, typically previously unseen, examples. In some cases, the aim is to determine the most likely class given the new example's inputs. In other cases, one may measure the probabilities of each class given the example's inputs and use this as a measure of

---

* In this chapter, we will assume that each individual example belongs to only one class; however, we will not assume that two examples with the same inputs must be in the same class. That is, we will allow for some noise in the input and output generation processes.

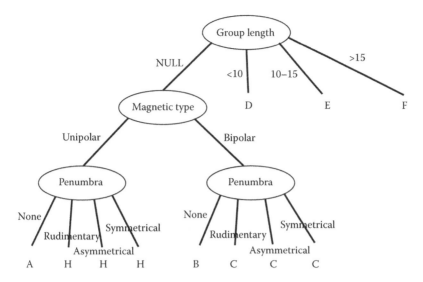

FIGURE 23.1 Example decision tree.

confidence—if the highest probability minus the second highest probability is high, then the model is quite confident in the highest-probability class. If this probability difference is low, then the confidence is low. Regardless of the problem, one needs the probability distribution of the class given the inputs, which is $P(Y|X)$ where X represents the new example and Y represents the vector of possible classes. Because we have a training set that is used to determine the probabilities, we are actually interested in $P(Y|X, D)$, where D represents the training data.

One example of a machine learning technique that estimates $P(Y|X, D)$ is nearest-neighbor classification [15]. Given a new example X, nearest-neighbor classification identifies one or more neighbors within D and chooses the class for X based on the classes of the neighbors. One possibility is to find the nearest neighbor and assign to X the class of that nearest neighbor. Another possibility is to find some number K nearest neighbors and assign the class most frequently seen among those neighbors or perhaps each class can get a weighted vote, with nearer neighbors getting stronger votes. One difficulty with this scheme is that the memory requirements for nearest-neighbor-based techniques grow with the size of the training set. Given how large modern datasets are in many disciplines, including astronomy, nearest-neighbor-based techniques are mostly impractical. One needs a machine learning technique that can compress the information from the training set into a form that is of constant size regardless of the size of the training set but is nevertheless able to use information learned from the training set to classify new examples.

The decision tree example given above is a good example of such a model. Decision trees with categorical features* cannot grow to a depth greater than the number of input features and cannot have a total number of nodes beyond $V^{|F|}$, where V is the maximum number

---

* A categorical feature can only be used once. An ordered discrete or continuous feature can be used as many times as it has unique values in the training set minus one.

of values that any feature can have and F is the set of features. This is independent of the number of training examples.

As explained above, classifier learning algorithms take a training set as input and return a model as output. However, this model is not of an arbitrary form. The model is drawn from a family of possible models by finding which model is best according to some predefined criteria. For example, the decision tree learning algorithm that we will describe later may have returned the decision tree shown in Figure 23.1. However, the set of possible models that the learning algorithm can return is limited. In particular, this learning algorithm will always return a decision tree and never another type of model such as a neural network. Additionally, as mentioned above, the decision tree will have depth that is no greater than the number of input features because each input feature can be used at most once on any path from the root to a leaf in the tree. Other learning algorithms may have different restrictions on the models that they may return. For example, a decision stump learning algorithm run on sunspot data may return only the top node ("Group Length") shown in Figure 23.1, the three leaves shown on the right with classes "D," "E," and "F," and the subtree rooted by the feature "Magnetic Type" replaced by the class most frequently seen among the training examples that have "Group Length" as "NULL." The set of possible models that a learning algorithm may return is typically called a *model family*.

More formally, instead of merely returning $P(Y|X,D)$, we often condition on a model family H and end up deriving the following:

$$P(Y|X,D) = \sum_{h=1}^{|H|} P(Y|X,H_h,D)P(H_h|D) \tag{23.1}$$

$$= \sum_{h=1}^{|H|} P(Y|X,H_h)P(H_h|D) \tag{23.2}$$

$$= \sum_{h=1}^{|H|} \frac{P(Y|X,H_h)P(D|H_h)P(H_h)}{P(D)}. \tag{23.3}$$

In Equation 23.1, we replace $P(Y|X,H_h,D)$ with $P(Y|X,H_h)$ to get Equation 23.2 because H_h is meant to contain all the information that we would ordinarily get from D. For example, if H_h is a decision tree learned from the training data D, we no longer need D but rather can just use H_h to return the predicted class or the probability of each possible class. Equation 23.2 conditions on the set of models, and returns a sum of probabilities $P(Y|X,H_h)$, each of which is weighted by $P(H_h|D)$.

Even using Equation 23.2 seems impractical because H can be a very large family. For example, H could be the entire set of decision trees over the set of features F. If each of the features is binary, then the total number of decision trees that could be returned by the learning algorithm is

$$\sum_{f=1}^{|F|} (|F|-f+1)^{2^{f-1}}$$

which can become large very quickly. For this reason, typical learning algorithms, such as the decision tree learning algorithm, choose one model from within H that is best according to some criterion. If the criterion is maximizing $P(H_h|D)$, then the chosen model is called the *maximum a posteriori* model, because it is the model with the highest posterior probability given the training data. Equation 23.3 is constructed by replacing $P(H_h|D)$ using Bayes's rule with $P(D|H_h)P(H_h)/P(D)$. Since $P(D)$ is the same for all models and often the prior probabilities of the models $P(H_h)$ are assumed to be the same, some learning methods choose the model that maximizes $P(D|H_h)$, which is the probability that the training data was generated by model H_h and is often called the *likelihood* of the data. The model that maximizes the likelihood is called the *maximum likelihood* model.

Clearly, using one model has risks—that model, although perhaps the single best-performing model on average, may perform poorly on some parts of the input space. Also, the model chosen is derived from that particular training dataset. Given a slightly different training set, a different model may be more appropriate. There is another possible method of performing *model selection* rather than either using all possible models or one model. One can use some intermediate number of models, all of which perform reasonably well (e.g., have relatively high values of $P(H_h|D)$), but which are different from one another so that they cover more of the input space or more of the space of models optimal over the possible training sets. Methods that return multiple such models are referred to as *ensembles*, and are the subject of Chapter 26.

The alert reader may have been inspired by the above use of Bayes's rule and noted that

$$P(Y|X) = \frac{P(X|Y)P(Y)}{P(X)}.$$

This may make one wonder whether there are methods that, instead of attempting to learn $P(Y|X)$ or a similar function that generates the predicted class or probability distribution over classes given the input, attempt to learn $P(X|Y)$, $P(Y)$, $P(X)$. There are methods that attempt to learn $P(X|Y)$ and $P(Y)$. Learning $P(X)$ separately is unnecessary because

$$P(X) = \sum_{y=1}^{|C|} P(X|Y = y)P(Y = y)$$

where C is the set of possible classes. Note that every factor in the above summand is either of the form $P(X|Y)$ or $P(Y)$, which are already being calculated.

Methods that learn an estimate of $P(Y|X)$ are called *discriminative* methods because they attempt to find a model that discriminates between multiple classes using the inputs. On the other hand, methods that learn an estimate of $P(X|Y)$ and $P(Y)$ are called *generative* methods because they attempt to learn a model of the process that generates the data. One assumes in this case that, in the true data generating process, a class is drawn at random from some prior distribution $P(Y)$ and then, given that class, a set of features is generated using the distribution $P(X|Y)$. We will give examples of generative models, such as Naïve Bayes classifiers, later in this chapter.

Sometimes, instead of directly using criteria like the posterior probability or likelihood to try to identify the best model, some other, more convenient criterion is used. One example criterion is classification error on the training set—the fraction of training examples for which the decision tree returns the incorrect class even after learning. The most commonly used decision tree learning algorithms—ID3 [15,18] and C4.5 [17]—use this criterion. There are some models that can return the posterior probability of an example being in each class. In this case, mean-squared error may be a better error criterion. So, if $\hat{y}_{i,c}$ is the model's prediction of the posterior probability of class c ($c \in \{1, 2, \ldots, |C|\}$) for training input $\mathbf{x}_i$, while $y_{i,c}$ is 1 if $\mathbf{x}_i$ is an example of class c and 0 otherwise, then the mean-squared error criterion is

$$\frac{1}{|T||C|} \sum_{i=1}^{|T|} \sum_{c=1}^{|C|} (y_{i,c} - \hat{y}_{i,c})^2.$$

Clearly, the more confident the classifier is in the correct class, the lower the value of the mean-squared error, and similarly the more confident the classifier is in an incorrect class, the higher the value of the mean-squared error.

Sometimes the error function is a sum of several criteria instead of just one criterion. Typically one criterion is an error term such as the mean-squared error, and the other one is a criterion designed to reduce the complexity of the model so that it does not *overfit* the training data—fit the training data so well that it fits the noise in the training data in addition to the signal and then perform poorly on the test set. For example, one way to guarantee that decision tree learning will fit the training data at least as well as any other decision tree is to generate a tree that is as deep as needed to make the leaves as homogeneous as possible (maximally skewed class distributions at each leaf). However, constructing a decision tree in this manner may reduce the feature selection benefit of decision tree learning and instead may include many irrelevant features. This may lead to lower performance on the test set.

We now describe some of the most popular classification learning algorithms.

23.3 DECISION TREES AND DECISION STUMPS

A decision tree is a tree structure consisting of nonterminal nodes, leaf nodes, and arcs. An example of a decision tree that may be produced in our example sunspot domain is depicted in Figure 23.1. Each nonterminal node represents a test on a feature value. In the example decision tree, the top node (called the *root* node) tests the value for "Group length." If the group length is either "< 10," "10 − 15," or "> 15," then the classification returned is **D**, **E**, or **F**, respectively. If the group length is "NULL," then instead of returning a class, an additional feature, "Magnetic Type," is checked. If the type is "Unipolar," then the "Penumbra" is checked. If the "Penumbra" is "None," then the class returned is **A**. If the "Penumbra" is "Rudimentary," "Asymmetrical," or "Symmetrical," the class returned is **H**. If the value of "Magnetic Type" is "Bipolar," then the feature "Penumbra" is checked as well. This time though, if the value of "Penumbra" is "None," then class **B** is returned, and otherwise class **C** is returned. There may have been other features measured for each example; however, none of them was used.

Decision_Tree_Learning (T,F,C)
   if $T_{i,C}$ is the same for all $i \in \{1, 2, \ldots, |T|\}$,
      return a leaf node labeled $T_{1,C}$.
   else if $|F| = 0$,
      return a leaf node labeled $\text{argmax}_c \sum_{i=1}^{|T|} I(T_{i,C} = c))$.
   else
      $f = \text{Choose\_Best\_Feature }(T, F)$
      Set tree to be a nonterminal node with test f.
      for each value v of feature f,
        $T_{f=v} = \phi$
      for each example $T_i \in T$,
        $v = T_{i,f}$
        Add example T_i to set $T_{f=v}$.
      for each value v of feature f,
        subtree = Decision_Tree_Learning $(T_{f=v}, F - f, C)$
        Add a branch to tree labeled f with subtree subtree.
  return tree.

FIGURE 23.2 Decision tree learning algorithm. This algorithm takes a training set T, feature set F, and class feature C, as inputs and returns a decision tree. T_i denotes the ith training example, $T_{i,f}$ denotes example i's value for feature f, and $T_{i,C}$ denotes example i's value for the class feature C.

    Decision trees are constructed in a top-down manner. One decision tree learning algorithm (ID3) is shown in Figure 23.2. If all the examples are of the same class, then the algorithm just returns a leaf node of that class. If there are no features left with which to construct a nonterminal node, then the algorithm has to return a leaf node. It returns a leaf node labeled with the default class (often chosen to be the class most frequently seen in the training set). If none of these conditions is true, then the algorithm finds the one feature value test that comes closest to splitting the entire training set into groups such that each group only contains examples of one class (we discuss this in more detail in the next paragraph). When such a feature is selected, the training set is split according to that feature. That is, for each value v of the feature, a training set $T_{f=v}$ is constructed such that all the examples in $T_{f=v}$ have value v for the chosen feature. The learning algorithm is called recursively on each of these training sets.

    We have yet to describe the function *Choose_Best_Feature* in Figure 23.2. The feature with the highest *information gain* is commonly used. Information gain is defined as follows:

$$\text{Gain}(T,f) = \text{Entropy}(T) - \sum_{v \in \text{values}(f)} \frac{|T_{f=v}|}{|T|} \text{Entropy}(T_{f=v}),$$

$$\text{Entropy}(T) = \sum_{i \in \text{values}(C)} -p_i \log_2 p_i.$$

    Here, *values*(f) is the set of possible values of feature f if f is a categorical feature. If f is an ordered feature (whether discrete or continuous), then different binary split points are considered (e.g., tests of the form $f < v$ that split the training set into two parts). T is the training set, $T_{f=v}$ is the subset of the training set having value v for feature f, C is the class feature, and p_i is the fraction of the training set having class C_i. The information gain is the entropy of the training set minus the sum of the entropies of the subsets of the training set that result from separating the training set using feature f. The more homogeneous the training set in terms of class values, the lower the entropy. Therefore, information gain measures how much a feature improves the separation of the training set into subsets of homogeneous classes.

A decision stump is a decision tree that is restricted to having only one nonterminal node. That is, the decision tree algorithm is used to find the one feature test that comes closest to splitting all the training data into groups of examples of the same class. One branch is constructed for each value of the feature, and a leaf node is constructed at the end of that branch. The training set is split into groups according to the value of that feature and each group is attached to the appropriate leaf. The class label most frequently seen in each group is the class label assigned to the corresponding leaf. The reader would be justified in wondering when a model so seemingly simplistic as the decision stump would be useful. Decision stumps are useful as part of certain ensemble learning algorithms in which multiple decision stumps are constructed to form a model that is more accurate than any single decision stump and often more accurate than much more complicated models.

We chose decision trees as the first machine learning model to describe in this chapter because the decision tree is one of the simplest and most intuitive models. It is rather similar to a set of if–then statements. In many application domains, domain experts need to understand how the model makes its classification decisions to justify its use. Decision trees are among the most transparent and easy-to-understand models, and are able to handle continuous and discrete data. Decision tree learning also includes feature selection—that is, it only uses the features that it needs to separate the training set into groups that are homogeneous or nearly homogeneous in terms of the class variable. Decision tree learning is relatively fast even on large datasets and can also be altered to run even faster on parallel computers quite easily [22].

Decision trees have drawbacks that lead us to mention other machine learning models. The tests within decision trees are based on single attributes only. Decision tree learning algorithms are also unable to learn arithmetic combinations of features that are effective for classifications. Additionally, they are not designed to represent probabilistic information. The fraction of examples of different classes at each leaf is often used as a representation of the probability, but this is a heuristic measure. The number of examples at a leaf node, or even at nonterminal nodes deep in the tree, may be relatively low sometimes. Therefore, *pruning* is often used to remove such parts of the decision tree. This often helps to alleviate *overfitting*. However, pruning also leads to relatively few features being considered in cases where there is a small amount of available training data. Additionally, decision trees are sensitive to small changes in the training set. This makes them very useful as part of ensemble learning methods where the diversity in decision trees enables their combination to perform better, but may lead to difficulties when used by themselves.

23.4 NAÏVE BAYES

We earlier discussed Bayes's rule as a way of motivating generative classifiers. In particular,

$$P(Y = y | X = \mathbf{x}) = \frac{P(Y = y)P(X = \mathbf{x} | Y = y)}{P(X = \mathbf{x})}.$$

Bayes's theorem tells us that to optimally predict the class of an example $\mathbf{x}$, we should predict the class y that maximizes the two expressions in the above equation.

Define F to be the set of features. If all the features are independent given the class, then we can rewrite $P(X = \mathbf{x} | Y = y)$ as $\prod_{f=1}^{|F|} P(X_f = \mathbf{x}_{(f)} | Y = y)$, where $\mathbf{x}_{(f)}$ is the fth feature value of example $\mathbf{x}$. The probabilities $P(Y = y)$ and $P(X_f = \mathbf{x}_{(f)} | Y = y)$ for all classes Y and all possible values of all features X_f are estimated from a training set. For example, $P(X_f = \mathbf{x}_{(f)} | Y = y)$ is the fraction of class-y training examples that have $\mathbf{x}_{(f)}$ as their fth feature value. Estimating $P(X = \mathbf{x})$ is unnecessary because it is the same for all classes; therefore, we ignore it. To classify a new example, we can return the class that maximizes

$$P(Y = y) \prod_{f=1}^{|F|} P(X_f = \mathbf{x}_{(f)} | Y = y). \tag{23.4}$$

This is known as the Naïve Bayes classifier because it operates under the naïve assumption that the features are independent given the class. One simple algorithm for learning Naïve Bayes classifiers is shown in Figure 23.3. For each training example, we just increment the appropriate counts: N is the number of training examples seen so far, N_y is the number of examples in class y, and $N_{y, \mathbf{x}_{(f)}}$ is the number of examples in class y having $\mathbf{x}_{(f)}$ as their value for feature f. $P(Y = y)$ is estimated by N_y / N and, for all classes y and feature values $\mathbf{x}_{(f)}$, $P(X_f = \mathbf{x}_{(f)} | Y = y)$ is estimated by $N_{y, \mathbf{x}_{(f)}} / N_y$. The algorithm returns a classification function that returns, for an example $\mathbf{x}$

$$\operatorname{argmax}_{y \in C} \frac{N_y}{N} \prod_{f=1}^{|F|} \frac{N_{y, \mathbf{x}_{(f)}}}{N_y}.$$

Naive-Bayes-Learning (T, F)
　　for each training example $(x, y) \in T$,
　　　　Increment N
　　　　Increment N_y
　　　　for $f \in \{1, 2, \dots, |F|\}$
　　　　　　Increment $N_{y, \mathbf{x}_{(f)}}$
　　return $h(x) = \operatorname{argmax}_{y \in Y} \dfrac{N_y}{N} \prod_{f=1}^{|F|} \dfrac{N_{y, \mathbf{x}_{(f)}}}{N_y}$

FIGURE 23.3 Naïve Bayes Learning Algorithm. This algorithm takes a training set T and feature set F as inputs and returns a Naïve Bayes classifier h. N is the number of training examples seen so far, N_y is the number of examples in class y, and $N_{y, \mathbf{x}_{(f)}}$ is the number of examples in class y that have $\mathbf{x}_{(f)}$ as their value for feature f.

Every factor in this equation estimates its corresponding factor in Equation 23.4. In spite of the naïvety of Naïve Bayes classifiers, they have performed quite well in many experiments [14].

Note that, in general,

$$P(X = \mathbf{x}|Y = y) = \prod_{f=1}^{|F|} P(X_1 = \mathbf{x}_{(1)}, \ldots, X_f = \mathbf{x}_{(f)}|Y = y)$$

where we derived $P(X = \mathbf{x}|Y = y) = \prod_{f=1}^{|F|} P(X_f = \mathbf{x}_{(f)}|Y = y)$ through the assumption of features being conditionally independent given the class. One can introduce other less restrictive assumptions of conditional independence, such as two groups of features being conditionally independent, to derive other types of Bayesian classifiers. Probabilistic networks [13,21] include Bayesian classifiers of this type.

Naïve Bayes classifiers have several advantages that make them quite popular. They are very easy and fast to train, are robust to missing values, and have a clear probabilistic semantics. In spite of their rather strong assumption that the input features are independent given the class, they perform quite well in situations where this assumption is not true, and there is some theoretical evidence to justify this [12]. Another advantage is that the probability distributions that are estimated involve only one variable each. This alleviates the *curse of dimensionality*, which is the problem that, as the dimensionality of the input space increases, the training set required to fill the space densely enough to derive accurate models grows exponentially. Naïve Bayes classifiers are also relatively insensitive to small variations in the training set.

The simplicity of Naïve Bayes models also has the drawback that it cannot represent situations in which features interact to classify data. For example, with decision trees, we saw that certain features are only used when other features have particular values. Such interactions cannot be represented in a Naïve Bayes classifier. Additionally, the strong performance of Naïve Bayes even with violations of the assumption of feature independence comes about because correct classification only requires the correct class's posterior probability to be higher than the others—the probabilities do not actually have to be correct. Naïve Bayes classifiers may not necessarily give the correct probabilities.

23.5 NEURAL NETWORKS

The multilayer perceptron is the most common neural network representation. It is often depicted as a directed graph consisting of nodes and arcs—an example is shown in Figure 23.4. Each column of nodes is a *layer*. The leftmost layer is the *input layer*. The inputs or features of an example to be classified are entered into the input layer. The second layer is the *hidden layer*. The third layer is the *output layer*, which, in classification problems, typically consists of as many outputs as classes. Information flows from the input layer to the hidden layer to the output layer via a set of arcs. Note that the nodes within a layer are not directly connected. In our example, every node in one layer is connected to every node in the next layer, but this is not required in general. Also, a neural network can have more or less than one hidden layer and can have any number of nodes in each hidden layer.

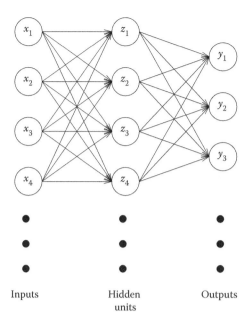

FIGURE 23.4 An example of a multilayer feedforward perceptron.

Each noninput node, its incoming arcs, and its single outgoing arc constitute a *neuron*, which is the basic computational element of a neural network. Each incoming arc multiplies the value coming from its origin node by the weight assigned to that arc and sends the result to the destination node. The destination node adds the values presented to it by all the incoming arcs, transforms it with a nonlinear activation function (to be described later in this chapter), and then sends the result along the outgoing arc. For example, the output of a hidden node z_j in our example neural network is

$$z_j = g \left(\sum_{i=1}^{|F|} w_{i,j}^{(1)} \mathbf{x}_{(f)} \right)$$

where $|F|$ is the number of input units, which is the same as the number of input features; $w_{i,j}^{(k)}$ is the weight on the arc in the kth layer of arcs that goes from unit i in the kth layer of nodes to unit j in the next layer (so $w_{i,j}^{(1)}$ is the weight on the arc that goes from input unit i to hidden unit j); and g is a nonlinear activation function. A commonly used activation function is the sigmoid function:

$$g(a) \equiv \frac{1}{1 + exp(-a)}.$$

The output of an output node y_j is

$$y_j = g \left(\sum_{i=1}^{Z} w_{i,j}^{(2)} z_i \right),$$

where Z is the number of hidden units. The outputs are clearly nonlinear functions of the inputs. Neural networks used for classification problems typically have one output per class. The example neural network depicted in Figure 23.4 is of this type. The outputs lie in the range $[0, 1]$. Each output value is a measure of the network's confidence that the example presented to it is a member of that output's corresponding class. Therefore, the class corresponding to the highest output value is returned as the prediction.

The most widely used method for setting the weights in a neural network is the backpropagation algorithm [6,15,20]. For each of the training examples in the training set T, its inputs are presented to the input layer of the network and the predicted outputs are calculated. The difference between each predicted output and the corresponding target output is calculated. The total error of the network is

$$E = \frac{1}{2} \sum_{t=1}^{|T|} (y_i - \hat{y}_i)^2, \tag{23.5}$$

where y_i and $\hat{y}_i$ are the true and predicted outputs, respectively, for the ith training example. In classification, neural networks are normally set up to have one output per class, so that y_i and $\hat{y}_i$ become vectors $\mathbf{y}_i$ and $\hat{\mathbf{y}}_i$. In the training set, $\mathbf{y}_{i,c} = 1$ if the ith training example is of class c and $\mathbf{y}_{i,c} = 0$ otherwise. By training the neural network with the mean-squared error criterion (Equation 23.5), the network attempts to estimate the posterior probability $P(Y|X)$ for every class Y. A separate value of E can be calculated for each class output. We can write E in terms of the parameters in the network as follows:

$$E = \frac{1}{2} \sum_{n=1}^{N} \left(y_n - g \left(\sum_{j=1}^{Z} w_{i,j}^{(2)} z_i \right) \right)^2,$$

$$= \frac{1}{2} \sum_{n=1}^{N} \left(y_n - g \left(\sum_{j=1}^{Z} w_{i,j}^{(2)} g \left(\sum_{i=1}^{|A|} w_{i,j}^{(1)} x_i \right) \right) \right)^2.$$

In order to adjust the weights to reduce the error, we calculate the derivative of E with respect to each weight and change the weight accordingly. The derivatives are

$$\frac{\partial E}{\partial w_{i,j}^{(2)}} = - \sum_{n=1}^{N} \left(y_n - g \left(\sum_{j=1}^{Z} w_{i,j}^{(2)} z_i \right) \right) g' \left(\sum_{j=1}^{Z} w_{i,j}^{(2)} z_i \right) z_i$$

$$\frac{\partial E}{\partial w_{i,j}^{(1)}} = \sum_{n=1}^{N} \left(y_n - g \left(\sum_{j=1}^{Z} w_{i,j}^{(2)} g \left(\sum_{i=1}^{|A|} w_{i,j}^{(1)} x_i \right) \right) \right)$$

$$\times g' \left(\sum_{j=1}^{Z} w_{i,j}^{(2)} g \left(\sum_{i=1}^{|A|} w_{i,j}^{(1)} x_i \right) \right) g' \left(\sum_{i=1}^{|A|} w_{i,j}^{(1)} x_i \right) x_i$$

Note that many factors in the derivatives are redundant and, if g is a sigmoid function, then $g' = g(1-g)$, so these derivatives can be calculated quickly. The weights on the arcs of the networks are adjusted according to these derivatives so that if the training example is presented to the network again, then the error would be less. The learning algorithm typically cycles through the training set many times—each cycle is called an *epoch* in the neural network literature.

Neural networks have performed well in a variety of domains for nearly 50 years. They are able to represent complicated interactions among features when necessary to derive a classification. They are also *universal approximators*—given a single hidden layer and an arbitrary number of hidden units, they can approximate any continuous function, and given two hidden layers and an arbitrary number of hidden units, they can approximate any function to arbitrarily high accuracy.

The universal approximator property of neural networks is nice in theory; however, in practice, because neural network learning is computationally intensive, reaching a small error can take excessive time and can lead to overfitting. Determining the best number of hidden units is more art than science. Neural networks are learned using gradient descent methods, which use partial derivatives of the error with respect to model parameters as shown above to adjust the parameters to improve the accuracy. However, gradient descent methods applied to nonlinear models such as neural networks are guaranteed to reach solutions that are locally optimal but not necessarily globally optimal. Additionally, the error as a function of the neural network's weights seems to be relatively complicated, because starting the learning from different initial weights tends to lead to very different final weights and the resulting networks have different errors in spite of learning with the same training set. This indicates that there are many locally optimal solutions. Neural networks are also nearly impossible for domain experts and even machine learning experts to interpret for the purpose of understanding how they arrive at their classifications.

23.6 SUPPORT VECTOR MACHINES

Support vector machines (SVM) were developed by Vapnik in 1979 (see [7] for a tutorial). SVMs are learned using a statistical learning method based on structural risk minimization (SRM). In SRM, a set of classification functions that classify a set of data are chosen in such a way that minimizing the training error (what Vapnik refers to as the empirical risk) yields the minimum upper bound on the test error (what Vapnik refers to as the actual risk).

The simplest example of an SVM is a linear hyperplane trained on data that are perfectly separable as shown on the left side of Figure 23.5. Given a set of input vectors, $\mathbf{x}_i \in \mathcal{R}^d$, and labels, $y_i \in \{-1, 1\}^*$, SVM finds a hyperplane described by its normal vector, $\mathbf{w}$, and distance from the origin, b, that divides the data perfectly and is equidistant from at least one point in each class that is closest to the hyperplane (this distance is shown as $m/2$ in Figure 23.5—the total distance for the two classes, m, is called the *margin*). This hyperplane

---

[*] For the case where the number of classes C is greater than two, typically the problem is split into $|C|$ two-class problems where, for each class $c \in C$, the corresponding class-c SVM learns to predict whether the example is in class c or not.

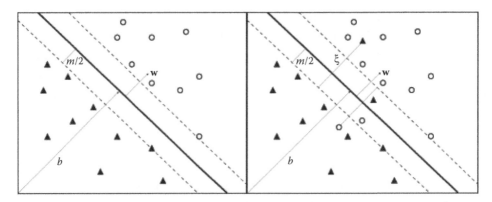

FIGURE 23.5 Optimum hyperplane for separable and nonseparable case.

is a decision boundary and the classification of an unknown input vector is determined by the sign of the vector operation

$$\mathbf{x}_i \cdot \mathbf{w} - b = d \qquad (23.6)$$

If $d \geq 0$ ($d < 0$) then the input is predicted to be in the class $y = +1$ ($y = -1$).

Learning an SVM requires using the training data to find the best value of $\mathbf{w}$. The SVM is constructed by solving the optimization problem

$$\min \| \mathbf{w} \|$$

$$\text{s.t.} \quad \mathbf{x}_i \cdot \mathbf{w} + b \geq 1 \quad \text{for } y_i = +1$$

$$\mathbf{x}_i \cdot \mathbf{w} + b \leq -1 \quad \text{for } y_i = -1,$$

where the last two equations can be combined into one as follows:

$$y_i(\mathbf{x}_i \cdot \mathbf{w} + b) - 1 \geq 0 \ \forall i \in \{1, 2, \ldots, |T|\}.$$

The objective function is chosen because we would like to maximize the margin, which turns out to be $2/ \| \mathbf{w} \|$, and so this is equivalent to minimizing $\| \mathbf{w} \|$.

If the data are not perfectly separable this method can be adapted to compensate for instances that occur on the wrong side of the hyperplane. In that case slack variables, ξ_i (one for each training example), are introduced that measure the error of misclassified instances of the training data—in particular, the slacks measure how far on the wrong side of the hyperplane these misclassified instances are (see the right side of Figure 23.5). SVMs find a hyperplane that best separates the data and minimizes the sum of the errors ξ_i by changing the optimization problem to the following:

$$\min \| \mathbf{w} \| + B \sum_{i=1}^{|T|} \xi_i \qquad (23.7)$$

$$\text{s.t.} \quad \mathbf{x}_i \cdot \mathbf{w} + b \geq 1 - \xi_i \quad \text{for } y_i = +1, \qquad (23.8)$$

$$\mathbf{x}_i \cdot \mathbf{w} + b \leq -1 + \xi_i \quad \text{for } y_i = -1, \tag{23.9}$$

$$\xi_i \geq 0 \qquad \forall i \in \{1, 2, \ldots, |T|\}, \tag{23.10}$$

where B is a user-defined weight on the slack variables. If B is large, the learning algorithm puts a large penalty on errors and will devise a more complicated model. If B is small, then the classifier is simpler but may have more errors on the training set. If the datasets are not balanced it is sometimes necessary to give the errors of one class more weight than the other. Additional parameters, μ_y (one for each class), can be added to weigh one class error over the other by replacing the objective in Equation 23.7 with

$$\min \| \mathbf{w} \| + B \sum_{i=1}^{|T|} \xi_i \mu_{y_i}.$$

We can write the Wolfe dual of the convex optimization problem shown in Equation 23.7 as follows:

$$\max \sum_{i=1}^{|T|} \alpha_i - \frac{1}{2} \sum_{i=1}^{|T|} \sum_{j=1}^{|T|} \alpha_i \alpha_j y_i y_j (\mathbf{x}_i \cdot \mathbf{x}_j) \tag{23.11}$$

$$\text{s.t.} \quad 0 \leq \alpha_i \leq B, \tag{23.12}$$

$$\sum_{i=1}^{|T|} \alpha_i y_i = 0. \tag{23.13}$$

This decision hyperplane can only be linear which is not suitable for many types of data. To overcome this problem a function can be used to map the data into a higher- or infinite-dimensional space and run SVM in this new space.

$$\Phi : \mathbf{R}^d \mapsto \mathcal{H}$$

Because of the "kernel trick," $\mathcal{H}$ may have infinite dimension while still making the learning algorithm practical. In particular, one never works directly with Φ but rather with a kernel function K such that

$$K(\mathbf{x}_i, \mathbf{x}_j) = \Phi(\mathbf{x}_i) \cdot \Phi(\mathbf{x}_j),$$

so that the dual formulation shown in Equation 23.11 becomes

$$\max \sum_{i=1}^{|T|} \alpha_i - \frac{1}{2} \sum_{i=1}^{|T|} \sum_{j=1}^{|T|} \alpha_i \alpha_j y_i y_j K(\mathbf{x}_i, \mathbf{x}_j)$$

$$\text{s.t.} \quad 0 \leq \alpha_i \leq B,$$

$$\sum_{i=1}^{|T|} \alpha_i y_i = 0.$$

There are many possible kernel functions available and new kernels can be built from the data itself. The kernels only need to meet Mercer's conditions [8] to be used in SVM. One example kernel function is a radial basis function,

$$K(\mathbf{x}_i, \mathbf{x}_j) = e^{-\gamma \|\mathbf{x}_i - \mathbf{x}_j\|^2},$$

where γ is a parameter entered by the user.

For a new example $\mathbf{x}$, the predicted class is the sign of

$$f(\mathbf{x}) = \sum_{i=1}^{|T|} \alpha_i y_i K(\mathbf{x}_i, \mathbf{x}) + b.$$

However, this can be simplified. The summation actually does not need to be calculated over all the training examples, but rather only those training examples whose distance from the hyperplane is exactly half the margin and those which are on the wrong side of the hyperplane. These are referred to as *support vectors*, and are the only training points having $\alpha_i \neq 0$. If V is the subset of T containing the support vectors, then the above classification function can be simplified to

$$f(\mathbf{x}) = \sum_{(\mathbf{s}, y) \in V} \alpha_i y K(\mathbf{s}, \mathbf{x}) + b.$$

As mentioned earlier, the sign of $f(\mathbf{x})$ indicates the predicted class. $|f(\mathbf{x})|$ indicates how confident the SVM is in its class. $|f(\mathbf{x})|$ does not correspond directly to a probability, but can nevertheless can be used as a measure of confidence.

SVMs are generated by solving a convex optimization problem, which means that the solution is guaranteed to be optimal for that training set, even though SVMs can be nonlinear in the original data space. They are also usable in problems such as text mining where a very large number of input features is present. They have been shown to perform very well experimentally in many domains. SVMs are applicable to many domains because kernels over many different types of inputs—discrete, continuous, graphs, sequences, and so on—have been derived. Multiple kernels can even be used at the same time [1,10,11]. These properties have led SVMs to be one of the most popular machine learning models in use today among machine learning researchers and practitioners, as can be seen in the many papers on SVMs in machine learning conferences over the last 10 years.

SVMs have the disadvantage, just like neural networks, of being impenetrable by domain experts wishing to understand how they arrive at their classifications. Learning SVMs is also computationally intensive, although there have been substantial efforts to reduce their running time (such as in Ref. [9]). They are also designed to solve two-class problems. As mentioned earlier, they can be set up to solve $|C|$-class problems by setting up $|C|$ SVMs, each of which solves a two-class problem. However, this means $|C|$ times the computation of a two-class problem. Also, if the set of classes is mutually exclusive and if more than one SVM predicts that the example is in its corresponding class, then it is unclear how to determine which class is the correct one.

23.7 FURTHER READING

We gave several references throughout the chapter for more information on the machine learning algorithms discussed here. There have also been several attempts at applying classification methods to astronomy data. Classification and other forms of analysis on data representing different solar phenomena are described in Ref. [4]. Classification of stars and galaxies in various sky surveys such as the Sloan Digital Sky Survey are described in Refs. [3,5]. See Ref. [19] for a description of the use of classification methods for matching objects recorded in different catalogues. Classification of sunspot images obtained by processing some NASA satellite images is described in Ref. [16]. A recent, general survey of the use of data mining and machine learning in astronomy is Ref. [2].

REFERENCES

1. F. R. Bach, G. R. G. Lanckriet, and M. I. Jordan. Multiple kernel learning, conic duality, and the SMO algorithm. In *International Conference on Machine Learning*, pp. 41–48, Banff, Canada, 2004.
2. N. M. Ball and R. J. Brunner. Data mining and machine learning in astronomy. *International Journal of Modern Physics D*, 19(7):1049–1106, 2009.
3. N. M. Ball, R. J. Brunner, and A. D. Myers. Robust machine learning applied to terascale astronomical datasets. In R. W. Argyle, P. S. Bunclark, and J. R. Lewis, editors, *Astronomical Data Analysis Software Systems (ADASS) XVII, ASP Confererence Series*, pp. 201–204, National Optical Astronomy Observatory, Tucson, Arizona, 2008.
4. J. M. Banda and R. Anrgyk. Usage of dissimilarity measures and multidimensional scaling for large scale solar data analysis. In *Proceedings of the 2010 Conference on Intelligent Data Understanding*, pp. 189–203, National Aeronautics and Space Administration, Mountain View, California, 2010.
5. K. D. Borne. A machine learning classification broker for petascale mining of large-scale astronomy sky survey databases. In *National Science Foundation Symposium on Next Generation of Data Mining and Cyber-Enabled Discovery for Innovation (NGDM-07)*. National Science Foundation, Baltimore, Maryland, 2007.
6. A. E. Bryson and Y.-C. Ho. *Applied Optimal Control*. Blaisdell Publishing Co., Waltham, Massachusetts, 1969.
7. C. J. C. Burges. A tutorial on support vector machines for pattern recognition. *DMKD*, 2:121–167, 1998.
8. N. Cristianini and J. Shawe-Taylor. *An Introduction to Support Vector Machines and Other Kernel-Based Learning Methods*. Cambridge University Press, Cambridge, UK, 2000.
9. S. Das, K. Bhaduri, N. Oza, and A. Srivastava. Nu-anomica: A fast support vector based anomaly detection technique. In *Proceedings of the IEEE International Conference on Data Mining (ICDM)*. Institute for Electrical and Electronics Engineers, pp. 101–109, Miami Beach, Florida, 2009.
10. S. Das, B. Matthews, K. Bhaduri, N. Oza, and A. Srivastava. Detecting anomalies in multivariate data sets with switching sequences and continuous streams. In *NIPS 2009 Workshop: Understanding Multiple Kernel Learning Methods*, Whistler, BC, Canada, 2009.
11. S. Das, B. Matthews, A. Srivastava, and N. C. Oza. Multiple kernel learning for heterogeneous anomaly detection: Algorithm and aviation safety case study. In B. Rao, B. Krishnapuram,

A. Tomkins, and Q. Yang, editors, *The Sixteenth ACM SIGKDD International Conference on Knowledge Discovery and Data Mining*, pp. 47–56, ACM Press, New York, NY, 2010.

12. P. Domingos and M. Pazzani. On the optimality of the simple bayesian classifier under zero-one loss. *Machine Learning*, 29:103–130, 1997.

13. D. Koller and N. Friedman. *Probabilistic Graphical Models: Principles and Techniques*. MIT Press, Cambridge, Massachusetts, 2009.

14. D. Michie, D. J. Spiegelhalter, and C. C. Taylor. *Machine Learning, Neural and Statistical Classification (edited collection)*. Ellis Horwood, New York, 1994.

15. T. Mitchell. *Machine Learning*. McGraw-Hill, Columbus, Ohio, 1997.

16. T. T. Nguyen, C. P. Willis, D. J. Paddon, S. H. Nguyen, and Hung Son Nguyen. Learning sunspot classification. *Fundamenta Informaticae*, 72(1):295–309, 2006.

17. J. R. Quinlan. *C4.5: Programs for Machine Learning*. Morgan Kaufman, San Mateo, California, 1992.

18. R. Quinlan. Induction of decision trees. *Machine Learning*, 1(1):81–106, 1986.

19. D. Rohde, M. Drinkwater, M. Gallagher, T. Downs, and M. Doyle. Machine learning for matching astronomy catalogues. In Z. R. Yang, R. Everson, and H. Yin, editors, *Intelligent Data Engineering and Automated Learning (IDEAL)*, pp. 702–707. Springer-Verlag, Berlin, Germany, 2004.

20. D. E. Rumelhart, G. E. Hinton, and R. J. Williams. Learning internal representations by error propagation. In D. E. Rumelhart and J. L. McClelland, editors, *Parallel Distributed Processing: Explorations in the Microstructure of Cognition*, pp. 318–362. Bradford Books/MIT Press, Cambridge, MA, 1986.

21. S. Russell and P. Norvig. *Artificial Intelligence: A Modern Approach*, 3rd Edition, Prentice-Hall, Englewood Cliffs, NJ, 2010.

22. A. Srivastava, E. Han, V. Kumar, and V. Singh. Parallel formulations of decision-tree classification algorithms. *Data Mining and Knowledge Discovery*, 3(3):237–261, 1999.

CHAPTER 24

On the Shoulders of Gauss, Bessel, and Poisson

Links, Chunks, Spheres, and Conditional Models

William D. Heavlin
Google, Inc.

CONTENTS

24.1 Introduction 524
    24.1.1 Notation for Observations 524
    24.1.2 Probability Models 525
    24.1.3 Statistical Models 525
    24.1.4 Maximum Likelihood Estimation 526
24.2 Chunks and Conditional Models 527
    24.2.1 Bessel's Personal Equations 527
    24.2.2 Nuisance Effects and Attenuation 527
    24.2.3 Chunking and Conditional Likelihood 528
        24.2.3.1 Kinds of Chunking 528
        24.2.3.2 Conditioning on the Within-Chunk Response 529
        24.2.3.3 Chunks of Size 2 530
24.3 The Spherical Approximation 530
    24.3.1 Chunks of Size 3 530
    24.3.2 The von Mises–Fisher Distribution 531
    24.3.3 The Spherical Likelihood Gradient 532
    24.3.4 The Spherical Throttling Ratio 532
    24.3.5 Some Normal Equations 534
    24.3.6 The Chunk-Specific Correlation Coefficient 534
    24.3.7 Assessing the Spherical Approximation 535
24.4 Related Methods 537
    24.4.1 Iteratively Reweighted Least Squares 537
    24.4.2 Marginal Regression 537

24.4.3 A Generalization to Many Features 538
24.4.4 Multiple-Stage Models 538
24.5 Conclusions 540
References 540

24.1 INTRODUCTION

The foregoing chapters illustrate how data processing and statistical technologies can be applied to issues of scientific interest to astronomers. Here I focus on an inverse mapping: applying the physical intuition that astronomers possess to advance the science of statistical modeling. I found this remarkably easy to do, and I hope you enjoy the perspective.

I address the topic of generalized linear models [6], for which the literature has emerged in the last 20–30 years. The particular ideas I present lie on the boundary between those established and those still unfolding. The result is a rather personal perspective, for which I hope to pull you all along.

We begin first by setting some notation, by specifying the models, and by describing key data structures. The main novelty in what follows is the spherical approximation introduced in Section 24.3. In that framework, we derive some interesting, slightly new and intriguingly familiar algorithms.

24.1.1 Notation for Observations

Consider a set of independent observations $(y_i, \mathbf{x}_i)$, $i = 1, 2, \ldots, n$. The y_i are responses, that is, the causal consequences of the system we observe, while the $\mathbf{x}_i$ are the associated predictors, features, and/or causes. With little loss of generality, we take each y_i to be a scalar, while each $\mathbf{x}_i$ is a J-dimensional vector. The matrix version of this notation denotes the responses by the n-element vector $\mathbf{y} = (y_1, y_2, \ldots, y_n)$ and the associated features are collated into an $(n \times J)$-matrix $\mathbf{X}$; the ith row of $\mathbf{X}$ is $\mathbf{x}_i^{\mathsf{T}}$.

Alternately, this set of observations can be (and typically is) grouped or *chunked* into subsets indexed by g. To support this case, we doubly index observations $(y_{gi}, \mathbf{x}_{gi} : g = 1, 2, \ldots, G; i = 1, \ldots, n_g)$. The corresponding vector and matrix notation denotes the responses of the gth chunk by the n_g-vector $\mathbf{y}_g = (y_{g1}, y_{g2}, \ldots, y_{gn_g})$ and the feature vectors of the gth chunk by the $(n_g \times J)$-matrix $\mathbf{X}_g$; the ith row of $\mathbf{X}_g$ is $\mathbf{x}_{gi}$.

Therefore, we index observations by either i or gi, and we reserve j to index features, that is, to reference the columns of $\mathbf{X}$ and $\mathbf{X}_g$.

The chunking index g allows us to recognize that datasets, particularly large ones, are made up of the combination many subsets, each of which consists of data observed in close proximity to one another, so the observations within a chunk are in some way correlated. Chunks can be coarse, corresponding to an observatory or observatory-day, or quite fine grained, limited to double fields of view or even just a few pixels. Laboratory scientists easily associate a chunk with a common calibration sample, one chunk per control or per calibration run.

A compact way to represent within-chunk correlation is with this pair of assumptions: (a) the chunks are mutually independent, and (b) the observations within a chunk are conditionally independent given the chunk.

24.1.2 Probability Models

Consider a response y. When the domain of y is the nonnegative integers $0,1,2,\ldots$, a common probability model is that of Poisson:

$$\Pr\{y|\eta\} = \frac{e^{-\eta}\eta^y}{y!}, \quad y = 0,1,2,\ldots \tag{24.1}$$

When the domain of y is the real line, $-\infty < y < \infty$, the Gaussian probability model has attractive properties:

$$\Pr\{y|\eta,\sigma\} = \frac{\exp\{-(y-\eta)^2/(2\sigma^2)\}}{\sqrt{2\pi}\sigma}, \quad -\infty < y < \infty. \tag{24.2}$$

To complete the common cases, when y is constrained to two values, 0 and 1, the logistic distribution is conceptually convenient:

$$\Pr\{y|\eta\} = \frac{\exp\{y\eta\}}{1 + \exp\{\eta\}}, \quad y = 0,1. \tag{24.3}$$

(The logistic distribution can also be derived from the Poisson by ignoring any counts of two and larger.)

A useful abstraction recognizes a common form to Equations 24.1, 24.2, and 24.3, the *exponential family* (EXP). Consider this form:

$$\Pr\{y_i|\eta_i\} = \exp\{y_i h_0(\eta_i) + h_1(y_i) + h_2(\eta_i)\}. \tag{24.4}$$

The latter encompasses the Poisson, Gaussian, and logistic distributions. The function h_1 manages the domain over which the distribution is defined, while the function h_2 ensures the sum or integration of all probabilities over this domain equals unity.

24.1.3 Statistical Models

Let us now integrate the data notation with the above probability models.

We associate each response y_i with an intensity parameter η_i, which in turn is a function of the features $\mathbf{x}_i$. A key simplification is to take η_i as a linear function of $\mathbf{x}_i$,

$$\eta_i = f(\mathbf{x}_i^{\mathrm{T}}\beta), \tag{24.5}$$

where $\mathbf{x}_i^{\mathrm{T}}\beta \equiv \sum_j x_{ij}\beta_j$. We refer to Equation 24.5 as *latent feature linearity* (LFL): the under-lying intensity η_i is unobservable, or latent, yet is nonetheless a linear combination of the available features. Since this form allows, say, polynomials of the features $\mathbf{x}_i$, and preserves the rather broad flexibility of specifying the features $\mathbf{x}$ themselves, LFL is not overly restrictive. Further, many of the models that we consider embed $\mathbf{x}^{\mathrm{T}}\beta$ in expressions like $\exp\{\mathbf{x}^{\mathrm{T}}\mathbf{b}\} = e^{x_1\beta_1}e^{x_2\beta_2}\ldots e^{x_J\beta_J}$; the multiplications implicit in the latter align reasonably well with physical and empirical laws of proportionality, and their generalization, the power laws.

As anticipated, we combine the assumption EXP of Equation 24.4 with that of LFL of Equation 24.5 to achieve this form:

$$\Pr\{y_i|\mathbf{x}_i^{\mathrm{T}}\beta\} = \exp\{y_i h_0(\mathbf{x}_i^{\mathrm{T}}\beta) + h_1(y_i) + h_2(\mathbf{x}_i^{\mathrm{T}}\beta)\}. \tag{24.6}$$

We make one further simplification, which is to linearize h_0. This is consistent with the Poisson, Gaussian, and logistic distributions, but does exclude certain distributions, for example, the gamma, that are part of the EXP but without a linear h_0. The result is that Equation 24.6 is transformed into

$$\Pr\{y_i|\mathbf{x}_i^{\mathrm{T}}\beta\} = \exp\{y_i \mathbf{x}_i^{\mathrm{T}}\beta + h_1(y_i) + h_2(\mathbf{x}_i^{\mathrm{T}}\beta)\}. \tag{24.7}$$

Taking h_0 as the identify function is sometimes called the *canonical link.*

24.1.4 Maximum Likelihood Estimation

For mutually independent observations, we can write the joint probability of observing our responses y_i, $i = 1, 2, \ldots, n$, conditional on the features $\mathbf{x}_i$, $i = 1, 2, \ldots, n$:

$$\prod_{i=1}^{n} \Pr\{y_i|\mathbf{x}_i^{\mathrm{T}}\beta\}$$

As a function of the observations $\{(y_i|, \mathbf{x}_i) : i = 1, \ldots, n\}$ but assuming that the parameter β is known, the latter is the *joint probability* of observing the facts $\{y_i : i = 1\ldots, n\}$, conditional on $\{\mathbf{x}_i : i = 1\ldots, n\}$. As a function of the unknown parameters β, but implicitly with knowledge of the observations $(y_i, \mathbf{x}_i)$, $i = 1, 2, \ldots, n$, the latter expression is called a *likelihood* (or the *likelihood of* β). Taking logarithms allows us to deal with the likelihood construct as a sum:

$$\sum_{i=1}^{n} \log \Pr\{y_i|\mathbf{x}_i^{\mathrm{T}}\beta\} \equiv L(\beta), \tag{24.8}$$

the latter expression referred to as the log-likelihood. Substituting (Equations 24.7 into 24.8) offers this simplified form:

$$L(\beta) = \sum_{i=1}^{n} \left\{ y_i \mathbf{x}_i^{\mathrm{T}}\beta + h_1(y_i) + h_2(\mathbf{x}_i^{\mathrm{T}}\beta_i) \right\}. \tag{24.9}$$

A plausible approach to estimating the unknown parameters β is to determine those that maximize the likelihood $\exp\{L(\beta)\}$, or, equivalently, that maximize $L(\beta)$. This motivates these estimating equations:

$$\nabla_\beta L(\beta) = \sum_{i=1}^{n} [y_i + h_2'(\mathbf{x}_i^{\mathrm{T}}\beta_i)]\mathbf{x}_i = 0 \tag{24.10}$$

Estimates derived by solving (Equation 24.10) are known to have good statistical properties, including consistency and asymptotic efficiency. The absence of the h_1-term suggests some underlying elegance. Deeper theory shows that $-h_2'(\mathbf{x}_i^T\boldsymbol{\beta})$ acts as a prediction of y_i, hence (Equation 24.10) strives to make uncorrelated with $\mathbf{x}_i$ the difference between the observed y_i and this predicted value.

24.2 CHUNKS AND CONDITIONAL MODELS

24.2.1 Bessel's Personal Equations

Nineteeth-century astronomers recorded stellar transits in the following way: (1) They noted the time on a clock to the nearest second. (2) Then they counted pendulum beats as the star crossed the telescopic field. (3) Against this regular progression of integer counts, they noted the positions of the star with respect to threads strung across the telescopic field. The resulting method, involving the coordination of perceptions by the ear and eye, proved unexpectedly variable from one astronomer to another. Over the years, astronomers refined their techniques, by reducing their lack of sleep, avoiding observations after eating, and so on [11].

In 1826, Friedrich Bessel performed a series of experiments that documented some systematic differences in observation among his peer astronomers [12]. A subset of his calculations are presented in Table 24.1.

Over the next few decades, this phenomenon, whereby each observer has his or her own *personal equation*, became part of standard astronomical practice. Also, for the nascent science of experimental psychology, the topic offered an early area of investigation.

24.2.2 Nuisance Effects and Attenuation

Bessel's personal equations are the first documented case of a larger and persistent statistical concept, for which the modern term is *nuisance effects*[6]. These are (a) effects in the sense that they genuinely represent some aspect of the observational process, and (b) nuisances in the sense that, unrelated to the phenomenon of interest, they do not properly belong as part of any prediction system.

From the latter point (b), the temptation is always to construct a predictive model that simply ignores nuisance effects. The alternative would be to define a set of chunks for which the effect is constant. (In Bessel's case, this would be the set of observations by a single astronomer.) Such a model typically takes the following form:

$$\eta_{gi} = \mu_g + \mathbf{x}_{gi}^T\boldsymbol{\beta}, \tag{24.11}$$

TABLE 24.1   Astronomers' Personal Equations

in 1820–21	B−W= −1.041
	S−W= −0.242
hence	B−S= −0.799
in 1823	B−A= −1.223
	S−A= −0.202
giving	B−S= −1.021

Note: A = Argelander,   B = Bessel,   S = Struve, W = Walbeck

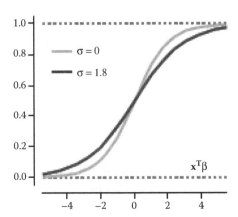

FIGURE 24.1 Equation 24.12 for two values of σ.

the nuisance effect for chunk g being represented by the parameter μ_g.

Nuisance effects cannot be ignored without hazard. Consider a random chunk effect in a logistic model:

$$\Pr\{y|\mathbf{x}^{\mathsf{T}}\beta\} = \int_{-\infty}^{\infty} \frac{\exp\{\sigma z + \mathbf{x}^{\mathsf{T}}\beta\}}{1 + \exp\{\sigma z + \mathbf{x}^{\mathsf{T}}\beta\}} f(z)\,dz. \tag{24.12}$$

The nuisance effect z appears here as a random effect for which the parameter σ modulates its magnitude. As Figure 24.1 illustrates, near the origin, the ratio of slopes of ($\sigma = 0$) to ($\sigma = 1.8$) is 1.5; the result of larger σ is diminished precision—*attenuation*.

Attenuation issues abound in machine learning, and may even be thought of as the inevitable consequence of the flat or rectangular $n \times J$ array as a data structure. As we see below, chunking is a primary strategy for reducing attenuation effects.

24.2.3 Chunking and Conditional Likelihood

24.2.3.1 Kinds of Chunking

In practice, chunking occurs in one of three ways. The first is a consequence of the data collection process itself. For example, in the case of Bessel's personal equations, described above, observations from each astronomer, and perhaps even from each astronomer and observation-day (more properly, observation-*night*), can define a chunk. By chunking in this way, one assesses the changes in the night sky as would be perceived by a single observer (a single astronomer, say). This approach is self-levelling: the systematic effects that define the chunking are separated from the differences recorded within the chunk, the atmospheric nuisance effects of a given night, its humidity, temperature, and transparency, for example.

Alternately, one can define a chunk by fixing the target of the observation. Thus, a single field of view, for example, one centered on Cygnus, measured over time can also define an effective chunk. Such chunking is sensitized to the changes among the observation times, for instance the changes within a given coordinate of the observation field. Image subtraction, whereby a bright spot stands out in the midst of grayscale background to highlight a nova, provides the canonical example of this kind of chunking.

The third reason for chunking is to assist with computation and parallelization (and so the associated scientific issues reside more in the background). In this scheme, a data chunk is scaled to the size of the available memory cache, and thereby helps to make parallel computation tractable. As we see in Section 24.3.2, conditional likelihoods are free of intercept terms; without a common intercept, within-chunk calculations do not need to communicate (in an algorithmic sense) with other chunks. In particular, both gradients (and any second derivative hessians) can be calculated within chunk; the overall gradient (hessian) is then assembled by simple summing across chunks.

In view of modern computational capabilities, this latter rationale might seem negligible, but it can provide the key enabler for parallel computing using graphics processing units (GPUs), limited as GPUs are by their local memory sizes. This implies working backward from the GPU memory size to determine the maximum feasible chunk size.

24.2.3.2 Conditioning on the Within-Chunk Response

In Equation 24.11, μ_g parametrizes the chunk effect; larger values of μ_g induce larger values of η_g. In this way, the overall magnitude of $\mathbf{y}_g = (y_{g1}, y_{g2}, \ldots)$ is closely connected to estimates of the size μ_g. As it happens, if we condition on the size or sizes of $\mathbf{y}_g$, we can form a likelihood free of μ_g[1].

Denote the (shuffled) set of responses $\{y_{g,\tau[i]} : i = 1, 2, \ldots, n\}$ for an unknown permutation τ of indices by $(\tau[1], \tau[2], \ldots, \tau[n])$. This is compactly represented by a permutation matrix $\mathbf{P}_\tau$, such that $\mathbf{y}[\tau[1, 2, \ldots, n]] = \mathbf{P}_\tau \mathbf{y}$, that is, the permutation τ that is induced by the left matrix multiplication of $\mathbf{P}_\tau$ upon the vector $\mathbf{y}$.

For group g, the conditional likelihood given the histogram of $\mathbf{y}_g$, that is, given $\{y_{g,\tau[i]} : i = 1, 2, \ldots, n\}$ for τ unknown, is

$$\exp\{L_{cg}(\beta)\} \equiv \frac{\prod_{i=1}^{n} Pr\{y_{gi}|\mathbf{x}_{gi}^{\mathbf{T}}\beta\}}{\sum_{\forall \tau} \prod_{i=1}^{n} Pr\{y_{g,\tau[i]}|\mathbf{x}_{gi}^{\mathbf{T}}\beta\}}$$

$$= \frac{\prod_{i=1}^{n} \exp\{y_{gi}(\mu_g + \mathbf{x}_{gi}^{\mathbf{T}}\beta)\}}{\sum_{\forall \tau} \prod_{i=1}^{n} \exp\{y_{g\tau[i]}(\mu_g + \mathbf{x}_{gi}^{\mathbf{T}}\beta)\}} = \frac{\exp\{\mathbf{y}_g^{\mathbf{T}}\mathbf{X}_g\beta\}}{\sum_{\forall \tau} \exp\{\mathbf{y}_g^{\mathbf{T}}\mathbf{P}_\tau^{\mathbf{T}}\mathbf{X}_g\beta\}}, \qquad (24.13)$$

recalling $\mathbf{X}_g$ is the $n_g \times J$ matrix whose ith row is $\mathbf{x}_{gi}$. Note that the rightmost expression of Equation 24.13 is free of μ_g. The overall conditional likelihood is thus the product $\exp\{L_c(\beta)\} \equiv \prod_{g=1}^{G} \exp\{L_{cg}(\beta)\}$, and the log-conditional likelihood is $L_c(\beta) \equiv \sum_{g=1}^{G} L_{cg}(\beta)$, with a simplification analogous to Equation 24.9. The gradient with respect to β likewise resembles (Equation 24.10):

$$\nabla_\beta L_c(\beta) = \sum_g \left[\mathbf{X}_g^{\mathbf{T}}\mathbf{y}_g - \frac{\sum_{\forall \tau} \omega_g(\tau)\mathbf{X}_g^{\mathbf{T}}\mathbf{P}_\tau\mathbf{y}_g}{\sum_{\forall \tau} \omega_g(\tau)} \right], \qquad (24.14)$$

where $\omega_g(\tau) = \exp\{\mathbf{y}_g^{\mathbf{T}}\mathbf{P}_\tau^{\mathbf{T}}\mathbf{X}_g\beta\}$, the rightmost term acting like predictions of $\mathbf{X}_g^{\mathbf{T}}\mathbf{y}_g$.

Note that Equation 24.14 is invariant to additive shifts in $\mathbf{y}_g$: replacing $\mathbf{y}_g$ with $\mathbf{y}_g + c$ for any constant c gives the same value of $L_{cg}(\beta)$. Likewise, $\mathbf{X}_g$ is also so invariant. Therefore, without loss of generality, we can enforce these constraints by directly requiring

$$\mathbf{1}^{\mathbf{T}}\mathbf{y}_g = 0; \quad \mathbf{1}^{\mathbf{T}}\mathbf{X}_g = \mathbf{0}^{\mathbf{T}}. \tag{24.15}$$

24.2.3.3 Chunks of Size 2

For a given g, when $y_{gi} = $ constant, $i = 1, \ldots, n_g$, $\exp\{L_{cg}(\beta)\}$ is likewise constant in β. Therefore, only chunks for which the y_{gi} vary are of interest. For relatively rare events, this simple fact often allows us to reduce the volume of data by one or more orders of magnitude.

Still, the number of terms in the denominators of $\exp\{L_{cg}(\beta)\}$ is potentially $n_g!$, so conditional models carry an issue of combinatorial complexity that unconditional likelihoods do not. The simplest, and least combinatorial, case involves chunks of size two, $n_g = 2$. In this case, the entire EXP simplifies uniformly to this doubly differenced form,

$$\frac{1}{1 + \exp\{-(y_{g2} - y_{g1})(x_{g2} - x_{g1})^{\mathbf{T}}\beta\}}.$$

Note that for the special case with binary responses where $y_{g2} = 0$, $x_{g2} = 0$, this reduces to (unchunked, hence unconditonal) logistic regression [4]. (y_{g2}, x_{g2}) thus can be interpreted as the control or background observation particular to chunk g. This same observation generalizes to chunk sizes $n_g \geq 2$: observations (y_{gi}, x_{gi}) only offer information relative to other observations within chunk g.

Just as differential equations naturally describe many physical laws, so too do conditional likelihoods, expressible as above by differenced observations, carry forward a physical intuition of causes observed in response to particular changes injected above and beyond a control-measured background. Extending the analogy, the intercept term corresponds to a system's initial conditions. Cosmological origins research notwithstanding, these are rather more difficult to investigate with any degree of satisfaction.

Conditional likelihoods, including those with $n_g = 2$ as above, do not support an intercept term. This is the same property that frees us of the nuisance parameters μ_g. Middle–school science classes emphasize the importance of establishing localized control conditions, and assessing careful changes only with respect to such controls. The same principle applies here: the chunked data structure, the differential contrast, and the conditional model form all represent this core scientific concept both compactly and well. In contrast, statistical models without explicit chunking find it difficult to introduce localized control-group-type data structures.

24.3 THE SPHERICAL APPROXIMATION

24.3.1 Chunks of Size 3

The critical computational issue for conditional likelihoods $L_{cg}(\beta)$ lies with the combinatorial denominator $\sum_{\forall \tau} \exp\{\mathbf{y}_g^{\mathbf{T}}\mathbf{P}_\tau^{\mathbf{T}}\mathbf{X}_g\beta\}$. One way to re-express this is as an average:

$$\sum_{\forall \tau} \exp\{\mathbf{y}_g^{\mathbf{T}}\mathbf{P}_\tau^{\mathbf{T}}\mathbf{X}_g\beta\} = n_g! \text{ave}\{\exp\{\mathbf{y}_g^{\mathbf{T}}\mathbf{P}_\tau^{\mathbf{T}}\mathbf{X}_g\beta\}\}, \tag{24.16}$$

where the permutations indexed by τ are the quantities being averaged over.

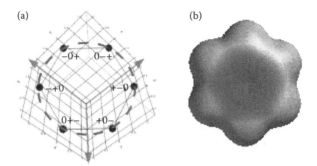

FIGURE 24.2 (a) The six permutations of the three-element vector $(-1, 0, +1)$. (b) Modeled electrostatic potential of benzene.

Figure 24.2 helps us visualize this for the case of chunk size $n_g = 3$. There are $3! = 6$ possible permutations. As noted by constraints (Equation 24.15), these permutations all coexist in $n_g - 1 = 2$ dimensions, and in this case form vertices of a hexagon. Further, the distance of these vertices from the origin is uniform, points on a circle of radius $\sqrt{2} = \sqrt{1^2 + (-1)^2 + 0^2}$. This is true in general: if $\mathbf{1}^T \mathbf{y}_g = 0$, then $\mathbf{1}^T \mathbf{P}_\tau \mathbf{y}_g = \mathbf{1}^T \mathbf{y}_g = 0$, and likewise they share a common (squared) radius, $\mathbf{y}_g^T \mathbf{P}_\tau^T \mathbf{P}_\tau \mathbf{y}_g = \mathbf{y}_g^T \mathbf{I} \mathbf{y}_g = \mathbf{y}_g^T \mathbf{y}_g$.

24.3.2 The von Mises–Fisher Distribution

The symmetry viewed in Figure 24.2a and illustrated by the common center and radius should not too surprising. Permutations are well recognized as a (discrete) group, and indeed, as a subgroup of the larger (continuous) orthogonal group of rigid rotations. Therefore, it seems plausible to approximate (Equation 24.16) by replacing the permutation group with the orthogonal one [8]:

$$\text{ave}\{\exp\{\mathbf{y}_g^T \mathbf{P}_\tau^T \mathbf{X}_g \beta\} | \tau \in \text{permutations}\} \approx \text{ave}\{\exp\{\mathbf{y}_g^T \mathbf{P}_\tau^T \mathbf{X}_g \beta\} | \tau \in \text{orthogonal}\} \quad (24.17)$$

Computationally, this turns out to be quite attractive. Whereas the left-hand side involves a calculation of combinatorial complexity, the implied integration on the right-hand side is tractable, comparable to the matrix multiplications implied by computing $\mathbf{y}_g^T \mathbf{X}_g \beta$. Although Watson and Williams [14] supplied the generalization from circles (von Mises) and spheres (Fisher) to p-dimensional hyperspheres, the literature has followed their lead and continues to call the distribution von Mises–Fisher [7].

We noted above that the (permutation-based) conditional likelihood (Equation 24.13) conditions on the histogram of $\mathbf{y}_g$. Conditioning on the likelihood associated with Equation 24.17 corresponds to conditioning on the variance of $\mathbf{y}_g$, $s_g^2 = \mathbf{y}_g^T \mathbf{y}_g / n_g$. (Recall the mean $\mathbf{1}^T \mathbf{y}_g / n_g$ is constrained by Equation 24.15 to be zero.) Thus, the statistical sense of the spherical approximation replaces the information of a histogram with zero mean with an alternative conditioning on the same histogram's standard deviation.

Section 24.3.7 investigates this approximation in some detail, but some physical intuition can be gleaned from considering the electrostatic potential curves of benzene, illustrated in Figure 24.2b [16]: For the smaller radii, the contours strongly resemble circles; at larger radii, the structure associated with the six atoms emerge more distinctly. A Taylor series approximation reinforces this point of view: everything else being equal, the approximation should be better for smaller radii.

The von Mises–Fisher distribution corresponds to a circular error distribution on the unit sphere in p dimensions. Its typical parametrization has the vector-valued mean direction μ consisting of a point on the hypersphere's surface, modulated by the scalar intensity parameter κ that guides the concentration of the distribution about this point. In this parametrization, the von Mises–Fisher distribution has this standard form:

$$\Pr\{\mathbf{u}|\mu,\kappa\} = \frac{\exp\{\kappa \mathbf{u}^\mathbf{T}\mu\}}{C(\kappa,\mu)}, C(\kappa,\mu) = \left(\frac{2\pi}{\kappa}\right)^p \kappa I_{p/2-1}(\kappa), \tag{24.18}$$

where $I_\nu(\kappa)$ is the modified Bessel function of order ν, and $\mathbf{u}$ represents a p-vector on the unit hypersphere centered at the origin.

24.3.3 The Spherical Likelihood Gradient

A change in notation facilitates applying von Mises–Fisher distribution to approximating conditional likelihoods. Define the mean direction unit vector μ_g as $\mathbf{X}_g\beta/||\mathbf{X}_g\beta||$, and define these scalar magnitudes: $\kappa_{g1}^2 = \mathbf{y}_g^\mathbf{T}\mathbf{y}_g$, $\kappa_{g2}^2(\beta) = \beta^\mathbf{T}\mathbf{X}_g^\mathbf{T}\mathbf{X}_g\beta$, $r_g^2(\beta) = \kappa_{g2}^2(\beta)/\kappa_{g1}^2$, and also $\kappa_g = \kappa_{g1}\kappa_{g2}$.

As is standard notation in directional statstics, let us further define the von Mises–Fisher direction mean as $\rho_\nu(\kappa)\mu$, where $\rho_\nu(\kappa) = I_\nu(\kappa)/I_{\nu-1}(\kappa)$.

Of particular interest is the gradient of any likelihood with respect to the parameters β:

$$\nabla_\beta L_{sg}(\beta) = \mathbf{X}_g^\mathbf{T}\left[\mathbf{y}_g - \frac{\kappa_{g1}}{\kappa_{g2}(\beta)}\rho(\kappa_{g1}\kappa_{g2}(\beta))\mathbf{X}_g\beta\right] \tag{24.19}$$

For compactness, in much of what follows, we suppress the β argument to κ_{g2} and the ν subscript of the radius function ρ.

24.3.4 The Spherical Throttling Ratio

In Equation 24.19, we see emerge the critical quantity $\kappa_{g1}/\kappa_{g2}\rho(\kappa_{g1}\kappa_{g2})$. This ratio goes to zero when κ_{g1} does so, that is, it acts to remove chunks with low amounts of information. For this reason, we term it as a *throttling ratio*.

In Figure 24.3, we attempt to characterize this throttling ratio. Taking the case of the logistic distribution, panels (a) and (b) display this ratio as a function of κ_{g1}; the pattern strongly resembles the sample variance s_g^2, defined in Section 24.3.2 above. Panel (c) plots this same ratio as a function of κ_{g2}. Here, with the exception of a chunk size of $n_g = 2$, the impact of changing κ_{g2} is negligible. This observation, the weak dependence on κ_{g2} and therefore β, foreshadows a valuable algorithmic property.

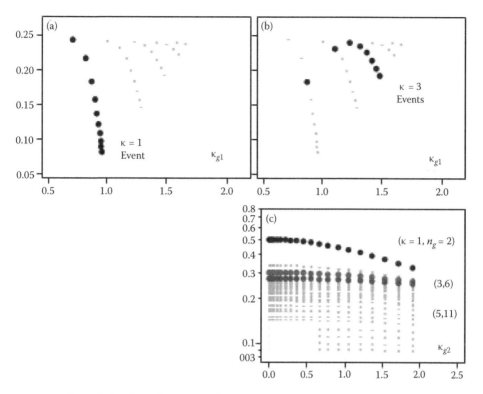

FIGURE 24.3 Plots of the throttling ratio. (a) vs κ_{g1} where $\mathbf{y}_g$ consists of a single 1 and the rest zeros; (b) where $\mathbf{y}_g$ consists of three 1s. (c) The logarithm of the throttling ratio as a function of κ_{g2}.

As defined above, $r_g^2(\beta) = \kappa_{g2}^2(\beta)/\kappa_{g1}^2$, and κ_{g1}^2 may be usefully reexpressed as $\kappa_{g1}^2 = [s_g^2 = \mathbf{y}_g^T\mathbf{y}_g/n_g] \times n_g$. This allows for a certain simplification of Equation 24.19:

$$\nabla_\beta L_{sg}(\beta) = \mathbf{X}_g^T \left[\mathbf{y}_g - \frac{\rho(r_g n_g s_g^2)}{r_g}\mathbf{X}_g\beta \right] \tag{24.20}$$

As noted just before, in this form the throttling ratio depends weakly on $r_g^2 = r_g^2(\beta)$ while acting to throttle chunks with small s_g^2.

From detailed numerical analysis, one can derive these bounds:

$$\frac{s_g^2}{1 + r_g^2 s_g^4} \le \frac{\rho(r_g n_g s_g^2)}{r_g} \le \frac{s_g^2}{1 + r_g^2 s_g^4 t}, \text{ where } t = v_g/(v_g + 1) \text{and } v_g = n_g/2. \tag{24.21}$$

Thus, the dependence of the throttling ratio on sample size is relatively weak, especially for the chunks with larger sizes n_g.

24.3.5 Some Normal Equations

If we set $\nabla_\beta \sum_g L_{sg}(\beta)$ equal to zero, from Equation 24.21 we can derive these equations for β:

$$\left[\sum_g \frac{\rho(r_g n_g s_g^2)}{r_g} \mathbf{X}_g^\mathsf{T} \mathbf{X}_g \right] \hat{\beta} = \mathbf{X}_g^\mathsf{T} \mathbf{y}_g \qquad (24.22)$$

These are strongly similar to Gauss' normal equations, which he derived to estimate β by least squares.

Several comments are now in order:

- The estimate $\hat{\beta}$ in Equation 24.22 is strongly similar to Gauss's original normal equations, which he derived by least squares.

- A key difference is that r_g depends on β, so (Equation 24.22) suggests or requires some fixed-point iteration.

- Indeed, $\rho(r_g n_g s_g^2)/r_g$ depends only weakly on $\hat{\beta}$ through $r_g = r_g(\hat{\beta})$. Therefore, (Equation 24.22) converges rapidly.

- $\rho(r_g n_g s_g^2)/r_g \approx s_g^2$ reweights chunks according to their information. In particular, chunks with no variation in their responses (so $s_g^2 = 0$) are simply eliminated from the likelihood gradient.

24.3.6 The Chunk-Specific Correlation Coefficient

In Equation 24.22, the factor r_g or r_g^2 is strongly suggested by the underlying mathematics. Its definition, $r_g^2(\beta) = \beta^\mathsf{T} \mathbf{X}_g^\mathsf{T} \mathbf{X}_g \beta / \mathbf{y}_g^\mathsf{T} \mathbf{y}_g$ is directly analogous to the multiple correlation coefficient, a construct here specific to chunk g. As such, it measures how well a model explains a given chunk. (Like the standard multiple correlation coefficient, $r_g^2 \geq 0$; unlike it, r_g^2 is not constrained to be uniformly below 1.)

Chunking provides us with a data structure by which to collate datasets. For definiteness, let us sort the chunks by the sizes of their values $\mathbf{y}_g^\mathsf{T} \mathbf{y}_g$, such that $g_1 < g_2$ implies $\mathbf{y}_{g_1}^\mathsf{T} \mathbf{y}_{g_1} \leq \mathbf{y}_{g_2}^\mathsf{T} \mathbf{y}_{g_2}$. Define a series of cumulative numerators by $N[g] = N[g-1] + \beta^\mathsf{T} \mathbf{X}_g^\mathsf{T} \mathbf{X}_g \beta$ and cumulative denominators by $D[g] = D[g-1] + \mathbf{y}_g^\mathsf{T} \mathbf{y}_g$. The numerator–denominator plot [9] of $N[g]$ versus $D[g]$ (Figure 24.4) has the following properties:

- The points are increasing and to the right, a consequence of the summands $\beta^\mathsf{T} \mathbf{X}_g^\mathsf{T} \mathbf{X}_g \beta$ and $\mathbf{y}_g^\mathsf{T} \mathbf{y}_g$ being both nonnegative. (In practice, the nonnegativity of the denominator summands is sufficient for such plots to be useful.)

- *The global slope property:* The slope from the origin to the last point estimates the average within-chunk multiple correlation coefficient R^2.

- *The local slope property:* The slope between any pair of points estimates the average within-chunk multiple correlation for that corresponding subset of chunks.

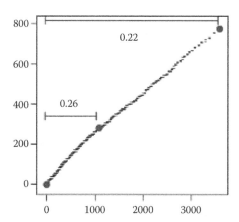

FIGURE 24.4 Numerator–denominator plot using the components of r_g^2.

In Figure 24.4, the chunks are sorted by $\mathbf{y}_g^T\mathbf{y}_g$. The larger initial slope implies better relative predictions for low-information chunks than for the higher-information ones. Note that were the fit dominated by a few outliers, the reverse would occur: the high-information chunks would have the higher r_g^2, as the outliers would work to distort the fit in their favor. In this sense, Figure 24.4 suggests a certain kind of robustness.

Finally, we note in passing that r_g^2 provides an avenue for model refinement: One can increase attention on modeling well those chunks with lower r_g^2 values. Boosting [10] shares a similar point of view.

24.3.7 Assessing the Spherical Approximation

From Equation 24.17, observe that to the extent that the spherical approximation is imperfect, it deviates as a function of this ratio:

$$Q(\kappa, \mathbf{u}_1, \mathbf{u}_2) = \frac{\text{ave}\{\exp\{\kappa\mathbf{u}_1^T\mathbf{P}_\tau^T\mathbf{u}_2\}|\tau \in \text{permutations}\}}{\text{ave}\{\exp\{\kappa\mathbf{u}_1^T\mathbf{P}_\tau^T\mathbf{u}_2\}|\tau \in \text{orthogonal}\}}, \tag{24.23}$$

where $\mathbf{u}_1$, $\mathbf{u}_2$ are unit vectors and κ is the intensity parameter of the von Mises–Fisher distribution encountered in Equation 24.18. Indeed, because the critical behavior is how $\log Q$ behaves when differentiated, the critical issue is the constancy of $\frac{\partial}{\partial \kappa} \log Q$. For various dimensions n_g and directions $\mathbf{u}_1$, $\mathbf{u}_2$, this can be investigated numerically. $8! = 40,320$, so this forms the upper limit of comfortable calculations; we can take the directions $\mathbf{u}_1$, $\mathbf{u}_2$ as random spherical radii.

A subset of results in presented in Figure 24.5. In panel (c) the phenomenon we see is straightforward: the spherical approximation works better and at larger radii κ for larger chunk sizes n_g. This stands to reason. Numerical integrations on a sphere involve sums of large numbers of values; as $n_g!$ approaches or exceeds standard Monte Carlo simulation counts ($10^3 - 10^4$), one expects the more symmetric permutation-based calculation to blend into any Monte Carlo-based estimate.

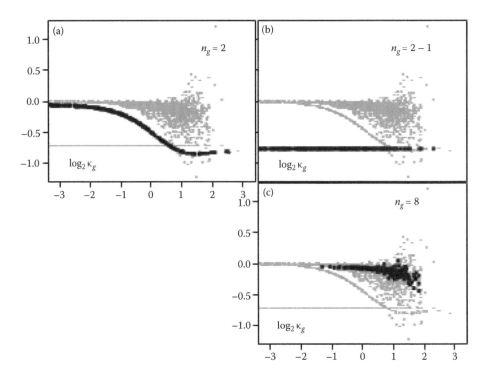

FIGURE 24.5 Numerically calculated values of $\frac{\partial}{\partial}\kappa \log Q$ as a function of radius κ. (a) Highlights results for $n_g = 2$. (b) Highlights the results for $n_g = 2$, modified so that the Bessel function order is reduced by $-1/2$. (c) Highlights the results for $n_g = 8$.

The pattern for $n_g = 2$ is more subtle. In Figure 24.5 panel (a), as a function of the radius κ, $\partial/\partial\kappa \log Q$ is not constant, but for each value of κ the offset is constant with respect to $\mathbf{u}_1$, $\mathbf{u}_2$. This suggests that a flattening correction is available; the result is panel (b), which consists of reducing the order of the Bessel functions in $\rho_v(\kappa) = I_v(\kappa)/I_{v-1}(\kappa)$ from $v = n_g/2 - 1$ to $v = (n_g - 1)/2 - 1$; such an adjustment is motivated by the fact that the centering implied by $\mathbf{1}^T\mathbf{y}_g = 0$ confines $\mathbf{y}_g$ to an $(n_g - 1)$-dimensional linear space.

To summarize, (Equation 24.1) the spherical approximation always works well for the smaller radii κ. This can occur when the response is relatively rare, when the within-chunk variation is small, and when the features or predictors are relatively weak—the case of the so-called weak learners. This intuition is captured in Figure 24.2b alluded to above. (2) As the chunk size n_g grows, the radius at which the spherical approximation remains viable grows also; the exact relationship is yet to be pinned down. (3) If the spherical approximation improves with increasing n_g, then it follows that the case of $n_g = 2$ should be the worst case. However, a second-order correction, consisting of replacing the Bessel order v with $v - 1/2$ enables an exact estimate in that case. (4) Therefore, at the moment of this writing, the actual worst case for the spherical approximation is with chunk sizes $n_g = 3$. Note that the six permutations of $(-1,0,+1)$ generate a regular hexagon in 3-space (Figure 24.2a). In contrast, the six permutations of $(1,0,0)$ result in three unique patterns, forming an equilateral triangle (and a degenerate hexagon). The spherical approximation is worse for the latter and better for the former.

24.4 RELATED METHODS

In the preceeding paragraphs, we have described connections between the spherical approximation and its algorithm (Equation 24.22), numerator–denominator plots, and boosting. In this section, we point out three related algorithms—iteratively reweighted least squares, marginal regression, and multiple-stage models. This sets the stage for us to propose an extension appropriate when the number of features J is large.

24.4.1 Iteratively Reweighted Least Squares

For unchunked data structures, the maximum likelihood estimates resemble those of Equation 24.22, taking this form:

$$\mathbf{X}^T \mathbf{D_w} \mathbf{X} \boldsymbol{\beta} = \mathbf{X}^T \mathbf{D_w} \mathbf{y}, \tag{24.24}$$

where $\mathbf{w} = \mathbf{w}(\boldsymbol{\beta})$ is a set of weights that in general depend on the parameters $\boldsymbol{\beta}$; usually $\mathbf{w}$ is the predicted variance of the corresponding element of $\mathbf{y}$ [5]. Because the dependence on $\mathbf{w}$ is relatively strong, the convergence of $\boldsymbol{\beta}$ in Equation 24.24 is relatively slow.

Some intuition for why this might be so can be gained by considering the direct estimation of Equation 24.11. Such direct estimation requires estimating the chunk-specific intercepts $\{\mu_g\}$; any iteration that perturbs any of these μ_g requires an adjustment to $\boldsymbol{\beta}$; during estimation, the intercepts $\{\mu_g\}$ and the feature coefficients $\boldsymbol{\beta}$ "communicate." For the related estimation of a single intercept μ, an analogous communication issue likewise slows convergence.

24.4.2 Marginal Regression

A primary challenge in machine learning is dealing with and selecting wisely from J features, J perhaps in the range of 10^3–10^9. In this sense, positing as it does a $J \times J$ matrix and the simultaneous solution of J Equations, (Equation 24.22) looks rather old-fashioned.

Most machine learning algorithms therefore avoid formulating estimation by systems of linear equations. Rather the contemporary focus is on gradient-based methods, including penalties of the form $L_p(\boldsymbol{\beta}) = \left[\sum_{j=1}^{J} |\beta_j|^p \right]^{1/p}$ to the maximum likelihood form. This general strategy is called *regularization*. The literature tends to emphasize L_1 and L_2: the former (often called the *lasso*[13]) acts to include only terms that offer benefit above a threshold, while the latter (sometimes termed *ridge regression* [3]) provides some numerical stability to solve linear equations.

An idea distinct from the gradient-and-penalty approach of L_p regularization is that of marginal regression [2]. Marginal regression, investigated mainly in the context of least squares, consists of estimating in parallel J single-feature models, then selecting those features for which the coefficients were largest. Marginal regression therefore consists of these two steps:

$$\tilde{\alpha}_{MR} = [\text{diag}(\mathbf{X}^T \mathbf{X})]^{-1} \mathbf{X}^T \mathbf{y},$$
$$\tilde{\beta}_{MR} = \tilde{\alpha}_{MR} \times \mathbf{1}\{|\tilde{\alpha}_{MR}| \geq \theta\}, \tag{24.25}$$

for some positive threshold θ. The analog for chunked data and the spherical approximation is with this pair of equations:

$$\tilde{\alpha}_{\text{sphere}} = \left[\sum_g \frac{\rho_g}{r_g} \text{diag}(\mathbf{X}_g^T \mathbf{X}_g) \right]^{-1} \sum_g \mathbf{X}_g^T \mathbf{y}_g,$$

$$\tilde{\beta}_{\text{sphere}} = \tilde{\alpha}_{\text{sphere}} \times \mathbf{1}\{|\tilde{\alpha}_{\text{sphere}}| \geq \theta\}. \tag{24.26}$$

Because the spherical approximation is highly linearized, this suggests an approach to large-J estimation, developed in the next section.

24.4.3 A Generalization to Many Features

Algorithm (Equation 24.25) creates both predictions $\hat{\mathbf{y}}_g = \mathbf{X}_g \beta_{\text{sphere}}$ and residuals $\mathbf{y}_g - \hat{\mathbf{y}}_g$. This property suggests the iterative adaptation of Equation 24.26 portrayed in Table 24.2. (A related algorithm replaces step 8 with Fisher score tests.)

For those familiar with boosting [10], the algorithm in Table 24.2 shares these similarities: the refinement over iterations, the affinity for so-called weak learners, and the building up of weak features into stronger ones. As per Section 24.3.7, we repeat that the spherical approximation works especially well when used in conjunction with weak learners.

24.4.4 Multiple-Stage Models

For several reasons, conditional models usually involve some postprocessing. The most concrete reason is that conditional models are unable to estimate an intercept, the coefficient of the always-present feature: because this feature 0 is such that $\mathbf{x}_{g0} = \mathbf{1}$, which the process of centering sets to zero: $\mathbf{1}^T \mathbf{x}_{g0} = 0$ implies $\mathbf{x}_{g0} = \mathbf{0}$ if $\mathbf{x}_{g0}$ is constant. At the minimum, any conditional model needs to recover an estimate of the intercept by postprocessing.

For example, for Poisson response y_i, $\Pr\{y_i | x_i\} = \lambda_i^{y_i} \exp\{-\lambda_i\}/y_i!$, the standard regression link function $\lambda_i = \exp\{\mu + \mathbf{x}_i^T \beta\}$. Conditional models can estimate β, hence $z_i = \mathbf{x}_i^T \beta$. The maximum likelihood estimate of μ given z_i can be expressed as

$$e^{\hat{\mu}} = \sum_i y_i / \sum_i e^{z_i}. \tag{24.27}$$

(The associated numerator–denominator plot consists of plotting the cumulative sums of y_i against the cumulative sums of $\exp\{z_i\}$; the global slope of the last point from the origin is (Equation 24.27), while the local slopes consist of similar estimates for that corresponding chunk of data.)

For the logistic model, the intercept estimate also makes use of $z_i = \mathbf{x}_i^T \beta$ and the auxiliary pseudo-probability $\pi_i = \exp\{z_i\}/(1 + \exp\{z_i\})$. The intercept (actually, the base-rate odds) can then be estimated as

$$e^{\hat{\mu}} = \sum_i y_i(1 - \pi_i) / \sum_i (1 - y_i)\pi_i. \tag{24.28}$$

For the Gaussian likelihood, the intercept term can be estimated as $\hat{\mu} = \sum_i (y_i - z_i)/n$.

TABLE 24.2 Algorithm, Modified from Marginal Regression, for Identifying Features When the Feature Count J Is Large

0	initialize	center $\mathbf{y}_g$, $\mathbf{X}_g$ such that $\mathbf{y}^T_g\mathbf{1} = 0, \mathbf{X}^T_g\mathbf{1} = \mathbf{0}$	used in steps 5, 6, 12, 16
1		$\hat{\mathbf{y}}_{g0} \leftarrow \tilde{\mathbf{y}}_{g0} \leftarrow \mathbf{1}$	for centering in step 5–6
2		$r^2_{g0} \leftarrow 0$	used on RHS of step 7
3	iterate $k=0,\ldots,K$	$\mathbf{u}_{gk} \leftarrow \hat{\mathbf{y}}_{gk}/\|\hat{\mathbf{y}}_{gk}\|$	unit vector form of pseudo-feature
4		$\mathbf{M}_{gk} \leftarrow \mathbf{I} - \mathbf{u}_{gk}\mathbf{u}^T_{gk}$	projection matrix $\perp \hat{\mathbf{y}}_{gk}$
5		$\mathbf{z}_{gk} \leftarrow \mathbf{M}_{gk}\mathbf{y}_g$	residual response $\perp \hat{\mathbf{y}}_{gk}$
6		$\mathbf{X}_{gk} \leftarrow \mathbf{M}_{gk}\mathbf{X}_g$	residual predictors $\perp \hat{\mathbf{y}}_{gk}$
7	a	$t_g \leftarrow n_g/(1 + n_g)$	motivated by Equation 24.21
	b	$s^2_{gk} \leftarrow \mathbf{z}^T_{gk}\mathbf{z}_{gk}/n_g$	within-chunk variance
	c	$\rho_{gk}/r_{gk} \leftarrow s^2_{gk}/\left[1 + r^2_{gk}s^4_{gk}t_g\right]$	and throttling factor
8		$\tilde{\alpha}_k \leftarrow \left[\sum_g \frac{\rho_{gk}}{r_{gk}}\text{diag}(\mathbf{X}^T_{gk}\mathbf{X}_{gk})\right]^{-1}\sum_g \mathbf{X}^T_{gk}\mathbf{z}_g$	marginal regression step:
9		$\tilde{\beta}_k \leftarrow \tilde{\alpha}_k \times \mathbf{1}\{|\tilde{\alpha}_k| \geq \theta\}$	selecting largest $\tilde{\alpha}_k$
10	…update	$\tilde{\mathbf{y}}_{gk} \leftarrow \mathbf{X}_g\tilde{\beta}_k$	new one-dimensional composite
11		$\tilde{\mathbf{X}}_{gk} \leftarrow [\hat{\mathbf{y}}_{gk}, \tilde{\mathbf{y}}_{gk}]$	old pseudo + new composite
12		$\tilde{\gamma}_k \leftarrow \left[\sum_g \frac{\rho_{gk}}{r_{gk}}\tilde{\mathbf{X}}^T_{gk}\tilde{\mathbf{X}}_{gk}\right]^{-1}\sum_g \mathbf{X}^T_{gk}\mathbf{y}_g$	two blending coefficients
13		$\hat{\mathbf{y}}_{gk} \leftarrow \tilde{\mathbf{X}}_{gk}\tilde{\gamma}_k$	blended new pseudo-feature for step 3
14		$r^2_{gk} \leftarrow \hat{\mathbf{y}}^T_{gk}\hat{\mathbf{y}}_{gk}/\mathbf{y}^T_{gk}\mathbf{y}_{gk}$	for RHS of step 7; go to step 3
15	finish	$\tilde{\mathbf{X}}_g \leftarrow [\tilde{\mathbf{y}}_{g0}, \tilde{\mathbf{y}}_{g1}, \tilde{\mathbf{y}}_{g2}, \ldots]$	K composite features
16		$\tilde{\eta} \leftarrow \left[\sum_g \frac{\rho_{gK}}{r_{gK}}\tilde{\mathbf{X}}^T_g\tilde{\mathbf{X}}_g\right]^{-1}\sum_g \tilde{\mathbf{X}}^T_g\mathbf{y}_g$	fit original $\mathbf{y}_g$ versus composite features

This latter estimator, as well as those of Equations 24.27 and 24.28, all support numerator–denominator plots. These allow us to detect level shifts in $\hat{\mu}$ associated with whatever is chosen as the order of cumulative summing. In contrast, with μ estimated jointly with the coefficient vector β, such level shifts would induce attentuation, both in β and in any model predictions.

For intercepts, then, conditional models require a second stage of model fitting. A modest generalization of this would estimate both an intercept μ and a slope γ on the one-dimensional pseudo-feature $\tilde{\mathbf{x}}_g = \mathbf{X}_g\beta$. For example, for the Poisson model, one would estimate the two-tuple (μ, γ) in this model:

$$\Pr\{y_{gi}| \tilde{x}_{gi}\} = \lambda_{gi}^{y_{gi}} \exp\{-\lambda_{gi}\}/y_{gi}!, \quad \lambda_{gi} = \exp\{\mu + \gamma\tilde{x}_{gi}\} \tag{24.29}$$

Estimating the intercept μ alone is analogous to one-point calibration, and estimating (μ, γ) as in Equation 24.29 corresponds to two-point calibration. There are four distinct reasons to do this: (1) As developed here, the joint second-stage estimation of (μ, γ) is a generalization of the always-necessary second-stage estimation of the intercept μ. In this context, estimating γ and, in particular, finding values significantly different from 1 protect us from attenuation effects that might otherwise go undetected. (2) The standard Gaussian link has also a scale parameter σ. For the Gaussian model, estimating γ corresponds to estimating $1/\sigma$ (sometimes called the precision or resolution). (3) Certain algorithms focus sufficiently on feature construction that their estimates of their coefficients are not calibrated with any certainty. (One such example is step 9 of the algorithm in Table 24.2, necessitating step 12.) (4) For reasons of data processing architecture, certain features are not conveniently available during earlier model fitting, but become so later. Sometimes it is convenient to add these secondary features as a second or third stage of model fitting.

The point is this: multiple-stage estimation should not be viewed as ad hoc or unaesthetic. Rather they are necessary manifestations of a larger theory, one closely connected to conditional models. In this sense, multiple-stage models have an intrinsic role in modern, production-scale machine learning algorithms.

24.5 CONCLUSIONS

Astronomers Gauss and Poisson are credited with the probability distributions that bear their respective names. This pair of distributions anchor the larger EXP, as particular *link functions* in the theory of generalized linear models.

Friedrich Bessel first investigated systematically observer-specific effects. He termed these as *personal equations*; contemporary parlance has come to call them *nuisance effects*. This core concept continues to animate statistical theory today [15]. To mitigate such nuisance effects, we propose *chunking* observations into groups. Chunking occurs in three ways: (a) chunking observations that occur together in the measurement process, (b) chunking observations that are made of the same phenomenon, the same galaxy, field of view, and so on, and (c) chunking observations that allows for efficient (usually parallel) computation.

In the field of statistics, Gauss also derived the least squares method of estimation and the so-called normal equations. The direct analog of such equations, represented here in Equation 24.22, reappears in this modern setting.

Here, we inject the spherical approximation into the theory of generalized linear models. The approximation itself is motivated by similar physical intuitions: that large numbers of discrete objects can be represented well by appropriate fields and continua. The resulting von Mises–Fisher distribution is familiar to the astronomy and planetary science community, being as they are the primary consumers of directional statistics. To my particular delight, this area rests on foundations laid by Bessel and other physicists.

The core framework here, that of conditional models, also grounds multiple-stage estimation in a larger theory. As such computational approaches take hold in astrostatistics, we can reflect on how the foundational contributions developed by astronomers have come full circle to benefit their field anew.

REFERENCES

1. Andersen, E. 1970. Asymptotic properties of conditional maximum-likelihood estimators. *Journal of the Royal Statistical Society, Series B*, 32, 283–301.

2. Fan, J. and Lv, J. 2008. Sure independence screening for ultra-high dimensional feature space (with discussion). *Journal of the Royal Statistical Society, Series B*, 70, 849–911.

3. Hoerl, A. E. 1962. Application of ridge analysis to regression problems, *Chemical Engineering Progress*, 58, 54–59.

4. Hosmer Jr., D. W. and Lemeshow, S. 2000. *Applied Logistic Regression*, 2nd edition, John Wiley & Sons: New York.

5. Jorgensen, M. 2006. Iteratively reweighted least squares, *Encyclopedia of Environmetrics*. John Wiley & Sons: New York.

6. McCullagh, P. and Nelder, J. A. 1989. *Generalized Linear Models*, 2nd edition, Chapman & Hall/CRC Press: Boca Raton, Florida.

7. Mardia, K. V. and Jupp, P. E. 2000. *Directional Statistics*, 2nd edition, John Wiley & Sons: New York.

8. Plis, S. M., Lane, T., and Calhoun, V. D. 2010. Permutations as angular data: Efficient inference in factorial spaces, *IEEE International Conference on Data Mining*, 403–410.

9. Page, E. S. 1954. Continuous inspection schemes, *Biometrika*, 41, 100–115.

10. Schapire, R. E. and Singer, Y. 1999. Improved boosting algorithms using confidence-rated predictors, *Machine Learning*, 37, 297–336.

11. Schaffer, S. 1988. Astronomers mark time: Discipline and the personal equation, *Science in Context*, 2, 101–131.

12. Stigler, S. M. 1986. *The History of Statistics: The Measurement of Uncertainty before 1900.* Harvard University Press: Cambridge, Massachusetts.

13. Tibshirani, R. 1996. Regression shrinkage and selection via the lasso, *Journal of the Royal Statistical Society, Series B*, 58 (1), 267–288.

14. Watson, G. S. and Williams, E. J. 1956. On the construction of significance tests on the circle and the sphere, *Biometrika*, vol. 43. pp. 344–352.

15. Wu, C. F. J and Hamada, M. 2002. *Experiments: Planning, Analysis, and Parameter Design Optimization.* John Wiley & Sons: New York.

16. Barron, A. 2010. Boron compounds with nitrogen donors. Connexions, January 25, 2010. http://cnx.org/content/m32856/1.2/.

CHAPTER 25

Data Clustering

Kiri L. Wagstaff

California Institute of Technology

CONTENTS

25.1 Introduction	543
25.2 *k*-Means Clustering	545
25.2.1 Compression via Clustering	547
25.2.2 Kernel *k*-Means	548
25.2.3 *k*-Means Limitations	548
25.3 EM Clustering	549
25.4 Hierarchical Agglomerative Clustering	552
25.5 Spectral Clustering	553
25.6 Constrained Clustering	555
25.7 Applications to Astronomy Data	556
25.8 Summary	557
25.9 Glossary	558
Acknowledgments	559
References	559

25.1 INTRODUCTION

On obtaining a new data set, the researcher is immediately faced with the challenge of obtaining a high-level understanding from the observations. What does a typical item look like? What are the dominant trends? How many distinct groups are included in the data set, and how is each one characterized? Which observable values are common, and which rarely occur? Which items stand out as anomalies or outliers from the rest of the data? This challenge is exacerbated by the steady growth in data set size [11] as new instruments push into new frontiers of parameter space, via improvements in temporal, spatial, and spectral resolution, or by the desire to "fuse" observations from different modalities and instruments into a larger-picture understanding of the same underlying phenomenon.

Data clustering algorithms provide a variety of solutions for this task. They can generate summaries, locate outliers, compress data, identify dense or sparse regions of feature space, and build data models. It is useful to note up front that "clusters" in this context refer to groups of items within some descriptive feature space, not (necessarily) to "galaxy clusters" which are dense regions in physical space. The goal of this chapter is to survey a variety of data clustering methods, with an eye toward their applicability to astronomical data analysis. In addition to improving the individual researcher's understanding of a given data set, clustering has led directly to scientific advances, such as the discovery of new subclasses of stars [14] and γ-ray bursts (GRBs) [38].

All clustering algorithms seek to identify groups within a data set that reflect some observed, quantifiable structure. Clustering is traditionally an *unsupervised* approach to data analysis, in the sense that it operates without any direct guidance about which items should be assigned to which clusters. There has been a recent trend in the clustering literature toward supporting *semisupervised* or *constrained* clustering, in which some partial information about item assignments or other components of the resulting output are already known and must be accommodated by the solution. Some algorithms seek a *partition* of the data set into distinct clusters, while others build a *hierarchy* of nested clusters that can capture taxonomic relationships. Some produce a single optimal solution, while others construct a probabilistic model of cluster membership.

More formally, clustering algorithms operate on a data set X composed of items represented by one or more features (dimensions). These could include physical location, such as right ascension and declination, as well as other properties such as brightness, color, temporal change, size, texture, and so on. Let D be the number of dimensions used to represent each item, $x_i \in \mathcal{R}^D$. The clustering goal is to produce an organization P of the items in X that optimizes an objective function $f : P \to \mathcal{R}$, which quantifies the quality of solution P. Often f is defined so as to maximize similarity within a cluster and minimize similarity between clusters. To that end, many algorithms make use of a measure $d : X \times X \to \mathcal{R}$ of the distance between two items. A partitioning algorithm produces a set of clusters $P = \{c_1, \ldots, c_k\}$ such that the clusters are nonoverlapping ($c_i \cap c_j = \emptyset, i \neq j$) subsets of the data set ($\bigcup_i c_i = X$). Hierarchical algorithms produce a series of partitions $P = \{p_1, \ldots, p_{n'}\}$. For a complete hierarchy, the number of partitions $n' = n$, the number of items in the data set; the top partition is a single cluster containing all items, and the bottom partition contains n clusters, each containing a single item. For model-based clustering, each cluster c_j is represented by a model m_j, such as the cluster center or a Gaussian distribution.

The wide array of available clustering algorithms may seem bewildering, and covering all of them is beyond the scope of this chapter. Choosing among them for a particular application involves considerations of the kind of data being analyzed, algorithm runtime efficiency, and how much prior knowledge is available about the problem domain, which can dictate the nature of clusters sought. Fundamentally, the clustering method and its representations of clusters carries with it a definition of what a cluster is, and it is important that this be aligned with the analysis goals for the problem at hand. In this chapter, I emphasize this point by identifying for each algorithm the cluster representation as a model, m_j, even for algorithms that are not typically thought of as creating a "model."

This chapter surveys a basic collection of clustering methods useful to any practitioner who is interested in applying clustering to a new data set. The algorithms include k-means (Section 25.2), EM (Section 25.3), agglomerative (Section 25.4), and spectral (Section 25.5) clustering, with side mentions of variants such as kernel k-means and divisive clustering. The chapter also discusses each algorithm's strengths and limitations and provides pointers to additional in-depth reading for each subject. Section 25.6 discusses methods for incorporating domain knowledge into the clustering process. This chapter concludes with a brief survey of interesting applications of clustering methods to astronomy data (Section 25.7).

The chapter begins with k-means because it is both generally accessible and so widely used that understanding it can be considered a necessary prerequisite for further work in the field. EM can be viewed as a more sophisticated verison of k-means that uses a generative model for each cluster and probabilistic item assignments. Agglomerative clustering is the most basic form of hierarchical clustering and provides a basis for further exploration of algorithms in that vein. Spectral clustering permits a departure from feature-vector-based clustering and can operate on data sets instead represented as affinity, or similarity matrices—cases in which only pairwise information is known.

The list of algorithms covered in this chapter is representative of those most commonly in use, but it is by no means comprehensive. There is an extensive collection of existing books on clustering that provide additional background and depth. Three early books that remain useful today are Anderberg's *Cluster Analysis for Applications* [3], Hartigan's *Clustering Algorithms* [25], and Gordon's *Classification* [22]. The latter covers basics on similarity measures, partitioning and hierarchical algorithms, fuzzy clustering, overlapping clustering, conceptual clustering, validations methods, and visualization or data reduction techniques such as principal components analysis (PCA), multidimensional scaling, and self-organizing maps. More recently, Jain et al. provided a useful and informative survey [27] of a variety of different clustering algorithms, including those mentioned here as well as fuzzy, graph-theoretic, and evolutionary clustering. Everitt's *Cluster Analysis* [19] provides a modern overview of algorithms, similarity measures, and evaluation methods.

25.2 k-MEANS CLUSTERING

The widely used k-means algorithm [36] provides a simple way to construct a collection of k clusters, $c_1, \ldots, c_k$. It begins with a random guess at the clusters and then iteratively improves their fit to the observed data. The model m_j for each cluster c_j is a point in feature space ("mean"), and each item is associated with its closest cluster mean, as dictated by the distance metric d.

k-Means seeks to minimize the following objective function:

$$f_{KM} = \frac{1}{n} \sum_{i=1}^{n} d(x_i, c_{a_i}), \tag{25.1}$$

where n is the number of items, a_i is the cluster to which item i is assigned, and $d(x_i, c)$ is the distance from item x_i to cluster c, which is the distance to c's model (mean), m. This objective function is referred to as the *variance* or *distortion* of a given clustering solution. The distance measure $d()$ is typically Euclidean distance, assigning equal weight to all dimensions, but it can be tailored for a particular application.

Algorithm 25.1   *k*-Means clustering [36]

1: **Inputs:** Data set X containing n items, number of clusters k
2: **Outputs:** Cluster models $M = \{m_j, j = 1 \dots k\}$, assignments $A = \{a_i, i = 1 \dots n\}$
3: Initialize the k cluster models with randomly chosen $x_i \in X$
4: **repeat**
5:     **for** $i = 1$ **to** n **do**
6:         Let c_j be x_i's closest cluster. Set a_i to j. // Item Assignment
7:     **end for**
8:     **for** $j = 1$ **to** k **do**
9:         Set cluster model m_j to be the mean of $\{x_i | a_i = j\}$. // Cluster update
10:    **end for**
11: **until** A does not change

The details of the *k*-means algorithm are given in Algorithm 25.1. The data set X and desired number of clusters are specified as inputs. The clustering process yields a set of cluster models M and a list A indicating the cluster assignment for each item in the data set. The cluster models are initialized with randomly chosen items from the data set (line 3). The algorithm then alternates between an assignment step (line 6), in which each item is assigned to its closest cluster (model), and a cluster update step (line 9), in which each cluster model is updated to be the mean of its currently assigned items. In this way, the cluster models are iteratively refined until the algorithm converges to a stable solution, which is when none of the item assignments change.

Figure 25.1 shows an example of the result of performing *k*-means clustering on a simple artificial data set. The data consist of 300 items, with 100 belonging to each of three clusters. Each item is described by two observed features, which correspond to their x and y coordinates in the figure. In the left panel, items are assigned symbols according to their true cluster membership. One cluster is well separated from the other two, while two of the clusters overlap slightly. On the right is shown the result of performing *k*-means clustering, with

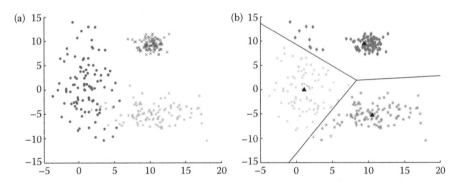

FIGURE 25.1   *k*-Means clustering on an artificial data set ($N = 300, k = 3$). The resulting cluster centers (M) are shown with triangular markers, and solid black lines indicate the associated Voronoi diagram cluster boundaries. (a) True cluster memberships. (b) *k*-Means output, $k = 3$.

k set to 3, on this data set. Here, items are assigned symbols according to their final assignment (A). The final cluster centers (M) are shown with triangular markers. While the cluster centers appear visually reasonable, some of the items from the leftmost cluster have been incorrectly associated with the topmost cluster (diamond markers). Also, k-means cannot model overlapping clusters, so it makes a hard division between the other two clusters.

The set of cluster models M can do more than just indicate which of the clusters best represents each data item. Using the assumption that any item will be assigned to its closest cluster center, the models can also predict how new items should be assigned. In fact, for any location in the feature space, M can indicate which cluster dominates. The boundaries between these regions can be computed as a Voronoi diagram (tessellation) of the space, shown by the solid black lines in Figure 25.1.

25.2.1 Compression via Clustering

The cluster models M can also provide a form of data compression. Rather than representing each item with its full list of observed features, one can instead simply record the models M and the assignments A. In the artificial data case, assuming that each item consists of two floating-point (4-byte) values, the space consumed by the data set shrinks by a factor of almost 25. The original data set consumes $300 \times 8 = 2400$ bytes, while the compressed data set consumes $3 \times 8 = 24$ bytes for M and 300×2 bits $= 74$ bytes for A, for a total of just 98 bytes. Of course, a lot of fidelity is lost, but if disk space is limited, or if the data must be transmitted across a bandwidth-constrained link, this lossy compression may be useful.

Figure 25.2 shows an example of k-means-based data compression for an image. The figure shows a grayscale version, but the image file is stored in color format. Each pixel in this 220×247 image ($N = 54,340$) is represented by three values (its red, blue, and green intensities). Clustering the pixels into $k = 4$ groups and replacing each pixel with its assigned m_j yields a lower-quality, yet recognizable version, of the image. Using JPEG encoding, the

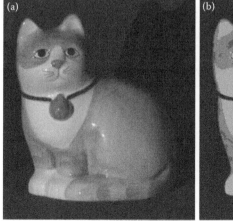

FIGURE 25.2 Clustering for data compression (vector quantization). The pixels in the original image were clustered with $k = 4$, and then each pixel was replaced with the mean of its assigned cluster, yielding a much smaller file. (a) Original image (29 KB JPEG). (b) Recoded image (12 KB JPEG).

file size is reduced by a factor of 2. Using an indexed encoding akin to that described above (e.g., GIF), which would store the four unique colors in M and use just two bits to encode each pixel, would yield an even smaller file size (2.6 KB, a factor of 10 reduction).

25.2.2 Kernel k-Means

Because k-means assigns each item to its closest cluster center, and partitions the feature space accordingly, it cannot model overlapping clusters or any nonlinear separation between clusters. Some of this expressivity can be achieved via *kernel* methods, which implicitly map the items into a higher-dimensional feature space [46]. A linear separation in the new feature space can map back down into a nonlinear separation in the original space. Kernel clustering works in the new feature space by reexpressing its objective function in terms of such a kernel rather than distance [21]. A popular kernel function is the radial basis function (RBF) kernel:

$$k(x_i, x_j) = k_{ij} = \exp\left(-\frac{d(x_i, x_j)^2}{2\sigma^2}\right), \tag{25.2}$$

where σ is a scaling factor. The value k_{ij} can be thought of as the similarity between items x_i and x_j. The corresponding objective function captures the intercluster similarity of the solution, which kernel k-means seeks to maximize:

$$f_{KKM} = \sum_{l=1}^{k} \frac{|C_l|}{n} \frac{1}{|C_l|^2} \sum_{x_i \in C_l, x_j \in C_l} k_{ij} \tag{25.3}$$

$$= \sum_{l=1}^{k} \frac{1}{n|C_l|} \sum_{x_i \in C_l, x_j \in C_l} k_{ij}, \tag{25.4}$$

where $|C_l|$ is the number of items assigned to cluster l. One challenge, however, is that the resulting cluster models M exist in the new feature space, not the original feature space, and it may be difficult (or impossible) to map them back into the original (data) space. Therefore, it may be desirable to optimize the kernelized objective function in another way, such as via deterministic annealing [21], or by using a one-class support vector machine to model the data [8,12].

25.2.3 k-Means Limitations

The k-means algorithm has some drawbacks or limitations that are important to consider when analyzing a new data set. First, it can be inefficient for large data sets. Although each iteration of the algorithm has a complexity that is linear in the number of items being clustered, the number of iterations is unpredictable and can grow extremely large for large data sets. One way to address this issue is to terminate before complete convergence; often the final iterations result in changes to only a small fraction of the items so omitting them may not affect the final cluster centers much, if at all. For tackling very large data sets, several modified versions of k-Means that employ sampling strategies to reduce its runtime have been developed, such as BIRCH [53] and CURE [23]. K-Means can also be accelerated by constructing a kd-tree over the feature space and then traversing the kd-tree to assign clusters to groups of items at a time [2,29,41]. For problems with high dimensionality, kd-trees are less effective. More generally, faster k-means performance can be achieved using distance bounds and the triangle inequality to skip unnecessary cluster–item distance calculations [24].

Second, k-means by its nature is sensitive to the choice of initial cluster models M. Random initialization is the most common strategy, but this means that the results can vary across runs of the algorithm, and some may be significantly better than others. The k-means++ algorithm [4] adopts a probabilistic sampling strategy to identify good starting centers for M that yields significant speed improvements. If any prior knowledge is available that can assist in choosing the cluster centers intelligently ("seeding"), then better results can also be obtained more quickly [5].

Finally, k-means requires the specification of the desired number of clusters. The objective function value of the resulting partition cannot be used directly to determine the best k value, since it can always be minimized by setting $k = N$, yielding $f_{KM} = 0$. A vast literature has sprung up proposing methods for automatically selecting the best k value, including the Akaike information criterion [1], the Bayes information criterion [43], and so on. Most such measures seek a trade-off between a tight fit to (good model of) the data versus the complexity of the learned model. Fundamentally, however, it may not be possible to select the best value for k without knowing how the clusters will be used, or what knowledge is sought. Increasing k yields cluster distinctions at finer levels of granularity, and being able to see partitions at a variety of levels can yield additional insights. This is the motivation behind agglomerative clustering (see Section 25.4). Consequently, any method that promises to automatically select k should be taken with a grain of salt (or a dose of domain knowledge).

For more extensive coverage of the history and evolution of the k-means algorithm, consult Anil Jain's retrospective [26].

25.3 EM CLUSTERING

The expectation-maximization (EM) algorithm provides a strategy for conducting the alternating optimization approach of k-means in a probabilistic setting [16]. It is a general method for estimating any model's parameters, but is widely used specifically for constructing clustering models. In contrast to the k-means algorithm, in which each cluster model m_j is a location in the feature space, EM employs probabilistic cluster models such as Gaussian or normal distributions. Another name for EM using Gaussian cluster models is *Gaussian mixture modeling*.

In the ensuing explanation, for simplicity I assume that clusters are equivalent to the mixtures being modeled. However, depending on the application, there may or may not be a one-to-one mapping between clusters and mixtures; for example, a single cluster could be composed of several mixtures. A determination of how to interpret the resulting clusters or mixtures should rely on knowledge of the problem domain.

EM clustering adopts a *generative* approach to modeling the data set. That is, the items in the data set are assumed to have come from a collection of k (noisy) sources. Each cluster is a sample of data drawn from one of those sources. The clusters are each modeled with a

Gaussian distribution. In one dimension, the probability of observing an item with value x is

$$\mathcal{N}_{\mu,\sigma}(x) = \frac{1}{\sqrt{2\pi\sigma^2}}\exp\left(-\frac{(x-\mu)^2}{2\sigma^2}\right), \tag{25.5}$$

where μ is the mean (center) of the distribution and σ is the standard deviation (σ^2 is the variance) and captures the amount of "spread" or noise. The model for a cluster in multiple dimensions is parameterized by a (μ_d, σ_d) pair for each dimension d; for convenience, let μ_j, σ_j refer to all d values for cluster j, and μ, σ refer collectively to the parameters for all clusters.

To identify the k cluster models, the EM algorithm estimates μ_j and σ_j for each cluster j. The best solution is one that maximizes the *likelihood* of the model given the data:

$$f_{EM} = \mathcal{L}(\mu, \sigma | X) = \prod_i^n p(x_i | \mu_j, \sigma_j). \tag{25.6}$$

Like k-means, EM alternates between two steps that correspond to item–cluster assignments and cluster model updates (see Algorithm 25.2). Initialization occurs randomly, as with k-means, or possibly uses a quick run of k-means to find reasonable starting points (line 3). The assignment step in EM is the expectation step, in which expected likelihood of the current solution is calculated (line 5), and the probability of membership in each cluster can be computed for each item (line 8). We can rewrite Equation 25.6 as

$$\mathcal{L}(\mu, \sigma | X) = \prod_i^n \sum_j^k \mathbb{I}(a_i = j)p(a_i = j)p(a_i = j | \mu_j, \sigma_j) \tag{25.7}$$

Algorithm 25.2 EM clustering [16]

1: **Inputs:** Data set X containing n items, number of clusters k
2: **Outputs:** Cluster models $M = \{m_j = (\mu_j, \sigma_j), \; j = 1 \ldots k\}$, assignments $A = \{a_i, \; i = 1 \ldots n\}$
3: Initialize the k cluster models with randomly chosen $x_i \in X$ (or k-means).
4: **repeat**
5:     Estimate likelihood $\mathcal{L}$ using Equation 25.8. // Expectation
6:     **for** $i = 1$ **to** n **do**
7:         **for** $j = 1$ **to** k **do**
8:             Update cluster memberships $p(a_i = j | \mu_j, \sigma_j)$ using Equation 25.9.
9:         **end for**
10:    **end for**
11:    **for** $j = 1$ **to** k **do**
12:         Update model params μ_j, σ_j with Equations 25.10 and 25.11. // Maximization
13:         Update prior cluster probabilities $p(a_i = j)$ using Equation 25.12.
14:    **end for**
15: **until** A does not change // Convergence

$$= \prod_{i}^{n} \sum_{j}^{k} \mathbb{I}(a_i = j) p(a_i = j) \mathcal{N}_{\mu_j, \sigma_j}(x), \qquad (25.8)$$

where $\mathbb{I}$ is the indicator function (1 if the expression $a_i = j$ is true, 0 otherwise), and $p(a_i = j)$ is the prior probability that any item belongs to cluster j. Each item's cluster membership probabilities must sum to 1; each such probability is the (weighted) probability that cluster j generated the item normalized by the probability sum over all clusters:

$$p(a_i = j | \mu_j, \sigma_j) = \frac{p(a_i = j) \mathcal{N}_{\mu_j, \sigma_j}(x_i)}{\sum_{l}^{k} p(a_i = l) \mathcal{N}_{\mu_l, \sigma_l}(x_i)}. \qquad (25.9)$$

The assignments a_i are set to $\arg \max_j p(a_i = j | \mu_j, \sigma_j)$.

The cluster model update step in EM is the maximization step (lines 12–13), during which the μ and σ values are refined to maximize the likelihood, and the prior cluster probabilities $p(a_i = j)$ are updated given the new assignments.

$$\mu_j = \frac{\sum_{i}^{n} p(a_i = j | \mu_j, \sigma_j) x_i}{\sum_{i}^{n} p(a_i = j | \mu_j, \sigma_j)} \qquad (25.10)$$

$$\sigma_j = \sqrt{\frac{\sum_{i}^{n} p(a_i = j | \mu_j, \sigma_j)(x_i - \mu_j)^2}{\sum_{i}^{n} p(a_i = j | \mu_j, \sigma_j)}} \qquad (25.11)$$

$$p(a_i = j) = \frac{1}{n} \sum_{i}^{n} \frac{p(a_i = j | \mu_j, \sigma_j)}{\sum_{l}^{k} p(a_i = l | \mu_l, \sigma_l)} \qquad (25.12)$$

Figure 25.3 shows the output of EM clustering on the same artificial data set with $k = 2$ and $k = 3$. The cluster models M are represented with ellipses dictated by μ, σ. For three clusters, in contrast to the k-means result, the upper right cluster is compactly modeled and does not include items from the leftmost cluster. The boundary between the other two clusters also

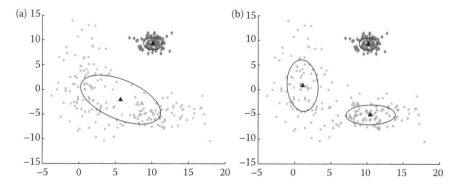

FIGURE 25.3 EM clustering on an artificial data set ($N = 300, k = 3$). The cluster centers (M) are shown with triangular markers, and the black ellipses mark illustrate the variance of each cluster model. (a) EM output, $k = 2$. (b) EM output, $k = 3$.

differs. In general, k-means is best suited for data with clusters with roughly the same size and extent, while EM can produce models that stretch or shrink to accommodate item density.

Although each item is shown with a symbol according to its cluster assignment a_i, the probabilistic modeling yields a "soft" assignment, $p(a_i = j|\mu_j, \sigma_j)$. Therefore, EM provides a posterior probability of each item's membership, as well as partial membership in each of the k clusters. For some applications, this can provide useful information beyond simply the most probable cluster for each item.

For more details about EM clustering, consult the excellent tutorial by Jeff Bilmes [10].

25.4 HIERARCHICAL AGGLOMERATIVE CLUSTERING

As noted earlier, there are cases in which it is useful to create a series of partitions of the data set, yielding a hierarchy of cluster organization at multiple levels of granularity. The resulting structure of nested partitions is called a *dendrogram*. Hierarchical agglomerative clustering (see Algorithm 25.3) usually proceeds in a bottom-up fashion, beginning with the partition in which each item is assigned its own cluster (line 3). For notational ease, the partitions are numbered from n (at the bottom of the hierarchy) to 1 (at the top) so that each partition's index also indicates the number of clusters it contains. Clusters are then iteratively merged to create the next higher partition in the hierarchy (line 5). The "model" of each cluster is emergent rather than explicit; the cluster consists of all items assigned to it, and each cluster is connected through the hierarchy to its parent and children.

Different definitions of cluster distance lead to different dendrogram structures [33]. Single-link or nearest-neighbor clustering defines the distance between two clusters C_i, C_j as the distance between their closest members:

$$d(C_i, C_j) = \min_{x_a \in C_i, x_b \in C_j} d(x_a, x_b). \tag{25.13}$$

Complete-link or furthest-neighbor clustering relies on the items with the largest separation:

$$d(C_i, C_j) = \max_{x_a \in C_i, x_b \in C_j} d(x_a, x_b). \tag{25.14}$$

Finally, average-link clustering computes all pairwise distances between items in the two clusters:

$$d(C_i, C_j) = \frac{1}{(n_i + n_j)(n_i + n_j - 1)} \sum_{x_a \in C_i} \sum_{x_b \in C_j} d(x_a, x_b). \tag{25.15}$$

Algorithm 25.3   Agglomerative clustering

1: **Inputs:** Data set X containing n items
2: **Outputs:** Partition hierarchy $H = \{p_1, \ldots, p_n\}, p_i = \{c_1, \ldots, c_i\}$.
3: Let p_n consist of one cluster per item, yielding n clusters.
4: **for** $i = n - 1$ **to** 1 **do**
5:    Create p_i by merging the two closest clusters in p_{i+1}.
6: **end for**

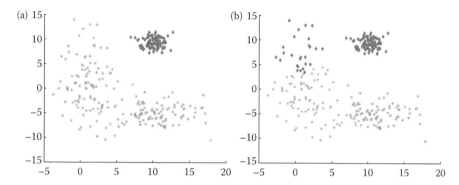

FIGURE 25.4 Agglomerative clustering on an artificial data set ($N = 300, k = 3$). (a) Single linkage. (b) Complete linkage.

Single-link clustering tends to form long "chains" of items as it links neighbors together; complete-link clustering is very sensitive to the presence of outliers, which will dominate the intercluster distance calculations; and average-link clustering is a compromise between the two.

A partition of any desired granularity can be selected from the hierarchy. Figure 25.4 shows the p_3 result of agglomerative clustering on the artificial data set with single-link and complete-link cluster merging. The single-link result does contain three clusters, but one cluster consists only of a single item (in the bottom right region), because its nearest-neighbor distance is so large that it is not merged with other items until p_2. The single-link process merged all the other items into their clusters first. The complete-link result spreads cluster membership more evenly, but it still yields an unintuitive result. These behaviors occur because agglomerative clustering makes very local decisions without modeling item density.

Unlike k-means or EM, this process is deterministic and always yields the same result for the same data set. In its naive form it is computationally expensive: computing all pairwise distances alone takes $\mathcal{O}(n^2)$ operations, and then building the hierarchy requires $\mathcal{O}(n^3)$ additional operations, to find the smallest such distance at each of n levels.

A hierarchy can also be built from the top down, using *divisive* clustering. One strategy for doing this is to create a minimum spanning tree (MST) across the data set in which links between items are weighted by their pairwise distance. Clustering proceeds by iteratively remove links, starting with the heaviest (largest-distance) link. Removing a single link from an MST increases the number of connected components (which correspond to clusters) by one, so the result is a nested hierarchy of partitions built from the top down.

25.5 SPECTRAL CLUSTERING

Spectral clustering departs from the preceding methods by examining the similarity between items, rather than their distance, and by basing clustering decisions on spectral properties of the associated data affinity matrix W. This process calculates the normalized graph Laplacian derived from W, then performs a regular k-means clustering in the space defined by the top

---

Algorithm 25.4   Spectral clustering [39]

1: **Inputs:** Data set X containing n items, number of clusters k

2: **Outputs:** Cluster models $M = \{m_j = (\mu_j, \sigma_j), j = 1 \ldots k\}$, assignments $A = \{a_i, i = 1 \ldots n\}$

3: Compute affinity matrix W using Equation 25.16.

4: Compute diagonal matrix D where $d_{ii} = \sum_j w_{ij}$.

5: Compute Laplacian matrix $L = D - W$ (unnormalized) or $L = D^{-\frac{1}{2}} W D^{-\frac{1}{2}}$ (normalized).

6: Compute $u_i \in \mathcal{R}^n$, the top k eigenvectors (smallest eigenvalues) of L.

7: Let $U \in \mathcal{R}^{n \times k} = [u_1, \ldots, u_k]$.

8: Normalize each u_i by dividing by $\sqrt{\sum_j u_{ij}^2}$.

9: Let y_i be row i in U, $i = 1 \ldots n$.

10: Apply k-means clustering to $Y = \{y_i\}$ to obtain k cluster models m_j and assignments A.

---

k eigenvectors [37,39]. In this sense, the spectral transformation or mapping is applied to enhance separability between the clusters.

In more detail, the first step is to compute the affinity matrix W (see Algorithm 25.4, line 3). A common way to calculate the similarity w_{ij} between two items x_i, x_j is to convert their distance using the RBF kernel described above (Equation 25.2):

$$w_{ij} = \exp\left(-\frac{d(x_i, x_j)^2}{2\sigma^2}\right), \tag{25.16}$$

where σ is a scale parameter. Larger distances translate to smaller affinities (similarities). To compute the graph Laplacian, the algorithm first computes the matrix D which is zero everywhere but on the diagonal; diagonal entries $d_{ii} = \sum_j w_{ij}$ (line 4) and then $L = D - W$ (line 5). It is usually better to compute a normalized Laplacian $L = D^{-\frac{1}{2}} W D^{-\frac{1}{2}}$. The matrix L is decomposed into its top k eigenvectors, which are associated with the k smallest eigenvalues (line 6). After normalizing the eigenvectors (lines 7–8), clustering of the data proceeds in their eigenspace (lines 9–10).

Figure 25.5 shows the result of spectral clustering on the artificial data set, using the approach of Ng et al. [39], as implemented in MATLAB® by Asad Ali. The tight clustering (high affinity) within the compact upper-right cluster is evident from the bright values in the lower–right portion of the affinity matrix. The remaining two clusters are more spread out, leading to fainter affinities, and the overlap between the clusters results in the bright affinity values that appear outside the block diagonal within-cluster regions.

The result of k-means clustering ($k = 3$) in the resulting (normalized) eigenspace is also shown. It is similar to the result obtained by EM clustering (Figure 25.3b) but did not require the assumption of Gaussian models for each cluster. Since k-means was used to obtain the clusters, the only cluster models involved are the means for each cluster, which exist in the

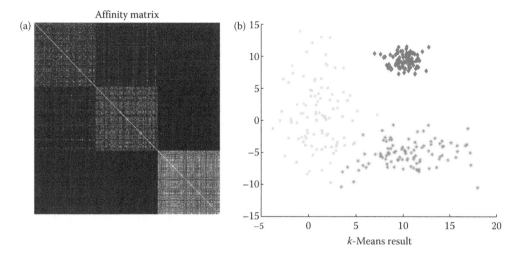

FIGURE 25.5 Spectral clustering on an artificial data set ($N = 300, k = 3$). (a) Affinity matrix. (b) k-Means result.

transformed eigenspace. The power of this model-free approach is illustrated more clearly for clusters that clearly are not Gaussian in structure.

For more details about spectral clustering, consult Ulrike von Luxburg's tutorial [48]. Also, the connection between kernel k-means and spectral clustering [17] is of particular interest.

25.6 CONSTRAINED CLUSTERING

Traditionally, clustering proceeds as an unsupervised process, that is, without any external guidance as to the kind of solution sought. It automatically identifies cluster models or regions of high density based solely on the information contained in the feature vector description of the items in the data set. However, most data analysis does not operate in a vacuum, and the investigator may have access to complementary information about the items or their relationships with each other that can guide the clustering process toward relevant solutions. The field of *constrained clustering* encompasses new clustering algorithms that can accommodate this kind of side information or domain knowledge.

This domain knowledge can take a wide variety of forms. It might consist of known cluster labels for a subset of the data, for example, if the goal is to cluster a collection of light curves with all known supernovae labeled but the rest of the items unlabeled. It might consist of knowledge about pairwise relationships between items, such as a statement that two items should, or should not, be assigned to the same cluster. These statements can capture some knowledge about the desired solution even when the particular clusters to which the items belong are not known (e.g., "These two light curves are too different to ever be clustered together, even if we don't know what types of sources they originate from" or "These two light curves come from sources with known x-ray characteristics that indicate they are the same type of source"). Domain knowledge can also provide guidance about the desired clusters, such as their location in feature space, their minimum or maximum sizes, the type of distribution used to model their assigned items, and so on.

Existing (partial) item labels can be used to "seed" the k-means clustering process to bias it toward relevant solutions [5]. Pairwise relationships are usually encoded as *must-link* and *cannot-link* constraints [49]. They can be incorporated into k-means straightforwardly by disallowing any assignments that would violate a constraint, via the cop-k means algorithm [50]. For cases in which a confidence can be associated with each constraint, they can be treated as soft constraints to be accommodated according to their confidence [51].

Algorithms have also been developed to incorporate pairwise constraints into EM clustering by requiring that the solution at each iteration remain consistent with the constraints [44] or by interpreting the constraints as prior probabilities on pairwise memberships in penalized probabilistic clustering [35]. Spectral clustering can incorporate constraints by modifying the affinity matrix W [28]. Constraint information can also be used to modify (tune) the distance metric, bringing must-linked pairs closer together and pushing cannot-linked pairs farther apart [32]. Any regular clustering algorithm can then proceed with the new metric and no further modifications. Bilenko et al. proposed a unified approach that enables k-means to accommodate both the constraints themselves and to learn an appropriate metric while clustering, showing this to be superior to doing just one or the other [9].

For more coverage of the ongoing developments in constrained clustering, consult the book edited by Basu et al. [6].

25.7 APPLICATIONS TO ASTRONOMY DATA

Clustering algorithms have a long history of being used to investigate and understand astronomical data sets.

One such area of investigation has been to determine how many fundamental types of GRBs there are, and how to describe these populations. Traditionally, GRBs were separated into two groups based on their duration ($>$ or <2 s). Mukherjee et al. [38] applied both EM clustering and agglomerative clustering to 797 bursts from the third Burst and Transient Source Experiment catalog. Both methods indicated that there were three clusters present, including one that had not been previously described (intermediate duration, intermediate brightness, and soft spectra). The existence of three GRB groups was confirmed in a later analysis using k-means and Dirichlet mixture modeling [13]. A more recent study by Veres et al. [47] identified a possible interpretation for this third group as x-ray flash signatures. Similarly, Cheeseman et al. [14] employed a Bayesian clustering algorithm (AutoClass) and discovered a new class of stars in the Infrared Astronomical Satellite low-resolution spectral catalog. Feitzinger and Galinski applied k-means clustering recursively to estimate the fractal dimension of star-forming sites in galaxies [20].

Clustering can also be used to identify anomalous items within a data set. The periodic curve anomaly detection (PCAD) algorithm uses a modified k-means clustering method to identify anomalous light curves and was able to identify interesting outliers in observations of periodic variable stars [42]. Relatedly, the DEMAC system conducts a PCA to identify outliers, albeit without explicitly identifying clusters in the data [18].

Constrained clustering can be used to cluster data with missing values, as has been demonstrated on Sloan Digital Sky Survey star/galaxy data [52]. Traditional approaches for dealing with missing values include marginalization, which removes either all features or all items that have missing values, and imputation, which fills in missing values with a best-guess value. The drawback of marginalization is that it may discard a significant amount of good data along with the missing values. Imputation, on the other hand, may insert unreliable values if it cannot be assumed that values are "missing at random." For example, a shape feature can be computed for sources that span multiple pixels in the sky, but is not well defined for a single-pixel source. In such a case, the fact that the feature is missing is itself informative, and there is no sensible replacement value. That is, imputation is appropriate when the missing value was simply unobserved, but not appropriate when the value was unobservable. In contrast, constrained clustering uses all of the available values without imputing any additional ones. The fully observed features are used as the data's feature space, and the remaining partially observed features are used to generate pairwise constraints between items based on their similarity in those features.

One challenge of clustering astronomy data sets is that they may be so large that scalability becomes a real concern. As noted above, the BIRCH [53] and CURE [23] algorithms employ sampling to speed up clustering for large data sets. It is also possible to split up the data set, separately cluster each subset, then combine the results (sometimes referred to as collaborative [40] or cooperative [30] clustering). If each subset can be clustered by a different processor, then this provides a natural way to parallelize the clustering process without modifying the clustering algorithm used. It is also possible to parallelize the k-means algorithm directly by dividing up the item-to-cluster-center distance calculations among many processors [31,54]; the EM algorithm has been parallelized both for computing the cluster memberships (distributing across processors by items) and the mixture model parameters (distributing by cluster) [34]. Spectral clustering can be parallelized by distributing the data set between processors and employing a parallel eigensolver [45]. There are also map-reduce versions of k-means [55] and EM [15] clustering.

Another key concern is the repeatability or stability of the clustering results, especially if they will be used (for example) to conclude the existence of a new category of sky objects, or to identify a particular object as anomalous. Algorithms such as k-means (or EM, as it is often initialized with a k-means run) produce results that are locally, but not necessarily globally, optimal. The solution produced by a given run is determined by the initialization of the algorithm, and the quality of these solutions can vary dramatically. Agglomerative, divisive, and spectral clustering algorithms are deterministic and so do not exhibit this sensitivity. Due to the variation in output, quantitative results for k-means are often reported as the average across a large number of trials. However, if information about key trends or outliers is sought, then the quality of a given solution, not the average one, is needed. One way to measure this is by the solution's *stability*, which can be estimated by running the algorithm several times and measuring the frequency with which each solution is found, or to perturb the data set slightly and recluster it to determine if the results change [7]. Stability estimates for a given solution can inform further interpretation of the clustering results.

25.8 SUMMARY

Data clustering algorithms can serve as useful tools for investigating new data sets, particularly when those data sets are too large for manual review of each item to be feasible.

Clustering methods can report descriptions of typical items, characterize populations defined as dense regions in feature space, identify anomalies in the data set, and more. A wide array of clustering algorithms exists. This chapter described four representative clustering algorithms: k-means, EM, agglomerative, and spectral clustering. These provide a working basis for current clustering research and applications. Interested readers are encouraged to investigate the additional reading noted in each section, as well as the pointers to other clustering algorithms.

To date, clustering algorithms have aided astronomy investigations by identifying previously unrecognized subgroups within data sets, detecting anomalous observations, and enabling the analysis of data that contains missing features. These methods are a useful addition to any investigator's toolbox and are expected to assist in further discoveries in the future.

25.9 GLOSSARY

Agglomerative clustering: Hierarchical clustering starting with one cluster per item and iteratively merging the two closest clusters.

Cannot-link constraint: Prior knowledge about two items that they should not be grouped into the same cluster.

Constrained clustering: Clustering in the presence of constraints on cluster memberships, commonly pairwise constraints (must-link and cannot-link).

EM: Expectation-maximization, a generic optimization technique that alternates between computing the expected likelihood of a given model with respect to the observed data, and then maximizing the model parameters given the data. EM clustering uses this strategy to optimize cluster models.

Gaussian mixture modeling: The use of Gaussian distributions to model each cluster within a data set. Typically, the parameters of the Gaussian distributions (mean and standard deviation) are identified using the EM algorithm.

Kernel clustering: The use of a kernel function while clustering, usually in place of a distance metric, which permits the clustering to proceed in a higher-dimensional feature space.

Kernel function: A function that takes in two data items and returns a value that represents their similarity.

k-Means clustering: An iterative algorithm that starts with a random guess at the best k cluster centers, then alternates between assigning all items to their closest clusters and updating the centers to reflect their current constituents.

Must-link constraint: Prior knowledge about two items that they should be grouped into the same cluster.

Semisupervised clustering: Clustering unlabeled data when some additional information is available (e.g., labeled items, constraints, preferences, previously built cluster models).

Spectral clustering: Using spectral methods to aid in identifying clusters in the data set, for example by mapping the items into the eigenspace defined by the top k eigenvectors of the similarity matrix, then performing clustering in this new space.

Unsupervised learning: Analyzing a data set to build or train a model in the absence of any labeled examples.

Vector quantization: Identifying the k most useful feature vectors within a data set (e.g., the k centers identified by k-means) and then replacing each item in the data set with its closest such vector, thereby greatly compressing the data set (in a lossy fashion).

ACKNOWLEDGMENTS

This chapter was prepared with the support of NSF grant #ITR-0325329, at the Jet Propulsion Laboratory, California Institute of Technology, under a contract with the National Aeronautics and Space Administration. Asad Ali authored the spectral clustering implementation used to generate Figure 25.5. © 2010, California Institute of Technology. Agreement says Taylor & Francis has the copyright.

REFERENCES

1. H. Akaike. A new look at the statistical model identification. *IEEE Transactions on Automatic Control*, 19(6):716–723, 1974.
2. K. Alsabti, S. Ranka, and V. Singh. An efficient k-means clustering algorithm. In *Proceedings of the 1st Workshop on High Performance Data Mining*, Orlando, FL, 1998.
3. M. Anderberg. *Cluster Analysis for Applications*. Academic Press, New York, 1973.
4. D. Arthur and S. Vassilvitskii. k-Means++: The advantages of careful seeding. In *Proceedings of the 18th Annual ACM-SIAM Symposium on Discrete Algorithms*, pp. 1027–1035, 2007.
5. S. Basu, A. Bannerjee, and R. Mooney. Semi-supervised clustering by seeding. In *Proceedings of the Nineteenth International Conference on Machine Learning*, pp. 19–26, Sydney, Australia, 2002.
6. S. Basu, I. Davidson, and K. L. Wagstaff, editors. *Constrained Clustering: Advances in Algorithms, Theory, and Applications*. CRC Press, Boca Raton, FL, 2008.
7. A. Ben-Hur, A. Elisseeff, and I. Guyon. A stability based method for discovering structure in clustered data. In *Proceedings of the Pacific Symposium on Biocomputing*, pp. 6–17, Lihue, HI, 2002.
8. A. Ben-Hur, D. Horn, H. T. Siegelmann, and V. Vapnik. Support vector clustering. *Journal of Machine Learning Research*, 2:125–137, 2001.
9. M. Bilenko, S. Basu, and R. J. Mooney. Integrating constraints and metric learning in semi-supervised clustering. In *Proceedings of the Twenty-First International Conference on Machine Learning*, pp. 11–18, Banff, Canada, 2004.
10. J. Bilmes. A gentle tutorial of the EM algorithm and its application to parameter estimation for Gaussian mixture and hidden Markov models. Technical Report TR-97-021, ICSI, 1997.
11. R. J. Brunner, S. G. Djorgovski, T. A. Prince, and A. S. Szalay. In J. Abello, P. M. Pardalos, and M. G. C. Resende, editors, *Handbook of Massive Data Sets*, Massive Datasets in Astronomy, pp. 931–979. Kluwer Academic Publishers, Dordrecht, The Netherlands, 2002.
12. F. Camastra and A. Verri. A novel kernel method for clustering. *IEEE Transactions on Pattern Analysis and Machine Intelligence*, 27(5):801–805, 2005.
13. T. Chattopadhyay, R. Misra, A. K. Chattopadhyay, and M. Naskar. Statistical evidence for three classes of gamma-ray bursts. *The Astrophysical Journal*, 667(2):1017–1023, 2007.
14. P. Cheeseman, J. Stutz, M. Self, W. Taylor, J. Geobel, K. Volk, and H. Walker. Automatic classification of spectra from the infrared astronomical satellite (IRAS). NASA Reference Publication 1217, National Technical Information Service, Springfield, VA, 1989.
15. A. Das, M. Datar, A. Garg, and S. Rajaram. Google news personalization: Scalable online collaborative filtering. In *Proceedings of the 16th International World Wide Web Conference*, pp. 271–280, Banff, Canada, 2007.

16. A. P. Dempster, N. M Laird, and D. B. Rubin. Maximum likelihood from incomplete data via the EM algorithm. *Journal of the Royal Statistical Society*, 39(1):1–38, Series B 1977.

17. I. S. Dhillon, Y. Guan, and B. Kulis. Kernel k-means, spectral clustering, and normalized cuts. In *Proceedings of the Tenth ACM SIGKDD International Conference on Knowledge Discovery and Data Mining*, pp. 551–556, Seattle, WA, 2004.

18. H. Dutta, C. Giannella, K. D. Borne, and H. Kargupta. Distributed top-k outlier detection from astronomy catalogs using the DEMAC system. In *Proceedings of the Seventh SIAM International Conference on Data Mining*, pp. 473–478, Minneapolis, MN, 2007.

19. B. S. Everitt, S. Landau, M. Leese, and D. Stahl. *Cluster Analysis. Probability and Statistics*, 5th edition. Wiley, Chichester, UK, 2011.

20. J. V. Feitzinger and T. Galinksi. The fractal dimension of star-forming sites in galaxies. *Astronomy and Astrophysics*, 179(1–2):249–254, 1987.

21. M. Girolami. Mercer kernel-based clustering in feature space. *IEEE Transactions on Neural Networks*, 13(3):780–784, 2002.

22. A. D. Gordon. *Classification*. Chapman & Hall, New York, NY, 1981.

23. S. Guha, R. Rastogi, and K. Shim. CURE: An efficient algorithm for clustering large databases. In *Proceedings of the ACM-SIGMOD International Conference on Management of Data*, pp. 73–84, Seattle, WA, 1998.

24. G. Hamerly. Making k-means even faster. In *Proceedings of the 2010 SIAM International Conference on Data Mining*, pp. 130–140, Columbus, OH, 2010.

25. J. A. Hartigan. *Clustering Algorithms*. Wiley, New York, NY, 1975.

26. A. K. Jain. Data clustering: 50 years beyond k-means. *Pattern Recognition Letters*, 31:651–666, 2010.

27. A. K. Jain, M. N. Murty, and P. J. Flynn. Data clustering: A review. *ACM Computing Surveys*, 31(3):264–323, 1999.

28. S. D. Kamvar, D. Klein, and C. D. Manning. Spectral learning. In *Proceedings of the 18th International Joint Conference on Artificial Intelligence*, pp. 561–566, Acapulco, Mexico, 2003.

29. T. Kanungo, D. M. Mount, N. S. Netanyahu, C. Piatko, R. Silverman, and A. Y. Wu. Computing nearest neighbors for moving points and application to clustering. In *Proceeding of the Tenth ACM-SIAM Symposium on Discrete Algorithms*, pp. S931–S932, Baltimore, MD, 1999.

30. R. Kashef and M. S. Kamel. Cooperative clustering. *Pattern Recognition*, 43(6):2315–2329, 2010.

31. K. Kerdprasop and N. Kerprasop. A lightweight method to parallel k-means clustering. *International Journal of Mathematics and Computers in Simulation*, 4:144–153, 2010.

32. D. Klein, S. D. Kamvar, and C. D. Manning. From instance-level constraints to space-level constraints: Making the most of prior knowledge in data clustering. In *Proceedings of the Nineteenth International Conference on Machine Learning*, pp. 307–313, Sydney, Australia, 2002.

33. G. N. Lance and W. T. Williams. A general theory of classificatory sorting strategies, 1. Hierarchical systems. *The Computer Journal*, 9:373–380, 1967.

34. P. E. López-de Teruel, J. M. García, and M. Acacio. The parallel EM algorithm and its applications in computer vision. In *Proceedings of Parallel and Distributed Processing Techniques and Applications*, pp. 571–578, Las Vegas, NV, 1999.

35. Z. Lu and T. K. Leen. Semi-supervised learning with penalized probabilistic clustering. In L. K. Saul, Y. Weiss, and L. Bottou, editors, *Advances in Neural Information Processing Systems 17*, pp. 849–856, MIT Press, Cambridge, MA, 2004.

36. J. B. MacQueen. Some methods for classification and analysis of multivariate observations. In *Proceedings of the Fifth Symposium on Math, Statistics, and Probability*, Vol. 1, pp. 281–297, Berkeley, CA, 1967. University of California Press.

37. M. Meilă and J. Shi. A random walks view of spectral segmentation. In *Proceedings of the Eighth International Workshop on Artificial Intelligence and Statistics*, Key West, FL, 2001.

38. S. Mukherjee, E. D. Feigelson, G. J. Babu, F. Murtagh, C. Fraley, and A. Raftery. Three types of gamma-ray bursts. *The Astrophysical Journal*, 508(1):314–327, 1998.

39. A. Y. Ng, M. I. Jordan, and Y. Weiss. On spectral clustering: Analysis and an algorithm. In T. G. Dietterich, S. Becker, and Z. Ghahramani, editors, *Proceedings of Advances in Neural Information Processing Systems 14*, pp. 849–856, MIT Press, Cambridge, MA, 2001.

40. W. Pedrycz. Collaborative fuzzy clustering. *Pattern Recognition Letters*, 23(14):1675–1686, 2002.

41. D. Pelleg and A. Moore. Accelerating exact k-means algorithms with geometric reasoning. In *Proceedings of the Fifth International Conference on Knowledge Discovery in Database*, pp. 277–281, San Diego, CA, 1999.

42. U. Rebbapragada, P. Protopapas, C. E. Brodley, and C. Alcock. Finding anomalous periodic time series: An application to catalogs of periodic variable stars. *Machine Learning*, 74(3):281–313, 2009.

43. G. E. Schwarz. Estimating the dimension of a model. *Annals of Statistics*, 8(2):461–464, 1978.

44. N. Shental, A. Bar-Hillel, T. Hertz, and D. Weinshall. Computing Gaussian mixture models with EM using equivalence constraints. In S. Thrun, L. Saul, and B. Schoelkopf, editors, *Advances in Neural Information Processing Systems 16*, MIT Press, Cambridge, MA, 2003.

45. Y. Song, W.-Y. Chen, H. Bai, C.-J. Lin, and E. Chang. Parallel spectral clustering. In W. Daelemans, B. Goethals, and K. Morik, editors, *Machine Learning and Knowledge Discovery in Databases*, Lecture Notes in Computer Science, Vol. 5212, pp. 374–389. Springer, Berlin/Heidelberg, 2008.

46. V. N. Vapnik. *Statistical Learning Theory*. Wiley, New York, 1998.

47. P. Veres, Z. Bagoly, I. Horváth, A. Mészáros, and L. G. Balázs. A distinct peak-flux distribution of the third class of gamma-ray bursts: A possible signature of x-ray flashes? *The Astrophysical Journal*, 725(2):1955–1965, 2010.

48. U. von Luxburg. A tutorial on spectral clustering. Technical Report TR-149, Max Planck Institute for Biological Cybernetics, 2006.

49. K. Wagstaff and C. Cardie. Clustering with instance-level constraints. In *Proceeding of the Seventeeth International Conference on Machine Learning*, pp. 1103–1110, Stanford, CA, 2000.

50. K. Wagstaff, C. Cardie, S. Rogers, and S. Schroedl. Constrained k-means clustering with background knowledge. In *Proceedings of the Eighteenth International Conference on Machine Learning*, pp. 577–584, Williamstown, MA, 2001.

51. K. L. Wagstaff. Intelligent clustering with instance-level constraints. PhD thesis, Cornell University, August 2002.

52. K. L. Wagstaff and V. G. Laidler. Making the most of missing values: Object clustering with partial data in astronomy. In *Proceedings of Astronomical Data Analysis Software and Systems XIV*, Vol. 347, pp. 172–176, Pasadena, CA, 2005.

53. T. Zhang, R. Ramakrishnan, and M. Livny. BIRCH: An efficient data clustering method for very large databases. In *Proceedings of the ACM-SIGMOD International Conference on Management of Data*, pp. 103–114, Montreal, Canada, 1996.

54. Y. Zhang, Z. Xiong, J. Mao, and L. Ou. The study of parallel k-means algorithm. In *Proceedings of the Sixth World Congress on Intelligent Control and Automation*, pp. 5868–5871, Dalian, China, 2006.

55. W. Zhao, H. Ma, and Q. He. Parallel k-means clustering based on MapReduce. In M. Jaatun, G. Zhao, and C. Rong, editors, *Cloud Computing*, Lecture Notes in Computer Science, Vol. 5931, pp. 674–679. Springer, Berlin/Heidelberg, 2009.

CHAPTER 26

Ensemble Methods

A Review

Matteo Re and Giorgio Valentini

Università degli Studi di Milano

CONTENTS

26.1 Introduction 563
26.2 Theoretical and Practical Reasons for Combining Classifiers 565
26.3 Taxonomies of Ensemble Methods 567
26.4 Nongenerative Ensembles 570
    26.4.1 Ensemble Fusion Methods 570
    26.4.2 Ensemble Selection Methods 572
26.5 Generative Ensembles 573
    26.5.1 Resampling Methods 573
    26.5.2 Feature Selection/Extraction Methods 575
    26.5.3 Mixture of Experts 576
    26.5.4 OC Methods 576
    26.5.5 Randomized Ensemble Methods 577
26.6 Ensemble Methods in Astronomy and Astrophysics 578
26.7 Conclusions 582
Acknowledgments 582
References 582

26.1 INTRODUCTION

Ensemble methods are statistical and computational learning procedures reminiscent of the human social learning behavior of seeking several opinions before making any crucial decision. The idea of combining the opinions of different "experts" to obtain an overall "ensemble" decision is rooted in our culture at least from the classical age of ancient Greece, and it has been formalized during the Enlightenment with the *Condorcet Jury Theorem* [45]), which proved that the judgment of a committee is superior to those of individuals, provided the individuals have reasonable competence.

Ensembles are sets of learning machines that combine in some way their decisions, or their learning algorithms, or different views of data, or other specific characteristics to obtain more reliable and more accurate predictions in supervised and unsupervised learning problems [48,116]. A simple example is represented by the *majority vote* ensemble, by which the decisions of different learning machines are combined, and the class that receives the majority of "votes" (i.e., the class predicted by the majority of the learning machines) is the class predicted by the overall ensemble [158].

In the literature, a plethora of terms other than ensembles has been used, such as fusion, combination, aggregation, and committee, to indicate sets of learning machines that work together to solve a machine learning problem [19,40,56,66,99,108,123], but in this chapter we maintain the term *ensemble* in its widest meaning, in order to include the whole range of combination methods.

Nowadays, ensemble methods represent one of the main current research lines in machine learning [48,116], and the interest of the research community on ensemble methods is witnessed by conferences and workshops specifically devoted to ensembles, first of all the *multiple classifier systems (MCS)* conference organized by Roli, Kittler, Windeatt, and other researchers of this area [14,62,85,149,173].

Several theories have been proposed to explain the characteristics and the successful application of ensembles to different application domains. For instance, Allwein, Schapire, and Singer interpreted the improved generalization capabilities of ensembles of learning machines in the framework of large margin classifiers [4,177], Kleinberg in the context of stochastic discrimination theory [112], and Breiman and Friedman in the light of the bias–variance analysis borrowed from classical statistics [21,70].

Empirical studies showed that both in classification and regression problems, ensembles improve on single learning machines, and moreover large experimental studies compared the effectiveness of different ensemble methods on benchmark data sets [10,11,49,188]. The interest in this research area is motivated also by the availability of very fast computers and networks of workstations at a relatively low cost that allow the implementation and the experimentation of complex ensemble methods using off-the-shelf computer platforms. However, as explained in Section 26.2 there are deeper reasons to use ensembles of learning machines, motivated by the intrinsic characteristics of the ensemble methods.

The main aim of this chapter is to introduce ensemble methods and to provide an overview and a bibliography of the main areas of research, without pretending to be exhaustive or to explain the detailed characteristics of each ensemble method. The paper is organized as follows. In the next section, the main theoretical and practical reasons for combining multiple learners are introduced. Section 26.3 depicts the main taxonomies on ensemble methods proposed in the literature. In Section 26.4 and 26.5, we present an overview of the main supervised ensemble methods reported in the literature, adopting a simple taxonomy, originally proposed in Ref. [201]. Applications of ensemble methods are only marginally considered, but a specific section on some relevant applications of ensemble methods in astronomy and astrophysics has been added (Section 26.6). The conclusion (Section 26.7) ends this paper and lists some issues not covered in this work.

26.2 THEORETICAL AND PRACTICAL REASONS FOR COMBINING CLASSIFIERS

There is no unified theory underlying ensemble methods, and several authors outlined that consistent and theoretically sound explanations of the success of classifier ensembles are not available [116] or are incomplete or assumption bound [90]. This is not surprising, considering the variety of the proposed approaches and the relative youngness of this research area.

Theories on ensemble methods. Despite these negative considerations, we would like to cite at least three theories able to explain the effectiveness of some of the most widely known and used supervised ensemble methods. The first one considers the ensembles in the framework of large margin classifiers [133], showing that ensembles enlarge the margins, enhancing the generalization capabilities of output coding (OC) [4] and boosting-based ensemble algorithms [177]. This interpretation is strictly related to Vapnik's statistical learning theory [203], which is likely the most accredited theory within the machine learning community.

The second is based on the classical bias–variance decomposition of the error [76], and it shows that ensembles can reduce variance [20,124] or both bias and variance [22,113,176]. Recently, Domingos proved that these two theories are two faces of the same coin. Indeed Schapire's notion of margins [177] can be expressed in terms of bias and variance and vice versa, and hence Schapire's bounds of ensemble's generalization error can be equivalently expressed in terms of the distribution of the margins or in terms of the bias–variance decomposition of the error, showing the equivalence of margin-based and bias–variance-based approaches [52,53].

Another general theory about ensemble methods has been proposed by Kleinberg [110,111]. His stochastic discrimination theory is founded on a set-theoretic abstraction to remove all the algorithmic details of classifiers and training procedures. By this abstraction the classifiers are considered as a combination of subsets of points of the feature space underlying a given problem, classifiers' decision regions are considered only in form of point sets, and the set of classifiers is just a sample into the power set of the feature space. A rigorous mathematical treatment starting from the "representativeness" of the examples used in machine learning problems leads to the design of ensemble of weak classifiers, whose accuracy is governed by the law of large numbers [38,109].

Statistical, representational, and computational reasons for combining multiple learners. Without pretending to depict a general theory on ensemble methods, Thomas Dietterich suggested three main reasons why an ensemble of classifiers might be better than a single classifier [48].

The first one is statistical. Indeed, learning algorithms try to find an hypothesis in a given space $\mathcal{H}$ of hypotheses, and in many cases if we have sufficient data they can find the optimal one for a given problem, but in real cases we have only limited data sets and sometimes only few examples are available. The "subregion" $\mathcal{S} \subset \mathcal{H}$ of the "optimal" hyphotheses with respect to the training error, may correspond to classifiers having different generalization performances. We could in principle try to select among them the simplest or the one with the lowest capacity, but in practice this is difficult: we can avoid this problem by averaging or combining the base classifiers to get a good approximation of the unknown true hypothesis.

Continuing to follow Dietterich's analysis, the second reason for combining multiple learners arises from the limited representational capability of learning algorithms. In many cases the unknown function to be approximated is not present in $\mathcal{H}$, but a combination of hypotheses drawn from $\mathcal{H}$ can expand the space of representable functions, possibly embracing also the true one. It is well known that many learning algorithms enjoy universal approximation properties [94,151], but these asymptotic features do not hold with finite data sets, since the effective space of hypotheses explored by the learning algorithm with small-sized data can be significantly smaller than the virtual $\mathcal{H}$ considered in the asymptotic case. From this standpoint ensembles can enlarge the effective hypotheses coverage, expanding the space of representable functions.

The third reason is computational, in the sense that training algorithms may get stuck in local optima. For instance, multilayer perceptrons apply gradient descent techniques to minimize an error function over the training data, and inductive decision trees employ a greedy local optimization approach, both resulting in suboptimal solutions due to multiple local minima of the underlying error function to be minimized. As a consequence, even if the learning algorithm could in principle find the best hypothesis, we actually may not be able to find it. An ensemble merging different local suboptimal solutions may achieve a better approximation, at least avoiding the worst local minima solutions.

The accuracy–diversity trade-off. A simple example, due another time to Dietterich [48], is useful to introduce another open theoretical issue about ensemble methods: the so-called accuracy–diversity trade-off [121]. Having a set of L classifiers whose error is lower than random guessing for a two-class classification problem (that is an error <0.5), it is easy to see that the overall error of the majority voting ensemble, given by the area under the binomial distribution where more than $L/2$ hypotheses are wrong, is significantly lower than the error of the base classifier. It is worth noting that this result is known since the end of the eighteenth century in the context of social sciences: in fact the *Condorcet Jury Theorem* [45]) proved that the judgment of a committee is superior to those of individuals, provided the individuals have reasonable competence (that is, on the assumption that their probability of being correct is >0.5). Nevertheless, this result holds only if the base classifiers (individuals) are independent: if their decisions are dependent, that is similar for a given input data set, we have not a series of independent Bernoulli experiments, and we have no guarantee of a reduced error when a majority voting ensemble is applied. This fact is a particular case of the more general problem of the relationships between accuracy and diversity of base learners within an ensemble: the performance of an ensemble depends not only on the accuracy of the component base learners, but also on their diversity, that is on their capability of responding differently to the same input. On the one hand, if each base learner provides the same predictions, there is no utility in combining their outputs, but on the other hand, if the base learners are maximally accurate, they generate the same correct predictions, with no diversity between them. The resulting trade-off between accuracy and diversity has been actively studied, starting from the pioneering work of Tumer and Gosh, who showed how ensemble error decreases as base model error decreases and diversity increases [195]. Nevertheless, Kuncheva studies showed that the relationships between accuracy and diversity of base learners are more complex, showing, for example, that the way the ensemble method generates base learners and the behavior of the base learning algorithms play a crucial role in determining the characteristics of the accuracy–diversity trade-off [120,121].

Empirical reasons for combining multiple learners. There are also practical reasons for using ensemble methods, as witnessed by their successful applications in several domains [143,144]. Indeed employing multiple learners can derive from the application context, such as when multiple sources data are available, inducing a natural decomposition of the problem. In more general cases we can dispose of different training sets, collected at different times, having eventually different features and we can use different specialized learning machine for each different item. Predictive performances of single models have been improved by the ensemble methodology in several application fields, ranging from information security [139], astronomy and astrophysics [13], geography and remote sensing [17], image retrieval [190], finance [128], to medicine [160], bioinformatics [166], and chemioinformatics [140].

26.3 TAXONOMIES OF ENSEMBLE METHODS

Considering the variety of ensemble techniques and the large number of combination schemes proposed in the literature, it is not surprising that a very large number of ensemble methods and algorithms are now available to the research community. To help the researchers and practitioners to get their bearings and to develop new methods and techniques, several taxonomies of ensemble methods have been proposed. Indeed, combination techniques can be grouped and analyzed in different ways, depending on the main classification criterion adopted. For instance, if we consider the representation of the input patterns as the main criterion, we can identify two distinct large groups, one that uses the same and one that uses different representations of the inputs [107,108].

Another criterion distinguishes between strictly speaking ensemble systems and modular systems: the latter is characterized by component learners devoted to different tasks by which the original problem has been decomposed, the former by a combination of a set of classifiers, each of which solves the same original task [122,184]. A taxonomy can also be based on the way diversity between base learners is achieved, that is implicitly between randomization methods like bagging and random subspace techniques, or by methods that explicitly improves diversity through a proper metric [28]. The differentiation between trainable and nontrainable ensembles represents another key to classify ensemble methods: nontrainable ensembles do not need training after the base learners have been induced (they apply "fixed" rules to combine base classifiers), while trainable ensembles imply the training of the combiner module, either during or after the base learners have been trained [58,116].

Sharkey [182] proposes a multidimensional taxonomy, founded on three dichotomies: (1) selection or combination of the multiple base learners; (2) methods based or not on the direct combination of base learner outputs; (3) methods based on ensembles or modular systems. Extending this approach, a more complex five-dimensional taxonomy has been proposed by Rokach, based on combiner usage, classifier dependency, diversity generation, ensemble size and the capability of ensemble methods to be applied with different base learning algorithms [171].

In her fundamental book on ensemble methods, Kuncheva proposes a four–level taxonomy based on the way ensembles are constructed [116]. At a first level, the author highlights the combination rules to assemble multiple classifiers, distinguishing between fusion methods that combine in some way the outputs of the base learners and selection methods, by which a single classifier is selected among the set of available base classifiers. At a second level we may consider different models, and we may design base learners for specific ensemble methods. At feature level, different subsets of features can be used for the classifiers. Finally, different data subsets, so that each base classifier in the ensemble is trained on its own data, can be used to build up the committee of learning machines.

In this survey we adopt the classification scheme originally proposed in [201] (with some minor modifications borrowed from [116]), not because we consider this taxonomy better than others, but simply because it is quite simple and clean and facilitates us to introduce the main ensemble methods presented in the literature. Indeed ensemble methods are characterized by two basic features: (1) the algorithms by which different base learners are combined; (2) the techniques by which different and diverse base learners are generated.

Our proposed taxonomy basically distinguishes between *nongenerative* ensemble methods that mainly rely on the former feature of ensemble methods, and *generative* ensembles that mainly focus on the latter. It is worth noting that the "combination" and the "generation" of base learners are somehow both present in all ensemble methods: the distinction between these two large classes depends on the predominance of the combination or of the generation component of the ensemble algorithm.

More precisely, *nongenerative* ensemble methods confine themselves to combine a set of possibly well-designed base classifiers: they do not actively generate new base learners but try to combine in a suitable way a set of existing base classifiers. Examples are methods that combine the output of a set of base learners by majority voting [158], or methods that select the best subset of base learners on the basis of their accuracy [156], or methods that combine the probabilistic output of a set of classifiers according to the Bayes rule [57]. Note that in all these cases the emphasis is placed on the way the base learners are combined or selected, and not on the way different and diverse classifiers are generated.

On the contrary, *generative* ensemble methods generate sets of base learners acting on the base learning algorithm or on the structure of the data set to try to actively improve diversity and accuracy of the base learners. In this case the emphasis is placed on the way diverse base learners are constructed, while the combination technique does not represent the main issue of the ensemble algorithm. Examples are resampling methods, that train base learners on different bootstrap replicates of the data [20], or random subspace algorithms that generate diverse base learners by using different randomly selected subsets of features [87], or mixture of experts methods, where a gating network performs the division of the input space and an ensemble of neural networks perform the effective calculation at each assigned region separately [103].

Table 26.1 provides a high-level scheme of the taxonomy of ensemble methods proposed in this paper. *Nongenerative* methods (Section 26.4) are partitioned in *ensemble fusion* (Section 26.4.1) and *ensemble selection* (Section 26.4.2) methods, while the other high–level and more heterogeneous branch of the taxonomy, that is *generative* ensembles (Section 26.5), is subdivided in *resampling* (Section 26.5.1), *feature selection* (Section 26.5.2), *mixture of experts* (Section 26.5.3), OC (Section 26.5.4), and *randomized ensembles* (Section 26.5.5) methods.

TABLE 26.1   A Taxonomy for Ensemble Methods

Nongenerative ensembles	Ensemble fusion methods	– Majority voting – Naive Bayes rule – Behavior–knowledge–space – Algebraic operators fusion – Fuzzy fusion – Decision template – Meta learning – Multilabel hierarchical methods
	Ensemble selection methods	– Test and select – Cascading classifiers – Dynamic classifier selection – Clustering–based selection – Pruning by statistical tests – Pruning by semidef. programming – Forward/Backward selection
Generative ensembles	Resampling methods	– Bagging – Boosting – Arcing – Cross-validated committees
	Feature selection and extraction methods	– Random subspace – Similarity-based selection – Input decimation – Feature subset search – Rotation forests
	Mixture of experts	– Gating network selection – Hierarchical mixture of experts – Hybrid experts
	Output coding methods	– One-per-class – Pairwise and correcting classifiers – ECOC – Data–driven ECOC
	Randomized methods	– Randomized decision trees – Random forests – Pasting small votes

It is worth noting that semisupervised and unsupervised ensemble methods have been recently proposed. Unfortunately, for lack of space we do not discuss these topics, and we refer the reader to Section 8.3 of Kuncheva's book [116], or to the brief review provided in Ref. [77]. However, some examples of the application of unsupervised ensemble methods to astronomy and astrophysics problems are described in Section 26.6 of this chapter.

26.4 NONGENERATIVE ENSEMBLES

Following the taxonomy proposed in Ref. [116], we can subdivide nongenerative strategies into *ensemble fusion* and *ensemble selection* methods. Both these approaches share the very general common property of using a predetermined set of learning machines previously trained with suitable algorithms. The base learners are then put together by a combiner module that may vary depending on the requirement of the output of the individual learning machines: for instance, the combiner may need the labels of the classes, or a ranking of the classes, or a support (e.g., the *a posteriori* probability estimation) for each class [210]. Moreover, we may distinguish between combiners that are trainable and not trainable [58]. Very schematically ensemble fusion methods combine all the outputs of the base classifiers, while ensemble selection methods try to choose the "best classifiers" among the set of the available base learners.

26.4.1 Ensemble Fusion Methods

The most popular ensemble fusion method is represented by the *majority voting* ensemble, by which each base classifier "votes" for a specific class, and the class that collects the majority of votes is predicted by the ensemble [106,124,158]. By generalizing this approach, Xu et al. proposed a *thresholded plurality vote*: by imposing a threshold on the number of votes to select the class, we may move from an *unanimity vote* rule, by which we choose a class only if all the base classifiers agree on the corresponding label, to the simple *majority voting* rule, by which it sufficient to achieve the majority of votes to select the class; intermediate cases can be considered by moving the threshold of votes, at the expenses of some possible unclassified example [210]. This approach can be refined assigning different weights to each classifier to optimize the performance of the combined classifier, according to the base learner accuracy estimated on a validation set [123,147]. By generalizing the analysis to linear combiners, Fumera and Roli showed the reasons why weighted average improves on the simple average combining rule [73].

Assuming conditional independence between classifiers, a *naive Bayes decision rule* selects the class with the highest posterior probability computed through the estimated class conditional probabilities and the Bayes' formula [54,57]. A Bayesian approach has also been used in *consensus–based classification* of multisource remote sensing data [15,16,27], outperforming conventional multivariate methods for classification. To overcome the problem of the independence assumption (that is unrealistic in most cases), the behavior–knowledge space method [96] considers each possible combination of class labels, filling a look-up table using the available data set, but this technique requires a huge volume of training data.

Other simple operators such as *minimum, maximum, average, product,* and *ordered weight averaging* have been applied to combine multiple classifiers [26,114,175]. On the basis of a common Bayesian framework, Kittler provided a theoretical underpinning of many existing classifier combination schemes based on the product and the sum rule, showing also that the sum rule is less sensitive to the errors of subsets of base classifiers [108].

Fuzzy set theory has also been applied to combine multiple learners through proper fuzzy aggregation connectives [39,40,105,205,207]. In particular, the fuzzy integral has been reported to give excellent results as a classifier combiner. Its effectiveness comes from the fact that it measures the "strength" of every subset of classifiers, and not only the strength of each individual classifier: as a consequence, for each example to be classified, the decision of the ensemble is based on the competence of every subset of base learners [121]. If the classifier outputs are possibilistic, *Dempster–Schafer* combination rules can be applied [169].

A valuable approach that takes into account the prediction scores (e.g., the support) of the base learners for each class to be predicted is represented by the *decision templates* (DT) [117]. The main idea behind DT consists in comparing a "prototypical answer" of the ensemble for the examples of a given class (the template), to the current answer of the ensemble to a specific example whose class needs to be predicted (the decision profile). Different similarity measures can be used to evaluate the matching between the matrix of classifier outputs for a given input, that is the decision profiles, and the matrix templates (one for each class) found as the class means of the classifier outputs. This approach can be easily applied to combine multiple set of features or sources of data to improve predictions [47,164]. In Ref. [165], the authors analyzed the tolerance of DT and other ensemble methods to noisy data, and showed that DT are particularly resistant to the addition of noisy data sets.

Ensemble fusion can also be performed by second-level trainable combiners, through *meta-learning* techniques [59]. For instance, in *stacking* methods the outputs of the base learners are interpreted as features in an intermediate space: the outputs are fed into a second-level machine to perform a trained combination of the base learners [208]. By extending this approach a new method based on multiresponse linear regression have been shown to outperform the original Wolpert's stacking approach [60]. Stacking requires a careful training of the base learners and of the combiner: if we use the same training data for both, it is likely to incur in overfitting. To avoid this problem, L_2 norm (ridge regression) and L_1 norm (Lasso regression) penalization have been introduced in linear models for stacked generalization [167]. Another type of meta-learning ensemble is represented by methods that use an arbiter or a combiner to decide recursively in a hierarchically structured input space on the basis of the predictions made by the base learners. The aim of this strategy is, respectively, to provide an alternate classification when the base learners disagree (*arbiter trees*) [34] or to combine the outputs of the base classifiers by learning their relationships with the correct labels (*combiner trees*) [35,95].

Several ensemble methods have been devoted to the *classification of hierarchically structured classes*, as those related to the classification of texts in the web or to the classification of genes in functional genomics [3,100,204]. Different ensemble–based algorithms have been proposed ranging from methods restricted to multilabels with single and no partial paths [46] to methods extended to multiple and also partial paths [31]. In this context, hierarchical ensemble methods are in general characterized by a two-step strategy: (a) flat learning of the classes as a set of independent classification problems; (b) combination of the predictions by exploiting the relationships between classes that characterize the hierarchy. Some recently published works clearly demonstrated that this approach ensures an increment in precision with respect to "flat" methods, but this comes at expenses of the overall recall [33,83,142]. A recently proposed hierarchical ensemble approach, proposed in the context of functional genomics, partially overcomes this problem, and can be, in principle, extended to other application domains characterized by the unbalance of the classes and hierarchically structured relationships between classes [198]. Hierarchically structured ensembles, originally proposed in Refs. [33,198] have also been successfully combined with majority voting ensembles and kernel fusion methods to integrate multiple sources of data in the context of gene function prediction problems [32]. Finally, a different method, based on ensembles of hierarchical multilabel decision trees, developed for the prediction of gene functions, is generally enough to be adapted to other classification problems with classes structured according to a direct acyclic graph [179].

26.4.2 Ensemble Selection Methods

This general approach tries to identify the "best" base classifier among the set of base learners for a specified input, and the output of the ensemble is the output of the selected best classifier. From a more general standpoint also, a subset of base classifiers can be chosen. In this case we need to decide to pick one of the selected outputs as the ensemble output, or to combine the output of the base learners, according, for example, to one of the ensemble fusion methods described in the previous section. To design an ensemble selection method we need to decide how to build the individual classifiers, how to evaluate the competence of each classifier on a specific input, and what selection strategy to use [116].

The *test and select* methodology relies on a greedy approach, by which a new learner is added to the ensemble only if the resulting squared error is reduced [158], but in principle any optimization technique can be used to select the "best" component of the ensemble, including genetic algorithms [125,146].

Another possible approach is represented by *cascading classifiers*. The base learners are applied sequentially, and if the confidence of the first classifier is high, its prediction is taken, otherwise the prediction is recursively demanded to the next classifier and so on. This "cascade" model is useful especially with real-time systems, since most of the inputs need to be processed by a few classifiers [6,74].

The competence of each base classifier can be estimated dynamically (that is during the operational phase) by using only *a priori* information on the features of the input pattern (with no knowledge about its predicted labels) or by using *a posteriori* information on the labels predicted by all the classifiers, by applying for example, k nearest neighbors estimates [209]. This approach is also known as *dynamic classifier selection* [80,91,209] and it is based on the definition of a function selecting for each pattern the classifier which is likely the most accurate, estimating, for instance, the accuracy of each classifier in a local region of the feature space surrounding an unknown test pattern [80].

Nevertheless, this dynamical approach might be too computationally intensive and a less demanding preestimation of the competence region can be applied. To this end, the region of competence for the base classifiers can be determined through clustering methods [115,130]. Clustering algorithms have also been employed to discover groups of base learners that make similar predictions, then models are picked from each cluster to both select a subset of the available base learners and to improve the diversity of the ensemble [81,82,127].

Other greedy search methods are based on a ranking that gives preference to classifiers that are able to correct the incorrect predictions of the ensemble, as in *orientation ordering* ensembles, thus assuring the selection of base learners able to improve the prediction of the overall ensemble [132]. Ensembles of heterogeneous classifiers can also be pruned by statistical procedures that select only those classifiers with significantly better performances, combined in a second step through majority voting [191,192].

Ensemble pruning can also be formulated as a *semidefinite programming* problem that minimizes misclassification and maximize diversity, with the constraints of selecting only a subset of the available classifiers: by setting the number of base classifiers to be selected as an input parameter Zhang et al. approximating solve the resulting quadratic integer programming problem in polynomial time [213].

Algorithms borrowed from the feature selection literature or for the solution of complex optimization tasks (tabu search) are proposed in Ref. [172]. Forward selection [30] and backward elimination [9] ensemble selection algorithms, adapted from the corresponding feature selection literature, respectively add or remove base classifiers selected according to the minimization of an objective function. Another valuable approach is represented by a greedy search that takes into account both ensemble decision strength and the diversity of the base learners to select the base learners [154]. Recently Partalas, Tsoumakas, and Vlahavas proposed a new ensemble pruning method via directed hill climbing: they introduced a new measure to select models that takes into account the uncertainty of the decision of the ensemble. Through the proposed *uncertainty weighted accuracy* measure, the authors select a small subset of base classifiers, achieving state-of-the art results on a large set of benchmark data sets [155].

26.5 GENERATIVE ENSEMBLES

In this section we introduce ensemble methods able to generate base learners by acting on the base learning algorithm or on the structure of the data set to try to actively boost diversity and accuracy of the base learners. These methods can perturb the structure and the characteristics of the available input data, as in *resampling* methods or in *feature selection/subsampling* methods, or can manipulate the aggregation and the coding of the classes (OC methods), or can select base learners specialized for a specific input region (*mixture of experts* methods). They can also randomly modify the base learning algorithm, or apply randomized procedures to the learning processes to improve the diversity or to avoid local minima of the error (*randomized* methods).

26.5.1 Resampling Methods

Resampling techniques can be used to generate different hypotheses. For instance, bootstrapping techniques [61] may be used to generate different training sets and a learning algorithm can be applied to the obtained subsets of data in order to produce multiple hypotheses. These techniques are effective especially with unstable learning algorithms, which are algorithms very sensitive to small changes in the training data, such as neural networks and decision trees [64].

Bagging (an acronym for *bootstrap aggregating*) builds up the ensemble by making bootstrap replicates of the training sets, and the multiple hypotheses, resulting from the application of a suitable learning algorithm to the perturbed data, are used to get an aggregated predictor [20]. The aggregation can be performed averaging the outputs in regression or by majority or weighted voting in classification problems [185,186].

By applying procedures to estimate the bias and variance of each base learner, an enhanced version of bagging, tailored to the characteristics of support vector machines is proposed in Ref. [199]: considering that bagging is a variance-reduction method, by selecting the support vector machines (SVMs) with a low bias, we may obtain ensembles that lower both the variance and the bias component of the error. This technique comes from a more general approach to design ensembles based on the bias–variance decomposition of the error, to exploit the specific learning characteristics of the base learners [200].

Another variant of bagging, based on a nonuniform probability to extract examples from the training set is *wagging*: while in bagging each example is drawn with equal probability from the available training data, in wagging each example is extracted according to a weight stochastically assigned [11].

Extending this approach, in *boosting* methods the learning algorithm at each iteration uses a different distribution or weighting over the training examples [55,56,67,174–176], according to the errors of the base learners. The most known algorithm in this family is surely *AdaBoost* [68,69]. This technique places the highest weight on the examples most often misclassified by the previous base learner: in this way the base learner focuses its attention on the hardest examples. Then the boosting algorithm combines the outputs of the base learners by weighted majority voting. Schapire and Singer showed that the training error exponentially drops down with the number of iterations [178] and Schapire et al. [177] proved that boosting enlarges the margins of the training examples, showing also that this fact translates into a superior upper bound on the generalization error. It is worth noting that this ensemble method is one of the most studied, with a solid theoretical background, and largely applied in several application domains.

Breiman proposed an algorithm similar to the AdaBoost algorithm, which he named *arcing* (adaptive resampling and combining) to investigate whether the success of AdaBoost is due to technical details or to the resampling scheme adopted [22,24]. His conclusions showed that AdaBoost is a well-theoretically founded algorithm, while arcing is basically a heuristic with empirical results comparable to AdaBoost [22]. Different variants of boosting algorithms for multiclass problems, or real-valued classifier outputs, have been proposed [5,178].

Relationships of boosting with logistic regression have been analyzed in Refs. [43,71], giving raise to a parameterized family of iterative algorithms [43] and to the *LogitBoost* algorithm, that casts AdaBoost in a statistical framework, by applying the cost functional of logistic regression and reinterpreting boosting as a generalized additive model [71]. Instead of reweighting, a new boosting by resampling technique can be adopted: a local error for each training example is computed and then used to update the probability of drawing the example at the next iteration of the algorithm [212].

Experimental work showed that bagging is effective with noisy data, while boosting, concentrating its efforts on noisy data, is quite sensitive to noise [49,163]. Nevertheless, boosting algorithms designed for noisy data partially overcome this problem [148,189]. Finally, another approach based on subsampling to achieve diversity consists in constructing training sets by leaving out disjoint subsets of the training data as in *cross-validated committees* [152,153] or sampling without replacement [183].

26.5.2 Feature Selection/Extraction Methods

Reducing the number of input features of the base learners, we can contrast the effects of the classical curse of dimensionality problem that characterize high-dimensional and sparse data [72]. For instance, by applying feature selection algorithms or subsampling methods to draw subsets of features from the available data, we can construct sets of diverse base classifiers that can be combined through appropriate ensemble fusion techniques.

A random strategy can be applied to select sets of features. In *random subspace* methods [87,120], a subset of features is randomly selected and assigned to an arbitrary learning algorithm: a random subspace of the original feature space is obtained, and classifiers are constructed inside this reduced subspace. The aggregation is usually performed using weighted voting on the basis of the base classifiers accuracy, but other techniques could be in principle applied. It has been shown that this method is effective for classifiers having a decreasing learning curve constructed on small and critical training sample sizes [187].

By using a dissimilarity representation of the objects, for example, the distances between the pairs of examples in the training set, we can construct a "similarity-based" feature space that resembles an approach similar to kernel methods [180]. By adopting this approach a linear discriminant classifier applied on subsets of randomly selected dissimilarity features has been proposed [157].

An open problem in random subspace methods is represented by the choice of the dimension of the projected subspace. Ho suggested a dimension about equal to the half of the available features [87], and in Ref. [29] a method based on a random search in the feature subset spaces is proposed. However, both methods are heuristics, even if supported by empirical evidence of their effectiveness.

Following a different approach, sets of features can also be chosen by nonrandom selection methods. For instance, the *input decimation* approach [104,150] reduces the correlation among the errors of the base classifiers, decoupling the base classifiers by training them with different subsets of the input features. It differs from the previous random subspace method, since for each class the correlation between each feature and the output of the class is explicitly computed, and the base classifier is trained only on the most correlated subset of features.

Extending this approach, various criteria instead of the simple correlation between features and class labels have been introduced; moreover, base learner selection is accomplished by checking both the accuracy and the diversity of the base learners [193,194]. Gunter and Bunke apply different feature subset search algorithms to find different subsets of features; to incrementally select the base learners, they take into account the diversity and the accuracy of the overall ensemble [84]. Genetic and evolutionary techniques have also been applied to construct ensembles based on feature subset selection [118,145,170].

Another effective method that relies on feature extraction techniques is represented by the *rotation forests* [168]. Features are randomly split into n subsets, and n axis rotation are performed to encourage simultaneously individual accuracy and diversity within the ensemble. In a comparative experimental study rotation forests achieve better results than those obtained with bagging, boosting, and random forests [119].

26.5.3 Mixture of Experts

A general approach similar to ensemble selection is represented by the *mixture of experts* methods [97,98]. It differs from selection methods by the fact that the recombination of the base learners is governed by a supervisor learning machine, which selects the most appropriate element of the ensemble on the basis of the available input data: a gating network performs the division of the input space and an ensemble of neural networks perform the effective calculation at each assigned region separately. The output of the gating network are interpreted as probabilities for selecting the expert responsible for the prediction on a given input. These probabilities can be used to stochastically select the expert, or to choose the expert according to a winner-takes-all paradigm, or as weights to combine the outputs of the multiple base learners (experts). Through this type of ensemble methods both the selector (the gating network) and the base classifiers can be trained with standard learning algorithms: the standard back-propagation algorithm or the expectation-maximization method [103]. An extension of this model is the *hierarchical mixture of experts* method, where the outputs of the different experts are nonlinearly combined by different supervisor gating networks hierarchically organized [97,101,102].

Cohen and Intrator extended the idea of constructing local simple base learners for different regions of input space, searching for appropriate architectures that should be locally used and for a criterion to select a proper unit for each region of the input space [41,42]. They proposed a hybrid multi layer perceptron (MLP)/radial basis function (RBF) network by combining RBF and perceptron units in the same hidden layer and using a forward selection procedure to add units until the error drops to a given threshold. Although the resulting *hybrid perceptron/radial network* is not in a strict sense an ensemble, the way by which the regions of the input space and the computational units are selected and tested could be, in principle, extended to ensembles of learning machines.

26.5.4 OC Methods

By manipulating the coding of classes in multiclass classification problems, we can construct ensembles able to partially correct errors committed by the base learners, exploiting the redundancy in the bit-string representation of the classes [48,51,138]. More precisely, OC methods decompose a multiclass classification problem in a set of two-class subproblems, and then recompose the original problem combining them to achieve the class label. An equivalent way of thinking about these methods consists in encoding each class as a bit string (named codeword), and in training a different two-class base learner (dichotomizer) in order to separately learn each codeword bit. When the dichotomizers are applied to classify new points, a suitable measure of dissimilarity between the codeword computed by the ensemble and the codeword classes is used to predict the class (e.g., Hamming distance) [50].

Different *decomposition schemes* have been proposed in the literature: In the One-Per-Class decomposition [8], each dichotomizer separates a single class from all others; in the *pairwise coupling* (PWC) decomposition [86], the task of each dichotomizer consists in separating a pair of classes, ignoring all other classes; the *correcting classifiers* (CC) and the PWC-CC are variants of the PWC decomposition scheme, which reduce the noise originated in the PWC scheme due to the processing of nonpertinent information performed by the PWC dichotomizers [141].

Error correcting output coding (ECOC) [50,51] is the most studied OC method, and has been successfully applied to several classification problems [2,18,78,196,211]. This decomposition method tries to improve the error-correcting capabilities of the codes generated by the decomposition through the maximization of the minimum distance between each pair of codewords [113]. This goal is achieved by means of the redundancy of the coding scheme [202].

The trade-off between error-recovering capabilities and complexity/learnability of the dichotomies induced by the decomposition scheme has been studied in Ref. [4]. The effectiveness of ECOC decomposition methods depends mainly on the design of the learning machines implementing the decision units, on the similarity of the ECOC codewords, on the accuracy of the dichotomizers, on the complexity of the multiclass learning problem, and on the correlation of the codeword bits [134–137].

The design of ECOC codes tuned to the characteristics of the data, in order to obtain codes and dichotomies that are both "simple" and able to recover errors of the base learners, is another interesting issue considered in several works [7,138]. In Ref. [44], it is shown that given a set of dichotomizers the problem of finding an optimal decomposition matrix is nondeterministic polynomial-complete (NP-complete): by introducing continuous codes and casting the design problem of continuous codes as a constrained optimization problem, we can achieve an optimal continuous decomposition using standard optimization methods.

Data-driven ECOC analyzes the distribution of data classes to optimize both the composition and the number of the base learners [214]. Compact ECOC codes, able to code classes using very compact but effective codes, well suited for problems characterized by a very large number of classes have been proposed in Ref. [161], and adapted to the characteristics and the distribution of the data [12]. Recently, ternary ECOC codes (adding a "don't care" bit) have been extensively studied and a taxonomy of both binary and ternary ECOC codes is proposed in Ref. [63].

26.5.5 Randomized Ensemble Methods

Randomness plays a central role to generate a set of diverse base classifiers: for instance, we can randomly draw examples (bagging, Section 26.5.1) or features (random subspace, Section 26.5.2) from the training data to construct ensembles of diverse classifiers. Moreover, several experimental results showed that randomized learning algorithms used to generate base elements of ensembles improve the performances of single nonrandomized classifiers.

For instance, in [49] randomized decision tree ensembles outperform single C4.5 decision trees [162], and adding Gaussian noise to the data inputs, together with bootstrap and weight regularization can result in large improvements in classification accuracy [163].

By extending this approach, Leo Breiman proposed a general class of ensembles, the *random forests* [25], using decision trees as base classifiers. A random forest can be constructed by randomly sampling from the data set, or by sampling from the feature set, or from both. For instance, along with selecting examples bootstrapped from the available training data, a random subset of features is drawn at each node of the tree and the best one is selected among this set to split the nodes. Random forests are also implicitly able to select the most relevant features associated to the classification problem they are applied to [25].

Randomness plays a role also when ensembles are built to deal with very large or distributed data sets. Indeed, in these situations, ordinary learning algorithms cannot directly process the data set as a whole. To this end, *pasting small votes* techniques, by which individual classifiers are trained on relatively small subsets of the available data, have been proposed [23]. By this method, training sets are sampled from a large data set either randomly (*Rvotes*, similar to bagging), or taking into account their importance for the classification task (*Ivotes*, similar to boosting). Breiman [23] and Chawla et al. [36] showed that importance small sampling-based ensembles such as *Ivotes* and their distribute counterpart *DI-votes* may also obtain better results with respect to single learners trained on the entire available learning set. In particular, Chawla et al. showed that this ensemble approach may improve accuracy by enhancing diversity between base learners, even if stable classifiers, such as naive Bayes, are used [37]. Through a comparative experimental analysis, the reasons why voting many classifiers built on small subsets of data work successfully are interpreted in the context of bias–variance analysis of the generalization error: the success of this approach is due to the very significant variance reduction, while bias remains substantially unchanged [197].

26.6 ENSEMBLE METHODS IN ASTRONOMY AND ASTROPHYSICS

The exponential raise in the available amount of astronomical and astrophysical data observed in the last years resulted into an ever–increasing gap between the availability of information about the natural objects (and/or phenomena) under investigation and our ability to extract useful knowledge from them. Astronomy has been among the first scientific disciplines to experience this flood of data. The sheer volume of data routinely analyzed in astronomical and astrophysical research projects needs the application of a multidisciplinary approach often involving the concurrent use of data mining, statistical, and machine learning techniques in order to effectively tackle the high levels of noise usually present in the data produced by large–scale astronomical projects.

The primary goal of this section dedicated to the application of ensemble methods in astronomy and astrophysics is not to present an exhaustive list of all the scientific papers recently published, but rather to put in light the potential benefits introduced by the application of the ensemble learning approach to common problems investigated in these rapidly growing research areas.

In a very broad sense the aim of ensemble learning, as in the case of all the other branches of *machine learning*, is to leverage a computational machine able to extract patterns from the data and then to translate these patterns into novel (and hopefully useful) knowledge. Indeed, a common problem faced by scientific investigators is not only to classify objects according to a preexisting classification scheme but also to highlight the eventual existence of relationships between uncategorized (or unlabeled) and more characterized objects. From this point of view, the choice of the ensemble method to be applied to the problem at hand is of paramount importance. We finally would like to stress that the described ensemble methods are not out-of-the-box solutions but rather *tools* that, if applied correctly, have strong potential for the production of interesting scientific results and are able to provide inspiration for new ideas and applications.

Supervised ensemble methods have been applied to several astronomical problems. They have proven to be effective in the automated annotation of the content of large publicly available catalogs. A good example of this class of problems is the automated morphological classification of galaxies.

In Ref. [13], the authors predicted the morphological class of 800 examples with both a single classifier and an ensemble of classifiers trained on bootstrap replicates of the training set (bootstrap aggregating or bagging, see Section 26.5.1). Performance were collected in the form of averaged classification errors obtained by means of a canonical 10-fold classification scheme in order to provide a good estimate of the generalization capabilities of the evaluated approaches. As component classifiers the authors evaluated artificial neural networks (commonly applied in astronomical classification problems) trained with backpropagation, a pruned decision tree and the naive Bayes classifier.

The collected experimental results clearly demonstrated the ability of the ensemble systems to reduce the overall classification error but with different patterns. The authors observed that the decrement in the classification error due to the application of the ensemble method was different according to the type of the component classifier and also that the error reduction effect was inversely related to the number of the output classes. Apart from the expected classification error reduction, a really crucial point emerging from the results presented in this work is that the same ensemble scheme may produce different results according to the nature of its component classifiers. Unfortunately, there is no way to know *a priori* which is the best type of classifier to be used to solve a particular problem being this strictly dependent not only on the nature of the problem at hand but also on the dataset to be evaluated. The only way to face this problem is to perform experiments involving ensemble methods based on different types of base learners. With respect to this problem an interesting feature of the ensemble methods is that their flexible nature allows the formation of the classifiers committee not only using instances of the same algorithm but also using committees composed by instances of different algorithms trained on the same datasets resulting into a sort of mixture of experts able to exploit the strength of different classifiers.

The combination of bagging and random selection of a small subset of features for splitting at each node is known as a random forest (see Section 26.5.5). Random forests have proven to be effective in the identification of quasars from the Faint Images of the Radio Sky at Twenty-cm (FIRST) survey [65]. The random forest ensemble method has also been applied in multiwavelength classification problems of data collected from different databases including the Sloan Digital Sky Survey (SDSS), United States Naval Observatory (USNO),

FIRST, and ROentger SATellite (ROSAT) mission [75]. In this experiment, the authors not only demonstrated that the investigated ensemble method is effective in astronomical objects classification, but also investigated the feature selection and feature weighting capabilities of this method when applied to data constituted by samples from optical and radio bands. The authors also noted the robustness of random forest in outlier detection.

In Ref. [79], the authors applied machine-learning methods to the automatic geomorphic mapping of planetary surfaces. In this experiment, the authors casted remotely sensed topographic data into semantically meaningful maps of landforms producing maps that are valuable research tools for planetary science. The proposed framework is composed by two distinct steps: mapping of the available topographic data (achieved by means of scene segmentation) followed by supervised classification of segments. The method was applied to six test sites on Mars. The collected experimental results showed that a combination of K-means-based agglomerative segmentation and both SVM with a quadratic kernel and bagging with C4.5 produce the best maps. This work demonstrates that the bagged ensemble of decision trees perform comparably with the support vector machine in the classification step, and this raises a crucial question: why should an investigator choose a classification scheme instead of another? As we discussed previously, there is not an easy way to answer this question (being it strictly related to the intrinsic nature of the investigated problem), but while the application of an ensemble method does not ensure an increment in the classification performance, it is granted to improve the generalization level, meaning that an ensemble system is expected to be more robust w.r.t. previously unseen data and this can make the difference in research fields in which the amount of the publicly available data and the acquisition rate of novel data is constantly growing (as in the case of astronomy and astrophysics).

Supervised methods can be applied only in the presence of *a priori* knowledge available in the form of a set of labels, but relevant problem is astronomy and astrophysics need an unsupervised approach.

A common problem faced by astronomers is the classification of celestial objects on the basis of their spectral emissions. A large volume of noisy, multidimensional data have been recently produced by using charge-coupled device (CCD) imaging spectrometers and made available to the scientific community by large–scale astronomical projects (data collected by the Chandra X-Ray Observatory and the X-Ray Multimirror Mission Newton). The sheer volume of these datasets raised the need to develop methods to classify and characterize the vast library of X-Ray spectra in an unsupervised fashion. This is of paramount importance in order to create a counterpart to the current parametric model fits, usually employed to classify X-Ray spectral data, and to provide the investigators with a family of tools able to produce an opinion originated by a different classification paradigm (the unsupervised one) w.r.t. the one upon which classical spectral classification models are based.

In a recent work [92], the authors applied an ensemble classifier consisting of agglomerative hierarchical clustering and k-means clustering applied to x-ray spectral classification. This method does not need spectral source models and can operate without information about the sources. It is also able to deal with a massive amount of data, a feature making it attractive for the analysis of the data produced by large–scale astrophysical investigations. This approach employs principal component analysis for dimensional reduction of the spectral bands followed by clustering. While PCA provides a means to automatically define optimal spectral band definitions from the data–set itself, the ensemble clustering method groups similar sources of x-ray emission by placing them in a three-dimensional spectral sequence and then grouping the ordered sources into clusters based on their spectra. The statistical issues behind this method are discussed in Ref. [93].

The ensemble learning paradigm has also been applied in a popular astronomical research area: galaxy spectra modeling. In Ref. [131], a statistical approach, ensemble learning for independent component analysis (ICA), was applied to the analysis of a synthetic galaxy spectral library. The authors found that the proposed method was able to reduce the data of the spectral library to six nonnegative independent components (ICs). They also found that the identified ICs were good templates for modeling normal galaxy spectra, as the ones contained in the SDSS. In this case, ensemble learning was applied, according to what originally suggested in Ref. [126], in order to provide a sort of heuristic able to efficiently estimate the nonlinear parameters of an ICA model. The approach proposed in Ref. [131] is not only able to solve the problem of data compression characterizing the first step of a spectra modeling task, but is also expected to manage effectively the risk of overfitting the training data.

As a last example of application of ensemble methods in astronomy/astrophysics, we would like to highlight the potential benefits introduced by the usage of ensemble systems in classification tasks involving missing data. Any classification algorithm independently by its supervised or unsupervised nature, assumes that each instance is associated with a complete set of values but real observational data often contain missing features. A common approach to tackle the problems due to the presence of incomplete instances is to simply remove them from the dataset under investigation before to start its analysis. This approach is suboptimal when a large fraction of the data points have missing features. A common technique to avoid this "filtering" approach is to impute the missing value, but often imputation techniques are based on the assumption that missing values occur by chance which is not always the case. In particular, in astronomy the absence of a value could have a physical meaning. In order to avoid both the prefiltering of incompletely described instances and the imputation of missing values, a solution based on an adapted clustering approach was proposed in Ref. [206]. The proposed method [k-means with Soft Constraints (KSC)] is based on soft constraints induced by the fully described instances to assist in the grouping of the incomplete ones. This approach is suitable only for exploratory (unsupervised) investigations but cannot be applied to supervised classification problems. In Ref. [159], the author proposed a solution based on the creation of an ensemble of classifiers, each trained with a random subset of the features, so that an instance with missing features can still be classified using learners whose training data did not include those attributes. The main parameters affecting the performance of the classifier are the number of random features used in the training of the component classifiers, and the total number of component learners to be generated in order to create the ensemble committee.

26.7 CONCLUSIONS

In this chapter, we provided an overview and a bibliography of ensemble methods. Despite our efforts, we are sure that we missed important research works and may be important research areas, since this field of machine learning is continuously growing, and new methods and innovative applications able to stimulate the development of new ensemble methods are objects of intensive research. To expand on this subject, we would like to mention an excellent book devoted to ensemble methods, the fundamental *Combining Pattern Classifiers* by Kuncheva [116], as well as the cited proceedings of the MCS conference, especially for the discussion of recent advanced topics on ensemble systems [14,62].

Our overview focused on supervised ensemble methods, since historically these were the first to be studied and applied to several application domains. More precisely, in this chapter, a general taxonomy, distinguishing between *generative* and *nongenerative* ensemble methods, has been proposed, considering the different ways supervised base learners can be generated or combined together. Nevertheless, ensemble methods have been also developed in the context of semisupervised and unsupervised ensemble methods, as witnessed by recent research works on the unsupervised exploratory analysis of data [77,143,144].

Several important issues have not been discussed in this paper. In particular, the theoretical problems behind ensemble methods need to be reviewed and discussed more in detail, and the discussion on the application of ensemble methods to real-world problems has been limited to some relevant problems in astronomy and astrophysics, since a discussion extended to all the application domains of ensemble systems if far beyond the scope of this chapter. To gain a general overview of the applications of ensemble methods, a good starting point could be the quite recent special issue on *Applications of Ensemble Methods of the Information Fusion Journal* [1].

Other open problems not covered in this chapter are the relationships between ensemble methods and data complexity [88,89,129], a systematic research of hidden commonalities among all the combination approaches despite their superficial differences, and a general analysis of the applications (and of the applicability) of these methods to supervised tasks such as active learning [181], or to semi–supervised learning [215].

ACKNOWLEDGMENTS

We would like to thank the anonymous reviewers for their comments and suggestions. The authors gratefully acknowledge partial support by the PASCAL2 Network of Excellence under EC grant no. 216886. This publication only reflects the authors' views.

REFERENCES

1. AA.VV. *Special Issue on Applications of Ensemble Methods*, volume 9 (1) of *Information Fusion Journal*. Elsevier, 2008.
2. D.W. Aha and R. Bankert. Cloud classification using error-correcting output codes. *Artificial Intelligence Applications: Natural Resources, Agriculture, and Environmental Science*, 11(1): 13–28, 1997.
3. N. Alaydie, C.K. Reddy, and F. Fotouhi. Hierarchical multi-label boosting for gene function prediction. In *Proceedings of the International Conference on Computational Systems Bioinformatics (CSB)*, pp. 14–25, Stanford, CA, 2010.

4. E.L. Allwein, R.E. Schapire, and Y. Singer. Reducing multiclass to binary: A unifying approach for margin classifiers. *Journal of Machine Learning Research*, 1:113–141, 2000.

5. E.L. Allwein, R.E. Schapire, and Y. Singer. Reducing multiclass to binary: A unifying approach for margin classifiers. In *Proceedings of the ICML'2000, The Seventeenth International Conference on Machine Learning*, pp. 113–141, 2000.

6. E. Alpaydin and C. Kaynak. Cascading classifiers. *Kybernetika*, 34(4):369–374, 1998.

7. E. Alpaydin and E. Mayoraz. Learning error-correcting output codes from data. In *ICANN'99*, pp. 743–748, Edinburgh, UK, 1999.

8. R. Anand, G. Mehrotra, C.K. Mohan, and S. Ranka. Efficient classification for multiclass problems using modular neural networks. *IEEE Transactions on Neural Networks*, 6:117–124, 1995.

9. R.E. Banfield, L.O. Hall, K.W. Bowyer, and P. Kegelmeyer. Ensemble diversity measure and their application to thinning. *Information Fusion*, 6(1):49–62, 2005.

10. R.E. Banfield, L.O. Hall, O. Lawrence, K.W. Bowyer, W. Kevin, and P. Kegelmeyer. A comparison of decision tree ensemble creation techniques. *IEEE Transactions on Pattern Analysis and Machine Intelligence*, 29(1):173–180, 2007.

11. E. Bauer and R. Kohavi. An empirical comparison of voting classification algorithms: Bagging, boosting and variants. *Machine Learning*, 36(1/2):105–139, 1999.

12. M.A. Bautista, X. Baro, O. Pujol, P. Radeva, J. Vitria, and S. Escalera. Compact evolutive design of error-correcting output codes. In O. Okun, M. Re, and G. Valentini, (eds), *ECML-SUEMA 2010 Proceedings*, pp. 119–128, Barcelona, Spain, 2010.

13. D. Bazell and W.D. Aha. Ensemble of classifiers for morphological galaxy classification. *The Astrophysical Journal*, 548:219–223, 2001.

14. J. Benediktsson, F. Roli, and J. Kittler. *Multiple Classifier Systems, 8th International Workshop, MCS2009*, volume 5519 of Lecture Notes in Computer Science. Springer-Verlag, Berlin, 2009.

15. J. Benediktsson, J. Sveinsson, O. Ersoy, and P. Swain. Parallel consensual neural networks. *IEEE Transactions on Neural Networks*, 8:54–65, 1997.

16. J. Benediktsson and P. Swain. Consensus theoretic classification methods. *IEEE Transactions on Systems, Man and Cybernetics*, 22:688–704, 1992.

17. J.A. Benediktsson, J. Chanussot, and M. Fauvel. Multiple classifier systems in remote sensing: From basics to recent developments. In M. Haindl, J. Kittler, and F. Roli, (eds), *Multiple Classifier Systems. Seventh International Workshop, MCS 2007, Prague, Czech Republic*, volume 4472 of Lecture Notes in Computer Science, pp. 511–512, Springer, 2007.

18. A. Berger. Error correcting output coding for text classification. In *IJCAI'99: Workshop on Machine Learning for Information Filtering*, Stockholm, Sweden, 1999.

19. C. M. Bishop. *Neural Networks for Pattern Recognition*. Clarendon Press, Oxford, 1995.

20. L. Breiman. Bagging predictors. *Machine Learning*, 24(2):123–140, 1996.

21. L. Breiman. Bias, variance and arcing classifiers. Technical Report TR 460, Statistics Department, University of California, Berkeley, CA, 1996.

22. L. Breiman. Arcing classifiers. *Annals of Statistics*, 26(3):801–849, 1998.

23. L. Breiman. Pasting small votes for classification in large databases and on-line. *Machine Learning*, 36:85–103, 1999.

24. L. Breiman. Prediction games and arcing classifiers. *Neural Computation*, 11(7):1493–1517, 1999.

25. L. Breiman. Random forests. *Machine Learning*, 45(1):5–32, 2001.

26. M. van Breukelen, R.P.W. Duin, D. Tax, and J.E. den Hartog. Combining classifiers for the recognition of handwritten digits. In *Ist IAPR TC1 Workshop on Statistical Techniques in Pattern Recognition*, pp. 13–18, Prague, Czech Republic, 1997.

27. G.J. Briem, J.A. Benediktsson, and J.R. Sveinsson. Boosting. Bagging and consensus based classification of multisource remote sensing data. In J. Kittler and F. Roli, (eds), *Multiple

Classifier Systems. Second International Workshop, MCS 2001, Cambridge, UK, volume 2096 of Lecture Notes in Computer Science, pp. 279–288, Springer-Verlag, 2001.

28. G. Brown, J. Wyatt, R. Harris, and X. Yao. Diversity creation methods: A survey and categorisation. *Information Fusion*, 6(1):5–20, 2005.

29. R. Bryll, R. Gutierrez-Osuna, and F. Quek. Attribute bagging: Improving accuracy od classifier ensembles by uing random feature subsets. *Pattern Recognition*, 36:1291–1302, 2003.

30. R. Caruana, A. Niculescu-Mizil, G. Crew, and A. Ksikes. Ensemble selction from libraries of models. In *21th International Conference on Machine Learning, ICML 2004*, pp. 18, ACM Press, 2004.

31. N. Cesa-Bianchi, C. Gentile, and L. Zaniboni. Hierarchical classification: Combining Bayes with SVM. In *Proceedings of the 23rd International Conference on Machine Learning*, pp. 177–184, ACM Press, 2006.

32. N. Cesa-Bianchi, M. Re, and G. Valentini. Functional inference in FunCat through the combination of hierarchical ensembles with data fusion methods. In *ICML-MLD 2nd International Workshop on Learning from Multi-Label Data*, pp. 13–20, Haifa, Israel, 2010.

33. N. Cesa-Bianchi and G. Valentini. Hierarchical cost-sensitive algorithms for genome-wide gene function prediction. *Journal of Machine Learning Research, W&C Proceedings, Machine Learning in Systems Biology*, 8:14–29, 2010.

34. P. Chan and S. Stolfo. A comparative evaluation of voting and meta-learning on partitioned data. In *Proceedings 12th ICML*, pp. 90–98, Tahoe City, California, USA, 1995.

35. P. Chan and S. Stolfo. On the accuracy of meta-learning for scalable data mining. *Journal of Intelligent Information Systems*, 8:5–28, 1997.

36. N.V. Chawla, L.O. Hall, K.W. Bowyer, and W.P. Kegelmeyer. Learning ensembles from bites: A scalable and accurate approach. *Journal of Machine Learning Research*, 5:421–451, 2004.

37. N.V. Chawla, L.O. Hall, K.W. Bowyer, T.E. Moore, and W.P. Kegelmeyer. Distributed pasting of small votes. In *Multiple Classifier Systems. Third International Workshop, MCS2002, Cagliari, Italy*, volume 2364 of Lecture Notes in Computer Science, pp. 52–61, Springer-Verlag, 2002.

38. D. Chen. Statistical estimates for Kleinberg's method of stochastic discrimination. PhD thesis, The State University of New York, Buffalo, USA, 1998.

39. S. Cho and J. Kim. Combining multiple neural networks by fuzzy integral and robust classification. *IEEE Transactions on Systems, Man and Cybernetics*, 25:380–384, 1995.

40. S. Cho and J. Kim. Multiple network fusion using fuzzy logic. *IEEE Transactions on Neural Networks*, 6:497–501, 1995.

41. S. Cohen and N. Intrator. A hybrid projection based and radial basis function architecture. In J. Kittler and F. Roli, (eds), *Multiple Classifier Systems. First International Workshop, MCS 2000, Cagliari, Italy*, volume 1857 of Lecture Notes in Computer Science, pp. 147–156, Springer-Verlag, 2000.

42. S. Cohen and N. Intrator. Automatic model selection in a hybrid perceptron/radial network. In *Multiple Classifier Systems. Second International Workshop, MCS 2001, Cambridge, UK*, volume 2096 of Lecture Notes in Computer Science, pp. 349–358, Springer-Verlag, 2001.

43. M. Collins, R.E. Schapire, and Y. Singer. Logistic regression, AdaBoost and Bregman distances. *Machine Learning*, 48:31–44, 2002.

44. K. Crammer and Y. Singer. On the learnability and design of output codes for multiclass problems. In *Proceedings of the Thirteenth Annual Conference on Computational Learning Theory*, pp. 35–46, Palo Alto, California, USA, 2000.

45. N.C. de Condorcet. *Essai sur l' application de l' analyse à la probabilité des decisions rendues à la pluralité des voix*. Imprimerie Royale, Paris, 1785.

46. O. Dekel, J. Keshet, and Y. Singer. Large margin hierarchical classification. In *Proceedings of the 21st International Conference on Machine Learning*, pp. 209–216, Omnipress, 2004.

47. C. Dietrich, G. Palm, and F. Schwenker. Decision templates for the classification of bioacoustic time series. *Information Fusion*, 4(2):101–109, 2003.

48. T.G. Dietterich. Ensemble methods in machine learning. In J. Kittler and F. Roli, (eds), *Multiple Classifier Systems. First International Workshop, MCS 2000, Cagliari, Italy*, volume 1857 of Lecture Notes in Computer Science, pp. 1–15, Springer-Verlag, 2000.

49. T.G. Dietterich. An experimental comparison of three methods for constructing ensembles of decision trees: Bagging, boosting and randomization. *Machine Learning*, 40(2):139–158, 2000.

50. T.G. Dietterich and G. Bakiri. Error-correcting output codes: A general method for improving multiclass inductive learning programs. In *Proceedings of AAAI-91*, pp. 572–577, AAAI Press/MIT Press, 1991.

51. T.G. Dietterich and G. Bakiri. Solving multiclass learning problems via error-correcting output codes. *Journal of Artificial Intelligence Research*, (2):263–286, 1995.

52. P. Domingos. A unified bias-variance decomposition and its applications. In *Proceedings of the Seventeenth International Conference on Machine Learning*, pp. 231–238, Morgan Kaufmann Stanford, CA, 2000.

53. P. Domingos. A unified bias-variance decomposition for zero-one and squared loss. In *Proceedings of the Seventeenth National Conference on Artificial Intelligence*, pp. 564–569, Austin, TX, 2000. AAAI Press.

54. P. Domingos and M. Pazzani. On the optimality of the simple Bayesian classifier under zero-one loss. *Machine Learning*, 29:103–130, 1997.

55. H. Drucker and C. Cortes. Boosting decision trees. In D. Touretsky, M. Mozer, and M. Hasselmo (eds), *Advances in Neural Information Processing Systems*, Vol. 8, pp. 479–485. MIT Press, Cambridge, MA, 1996.

56. H. Drucker, C. Cortes, L. Jackel, Y. LeCun, and V. Vapnik. Boosting and other ensemble methods. *Neural Computation*, 6(6):1289–1301, 1994.

57. R.O. Duda, P.E. Hart, and D.G. Stork. *Pattern Classification*, 2nd edition Wiley & Sons, New York, 2001.

58. R. Duin. The combining classifier: To train or not to train? In *Proceedings of the 16th International Conference on Pattern Recognition, ICPR'02*, pp. 765–770, Canada, 2002.

59. R.P.W. Duin and D.M.J. Tax. Experiments with classifier combination rules. In J. Kittler and F. Roli, (eds), *Multiple Classifier Systems. First International Workshop, MCS 2000, Cagliari, Italy*, volume 1857 of Lecture Notes in Computer Science, pp. 16–29, Springer-Verlag, 2000.

60. S. Dzeroski and B. Zenko. Is combining classifiers with stacking better than selcting the best one? *Machine Learning*, 54(3):255–273, 2004.

61. B. Efron and R. Tibshirani. *An Introduction to the Bootstrap*. Chapman and Hall, New York, 1993.

62. N. El Gayar, F. Roli, and Kittler. *Multiple Classifier Systems, 9th International Workshop, MCS2010*, volume 5997 of Lecture Notes in Computer Science. Springer-Verlag, Berlin, 2010.

63. S. Escalera, O. Pujol, and P. Radeva. On the decoding process in ternary error-correcting output codes. *IEEE Transactions on Pattern Analysis and Machine Intelligence*, 32(1):120–134, 2010.

64. T. Evgeniou, M. Pontil, and A. Elisseeff. Leave one out error, stability, and generalization of voting combinations of classifiers. *Machine Learning*, 55(1):71–97, 2004.

65. E.D. Feigelson, G. Jogesh Babu, L. Breiman, M. Last, and J. Rice. Random forests: Finding quasars. In E.D. Feigelson and G. Jogesh Babu (eds), *Statistical Challenges in Astronomy*, pp. 243–254, Springer, New York, 2003.

66. E. Filippi, M. Costa, and E. Pasero. Multi-layer perceptron ensembles for increased performance and fault-tolerance in pattern recognition tasks. In *IEEE International Conference on Neural Networks*, pp. 2901–2906, Orlando, Florida, 1994.

67. Y. Freund. Boosting a weak learning algorithm by majority. *Information and Computation*, 121(2):256–285, 1995.

68. Y. Freund and R. Schapire. A decision-theoretic generalization of on-line learning and an application to boosting. *Journal of Computer and Systems Sciences*, 55(1):119–139, 1997.

69. Y. Freund and R.E. Schapire. Experiments with a new boosting algorithm. In *Proceedings of the 13th International Conference on Machine Learning*, pp. 148–156, Morgan Kauffman, 1996.

70. J. Friedman and P. Hall. On bagging and nonlinear estimation. Technical Report Technical Report, Statistics Department, University of Stanford, CA, 2000.

71. J. Friedman, T. Hastie, and R. Tibshirani. Additive logistic regression: A statistical view of boosting. *Annals of Statistics*, 38(2):337–374, 2000.

72. J.H. Friedman. On bias, variance, 0/1 loss and the curse of dimensionality. *Data Mining and Knowledge Discovery*, 1:55–77, 1997.

73. G. Fumera and F. Roli. A theoretical and experimental analysis of linear combiners for multiple classifer systems. *IEEE Transactions on Pattern Analysis and Machine Intelligence*, 27(6):942–956, 2005.

74. J. Gamma and P. Brazdil. Cascade generalization. *Machine Learning*, 41(3):315–343, 2000.

75. Y. Zhang, D. Gao and Y. Zhao. Random forest algorithm for classification of multiwavelength data. *Research in Astronomy Astrophysics*, 9(2):220–226, 2009.

76. S. Geman, E. Bienenstock, and R. Doursat. Neural networks and the bias-variance dilemma. *Neural Computation*, 4(1):1–58, 1992.

77. R. Ghaemi, M. Sulaiman, H. Ibrahim, and N. Mustapha. A survey: clustering ensembles techniques. In *World Academy of Science, Engineering and Technology 50*, pp. 636–645, 2009.

78. R. Ghani. Using error correcting output codes for text classification. In *ICML 2000: Proceedings of the 17th International Conference on Machine Learning*, pp. 303–310, Morgan Kaufmann Publishers, San Francisco, US, 2000.

79. S. Ghosh, T.F. Stepinski, and R. Vilalta. Automatic annotation of planetary surfaces with geomorphic labels. *IEEE Transactions on Geoscience and Remote Sensing*, 48(1):175–185, 2010.

80. G. Giacinto and F. Roli. Dynamic classifier fusion. In J. Kittler and F. Roli (eds), *Multiple Classifier Systems. First International Workshop, MCS 2000, Cagliari, Italy*, volume 1857 of Lecture Notes in Computer Science, pp. 177–189, Springer-Verlag, 2000.

81. G. Giacinto and F. Roli. An approach to the automatic design of multiple classifier systems. *Pattern Recognition Letters*, 22(1):25–33, 2001.

82. G. Giacinto, F. Roli, and G. Fumera. Design of effective multiple classifier systems by clustering of classifiers. In *15th International Conference on Pattern Recognition ICPR 2000*, pp. 160–163, Barcelona, Spain, 2000.

83. Y. Guan, C.L. Myers, D.C. Hess, Z. Barutcuoglu, A. Caudy, and O.G. Troyanskaya. Predicting gene function in a hierarchical context with an ensemble of classifiers. *Genome Biology*, 9(suppl 1):S3, 2008.

84. S. Gunter and H. Bunke. Feature selection algorithms for the generation of multiple classifier systems and their application to handwritten word recognition. *Pattern Recognition Letters*, 25:1323–1336, 2004.

85. M. Haindl, F. Roli, and Kittler. *Multiple Classifier Systems, 7th International Workshop, MCS2007*, volume 4472 of Lecture Notes in Computer Science. Springer-Verlag, Berlin, 2007.

86. T. Hastie and R. Tibshirani. Classification by pairwise coupling. *Annals of Statistics*, 26(1):451–471, 1998.

87. T.K. Ho. The random subspace method for constructing decision forests. *IEEE Transactions on Pattern Analysis and Machine Intelligence*, 20(8):832–844, 1998.

88. T.K. Ho. Complexity of classification problems ans comparative advantages of combined classifiers. In J. Kittler and F. Roli, (eds), *Multiple Classifier Systems. First International*

Workshop, MCS 2000, Cagliari, Italy, volume 1857 of Lecture Notes in Computer Science, pp. 97–106, Springer-Verlag, 2000.

89. T.K. Ho. Data complexity analysis for classifiers combination. In J. Kittler and F. Roli, (eds), *Multiple Classifier Systems. Second International Workshop, MCS 2001, Cambridge, UK*, volume 2096 of Lecture Notes in Computer Science, pp. 53–67, Springer-Verlag, Berlin, 2001.

90. T.K. Ho. Multiple classifier combinations: Lessons and the next steps. In A. Kandel and K. Bunke, (eds), *Hybrid Methods in Pattern Recognition*, pp. 171–198, World Scientific, Hackensack, NJ, USA, 2002.

91. T.K. Ho, J.J. Hull, and S.N. Srihari. Decision combination in multiple classifiers. *IEEE Transactions on Pattern Analysis and Machine Intelligence*, 19(4):405–410, 1997.

92. S.M. Hojnacki, J.H. Kastner, G. Micela, E.D. Feigelson, and S.M. LaLonde. An x-ray spectral classification algorithm with application to young stellar clusters. *Astrophysical Journal*, 659:659, 2007.

93. S. Hojnacki, G. Micela, S. Lalonde, E. Feigelson, and J. Kastner. An unsupervised, ensemble clustering algorithm: A new approach for classification of x-ray sources. *Statistical Methodology*, 5:350–360, 2008.

94. K. Hornik. Approximation capabilities of multilayer feedforward networks. *Neural Networks*, 4:251–257, 1991.

95. T. Hothorn and B. Lausen. Bundling classifers by bagging trees. *Computational Statistics and Data Analysis*, 49:1068–1078, 2005.

96. Y.S. Huang and C.Y. Suen. Combination of multiple experts for the recognition of unconstrained handwritten numerals. *IEEE Transactions on Pattern Analysis and Machine Intelligence*, 17:90–94, 1995.

97. R.A. Jacobs. Methods for combining experts probability assessment. *Neural Computation*, 7:867–888, 1995.

98. R.A. Jacobs, M.I. Jordan, S.J. Nowlan, and G.E. Hinton. Adaptive mixtures of local experts. *Neural Computation*, 3(1):125–130, 1991.

99. A. Jain, R. Duin, and J. Mao. Statistical pattern recognition: A review. *IEEE Transactions on Pattern Analysis and Machine Intelligence*, 22:4–37, 2000.

100. X. Jiang, N. Nariai, M. Steffen, S. Kasif, and E. Kolaczyk. Integration of relational and hierarchical network information for protein function prediction. *BMC Bioinformatics*, 9:350, 2008.

101. M. Jordan and R. Jacobs. Hierarchies of adaptive experts. In J. Moody, S. Hanson, and R. Lippmann (eds), *Advances in Neural Information Processing Systems*, Vol. 4, pp. 985–992, Morgan Kauffman, San Mateo, CA, 1992.

102. M.I. Jordan and R.A. Jacobs. Hierarchical mixture of experts and the em algorithm. *Neural Computation*, 6:181–214, 1994.

103. M.I. Jordan and L. Xu. Convergence results for the EM approach to mixture of experts architectures. *Neural Networks*, 8:1409–1431, 1995.

104. K. Tumer and C.N. Oza. Input decimated ensembles. *Pattern Analysis and Applications*, 6:65–77, 2003.

105. J.M. Keller, P. Gader, H. Tahani, J. Chiang, and M. Mohamed. Advances in fuzzy integration for pattern recognition. *Fuzzy Sets and Systems*, 65:273–283, 1994.

106. F. Kimura and M. Shridar. Handwritten numerical recognition based on multiple algorithms. *Pattern Recognition*, 24(10):969–983, 1991.

107. J. Kittler. Combining classifiers: A theoretical framework. *Pattern Analysis and Applications*, 1:18–27, 1998.

108. J. Kittler, M. Hatef, R.P.W. Duin, and J. Matas. On combining classifiers. *IEEE Transactions on Pattern Analysis and Machine Intelligence*, 20(3):226–239, 1998.

109. E.M. Kleinberg. Stochastic discrimination. *Annals of Mathematics and Artificial Intelligence*, 1:207–239, 1990.

110. E.M. Kleinberg. An overtraining-resistant stochastic modeling method for pattern recognition. *Annals of Statistics*, 4(6):2319–2349, 1996.

111. E.M. Kleinberg. A mathematically rigorous foundation for supervised learning. In J. Kittler and F. Roli (eds), *Multiple Classifier Systems. First International Workshop, MCS 2000, Cagliari, Italy*, volume 1857 of Lecture Notes in Computer Science, pp. 67–76, Springer-Verlag, 2000.

112. E.M. Kleinberg. On the algorithmic implementation of stochastic discrimination. *IEEE Transactions on Pattern Analysis and Machine Intelligence*, 22(5):473–490, 2000.

113. E. Kong and T.G. Dietterich. Error-correcting output coding correct bias and variance. In *The XII International Conference on Machine Learning*, pp. 313–321, Morgan Kauffman, San Francisco, CA, 1995.

114. L.I. Kuncheva. An application of OWA operators to the aggregation of multiple classification decisions. In *The Ordered Weighted Averaging Operators. Theory and Applications*, pp. 330–343, Kluwer Academic Publisher, USA, 1997.

115. L.I. Kuncheva. Switching between selection and fusion in combining classifiers: An experiment. *IEEE Transactions on Systems, Man and Cybernetics*, 32(2):146–156, 2002.

116. L.I. Kuncheva. *Combining Pattern Classifiers: Methods and Algorithms*. Wiley-Interscience, New York, 2004.

117. L.I. Kuncheva, J.C. Bezdek, and R.P.W. Duin. Decision templates for multiple classifier fusion: An experimental comparison. *Pattern Recognition*, 34(2):299–314, 2001.

118. L.I. Kuncheva and L.C. Jain. Designing classifier fusion systems by genetic algorithms. *IEEE Trnsactions on Evolutionary Computation*, 4(4):327–336, 2000.

119. L.I. Kuncheva and J. Rodriguez. An experimental study on rotation forest ensembles. In M. Haindl, J. Kittler, and F. Roli (eds), *Multiple Classifier Systems. Seventh International Workshop, MCS 2007, Prague, Czech Republic*, volume 4472 of Lecture Notes in Computer Science, pp. 459–468, Springer, 2007.

120. L.I. Kuncheva, F. Roli, G.L. Marcialis, and C.A. Shipp. Complexity of data subsets generated by the random subspace method: An experimental investigation. In J. Kittler and F. Roli (eds), *Multiple Classifier Systems. Second International Workshop, MCS 2001, Cambridge, UK*, volume 2096 of Lecture Notes in Computer Science, pp. 349–358, Springer-Verlag, 2001.

121. L.I. Kuncheva and C.J. Whitaker. Measures of diversity in classifier ensembles. *Machine Learning*, 51:181–207, 2003.

122. L. Lam. Classifier combinations: Implementations and theoretical issues. In *Multiple Classifier Systems. First International Workshop, MCS 2000, Cagliari, Italy*, volume 1857 of Lecture Notes in Computer Science, pp. 77–86, Springer-Verlag, 2000.

123. L. Lam and C. Sue. Optimal combination of pattern classifiers. *Pattern Recognition Letters*, 16:945–954, 1995.

124. L. Lam and C. Sue. Application of majority voting to pattern recognition: An analysis of its behavior and performance. *IEEE Transactions on Systems, Man and Cybernetics*, 27(5):553–568, 1997.

125. W.B. Langdon and B.F. Buxton. Genetic programming for improved receiver operating characteristics. In J. Kittler and F. Roli (eds), *Second International Conference on Multiple Classifier System*, volume 2096 of LNCS, pp. 68–77, Springer-Verlag, Cambridge, 2001.

126. H. Lapplainen. Nonlinear independent component analysis using ensemble learning: Theory. In *Proceedings of the 1st International Workshop on Independent Component Analysis and Blind Signal Separation*, p. 7, 1998.

127. A. Lazarevic and Z. Obradovic. Effective pruning of neural network classifiers. In *Proceedings of the IEEE International Joint Conference on Neural Networks IJCNN'01*, pp. 796–801, IEEE, Washington, DC, USA, 2001.

128. W. Leigh, R. Purvis, and J.M. Ragusa. Forecasting the NYSE composite index with technical analysis, pattern recognizer, neural networks and genetic algorithm: A case study. *Decision Support Systems*, 32(4):361–377, 2002.

129. M. Li and P. Vitanyi. *An Introduction to Kolmogorov Complexity and Its Applications*. Springer-Verlag, Berlin, 1993.

130. R. Liu and B. Yuan. Multiple classifier combination by clustering and selection. *Information Fusion*, 2:163–168, 2001.

131. H. Lu, H. Zhou, J. Wang, T. Wang, X. Dong, Z. Zhuang, and C. Li. Ensemble learning for independent component analysis of normal galaxy spectra. *Astronomical Journal*, 131:790–805, 2006.

132. G. Martinez-Muniz and A. Suarez. Pruning in ordered bagging ensembles. In *23th International Conference on Machine Learning, ICML 2006*, pp. 609–616, ACM Press, 2006.

133. L. Mason, P. Bartlett, and J. Baxter. Improved generalization through explicit optimization of margins. *Machine Learning*, 38(3):243–255, 2000.

134. F. Masulli and G. Valentini. Effectiveness of error correcting output codes in multiclass learning problems. In Lecture Notes in Computer Science, volume 1857, pp. 107–116, Springer-Verlag, Berlin, Heidelberg, 2000.

135. F. Masulli and G. Valentini. Quantitative evaluation of dependence among outputs in ECOC classifiers using mutual information based measures. In K. Marko and P. Webos (eds), *Proceedings of the International Joint Conference on Neural Networks IJCNN'01*, volume 2, pp. 784–789, Piscataway, NJ, USA, IEEE, 2001.

136. F. Masulli and G. Valentini. Effectiveness of error correcting output coding decomposition schemes in ensemble and monolithic learning machines. *Pattern Analysis and Application*, 6:285–300, 2003.

137. F. Masulli and G. Valentini. An experimental analysis of the dependence among codeword bit errors in ecoc learning machines. *Neurocomputing*, 57:189–214, 2004.

138. E. Mayoraz and M. Moreira. On the decomposition of polychotomies into dichotomies. In *The XIV International Conference on Machine Learning*, pp. 219–226, Nashville, TN, July 1997.

139. E. Menahem, A. Shabtai, L. Rokach, Y. Elovici, and A. Troiha. Improving malware detection by applying multi-iducer ensemble. *Computational Statistics and Data Analysis*, 53(4):1483–1494, 2009.

140. C. Merkwirth, H. Mauser, T. Schulz-Gasch, O. Roche, M. Stahl, and T. Lengauer. Ensemble methods for classification in cheminformatics. *Journal of Chemical Information and Modeling*, 44(6):1971–1978, 2009.

141. M. Moreira and E. Mayoraz. Improved pairwise coupling classifiers with correcting classifiers. In C. Nedellec and C. Rouveirol (eds), *Lecture Notes in Artificial Intelligence*, Volume 1398, pp. 160–171, Berlin, Heidelberg, New York, 1998.

142. G. Obozinski, G. Lanckriet, C. Grant, M. Jordan, and W.S. Noble. Consistent probabilistic output for protein function prediction. *Genome Biology*, 9(supp. 1), 2008.

143. O. Okun and G. Valentini (eds.). *Supervised and Unsupervised Ensemble Methods and Their Applications*, volume 126 of Studies in Computational Intelligence. Springer-Verlag, Berlin, 2008.

144. O. Okun and G. Valentini (eds.). *Applications of Supervised and Unsupervised Ensemble Methods*, volume 245 of Studies in Computational Intelligence. Springer-Verlag, Berlin, 2009.

145. D.W. Opitz. Feature selection for ensembles. In *Proceedings of the 16th National Conference on Artificla Intelligence, AAAI*, pp. 379–384, 1999.

146. D.W. Opitz and J.W. Shavlik. Actively searching for an effective neural network ensemble. *Connection Science*, 8(3/4):337–353, 1996.

147. D.W. Opitz and J.W. Shavlik. Generating accurate and diverse members of a neural-network ensemble. In D. Touretzky, M. Mozer, and M. Hasselmo (eds), *Advances in Neural Information Processing Systems*, volume 8, pp. 535–541, MIT Press, Cambridge, MA, 1996.

148. N.C. Oza. Aveboost2: Boosting for noisy data. In *Multiple Classifier Systems. Fifth International Workshop, MCS 2004, Cagliari, Italy*, volume 3077 of Lecture Notes in Computer Science, pp. 31–40, Springer-Verlag, 2004.

149. N.C. Oza, R. Polikar, F. Roli, and Kittler. *Multiple Classifier Systems, 6th International Workshop, MCS2005*, volume 3541 of Lecture Notes in Computer Science. Springer-Verlag, Berlin, 2005.

150. N.C. Oza and K. Tumer. Input decimation ensembles: Decorrelation through dimensionality reduction. In J. Kittler and F. Roli (eds), *Multiple Classifier Systems. Second International Workshop, MCS 2001, Cambridge, UK*, volume 2096 of Lecture Notes in Computer Science, pp. 238–247, Springer-Verlag, 2001.

151. J. Park and I.W. Sandberg. Approximation and radial basis function networks. *Neural Computation*, 5(2):305–316, 1993.

152. B. Parmanto, P. Munro, and H. Doyle. Improving committe diagnosis with resampling techniques. In D.S. Touretzky, M. Mozer, and M. Hesselmo (eds), *Advances in Neural Information Processing Systems*, volume 8, pp. 882–888, MIT Press, Cambridge, MA, 1996.

153. B. Parmanto, P. Munro, and H. Doyle. Reducing variance of committee predition with resampling techniques. *Connection Science*, 8(3/4):405–416, 1996.

154. I. Partalas, G. Tsoumakas, and I. Vlahavas. Focused ensemble selection: A diversity-based method for greedy ensemble selection. In *Proceeding of the 2008 Conference on ECAI 2008: 18th European Conference on Artificial Intelligence*, pp. 117–121, IOS-Press, 2008.

155. I. Partalas, G. Tsoumakas, and I. Vlahavas. An ensemble uncertainty aware measure for directed hill climbing ensemble pruning. *Machine Learning*, 81(3):257–282, 2010.

156. D. Partridge, and W.B Yates. Engineering multiversion neural-net systems. *Neural Computation*, 8:869–893, 1996.

157. E. Pekalska, M. Skurichina, and R.P.W. Duin. Combining Fisher linear discriminant for dissimilarity representations. In J. Kittler and F. Roli (eds), *Multiple Classifier Systems. First International Workshop, MCS 2000, Cagliari, Italy*, volume 1857 of Lecture Notes in Computer Science, pp. 230–239, Springer-Verlag, 2000.

158. M.P. Perrone and L.N. Cooper. When networks disagree: Ensemble methods for hybrid neural networks. In R.J. Mammone (ed.), *Artificial Neural Networks for Speech and Vision*, pp. 126–142, Chapman & Hall, London, 1993.

159. R. Polikar, J. DePasquale, H. Syed Mohammed, G. Brown, and L.I. Kuncheva. Learn++.mf: A random subspace approach for the missing feature problem. *Pattern Recognition*, 43:3817–3832, 2010.

160. R. Polikar, A. Topalis, D. Parikh, D. Green, J. Frymiare, J. Kounios, and C. Clark. An ensemble based data fusion approach for early diagnosis of alzheimer disease. *Information Fusion*, 9:83–95, 2008.

161. O. Pujol, P. Radeva, and J. Vitria. Discriminant ECOC: A heuristic method for application dependent design of error correcting output codes. *IEEE Transactions on Pattern Analysis and Machine Intelligence*, 28:1001–1007, 2006.

162. J.R. Quinlan. *C4.5 Programs for Machine Learning*. Morgan Kauffman, London, UK, 1993.

163. Y. Raviv and N. Intrator. Bootstrapping with noise: An effective regularization technique. *Connection Science*, 8(3/4):355–372, 1996.

164. M. Re and G. Valentini. Integration of heterogeneous data sources for gene function prediction using decision templates and ensembles of learning machines. *Neurocomputing,* 73(7–9):1533–1537, 2010.

165. M. Re and G. Valentini. Noise tolerance of multiple classifier systems in data integration-based gene function prediction. *Journal of Integrative Bioinformatics,* 7(3):139, 2010.

166. M. Re and G. Valentini. Simple ensemble methods are competitive with state-of-the-art data integration methods for gene function prediction. *Journal of Machine Learning Research, W&C Proceedings, Machine Learning in Systems Biology,* 8:98–111, 2010.

167. S. Reid and G. Grudic. Regularized linear models in stacked generalization. In J. Kittler, J. Benediktsson, and F. Roli, (eds), *Multiple Classifier Systems. Eighth International Workshop, MCS 2009, Reykjavik, Iceland,* volume 5519 of Lecture Notes in Computer Science, pp. 112–121, Springer, 2009.

168. J. Rodriguez and L.I. Kuncheva. Rotation forest: A new classifier ensemble method. *IEEE Transactions on Pattern Analysis and Machine Intelligence,* 28(10):1619–1630, 2006.

169. G. Rogova. Combining the results of several neural neetworks classifiers. *Neural Networks,* 7:777–781, 1994.

170. L. Rokach. Genetic algorithm-based feature set partitioning for classifiaction problems. *Pattern Recognition,* 41(5):1676–1700, 2008.

171. L. Rokach. Taxonomy for characterizing ensemble methods in classification asks: A reveiw and annotated bibliography. *Computational Statistics and Data Analysis,* 53:4046–4072, 2009.

172. F. Roli, G. Giacinto, and G. Vernazza. Methods for designing multiple classifier systems. In J. Kittler and F. Roli (eds), *Multiple Classifier Systems. Second International Workshop, MCS 2001, Cambridge, UK,* volume 2096 of Lecture Notes in Computer Science, pp. 78–87, Springer-Verlag, 2001.

173. F. Roli, J. Kittler, and T. Windeatt. *Multiple Classifier Systems, Fifth International Workshop, MCS2004,* volume 3077 of Lecture Notes in Computer Science. Springer-Verlag, Berlin, 2004.

174. R. Schapire and Y. Singer. Boostexter: A boosting-based system for text categorization. *Machine Learning,* 39(2/3):135–168, 2000.

175. R.E. Schapire. The strenght of weak learnability. *Machine Learning,* 5(2):197–227, 1990.

176. R.E. Schapire. A brief introduction to boosting. In T. Dean (ed.), *16th International Joint Conference on Artificial Intelligence,* pp. 1401–1406, Morgan Kauffman, 1999.

177. R.E. Schapire, Y. Freund, P. Bartlett, and W. Lee. Boosting the margin: A new explanation for the effectiveness of voting methods. *Annals of Statistics,* 26(5):1651–1686, 1998.

178. R.E. Schapire and Y. Singer. Improved boosting algorithms using confidence-rated predictions. *Machine Learning,* 37(3):297–336, 1999.

179. L. Schietgat, C. Vens, J. Struyf, H. Blockeel, and S. Dzeroski. Predicting gene function using hierarchical multi-label decision tree ensembles. *BMC Bioinformatics,* 11:2, 2010.

180. B. Scholkopf and A. Smola. *Learning with Kernels.* MIT Press, Cambridge, MA, 2002.

181. B. Settles. Active learning literature survey. Technical Report Computer Sciences Technical Report 1648, University of Wisconsin, Madison, 2010.

182. A. Sharkey. Types of multi-net systems. In F. Roli and J. Kittler (eds), *Multiple Classifier Systems, Third International Workshop, MCS2002,* volume 2364 of Lecture Notes in Computer Science, pp. 108–117, Springer-Verlag, 2002.

183. A. Sharkey, N. Sharkey, and G. Chandroth. Diverse neural net solutions to a fault diagnosis problem. *Neural Computing and Applications,* 4:218–227, 1996.

184. A. Sharkey (ed.). *Combining Artificial Neural Nets: Ensemble and Modular Multi-Net Systems.* Springer-Verlag, London, 1999.

185. M. Skurichina and R.P.W. Duin. Bagging for linear classifiers. *Pattern Recognition,* 31(7):909–930, 1998.

186. M. Skurichina and R.P.W. Duin. Bagging and the random subspace method for redundant feature spaces. In *Multiple Classifier Systems. Second International Workshop, MCS 2001, Cambridge, UK*, volume 2096 of Lecture Notes in Computer Science, pp. 1–10, Springer-Verlag, 2001.

187. M. Skurichina and R.P.W. Duin. Bagging, boosting and the random subspace method for linear classifiers. *Pattern Analysis and Applications*, 5(2):121–135, 2002.

188. S.Y. Sohna and H.W. Shinb. Experimental study for the comparison of classifier combination methods. *Pattern Recognition*, 40:33–40, 2007.

189. Y. Sun, S. Todorovic, and L. Li. Reducing the overfitting of adaboost by controlling its data distribution skewness. *International Journal of Pattern Recognition and Artificial Intelligence*, 20(7):1093–1116, 2006.

190. D. Tao, X. Tang, X. Li, and X. Wu. Asymmetric bagging and random subspace for support vector machine-based relevance feedback in image retrieval. *IEEE Transactions on Pattern Analysis and Machine Intelligence*, 28(7):1088–1099, 2006.

191. G. Tsoumakas, L. Angelis, and I. Vlahavas. Selective fusion of heterogeneous classifiers. *Intelligent Data Analysis*, 9(6):511–525, 2005.

192. G. Tsoumakas, I. Katakis, and I. Vlahavas. Effective voting of heterogeneous classifiers. In *Proceedings of the 15th European Conference on Machine Learning, ECML 2004*, pp. 465–476, Pisa, Italy, 2004.

193. A. Tsymbal, M. Pechenizkiy, and P. Cunningham. Diversity in search strategies for ensemble feature selection. *Information Fusion*, 6:83–98, 2006.

194. A. Tsymbal, S. Puuronen, and D.W. Patterson. Ensemble feature selection with the simple bayesian classifiaction. *Information Fusion*, 4:87–100, 2003.

195. K. Tumer and J. Ghosh. Error correlation and error reduction in ensemble classifiers. *Connection Science*, 8(3/4):385–404, 1996.

196. G. Valentini. Gene expression data analysis of human lymphoma using support vector machines and output coding ensembles. *Artificial Intelligence in Medicine*, 26(3):283–306, 2002.

197. G. Valentini. An experimental bias-variance analysis of SVM ensembles based on resampling techniques. *IEEE Transactions on Systems, Man and Cybernetics, Part B: Cybernetics*, 35(6):1252–1271, 2005.

198. G. Valentini. True path rule hierarchical ensembles for genome-wide gene function prediction. *IEEE ACM Transactions on Computational Biology and Bioinformatics*, 8(3), 2011.

199. G. Valentini and T.G. Dietterich. Low bias bagged support vector machines. In T. Fawcett and N. Mishra (eds), *Machine Learning, Proceedings of the Twentieth International Conference (ICML 2003)*, pp. 752–759, AAAI Press, Washington D.C., USA, 2003.

200. G. Valentini and T.G. Dietterich. Bias–variance analysis of support vector machines for the development of SVM-based ensemble methods. *Journal of Machine Learning Research*, 5:725–775, 2004.

201. G. Valentini and F. Masulli. Ensembles of learning machines. In *Neural Nets WIRN-02*, volume 2486 of Lecture Notes in Computer Science, pp. 3–19, Springer-Verlag, 2002.

202. J. Van Lint. *Coding Theory*. Springer Verlag, Berlin, 1971.

203. V. N. Vapnik. *Statistical Learning Theory*. Wiley, New York, 1998.

204. C. Vens, J. Struyf, L. Schietgat, S. Dzeroski, and H. Blockeel. Decision trees for hierarchical multi-label classification. *Machine Learning*, 73(2):185–214, 2008.

205. A. Verikas, A. Lipnickas, K. Malmqvist, M. Bacauskiene, and A. Gelzinis. Soft combination of neural classifiers: A comparative study. *Pattern Recognition Letters*, 20:429–444, 1999.

206. K.L. Wagstaff and V.G. Laidler. Making the most of missing values: Object clustering with partial data in astronomy. In *Astronomical Data Analysis Software and Systems XIV*, ASP Conference Series, Vol. 347, Proceedings of the Conference held 24–27 October, 2004 in Pasadena, California, USA, p. 172, 2005.

207. D. Wang, J.M. Keller, C.A. Carson, K.K. McAdoo-Edwards, and C.W. Bailey. Use of fuzzy logic inspired features to improve bacterial recognition through classifier fusion. *IEEE Transactions on Systems, Man and Cybernetics*, 28B(4):583–591, 1998.

208. D.H. Wolpert. Stacked generalization. *Neural Networks*, 5:241–259, 1992.

209. K. Woods, W.P. Kegelmeyer, and K. Bowyer. Combination of multiple classifiers using local accuracy estimates. *IEEE Transactions on Pattern Analysis and Machine Intelligence*, 19(4):405–410, 1997.

210. L. Xu, C. Krzyzak, and C. Suen. Methods of combining multiple classifiers and their applications to handwritting recognition. *IEEE Transactions on Systems, Man and Cybernetics*, 22(3):418–435, 1992.

211. C. Yeang, S. Ramaswamy, P. Tamayo, S. Mukherjee, R.M. Rifkin, M. Angelo, M. Reich, E. Lander, J. Mesirov, and T. Golub. Molecular classification of multiple tumor types. In *ISMB 2001, Proceedings of the 9th International Conference on Intelligent Systems for Molecular Biology*, pp. 316–322, Oxford University Press, Copenaghen, Denmark, 2001.

212. C.X. Zhang and J.S. Zhang. A local boosting algorithm for solving classification problems. *Computational Statistics and Data Analysis*, 52(4):1928–1941, 2008.

213. Y. Zhang, S. Burer, and W.N. Street. Ensemble pruning via semi-definite programming. *Journal of Machine Learning Research*, 7:1315–1338, 2006.

214. J. Zhou, H. Peng, and C. Suen. Data-driven decompositon for multi-class classification. *Pattern Recognition*, 41(1):67–76, 2008.

215. X. Zhu and A. Goldberg. *Introduction to Semi-Supervised Learning. Synthesis Lectures on Artificial Intelligence and Machine Learning*. Morgan & Claypool, 2009.

CHAPTER 27

Parallel and Distributed Data Mining for Astronomy Applications

Kamalika Das and Kanishka Bhaduri
NASA Ames Research Center

CONTENTS

27.1	Introduction	595
27.2	Parallel Algorithms for Data Mining	598
	27.2.1 Parallel Data Clustering Algorithms	598
	27.2.2 Parallel Data Classification Algorithms	600
	27.2.3 MapReduce Framework	602
27.3	Distributed Algorithms for Data Mining	604
	27.3.1 Distributed Clustering	604
	27.3.2 Distributed Classification	606
	27.3.3 Data Mining in P2P Environments	607
	27.3.3.1 Approximate Algorithms for P2P Data Mining	607
	27.3.3.2 Exact Algorithms for P2P Data Mining	608
	27.3.4 Message Passing Interface	608
27.4	Data Mining in GRID Environments	609
27.5	DDM for Astronomy	610
27.6	Conclusion	611
Acknowledgment		611
References		611

27.1 INTRODUCTION

Due to advances in data collection capabilities, storage, and computing technologies, astronomy has become a data-rich discipline. Over the last few years, numerous surveys have systematically looked at the entire sky using different wavelengths of light and sophisticated instruments. The data generated from many of these surveys exceed several terabyte, and often reach peta byte scales. Brunner et al. [18] present an overview of some of these massive astronomy datasets available to the researchers. All these data are meaningless unless they are analyzed for new and interesting astronomical discoveries.

Data mining is playing an increasingly important role in astronomy research [3,11,51] involving very large sky surveys such as Sloan Digital Sky Survey (SDSS), the 2-Micron All-Sky Survey, and many others mentioned in Ref. [18]. These sky surveys are offering a new way to study and analyze the behavior of the astronomical objects. However, handling and processing such large volumes of stored datasets for scientific discovery pose significant computational challenges. Traditional analysis techniques do not scale to this volume of data on a single computer. So we need a fundamentally different way of analyzing these massive datasets.

The next generation of sky surveys are poised to take a step further by incorporating sensors that will collect large volumes of data and stream them in real time back to the data storage centers. For example, the Large Synoptic Survey Telescopes (LSST) will take repeated images of the night sky every 20 s. This will generate 30 terabytes of calibrated imagery every night that will need to be coanalyzed with other astronomical data stored at different locations around the world. Event identification and classification in such data sets may provide useful insights in to unique astronomical phenomenon displaying astrophysically significant variations: quasars, supernovae, variable stars, and potentially hazardous asteroids. Analyzing such datasets is challenging not only because of their large size (several petabytes) but also because the data may be streaming.

Existing workarounds for solving these massive data analysis problems can be divided into two categories: (1) to scale up the algorithms using approximation techniques where the final result of the data mining algorithm is within some ϵ-bound of the true or actual result, or (2) to develop decomposable versions of the problem that can be solved either exactly or approximately by leveraging the computing power of multiple computers.

In this chapter, we focus on the discussion of scaling up data mining algorithms using the second type of technique, that is, using multiple computers. We distinguish, within this category, between two related but different frameworks for scaling up data mining algorithms, namely, the parallel computing framework and the distributed computing framework. Distributed computing is a generic term referring to all scenarios where the computing system consists of multiple computers (or autonomous processors) communicating via a computer network in order to achieve a common goal. Parallel computing is considered to be a special instance of distributed computing when the processes communicate via shared memory. However, in this chapter we distinguish between parallel and distributed computing in the following way. A parallel computing framework is one in which there exists a high-performance shared memory multiprocessor computer or a cluster of computers. These systems are tightly coupled and may have one master processor coordinating the actions of all the slaves. The processes are highly synchronized based on a single-system clock. A distributed computing framework, on the other hand, uses large-scale autonomous distributed computers connected via a communication infrastructure with or without the existence of a master–slave architecture. The systems are loosely coupled and are often asynchronous in nature with each autonomous computer maintaining its

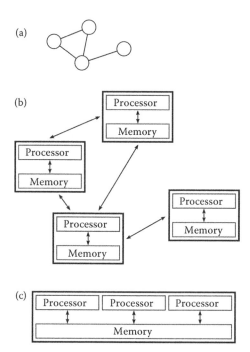

FIGURE 27.1 Conceptual diagram of distributed and parallel computing. (a)–(b) Distributed computing. (c) Parallel computing. (Adapted from http://en.wikipedia.org/wiki/File: Distributed-parallel.svg.)

own clock. Figure 27.1 shows a diagrammatic representation of parallel and distributed computing frameworks. Figure 27.1a shows that distributed computing infrastructure can be viewed as a graph with the nodes representing computers and the links between them representing communication pathways. In Figure 27.1b, each processor is shown explicitly having its own memory. Figure 27.1c shows a parallel computing framework in which the processors communicate via shared memory architecture. The *parallel computing* framework becomes very useful when the data are initially at one central location. When such a central dataset is too huge to be analyzed by a single off-the-shelf computer, it is split into multiple partitions and each partition is assigned to one processor within a cluster. Once the processors finish their computation, the results are combined by the master computer to produce the final output. For this type of computation, it is important to note that the tasks executed on each of the data partitions are independent, with no or minimal requirement for interprocess communication. This helps the parallel data mining algorithms to scale up using less resources. The *distributed computing* framework on the other hand, is useful when the data are inherently distributed to start with and the goal is to build a global model without centralizing all of the data. Distributed computing techniques generally require communication among the processes. While a parallel algorithm is measured by its speed-up with respect to a comparable centralized algorithm, a distributed algorithm is often measured by the fraction of communication (with respect to complete data centralization) it needs, in order to output the correct result.

The rest of the chapter is organized as follows. In the next section (Section 27.2), we discuss parallel data clustering and classification algorithms. In the same section we discuss the MapReduce framework which is amenable to large-scale parallel processing with massive datasets. In Section 27.3, we discuss distributed techniques for data classification and clustering along with the message passing interface (MPI) for implementing distributed data mining (DDM) algorithms. We also discuss another emerging area of DDM which is known as peer-to-peer (P2P) data mining. We briefly present some data mining techniques on the Grid infrastructure in Section 27.4. In Section 27.5, we discuss applications of the above techniques for astronomical applications. Finally, we conclude the chapter in Section 27.6.

27.2 PARALLEL ALGORITHMS FOR DATA MINING

In most of today's applications, data are being collected at a high rate due to easier availability and better reliability of data collection sensors. This is true not only for almost all disciplines of science such as astronomy, physics, chemistry, and biology but also for human-engineered systems such as commercial aircraft, automobiles, and communication devices. Data mining/knowledge discovery from databases plays a vital role in analyzing these datasets whose sizes often exceed tens or thousands of petabytes. The sizes of these datasets are well beyond the scope of a single commodity computer to perform the analysis. Either the computational complexity of the algorithm becomes the bottleneck or the data are too huge to fit in memory. Depending on the algorithm, one solution is to read blocks of data into main memory, perform the required computation, then save the result and repeat with the other blocks. However, such disk-based solutions may be too slow for a large number of blocks. It is appealing to develop solutions in which a cluster of computers can be used for both memory and computational needs. To this end, several researchers have proposed algorithms and systems which first split the data and then let each computer in a cluster work on a data split. Finally, the results from each of these splits are combined to derive the final result of the overall computation. In this section, we provide several examples of data mining algorithms and systems which are specifically designed for such tightly coupled cluster of computers.

27.2.1 Parallel Data Clustering Algorithms

Data clustering is one of the basic tasks in exploratory data analysis. Clustering of the data not only helps one to understand the characteristics of the data, but is also often used as a preprocessing step to identify different modes or regimes of the data so that they can be analyzed separately. Clustering algorithms often suffer from high computational and storage costs. Parallelizing is thus a natural choice by which large datasets can be clustered in a reasonable time. Clustering may be important for astronomy datasets to identify groups of similarly behaved galaxies, planets or other astrophysical objects.

Olson [74] presents several parallel algorithms for agglomerative hierarchical clustering. They consider two types of distance metrics: (1) *graph based* in which the distance between two clusters is determined by computing the minimum, average or maximum length of the edges between the members of the cluster, and (2) geometric metrics in which the distance between two clusters is defined as the centroid, median, or minimum variance distance between the members in the two groups. Both graph-based and geometric metric-based hierarchical algorithms have a computational complexity of $O(n^2)$, where n is the number of points in the dataset [34]. The most computationally expensive operation in hierarchical clustering is in finding the closest distance between two clusters for each iteration. The paper suggests several algorithms for speeding up this computation using both concurrent read, concurrent write (CRCW) model and exclusive read, exclusive write (EREW) model. The idea is for each processor to be in charge of one cluster and its associated nearest-neighbor distances so that the task of agglomerating the clusters is distributed across all the processors. These algorithms run in linear time with respect to the number of points n, for most of the distance metrics and CRCW architectures. For the EREW model, the algorithms run in $O(n \log n)$ time.

Another popular clustering technique is the k-means algorithm which is a partitioning-based clustering method. In this method, the number of clusters is provided before the algorithm starts. Initially, each data point is assigned to one cluster (based on some notion of distance to cluster centroids) and at each successive iteration points may be reassigned to other clusters and new cluster centroids are computed. Although the advantage of k-means algorithm over other clustering methods is its linear computational complexity, still the computational burden is in finding the distance of each point from each cluster centroid once per iteration. Fortunately, this process can easily be parallelized noting that, for each cluster, the distance of all points in that cluster to the cluster centroids only depends on the points and the cluster centroids themselves. Dhillon and Modha [36] present a parallel version of this algorithm which aims to handle this problem elegantly. In the first step, the data are distributed across the processors and a random choice of the k centroids are also broadcast. Then the algorithm proceeds as follows:

- Compute the distance of all points in that partition (points assigned to one processor) to all the centroids since the latter are available to all the processors.

- For all points in that partition, find the cluster assignment based on the minimum distance over all the clusters.

- Update the cluster centroids at each processor independently.

- Recompute the cluster centroids by requesting the centroids from the other processors.

The steps mentioned above are executed iteratively till the algorithm converges. The authors have implemented the algorithm in the MPI which is a standardized, portable, and widely available message-passing system [49]. Several experiments conducted by the authors show that the algorithm demonstrates linear scale-up (with respect to the number of processors).

In some cases, k-means may produce unsatisfactory results if the data itself are not grouped into convex regions. An alternative approach, proposed first by Shi and Malik [79] is to embed the data first into a k-dimensional eigen-space and then perform clustering in that space. This technique is known as spectral clustering and several algorithms have been proposed for that. As shown by Ng et al. [28], spectral clustering often produces more meaningful results compared to simple distance-based k-means when the data are, for example, arranged in an annular shape. For many applications such as astronomy, this may be interesting for clustering nonconvex-shaped galaxies/galactic objects. Given a dataset $\mathbf{S} = \{\mathbf{x}_1, \ldots, \mathbf{x}_n\}$, where $\mathbf{x}_i \in \mathbb{R}^d$, if we want to group the data into k clusters using a spectral clustering algorithm, then the steps are as follows:

- Compute weighted adjacency matrix (a similarity matrix) $\mathbf{W} \in \mathbb{R}^{d \times d}$, where $W_{ij} = \exp(-||\mathbf{x}_i - \mathbf{x}_j||^2 / 2\sigma^2)$ if $i \neq j$ and $W_{ij} = 0$ otherwise.

- Define a diagonal matrix $\mathbf{D}$ whose (i, i)th element is the degree of the ith node in the graph and construct the Laplacian matrix as $\mathbf{L} = \mathbf{D}^{-1/2} \mathbf{W} \mathbf{D}^{-1/2}$.

- Compute the first k eigenvectors of $\mathbf{L}$.

- Let $\mathbf{E} \in \mathbb{R}^{n \times k}$ be the matrix in which each column is an eigenvector.

- Run k-means algorithm on $\mathbf{E}$ taking each input vector as a row of $\mathbf{E}$.

- Assign a cluster label to each of the original n points.

The above algorithm works quite well if the entire data similarity matrix $\mathbf{W}$ fits in memory and the eigenvectors can be computed efficiently. Spectral clustering becomes too expensive and eventually impossible to compute if the data are too big. To solve this problem, Chen et al. [25] have proposed a parallel spectral clustering (PSC) algorithm that can run on multiple computers. Their main idea is to first split the dataset among p processors such that each processor has n/p data points. To reduce storage complexity at each node, instead of computing the full dense similarity matrix containing all the points, the authors in Ref. [25] propose to compute the t most similar items (using an efficient max-heap) for each point under the assumption that the rest of the distances are quite small and can safely be ignored. This is then stored in a sparse format to save space. In the next step, a parallel eigensolver is used to find the top k eigenvectors of the Laplacian matrix such that the eigenvectors are stored in a distributed fashion at all the nodes. Finally, a parallel k-means clustering algorithm is used to cluster the eigenvectors into k groups. Experimental results reported in Ref. [25] on image databases of size close to 200,000 show good efficiency and scalability using 64 computers. The source code of the PSC is available at http://code.google.com/p/pspectralclustering/.

27.2.2 Parallel Data Classification Algorithms

In this section, we discuss supervised techniques for data analysis. Classification is one of the most popular supervised techniques in which the task is to learn a function mapping the input to the output. Once this function is learned based on a training set, the goal is to predict the outputs of some test instances. In classification problems, the inputs can be continuous or discrete while the outputs need to be discrete. Moreover, unlike clustering, classification algorithms typically need a separate training and test set to learn the function and test its generalization power on unseen examples, respectively. Classification algorithms require labeled instances for the training set. Classification is a natural choice where the task is to distinguish objects of different classes in an automated fashion. Identification of different types of galaxies, stars, and planets can be accomplished using data classification algorithms. Unfortunately, many of these algorithms are resource intensive and may require high computational power to learn the mapping function. In this section, we discuss algorithms for speeding up classification algorithms using parallel computing methodologies.

Several parallel implementations have been developed for some very popular data classification algorithms such as support vector machines, decision trees, and so on Chang et al. [23] have developed a parallel support vector machine (PSVM) algorithm which reduces the memory requirement from $O(n^2)$ to $O(np/m)$ and the running time from $O(n^3)$ to $O(np^2/m)$, where n is the number of data instances, m is the number of computers available, and p is the reduced matrix dimension of the incomplete Cholesky factorization [46] of the input data. It has been empirically verified by the authors in Ref. [23] that p can be set to $\sqrt{n}$ without degrading the quality of the matrix reconstruction. Solution of SVM in the dual domain requires one to solve a convex quadratic programming problem with linear constraints. This is most efficiently done using the interior point methods. However, the computational complexity of this is $O(n^3)$. PSVM circumvents both the memory requirement and the computation requirement by loading the training data into the m computers in a round-robin fashion, thereby reducing the memory requirement from $O(n^2)$ to $O(nd/m)$ per computer, where d is the dimensionality of the data. In the second step, each computer computes an incomplete cholesky factorization of the loaded data matrix and stores only the factorized data, thereby reducing the memory requirement further to $O(np/m)$, where $p/m(\ll n)$ is the column dimension of the factorized matrix. To reduce the computational overhead, the authors propose a parallel interior point algorithm (IPM) which coordinates the solution of the dual SVM formulation using the interior point method among all the computers. This parallel IPM reduces the computational cost from $O(n^3)$ to $O(np^2/m)$. Experimental results on publicly available datasets show that the algorithm achieves comparable accuracy to LIBSVM, a state-of-the-art centralized SVM routine. The technique also shows linear scale-up in the computational time when the number of computers is increased linearly. The open-source software can be downloaded from http://code.google.com/p/psvm/.

Decision trees is another popular classification technique. There have been several attempts in making them scale to large datasets by parallelizing their construction over a network of computers. The major drawback of decision trees is the need to sort the instances in order to figure out where to split a tree node. The various techniques for parallel tree building can roughly be divided into three groups [1]: (1) task-parallel, in which the tree nodes are distributed across the computers and each computer contains a small subset of all the tree nodes, (2) data parallel, in which the data are split among the computers and each computer is responsible in contributing to the tree based on only its share, and (3) hybrid, in which data parallelism is exploited toward the first stages followed by task parallelism toward the end. Parallel decision tree algorithms accommodate the need for data sorting by presorting, distributed sorting or approximations. Approximations are carried out through compact data representations such as histograms, counts, and so on [5,78].

In the meta-learning literature, a decision tree is learned independently at each of the data partitions [12,21]. A test tuple is classified by taking a majority vote of all the individual classifiers. While this approach may be very simple to implement and easy to use, performance degrades if the data partitions are much different. The core of the decision tree induction algorithm is the evaluation of impurity gains for determining the splitting attribute. Most of the impurity gain calculations (e.g., Gini [15], information gain, misclassification gain) can be expressed as sufficient statistics (counts of how many tuples belong to a particular class for each attribute value) over the datasets. For each level of the tree, a computer can build histogram counts of the attribute-label values and these are then broadcast to all the other computers. At the end of this step, each computer can compute the correct impurity measure for each attribute and select the splitting node as the one which minimizes this measure. This process is then repeated recursively for each level of the tree until some stopping criterion (such as depth or required fraction of purity in the leaf nodes) is reached. Based on minor variations of this technique, several parallel decision tree induction algorithms have been proposed such as Refs. [5,55,78].

27.2.3 MapReduce Framework

One of the more popular frameworks which lends itself nicely to massively parallel computations is the MapReduce framework proposed by Dean and Ghemawat [35]. A given dataset is first broken into multiple smaller sets. A map function is invoked for each of these smaller sets independently. The task of the map function is to take in a pair (#key, #value) in one domain and output a pair (#key, #value) in another domain. This map function is applied to every item in the input dataset, thereby converting the dataset to a collection of key-value pairs. The framework then groups and stores all these intermediate key-value pairs such that all the values for each key is stored as a group. The reduce function is then applied in parallel to each group which collects and outputs a single value for each group. The following is a schematic description of the map and reduce functions:

$$\text{Map}(k_1, v_1) \rightarrow \text{list}(k_2, v_2)$$

$$\text{Reduce}(k_2, \text{list}(v_2)) \rightarrow \text{list}(v_3)$$

In the above, a mapper takes in a key–value pair k_1, v_1 and outputs a list of key–value pairs (k_2, v_2) in a different domain. The reducer then takes all the values having the same key k_2 and outputs another list of values v_3 which are in the same domain as the original data. There are many open-source MapReduce implementations available today. Apache Hadoop[*] is one such working system implemented in Java. It should be noted that MapReduce is good for noniterative algorithms in which the individual processes are independent and communication is only needed while going from the map phase to the reduce phase. Once the MapReduce framework is setup, it is fairly easy to implement advanced data mining algorithms. For data mining algorithms to be amenable to MapReduce, one must be able to

---

[*] http://hadoop.apache.org/

express it in a decomposable form, that is, computations should be able to be carried out at each of the partitions independently and then combined to form the global solution.

As an example, let us see how a multivariate regression model can be evaluated using MapReduce. In regression, the goal is to model a function from input to output, but the output in this case is real valued instead of discrete (as in classification). Consider a data matrix $\mathbf{D}$ having m rows (instances) and d columns (features) such that $\mathbf{D} = [(\mathbf{x}_1, y_1), (\mathbf{x}_2, y_2), \ldots, (\mathbf{x}_m, y_m)]^T$, where $\mathbf{x}_j = [x_{j.1} \ldots x_{j.(d-1)}] \in \mathbb{R}^{d-1}$ and $y_j \in \mathbb{R}$. Every data tuple can be viewed as an input and output (target) pair. Let the input–output be related linearly as follows: $\widehat{y}_j = f(\mathbf{x}_j) = w_0 + w_1 x_{j.1} + w_2 x_{j.2} + \cdots + w_{d-1} x_{j.(d-1)} = \mathbf{w} \widetilde{\mathbf{x}}_j^T$, where $\mathbf{w} = [w_0 w_1 \ldots w_{d-1}]$ and $\widetilde{\mathbf{x}}_j = [1 \quad \mathbf{x}_j]$. Linear least squares method finds the optimal values of $\mathbf{w}$ by minimizing the squared error between $\widehat{y}$ and y. First, we rewrite $\mathbf{D} = [\mathbf{X} \quad \mathbf{y}]$, where $\mathbf{X} = [\widetilde{\mathbf{x}}_1, \widetilde{\mathbf{x}}_2, \ldots, \widetilde{\mathbf{x}}_m]^T$ is the set of all inputs and $\mathbf{y} = [y_1, y_2, \ldots, y_m]^T$ is the set of all outputs. Following the least squares technique,

$$\mathbf{w} = (\mathbf{X}^T \mathbf{X})^{-1} (\mathbf{X}^T \mathbf{y}).$$

Now assume that we need to do the same computation using MapReduce. Our first task is to split $\mathbf{D}$ into m partitions $\mathbf{D}_1, \mathbf{D}_2, \ldots, \mathbf{D}_m$ such that $\mathbf{D} = \cup_{i=1}^{m} \mathbf{D}_i$. Using the same notations for $\mathbf{X}_i$ and $\mathbf{y}_i$ for each partition, it can be easily verified that

$$\mathbf{X}^T \mathbf{X} = \sum_{i=1}^{m} \mathbf{X}_i^T \mathbf{X}_i, \quad \mathbf{X}^T \mathbf{y} = \sum_{i=1}^{m} \mathbf{X}_i^T \mathbf{y}_i.$$

Once these two matrices are known, $\mathbf{w}$ is given by

$$\mathbf{w} = \left(\sum_{i=1}^{m} \mathbf{X}_i^T \mathbf{X}_i \right)^{-1} \left(\sum_{i=1}^{m} \mathbf{X}_i^T \mathbf{y}_i \right)$$

The above expression suggests that we use two sets of mappers for the distributed computation, each containing m mappers. The first set will compute the value of $\sum_{i=1}^{m} \mathbf{X}_i^T \mathbf{X}_i$ and the second set will compute the value of $\sum_{i=1}^{m} \mathbf{X}_i^T \mathbf{y}_i$. All the outputs of the first mappers will go into a single reducer which will simply sum the values of $\mathbf{X}_i^T \mathbf{X}_i$ while the output of the second set will be reduced by a single reducer to compute $\mathbf{X}_i^T \mathbf{y}_i$. The specifications for the first set of mappers and reducers will be

$$\text{Map}(i, \mathbf{D}_i) \rightarrow list(1, \mathbf{X}_i^T \mathbf{X}_i)$$

$$\text{Reduce}(1, list(\mathbf{X}_i^T \mathbf{X}_i)) = \sum_{i=1}^{m} \mathbf{X}_i^T \mathbf{X}_i$$

Similarly, for the second set, the input–output of the mappers will be

$$\text{Map}(i, \mathbf{D}_i) \rightarrow list(1, \mathbf{X}_i^T \mathbf{y}_i)$$

$$\text{Reduce}(1, list(\mathbf{X}_i^T \mathbf{y}_i)) = \sum_{i=1}^{m} \mathbf{X}_i^T \mathbf{y}_i$$

There are several other machine learning algorithms that can be adapted to the MapReduce framework. They include the naive Bayes, k-means, SVM, principal components analysis, decision trees, expectation maximization, and more. Chu et al. [27] and Panda et al. [75] provide an introductory analysis on how to formulate these algorithms in the MapReduce framework.

In many astronomy applications, however, it is not the case that the entire data is stored in a single location. For example, the very large sky surveys such as GALEX [42], ROSAT [77], and so on produce enormous geographically distributed catalogs of astronomical objects and analyzing them by downloading them all to a single computer can pose serious scalability issues. This motivates the need to develop communication efficient DDM techniques which are discussed in the next section.

27.3 DISTRIBUTED ALGORITHMS FOR DATA MINING

In this section, we present a brief overview of the existing work on DDM. DDM, as the name suggests, deals with the problem of data analysis in environments with distributed data, computing nodes, and users. In DDM, the data are inherently distributed and it is difficult to centralize the data due to bandwidth constraints. DDM aims to solve this problem by developing data analysis algorithms which can work with the data *in situ*. For an introduction to the area, interested readers are referred to the books by Kargupta et al. [59,60], by Ghosh [43], and several surveys [84,85].

Depending on how the data are distributed across the various sites, a natural way to categorize the DDM algorithms is as follows:

- *Horizontally partitioned*—in which all the features are present at all the sites, but the samples/data tuples are distributed across the sites.

- *Vertically partitioned*—in which the features are distributed across all the sites (may be overlapping or disjoint).

DDM algorithms have been proposed for many popular data analysis tasks. In the next two sections, we present examples of some popular DDM algorithms.

27.3.1 Distributed Clustering

As stated earlier, data clustering deals with an unsupervised way of grouping data items based on some notion of similarity among those items. While the problem of clustering has been well studied when the data are available at one location, developing algorithms when the data are stored at different locations is challenging. In the remainder of this section, we present several algorithms which have been developed for data clustering from distributed databases.

Kargupta et al. [61] have developed a principal component analysis (PCA)-based clustering technique on the collective data mining (CDM) framework in which the dataset is such that each site contains only a subset of variables or features. In DDM parlance, this is commonly referred to as heterogeneously data-distributed scenario. Each local site performs PCA, projects the local data along the principal components, and applies a known clustering algorithm. The communication complexity of such a technique is much smaller than centralizing all the data.

Klusch et al. [65] consider the problem of distributed clustering over homogenously distributed data using kernel-density estimates (KDE). KDE is a nonparametric technique for estimating the probability density function of a random variable. They use a local definition of density-based cluster-points that have the same density (according to some kernel in KDE) are put in the same cluster. They build local clusters of the data and transmit these to a central site. The central site then combines these clusters.

k-Means clustering of data is a popular data mining technique. Distributed versions of k-means clustering algorithms have been proposed by various researchers till date. Eisenhardt et al. [38] propose a distributed method for document clustering using k-means. Their technique is enhanced with a "probe and echo" mechanism for updating cluster centroids. The algorithm is synchronized and each round corresponds to a k-means iteration. There is one designated initiator node. It launches a probe message to all its neighbors and sends its local centroids and weights. Whenever a node P_j receives a probe message from its neighbor P_i, it first updates its current centroids with the ones received from P_i and then forwards the probe to all its neighbors (except for P_i). When a site has received a probe from everyone, it forwards the message to the one from which it first received the probe. This process continues until a termination criterion stops the iterations. Based on this technique, several extensions have been proposed. Bandyopadhyay et al. [4] have proposed an algorithm for k-means clustering based on sampling. In their approach, only the centroids of the local sites need to be communicated. The paper presents a theoretical proof of quality and convergence (in some restricted cases, e.g., when the data is sampled uniformly from the network). Datta et al. [33] later enhanced this algorithm to work in an asynchronous environment. Unlike the algorithm proposed in Ref. [4], the algorithm proposed by Datta et al. does not require the sites to synchronize their centroids after every round of computation. While a formal proof of correctness cannot be provided for this general case, the authors empirically show that their algorithm is quite accurate.

Johnson and Kargupta [56] have proposed a hierarchical clustering algorithm on heterogeneously distributed data. The idea is to generate local dendograms (structure which denotes the hierarchical clusters) and then combine them at a central site. Lazarevic et al. [68] considers the problem of combining spatial clusterings to produce a global regression-based classifier on a homogenously distributed data. Several other techniques for distributed heterogenous data clustering have been proposed using ensemble-based approach such as Refs. [81] and [40].

27.3.2 Distributed Classification

In this section we discuss another popular data mining technique, viz., classification. As in the case of clustering, data classification becomes a challenging task if the data are not at one location. In this section, we show how classification models can be derived from distributed data by minimizing the communication among the data centers.

Ensemble-based classifier learning is well suited to DDM framework. In ensemble techniques such as bagging [16], boosting [41], and random forests [17], several weak classifiers are induced from different partitions of the data and then they are combined using voting techniques or otherwise to produce the output. These techniques can be adopted for distributed learning by learning a weak classifier from each distributed data site and then combining them at a central location. In the literature of DDM, this is popularly known as the meta-learning framework [21,22]. Several strategies for combining the classifiers have been proposed such as voting, arbitration, and combiner. The meta-learning framework for horizontally partitioned datasets is implemented as part of the JAM system [80] which is a Java-based multiagent framework for distributed databases.

The problem of learning from heterogeneously partitioned data has been addressed by several researchers. Park et al. [76] have developed algorithms for learning from heterogenous datasets using an evolutionary technique based on the gene expression messy genetic algorithm. Their work first builds local classifiers and then identifies a selection of tuples that none of these local classifiers can correctly classify. These tuples are centralized and a new classifier is built on these tuples. The classification of a new tuple is done by taking a majority vote of the local classifiers and the centralized classifier.

Caragea et al. [20] present a decision tree induction algorithm for both distributed homogenous and heterogenous environments. Noting that the crux of any decision tree algorithm is the use of an effective splitting criterion, the authors propose a method by which this criterion can be evaluated in a distributed fashion. More specifically, the paper shows that by only centralizing summary statistics such as counts of instances that satisfy specific constraints on the values of the attributes from each site to one location, there can be huge savings in terms of communication when compared to brute force centralization. Moreover, the distributed decision tree induced is the same compared to a centralized scenario.

A different approach is taken by Giannella et al. [45] for distributed classifier learning. They have used Gini information gain as the impurity measure and have shown that Gini between two attributes can be formulated as a dot product between two binary vectors. To cut down the communication complexity, the authors have evaluated the dot product after projecting the vectors in a random subspace. Instead of sending either the raw data or the large binary vectors, the distributed sites communicate only these projected low-dimensional vectors. The paper shows that using only 20% of the communication cost necessary to centralize the data, they can build trees which are at least 80% accurate compared to the trees produced by centralization.

The CDM framework by Kargupta et al. [58] proposes algorithms for data mining from heterogenous data sites using orthogonal basis functions. The basic idea is to represent the model to be built using an orthonormal basis set such as Fourier coefficients. These coefficients are then evaluated at each local site and they are transferred to a central location. Coefficients involving cross terms between the sites are evaluated at the central site after centralizing a sample of the data. Several data mining algorithms have been proposed using this technique. Bayesian networks from distributed heterogenous data [24], decision trees [62], distributed multivariate regression [50], distributed PCA, and data clustering [61] are some of the algorithms developed using the CDM framework.

27.3.3 Data Mining in P2P Environments

An emerging area of data mining is in the area of P2P networks. In this model, the data are distributed over a large number of partitions with the partitions being connected in an *ad hoc* fashion, much like a peer-to-peer network. The goal is to build data mining models in such environments. One of the primary differences between traditional DDM and P2P is the huge number of sites and the need for asynchronous computation. Datta et al. [32] present an overview of DDM algorithms for P2P networks.

27.3.3.1 Approximate Algorithms for P2P Data Mining

In approximate P2P data mining algorithms, the goal is to approximate the results of a similar data mining algorithm which can be executed on the entire dataset. It is desirable to have bounds on the error induced as a result of the approximation. We discuss several approximate P2P data mining algorithms next.

The first type of approximation algorithms are the sampling-based algorithms in which peers sample data using some variations of graph random walk from their own partition and those of several neighbors' and then build models assuming that these data are representative of that of the entire set of peers. Examples for these algorithms include the naive Bayes algorithm by Kowalczyk et al. [66], and more. The distributed k-means clustering algorithm by Bandyopadhyay et al. [4] and Datta et al. [33] which have been discussed earlier can also be used for approximate P2P data clustering. More recently, Das et al. [30] have developed an algorithm for identifying significant inner product entries in a P2P network in a horizontally partitioned data distribution scenario. The proposed algorithm uses a variant of Metropolis-Hastings random walk [26] to draw random samples from the network and using the results from the ordinal decision theory bounds the quality of the result and the communication complexity.

Another type of distributed algorithms are the gossip-based algorithms first introduced by Kempe et al. [64], in which they showed that each peer, by contacting a small number of nodes chosen at random, can get the result of the computation exponentially fast. Boyd et al. [13] enhanced the protocol for general graphs. The most important quality of gossip-based algorithms is that they provide probabilistic guarantees for the accuracy of the result. However, the first gossip-based algorithms required that the algorithm be executed from scratch if the data change in order to maintain those guarantees. This problem was later addressed by Mehyar et al. [73], and by Jelasity et al. [54]. Mehyar et al. [73] propose a graph Laplacian-based approach to compute the average of a set of points in a P2P network. Each peer has a number x_i and an estimate of the global average z_i. The paper shows that the rate at which $z_i \rightarrow \frac{1}{n} \sum_{i=1}^{n} x_i$ is exponential, thereby achieving fast convergence.

Gossip-based algorithms have been used for sum, average, minimum, maximum, and many other computations in large P2P networks. They have also been used or advanced tasks such as distributed spectral clustering [63].

27.3.3.2 Exact Algorithms for P2P Data Mining

Exact algorithms are an exciting paradigm of computation whereby the result generated by the distributed algorithms is exactly the same as if all the peers were given all the data without the (trivial) need to centralize all of the data. Thus, contrary to approximate techniques, these algorithms produce the correct result everytime they are executed.

Exact algorithms for P2P data mining include the majority voting and association rule mining protocol developed by Wolff and Schuster [83]. This algorithm and the ones discussed in this section guarantee *eventual correctness*—when the computation terminates, each node computes the correct result compared to a centralized setting. In the simplest form, majority rule protocol deals with the following computation: suppose each node has two real numbers x_i and y_i, and the goal is to find out if $\sum_{i=1}^{n} x_i \geq \sum_{i=1}^{n} y_i$, where there are n peers in the network. This protocol is eventually correct, fault tolerant, and robust to data and network changes. They are efficient and require far less resources compared to broadcast. Based on its variants, researchers have further proposed more complicated algorithms such as facility location [67], outlier detection [14], and meta-classification [71].

Based on the above protocol, Bhaduri et al. [9] have proposed a decision tree learning algorithm which can build the same tree on all the peers in an asynchronous fashion. First, the authors show that comparison of two features can be accomplished by concurrently running four majority votes. The next step is to choose top 1-out-of-k attributes and this can be easily accomplished by running the previous comparison per attribute pair. Finally, the tree can be built asynchronously by performing this 1-out-of-k comparison for each level of the tree. More recently, Bhaduri and Kargupta [7] have proposed an exact algorithm for multivariate regression in P2P networks. The idea is to use a two-step approach. A local algorithm is used to track the fitness of the current regression model and the data. If the data have changed such that the model no longer fits the data, a feedback loop is used and the model is rebuilt. Experimental results show the accuracy and low monitoring cost of the proposed technique. This has been refined in a more recent publication [6] in which the authors showed how coefficient of variation (R^2) can be used to find the goodness of fit in a distributed setting. Both soft clustering using EM [8] and its deterministic version [82] have been proposed for large distributed networks. Ping et al. [71] have proposed a meta-classification algorithm in which every peer computes a weak classifier based on its own data. Then, these classifiers are merged into a meta-classifier by computing—per test sample—the majority of the outcomes of the weak classifiers. The computation of weak classifiers requires no communication overhead at all, and the majority is computed using the majority voting protocol.

27.3.4 Message Passing Interface

A popular and widely accepted framework for distributed processing is the MPI [49]. MPI is a language-independent communication protocol used to communicate between processes running on different computers connected via a communication infrastructure. While there are many types of topologies and communication protocols supported by MPI, the following are the most popular ones:

- MPI_Send: allows one specified process to send a message to a second specified process in a point-to-point fashion.

- MPI_Bcast: fetches data from one node and broadcasts it to all processes in the process group.

- MPI_Reduce: collects data from all processes in a group, performs an operation, and stores the results on one node.

An MPI program is loaded onto the main memory of a computer which allows the same program to be executed on multiple data. Indeed, the processes can communicate with each other and synchronize. MPI is useful for iterative algorithms in which there is a clear dependence among the data partitions and interprocess communication is necessary for the algorithm to compute. Although MPI can also be used for implementing parallel algorithms, it is observed that MPI is better suited for distributed implementations of data mining algorithms. However, it should be noted that MPI is by no means the only interface for implementation of DDM algorithms. Other popular choices include Java RMI, JADE [53], JXTA [57], and many more.

27.4 DATA MINING IN GRID ENVIRONMENTS

A grid can be defined as "the ability, using a set of open standards and protocols, to gain access to applications and data, processing power, storage capacity and a vast array of other computing resources over the Internet" [47]. Grid computing has been popularized by Foster et al. in their seminal work [39]. A grid is a type of parallel and distributed system that enables the sharing, selection, and aggregation of resources distributed across multiple administrative domains based on the resources availability, capacity, performance, cost, and users'"quality-of-service requirements." Since many disciplines of science involve scavenging through massive volumes of data, computational grids offer a perfect platform for various data mining tasks. Grid systems differ from P2P systems in the sense that grid systems have a built-in centralized control structure and network infrastructure which assigns jobs and ensures optimal system performance. Since the grid uses resources from various sources, proper resource allocation to different jobs is a challenging job. Hoschek et al. [52] discuss the data management issues for grid data mining. Some interesting and ongoing grid projects involve data mining over the grid, for example, Papyrus [2], the Data Grid [31], and the Knowledge Grid [19].

Grid computing has gained popularity as a distributed computational resource for a plethora of massive computational tasks which is difficult to handle by a single computational resource. The grid data mining for astronomy (GRIST) project [48] is an attempt at using the national science foundation (NSF) Teragrid project in analyzing massive astronomy datasets from the national virtual observatories (NVOs). All the analysis tools are implemented as Web services and the system has the ability of harnessing the power of grid computers to execute the submitted jobs.

27.5 DDM FOR ASTRONOMY

In this section, we discuss some recent advances in analyzing large astronomy datasets using parallel/DDM techniques.

One of the earliest discussions on using DDM for astronomy applications can be found in the work by Borne [10] in which the author argues that with increasing sizes of astronomy catalogs stored at multiple locations, analyzing them in one machine is impossible. He suggests using DDM as an alternative means of computation, thereby allowing scientists to discover knowledge from these vast and geographically distributed data repositories.

McConnell and Skillicorn [72] apply ensemble technique (meta-learning) in order to learn models from distributed astronomy datasets. In their approach, they have used an ensemble of decision trees classifiers. The basic approach is to learn a decision tree at each site and then use simple voting or weighted voting in order to produce the final output. The algorithm has been tested on five astronomy datasets ranging from 1000 to 11,000 objects and 3 to 12 attributes. Two of these datasets have been downloaded from SDSS data release 3. The datasets have between 3 and 14 classes and the goal is to classify these datasets. As demonstrated in the experiments, the ensemble approach produces comparable results to the centralized decision tree with very low communication overhead. This implies that this approach can easily be extended to very large datasets.

The US NVO is a large-scale effort by the NSF to develop an information infrastructure by which users can access distributed astronomy data archives. It contains services for querying, storing, analyzing, and visualizing these astronomy catalogs. New data analysis modules can be added to the infrastructure. The DEMAC architecture described by Giannella et al. [44] and Dutta et al. [37] fits well with the NVO as a new service. Their framework sits as a Web service layer on top of the data layer and the users can access it through the functions of the NVO. This layer runs a DDM algorithm in the background using multiple data sources and the users are able to download the learned model instead of the raw data. There are two algorithms which are discussed in the papers. Both work under the assumption that the data at each site contains a subset of the variables (vertically partitioned) and the tuples have been cross-matched. The first algorithm is a distributed PCA algorithm for identifying the galactic fundamental plane (the imaginary plane formed by the sun at the center and aligned with the center of the Milky Way galaxy) and the second is the outlier detection algorithm based on the estimated covariance. Both these algorithms rely on random projections to project the local data in a low-dimensional subspace which is then shared among the sites. The projected data are then used to reconstruct the covariance matrix. It can be shown that, on average, the covariance matrix is well approximated using random projection [69]. Once the covariance matrix is known at each site, PCA calculation simply boils down to performing eigen analysis of this matrix which does not take any communication at all. Identifying the outliers from this dataset based on PCA requires one to project the dataset in the residual subspace spanned by the few most insignificant eigenvectors. Since the entire dataset is not available at one location, computing the top outliers needs distributed processing. It has been shown in Ref. [37], that this top-k computation can be reduced to a series of max computations for which the authors have proposed a distributed algorithm.

While the above algorithm works well in the static data scenario, a challenging task is to develop distributed algorithms which can deal with high-throughput data streams such as the next-generation LSST pipeline data [70]. The latter would produce repeated images of the sky and the goal is to determine changes in such images in real time. The detection algorithm needs to take into account not only the LSST pipeline data, but also other astronomy datasets distributed geographically. Das et al. [29] present an algorithm which can identify changes in the fundamental plane by tracking the principal components of the covariance matrix. In their approach, they assume that the data are distributed among a large number of sites horizontally and the sites are connected in an *ad hoc* fashion. Each of these sites gets a stream of tuples and the goal is to detect any changes in the eigenspace of the global covariance matrix. The proposed algorithm takes a two step approach. First, given an estimate of the principal eigenvector and eigenvalue (henceforth called the model), the algorithm at each site asynchronously checks if the data fit the model. This checking is done using a data-dependent set of rules which guarantee that if the model fits the global data, no communication is wasted and the algorithm converges. Only when the data do not fit the model at any site, they send messages which result in a globally correct termination condition. In the second phase, once the sites jointly discover that the model is outdated, it is rebuilt using network sampling. The eventual correctness and high scalability of this algorithm make it an ideal candidate for analyzing the data generated by the high-throughput data streams such as LSST.

27.6 CONCLUSION

Parallel and distributed data mining will continue to play an important role in analysis of modern astronomy datasets which are often too large and/or distributed at different geographical locations. In this chapter, we have discussed several strategies for scaling up data mining algorithms for handling large datasets using multiple computers. Parallel algorithms leverage the computation and storage power of many shared memory computers such as computing clusters. DDM algorithms, on the other hand, use a collection of user computers connected in an *ad hoc* fashion through a communication infrastructure and are ideal for citizen science. Both areas of parallel and distributed data mining are relatively new research areas and offer plenty of opportunities for both novel algorithm development and astronomical discoveries.

ACKNOWLEDGMENT

The authors acknowledge funding from the NASA ARMD SSAT project.

REFERENCES

1. N. Amado, J. Gama, and F. Silva. Parallel implementation of decision tree learning algorithms. In *Proceedings of the 10th Portuguese Conference on Artificial Intelligence on Progress in Artificial Intelligence, Knowledge Extraction, Multi-Agent Systems, Logic Programming and Constraint Solving*, pp. 6–13, Porto, Portugal, 2001.

2. S. Bailey, R. Grossman, H. Sivakumar, and A. Turinsky. Papyrus: A system for data mining over local and wide area clusters and super-clusters. In *Proceedings of CDROM'99*, p. 63, New York, USA, 1999.

3. N. M. Ball and R. J. Brunner. Data mining and machine learning in astronomy. *International Journal of Modern Physics D*, 19(7):1049–1106, 2010.

4. S. Bandyopadhyay, C. Giannella, U. Maulik, H. Kargupta, K. Liu, and S. Datta. Clustering distributed data streams in peer-to-peer environments. *Information Science*, 176(14):1952–1985, 2006.

5. Y. Ben-Haim and E. Tom-Tov. A streaming parallel decision tree algorithm. *Journal of Machine Learning Research*, 11:849–872, March 2010.

6. K. Bhaduri, K. Das, and C. Giannella. Distributed monitoring of the R^2 statistic for linear regression. In *SIAM International Conference on Data Mining*, pp. 438–449, Mesa, AZ, 2011.

7. K. Bhaduri and H. Kargupta. A scalable local algorithm for distributed multivariate regression. *Statistical Analysis and Data Mining J.*, 1(3):177–194, 2008.

8. K. Bhaduri and A. N. Srivastava. A local scalable distributed expectation maximization algorithm for large peer-to-peer networks. In *Proceedings of ICDM'09*, pp. 31–40, Miami, Florida, USA, 2009.

9. K. Bhaduri, R. Wolff, C. Giannella, and H. Kargupta. Distributed decision tree induction in P2P systems. *Statistical Analysis and Data Mining*, 1(2):85–103, 2008.

10. K. Borne. Distributed data mining in the national virtual observatory. *SPIE Data Mining and Knowledge Discovery: Theory, Tools, and Technology V*, 5098:211–218, 2003.

11. K. Borne. Scientific data mining in astronomy. In H. Kargupta, J. Han, P. S. Yu, R. Motwani, and V. Kumar, editors, *Next Generation of Data Mining*, Chapter 5, pp. 91–114. CRC Press, Boca Raton, FL, 2009.

12. K. W. Bowyer, L. O. Hall, T. Moore, N. Chawla, and W. P. Kegelmeyer. A parallel decision tree builder for mining very large visualization datasets. In *Proceedings of IEEE International Conference on Systems, Man, and Cybernetics, 2000*, Vol. 3, pp. 1888–1893, 2000.

13. S. Boyd, A. Ghosh, B. Prabhakar, and D. Shah. Gossip algorithms: Design, analysis and applications. In *Proceedings of Infocom'05*, pp. 1653–1664, Miami, FL, March 2005.

14. J. Branch, B. Szymanski, C. Giannella, R. Wolff, and H. Kargupta. In-network outlier detection in wireless sensor networks. In *Proceedings of ICDCS'06*, p. 51, Lisbon, Portugal, July 2006.

15. L. Breiman, J. Friedman, R. Olshen, and C. Stone. *Classification and Regression Trees*. Wadsworth and Brooks, Monterey, CA, 1984.

16. L. Breiman. Bagging predictors. *Machine Learning*, 24(2):123–140, 1996.

17. L. Breiman. Random forests. *Machine Learning*, 45(1):5–32, 2001.

18. R. Brunner, G. Djorgovski, T. Prince, and A. Szalay. Handbook of massive data sets. In J. Abello, P. M. Pardalos, and M. G. C. Resende, editors, *Massive Datasets in Astronomy*, pp. 931–979. Kluwer Academic Publishers, Norwell, MA, USA, 2002.

19. M. Cannataro and D. Talia. The knowledge grid. *Communication of ACM*, 46(1):89–93, 2003.

20. D. Caragea, A. Silvescu, and V. Honavar. A framework for learning from distributed data using sufficient statistics and its application to learning decision trees. *International Journal of Hybrid Intelligent Systems*, 1(1–2):80–89, 2004.

21. P. K. Chan and S. Stolfo. Experiments on multistrategy learning by meta-learning. In *Proceedings of the Second International Conference on Information and Knowledge Management*, Washington, DC, USA, pp. 314–323, 1993.

22. P. K. Chan and S. J. Stolfo. Toward parallel and distributed learning by meta-learning. In *Working Notes of AAAI Workshop on Knowledge Discovery in Databases*, Washington, DC, pp. 227–240, 1993.

23. E. Y. Chang, K. Zhu, H. Wang, H. Bai, J. Li, Z. Qiu, and H. Cui. PSVM: Parallelizing support vector machines on distributed computers. In *Proceedings of Neural Information Processing Systems*, 2007.

24. R. Chen, K. Sivakumar, and H. Kargupta. Collective mining of bayesian networks from distributed heterogeneous data. *Knowledge and Information Systems*, 6(2):164–187, 2004.

25. W. Chen, Y. Song, H. Bai, C. Lin, and E. Chang. Parallel spectral clustering in distributed systems. *IEEE Transactions on Pattern Analysis and Machine Intelligence*, 99(RapidPosts), pp. 568–586, 2010.

26. S. Chib and E. Greenberg. Understanding the Metropolis-Hastings algorithm. *The American Statistician*, 49(4):327–335, 1995.

27. C. Chu, S. Kim, Y. Lin, Y. Yu, G. Bradski, A. Ng, and K. Olukotun. Map-Reduce for machine learning on multicore. In *Proceedings of NIPS'06*, pp. 281–288, Vancouver, BC, Canada, 2006.

28. A. Ng, M. Jordan, and Y. Weiss. On spectral clustering: Analysis and an algorithm. In *Proceedings of Advances in Neural Information Processing Systems*, 849–856, Vancouver, BC, Canada, 2001.

29. T. Mahule, K. D. Borne, S. Dey, S. Arora, and H. Kargupta. PADMINI: A peer-to-peer distributed astronomy data mining system and a case study. *CIDU*, 243–257, 2010.

30. K. Das, K. Bhaduri, K. Liu, and H. Kargupta. Distributed identification of top-*l* inner product elements and its application in a peer-to-peer network. *IEEE Transactions on Knowledge and Data Engineering (TKDE)*, 20(4):475–488, 2008.

31. The DataGrid Project. http://eu-datagrid.web.cern.ch/eu-datagrid/default.htm.

32. S. Datta, K. Bhaduri, C. Giannella, R. Wolff, and H. Kargupta. Distributed data mining in peer-to-peer networks. *IEEE Internet Computing Special Issue on Distributed Data Mining*, 10(4):18–26, 2006.

33. S. Datta, C. Giannella, and H. Kargupta. K-Means clustering over large, dynamic networks. In *Proceedings of SDM'06*, pp. 153–164, Maryland, 2006.

34. W. Day and H. Edelsbrunner. Efficient algorithms for agglomerative hierarchical clustering methods. *Journal of Classification*, 1:7–24, 1984.

35. J. Dean and S. Ghemawat. MapReduce: Simplified data processing on large clusters. *Commun. ACM*, 51:107–113, 2008.

36. I. Dhillon and D. Modha. A data-clustering algorithm on distributed memory multiprocessors. In *Revised Papers from Large-Scale Parallel Data Mining, Workshop on Large-Scale Parallel KDD Systems, SIGKDD*, pp. 245–260, Boston, MA, 2000.

37. H. Dutta, C. Giannella, K. D. Borne, and H. Kargupta. Distributed top-K outlier detection from astronomy catalogs using the DEMAC system. In *SDM*, 2007.

38. M. Eisenhardt, W. Müller, and A. Henrich. Classifying documents by distributed P2P clustering. In *GI Jahrestagung*, pp. 286–291, 2003.

39. I. Foster and C. Kesselman. *The Grid: Blueprint for a New Computing Infrastructure*. Morgan Kaufmann, San Francisco, CA, USA, 2004.

40. A. Fred and A. Jain. Data clustering using evidence accumulation. In *Proceedings of ICPR'02*, p. 40276, Washington, DC, USA, 2002.

41. J. Friedman, T. Hastie, and R. Tibshirani. Additive logistic regression: A statistical view of boosting. *The Annals of Statistics*, 38(2):337–374, 2000.

42. The Galaxy Evolution Explorer. http://www.galex.caltech.edu/about/overview.html.

43. S. Ghosh. *Distributed Systems: An Algorithmic Approach*. CRC Press, Boca Raton, FL, 2006.

44. C. Giannella, H. Dutta, K. Borne, R. Wolff, and H. Kargupta. Distributed data mining for astronomy catalogs. In *Proceedings of Scientific Data Mining'06 Workshops*, Bethesda, MD, USA, 2006.

45. C. Giannella, K. Liu, T. Olsen, and H. Kargupta. Communication efficient construction of deicision trees over heterogeneously distributed data. In *Proceedings of ICDM'04*, pp. 67–74, Brighton, UK, 2004.

46. G. Golub and C. Van Loan. *Matrix Computations*, 3rd ed. Johns Hopkins University Press, Baltimore, MD, USA, 1996.

47. What is grid? http://www.eu-degree.eu/DEGREE/General%\20questions/copy_of_what-is-grid.

48. GRIST: Grid data mining for astronomy. http://grist.caltech.edu/index.html.

49. W. Gropp, E. Lusk, and A. Skjellum. *Using MPI: Portable Parallel Programming with the Message Passing Interface.* MIT Press, Cambridge, MA, 1996.

50. D. E. Hershberger and H. Kargupta. Distributed multivariate regression using wavelet-based collective data mining. *Journal of Parallel and Distributed Computing*, 61(3):372–400, 2001.

51. T. Hinke and J. Novotny. Data mining on NASA's information power grid. In *Proceedings of HPDC'00*, p. 292, Pittsburgh, PA, USA, 2000.

52. W. Hoschek, F. J. Jaén-Martínez, A. Samar, H. Stockinger, and K. Stockinger. Data management in an international data grid project. In *Proceedings of the First International Workshop on Grid Computing*, pp. 77–90, London, UK, 2000.

53. JADE: Java Agent DEvelopment Framework. http://jade.tilab.com/.

54. M. Jelasity, A. Montresor, and O. Babaoglu. Gossip-based aggregation in large dynamic networks. *ACM Transactions on Computer Systems (TOCS)*, 23(3):219–252, 2005.

55. R. Jin and G. Agrawal. Communication and memory efficient parallel decision tree construction. In *SDM*, pp. 119–129, San Francisco, CA, 2003.

56. E. L. Johnson and H. Kargupta. Collective, hierarchical clustering from distributed, heterogeneous data. In *Revised Papers from Large-Scale Parallel Data Mining, Workshop on Large-Scale Parallel KDD Systems, SIGKDD*, pp. 221–244, London, UK, 2000.

57. JXTA: Open source peer-to-peer protocol. https://jxta.dev.java.net/.

58. H. Kargupta, B. Park, D. Hershbereger, and E. Johnson. Collecttive data mining: A new perspective towards distributed data mining. In H. Kargupta and P. Chan, editors, *Advances in Distributed and Parallel Knowledge Discovery*. AAAI/MIT Press, pp. 133–184, Cambridge, MA, USA, 1999.

59. H. Kargupta and K. Sivakumar. *Existential Pleasures of Distributed Data Mining. Data Mining: Next Generation Challenges and Future Directions.* AAAI/MIT Press, Cambridge, MA, USA, 2004.

60. H. Kargupta and P. Chan, editors. *Advances in Distributed and Parallel Knowledge Discovery*. MIT Press, Cambridge, MA, USA, 2000.

61. H. Kargupta, W. Huang, K. Sivakumar, and E. L. Johnson. Distributed clustering using collective principal component analysis. *Knowledge and Information Systems*, 3(4):422–448, 2001.

62. H. Kargupta, Byung-Hoon Park, and H. Dutta. Orthogonal decision trees. *IEEE Transactions on Knowledge and Data Engineering*, 18(8):1028–1042, 2006.

63. D. Kempe and F. McSherry. A decentralized algorithm for spectral analysis. In *Proceedings of STOC'04*, pp. 561–568, Chicago, IL, USA, 2004.

64. D. Kempe, A. Dobra, and J. Gehrke. Computing aggregate information using gossip. In *Proceedings of FOCS'03*, Cambridge, 2003.

65. M. Klusch, S. Lodi, and G. L. Moro. Distributed clustering based on sampling local density estimates. In *Proceedings of IJCAI'03*, pp. 485–490, Mexico, August 2003.

66. W. Kowalczyk, M. Jelasity, and A. E. Eiben. Towards data mining in large and fully distributed peer-to-peer overlay networks. In *Proceedings of BNAIC'03*, pp. 203–210, University of Nijmegen, Nijmegen, Netherlands, 2003.

67. D. Krivitski, A. Schuster, and R. Wolff. A local facility location algorithm for large-scale distributed systems. *Journal of Grid Computing*, 5(4):361–378, 2007.
68. A. Lazarevic, D. Pokrajac, and Z. Obradovic. Distributed clustering and local regression for knowledge discovery in multiple spatial databases. In *Proceedings of ESANN'00*, pp. 129–134, Bruges, Belgium, April 2000.
69. K. Liu, H. Kargupta, and J. Ryan. Random projection-based multiplicative data perturbation for privacy preserving distributed data mining. *IEEE Transactions on Knowledge and Data Engineering*, 18:92–106, 2006.
70. LSST: Large Synoptic Survey Telescope. http://www.lsst.org/lsst.
71. P. Luo, H. Xiong, K. Lü, and Z. Shi. Distributed classification in peer-to-peer networks. In *Proceedings of SIGKDD'07*, pp. 968–976, San Jose, CA, USA, 2007.
72. S. M. McConnell and D. Skillicorn. Distributed data mining for astrophysical datasets. In *Proceedings of Astronomical Data Analysis Software and Systems XIV*, p. 360, 2005.
73. M. Mehyar, D. Spanos, J. Pongsajapan, S. H. Low, and R. Murray. Distributed averaging on peer-to-peer networks. In *Proceedings of CDC'05*, Spain, pp. 7446–7451, 2005.
74. C. F. Olson. Parallel algorithms for hierarchical clustering. *Parallel Computing*, 21:1313–1325, 1995.
75. B. Panda, J. Herbach, S. Basu, and R. Bayardo. Planet: Massively parallel learning of tree ensembles with mapreduce. *Proceedings of the VLDB Endowment*, 2:1426–1437, 2009.
76. B. Park, H. Kargupta, E. Johnson, E. Sanseverino, D. Hershberger, and L. Silvestre. Distributed, collaborative data analysis from heterogeneous sites using a scalable evolutionary technique. *Applied Intelligence*, 16(1):19–42, 2002.
77. The Röntgen Satellite. http://heasarc.gsfc.nasa.gov/docs/rosat/rosgof.html.
78. J. Shafer, R. Agrawal, and M. Mehta. SPRINT: A scalable parallel classifier for data mining. In *Proceedings of the 22th International Conference on Very Large Data Bases*, pp. 544–555, Bombay, India, 1996.
79. J. Shi and J. Malik. Normalized cuts and image segmentation. In *Proceedings of CVPR'97*, p. 731, San Juan, Puerto Rico, 1997.
80. S. J. Stolfo, A. L. Prodromidis, S. Tselepis, W. Lee, D. W. Fan, and P. K. Chan. JAM: Java agents for meta-learning over distributed databases. In *Proceedings of KDD'97*, pp. 74–81, Newport Beach, CA, 1997.
81. A. Strehl and J. Ghosh. Cluster ensembles—A knowledge reuse framework for combining multiple partitions. *Journal of Machine Learning Research*, 3:583–617, 2003.
82. R. Wolff, K. Bhaduri, and H. Kargupta. A generic local algorithm for mining data streams in large distributed systems. *TKDE*, 21(4):465–478, 2009.
83. R. Wolff and A. Schuster. Association rule mining in peer-to-peer systems. *IEEE Transactions on Systems, Man and Cybernetics—Part B*, 34(6):2426–2438, 2004.
84. M. J. Zaki. Parallel and distributed association mining: a survey. *IEEE Concurrency*, 7(4):14–25, 1999.
85. M. J. Zaki. Parallel and distributed data mining: An introduction. In *Revised Papers from Large-Scale Parallel Data Mining, Workshop on Large-Scale Parallel KDD Systems, SIGKDD*, pp. 1–23, London, UK, 2000.

CHAPTER 28

Pattern Recognition in Time Series

Jessica Lin, Sheri Williamson, and Kirk D. Borne
George Mason University

David DeBarr
Microsoft Corporation

CONTENTS

28.1 Introduction	617
28.2 Pattern Recognition Approaches	620
28.2.1 Pattern Recognition Algorithms	621
28.2.2 Semi-Supervised Learning	622
28.2.3 Unsupervised Learning	623
28.2.4 Time-Series Representations and Symbolic Aggregate Approximation	625
28.2.5 Similarity Measure	628
28.2.5.1 Shape-Based Similarity	629
28.2.5.2 Experimental Results	632
28.2.5.3 Structural Similarity	633
28.2.5.4 Bag-of-Patterns (BOP) Representation	635
28.3 Astronomical Applications: Current and Future	637
28.4 Summary	639
References	639

28.1 INTRODUCTION

Perhaps the most commonly encountered data types are time series, touching almost every aspect of human life, including astronomy. One obvious problem of handling time-series databases concerns with its typically massive size—gigabytes or even terabytes are common, with more and more databases reaching the petabyte scale. For example, in telecommunication, large companies like AT&T produce several hundred millions long-distance records per day [Cort00]. In astronomy, time-domain surveys are relatively new—these are surveys that cover a significant fraction of the sky with many repeat observations, thereby producing time series for millions or billions of objects. Several such time-domain sky surveys are now completed, such as the MACHO [Alco01], OGLE [Szym05], SDSS Stripe 82 [Bram08], SuperMACHO [Garg08], and Berkeley's Transients Classification Pipeline (TCP) [Star08] projects. The Pan-STARRS project is an active sky survey—it began in 2010, a 3-year survey covering three-fourths of the sky with ~60 observations of each field [Kais04]. The Large Synoptic Survey Telescope (LSST) project proposes to survey 50% of the visible sky repeatedly approximately 1000 times over a 10-year period, creating a 100-petabyte image archive and a 20-petabyte science database (http://www.lsst.org/). The LSST science database will include time series of over 100 scientific parameters for each of approximately 50 billion astronomical sources—this will be the largest data collection (and certainly the largest time-series database) ever assembled in astronomy, and it rivals any other discipline's massive data collections for sheer size and complexity.

More common in astronomy are time series of flux measurements. As a consequence of many decades of observations (and in some cases, hundreds of years), a large variety of flux variations have been detected in astronomical objects, including periodic variations (e.g., pulsating stars, rotators, pulsars, eclipsing binaries, planetary transits), quasi-periodic variations (e.g., star spots, neutron star oscillations, active galactic nuclei), outburst events (e.g., accretion binaries, cataclysmic variable stars, symbiotic stars), transient events (e.g., gamma-ray bursts (GRB), flare stars, novae, supernovae (SNe)), stochastic variations (e.g., quasars, cosmic rays, luminous blue variables (LBVs)), and random events with precisely predictable patterns (e.g., microlensing events). Several such astrophysical phenomena are wavelength-specific cases, or were discovered as a result of wavelength-specific flux variations, such as soft gamma ray repeaters, x-ray binaries, radio pulsars, and gravitational waves. Despite the wealth of discoveries in this space of time variability, there is still a vast unexplored region, especially at low flux levels and short time scales (see also the chapter by Bloom and Richards in this book). Figure 28.1 illustrates the gap in astronomical knowledge in this time-domain space. The LSST project aims to explore phenomena in the time gap.

In addition to flux-based time series, astronomical data also include motion-based time series. These include the trajectories of planets, comets, and asteroids in the Solar System, the motions of stars around the massive black hole at the center of the Milky Way galaxy, and the motion of gas filaments in the interstellar medium (e.g., expanding supernova blast wave shells). In most cases, the motions measured in the time series correspond to the actual changing positions of the objects being studied. In other cases, the detected motions indirectly reflect other changes in the astronomical phenomenon, such as light echoes reflecting across vast gas and dust clouds, or propagating waves.

For concreteness, we define time series below:

Definition 28.1 *Time Series*: A time series $T = t_1, \ldots, t_m$ is an ordered set of m real-valued variables.

Figure 28.2 shows examples of time-series data on several types of variable stars. (Reproduced from Rebbapragada et al. [Rebb09].)

Most classic data mining algorithms do not perform or scale well on time-series data. The intrinsic structural characteristics of time-series data such as the high dimensionality and feature correlation, combined with the measurement-induced noises that beset real-world

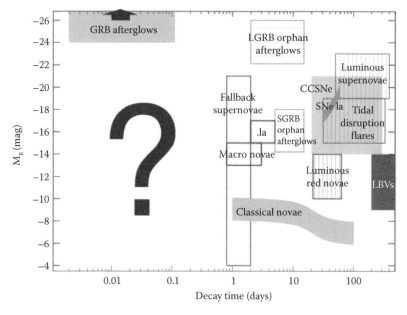

FIGURE 28.1 (**See color insert.**) Discovery space of astronomical variability. Numerous classes of flux variability have been identified through analysis of astronomical time-series observations. But there is still a vast unexplored region of this parameter space that the LSST survey aims to explore. The axes on this plot are time scale of variability (abscissa) and the negative logarithm of the flux (ordinate), measured in magnitudes, where magnitudes decrease as an object gets brighter. (From LSST Science Collaborations and LSST Project, *LSST Science Book*, version 2.0, arXiv:0912.0201, http://www.lsst.org/lsst/scibook, 2009 and adapted from Rau, A. et al. *Exploring the Optical Transient Sky with the Palomar Transient Factory*. Astronomical Society of the Pacific, 886, 1334–1351, 2009.)

time-series data, pose challenges that render classic data mining algorithms ineffective and inefficient for time series. As a result, time-series data mining has attracted enormous amount of attention in the past two decades. Some of the areas that have seen the majority of research in time-series data mining include indexing (query by content) [Agra93, Agra95, Came10, Chan99, Chen05, Ding08, Falo94, Keog01, Keog02b, Lin07, Yi00], clustering [Kalp01, Keog98, Keog04, Liao05, Lin04], classification [Geur01, Rado10, Wei06], anomaly detection [Dasg99, Keog02c, Keog05, Pres09, Yank08], and motif discovery [Cast10, Chiu03, Lin02, Minn06, Minn07, Muee09, Muee10, Tang08]. All tasks require that the notion of similarity be defined. As a result, similarity search has been an active research area in the last two decades.

In all of the aforementioned astronomical cases, pattern recognition and mining of the time-series data falls into the usual three categories: supervised learning (classification), unsupervised learning (pattern detection, clustering, class discovery, characterization, change detection, and Fourier, wavelet, or principal component decomposition), and semi-supervised learning (semi-supervised classification, outlier/surprise/novelty/anomaly detection). For each of these applications, scientific knowledge discovery is the goal. Approaches to pattern recognition tasks vary by representation for the input data, similarity/distance measurement function, and pattern recognition technique. In this chapter, we discuss the various supervised, semi-supervised, and unsupervised learning techniques, representations,

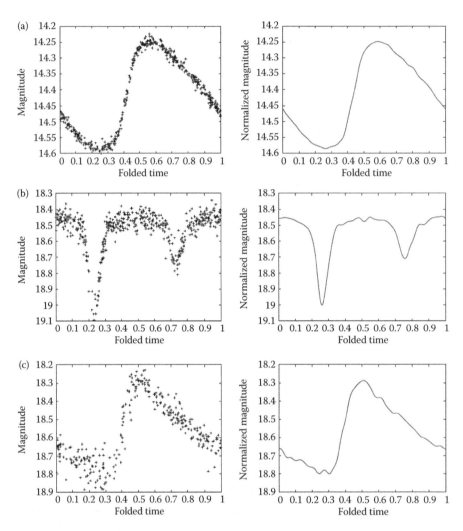

FIGURE 28.2 Examples of time-series data for three different types of variable stars—the left panel in each case is the measured data, and the right panel is the processed data (including smoothing, interpolation, and spike removal). (a) Cepheid. OGLE-LMC_SC4-53463. Period 5.4 days. (b) Eclipsing Binary. OGLE052209.11-694441.9. Period 7.7 days. (c) RR Lyrae. OGLE053520.04-703554.2. Period 0.34 days. (Data and processing results from U. Rebbapragada et al., *Machine Learning* 74(3): 281–313, 2009.)

and similarity measures commonly used for time series. While the majority of work concentrated on univariate time series, we will also discuss the applications of some of the techniques on multivariate time series.

28.2 PATTERN RECOGNITION APPROACHES

Previous work on time-series pattern recognition focuses on one of the three areas: pattern recognition algorithms, efficient time-series representations and dimensionality reduction techniques, and similarity measures for time-series data.

28.2.1 Pattern Recognition Algorithms

Pattern recognition is the process of automatically mapping an input representation for an entity or relationship to an output category. The recognition task is generally categorized based on how the learning procedure determines the output category. The learning procedure can be supervised (when a given pattern is assigned to one of the predefined classes, using labeled data to build a model or guide the pattern classification), unsupervised (when a given pattern is assigned to an unknown class), or semi-supervised (when a given pattern is assigned to one of the predefined classes, using both labeled and unlabeled data). For supervised learning, or classification, a functional model is often used to map observed inputs to output categories. A great deal of model construction techniques have been developed for this purpose [Duda00], including decision trees, rule induction, Bayesian networks, memory-based reasoning, support vector machines (SVMs), and neural networks.

Memory-based reasoning methods, such as nearest neighbor, have been successfully used for classification of time-series data. The Nearest Neighbor (NN) classification algorithm works by computing the distance between the object to be classified and each member of the training set [Han00]. The classification of the object to be classified is predicted to be the same as the classification of the nearest training set member. A common variation of this algorithm predicts the classification of the test object to be the most common classification found among the "k" nearest neighbors in the training set (k-NN). The 1-NN classifier, with leaving-one-out cross validation, has become the standard method used to compare and evaluate the utilities of time-series representations and similarity measures [Ding08].

Another common choice of classifier is decision tree. A decision tree [Han00] is classic machine learning tool that uses a flow-chart-like tree structure, in which an internal node denotes a test on an attribute, a branch denotes the outcome on a test, and a leaf node denotes the class label or class distribution. By placing the attribute that best distinguishes the data (i.e., the one resulting in the highest information gain) at the root, a decision tree induction process recursively partitions the data samples into subsets of samples based on the attribute splitting criterion. The resulting tree can be used to classify future incoming samples, and it can also be easily converted to a set of rules that generalize the behaviors of the data. While decision trees are defined for real data, attempting to classify time series using the raw data would clearly be a mistake, since the high dimensionality and noise levels would result in a deep, bushy tree with poor accuracy [Lin07]. In an attempt to overcome this problem, Geurts [Geur01] suggested representing the time series as a Regression Tree (RT), and training the decision tree directly on this representation. The technique shows great promise.

SVMs [Duda00, Rodr05] are commonly used for constructing classification models for high-dimensional representations. The observations from the training set which best define the decision boundary between classes are chosen as support vectors (they support/define the decision boundary). A nonzero weight is assigned to each support vector. The classification of a test object is predicted by computing a weighted sum of similarity function output between the new object and the support vectors for each class. The predicted class is the class with the largest sum, relative to a bias (offset) for the decision boundary. Kernel functions are used to measure the similarity between objects. Common kernel functions for this purpose include the cosine similarity function and the Gaussian RBF. SVMs have been used for time-series classification by several researchers. Chaovalitwongse and Pardalos used SVMs with dynamic time warping kernel for brain activity classification [Chao08]. Kampouraki et al. [Kamp09] extracted features from ECG time series using statistical methods and signal analysis techniques, and classified the data using SVMs. Eads et al. proposed to use genetic algorithms to extract time-series features, which are then classified using SVMs [Eads02].

Artificial neural networks (ANNs) [Lipp87], also known as neural networks or neural nets, are analogous to their namesake, biological neural networks, in that both receive multiple inputs and respond with a single output. ANNs are comprised of connected nodes, called neurons, which are partitioned into layers. Networks may contain an input layer, an output layer, and a hidden inner layer, and additional hidden layers may be added to increase the complexity of the network. Weights are assigned to each of the links between neurons, and they are updated as part of the learning process. A simple example of an ANN is a perceptron [Rose58], which is a two-class linear classifier that is composed of a single-layered neural network. During training, each example is fed through the network, and the weights are updated according to the difference between the actual and expected output. During testing, the weighted sum of the network (the dot product of the instance and the vector of weights) is compared to a bias parameter to classify each instance as either a positive or negative instance. ANNs generalize well to previously unseen instances because they map every possible input to some output. Additionally, they are robust because even if a node is faulty or is removed, since the work is distributed across the network, it is still possible to achieve good results. ANNs have been successfully applied to domains such as character recognition [Desa10], face detection [Aitk03], intrusion detection [Amin06], speech recognition [Dede10], autonomous driving [Pome89], and astronomy [Coll04, Wang09]. In recent years, research has shown that constructing an ensemble of neural networks, where the majority vote determines the classification, can improve the accuracy of ANNs [Kim07, Kras07, Mink08, Zhou02].

28.2.2 Semi-Supervised Learning

All of the techniques described so far are supervised learning techniques, for which training examples with known classification labels are used to construct a model to label test examples with unknown classification labels. Semi-supervised learning on the other hand involves using both the training examples with known classification labels and the test examples with unknown classification labels to construct a model to label the examples with unknown classification labels. While it is possible to construct a model for labeling future test examples, it is also possible to construct a model that focuses on the current test examples. Vapnik referred to this as transduction [Vapn98], suggesting that it is better to focus on the simpler problem of classifying the current test examples rather than trying to construct a model to future test examples. For time series, Wei and Keogh [Wei06] proposed a semi-supervised classification technique based on 1-NN. The algorithm starts by training the classifier using all labeled data. It then classifies unlabeled data, and adds the most confidently classified unlabeled data into the training set [Wei06]. This process is repeated until some stopping criterion is reached. This algorithm is an example of *self-training* algorithms [Zhu05]. The authors have shown that their algorithm requires only a handful of labeled data to construct an accurate classifier.

In addition to semi-supervised classification, semi-supervised learning is also applicable to anomaly detection and novelty discovery, since we often do not know what is anomalous in advance. In fact, anomaly detection is often regarded as a binary classification problem. A simple idea for detecting anomalous behavior in time series is to examine previously observed normal data and build a model of it. Data obtained in the future can be compared to this model and any lack of conformity can signal an anomaly [Dasg99]. In order to achieve this, Keogh et al. combined a statistically sound scheme with an efficient combinatorial approach [Keog02c]. The statistically scheme is based on Markov chains and normalization. Markov chains are used to model the "normal" behavior, which is inferred from the previously observed data. The time- and space-efficiency of the algorithm comes from the use of a suffix tree as the main data structure. Each node of the suffix tree represents a pattern. The tree is annotated with a score obtained comparing the support of a pattern observed in the new data with the support recorded in the Markov model. Model-free anomaly detection approaches have also been proposed. In a previous work, we defined time-series discords to be subsequences of a longer time series that are maximally different from all the rest of the subsequences [Keog06b]. They thus capture the sense of the most unusual subsequence within a time series. Time-series discords are superlative anomaly detectors, able to detect subtle anomalies in diverse domains. In fact, in a recent extensive empirical study, Chandola et al. concluded that "... *on 19 different publicly available data sets, comparing 9 different techniques, time series discord is the best overall technique among all techniques*" [Chan09].

28.2.3 Unsupervised Learning

Unlike supervised or semi-supervised learning, unsupervised learning (e.g., clustering) involves only unlabeled data. One of the most widely used clustering approaches is hierarchical clustering, due to the great visualization power it offers [Keog98]. Hierarchical clustering produces a nested hierarchy of similar groups of objects, according to a pairwise distance matrix of the objects. One of the advantages of this method is its generality, since the user does not need to provide any parameters such as the number of clusters. However, its application is limited to only small datasets, due to its quadratic (or higher order) computational complexity.

A faster method to perform clustering is k-Means [McQu67, Brad98]. Table 28.1 summarizes the algorithm.

TABLE 28.1   An Outline of the k-Means Algorithm

Algorithm k-Means
1   Decide on a value for k.
2   Initialize the k cluster centers (randomly, if necessary).
3   Decide the class memberships of the N objects by assigning them to the nearest cluster center.
4   Reestimate the k cluster centers, by assuming the memberships found above are correct.
5   If none of the N objects changed membership in the last iteration, exit. Otherwise goto 3.

The basic intuition behind k-Means (and in general, iterative refinement algorithms) is the continuous reassignment of objects into different clusters, so that the within-cluster distance is minimized. Therefore, if x are the objects and c are the cluster centers, k-Means attempts to minimize the following objective function:

$$F = \sum_{m=1}^{k} \sum_{i=1}^{N} \|x_i - c_m\| \tag{28.1}$$

The k-Means algorithm for N objects has a complexity of $O(kNrD)$ [McQu67], where k is the number of clusters specified by the user, r is the number of iterations until convergence, and D is the dimensionality of the points. The shortcomings of the algorithm are its tendency to favor spherical clusters, and its requirement for prior knowledge on the number of clusters, k. The latter limitation can be mitigated by attempting all values of k within a large range. Various statistical tests can then be used to determine which value of k is most parsimonious. Since k-Means is essentially a hill-climbing algorithm, it is guaranteed to converge on a local but not necessarily global optimum. In other words, the choices of the initial centers are critical to the quality of results. Nevertheless, in spite of these undesirable properties, for clustering large datasets of time-series, k-Means is preferable due to its faster running time.

Another well-known partitional clustering algorithm is the expectation–maximization (EM) algorithm. The EM algorithm with Gaussian Mixtures is very similar to k-Means. As with k-Means, the algorithm begins with an initial guess as to the location of the clusters. The steps are: (1) the data are apportioned into separate localized data elements weighted according to the cluster models (the "E" step); and then (2) the parameters of the cluster models (location, width, etc.) are fit to these data elements by maximizing some goodness-of-fit quantity (the "M" step); (3) steps (1) and (2) are repeated until the model converges. In some cases, convergence is slow or otherwise problematic.

The major distinction between EM and k-Means is that k-Means attempts to model the data as a collection of k spherical regions, with every data object belonging to exactly one cluster. In contrast, EM models the data as a collection of k Gaussians, with every data object having some degree of membership in each cluster (in fact, although Gaussian models are most common, other distributions are possible). The major advantage of EM over k-Means is its ability to model a much richer set of cluster shapes. This generality has made EM (and its many variants and extensions) the clustering algorithm of choice in data mining [Demp77] and bioinformatics [Lawr90].

Specialized clustering algorithms have been proposed for time series. For example, Lin et al. [Lin04] proposed an incremental and iterative version of the k-Means algorithm called the i-kMeans. The algorithm works by leveraging off the multiresolution property of wavelets, which mitigates the dilemma of initial centers selection for k-Means [Lin04]. They further extended the approach to a general framework, which works for any iterative refining clustering algorithms (such as EM), with any multiresolution decomposition methods. Rodriguez et al. [Rodr06] proposed a hierarchical clustering algorithm for time-series data streams. The algorithm incrementally constructs a tree-like hierarchy of clusters, using a correlation-based dissimilarity measure between time series [Rodr06].

28.2.4 Time-Series Representations and Symbolic Aggregate Approximation

With the rapid growth of storage technology, datasets from practical applications typically do not fit in main memory, and disk I/O tends to be the bottleneck for any data mining task. The sheer volume of time-series databases makes sequential scanning of the databases infeasible for any data mining and retrieval tasks. A simple generic framework for time-series data mining has thus emerged [Lin07]. It works by (1) approximating the time-series data with some representation that typically fits in the main memory, (2) approximately solving the problem at hand using the representation, and (3) validating the results by making (hopefully) a small number of disk accesses.

It is obvious that the quality of the framework heavily relies on the quality of the representation. If the representation is faithful to the original data, then the approximate solutions obtained from Step 2 will be close to the actual solutions. Towards this end, many high-level time-series representations have been proposed, including discrete Fourier transform (DFT) [Agra93], discrete wavelet transform (DWT) [Chan99], piecewise linear aggregate (PLA) and piecewise constant models (PAA) [Keog01], and adaptive piecewise constant approximation (APCA) [Keog01]. Figure 28.3 illustrates a hierarchy of the various representations in the literature.

These representations allow fast access to the "approximation" version of the data via some index structure, so as to relieve the burden associated with disk I/Os. In addition, since time-series data are typically very high dimensional and noisy, it is essential to reduce its dimensionality via one of these representations. In fact, even though most of the representations use only a few coefficients to represent the data, they often result in better accuracy than the raw data due to the implicit noise removal.

One important feature of most of the representations is that they are real valued. This limits the algorithms, data structures, and definitions available for them. Such limitations have led researchers to consider using a symbolic representation for time series. In addition, there is an enormous wealth of existing algorithms and data structures that allow the efficient manipulations of strings, for example, hashing, Markov models, suffix trees, decision trees, etc. Such algorithms have received decades of attention in the text retrieval community, and more recent attention from the bioinformatics community.

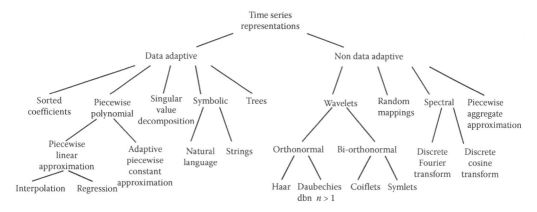

FIGURE 28.3 A hierarchy of all the various time-series representations in the literature.

While many symbolic representations of time series have been introduced over the past decades, they suffer from two fatal flaws: (1) The dimensionality of the symbolic representation is the same as the original data, and virtually all data mining and indexing algorithms degrade exponentially with dimensionality. (2) Although distance measures can be defined on the symbolic approaches, these distance measures have little correlation with distance measures defined on the original signals. In our previous work, we introduced a symbolic representation for time series called Symbolic Aggregate approXimation (SAX) [Lin07] that remedies the problems mentioned above. More specifically, SAX is both time- and space-efficient, and we can define a distance measure that guarantees the distance between two SAX strings to be no larger than the true distance between the original time series. This *lower-bounding* property is at the core of almost all algorithms in time-series data mining [Falo94]. It plays an important role in indexing and similarity search, since it is the essential key to guarantee no false dismissal of results. In general, without a lower-bounding distance measure, one cannot meaningfully compute the approximate solution in the representation space, since the approximate solution obtained may be arbitrarily dissimilar to the true solution obtained from the original data.

SAX performs the discretization by dividing a time series into w equal-sized segments. For each segment, the mean value for the points within that segment is computed. Aggregating these w coefficients forms the Piecewise Aggregate Approximation (PAA) representation of T. Each coefficient is then mapped to a symbol according to a set of breakpoints that divide the distribution space into α equiprobable regions, where α is the alphabet size specified by the user. If the symbols were not equiprobable, some of the symbols would occur more frequently than others. As a consequence, we would inject a probabilistic bias in the process. It has been noted that some data structures such as suffix trees produce optimal results when the symbols are of equiprobability [Croc94]. The discretization steps are summarized in Figure 28.4.

Preliminary results show that SAX is competitive compared to other time-series representations [Lin07]. SAX has had a large impact in both industry and academia. It has received a large number of references, and it has been used worldwide in many domains.[*] To name a few, Chen et al. convert palmprint to time series, then to SAX, and then perform biometric recognition [Chen05b]. Murakami et al. use SAX to find repeated patterns in robot sensors

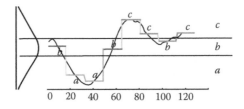

FIGURE 28.4 A time series is discretized by first obtaining a PAA approximation and then using predetermined breakpoints to map the PAA coefficients into SAX symbols. In the example above, with $n = 128$, $w = 8$ and $a = 3$, the time series is mapped to the word *baabccbc*.

---

[*] For a more complete list of papers that extend and use SAX, see http://www.cs.gmu.edu/~jessica/sax.htm

[Mura04]. Pouget et al. use SAX for the detection of multiheaded stealthy attack tools [Poug06]. McGovern et al. use SAX for the prediction of severe weather phenomena such as tornados, thunderstorms, hail, and floods [McGo10].

Many researchers have introduced various time-series data mining algorithms that use SAX as a representation. For example, Bakalov et al. use SAX to do spatiotemporal trajectory joins [Baka05]. Many motif discovery [Cast10, Chiu03, Lin02, Minn06, Minn07, Muee09, Muee10, Tang08] and anomaly detection [Dasg99, Keog02c, Keog05, Pres09, Yank08] algorithms have been proposed that utilize SAX as the representation.

While there have been dozens of applications of SAX to diverse problems, there has been surprisingly little work to augment or extend the SAX representation itself. We attribute this to the generality of the original framework; it simply works very well for most problems. Nevertheless, there have been some SAX extensions, which we consider next. Lkhagva et al. augment each SAX symbol by incorporating the minimum and maximum value in the range [Lkha06]. Thus each SAX segment contains a triplet of information, rather than a single symbol. Very preliminary results are presented which suggest that this may be useful in some domains. Bernard Hugueney [Hugu06] has recently suggested that SAX could be improved by allowing SAX symbols to adaptively represent different length sections of a time series. Just as SAX may be seen as a symbolic version of the PAA representation, Dr. Hugueneys' approach may be seen as a symbolic version of the APCA representation [Keog01]. In contrast to Hugueneys' idea of allowing segment lengths to be adaptive, work by Morchen and Ultsch has suggested that the breakpoints should be adaptive [Morc05]. The authors make a convincing case that in some datasets this may be better than the Gaussian assumption. Shieh and Keogh introduced iSAX [Shie08], the generalization of SAX that allows multiresolution indexing and mining of massive time-series data.

To see how SAX compares to Euclidean distance, we ran an extensive 1-NN classification experiment and compared the error rates on 22 publicly available datasets [Lin07]. Each dataset is split to training and testing parts. We use the training part to search the best value for SAX parameters w (number of SAX words) and a (size of the alphabet):

- For w, we search from 2 up to $n/2$ (n is the length of the time series). Each time we double the value of w.

- For a, we search each value between 3 and 10.

- If there is a tie, we use the smaller values.

The compression ratio (last column of Table 28.2) is calculated as: $w \times \lceil \log_2 a \rceil / n \times 32$, because for SAX representation we only need $\lceil \log_2 a \rceil$ bits per word, while for the original time series we need 4 bytes (32 bits) for each value.

Then we classify the testing set based on the training set using one nearest neighbor classifier and report the error rate. The results are shown in Table 28.2 [Lin07]. From this experiment, we can conclude that SAX is competitive with Euclidean distance, but requires far less space. More experiments that compare different time-series representations can be found in [Ding08].

TABLE 28.2   1-NN Classification Results between SAX and Raw Data Using Euclidean Distance

Name	Number of Classes	Size of Training Set	Size of Testing Set	Time-Series Length	1-NN EU Error	1-NN SAX Error	w	a	Compression Ratio (%)
Synthetic control	6	300	300	60	0.12	0.02	16	10	3.33
Gun-point	2	50	150	150	0.087	0.18	64	10	5.33
CBF	3	30	900	128	0.148	0.104	32	10	3.13
Face (all)	14	560	1690	131	0.286	0.330	64	10	6.11
OSU leaf	6	200	242	427	0.483	0.467	128	10	3.75
Swedish leaf	15	500	625	128	0.211	0.483	32	10	3.13
Trace	4	100	100	275	0.24	0.46	128	10	5.82
Two patterns	4	1000	4000	128	0.093	0.081	32	10	3.13
Wafer	2	1000	6174	152	0.0045	0.0034	64	10	5.26
Face (four)	4	24	88	350	0.216	0.170	128	10	4.57
Lightning-2	2	60	61	637	0.246	0.213	256	10	5.02
Lightning-7	7	70	73	319	0.425	0.397	128	10	5.02
ECG	2	100	100	96	0.12	0.12	32	10	4.17
Adiac	37	390	391	176	0.389	0.890	64	10	4.55
Yoga	2	300	3000	426	0.170	0.195	128	10	3.76
Fish	7	175	175	463	0.217	0.474	128	10	3.46
Plane	7	105	105	144	0.038	0.038	64	10	5.56
Car	4	60	60	577	0.267	0.333	256	10	5.55
Beef	5	30	30	470	0.467	0.567	128	10	3.40
Coffee	2	28	28	286	0.25	0.464	128	10	5.59
Olive oil	4	30	30	570	0.133	0.833	256	10	5.61

28.2.5 Similarity Measure

Another active area of research in time-series data mining is similarity measure, since all tasks, including classification and clustering, require that the notion of similarity be defined. As a result, many distance/similarity measures have been proposed for time-series data. We start by defining $Dist(A, B)$, the distance between two objects A and B.

Definition 28.2 *Distance*: *Dist* is a function that has two objects A and B as inputs and returns a nonnegative value R, which is said to be the distance from A to B.

Each time series is normalized to have a mean of zero and a standard deviation of one before calling the distance function, since it is well understood that in virtually all settings, it is meaningless to compare time series with different offsets and amplitudes [Keog02]. It should be noted that not all similarity measures are distance metrics. For a function calculating the similarity/dissimilarity between the time series A and B to qualify as a distance metric, it must satisfy the following conditions [Keog04b]:

1. Symmetry $D(A, B) = D(B, A)$
2. Constancy $D(A, A) = 0$
3. Positivity $D(A, B) = 0$ if and only if $A = B$
4. Triangle Inequality $D(A, B) <= D(A, C) + D(B, C)$

If a similarity measure satisfies the requirements to be a distance metric, then it can be used to index the time series in a database, which makes queries (e.g., searching for the nearest neighbor) more efficient.

Similarity measures can be categorized, based on how features are extracted and how similarity is determined, into shape-based similarity and structure-based similarity. The former determines the similarity of two time series by comparing their individual point values, whereas the latter looks at the higher-level structures. We discuss the two kinds of similarity in more details in the following sections.

28.2.5.1 Shape-Based Similarity

Given two time series Q and C, their shape-based similarity can be determined by comparing local point values. By far the most common distance measure for time series is the Euclidean distance. Assuming that Q and C are of the same length n, Equation 28.2 defines their Euclidean distance.

$$D(Q, C) \equiv \sqrt{\sum_{i=1}^{n}(q_i - c_i)^2} \tag{28.2}$$

The simplicity and efficiency of Euclidean distance makes it the most popular distance measure in data mining and it has the advantage of being a distance metric. However, a major drawback for Euclidean distance is that it is very brittle. It requires that both input sequences be of the same length, and it is sensitive to distortions, for example, shifting, along the time axis. Such a problem can generally be handled by elastic distance measures such as dynamic time warping (DTW) [Keog02b] and longest common subsequence (LCSS) [Vlac02].

DTW searches for the best alignment between two time series, attempting to minimize the distance between them. It is typically implemented using dynamic programming, and an example implementation is given in Table 28.3.

Figure 28.5 demonstrates the difference between Euclidean distance and DTW. Note that with Euclidean distance, the dips and peaks in the pair of time series are misaligned and therefore not matched, whereas with DTW, the dips and peaks are aligned with their corresponding points from the other time series.

TABLE 28.3 An Algorithm That Calculates the DTW Similarity between Two Time Series, A[0 ... n] and B[0 ... m], Using Some Distance Measure, D(x, y)

Algorithm DTW(A, B)
1   Initialize a matrix dtw[0 ... n][0 ... m] with all values 0
2   for $i = 1$ to n
3   for $j = 1$ to m
4   dtw[i][j] = D(A[i],B[j]) + min(dtw[$i-1$][j], dtw[i][$j-1$], dtw[$i-1$][$j-1$])
5   endfor
6   endfor
7   return dtw[n][m]

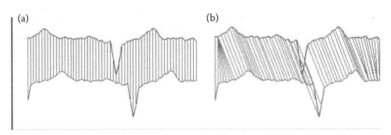

FIGURE 28.5 (a) Alignment for Euclidean distance between two sequence data. (b) Alignment for dynamic time warping distance between two sequence data.

Some global constraint on the warping path is typically specified to restrict the warping paths. The advantages of using a global constraint are two-fold: (1) it produces more intuitive alignment and (2) it speeds up the computation by narrowing the search space. A large warping window causes the search to become prohibitively expensive, as well as possibly allowing meaningless matching between points that are far apart. On the other hand, a small window might prevent us from finding the best solution. It has been shown by Ratanamahatana and Keogh [Rata04] that by learning the best size and shape of the global constraint for different datasets, higher accuracy can be achieved.

While DTW is a more robust distance measure than Euclidean distance, it is also a lot more computationally intensive than Euclidean distance, even with the presence of a global constraint. Keogh proposed an indexing scheme for DTW that allows faster retrieval [Keog02b]. Nevertheless, DTW is still at least several orders slower than Euclidean distance.

Notice that DTW requires a distance measure for comparing two observations, one from each time series; typically the L1 or L2 norm is used.

LCSS, as its name suggests, finds the length of the longest matching subsequence. Originally created for discrete values, it can be adapted to continuous values in time series by redefining a "match" to be whenever the difference between two observations in the time series is below a given threshold. Also typically implemented using dynamic programming, the algorithm is given in Table 28.4.

TABLE 28.4   An Algorithm That Calculates the LCSS between Two Time Series, $A[0 \ldots n]$ and $B[0 \ldots m]$, Using Some Distance Measure, $D(x, y)$, and a Threshold, δ

Algorithm LCSS(A, B)
1   Initialize a matrix $lcss[0 \ldots n][0 \ldots m]$ with all values 0
2   for $i = 1$ to n
3   for $j = 1$ to m
4   if $(D(A[i],B[j]) < \delta)$
5          $lcss[i][j] = lcss[i - 1][j - 1] + 1$
6   else
7          $lcss[i][j] = \max(lcss[i][j - 1], lcss[i - 1, j])$
8   endif
9   endfor
10   endfor
11   return $1 - lcss[n][m]/\min(n,m)$

Compared to Euclidean distance, DTW and LCSS are more elastic, supporting local time shifts and variations in lengths of pairs of time series, but they are also more expensive to compute. Of the three measures, LCSS is the least sensitive to noise because it includes a threshold to define a "match."

Note that the aforementioned definition of LCSS works well for univariate time series, where there are observations through time of only one variable. However, multivariate time series can encounter a situation where one threshold does not hold well across all variables. Thus, we introduce a variation on LCSS, which we call LCSS Relaxed, in our experimental results. In this variation, to constitute a "match" for multivariate time series, only a percentage of variables must "match" according to the given threshold.

Since DTW and LCSS do not satisfy the triangle inequality, they are not distance metrics. Other elastic distance metrics such as edit distance with real penalty (ERP) [Chen04] and time-warp edit distance (TWED) [Mart09] have been proposed. Both ERP and TWED have their roots in the classic string edit distance, which calculates the minimum number of insertions, deletions, and substitutions required to transform one string into another. ERP calls an inserted or deleted element to be a "gap" in the opposing time series and it defines a constant, g, to be used in calculating the penalty for gaps.

Given two time series, $A[0 \ldots n]$ and $B[0 \ldots m]$, and a constant g, ERP is determined by using Equation 28.3 as the recursive formula for dynamic programming.

$$ERP[i][j] = \begin{cases} \sum_{k=0}^{n} |A[k] - g| & : j = 0 \\ \sum_{k=0}^{m} |B[k] - g| & : i = 0 \\ min(ERP[i-1][j-1] + |A[i] - B[j]|, & \\ ERP[i-1][j] + |A[i] - g|, & \\ ERP[i][j-1] + |B[j] - g|) & : otherwise \end{cases} \tag{28.3}$$

A feature of particular interest for TWED is that it can account for time stamp differences between each observation. That is, each observation in the time series is taken at a different moment in time, and while other measures assume observations were taken at a consistent sampling rate, TWED can account for inconsistent observations. This makes it possible to compare time series with irregular or inconsistent sampling rates or missing values. In addition to using a similar gap penalty as ERP, TWED also defines a constant, λ, which controls the stiffness of elastic matching.

Using dynamic programming, TWED is determined by Equation 28.4, where $Dist$ is a distance measure (e.g., $Dist(a, b) = |a - b| + \lambda|t_a - t_b|$, where t_i is the timestamp for observing instance i in time series I)

$$TWED[i][j] = \begin{cases} 0 & : i = j = 0 \\ min(TWED[i-1][j-1] + D(A[i], B[j]), & \\ TWED[i-1][j] + D(A[i], A[i-1]) + g, & \\ TWED[i][j-1] + D(B[j], B[j-1]) + g) & : otherwise \end{cases} \tag{28.4}$$

28.2.5.2 Experimental Results

Several studies that compare time series similarity measures have been performed [Ding08, Keog02, Keog06a]. Most of them focus on univariate time series. In this chapter, we compare the various similarity measures discussed earlier on multivariate time-series classification. Five time-series datasets were chosen, and Table 28.5 shows the characteristics of the datasets.

For each dataset, classification was performed using the k-NN algorithm with $k = 3$, and the error rate (the percentage of incorrectly recorded instances) was recorded and is shown in Table 28.6. For each dataset, the numbers in bold face denote the best (smallest) error rate.

TABLE 28.5   Important Features of Datasets

	AUSLAN[a]	Japanese Vowels[b]	Wafer[c]	ECG[d]	Star Light Curve[e]
Number of variables	22	12	6	2	1
Length of time series	57 (avg)	7–29	104–198	39–152	1024
Number of classes	20 (subset of original)	9	2	2	3
Size of training set	540	270	1194	200	1024
Size of test set[f]	(use 9 fold cross-validation)	370	(use 10-fold cross-validation)	(use 10-fold cross-validation)	(use 10-fold cross-validation)

[a]   M. Kadous. Australian Sign Language Signs (High Quality) Data Set. The UCI KDD Archive, 26 February 2002. http://archive.ics.uci.edu/ml/databases/auslan2/

[b]   M. Kudo, J. Toyama, M. Shimbo. Japanese Vowels. The UCI KDD Archive, 13 June 2000. http://archive.ics.uci.edu/ml/databases/JapaneseVowels/

[c]   R. Olszewski. Wafer Database. Carnegie Mellon University, 8 October 2001. http://www.cs.cmu.edu/~bobski/

[d]   R. Olszewski. ECG Database. Carnegie Mellon University, 8 October 2001. http://www.cs.cmu.edu/~bobski/

[e]   Dragomir Yankov, Eamonn Keogh, and Umaa Rebbapragada. 2008. Disk aware discord discovery: finding unusual time series in terabyte sized datasets. *Knowledge and Information Systems* 17, 2 (November 2008), 241–262.

[f]   Note that some data sets already had the training and test sets divided. The others used cross validation (divide the data into groups and leave one group out for testing and train on the other groups).

TABLE 28.6   Error Rates for Each Similarity Measure by Dataset

	AUSLAN	Japanese Vowels	Wafer	ECG	Star Light Curve
Euclidean	0.1722	0.3838	0.08333	0.1778	0.1767
DTW (full)	0.1556	0.1063	0.0909	0.1889	0.1647
DTW (window)	0.1444	0.1063	0.0656	0.1722	0.1566
EDR	0.4167	0.1417	0.3131	0.2000	0.1606
ERP	0.2778	0.2970	0.0556	0.1944	0.3936
LCSS	0.4185	0.1226	0.1363	0.1278	0.1687
LCSS relaxed	0.2259	0.1172	0.1091	0.1278	0.1687
TWED	0.1741	0.1063	0.0318	0.1278	0.1606

Variations on the DTW and LCSS algorithms were compared. Specifically, DTW was performed both with and without a specified warping window size. Consistent with the observation made by Ratanamahatana and Keogh [Rata04], the results indicate that imposing a window constraint on the search for the best alignment not only increases the algorithm's efficiency but has the same or better performance. Additionally, the results for the variation on LCSS (LCSS Relaxed), described in the previous section, demonstrate that relaxing the definition of a "match" has beneficial results.

As it turns out, Euclidean distance is indeed just as good as other more complex measures, and it is the fastest, giving it an added advantage. In fact, it had similar performance to DTW (with a window constraint) for four out of five datasets. Thus, it is recommended as a valid and computationally inexpensive option for measuring similarity. Our results are consistent with those reported for univariate time series by other researchers [Keog06a, Ding08].

The error rates for DTW with a warping window were consistently less than or equal to the full DTW (no window). This shows that allowing any observation to match any other observation leads to misclassification. The intuition behind this is that while there may be time shifts between two similar time series, this time shift is local. That is, if the two time series are truly similar, then an observation at the beginning of the first should not be matching up with an observation at the end of the second. Otherwise, the two time series are not truly similar, but DTW (without a window) will still give these series an advantage by minimizing the penalty. Therefore, DTW should be used with a window constraint. Again, the results we obtained on multivariate time series are consistent with those reported by Ratanamahatana and Keogh [Rata04], for univariate time series.

For the similarity measures of the edit distance variety (i.e., EDR, ERP, and TWED), TWED appears to be the best measure. In fact, for multivariate time series, it appears that TWED is a better option than the nonconstraint DTW in most cases. This is most likely due to the fine-tuned penalty system TWED provides. However, constrained DTW performs just as well as TWED (best case for three out of five datasets).

28.2.5.3 Structural Similarity

Shape-based similarities work well for short signals or subsequences; however, for long time-series data (generally speaking, with a length of hundreds or more), they produce poor results. To illustrate this, we extracted subsequences of length 2048 from six different records on PhysioNet [Gold97], an online medical archive containing digital recordings of physiological signals. Signals #1–3 are measurements on respiratory rates, and signals #4–6 are ECG signals. Visually, these two vital sign readouts are readily distinguishable (Figure 28.6), however, if we try to cluster them using Euclidean distance as the distance measure, the result is disappointing. Figure 28.6 shows the hierarchical clustering results using Euclidean distance. The dendrogram illustrates that the only datasets that are correctly clustered are datasets #5 and #6 (i.e., they share a common parent node).

This is a surprising result, as these two sets of signals are visually very different. One reason for the poor clustering result could be due to the imperfect alignment of data points between signals from the same set. In addition, the presence of anomalous points, as shown in the beginning of dataset #4, could also throw off the distances computed. Dynamic time

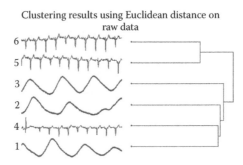

Clustering results using Euclidean distance on raw data

FIGURE 28.6 Result of hierarchical clustering using Euclidean distance on raw data. In fact, using DTW produces the same result.

warping may be used to mitigate the out-of-phase problem. Indeed, we were able to obtain perfect clustering result using DTW on this dataset. However, we would like to emphasize that DTW is computationally expensive—it is a quadratic algorithm, which makes it less desirable for long and large datasets.

A more appropriate alternative to determine similarity between long time series is to measure their similarity based on higher-level structures. Compared to the large amount of work on shape-based similarity, there is relatively little work on finding structural similarity. This is an unfortunate oversight, as the structural approach is particularly useful in applications where domain experts compare signals based on the arrangement of morphological events present in the signals [Olsz01c], for example, radar signal detection and speech recognition. In addition, while some of the shape-based approaches work well for short time series data, they typically fail to produce satisfactory results when the time series are long.

Several structure- or model-based similarity measures have been proposed that extract global features such as autocorrelation, skewness, and model parameters from data [Nano01, Wang06]. Deng et al. [Deng97] proposed learning ARMA model on the time series, and using the model coefficients as the feature. Ge and Smyth [Ge00] proposed a deformable Markov Model template for temporal pattern matching, in which the data is converted to a piecewise linear model. Nanopoulos et al. [Nano01] proposed extracting statistical features of time series, and classifying the data using multilayer perceptron (MLP) neural network.

Keogh et al. [Keog04] proposed a compression-based distance measure that compares the co-compressibility between datasets. Motivated by Kolmogorov Complexity [Keog04, Li97] and promising results shown in similar work in bioinformatics and computational theory, the authors devised a new dissimilarity measure called CDM (Compression-based Dissimilarity Measure). Given two datasets (strings) x and y, their compression-based dissimilarity measure can be formulated as follows:

$$CDM(x, y) = \frac{C(xy)}{C(x) + C(y)} \tag{28.5}$$

where $C(x)$ and $C(y)$ are the compressed sizes of the string x and y, respectively, and $C(xy)$ is the compressed size of the concatenated string $x + y$. The compression-based distance measure is shown to produce superior results compared to other existing structural similarity approaches [Keog04].

28.2.5.4 Bag-of-Patterns (BOP) Representation

In a previous work, we proposed a histogram-based representation that allows computation of similarity between data based on high-level structure, using a representation similar to the one widely used for text data [Lin09]. In the Vector Space Model [Salt75], a document is represented as a vector: each dimension of the vector corresponds to one word in the vocabulary, and its value is the (relative) frequency of occurrences for the corresponding word in the document. As a result, a p-by-q term-document matrix X is constructed, where p is the number of unique terms in the corpus, q is the number of documents, and each element $X_{i,j}$ is the frequency of the ith word occurring in the jth document. This "bag of words" representation works well for documents. It is able to capture the structure of a document, without knowing the exact locations or ordering of the words within.

We proposed to represent time-series data in a similar fashion, that is, as a combination of patterns. However, there are two challenges for representing real-valued time series as a "bag of patterns": (1) how to define patterns, and (2) how to find these patterns. Since time series are composed of real-valued data points, unlike textual data, there are no clear "delimiters" between patterns. In our work, we used a simple sliding-window scheme and the SAX representation to overcome these two challenges.

The algorithm works as follows. First, we construct the pattern "vocabulary" for our time-series database. The easiest way to achieve this is to extract subsequences of a fixed length n, normalize the subsequences so they have a mean of *zero* and standard deviation of *one*, and then convert them to SAX strings. As a result, we obtain a set of strings, each of which corresponds to a subsequence in the time series. (Recall that SAX has two parameters, α and w.) If we choose $\alpha = 4$ and $w = 8$, the resulting dictionary size is $\alpha^w = 4^8 = 65,536$. Despite its apparent size, the matrix is likely to be sparse, as with textual data. In our experiments, we find that only about 10% of all strings have some subsequence mapped to them [Lin09]. Therefore, we can eliminate words that never occur in any data, or store only the list of occurring SAX strings for each time series.

The construction of the bag-of-patterns matrix is straightforward. The matrix M is a SAX-sequence matrix, analogous to term-document matrices used for information retrieval and text mining. Each row i denotes a SAX string (i.e., a pattern) from the pattern vocabulary; each column j denotes a time series; and each $M_{i,j}$ stores the frequency of string i occurring in time series j. The matrix is a histogram of all pattern counts, which provides a summary for the time-series dataset. Once we build the matrix M, we can then use any applicable distance measures, typically Euclidean distance, or dimensionality reduction techniques to compute the similarity between them.

In Figure 28.6, we showed a simple example where using Euclidean Distance on the raw data fails to find the correct clusters, and noted that although DTW was able to cluster the data correctly, it is computationally expensive. Figure 28.7 shows the clustering results on the same datasets, using the bag-of-patterns approach, a linear time algorithm. Note we are now clustering on the transformed time series, or the "histogram" of the patterns. For clarity,

Clustering using bag-of-patterns approach

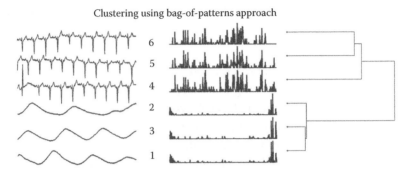

FIGURE 28.7 New clustering result on the same data shown in Figure 28.6. This time, we use our frequency-based, bag-of-patterns approach, and combine it with Euclidean distance. The two clusters are well separated.

the original, corresponding time series are also plotted on the left of the dendrogram. We can see clearly that the time series clustered together have similar pattern distribution, which explains why certain datasets are grouped together.

To see how the algorithm does on long time series, and on datasets with less diverse structures, Figures 28.8 and 28.9 show the hierarchical clustering results on 20 ECG records obtained from PhysioNet [Gold97]. This dataset, containing data from patients with various heart conditions, forms four clusters. Each record is of length 15,000. Figure 28.7 shows the clustering results using Euclidean distance. We also tried using DTW on the raw data, as well as using Euclidean distance on DFT coefficients, with DFT dimensionality ranging from 100 to 1000, and obtained similarly dissatisfactory results.

For comparison, the clustering results using the bag-of-patterns approach are shown in Figure 28.9. As the figure illustrates, all four clusters are cleanly and well separated. In addition, the histograms (of the pattern counts, shown in the middle of the plot) reveal the structure of the data, and provide insight for the clustering results. Note that Keogh et al. reported similar results using the compression-based algorithm, CDM [Keog04].

Clustering result using Euclidean distance on raw data

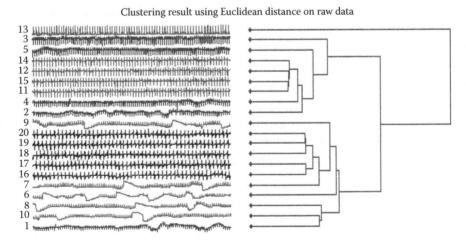

FIGURE 28.8 Clustering result on raw ECG data using Euclidean distance. Only nine datasets are correctly clustered (#11, #12, #14, #15, #16–#20).

Clustering result using the bag-of-patterns approach

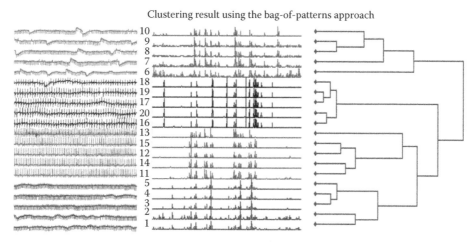

FIGURE 28.9 (**See color insert.**) Clustering result on the same ECG datasets, using our bag-of-patterns approach. All data are clustered correctly. This figure is best viewed in color.

TABLE 28.7   Classification Error Rates on Three Different Methods. Our Method Shows a Drastic Improvement

	Raw Data + Euclidean Distance	Raw Data + DTW	Bag of Patterns Representation + Euclidean Distance
Error rate	0.56	0.272	0.004

To demonstrate the effectiveness on 1-nearest-neighbor classification, we extracted 250 records from the PhysioNet archive [Gold97]. Each record contains 2048 points. These records are extracted from various databases containing different vital signs, or patients with different heart conditions. We separated the records into five classes, and labeled them according to the databases that they are extracted from. We use the leave-one-out cross validation, and count the number of misclassified objects, mc. The error rate is the ratio of mc and the total number of objects (i.e., 250). For this experiment, we also compared with DTW. The improvement is astounding. The error rate of 0.004 means that there is only one misclassified object out of 250 objects (Table 28.7).

The results shown above demonstrate that the structure-based similarity measure has the potential of distinguishing different datasets, even when the differences in structures are subtle. Unlike a distance measure that computes point-to-point distances, it captures the global structure of the data consisting of an unordered set of local features, rather than just local differences. This time-invariant feature is useful if we are interested in the overall structures of the time series. At the same time, some orderings are preserved within subsequences.

28.3 ASTRONOMICAL APPLICATIONS: CURRENT AND FUTURE

As described in Section 28.1, astronomical time-series mining includes supervised, unsupervised, and semi-supervised learning applications (see the chapter by Bloom and Richards in this book). Classification of stars and other phenomena into one of the known types of variability (described in Section 28.1) is a very common and traditional astronomy application [Bloo08]. Usually the first step is to determine the type of variability (e.g., periodic, transient, stochastic) and then to determine the specific astrophysical class of object within that variability type. For example, classification of supernovae of different types is a very important science objective of many large sky surveys, since distant supernovae in far away galaxies assist in the measurement of the Dark Energy content of the Universe and thereby contribute to our understanding of the origin and ultimate fate of the cosmos. Pattern recognition and detection are critical steps in identifying and classifying observed supernovae into known classes, since supernovae of a specific type (Type Ia) are most useful for the cosmological studies.

Unsupervised learning applications are also critical. These include early characterization and description of transients (e.g., early detection of an exploding supernova), discovery of new classes of variability, component decomposition and analysis, and changes in the stationarity of a source. Class discovery is one of the great promises of large astronomical sky surveys—the massive quantity of objects to be studied and of data to be collected should enable the discovery of a grand number of new classes of objects and of astrophysics behaviors. As in most unsupervised learning applications, the same is true here: similarity measures between different time series are crucial ingredients to the learning and discovery phase.

Semi-supervised learning is applicable to anomaly detection and novelty discovery. This is an enormously important aspect of the large sky surveys, especially the petascale surveys of the future (e.g., LSST), since these surveys will generate tens of billions of individual objects' time series. These enormous time-series databases will enable the detection and discovery of extremely rare one-in-a-billion, maybe one-in-a-trillion, types of phenomena. The huge discovery space that remains to be explored (Figure 28.1) bodes well for a vast number of new discoveries in time-domain astronomy.

The LSST will present an unprecedented time-series data mining challenge. The project anticipates the detection of approximately one million astronomical events each and every night for 10 years. These events include objects that have changed in brightness since their previous observation, or changed in position since their previous observation, or have appeared for the very first time in the night sky. The rapid characterization and probabilistic classification of each of these million events is not only critical to the success of the scientific mission of the LSST, but it is a scientific requirement—that is, the project is required to send an event notification to the worldwide astronomical research community within 60 s of detection of an event. Since there are roughly 1000 image pairs obtained nightly, one pair every 40 s, each of which contain tens of millions of sources, therefore the project must characterize and classify (roughly, probabilistically) about 1000 new events every 40 s throughout the night, on every night, for 10 years, while also analyzing the time series of all of the tens of millions of sources in each image (in order to determine if any of them have changed in brightness or position). The resulting characterizations and classifications are critical inputs to the next phase of scientific discovery—astronomical facilities worldwide will conduct follow-up observations on the most scientifically interesting objects in the LSST event stream. Since there are not enough astronomers or astronomical facilities in the world to follow up on each of the 1 million events each night, then intelligent sorting, filtering, and prioritizing are required. The application of informative and scalable time-series pattern detection and recognition algorithms is consequently essential to the success of the LSST project [Born08].

Finally, we note that the human eye and the human mind are very good at detecting patterns and anomalies in data, and time-series data are no exception. So, it is feasible to conceive of a Citizen Science project that addresses astronomical time-series data. Such a project has been conceived and it has been tentatively named "Light Curve Zoo" (after Galaxy Zoo; see the chapter by Fortson et al. in this book). In Light Curve Zoo, citizen science volunteers would be asked to identify known prespecified patterns (characterizations) or unusual features in light curves (i.e., time-series observations of an astronomical object). The database of characterizations and features aggregated from this citizen science project may then be subjected to mining and exploration by the research community. The application of the similarity-based mining and pattern recognition techniques presented in this chapter will then yield a rich harvest of astronomical discovery from the massive LSST event stream and its billions of individual source's time-series data.

28.4 SUMMARY

In this chapter, we discuss different approaches for time-series pattern recognition. Approaches to pattern recognition tasks vary by representation for the input data, similarity/distance measurement function, and pattern recognition technique. We discuss the various pattern recognition techniques, representations, and similarity measures commonly used for time series. While most existing time-series representations proposed in literature are real-valued, in the recent years, the discretization of time series, in particular, using SAX, has become a common practice. Many SAX-based algorithms have been proposed. We discuss two different types of similarity measures: shape-based and structure-based similarity. While the majority of work concentrated on univariate time series, we perform an experimental study on using different similarity measures to classify multivariate time series. We conclude that for classifying short time series, simple techniques, for example, k-NN as classifier, Euclidean distance as distance measure, and SAX as representation, are just as good as the other more complex approaches, and their efficiency and simplicity offer added advantages. For classifying long time series, a structure-based similarity measure such as CDM or BOP should be considered. We discuss the applicability of time-series pattern recognition techniques on astronomy data, as well as the current and future directions for data mining in astronomy.

REFERENCES

Agra93   R. Agrawal, C. Faloutsos, and A. N. Swami. Efficient similarity search in sequence databases. In *Proceedings of the 4th International Conference on Foundations of Data Organization and Algorithms (FODO '93)*, David B. Lomet (Ed.). Springer-Verlag, London, UK, pp. 69–84, 1993.

Agra95   R. Agrawal, G. Psaila, E. L. Wimmers, and M. Zait. Querying shapes of histories. In *Proceedings of the 21st International Conference on Very Large Databases*. Zurich, Switzerland. pp. 502–514, September 11–15, 1995.

Aitk03   M. J. Aitkenhead and A. J. S. McDonald, A neural network face recognition system, *Engineering Applications of Artificial Intelligence*, 16(3), 167–176, 2003.

Alco01   C. Alcock, R. A. Allsman, D. R. Alves, T. S. Axelrod, A. C. Becker, D. P. Bennett, K. H. Cook et al. The MACHO project: Microlensing detection efficiency. *ApJS*, 136, 439, 2001.

Amin06   M. Amini, R. Jalili, and H. R. Shahriari. RT-UNNID: A practical solution to real-time network-based intrusion detection using unsupervised neural networks, *Computers & Security*, 25(6), 459–468, 2006.

Baka05   P. Bakalov, M. Hadjieleftherious, and V. J. Tsotras. Time relaxed spatiotemporal trajectory. In *Proceedings of the ACM International Symposium on Advances in Geographic Information Systems*. Bremen, Germany. November 4–5, 2005.

Bloo08   J. S. Bloom, D. L. Starr, N. R. Butler, P. Nugent, M. Rischard, D. Eads, and D. Poznanski. Towards a real-time transient classification engine. *Astronomische Nachrichten*, 329(3), 284–287, 2008.

Born08   K. D. Borne. A machine learning classification broker for the LSST transient database. *Astronomische Nachrichten*, 329, 255, 2008.

Brad98   P. Bradley, U. Fayyad, and C. Reina. Scaling clustering algorithms to large databases. In *Proceedings of the 4th International Conference on Knowledge Discovery and Data Mining*. New York, NY, pp. 9–15, August 27–31, 1998.

Bram08   D. M. Bramich, S. Vidrih, L. Wyrzykowski, J. A. Munn, H. Lin, N. W. Evans, M. C. Smith et al. Light and motion in SDSS Stripe 82: The catalogues. *Monthly Notices of the Royal Astronomical Society*, 386, 887–902, 2008.

Came10   A. Camerra, T. Palpanas, J. Shieh, and E. Keogh, iSAX 2.0: Indexing and mining one billion time series, ICDM, *IEEE International Conference on Data Mining*, pp. 58–67, 2010.

Cast10   N. Castro and P. J. Azevedo. Multiresolution motif discovery in time series. *SDM*, 665–676, 2010.

Chan99   K. Chan and A. W. Fu. Efficient time series matching by wavelets. In *Proceedings of the 15th IEEE International Conference on Data Engineering*. Sydney, Australia. pp. 126–133, March 23–26, 1999.

Chan09   V. Chandola, D. Cheboli, and V. Kumar. *Detecting Anomalies in a Timeseries Database*. CS Technical Report 09-004, January 2009, Computer Science Department, University of Minnesota.

Chao08   W. A. Chaovalitwongse and P. M. Pardalos. On the time series support vector machine using dynamic time warping kernel for brain activity classification. *Cybernetics and Sys. Anal.*, 44(1), 125–138, 2008.

Chen04   L. Chen and R. Ng. On the marriage of Lp-norms and edit distance. In *Proceedings of the Thirtieth International Conference on Very Large Data Bases—Volume 30 (VLDB'04)*, pp. 792–803, 2004.

Chen05   L. Chen. Similarity search over time series and trajectory data. Ph.D. Dissertation. University of Waterloo, Waterloo, Ont., Canada, Canada, AAINR03008, 2005.

Chen05b   J. S. Chen, Y. S. Moon, and H. W. Yeung. Palmprint authentication using time series. In *Proceedings of the 5th International Conference on Audio- and Video-Based Biometric Person Authentication*. Hilton Rye Town, NY. July 20–22, 2005.

Chiu03   B. Chiu, E. Keogh, and S. Lonardi. Probabilistic discovery of time series motifs. In *Proceedings of the Ninth ACM SIGKDD International Conference on Knowledge Discovery and Data Mining* (Washington, DC, August 24–27, 2003). KDD '03. ACM Press, New York, pp. 493–498.

Coll04   A. A. Collister and O. Lahav. ANNz: Estimating photometric redshifts using artificial neural networks, *Publ. Astron. Soc. Pac.* 116:345–351, 2004.

Cort00   C. Cortes, K. Fisher, D. Pregibon, and A. Rogers. Hancock: A language for extracting signatures from data streams. In *ACM SIGKDD International Conference on Knowledge Discovery and Data Mining*, pp. 9–17, 2000.

Croc94   M. Crochemore, A. Czumaj, L. Gasjeniec, S. Jarominek, T. Lecroq, W. Plandowski, and W. Rytter. Speeding up two string-matching algorithms. *Algorithmica*, 12, 247–267, 1994.

Dasg99   D. Dasgupta and S. Forrest. Novelty detection in time series data using ideas from immunology. In *Proceedings of the 8th International Conference on Intelligent Systems*. Denver, CO, June 24–26, 1999.

Dede10   G. Dede and M. H. Sazli. Speech recognition with artificial neural networks. *Digital Signal Processing*, 20(3), 763–768, 2010.

Demp77   A. Dempster, N. Laird, and D. Rubin. Maximum likelihood from incomplete data via the EM algorithm. *Journal of the Royal Statistical Society, Series B*, 39(1), pp. 1–38, 1977.

Deng97   K. Deng, A. Moore, and M. Nechyba. Learning to recognize time series: Combining ARMA models with memory-based learning. *IEEE International Symposium on Computational Intelligence in Robotics and Automation*, 1, 246–250, 1997.

Desa10   A. A. Desai. Gujarati handwritten numeral optical character reorganization through neural network. *Pattern Recognition*, 43(7), 2582–2589, 2010.

Ding08   H. Ding, G. Trajcevski, P. Scheuermann, X. Wang, and E. Keogh. Querying and mining of time series data: Experimental comparison of representations and distance measures. *Proceedings of the VLDB Endowment*, 1(2), 1542–1552, 2008.

Duda00   R. O. Duda, P. E. Hart, and D. G. Stork. *Pattern Classification*, 2nd Edition, Wiley-Interscience, New York, NY.

Eads02   D. Eads, D. Hill, S. Davis, S. Perkins, J. Ma, R. Porter, and J. Theiler. Genetic algorithms and support vector machines for time series classification. *Proc. SPIE* 4787, pp. 74–85, July, 2002.

Falo94   C. Faloutsos, M. Ranganathan, and Y. Manolopulos. Fast subsequence matching in time-series databases. *SIGMOD Record*, 23, 419–429, 1994.

Garg08   A. Garg. Microlensing candidate selection and detection efficiency for the SuperMA-CHO Dark Matter search. PhD Thesis. Harvard University, 2008.

Ge00   X. Ge and P. Smyth. Deformable Markov model templates for time-series pattern matching. In *Proceedings of the 6th ACM SIGKDD*. Boston, MA, pp. 81–90, August 20–23, 2000.

Geur01   P. Geurts. Pattern extraction for time series classification. In *Proceedings of the 5th European Conference on Principles of Data Mining and Knowledge Discovery*. Freiburg, Germany, pp. 115–127, 2001.

Gold97   A. L. Goldberger, L. Amaral, L. Glass, J. M. Hausdorff, P. Ch. Ivanov, R. G. Mark, J. E. Mietus, G. B. Moody, C. K. Peng, and H. E. Stanley. PhysioBank, PhysioToolkit, and PhysioNet: Circulation, 101(23), e215–e220. *Discovery*, 1(3), 1997.

Han00   J. Han and M. Kamber. *Data Mining: Concepts and Techniques*. Morgan Kaufmann Publishers Inc., San Francisco, CA, USA, 2000.

Hugu06   B. Hugueney. Adaptive segmentation-based symbolic representation of time series for better modeling and lower bounding distance measures. In *Proceedings of the 10th European Conference on Principles and Practice of Knowledge Discovery in Databases*. Berlin, Germany, pp. 545–552, September 18–22, 2006.

Kado02   M. Kadous. Australian Sign Language Signs (High Quality) Data Set. The UCI KDD Archive, 26 February 2002.

Kais04   N. Kaiser. Pan-STARRS: A wide-field optical survey telescope array. *Proceedings of the SPIE*, 5489, 11–22, 2004.

Kamp09   A. Kampouraki, G. Manis, and C. Nikou. Heartbeat time series classification with support vector machines. *IEEE Transactions on Information Technology in Biomedicine*, 13(4), 512–518, 2009.

Kalp01   K. Kalpakis, D. Gada, and V. Puttagunta. Distance measures for effective clustering of ARIMA time-series. In *Proceedings of the 2001 IEEE International Conference on Data Mining*. San Jose, CA, pp. 273–280, November 29–December 2, 2001.

Keog98   E. Keogh and M. Pazzani. An enhanced representation of time series which allows fast and accurate classification, clustering and relevance feedback. In *Proceedings of the 4th International Conference on Knowledge Discovery and Data Mining*. New York, NY, pp. 239–241, August 27–31, 1998.

Keog04   E. Keogh, J. Lin, and W. Truppel. *Clustering of Time Series Subsequences is Meaningless: Implications for Past and Future Research*. Knowledge and Information Systems (KAIS), Springer-Verlag, New York.

Keog01   E. Keogh, K. Chakrabarti, and M. Pazzani. Locally adaptive dimensionality reduction for indexing large time series databases. In *Proceedings of ACM SIGMOD Conference on Management of Data*. Santa Barbara, pp. 151–162, May 21–24, 2001.

Keog02   E. Keogh and S. Kasetty. On the need for time series data mining benchmarks: A survey and empirical demonstration. In *Proceedings of the 8th ACM SIGKDD International Conference on Knowledge Discovery*. Edmonton, Alberta, Canada, pp. 102–111, 2002.

Keog02b   E. Keogh. Exact indexing of dynamic time warping. In *Proceedings of the 28th International Conference on Very Large Data Bases*. Hong Kong, China, August 20–23, 2002.

Keog02c   E. Keogh, S. Lonardi, and B. Chiu. Finding surprising patterns in a time series database in linear time and space. In *Proceedings of the 8th ACM SIGKDD International Conference on Knowledge Discovery and Data Mining*. Edmonton, Alberta, Canada, pp. 550–556, July 23–26, 2002.

Keog04   E. Keogh, S. Lonardi, and C. A. Ratanamahatana. Towards parameter-free data mining. In *Proceedings of the Tenth ACM SIGKDD International Conference on Knowledge Discovery and Data Mining*. Seattle, WA, USA, KDD'04, August 22–25, 2004.

Keog04b   E. Keogh. Tutorial in SIGKDD 2004. *Data Mining and Machine Learning in Time Series Databases*, http://www.cs.ucr.edu/~eamonn/tutorials.html.

Keog05   E. Keogh, J. Lin, and A. W. Fu. HOT SAX: Efficiently finding the most unusual time series subsequence. In *Proceedings of the 5th IEEE International Conference on Data Mining*. Houston, TX, pp. 226–233, November 27–30, 2005.

Keog06a   E. Keogh, X. Xi, L. Wei, and C. A. Ratanamahatana. The UCR Time Series Classification/Clustering Homepage, 2006, www.cs.ucr.edu/~eamonn/time_series_data/

Keog06b   E. Keogh, J. Lin, and A. Fu. Finding the most unusual time series subsequence: Algorithms and applications. *Knowledge and Information Systems (KAIS)*. Springer-Verlag, 2006.

Kim07   K-J. Kim and S-B. Cho. Evolutionary ensemble of diverse artificial neural networks using speciation. *Neurocomputing*, 71(7–9), *Progress in Modeling, Theory, and Application of Computational Intelligence—15th European Symposium on Artificial Neural Networks 2007, 15th European Symposium on Artificial Neural Networks 2007*, pp. 1604–1618, March 2008.

Kras07   V. M. Krasnopolsky. Reducing uncertainties in neural network Jacobians and improving accuracy of neural network emulations with NN ensemble approaches, *Neural Networks* 20(4), *Computational Intelligence in Earth and Environmental Sciences*, 454–461, 2007.

Lawr90   Lawrence, C. and Reilly, A. An expectation maximization (EM) algorithm for the iden-
          tification and characterization of common sites in unaligned biopolymer sequences.
          Proteins, 7, 41–51, 1990.

Li97   M. Li and P. Vitanyi. *An Introduction to Kolmogorov Complexity and Its Applications*,
          2nd Edition, Springer Verlag, New York, 1997.

Liao05   T. W. Liao. Clustering of time series data—A survey. *Pattern Recognition*, 38(11),
          1857–1874, 2005.

Lin02   J. Lin, E. Keogh, P. Patel, and S. Lonardi. *Finding Motifs in Time Series, the 2nd
          Workshop on Temporal Data Mining, the 8th ACM International Conference on
          Knowledge Discovery and Data Mining*. Edmonton, Alberta, Canada, pp. 53–68,
          2002.

Lin04   J. Lin, M. Vlachos, E. Keogh, and D. Gunopulos. Iterative incremental clustering of
          time series. *IX Conference on Extending Database Technology (EDBT)*, March 14–18,
          2004.

Lin07   J. Lin, E. J. Keogh, and L. Wei. Stefano lonardi: Experiencing SAX: A novel symbolic
          representation of time series. *Data Mining and Knowledge Discovery*, 15(2), 107–144,
          2007.

Lin09   J. Lin and Y. Li. Finding structural similarity in time series using bag-of-patterns
          representation. In *Proceedings of the 21st International Congress on Scientific and
          Statistical Database Management*. New Orleans, LA, June 2–4, 2009.

Lipp87   R. Lippmann. An introduction to computing with neural nets. *ASSP Magazine, IEEE*,
          4(2), 4–22, 1987.

Lkha06   B. Lkhagva, Y. Suzuki, and K. Kawagoe. New time series data representation ESAX
          for financial applications. In *Proceedings of the 22nd International Conference on Data
          Engineering Workshops*. Atlanta, GA, p. 115, April 3–8, 2006.

Lsst09   LSST Science Collaborations and LSST Project, LSST Science Book, version 2.0,
          arXiv:0912.0201, http://www.lsst.org/lsst/scibook, 2009.

Mart09   P. Marteau. Time warp edit distance with stiffness adjustment for time series
          matching. *IEEE Transactions on Pattern Analysis and Machine Intelligence*, 31(2),
          306–318, 2009.

McGo10   A. McGovern, D. H. Rosendahl, R. A. Brown, and K. K. Droegemeier. Identifying
          predictive multi-dimensional time series motifs: An application to severe weather
          prediction. *Data Mining and Knowledge Discovery*, 22(1), 232–258, 2011.

McQu67   J. McQueen, Some methods for classification and analysis of multivariate observa-
          tion. In *Proceedings of the 5th Berkeley Symposium on Mathematical Statistics and
          Probability*, L. Le Cam and J. Neyman (Eds.), University of California Press, Berkeley,
          CA. Vol. 1, pp. 281–297.

Mink08   F. L. Minku and Teresa B. Ludermir. Clustering and co-evolution to construct neural
          network ensembles: An experimental study. *Neural Networks*, 21(9), 2008.

Minn06   D. Minnen, T. Starner, I. Essa, and C. Isbell. Activity discovery: sparse motifs from
          multivariate time series. *Snowbird Learning Workshop*, Snowbird, Utah, April 4–7,
          2006.

Minn07   D. Minnen, C. L. Isbell, I. Essa, and T. Starner. Discovering multivariate motifs using
          subsequence density estimation and greedy mixture learning. *Twenty-Second Conf.
          on Artificial Intelligence (AAAI-07)*, Vancouver, B.C., July 22–26, 2007.

Morc05   F. Morchen and A. Ultsch. Optimizing time series discretization for knowledge
          discovery. In *Proceedings of the 11th ACM SIGKDD International Conference on
          Knowledge Discovery and Data Mining*. Chicago, IL, pp. 660–665, August 21–24,
          2005.

Muee09   A. Mueen, E. Keogh, Q. Zhu, S. Cash, and B. Westover. Exact discovery of time series motifs. In *Proceedings of the 2009 SIAM International Conference on Data Mining (SDM09)*, Sparks, NV, April 30–May 2, 2009.

Muee10   A. Mueen and E. Keogh. Online discovery and maintenance of time series motifs. In *Proceedings of the 16th ACM SIGKDD International Conference on Knowledge Discovery and Data Mining* (KDD '10), 2010.

Mura04   K. Murakami, Y. Yano, S. Doki, and S. Okuma. Behavior extraction from a series of obObserved Robot motion. In *Proceedings of JSME Conference on Robotics and Mechatronics*. Nagoya, Japan, June, 2004.

Nano01   A. Nanopoulos, R. Alcock, and Y. Manolopoulos. Feature-based classification of time-series data. In *Information Processing and Technology*, N. Mastorakis and S. D. Nikolopoulos (Eds.), Nova Science Publishers, Commack, NY, 49–61, 2001.

Olsz01c   R. Olszewski. *Generalized Feature Extraction for Structural Pattern Recognition in Time-Series Data*. PhD thesis, Carnegie Mellon University, Pittsburgh, PA, 2001.

Pres09   D. Preston, Pavlos Protopapas, Carla Brodley. Event discovery in time series. *SDM 2009*.

Pome89   D. A. Pomerleau. *ALVINN: An Autonomous Land Vehicle in a Neural Network*. Technical Report CMU-CS-89–107, Carnegie Mellon Univ., 1989.

Poug06   F. Pouget, G. Urvoy-Keller, and M. Dacier. Time signature to detect multi-headed stealthy attack tools. In *Proceedings of the 18th Annual FIRST Conference*. Baltimore, MD, June 25–30, 2006.

Rado10   M. Radovanovic, A. Nanopoulos, and M. Ivanovic. Time-series classification in many intrinsic dimensions. *SDM*, 677–688, 2010.

Rata04   C. A. Ratanamahatana and E. Keogh. Making time-series classification more accurate using learned constraints. In *Proceedings of SIAM International Conference on Data Mining (SDM '04)*, Lake Buena Vista, FL, April 22–24, 2004, pp. 11–22, 2004.

Rau09   A. Rau et al. Exploring the optical transient sky with the palomar transient factory. *Publications of the Astronomical Society of the Pacific*, 886, 1334–1351, 2009.

Rebb09   U. Rebbapragada, P. Protopapas, C. E. Brodley, and C. R. Alcock. Finding anomalous periodic time series. *Machine Learning*, 74(3), 281–313, 2009.

Rodr05   J. J. Rodríguez, C. J. Alonso, and J. A. Maestro. Support vector machines of interval-based features for time series classification. *Knowledge-Based Systems*, 18(4–5), 171–178, 2005.

Rodr06   P. P. Rodrigues, J. Gama, and J. P. Pedroso. ODAC: Hierarchical clustering of time series data streams. *Proc. Sixth SIAM International Conf. Data Mining*, 499–503, 2006.

Rose58   F. Rosenblatt. The perceptron: A probabilistic model for information storage and organization in the brain. *Psychological Review*, 65(6), 386–408, 1958.

Salt75   G. Salton, A. Wong, and C. S. Yang. A vector space model for automatic indexing. *Commun. ACM*, 19(11), 613–620, 1975.

Shie08   J. Shieh and E. Keogh. iSAX: Indexing and mining terabyte sized time series. In *Proceedings of the 14th ACM SIGKDD International Conference on Knowledge Discovery and Data Mining* (KDD '08). ACM, New York, NY, USA, 623–631, 2008.

Star08   D. Starr, J. Bloom, and J. Brewer. Realtime astronomical time-series classification and broadcast pipeline. In *Proceedings of the 7th Python in Science Conference*, 2008.

Szym05   M. K. Szymanski. The optical gravitational lensing experiment. Internet Access to the OGLE Photometry Data Set: OGLE-II BVI maps and I-band data. *Acta Astronomica*, 55, 43–57, 2005.

Tang08	H. Tang and S. S. Liao. Discovering original motifs with different lengths from time series. *Knowlege-Based Systems*, 21(7), 666–671, 2008.
Vapn98	V. Vapnik. Estimating the values of function at given points. *Statistical Learning Theory*, Wiley-Interscience, New York, 1998.
Vlac02	M. Vlachos, D. Gunopoulos, and G. Kollios. Discovering similar multidimensional trajectories. In *Proceedings of the 18th International Conference on Data Engineering*, (ICDE'02), 2002.
Wang06	X. Wang, K. Smith, and R. Hyndman. Characteristic-based clustering for time series data. *Data Mining and Knowledge Discovery*, 13(3), 335–364, 2006.
Wang09	T. Wang, J.-S. Huang, and Q.-S. Gu. Photometric redshifts of galaxies from SDSS and 2MASS. *Research in Astronomy and Astrophysics*, 9(4), 390–400, 2009.
Wei06	L. Wei and E. Keogh. Semi-supervised time series classification. In *Proceedings of the 12th ACM SIGKDD International Conference on Knowledge Discovery and Data Mining (KDD '06)*. ACM, New York, NY, USA, pp. 748–753, 2006.
Yank08	Dragomir Yankov, Eamonn Keogh, and Umaa Rebbapragada. 2008. Disk aware discord discovery: finding unusual time series in terabyte sized datasets. *Knowledge and Information Systems* 17, 241–262, November 2008.
Yi00	B. K. Yi and C. Faloutsos. Fast time sequence indexing for arbitrary Lp norms. In *Proceedings of the 26th Interenational Conference on Very Large Databases*. Cairo, Egypt, pp. 385–394, September 10–14, 2000.
Zhou02	Z-H. Zhou, J. Wu, and W. Tang. Ensembling neural networks: Many could be better than all. *Artificial Intelligence*, 137(1–2), 239–263, 2002.
Zhu05	X. Zhu. Semi-supervised learning literature survey. Technical Report, no. 1530, Computer Sciences, University of Wisconsin-Madison, 2005.

CHAPTER 29

Randomized Algorithms for Matrices and Data*

Michael W. Mahoney

Stanford University

CONTENTS

29.1	Introduction	648
29.2	Matrices in Large-Scale Scientific Data Analysis	649
	29.2.1 A Brief Background on Matrices	649
	29.2.2 Motivating Scientific Applications	650
	29.2.3 Randomization as a Resource	652
29.3	Randomization Applied to Matrix Problems	652
	29.3.1 Random Sampling and Random Projections	652
	29.3.2 Randomization for Large-Scale Matrix Problems	655
	29.3.3 A Retrospective and a Prospective	656
29.4	Randomized Algorithms for LS Approximation	657
	29.4.1 Different Perspectives on LS Approximation	657
	29.4.2 A Simple Algorithm for Approximating LS Approximation	658
	29.4.3 A Basic Structural Result	659
	29.4.4 Making This Algorithm Practical: In Theory	660
	29.4.5 Making This Algorithm Practical: In Practice	661
29.5	Randomized Algorithms for Low-Rank Matrix Approximation	662
	29.5.1 A Random Sampling Algorithm	662
	29.5.2 A Second Random Sampling Algorithm	664
	29.5.3 Several Random Projection Algorithms	666
29.6	Conclusion	668
	Acknowledgments	668
	References	669

---

* This chapter is an abridged version of a monograph of the same title that is in press in the NOW Publishers' "Foundations and Trends in Machine Learning" series.

29.1 INTRODUCTION

This chapter reviews recent work on *randomized matrix algorithms*. By "randomized matrix algorithms," we refer to a class of recently developed random sampling and random projection algorithms for ubiquitous linear algebra problems such as least-squares (LS) regression and low-rank matrix approximation. These developments have been driven by applications in large-scale data analysis—applications which place very different demands on matrices than traditional scientific computing applications. Thus, in this review, we will focus on highlighting the simplicity and generality of several core ideas that underlie the usefulness of these randomized algorithms in scientific applications such as genetics (where these algorithms have already been applied) and astronomy (where, hopefully, in part due to this review they will soon be applied).

The work we will review here had its origins within theoretical computer science (TCS). An important feature in the use of randomized algorithms in TCS more generally is that one must identify and then algorithmically deal with relevant "nonuniformity structure" in the data. For the randomized matrix algorithms to be reviewed here and that have proven useful recently in numerical linear algebra (NLA) and large-scale data analysis applications, the relevant nonuniformity structure is defined by the so-called *statistical leverage scores*. Defined more precisely below, these leverage scores are basically the diagonal elements of the projection matrix onto the dominant part of the spectrum of the input matrix. As such, they have a long history in statistical data analysis, where they have been used for outlier detection in regression diagnostics. More generally, these scores often have a very natural interpretation in terms of the data and processes generating the data. For example, they can be interpreted in terms of the leverage or influence that a given data point has on, say, the best low-rank matrix approximation; and this often has an interpretation in terms of high-degree nodes in data graphs, very small clusters in noisy data, coherence of information, articulation points between clusters, and so on.

Historically, the first generation of randomized matrix algorithms (to be described in Section 29.3) did *not* gain a foothold in NLA and only heuristic variants of them were used in machine learning and data analysis applications. In the second generation of randomized matrix algorithms (to be described in Sections 29.4 and 29.5) that *has* led to high-quality numerical implementations and useful machine learning and data analysis applications, two key developments were crucial.

- *Decoupling the randomization from the linear algebra.* This was originally implicit within the analysis of the second generation of randomized matrix algorithms, and then it was made explicit. By making this explicit, not only were improved quality-of-approximation bounds achieved, but also *much* finer control was achieved in the application of randomization. For example, it permitted easier exploitation of domain expertise, in both numerical analysis and data analysis applications.

- *Importance of statistical leverage scores.* Although these scores have been used historically for outlier detection in statistical regression diagnostics, they have also been crucial in the recent development of randomized matrix algorithms. Roughly, the best random sampling algorithms use these scores to construct an importance sampling distribution to sample with respect to; while the best random projection algorithms rotate to a basis where these scores are approximately uniform and thus in which uniform sampling is appropriate.

The remainder of this chapter will cover these topics in greater detail.

29.2 MATRICES IN LARGE-SCALE SCIENTIFIC DATA ANALYSIS

29.2.1 A Brief Background on Matrices

Matrices arise in machine learning and modern massive data set (MMDS) analysis in many guises. One broad class of matrices to which randomized algorithms have been applied is *object-feature matrices.*

- *Matrices from object-feature data.* An $m \times n$ real-valued matrix A provides a natural structure for encoding information about m objects, each of which is described by n features. In astronomy, for example, very small angular regions of the sky imaged at a range of electromagnetic frequency bands can be represented as a matrix. Similarly, in genetics, DNA single nucleotide polymorphism (SNP) or DNA microarray expression data can be represented in such a framework, with A_{ij} representing the expression level of the ith gene or SNP in the jth experimental condition or individual.

A fundamental property of matrices that is of broad applicability in both data analysis and scientific computing is the singular value decomposition (SVD). If $A \in \mathbb{R}^{m \times n}$, then there exist orthogonal matrices $U = [u^1 u^2 \dots u^m] \in \mathbb{R}^{m \times m}$ and $V = [v^1 v^2 \dots v^n] \in \mathbb{R}^{n \times n}$ such that $U^{\mathrm{T}} A V = \Sigma = \mathbf{diag}(\sigma_1, \dots, \sigma_\xi)$, where $\Sigma \in \mathbb{R}^{m \times n}$, $\xi = \min\{m, n\}$ and $\sigma_1 \geq \sigma_2 \geq \dots \geq \sigma_\xi \geq 0$. The σ_i are the singular values of A, the column vectors u^i, v^i are the ith left and the ith right singular vectors of A, respectively. If $k \leq r = \mathrm{rank}(A)$, then the SVD of $A = U \Sigma V^{\mathrm{T}}$ may be written as

$$A = \begin{bmatrix} U_k & U_{k,\perp} \end{bmatrix} \begin{bmatrix} \Sigma_k & \mathbf{0} \\ \mathbf{0} & \Sigma_{k,\perp} \end{bmatrix} \begin{bmatrix} V_k^{\mathrm{T}} \\ V_{k,\perp}^{\mathrm{T}} \end{bmatrix} = U_k \Sigma_k V_k^{\mathrm{T}} + U_{k,\perp} \Sigma_{k,\perp} V_{k,\perp}^{\mathrm{T}}. \tag{29.1}$$

Here, Σ_k is the $k \times k$ diagonal matrix containing the top k singular values of A, and $\Sigma_{k,\perp}$ is the $(r - k) \times (r - k)$ diagonal matrix containing the bottom $r - k$ nonzero singular values of A, V_k^{T} is the $k \times n$ matrix consisting of the corresponding top k right singular vectors,[*] and so on. By keeping just the top k singular vectors, the matrix $A_k = U_k \Sigma_k V_k^{\mathrm{T}}$ is the best rank-k approximation to A, when measured with respect to the spectral and Frobenius norm. Let $\|A\|_F^2 = \sum_{i=1}^m \sum_{j=1}^n A_{ij}^2$ denote the square of the Frobenius norm; let $\|A\|_2 = \sup_{x \in \mathbb{R}^n, \, x \neq 0} \|Ax\|_2 / \|x\|_2$ denote the spectral norm; and, for any matrix $A \in \mathbb{R}^{m \times n}$,

---

[*] In the text, we will sometimes overload notation and use V_k^{T} to refer to *any* $k \times n$ orthonormal matrix spanning the space spanned by the top-k right singular vectors (and similarly for U_k and the left singular vectors). The reason is that this basis is used only to compute the "importance sampling" probabilities—since those probabilities are proportional to the diagonal elements of the projection matrix onto the span of this basis, the particular basis does not matter.

let $A_{(i)}, i \in [m]$ denote the ith *row* of A as a row vector, and let $A^{(j)}, j \in [n]$ denote the jth *column* of A as a column vector.

29.2.2 Motivating Scientific Applications

To illustrate a few examples where randomized matrix algorithms have already been applied in scientific data analysis, recall that "the human genome" consists of a sequence of roughly 3 billion base pairs on 23 pairs of chromosomes, roughly 1.5% of which codes for ~20,000–25,000 proteins. There are numerous types of individual differences or polymorphic variations, but the most amenable to large-scale applications is the analysis of SNPs, which are known locations in the human genome where two alternate nucleotide bases (or alleles, out of A, C, G, and T) are observed in a nonnegligible fraction of the population. These SNPs occur quite frequently, ca. 1 b.p. per thousand, and thus they are effective genomic markers for the tracking of disease genes (i.e., they can be used to perform classification into sick and not sick) as well as population histories (i.e., they can be used to infer properties about population genetics and human evolutionary history).

In both cases, $m \times n$ matrices A naturally arise, either as a people-by-gene matrix, in which A_{ij} encodes information about the response of the jth gene in the ith individual/condition, or as people-by-SNP matrices, in which A_{ij} encodes information about the value of the jth SNP in the ith individual. Thus, matrix computations have received attention in these genetics applications [1–6]. Depending on the size of the data and the scientific problem under consideration, randomized algorithms can be useful in one or more of several ways.

For example, a common genetics challenge is to determine whether there is any evidence that the samples in the data are from a population that is structured, that is, are the individuals from a homogeneous population or from a population containing genetically distinct subgroups? To address this question, it is common to perform a procedure such as the following. Given an $m \times n$ data matrix A:

- Compute a full or partial SVD or perform a QR decomposition,[*] thereby computing the eigenvectors and eigenvalues of the correlation matrix AA^{T}.

- Appeal to a statistical model selection criterion to determine either the number k of principal components to keep in order to project the data onto or whether to keep an additional principal component as significant.

The computational bottleneck is typically computing the SVD or a QR decomposition. For small- to medium-sized data, this is not a problem. Thus, these methods can be quite fast even for very large data if one of the dimensions is small, for example, 10^2 individuals typed at 10^7 SNPs. On the other hand, if both m and n are large, for example, 10^3 individuals at 10^6 SNPs, or 10^4 individuals at 10^5 SNPs, then, for interesting values of the rank parameter k, the $O(mnk)$ running time of even the QR decomposition can be prohibitive. As we

---

[*] A *QR decomposition* of a matrix A is a decomposition of the matrix into an orthogonal matrix Q and an upper triangular matrix R [7]. It is often used to solve the linear LS problem; and it is the basis for the QR algorithm to calculate the eigenvalues and eigenvectors of a matrix.

will see below, however, by exploiting randomness inside the algorithm, one can obtain an $O(mn \log k)$ running time. Since interesting values for k are often in the hundreds, this improvement from $O(k)$ to $O(\log k)$ can be quite significant in practice, and thus one can apply the above procedure for identifying structure in DNA SNP data on much larger data sets than would have been possible with traditional deterministic methods [8].

More generally, a common *modus operandi* in applying NLA and matrix techniques such as principal component analysis (PCA) and the SVD to DNA microarray, DNA SNPs, and other data problems is:

- Model the people-by-gene or people-by-SNP data as an $m \times n$ matrix A.

- Perform the SVD (or related eigenmethods such as PCA or recently popular manifold-based methods that boil down to the SVD) to compute a small number of eigengenes or eigenSNPs or eigenpeople that capture most of the information in the data matrix.

- Interpret the top eigenvectors as meaningful in terms of underlying biological processes; or apply a heuristic to obtain actual genes or actual SNPs from the corresponding eigenvectors in order to obtain such an interpretation.

In certain cases, such reification may lead to insight and such heuristics may be justified. In general, however, the justification for interpretation comes from domain knowledge and not the mathematics [2,9–11]. The reason is that the eigenvectors themselves, being mathematically defined abstractions, can be calculated for any data matrix and thus are not easily understandable in terms of processes generating the data.

For this and other reasons, a common task in genetics and other areas of data analysis is as follows: given an input data matrix A and a parameter k, find the best subset of exactly k *actual* DNA SNPs or *actual* genes, that is, *actual* columns or rows from A, to use to cluster individuals, reconstruct biochemical pathways, reconstruct signal, perform classification, or inference, and so on. Unfortunately, common formalizations of this algorithmic problem lead to intractable optimization problems [12,13]. For example, consider the so-called column subset selection problem (CSSP) [14]: given as input an arbitrary $m \times n$ matrix A and a rank parameter k, choose the set of exactly k columns of A s.t. the $m \times k$ matrix C minimizes (over all $\binom{n}{k}$ sets of such columns) the error:

$$\min \|A - P_C A\|_\xi = \min \|A - CC^+ A\|_\xi, \tag{29.2}$$

where $\xi \in \{2, F\}$ represents the spectral or Frobenius norm of A, where C^+ represents the Moore–Penrose pseudoinverse of C, and where $P_C = CC^+$ is the projection onto the subspace spanned by the columns of C. As we will see below, however, by exploiting randomness inside the algorithm, one can find a small set of actual columns that is provably nearly optimal. Moreover, this algorithm and obvious heuristics motivated by it have already been applied successfully to problems of interest to geneticists such as genotype reconstruction in unassayed populations, identifying substructure in heterogeneous populations, and inference of individual ancestry [11,15–19].

Broadly similar issues arise in many other MMDS application areas. In astronomy, for example, PCA and the SVD have been used directly for spectral classification [20–23], to predict morphological types using galaxy spectra [24], to select quasar candidates from sky surveys [25–29], and so on. Size is an issue, but so too is understanding the data [30,31]; and many of these studies have found that principal components of galaxy spectra (and their elements) correlate with various physical processes such as star formation (via absorption and emission line strengths of, for example, the Hα line) as well as with galaxy color and morphology. Rank reduction techniques can be used in other ways, for example, they can be used with Gaussian process regression and applied in astronomy to the problem of photometric redshift prediction [32–34].

29.2.3 Randomization as a Resource

These examples illustrate two common reasons for using randomization in the design of matrix algorithms for large-scale data problems:

- *Faster Algorithms.* In some computation-bound applications, one simply wants *faster algorithms* that return more or less the exact answer. In many of these applications, one thinks of the rank parameter k as the numerical rank of the matrix, and thus one wants to choose the error parameter ϵ such that the approximation is precise of the order of machine precision.

- *Interpretable Algorithms.* In other analyst-bound applications, one wants *simpler algorithms or more interpretable output* in order to obtain qualitative insight in order to pass to a downstream analyst. In these cases, k is determined according to some domain-determined model selection criterion, in which case the difference between σ_k and σ_{k+1} may be small or it may be that $\sigma_{k+1} \gg 0$.[*] Thus, it is acceptable (or even desirable since there is noise in the data) if ϵ is chosen such that the approximation is much less precise.

Thus, randomization can be viewed as a computational resource to be exploited in order to lead to "better" algorithms, in a broad sense of the word.

29.3 RANDOMIZATION APPLIED TO MATRIX PROBLEMS

Before describing recently developed randomized algorithms for LS approximation and low-rank matrix approximation that underlie applications such as those described in Section 29.2, in this section we will provide a brief overview of the immediate precursors of that work.

29.3.1 Random Sampling and Random Projections

Given an $m \times n$ matrix A, it is often of interest to sample randomly a small number of actual columns from that matrix. A naïve way to perform this random sampling this would be to select those columns uniformly at random in i.i.d. trials. A more sophisticated and much

---

[*] Recall that σ_i is the ith singular value of the data matrix.

more powerful way to do this would be to construct an *importance sampling distribution* $\{p_i\}_{i=1}^n$, and then perform the random sample according to it.

To illustrate this importance sampling approach in a simple setting, consider the problem of approximating the product of two matrices. Given as input any arbitrary $m \times n$ matrix A and any arbitrary $n \times p$ matrix B:

- Compute the importance sampling probabilities $\{p_i\}_{i=1}^n$, where

$$p_i = \frac{\|A^{(i)}\|_2 \|B_{(i)}\|_2}{\sum_{i'=1}^n \|A^{(i')}\|_2 \|B_{(i')}\|_2}. \tag{29.3}$$

- Randomly select (and rescale appropriately—if the jth column of A is chosen, then scale it by $1/\sqrt{cp_j}$; see Ref. [35] for details) c columns of A and the corresponding rows of B (again rescaling in the same manner), thereby forming $m \times c$ and $c \times n$ matrices C and R, respectively.

This algorithm is described in more detail in Ref. [35], where it is shown that Frobenius norm bounds of the form

$$\|AB - CR\|_F \leq \frac{O(1)}{\sqrt{c}} \|A\|_F \|B\|_F \tag{29.4}$$

hold both in expectation and with high probability. (Issues associated with potential failure probabilities, big-O notation, etc. for this pedagogical example are addressed in Ref. [35]—these issues will be addressed in more detail for the algorithms of the subsequent sections.)

The analysis of the Frobenius norm bound (29.4) is quite simple [35], using very simple linear algebra and only elementary probability, and it can be improved. Most relevant for the randomized matrix algorithms of this review is the bound of [36,37], where much more sophisticated methods were used to show that if $B = A^T$ is an $n \times k$ orthogonal matrix Q (i.e., its k columns consist of k orthonormal vectors in $\mathbb{R}^n$),[*] then, under certain assumptions satisfied by orthogonal matrices, spectral norm bounds of the form

$$\|I - CC^T\|_2 = \|Q^T Q - Q^T SS^T Q\|_2 \leq O(1)\sqrt{\frac{k \log c}{c}}, \tag{29.5}$$

hold both in expectation and with high probability. In this and other cases below, one can represent the random sampling operation with a *random sampling matrix S*—for example, if the random sampling is implemented by choosing c columns, one in each of c i.i.d. trials,

---

[*] In this case, $Q^T Q = I_k$, $\|Q\|_2 = 1$, and $\|Q\|_F^2 = k$. Thus, the right-hand side of (29.4) would be $O(1)\sqrt{k^2/c}$. The tighter spectral norm bounds of the form (29.5) on the approximate product of two orthogonal matrices can be used to show that all the singular values of $Q^T S$ are nonzero and thus that rank is not lost—a crucial step in relative-error and high-precision randomized matrix algorithms.

then the $n \times c$ matrix S has entries $S_{ij} = 1/\sqrt{cp_i}$ if the ith column is picked in the jth independent trial, and $S_{ij} = 0$ otherwise—in which case $C = AS$.

Alternatively, given an $m \times n$ matrix A, one might be interested in performing a random projection by post multiplying A by an $n \times \ell$ *random projection matrix* Ω, thereby selecting ℓ linear combinations of the columns of A. There are several ways to construct such a matrix:

- Johnson and Lindenstrauss consider an orthogonal projection onto a random ℓ-dimensional space [38], where $\ell = O(\log m)$, and Ref. [39] considers a projection onto ℓ random orthogonal vectors. (In both cases, as well as below, the obvious scaling factor of $\sqrt{n/\ell}$ is needed.)

- Indyk and Motwani [40] and Dasgupta and Gupta [41] choose the entries of Ω as independent, spherically symmetric random vectors, the coordinates of which are ℓ i.i.d. Gaussian $N(0, 1)$ random variables. Achlioptas [42] chooses the entries of $n \times \ell$ matrix Ω as $\{-1, +1\}$ random variables and also shows that a constant factor—up to $2/3$—of the entries of Ω can be set to 0.

- Ailon and Chazelle and Matousek [43–45] choose $\Omega = DHP$, where D is a $n \times n$ diagonal matrix, where each D_{ii} is drawn independently from $\{-1, +1\}$ with probability $1/2$; H is an $n \times n$ normalized Hadamard transform matrix, defined below; and P is an $n \times \ell$ random matrix constructed as follows: $P_{ij} = 0$ with probability $1 - q$, where $q = O(\log^2(m)/n)$; and otherwise either P_{ij} is drawn from a Gaussian distribution with an appropriate variance, or P_{ij} is drawn independently from $\{-\sqrt{1/\ell q}, +\sqrt{1/\ell q}\}$, each with probability $q/2$.

As with random sampling matrices, postmultiplication by the $n \times \ell$ random projection matrix Ω amounts to operating on the columns—in this case, choosing linear combinations of columns; and thus premultiplying by Ω^{T} amounts to choosing a small number of linear combinations of rows.

An important technical point is that the last Hadamard-based construction is of particular importance for fast implementations (both in theory and in practice). Recall that the (nonnormalized) $n \times n$ matrix of the Hadamard transform H_n may be defined recursively as

$$H_n = \begin{bmatrix} H_{n/2} & H_{n/2} \\ H_{n/2} & -H_{n/2} \end{bmatrix}, \qquad \text{with} \qquad H_2 = \begin{bmatrix} +1 & +1 \\ +1 & -1 \end{bmatrix},$$

in which case the $n \times n$ normalized matrix of the Hadamard transform, to be denoted by H hereafter, is equal to $\frac{1}{\sqrt{n}}H_n$. Importantly, applying the *randomized Hadamard transform*, that is, computing the product xDH for any vector $x \in \mathbb{R}^n$ takes $O(n \log n)$ time (or even $O(n \log r)$ time if only r elements in the transformed vector need to be accessed). Applying such a *structured random projection* was first proposed in [43,44], it was first applied in the context of randomized matrix algorithms in [46,47], and there has been a great deal of research in recent years on variants of this basic structured random projection that are better in theory or in practice [45,47–52]. For example, one could choose $\Omega = DHS$, where

S is a random sampling matrix, as defined above, that represents the operation of uniformly sampling a small number of columns from the randomized Hadamard transform.

By now, the relationship between sampling algorithms and projection algorithms should be clear. Random sampling algorithms identify a coordinate-based nonuniformity structure, and they use it to construct an importance sampling distribution. For these algorithms, the "bad" case is when that distribution is extremely nonuniform, that is, when most of the probability mass is localized on a small number of columns. In that case, uniform sampling will perform poorly, while using an importance sampling distribution that provides a bias toward these columns will perform much better (at preserving distances, angles, subspaces, and other quantities of interest). On the other hand, random projections and randomized Hadamard transforms destroy or "wash out" or uniformize that coordinate-based nonuniformity structure by rotating to a basis where the importance sampling distribution is very delocalized and thus where uniform sampling is nearly optimal (but by satisfying the Johnson–Lindenstrauss (JL) lemma they too preserve metric properties of interest).

29.3.2 Randomization for Large-Scale Matrix Problems

Consider the following random projection algorithm that was introduced in the context of understanding the success of latent semantic analysis [53]. Given an $m \times n$ matrix A and a rank parameter k:

- Construct an $n \times \ell$, with $\ell \geq \alpha \log m/\epsilon^2$ for some constant α, random projection matrix Ω, as in the previous subsection.

- Return $B = A\Omega$.

This algorithm, which amounts to choosing uniformly a small number ℓ of columns in a randomly rotated basis, was introduced in Ref. [53], where it is proven that

$$\|A - P_{B_{2k}}A\|_F \leq \|A - P_{U_k}A\|_F + \epsilon\|A\|_F, \tag{29.6}$$

holds with high probability. (Here, B_{2k} is the best rank-$2k$ approximation to the matrix B; $P_{B_{2k}}$ is the projection matrix onto this $2k$-dimensional space; and P_{U_k} is the projection matrix onto U_k, the top k left singular vectors of A). The analysis of this algorithm boils down to the JL ideas of the previous subsection applied to the rows of the input matrix A.

Next, consider the following random sampling algorithm that was introduced in the context of clustering large datasets [54]. Given an $m \times n$ matrix A and a rank parameter k:

- Compute the importance sampling probabilities $\{p_i\}_{i=1}^n$, where $p_i = \|A^{(i)}\|_2^2/\|A\|_F^2$.

- Randomly select and rescale $c = O(k \log k/\epsilon^2)$ columns of A according to these probabilities to form the matrix C.

This algorithm was introduced in Ref. [54], although a more complex variant of it appeared in Ref. [55]. The original analysis was extended and simplified in Ref. [56], where it is proven that

$$\|A - P_{C_k}A\|_2 \leq \|A - P_{U_k}A\|_2 + \epsilon\|A\|_F \quad \text{and} \tag{29.7}$$

$$\|A - P_{C_k}A\|_F \leq \|A - P_{U_k}A\|_F + \epsilon\|A\|_F \tag{29.8}$$

hold with high probability. (Here, C_k is the best rank-k approximation to the matrix C, and P_{C_k} is the projection matrix onto this k-dimensional space). This additive-error column-based matrix decomposition, as well as heuristic variants of it, has been applied in a range of data analysis applications [15,57–60]. The analysis of this algorithm boils down to an approximate matrix multiplication result, in a sense that will be constructive to consider in some detail.

As an intermediate step in the proof of the previous results, that was made explicit in Ref. [56], it was shown that

$$\|A - P_{C_k}A\|_2^2 \leq \|A - P_{U_k}A\|_2^2 + 2\|AA^\mathsf{T} - CC^\mathsf{T}\|_2 \quad \text{and}$$

$$\|A - P_{C_k}A\|_F^2 \leq \|A - P_{U_k}A\|_F^2 + 2\sqrt{k}\|AA^\mathsf{T} - CC^\mathsf{T}\|_F.$$

These bounds decouple the linear algebra from the randomization in the following sense: they hold for *any* set of columns, and the effect of the randomization enters only through the "additional error" term. By using $p_i = \|A^{(i)}\|_2^2 / \|A\|_F^2$ as the importance sampling probabilities, this algorithm is effectively saying that the relevant nonuniformity structure in the data is defined by the Euclidean norms of the original matrix. In doing so, this algorithm can take advantage of (29.4) to provide additive-error bounds (so-called additive since the additional error is of the form ϵ times the scale $\|A\|_F$, as opposed to the scale $\|A - P_{U_k}A\|_F$) of the form (29.8). A similar thing was seen in the analysis of the random projection algorithm—since the JL lemma was applied directly to the columns of A, additive-error bounds of the form (29.6) were obtained.

29.3.3 A Retrospective and a Prospective

Much of the early work in TCS focused on randomly sampling columns according to an importance sampling distribution that depended on the Euclidean norm of those columns [35,37,54–56,61]. This had the advantage of being "fast," in the sense that it could be performed in a small number of "passes" over that data from external storage, and also that additive-error quality-of-approximation bounds could be proved. On the other hand, this had the disadvantage of being less immediately applicable to scientific computing and large-scale data analysis applications. At root, this is since the algorithms did not highlight "interesting" or "relevant" nonuniformity structure, which then led to bounds that were rather coarse. For example, columns are easy to normalize and are often normalized during data preprocessing.

Relatedly, bounds of the form (29.4) do not exploit the underlying vector space structure. This is analogous to how the JL lemma was applied in the analysis of the random projection algorithm—by applying the JL lemma to the actual rows of A, as opposed to some other more refined vectors associated with the rows of A, the underlying vector space structure was ignored and only coarse additive-error bounds were obtained. To obtain improvements and to bridge the gap between TCS, NLA, and data applications, much finer bounds that take into account the vector space structure in a more refined way were needed. To do so, it helped to identify more refined structural properties that decoupled the random matrix ideas from the underlying linear algebraic ideas—understanding this will be central to the next two sections.

29.4 RANDOMIZED ALGORITHMS FOR LS APPROXIMATION

In this section and the next, we will describe randomized matrix algorithms for the LS approximation and low-rank approximation problems.

29.4.1 Different Perspectives on LS Approximation

Consider the problem of finding a vector x such that $Ax \approx b$, where the rows of A and elements of b correspond to constraints and the columns of A and elements of x correspond to variables. In the very *overconstrained LS approximation problem*, where the $m \times n$ matrix A has $m \gg n$,[*] there is in general no vector x such that $Ax = b$, and it is common to quantify "best" by looking for a vector x_{opt} such that the Euclidean norm of the residual error is small, that is, to solve the LS approximation problem

$$x_{\mathrm{opt}} = \mathrm{argmin}_x \|Ax - b\|_2. \tag{29.9}$$

There are a number of different perspectives one can adopt on this LS problem. Two major perspectives of central importance in this review are the following.

- *Algorithmic perspective.* From an algorithmic perspective, the relevant question is: how long does it take to compute x_{opt}? The answer to this question is that is takes $O(mn^2)$ time [7]. Thus, from an algorithmic perspective, a natural next question to ask is: can the general LS problem be solved, *either exactly or approximately* in $o(mn^2)$ time, with no assumptions at all on the input data?

- *Statistical perspective.* From a statistical perspective, the relevant question is: when is computing this x_{opt} the right thing to do? The answer to this is that this LS optimization is the right problem to solve when the relationship between the "outcomes" and "predictors" is roughly linear and when the error processes generating the data are "nice" [62]. Thus, from a statistical perspective, a natural next question to ask is: what should one do when the assumptions underlying the use of LS methods are not satisfied or are only imperfectly satisfied?

When adopting the statistical perspective, it is common to check the extent to which the assumptions underlying the use of LS have been satisfied. To do so, it is common to assume that $b = Ax + \varepsilon$, where b is the response, the columns $A^{(i)}$ are the carriers, and ε is a "nice"

---

[*] In this section only, we will assume that $m \gg n$.

error process. Then, $x_{opt} = (A^T A)^{-1} A^T b$, and thus $\hat{b} = Hb$, where the projection matrix onto the column space of A,

$$H = A(A^T A)^{-1} A^T,$$

is the so-called *hat matrix*. It is known that H_{ij} measures the influence or statistical leverage exerted on the prediction $\hat{b}_i$ by the observation b_j [62–66]. Relatedly, if the ith diagonal element of H is particularly large then the ith data point is particularly sensitive or influential in determining the best LS fit, thus justifying the interpretation of the elements H_{ii} as *statistical leverage scores* [11]. These leverage scores have been used extensively in classical regression diagnostics to identify potential outliers by, for example, flagging data points with leverage score greater than 2 or 3 times the average value in order to be investigated as errors or potential outliers [62].

In order to compute these quantities exactly, recall that if U is *any* orthogonal matrix spanning the column space of A, then $H = P_U = UU^T$ and thus

$$H_{ii} = \|U_{(i)}\|_2^2,$$

that is, the statistical leverage scores equal the Euclidean norm of the *rows* of any such matrix U [11,67]. More generally, and of interest for the low-rank matrix approximation algorithms in Section 29.5, the *statistical leverage scores relative to the best rank-k approximation to A* are the diagonal elements of the projection matrix onto the best rank-k approximation to A. Thus, they can be computed from

$$(P_{U_k})_{ii} = \|U_{k,(i)}\|_2^2,$$

where $U_{k,(i)}$ is the ith row of any matrix spanning the space spanned by the top k left singular vectors of A (and similarly for the right singular subspace if columns rather than rows are of interest). In many diagnostic applications, for example, when one is interested in exploring and understanding the data to determine what would be the appropriate computations to perform, this computation time is not the bottleneck.

29.4.2 A Simple Algorithm for Approximating LS Approximation

Returning to the algorithmic perspective, consider the following random sampling algorithm for the LS approximation problem [67,68]. Given a very overconstrained LS problem, where the input matrix A and vector b are *arbitrary*, but $m \gg n$:

- Compute the normalized statistical leverage scores $\{p_i\}_{i=1}^m$, that is, compute $p_i = \|U_{(i)}\|_2^2/n$, where U is the $m \times n$ matrix consisting of the left singular vectors of A.[*]

---

[*] Stating this in terms of the singular vectors is a convenience, but it can create confusion. In particular, although computing the SVD is sufficient, it is by no means necessary—here, U can be *any* orthogonal matrix spanning the column space of A [11]. Moreover, these probabilities are robust, in that any probabilities that are close to the leverage scores will suffice; see Ref. [35]. Finally, as we will describe below, these probabilities can be approximated quickly, that is, more rapidly than the time needed to compute a basis exactly, or the matrix can be preprocessed quickly to make them nearly uniform.

- Randomly sample and rescale $r = O(n \log n/\epsilon^2)$ constraints, that is, rows of A and the corresponding elements of b, using these scores as an importance sampling distribution.

- Solve the induced subproblem $\tilde{x}_{\mathrm{opt}} = \mathrm{argmin}_x \| SAx - Sb \|_2$, where the $r \times m$ matrix S represents the sampling operation.

The induced subproblem can be solved using any appropriate direct or iterative LS solver [7] as a black box. As stated, this algorithm will compute all the statistical leverage scores exactly, and thus it will not be faster than the traditional algorithm for the LS problem—importantly, we will see below how to get around this problem.

Since this overconstrained LS algorithm samples constraints and not variables, the dimensionality of the vector $\tilde{x}_{\mathrm{opt}}$ that solves the subproblem is the same as that of the vector x_{opt} that solves the original problem. This algorithm is described in more detail in Refs. [47,67,68], where it is shown that relative-error bounds of the form

$$\| b - A\tilde{x}_{\mathrm{opt}} \|_2 \leq (1 + \epsilon) \| b - Ax_{\mathrm{opt}} \|_2 \text{ and} \tag{29.10}$$

$$\| x_{\mathrm{opt}} - \tilde{x}_{\mathrm{opt}} \|_2 \leq O(\epsilon) \| x_{\mathrm{opt}} \|_2 \tag{29.11}$$

hold. Of course, there is randomization inside this algorithm, and it is possible to flip a fair coin "heads" 100 times in a row. Thus, as stated, that is, with $r = O(n \log n/\epsilon^2)$, this algorithm might fail with a probability that is no greater than a constant (say $1/2$ or $1/10$ or $1/100$, depending on the (modest) constant hidden in the $O(\cdot)$ notation) that is independent of m and n. Of course, using standard methods [69], this can easily be improved to be an arbitrarily small δ. For example, if $r = O(n \log(n) \log(1/\delta)/\epsilon^2)$ in the above algorithm; alternatively, if one repeats the above algorithm $O(\log(1/\delta))$ times and keeps the best of the results.

29.4.3 A Basic Structural Result

What the above random sampling algorithm highlights is that the "relevant nonuniformity structure" that needs to be dealt with in order to solve the LS approximation problem is defined by the statistical leverage scores. To see "why" this works, it is helpful to identify a deterministic structural condition sufficient for relative-error approximation—doing so decouples the linear algebraic part of the analysis from the randomized matrix part. This condition, implicit in the analysis of [67,68], was made explicit in [47].

Consider preconditioning or premultiplying the input matrix A and the target vector b with some *arbitrary* matrix X, and consider the solution to the LS approximation problem

$$\tilde{x}_{\mathrm{opt}} = \mathrm{argmin}_x \| X(Ax - b) \|_2. \tag{29.12}$$

For notational simplicity, let $b^\perp = U_A^\perp U_A^{\perp \mathrm{T}} b$ denote the part of the right–hand–side vector b lying outside of the column space of A. Then, the following structural condition holds.

- *Structural condition underlying the randomized LS algorithm.* Under the assumption that X satisfies the following two conditions:

$$\sigma^2_{\min}(XU_A) \geq 1/\sqrt{2}; \text{ and} \tag{29.13}$$

$$\|U_A^T X^T X b^\perp\|_2^2 \leq \frac{\epsilon}{2}\|Ax_{\text{opt}} - b\|_2^2, \tag{29.14}$$

for some $\epsilon \in (0, 1)$, the solution vector $\tilde{x}_{\text{opt}}$ to the LS approximation problem (29.12) satisfies relative-error bounds of the form (29.10) and (29.11).

Several things should be noted about these two structural conditions:

1. Since $\sigma_i(U_A) = 1$, for all $i \in [n]$, Condition (29.13) indicates that the rank of XU_A is the same as that of U_A. Thus, one should think of Condition (29.13) as stating that XU_A is an approximate isometry.

2. Since before preprocessing by X, $b^\perp = U_A^\perp U_A^{\perp T} b$ is clearly orthogonal to U_A, Condition (29.14) simply states that after preprocessing $Xb^\perp$ remains approximately orthogonal to XU_A.

3. Condition (29.13) and (29.14) both boil down to the problem of approximating the product of two matrices, and thus the algorithmic primitives on approximate matrix multiplication from Section 29.3.1 will be useful, either explicitly or within the analysis.

It should be emphasized that there is no randomization in these two structural conditions—they are deterministic statements about an arbitrary matrix X that represent a structural condition sufficient for relative-error approximation. Thus, the effect of randomization enters only via X, and it is decoupled from the linear algebraic structure.

29.4.4 Making This Algorithm Practical: In Theory

In this section, we will describe *two* ways to speed up the random sampling algorithm of Section 29.4.2 so that it runs in $o(mn^2)$ time for arbitrary input.

First, consider the following structured random projection algorithm for approximating the solution to the LS approximation problem.

- Premultiply A and b with an $n \times n$ randomized Hadamard transform HD.
- Uniformly sample roughly $r = O\left(n(\log n)(\log m) + \frac{n(\log m)}{\epsilon}\right)$ rows from HDA and the corresponding elements from HDb.
- Solve the induced subproblem $\tilde{x}_{\text{opt}} = \text{argmin}_x\|SHDAx - SHDb\|_2$, where the $r \times m$ matrix S represents the sampling operation.

This algorithm, which first preprocesses the input with a structured random projection and then it solves the induced subproblem, as well as a variant of it that uses the original "fast" JL transform [43–45], was presented in [46,47] (where a precise statement of r is given and) where it is shown that relative-error bounds of the form (29.10) and (29.11) hold.

To understand this result, recall that premultiplication by a randomized Hadamard transform is a unitary operation and thus does not change the solution; and that from the SVD of A and of HDA it follows that $U_{HDA} = HDU_A$. Thus, the "right" importance sampling distribution for the preprocessed problem is defined by the diagonal elements of the projection matrix onto the span of HDU_A. Importantly, application of such a Hadamard transform tends to "uniformize" the leverage scores, in the sense that all the leverage scores associated with U_{HDA} are (up to logarithmic fluctuations) uniform [43,47]. Thus, uniform sampling probabilities are optimal, up to a logarithmic factor which can be absorbed into the sampling complexity. Overall, this relative-error approximation algorithm for the LS problem run in $o(mn^2)$ time [46,47]—essentially $O\left(mn \log(n/\epsilon) + \frac{n^3 \log^2 m}{\epsilon}\right)$ time, which is much less than $O(mn^2)$ when $m \gg n$.

Second, another way to make the above algorithm fast is to approximate the leverage scores in $o(mn^2)$ time and then call the random sampling algorithm of Section 29.4.2 using those approximate scores as the importance sampling distribution. Drineas et al. [70] presents a randomized algorithm (not described in detail here) that returns a relative-error approximation to *every m* statistical leverage score, that is, the approximation $\tilde{p}_i$ is within a multiplicative $1 \pm \epsilon$ factor of the exact leverage score probabilities, for every $i = 1, \ldots, m$. Within the analysis, this algorithm boils down to an *under*constrained LS problem, in which a structured random projection is carefully applied, in a manner somewhat analogous to the fast *over*constrained LS random projection algorithm above; and thus this algorithm runs in $o(mn^2)$ time—basically the same time as the fast random projection algorithm above.

Thus, the random sampling algorithm of Section 29.4.2, which naïvely needs $O(mn^2)$ to compute the importance sampling distribution can be sped up in two different ways. One can spend $o(mn^2)$ to uniformize the sampling probabilities and then sample uniformly; or one can spend $o(mn^2)$ time to approximately compute the sampling probabilities and then use those approximations as an importance sampling distribution. Both procedures take basically the same time (which should not be surprising since the approximation of the statistical leverage scores boils down to an underconstrained LS problem).

29.4.5 Making This Algorithm Practical: In Practice

Several "rubber-hits-the-road" issues need to be dealt with in order for the algorithms of the previous subsection to yield to high-precision numerical implementations that beat traditional numerical code, either in specialized scientific applications or when compared with popular numerical libraries. The following issues are most significant:

- *Awkward ϵ dependence.* The sampling complexity, that is, the number of columns or rows needed by the algorithm, scales as $1/\epsilon^2$ or $1/\epsilon$, which is the usual poor asymptotic complexity for Monte Carlo methods.

- *Numerical conditioning and preconditioning.* So far, nothing has been said about numerical conditioning issues, although it is well known that these issues are crucial when matrix algorithms are implemented numerically.

- *Forward error versus backward error.* The bounds above, for example, (29.10) and (29.11), provide so-called forward error bounds. The standard stability analysis in NLA is in terms of the backward error, where the approximate solution $\tilde{x}_{opt}$ is shown to be the exact solution of some slightly perturbed problem $x_{opt} = \text{argmin}_x \|(A + \delta A)x - b\|_2$, where $\|\delta A\| \leq \tilde{\epsilon}\|A\|$ for some small $\tilde{\epsilon}$.

All three of these issues can be dealt with by coupling the randomized algorithms of the previous subsection with traditional tools from iterative NLA algorithms. This was first done by Rokhlin and Tygert [49], and these issues were addressed in much greater detail by Avron and Maymounkov [52]. Both of these implementations take the following form:

- Premultiply A by a structured random projection, for example, $\Omega = SHD$ from Section 29.3.1, which represents uniformly sampling a few rows from a randomized Hadamard transform.

- Perform a QR decomposition on ΩA, so that $\Omega A = QR$.

- Use the R from the QR decomposition as a preconditioner for an iterative method.

In general, iterative algorithms compute an ϵ-approximate solution to a LS problem like (29.9) via $O(\kappa(A)\log(1/\epsilon))$ iterations, where $\kappa(A) = (\sigma_{max}(A))/(\sigma_{min}(A))$ is the condition number of the input matrix (which could be quite large, thereby leading to slow convergence of the iterative algorithm). In this case, by choosing the dimension of the random projection appropriately, for example, as discussed in the previous subsections, one can show that $\kappa(AR^{-1})$ is bounded above by a small data-independent constant. That is, by using the R matrix from a QR decomposition of ΩA, one obtains a very good preconditioner for the original problem (29.9), independent of course of any assumptions on the original matrix A. Overall, applying the structured random projection in the first step takes $o(mn^2)$ time; performing a QR decomposition of ΩA is fast since ΩA is much smaller than A; and one needs to perform only $O(\log(1/\epsilon))$ iterations, each of which needs $O(mn)$ time, to compute the approximation.

29.5 RANDOMIZED ALGORITHMS FOR LOW-RANK MATRIX APPROXIMATION

In this section, we will describe several related randomized algorithms for low-rank matrix approximation that underlie applications such as those described in Section 29.2. The algorithms build on those of Section 29.3.2 in very nontrivial ways that lead them to achieve improved worst-case bounds and to be useful in both numerical analysis and data analysis applications.

29.5.1 A Random Sampling Algorithm

Additive-error bounds (of the form proved for the low-rank algorithms of Section 29.3.2) are rather coarse, and the gold standard in TCS is to establish much finer relative-error bounds of the form provided in (29.15) below. To motivate the importance sampling probabilities used to achieve such relative-error guarantees, recall that if one is considering a matrix with $k - 1$ large singular values and one much smaller singular value, then the directional information of the kth singular direction will be hidden from the Euclidean norms of the input matrix. Intuitively, the reason is that, since $A_k = U_k \Sigma_k V_k^T$, the Euclidean norms of the columns of A are convolutions of "subspace information" (encoded in U_k and V_k^T) and "size-of-A information" (encoded in Σ_k). This *suggests* deconvoluting subspace information and size-of-A information by choosing importance sampling probabilities that depend on the Euclidean norms of the columns of V_k^T. Intuitively, this importance sampling distribution defines a nonuniformity structure over $\mathbb{R}^n$ that indicates *where* in the n-dimensional space the information in A is being sent, independent of *what* that (singular value) information is. More formally, these quantities are proportional to the diagonal elements of the projection matrix onto the span of V_k^T,[*] and thus they generalize the statistical leverage scores of Section 29.4.

This idea was suggested in [11,67], and it forms the basis for the algorithm from the TCS literature that achieves the strongest Frobenius norm bounds. Given an $m \times n$ matrix A and a rank parameter k:

- Compute the importance sampling probabilities $\{p_i\}_{i=1}^n$, where $p_i = \frac{1}{k}\|V_k^{T(i)}\|_2^2$, where V_k^T is *any* $k \times n$ orthogonal matrix spanning the top-k right singular subspace of A.

- Randomly select and rescale $c = O(k \log k / \epsilon^2)$ columns of A according to these probabilities.

A more detailed description of this basic random sampling algorithm may be found in Refs. [11,67], where it is proven that

$$\|A - P_{C_k}A\|_F \leq (1 + \epsilon)\|A - P_{U_k}A\|_F \tag{29.15}$$

holds. The analysis of this algorithm boils down to choosing a set of columns that are relative-error good at capturing the Frobenius norm of A, when compared with the basis provided by the top-k singular vectors. That is, it boils down to the randomized algorithm for the LS approximation problem from Section 29.4; see Refs. [11,67] for details. This algorithm and related algorithms that randomly sample columns and/or rows provide what is known as CX or CUR[†] matrix decompositions [11,61,67]; and this relative-error column-based matrix decomposition, as well as heuristic variants of it, has been applied in a range of data analysis applications, ranging from term-document data to DNA SNP data [11,15,16].

---

[*] Here, we are sampling columns and not rows, as in the algorithms of Section 29.4, and thus we are dealing with the right, rather than the left, singular subspace; but clearly the ideas are analogous.

[†] Note that the names CX and CUR are *not* acronyms, but instead they refer to the names of suggestively chosen matrices that constitute the decompositions. In that sense, the names are more analogous to QR decompositions (which refer to an orthogonal matrix Q and an upper triangular matrix R), as opposed to the SVD (which is an acronym for the Singular Value Decomposition). In the case of CX and CUR, C refers to a matrix of columns of the original matrix, R to a matrix of rows of the original matrix, and X and U are encoding matrices.

29.5.2 A Second Random Sampling Algorithm

The algorithm of the previous subsection randomly samples $O(k \log k/\epsilon^2)$ columns, and then in order to compare with the bound provided by the SVD, it "filters" those columns through an exactly rank-k space. In this subsection, we will describe a randomized algorithm for selecting *exactly* k columns from an input matrix. Importantly, this algorithm extends the ideas of the previous relative-error TCS algorithm to obtain bounds of the form used historically in NLA. Moreover, the main structural result underlying the analysis of this algorithm permits *much* finer control on the application of randomization, for example, for high-performance numerical implementation of both random sampling and random projection algorithms.

Consider the following more sophisticated version of a two-stage hybrid algorithm. Given an arbitrary $m \times n$ matrix A and rank parameter k:

- (Randomized phase) Compute the importance sampling probabilities $\{p_i\}_{i=1}^{n}$, where $p_i = \frac{1}{k}\|V_k^{T(i)}\|_2^2$, where V_k^T is *any* $k \times n$ orthogonal matrix spanning the top-k right singular subspace of A. Randomly select and rescale $c = O(k \log k)$ columns of V_k^T according to these probabilities.

- (Deterministic phase) Let $\tilde{V}^T$ be the $k \times O(k \log k)$ nonorthogonal matrix consisting of the down-sampled and rescaled columns of V_k^T. Run a deterministic QR algorithm on $\tilde{V}^T$ to select exactly k columns of $\tilde{V}^T$. Return the corresponding columns of A.

A more detailed description of this algorithm may be found in Refs. [14,71], where it is shown that with extremely high probability the following spectral[*] and Frobenius norm bounds hold:

$$\|A - P_C A\|_2 \leq O(k^{3/4} \log^{1/2}(k) n^{1/2})\|A - P_{U_k}A\|_2$$

$$\|A - P_C A\|_F \leq O(k \log^{1/2} k)\|A - P_{U_k}A\|_F.$$

Note that both the original choice of columns in the first phase, as well as the application of the QR algorithm in the second phase, involve the matrix V_k^T, that is, the matrix defining the relevant nonuniformity structure over the columns of A in the relative-error algorithm of Section 29.5.1. In particular, it is critical to the success of this algorithm that the QR procedure in the second phase be applied to the randomly sampled version of V_k^T, rather than of A itself. Since this algorithm may be viewed as postprocessing the relative-error random sampling algorithm of the previous subsection to remove redundant columns, it has been applied successfully to a range of data analysis problems; see, for example, [17–19] and [72–74].

With respect to running time, the computational bottleneck for this algorithm (as well as the relative-error algorithm of the previous subsection) is computing $\{p_i\}_{i=1}^{n}$, for which it

---

[*] Note that to establish the spectral norm bound, Refs. [14,71] used slightly more complicated (but still depending only on information in V_k^T) importance sampling probabilities, but this may be an artifact of the analysis.

suffices to compute *any* $k \times n$ matrix V_k^T that spans the top-k right singular subspace of A. (In particular, a full or partial SVD computation is *not* necessary.) Thus, this running time is of the same order as the running time of the QR algorithm used in the second phase when applied to the original matrix A, typically roughly $O(mnk)$ time, which represents one point of comparison for the CSSP. (Not surprisingly, one could also perform a random projection, such as those described in Section 29.5.3 below, to approximate this basis, and then use that approximate basis to compute approximate importance sampling probabilities, in which case similar bounds would hold but the running time would be improved to $O(mn \log k)$ time.) Moreover, this algorithm scales up to matrices with thousands of rows and millions of columns, whereas existing off-the-shelf implementations of traditional QR algorithms may fail to run at all.

As with the relative-error LS algorithm of Section 29.4, in order to see "why" this algorithm for the CSSP works, it is helpful to identify a structural condition that decouples the randomization from the linear algebra. This structural condition was first identified in Refs. [14,71], and it was subsequently improved by Halko et al. [75]. To identify it, consider preconditioning or postmultiplying the input matrix A by some *arbitrary* matrix X. Thus, for the above–randomized algorithm, the matrix X is a carefully constructed random sampling matrix, but it could be a random projection, or more generally *any* other arbitrary matrix X. Recall that if $k \leq r = \text{rank}(A)$, then the SVD of A may be written as

$$A = U_A \Sigma_A V_A^T = U_k \Sigma_k V_k^T + U_{k,\perp} \Sigma_{k,\perp} V_{k,\perp}^T,$$

where U_k is the $m \times k$ matrix consisting of the top k singular vectors, $U_{k,\perp}$ is the $m \times (r-k)$ matrix consisting of the bottom $r-k$ singular vectors, and so on. Then, the following structural condition holds.

- *Structural condition underlying the randomized low-rank algorithm.* If $V_k^T X$ has full rank, then for $\xi \in \{2, F\}$, that is, for both the Frobenius and spectral norms,

$$\|A - P_{AX}A\|_\xi^2 \leq \|A - A_k\|_\xi^2 + \|\Sigma_{k,\perp}(V_{k,\perp}^T X)(V_k^T X)^\dagger\|_\xi^2 \qquad (29.16)$$

holds, where P_{AX} is a projection onto the span of AX, and where the dagger symbol represents the Moore–Penrose pseudoinverse.

This structural condition characterizes the manner in which the behavior of the low-rank algorithm depends on the interaction between the right singular vectors of the input matrix and the matrix X. Note that the assumption that $V_k^T X$ does not lose rank is the generalization of Condition (29.14) of Section 29.4. Also, note that the form of this structural condition is the same for both the spectral and Frobenius norms.

As with the LS problem, given this structural insight, what one does with it depends on the application: one can compute the basis V_k^T exactly if that is not computationally prohibitive and if one is interested in extracting exactly k columns; or one can perform a random projection and ensure that with high probability the structural condition is satisfied.

Moreover, by decoupling the randomization from the linear algebra, it is easier to parameterize the problem in terms more familiar to NLA and scientific computing: for example, one can consider sampling $\ell > k$ columns and projecting onto a rank-k', where $k' > k$, approximation to those columns; or one can couple these ideas with traditional methods such as the power iteration method. Several of these extensions will be the topic of the next subsection.

29.5.3 Several Random Projection Algorithms

Consider the following basic random projection algorithm, which is the first of three random projection algorithms we will review in this subsection that draw on the ideas of the previous subsections in progressively finer ways. Given an $m \times n$ matrix A and a rank parameter k:

- Construct an $n \times \ell$, with $\ell = O(k/\epsilon)$, structured random projection matrix Ω, for example, $\Omega = DHS$ from Section 29.3.1, which represents uniformly sampling a few rows from a randomized Hadamard transform.

- Return $B = A\Omega$.

This algorithm, which amounts to choosing uniformly a small number ℓ of columns in a randomly rotated basis, was introduced in Ref. [46], where it is proven that

$$\|A - P_{B_k}A\|_F \leq (1 + \epsilon)\|A - P_{U_k}A\|_F \qquad (29.17)$$

holds with high probability. (Recall that B_k is the best rank-k approximation to the matrix B, and P_{B_k} is the projection matrix onto this k-dimensional space.) This algorithm runs in $O(Mk/\epsilon + (m + n)k^2/\epsilon^2)$ time, where M is the number of nonzero elements in A, and it requires two passes over the data from external storage. Although this algorithm is very similar to the additive-error random projection algorithm of [53] that was described in Section 29.3.2, this algorithm achieves much stronger relative-error bounds by performing a much more refined analysis. Basically, Sarlós [46] (and also the improvement [76]) modifies the analysis of the relative-error random sampling of [11,67] that was described in Section 29.5.1, which in turn relies on the relative-error random sampling algorithm for LS approximation [67,68] that was described in Section 29.4. In the same way that we saw in Section 29.4.4 that fast structured random projections could be used to uniformize coordinate-based nonuniformity structure for the LS problem, after which fast uniform sampling was appropriate, here uniform sampling in the randomly rotated basis achieves relative-error bounds. In showing this, Sarlós [46] also states a "subspace" analogue to the JL lemma, in which the geometry of an entire subspace of vectors (rather than just N pairs of vectors) is preserved. Thus, one can view the analysis of Ref. [46] as applying JL ideas, not to the rows of A itself, as was done by Papadimitriou et al. [53], but instead to vectors defining the subspace structure of A. Thus, with this random projection algorithm, the subspace information and size-of-A information are deconvoluted *within the analysis*, whereas with the random sampling algorithm of Section 29.5.1, this took place *within the algorithm* by modifying the importance sampling probabilities.

As with the randomized algorithms for the LS problem, several rubber-hits-the-road issues need to be dealt with in order for randomized algorithms for the low-rank matrix approximation problem to yield to high-precision numerical implementations that beat traditional deterministic numerical code. In addition to the issues described previously, the main issue here is the following.

- *Minimizing the oversampling factor.* In practice, choosing even $O(k \log k)$ columns, in either the original or a randomly rotated basis, even when the big-O notation hides only modest constants, can make it difficult for these randomized matrix algorithms to beat previously existing high-quality numerical implementations. Ideally, one could parameterize the problem so as to choose some number $\ell = k + p$ columns, where p is a modest additive oversampling factor, for example, 10 or 20 or k, and where there is no big-O constant.

When attempting to be this aggressive at minimizing the size of the sample, the choice of the oversampling factor p is more sensitive to the input than in the algorithms we have reviewed so far. For example, when parameterized this way, p could depend on the size of the matrix dimensions, the decay properties of the spectrum, and the particular choice made for the random projection matrix [48,75,77–79]. Moreover, for worst-case input matrices, such a procedure may fail.

That being said, running the risk of such a failure might be acceptable if, for example, one can efficiently couple to a diagnostic to check for such a failure, and if one can then correct for it by, for example, choosing more samples if necessary. The best numerical implementations of randomized matrix algorithms for low-rank matrix approximation do just this, and the strongest results in terms of minimizing p take advantage of Condition (29.16) in a somewhat different way than was originally used in the analysis of the CSSP [75]. For example, rather than choosing $O(k \log k)$ dimensions and then filtering them through *exactly* k dimensions, as the relative-error random sampling and relative-error random projection algorithm do, one can choose some number ℓ of dimensions and project onto a k'-dimensional subspace, where $k < k' \leq \ell$, while exploiting Condition (29.16) to bound the error, as appropriate for the computational environment at hand [75].

Next, consider a second random projection algorithm that will address this issue. Given an $m \times n$ matrix A, a rank parameter k, and an oversampling factor p:

- Set $\ell = k + p$.

- Construct an $n \times \ell$ random projection matrix Ω, either with i.i.d. Gaussian entries or in the form of a structured random projection such as $\Omega = DHS$ which represents uniformly sampling a few rows from a randomized Hadamard transform.

- Return $B = A\Omega$.

Although this is quite similar to the algorithms of [46,53], algorithms parameterized in this form were introduced in Refs. [48,77,78], where a suite of bounds of the form

$$\|A - Z\|_2 \lesssim 10\sqrt{\ell \min\{m, n\}}\|A - A_k\|_2$$

are shown to hold with high probability. Here, Z is a rank-k-or-greater matrix easily constructed from B.

Finally, consider a third random projection algorithm that will address the issue that the decay properties of the spectrum can be important when it is of interest to minimize the oversampling very aggressively. Given an $m \times n$ matrix A, a rank parameter k, an oversampling factor p, and an iteration parameter q:

- Set $\ell = k + p$.

- Construct an $n \times \ell$ random projection matrix Ω, either with i.i.d. Gaussian entries or in the form of a structured random projection such as $\Omega = DHS$ which represents uniformly sampling a few rows from a randomized Hadamard transform.

- Return $B = (AA^\mathsf{T})^q A\Omega$.

This algorithm (as well as a numerically stable variant of it) was introduced in Ref. [79], where it is shown that bounds of the form

$$\|A - Z\|_2 \lesssim (10\sqrt{\ell \min\{m, n\}})^{1/(4q+2)} \|A - A_k\|_2$$

hold with high probability. (This bound should be compared with that for the previous algorithm, and thus Z is a rank-k-or-greater matrix easily constructed from B.) Basically, this random projection algorithm modifies the previous algorithm by coupling a form of the power iteration method within the random projection step.

29.6 CONCLUSION

Randomization has had a long history in scientific applications. Within the last few decades, randomization has also proven to be useful in a very different way—as a powerful resource in TCS for establishing worst-case bounds for a wide range of computational problems. That is, in the same way that space and time are valuable resources available to be used judiciously by algorithms, it has been discovered that exploiting randomness as an algorithmic resource *inside the algorithm* can lead to better algorithms.

Perhaps since its original promise was oversold, and perhaps due to the greater-than-expected difficulty in developing high-quality numerically stable software for scientific computing applications, randomization *inside the algorithm* for common matrix problems was mostly "banished" from scientific computing and NLA in the 1950s. Thus, it is refreshing that within just the last few years, novel algorithmic perspectives from TCS have worked their way back to NLA, scientific computing, and scientific data analysis.

ACKNOWLEDGMENTS

I would like to thank the numerous colleagues and collaborators with whom these results have been discussed in preliminary form—in particular, Petros Drineas, with whom my contribution to the work reviewed here was made.

REFERENCES

1. O. Alter, P.O. Brown, and D. Botstein. Singular value decomposition for genome-wide expression data processing and modeling. *Proceedings of the National Academy of Sciences of the United States of America*, 97(18):10101–10106, 2000.

2. F.G. Kuruvilla, P.J. Park, and S.L. Schreiber. Vector algebra in the analysis of genome-wide expression data. *Genome Biology*, 3:research0011.1–0011.11, 2002.

3. Z. Meng, D.V. Zaykin, C.F. Xu, M. Wagner, and M.G. Ehm. Selection of genetic markers for association analyses, using linkage disequilibrium and haplotypes. *American Journal of Human Genetics*, 73(1):115–130, 2003.

4. B.D. Horne and N.J. Camp. Principal component analysis for selection of optimal SNP-sets that capture intragenic genetic variation. *Genetic Epidemiology*, 26(1):11–21, 2004.

5. Z. Lin and R.B. Altman. Finding haplotype tagging SNPs by use of principal components analysis. *American Journal of Human Genetics*, 75:850–861, 2004.

6. N. Patterson, A.L. Price, and D. Reich. Population structure and eigenanalysis. *PLoS Genetics*, 2(12):2074–2093, 2006.

7. G.H. Golub and C.F. Van Loan. *Matrix Computations*. Johns Hopkins University Press, Baltimore, 1996.

8. S. Georgiev and S. Mukherjee. Unpublished results, 2011.

9. S.J. Gould. *The Mismeasure of Man*. W. W. Norton and Company, New York, 1996.

10. P. Menozzi, A. Piazza, and L. Cavalli-Sforza. Synthetic maps of human gene frequencies in Europeans. *Science*, 201(4358):786–792, 1978.

11. M.W. Mahoney and P. Drineas. CUR matrix decompositions for improved data analysis. *Proceedings of the National Academy of Sciences of the United States of America*, 106:697–702, 2009.

12. A. Civril and M. Magdon-Ismail. On selecting a maximum volume sub-matrix of a matrix and related problems. *Theoretical Computer Science*, 410:4801–4811, 2009.

13. A. Civril and M. Magdon-Ismail. Column based matrix reconstruction via greedy approximation of SVD. Unpublished Manuscript, 2009.

14. C. Boutsidis, M.W. Mahoney, and P. Drineas. An improved approximation algorithm for the column subset selection problem. In *Proceedings of the 20th Annual ACM-SIAM Symposium on Discrete Algorithms*, pp. 968–977, 2009.

15. P. Paschou, M. W. Mahoney, A. Javed, J. R. Kidd, A. J. Pakstis, S. Gu, K. K. Kidd, and P. Drineas. Intra- and interpopulation genotype reconstruction from tagging SNPs. *Genome Research*, 17(1):96–107, 2007.

16. P. Paschou, E. Ziv, E.G. Burchard, S. Choudhry, W. Rodriguez-Cintron, M.W. Mahoney, and P. Drineas. PCA-correlated SNPs for structure identification in worldwide human populations. *PLoS Genetics*, 3:1672–1686, 2007.

17. P. Paschou, P. Drineas, J. Lewis, C.M. Nievergelt, D.A. Nickerson, J.D. Smith, P.M. Ridker, D.I. Chasman, R.M. Krauss, and E. Ziv. Tracing sub-structure in the European American population with PCA-informative markers. *PLoS Genetics*, 4(7):e1000114, 2008.

18. P. Drineas, J. Lewis, and P. Paschou. Inferring geographic coordinates of origin for Europeans using small panels of ancestry informative markers. *PLoS ONE*, 5(8):e11892, 2010.

19. P. Paschou, J. Lewis, A. Javed, and P. Drineas. Ancestry informative markers for fine-scale individual assignment to worldwide populations. *Journal of Medical Genetics*, page doi:10.1136/jmg.2010.078212, 2010.

20. A. J. Connolly, A. S. Szalay, M. A. Bershady, A. L. Kinney, and D. Calzetti. Spectral classification of galaxies: An orthogonal approach. *Astronomical Journal*, 110(3):1071–1082, 1995.

21. A. J. Connolly and A. S. Szalay. A robust classification of galaxy spectra: Dealing with noisy and incomplete data. *Astronomical Journal*, 117(5):2052–2062, 1999.

22. D. Madgwick, O. Lahav, K. Taylor, and the 2dFGRS Team. Parameterisation of galaxy spectra in the 2dF galaxy redshift survey. In *Mining the Sky: Proceedings of the MPA/ESO/MPE Workshop, ESO Astrophysics Symposia*, pp. 331–336, 2001.

23. C. W. Yip, A. J. Connolly, A. S. Szalay, T. Budavári, M. SubbaRao, J. A. Frieman, R. C. Nichol et al. Distributions of galaxy spectral types in the Sloan Digital Sky Survey. *Astronomical Journal*, 128(2):585–609, 2004.

24. S.R. Folkes, O. Lahav, and S.J. Maddox. An artificial neural network approach to the classification of galaxy spectra. *Monthly Notices of the Royal Astronomical Society*, 283(2):651–665, 1996.

25. C. W. Yip, A. J. Connolly, D. E. Vanden Berk, Z. Ma, J. A. Frieman, M. SubbaRao, A. S. Szalay et al. Spectral classification of quasars in the Sloan Digital Sky Survey: Eigenspectra, redshift, and luminosity effects. *Astronomical Journal*, 128(6):2603–2630, 2004.

26. N. M. Ball, J. Loveday, M. Fukugita, O. Nakamura, S. Okamura, J. Brinkmann, and R. J. Brunner. Galaxy types in the Sloan Digital Sky Survey using supervised artificial neural networks. *Monthly Notices of the Royal Astronomical Society*, 348(3):1038–1046, 2004.

27. T. Budavári, V. Wild, A. S. Szalay, L. Dobos, and C.-W. Yip. Reliable eigenspectra for new generation surveys. *Monthly Notices of the Royal Astronomical Society*, 394(3):1496–1502, 2009.

28. R. C. McGurk, A. E. Kimball, and Z. Ivezić. Principal component analysis of SDSS stellar spectra. *Astronomical Journal*, 139:1261–1268, 2010.

29. T. A. Boroson and T. R. Lauer. Exploring the spectral space of low redshift QSOs. Technical report. Preprint: arXiv:1005.0028, 2010.

30. R.J. Brunner, S.G. Djorgovski, T.A. Prince, and A.S. Szalay. Massive datasets in astronomy. In J. Abello, P.M. Pardalos, and M.G.C. Resende (eds), *Handbook of Massive Data Sets*, pp. 931–979. Kluwer Academic Publishers, Berlin, 2002.

31. N. M. Ball and R. J. Brunner. Data mining and machine learning in astronomy. *International Journal of Modern Physics D*, 19(7):1049–1106, 2010.

32. L. Foster, A. Waagen, N. Aijaz, M. Hurley, A. Luis, J. Rinsky, C. Satyavolu, M. J. Way, P. Gazis, and A. Srivastava. Stable and efficient Gaussian process calculations. *Journal of Machine Learning Research*, 10:857–882, 2009.

33. M. J. Way, L. V. Foster, P. R. Gazis, and A. N. Srivastava. New approaches to photometric redshift prediction via Gaussian process regression in the Sloan Digital Sky Survey. *Astrophysical Journal*, 706(1):623–636, 2009.

34. M. J. Way. Galaxy zoo morphology and photometric redshifts in the Sloan Digital Sky Survey. Technical report. Preprint: arXiv:1104.3758, 2011.

35. P. Drineas, R. Kannan, and M.W. Mahoney. Fast Monte Carlo algorithms for matrices I: Approximating matrix multiplication. *SIAM Journal on Computing*, 36:132–157, 2006.

36. M. Rudelson. Random vectors in the isotropic position. *Journal of Functional Analysis*, 164(1):60–72, 1999.

37. M. Rudelson and R. Vershynin. Sampling from large matrices: An approach through geometric functional analysis. *Journal of the ACM*, 54(4):Article 21, 2007.

38. W.B. Johnson and J. Lindenstrauss. Extensions of Lipshitz mapping into Hilbert space. *Contemporary Mathematics*, 26:189–206, 1984.

39. P. Frankl and H. Maehara. The Johnson-Lindenstrauss lemma and the sphericity of some graphs. *Journal of Combinatorial Theory Series A*, 44(3):355–362, 1987.

40. P. Indyk and R. Motwani. Approximate nearest neighbors: Towards removing the curse of dimensionality. In *Proceedings of the 30th Annual ACM Symposium on Theory of Computing*, pp. 604–613, 1998.

41. S. Dasgupta and A. Gupta. An elementary proof of a theorem of Johnson and Lindenstrauss. *Random Structures and Algorithms*, 22(1):60–65, 2003.

42. D. Achlioptas. Database-friendly random projections: Johnson-lindenstrauss with binary coins. *Journal of Computer and System Sciences*, 66(4):671–687, 2003.

43. N. Ailon and B. Chazelle. Approximate nearest neighbors and the fast Johnson-Lindenstrauss transform. In *Proceedings of the 38th Annual ACM Symposium on Theory of Computing*, pp. 557–563, 2006.

44. N. Ailon and B. Chazelle. The fast Johnson-Lindenstrauss transform and approximate nearest neighbors. *SIAM Journal on Computing*, 39(1):302–322, 2009.

45. J. Matousek. On variants of the Johnson–Lindenstrauss lemma. *Random Structures and Algorithms*, 33(2):142–156, 2008.

46. T. Sarlós. Improved approximation algorithms for large matrices via random projections. In *Proceedings of the 47th Annual IEEE Symposium on Foundations of Computer Science*, pp. 143–152, 2006.

47. P. Drineas, M.W. Mahoney, S. Muthukrishnan, and T. Sarlós. Faster least squares approximation. *Numerische Mathematik*, 117(2):219–249, 2010.

48. E. Liberty, F. Woolfe, P.-G. Martinsson, V. Rokhlin, and M. Tygert. Randomized algorithms for the low-rank approximation of matrices. *Proceedings of the National Academy Sciences of the United States of America*, 104(51):20167–20172, 2007.

49. V. Rokhlin and M. Tygert. A fast randomized algorithm for overdetermined linear least-squares regression. *Proceedings of the National Academy Sciences of the United States of America*, 105(36):13212–13217, 2008.

50. N. Ailon and E. Liberty. Fast dimension reduction using Rademacher series on dual BCH codes. In *Proceedings of the 19th Annual ACM-SIAM Symposium on Discrete Algorithms*, pp. 1–9, 2008.

51. E. Liberty, N. Ailon, and A. Singer. Dense fast random projections and lean Walsh transforms. In *Proceedings of the 12th International Workshop on Randomization and Computation*, pp. 512–522, 2008.

52. H. Avron, P. Maymounkov, and S. Toledo. Blendenpik: Supercharging LAPACK's least-squares solver. *SIAM Journal on Scientific Computing*, 32:1217–1236, 2010.

53. C.H. Papadimitriou, P. Raghavan, H. Tamaki, and S. Vempala. Latent semantic indexing: A probabilistic analysis. *Journal of Computer and System Sciences*, 61(2):217–235, 2000.

54. P. Drineas, A. Frieze, R. Kannan, S. Vempala, and V. Vinay. Clustering large graphs via the singular value decomposition. *Machine Learning*, 56(1–3):9–33, 2004.

55. A. Frieze, R. Kannan, and S. Vempala. Fast Monte-Carlo algorithms for finding low-rank approximations. *Journal of the ACM*, 51(6):1025–1041, 2004.

56. P. Drineas, R. Kannan, and M.W. Mahoney. Fast Monte Carlo algorithms for matrices II: Computing a low-rank approximation to a matrix. *SIAM Journal on Computing*, 36:158–183, 2006.

57. M.W. Mahoney, M. Maggioni, and P. Drineas. Tensor-CUR decompositions for tensor-based data. In *Proceedings of the 12th Annual ACM SIGKDD Conference*, pp. 327–336, 2006.

58. J. Sun, Y. Xie, H. Zhang, and C. Faloutsos. Less is more: Compact matrix decomposition for large sparse graphs. In *Proceedings of the 7th SIAM International Conference on Data Mining*, 2007.

59. H. Tong, S. Papadimitriou, J. Sun, P.S. Yu, and C. Faloutsos. Colibri: Fast mining of large static and dynamic graphs. In *Proceedings of the 14th Annual ACM SIGKDD Conference*, pp. 686–694, 2008.

60. F. Pan, X. Zhang, and W. Wang. CRD: Fast co-clustering on large datasets utilizing sampling-based matrix decomposition. In *Proceedings of the 34th SIGMOD International Conference on Management of Data*, pp. 173–184, 2008.

61. P. Drineas, R. Kannan, and M.W. Mahoney. Fast Monte Carlo algorithms for matrices III: Computing a compressed approximate matrix decomposition. *SIAM Journal on Computing*, 36:184–206, 2006.

62. S. Chatterjee and A.S. Hadi. *Sensitivity Analysis in Linear Regression*. John Wiley & Sons, New York, 1988.

63. D.C. Hoaglin and R.E. Welsch. The hat matrix in regression and ANOVA. *American Statistician*, 32(1):17–22, 1978.

64. S. Chatterjee and A.S. Hadi. Influential observations, high leverage points, and outliers in linear regression. *Statistical Science*, 1(3):379–393, 1986.

65. P.F. Velleman and R.E. Welsch. Efficient computing of regression diagnostics. *American Statistician*, 35(4):234–242, 1981.

66. S. Chatterjee, A.S. Hadi, and B. Price. *Regression Analysis by Example*. John Wiley & Sons, New York, 2000.

67. P. Drineas, M.W. Mahoney, and S. Muthukrishnan. Relative-error CUR matrix decompositions. *SIAM Journal on Matrix Analysis and Applications*, 30:844–881, 2008.

68. P. Drineas, M.W. Mahoney, and S. Muthukrishnan. Sampling algorithms for ℓ_2 regression and applications. In *Proceedings of the 17th Annual ACM-SIAM Symposium on Discrete Algorithms*, pp. 1127–1136, 2006.

69. R. Motwani and P. Raghavan. *Randomized Algorithms*. Cambridge University Press, New York, 1995.

70. P. Drineas, M. Magdon-Ismail, M. W. Mahoney, and D. P. Woodruff. Fast approximation of matrix coherence and statistical leverage. Technical Report. Preprint: arXiv:1109.3843, 2011.

71. C. Boutsidis, M.W. Mahoney, and P. Drineas. An improved approximation algorithm for the column subset selection problem. Technical report. Preprint: arXiv:0812.4293v2, 2008.

72. C. Boutsidis, M.W. Mahoney, and P. Drineas. Unsupervised feature selection for principal components analysis. In *Proceedings of the 14th Annual ACM SIGKDD Conference*, pp. 61–69, 2008.

73. C. Boutsidis, M.W. Mahoney, and P. Drineas. Unsupervised feature selection for the k-means clustering problem. In *Annual Advances in Neural Information Processing Systems 22: Proceedings of the 2009 Conference*, 2009.

74. B. Savas and I. Dhillon. Clustered low rank approximation of graphs in information science applications. In *Proceedings of the 11th SIAM International Conference on Data Mining*, 2011.

75. N. Halko, P.-G. Martinsson, and J. A. Tropp. Finding structure with randomness: Probabilistic algorithms for constructing approximate matrix decompositions. Technical report. Preprint: arXiv:0909.4061, 2009.

76. N.H. Nguyen, T.T. Do, and T.D. Tran. A fast and efficient algorithm for low-rank approximation of a matrix. In *Proceedings of the 41st Annual ACM Symposium on Theory of Computing*, pp. 215–224, 2009.

77. P.-G. Martinsson, V. Rokhlin, and M. Tygert. A randomized algorithm for the decomposition of matrices. *Applied and Computational Harmonic Analysis*, 30:47–68, 2011.

78. F. Woolfe, E. Liberty, V. Rokhlin, and M. Tygert. A fast randomized algorithm for the approximation of matrices. *Applied and Computational Harmonic Analysis*, 25(3):335–366, 2008.

79. V. Rokhlin, A. Szlam, and M. Tygert. A randomized algorithm for principal component analysis. *SIAM Journal on Matrix Analysis and Applications*, 31(3):1100–1124, 2009.

Index

Note: n = Footnote

A

a posteriori probability, 303, 304
a priori probability, 304
Abell 1882 system, 346
    Voronoi Tessellation diagram, 347
ACD. *See* Anticoincident detector (ACD)
Active galactic nuclei (AGN), 43, 345
    host galaxies, 223
    property characterization, 501
AdaBoost algorithm, 574
Adaptive piecewise constant approximation (APCA), 625
Adaptive resampling and combining algorithm (*Arcing* algorithm), 574
ADQA. *See* Automated data quality assessment (ADQA)
ADS. *See* Astrophysics Data System (ADS)
Advanced LIGO detector (advLIGO detector), 415–416
Advanced modeling for cross-identification, 123
    folding in photometry, 123–124
    stars with unknown proper motion, 124–127
    varying prior over sky, 124
advLIGO detector. *See* Advanced LIGO detector (advLIGO detector)
AGB stars. *See* Asymptotic gaint branch stars (AGB stars)
Agglomerative clustering, 545, 558. *See also* Data clustering
    hierarchical, 552–553
    k-means algorithm, 549
AGN. *See* Active galactic nuclei (AGN)
AID. *See* Automated interaction detection (AID)
All Nearest Neighbors (AllNN), 463, 465
    dual-tree algorithm, 465, 467
    experimental performance, 468
    kd-tree, 466
    naive approach, 465
    pruning parts of computation, 466
    runtime guarantees, 467
    single-tree algorithm, 467
    space-partitioning trees, 467
AllNN. *See* All nearest neighbors (AllNN)
ALMA. *See* Atacama Large Millimeter Array (ALMA)
AM. *See* Amplitude-modulation (AM)
Amplitude-modulation (AM), 498
ANN. *See* Artificial neural networks (ANN)
Anticoincident detector (ACD), 250, 251
    background rejection analysis, 257
    GamProb CTs, 258
    prefilters for, 255
    reason for failure, 257
    tiles in Hardware Trigger, 255, 256
APCA. *See* Adaptive piecewise constant approximation (APCA)
Approximation–approximation coefficients, 206, 207
Approximation–detail coefficients, 206, 207
Arcing algorithm. *See* Adaptive resampling and combining algorithm (Arcing algorithm)
Arcseconds, 114n
Artificial neural networks (ANN), 105, 622
    SOM, 107
    uses, 216
ASKAP. *See* Australian Square Kilometre Array Pathfinder (ASKAP)

ASP. *See* Automated science processing (ASP)

Astroinformatics, 164

Astronomers, 11. *See also* Astronomy
    armchair astronomer, 91
    celestial object classification problem, 580
    clustering scheme validation, 5
    friends-of-friends algorithm, 4
    using nonprobabilistic method, 14
    personal equation, 527
    responsibility, 33
    sky survey studies, 450

Astronomical research contexts, 454

Astronomical spectra analyses, 267. *See also*
        Circumstellar envelopes of AGB stars
    automated spectral analyses, 283
    line lists, 279
    molecular infrared spectroscopy, 268–270
    molecular modeling, 271–272
    NNLS spectral fits, 270–271
    PAH database and model, 274–277
    PAHs in space, 273–274
    ro-vibrational spectrum, 268–269

Astronomical unit (AU), 356

Astronomy, 33–34. *See also* General relativity; Pattern
        recognition; Time-series databases
    astronomical variability discovery space, 619
    automated neural networks, 7, 8
    class discovery, 638
    computation, 13–14
    conflicts, 11–12
    data mining for, 595
    DDM for, 610–611
    digital revolution, 24–25
    ensemble methods in, 578–581
    estimation, 14–15
    flux variations, 618
    flux-based time series, 618
    historical classifications, 3–4
    machine learning classification, 6, 7, 8–9
    motion-based time series, 618
    philosophical skepticism to Bayesian inference,
        15–16
    planetary theory, 12–13
    supervised classification, 5–6
    survey science, 163, 164
    time-domain surveys, 617–618
    time-series data, 617–620, 637–639
    unsupervised clustering in, 4–5

Astrophysics, 448, 618
    ensemble methods in, 581
    exoplanets, 355
    Kepler mission in, 368

    time-domain astronomy, 451
    variable optical sky exploration, 170

Astrophysics Data System (ADS), 452

Asymptotic gaint branch stars (AGB stars), 273, 385,
        387. *See also* Circumstellar envelopes of
        AGB stars
    dust emission, 282
    infrared spectrum of, 281, 282
    ISO-SWS spectrum, 281–282
    star O Cet, 282

Atacama Large Millimeter Array (ALMA), 168

AU. *See* Astronomical unit (AU)

Australian Square Kilometre Array Pathfinder
        (ASKAP), 94, 169, 423

Autoclass method, 106, 385

Automated data quality assessment (ADQA),
        173–175. *See also* Large Synoptic Survey
        Telescope (LSST)
    automated discovery, 176
    data complexity, 175–176
    data mining research, 178
    low false transient alert rate, 176, 178
    retuning of algorithm behavior, 175
    unsupervised learning algorithm, 177

Automated Interaction Detection (AID), 240

Automated neural networks, 7, 8

Automated science processing (ASP), 41–42
    Bayesian blocks reconstruction, 46, 47
    burst candidate direction estimation, 49
    clustering statistic computation, 48–49
    GRB refinement task, 45–46
    implementation and execution, 52–53
    LAT detection, 46–47
    LAT error contours, 46, 48
    signal types, 43–45

Autonomous repoints, 42

B

B component. *See* Magnetic component
    (B component)

B mode, 59, 301. *See also* E mode
    decomposition, 80, 300, 301
    second-order statistics, 313, 314

Background cosmic rays (BKG), 240, 258, 262
    CT-generated probability, 257
    probability distributions, 259, 260

Backpropagation algorithm, 516

Bagging. *See* Bootstrap aggregating (Bagging)

Bag-of-patterns representation (BOP), 635–637

Ball trees, 467

Balmer break, 324, 332

BAO. *See* Baryonic acoustic oscillations (BAO)

Baryonic acoustic oscillations (BAO), 173
BAT. *See* Burst Alert Telescope (BAT)
Bayes factor, 116, 117, 121
    all-sky Bayes factor, 120
    as angular separation function, 118
Bayes theorem, 69, 303, 513
Bayes' theory, 16
Bayesian approach, 65, 115, 309. *See also* Advanced
        modeling for cross-identification
    advantages, 121–123
    Bayes factor, 116, 121
    intrinsic stellar variability separation, 370–371
    limited sky coverage, 120
    MAP application to Kepler data, 373–377
    MAP approach, 371–373
    MAP implementation considerations, 377
    observational evidence, 116–118
    PowellSnake, 63–64
    probability of association, 118–120
    problem with PDC, 368–370
    for training set, 330
Bayesian filters, 303
    a posteriori probability, 303
    multiscale, 305, 306
Bayesian Network classification, 104
Bayesian statistics, 25
Bessel's personal equation, 13, 527–528, 540
Biparametric scale-adapter filter (BSAF), 65
BKG. *See* Background cosmic rays (BKG)
Black holes, 408. *See also* Gravitational wave
    binary system detection, 430
    emri events, 420
    feeding modes, 222–223
    growth, 223
    massive black-hole binaries, 430
    observing, 412
    supermassive, 43, 422, 423
    supernova explosion, 414
Blazars, 43, 45. *See also* Gamma ray bursts (GRBs)
Blind Source Separation problem (BSS problem),
        65-66, 68n
    GMCA, 71
    SMICA, 70
Bode's law, 17
BolPol, 72
Bonferroni adjustment, 24
BOOMERanG experiment, 56, 57
Boosting, 105, 243, 574
Bootstrap *aggregating* (*Bagging*), 574, 575
BOP. *See* Bag-of-patterns representation (BOP)
Borne data mining research project, 453
Borůvka's algorithm, 469

Boscovitch's procedure, 14
Brent's method, 64n
Brightest cluster galaxy algorithm (maxBCG
        algorithm), 344
Broadband photometry, 326
BSAF. *See* Biparametric scale-adapter filter (BSAF)
BSS problem. *See* Blind Source Separation problem
        (BSS problem)
Bump hunting, 30
Burst Alert Telescope (BAT), 45–46

C

CAL. *See* Calibration (CAL); Calorimeter (CAL)
Calibration (CAL), 357
Calorimeter (CAL), 250, 251
Cannot-link constraint, 556, 558
Canonical link, 526
Carnegie Supernova Project (CSP), 5
CART. *See* Classification and Regression Trees
        (CART)
Cartography experiments, 398
CAS. *See* Catalog Archive Server (CAS); Central
        Authentication Service (CAS)
Catalog Archive Server (CAS), 222
Categorical feature, 507n
Causality, 16
    assessment, 30–31
    twentieth-century suspicions, 26
Causation, 23
CBI. *See* Cosmic Background Imager (CBI)
CC. *See* Correcting classifiers (CC)
CCA. *See* Correlated component analysis (CCA)
CCD. *See* Charge-coupled device (CCD)
CDM. *See* Collective data mining (CDM);
        Compression-based Dissimilarity Measure
        (CDM)
CDM paradigm. *See* Cold Dark Matter paradigm
        (CDM paradigm)
CDPP. *See* Combined Differential Photometric
        Precision (CDPP)
Celestial objects, 3
Central Authentication Service (CAS), 227
Central-limit theorem, 14–15
CGRO. *See* Compton Gamma Ray Observatory
        (CGRO)
Chaining, 5
Charge-coupled device (CCD), 24, 119, 356n, 580
    Channel 2.1, 370
    Pan-STARRS, 166
    plotted *vs.* time, 163
    SkyMapper, 165
Chirp signals, 425

Chunking, 528, 540
    and conditional likelihood, 528
    conditioning on within-chunk response, 529–530
    for definiteness, 534
    kinds of, 528–529
Chunks, 527. *See also* Generalized linear models;
        Spherical approximation
    attenuation, 528
    Bessel's personal equations, 527
    and conditional models, 527
    nuisance effects, 527–528
    random chunk effect, 528
    size 2 chunks, 530
    size 3 chunks, 530-532
CIB. *See* Cosmic infrared background (CIB)
Circumstellar envelopes of AGB stars, 277
    column density, 280n
    emission from circumstellar dust, 282
    geometry and approximation for, 278
    grid searching, 280
    infrared spectrum of stars, 281, 282
    ray-tracing method, 279
    spectroscopic databases, 279
Citizen science projects, 230–232
Citizen scientist, 219, 224, 229
    volunteers community, 225
    in Zooniverse development, 228
Classification, astronomical object, 383. *See also*
        Miras identification with SVM; Neural
        networks
    feature space, 395
    machine learning classifications, 385
    microlensing event selection, 391–393
    periodic variability classification algorithms,
        384–385
    template light curves, 394
    transient variability classification, 393–397
Classification and Regression Trees (CART), 240
Classification learning algorithm, 506. *See also*
        Support vector machines (SVMs)
    categorical feature, 507n
    classification error on training set, 510
    decision stump, 512
    decision tree, 507, 508, 510–512
    discriminative method, 509
    ensembles, 509
    generative method, 509
    hyperplane, 517–519
    information gain, 511
    maximum a posteriori model, 509
    maximum likelihood model, 509
    naïve Bayes classifier, 513–514

    nearest neighbor classification, 507
    neural networks, 514–517
Classification learning task, 505
Classification Tree (CT), 105, 240, 241
    BKG and GAM, 240
    boosting, 243
    ensembles, 242, 243
    Fermi-LAT, 263
    Gini and *Entropy*, 240–241
    importance of independent variables, 245, 246
    leaf nodes, 241, 242
    MAGIC analysis, 263–264
    measuring performance, 243–245
    in practice, 246
    probabilities distributions, 242–243
    root node, 241, 242
    termination nodes, 241
    training bias minimization, 242, 246–250
Classifiers, 103, 565. *See also* Ensemble methods
    Bayes' classifier, 103
    combination schemes, 570
    competence estimation, 572
    dynamic classifier selection, 572
    Gaia spectral, 399
    Gaussian mixture, 104
    KNN, 104–105
    naïve Bayes, 513
    randomness, 577
    RF, 93
    SDSS SVM, 8
    SOM, 398
Clusters, 338, 543–544. *See also* Data clustering;
        Galaxy clusters
    expansion constant, 470
    modeling, 549–550
    SZ cluster extraction, 64–65
CMB. *See* Cosmic microwave background (CMB)
COBE. *See* Cosmic background explorer (COBE)
CoCoRaHS Network. *See* Community Collaborative
        Rain, Hail and Snow Network (CoCoRaHS
        Network)
Cold Dark Matter paradigm (CDM paradigm), 341
    for data mining, 606–607
    N-body simulations, 341–342
Collaborative clustering, 557. *See also* Data clustering
Collective data mining (CDM), 605, 606
Column density, 280n
Column subset selection problem (CSSP), 651
Combined Differential Photometric Precision
        (CDPP), 366, 367
Commander algorithm, 73

Community Collaborative Rain, Hail and Snow Network (CoCoRaHS Network), 233
Component separation in spherical harmonic domain
   MEM, 69–70
   SMICA, 66, 70
Compression-based Dissimilarity Measure (CDM), 634–635
Compton Gamma Ray Observatory (CGRO), 237
Compute Unified Device Architecture (CUDA), 130
Concurrent read, concurrent write model (CRCW model), 599
Conditional density, 331
Conditional models, 538, 539
Condorcet Jury Theorem, 563, 566
Constrained clustering, 544, 555–556, 558. *See also* Data clustering
Context-specific features, 103
Convergence, 293, 294, 299
   bispectrum, 314–315
   kurtosis, 315
   skewness, 315
   SOM, 398
   three-point correlation function, 315
   2D convergence power spectrum, 311
Convergence field, 306, 316. *See also* Shear field
   density contrast and, 308
   lensing potential and, 307
Cooperative clustering. *See* Collaborative clustering
Correcting classifiers (CC), 577
Correlated component analysis (CCA), 69
Correlation, 26
   chunk-specific correlation coefficient, 534–535
   n-point correlation functions, 463, 471
   time series, 367
Cosmic background explorer (COBE), 56, 134
   all sky CMB maps, 137
   sky cube pixelization scheme, 137–138
Cosmic Background Imager (CBI), 57
Cosmic infrared background (CIB), 453
Cosmic microwave background (CMB), 56, 134, 481
   angular power spectrum, 57
   celestial sphere pixelization, 137
   COBE, 56
   cube pixelization, 137–138
   ECP pixelization, 136
   HEALPix, 140–142
   HTM pixelization, 137
   icosahedron pixelization, 138–139
   igloo pixelization, 139–140
   map by WMAP, 57
   MF, 63

MHW, 63
   Planck Surveyor, 57–58
   polarization, 59, 79–82
   PowellSnake, 63–64
   radiation, 80n
   SExtractor, 64
   SZ cluster extraction, 64–65
   temperature power spectrum, 58, 59
   WMAP, 57
Cosmic microwave background (CMB) component separation, 65
   CCA, 69
   FastICA technique, 68
   GMCA, 71
   ILC, 67
   MEM, 69–70
   modeling sky emission, 65–66
   SMICA, 70
   template fitting, 66–67
   toward wavelet-based methods, 70–71
Cosmic microwave background (CMB) data analysis pipeline, 60
   map-making, 61
   multichannel maps to cosmological parameters, 61–62
   raw to time-ordered data, 60
Cosmic microwave background (CMB) map statistical analysis
   data map by WMAP team, 78
   non-Gaussian signatures, 78
   sparse inpainting, 79
Cosmic rays, 238
   incoming rates, 238–239
   LAT trigger use, 239
   target pixel time series scanning, 361
Cosmic shear amplitude, 295
Cosmic variance, 33–34
Cosmological model constraints, 310. *See also* Non-Gaussian statistics
   convergence and shear power spectra, 311–312
   E and B mode separation, 313–314
   methodology, 310–311
   second-order statistics, 311
   shear tomography, 313
   shear two-point correlation function, 312–313
   shear variance, 312
Cosmological parameters, 60. *See also* Cosmic microwave background (CMB)
   CMB power spectrum use, 59
   estimation, 76
   maximum likelihood method, 310–311
   MCMC, 76–77

Cosmological parameters (*continued*)
    from multichannel maps to, 61–62
    quantification, 311
    using second-order statistics, 311, 312
    shear power spectrum use, 297
Cosmological populations, 453
Cosmology population Monte Carlo (CosmoPMC),
    77
CosmoPMC. *See* Cosmology population Monte
    Carlo (CosmoPMC)
Cover trees, 467
CRCW model. *See* Concurrent read, concurrent
    write model (CRCW model)
cROMAster, 74
Cross-identification of sources, 114
    advanced modeling, 123–127
    Bayesian approach, 115–121
    crossmatching, 127–130, 455
    GALEX, 114, 115
    2MASS, 114, 115
    SDSS, 114, 115
Crossmatching, 127, 455
    locality and storage, 128
    matching in databases, 128
    parallel on GPUs, 129–130
    recursive evaluation, 127–128
Cross-matching. *See* Object cross-identification
CrossSpect, 74–75
Crude clustering algorithm, 49
CSP. *See* Carnegie Supernova Project (CSP)
CSSP. *See* Column subset selection problem (CSSP)
CT. *See* Classification tree (CT)
CTBCPFGamProb, 246
Cube pixelization, 137–138
CUDA. *See* Compute Unified Device Architecture
    (CUDA)
CUR matrix decompositions, 663n
Curl component. *See* Magnetic component
    (B component)
Curse of dimensionality, 103, 175, 176, 514
Curvelet transform on sphere, 187, 188
Custom CUDA kernel, 130
CX matrix decompositions. *See* CUR matrix
    decompositions
Cyberinfrastructure, 447
    for cyber-enabled scientific discovery, 164
    e-Science, 451

D

Dark energy, 172, 288
    DES, 168
    distant explosions of supernovae, 90
    effects, 173
    gravitational lensing use, 294
    LSST use, 173
    2-point correlation use, 481
Dark Energy Survey (DES), 5, 168
Dark matter, 288
    cluster detection, 343
    clusters of, 173
    distant explosions of supernovae, 90
    filtering use, 303
    gravitational lensing use, 294
    mass map, 306
    particle halo, 340
Dark matter 2D mapping, 299
    Bayesian methods, 303
    convergence, 299
    distribution map, 306
    E and B modes, 300–301
    entropy, 305
    filling masked regions, 301
    global inversion, 299–300
    linear filters, 303
    local inversion, 300
    MCA, 302
    multichannel MEM-ICF method, 305
    multiresolution representations, 301–302
    power spectrum measurement, 301
    sparse representations of signals, 305
    two-point statistics of cosmic shear field, 301
Dark matter 3D mapping, 306
    angular diameter distance, 307
    Karhunen–Loeve transform, 309
    linear filters, 308–309
    MEM filter, 309
    nonlinear filters, 309
    3D convergence field reconstruction, 307
    3D gravitational potential, 307
DASI. *See* Degree Angular Scale Interferometer
    (DASI)
Data clustering, 543, 557, 598. *See also* Data mining
    agglomerative clustering, 552–553
    algorithms, 543–545
    applications to astronomy data, 556–557
    constrained clustering, 555–556
    EM clustering, 549–552
    k-means clustering, 545–549
    marginalization, 556–557
    spectral clustering, 553–555
Data management center (DMC), 359
Data mining, 449, 596, 611. *See also* Distributed
    mining of data (DMD)
    assessing causality, 30–31

Borne data mining research project, 453
correlation *vs.* dependence, 31–32
cosmic variance, 34–35
detecting patterns, 28–30
distance metrics, 598
distributed algorithms for, 604
distributed classification, 606–607
distributed clustering, 604–605
of distributed data, 452–454
early planet hunting, 17–18
in general relativity, 20
in GRID environments, 609–610
k-means algorithm, 599
knowledge discovery in databases, 449
MapReduce framework, 602–604
MDD, 451
methods, 288
MPI, 608–609
nearest-neighbor identification, 455
object cross-correlation, 455
object cross-identification, 455
P2P environments, 607–608
parallel data classification algorithms, 600–602
parallel data clustering algorithms, 598–600
planetary theory from, 12–13
probability in science, 32–33
research for LSST project, 178
reverse, 22
statistical malfeasance, 34–35
surprise detection, 456
systematic data exploration, 455–456
VO, 451–452
Zooniverse results, 230
Data quality assessment (DQA), 173
Data reduction techniques, 545
Data validation module (DV module), 359, 366, 367
DDM. *See* Distributed data mining (DDM)
Decision templates (DT), 571
Decision tree, 506, 507, 510, 621. *See also* Pattern recognition
decision stump, 512
drawback of, 601
for GZ2, 224
induction algorithm, 606
learning algorithm, 508, 511, 608
in meta-learning literature, 602
Deep drilling fields, 167
Deep Space Network (DSN), 357
Degree Angular Scale Interferometer (DASI), 57
Dendrogram, 552, 633
Denoising algorithm, spherical MSVST 2D–1D wavelet, 209

DES. *See* Dark Energy Survey (DES)
Detail–approximation coefficients, 206, 207
Detail–detail coefficients, 206, 207
DFT. *See* Discrete Fourier transform (DFT)
Diffuse γ-ray emission, 199
Digital revolution, 24–25
Dipole moment, 412
Discovery informatics, 448
Discrete Fourier transform (DFT), 625
Discrete wavelet transform (DWT), 625
Dispersion, 413
Distance, 628
angular diameter, 307
elastic distance measures, 629, 631
Euclidean distance, 629
pairwise distance matrix, 102
Distributed computing framework, 596–597. *See also* Data mining; Parallel computing framework
Distributed data mining (DDM), 451, 452, 598
algorithms, 604
for astronomy, 610–611
data in, 604
meta-learning framework, 606
Distributed mining of data (DMD), 451, 456
correlation mining work, 457
elliptical galaxies, 456
local galaxy density, 457
PCA on SDSS–2MASS data set, 457–458
plane galaxy parameters, 459
DMC. *See* Data management center (DMC)
DMD. *See* Distributed mining of data (DMD)
Domain-based classification, 96. *See also* Feature-based classification
variability of stars across H–R diagram, 96, 97
variability selection of quasars, 98, 99
variable sources, 98
Double counting, 19
DQA. *See* Data quality assessment (DQA)
Dropout techniques, 330
DSN. *See* Deep Space Network (DSN)
DT. *See* Decision templates (DT)
DTW. *See* Dynamic time warping (DTW)
Dual-tree algorithms, 465, 467
guaranteeing error in, 474–475
for KDE problem, 473
DUALTREEBORUVKA, 469, 470
DV module. *See* Data validation module (DV module)
DWT. *See* Discrete wavelet transform (DWT)
Dynamic time warping (DTW), 629
advantages, 631

Dynamic time warping (*continued*)
 algorithm, 629
 alignment for, 630
 error rates for, 633
 global constraint, 630

E

E component. *See* Electric component
   (E component)
E mode, 301
 least square estimator, 301
 polarization spectrum, 57
E/S0 ridgeline, 338, 345
E–B coupling, 82
ECOC. *See* Error correcting output coding (ECOC)
ECP. *See* Equidistant cylindrical projection (ECP)
Edit distance with real penalty (ERP), 631
EGRET. *See* Energetic gamma-ray experiment
   telescope (EGRET)
Electric component (E component), 300
Electromagnetic waves, 411. *See also* Gravitational
   wave
 impacts on, 413
 monopole moment of a charge distribution, 412
 source, 412
ELTs. *See* Extremely Large Telescopes (ELTs)
EM algorithm. *See* Expectation-maximization
   algorithm (EM algorithm)
EM clustering. *See* Expectation-maximization
   clustering (EM clustering)
Empirical approaches, 324–325, 329
Emri events, 420
EMST. *See* Euclidean Minimum Spanning Tree
   (EMST)
Energetic gamma-ray experiment telescope
   (EGRET), 50, 184, 237, 238
Ensemble methods, 563, 582
 accuracy–diversity trade-off, 566–567
 in astronomy and astrophysics, 578–581
 characterization, 571–572
 classifiers, 565
 ensemble fusion methods, 570–572
 ensemble pruning, 573
 ensemble selection methods, 572–573
 feature selection methods, 575–576
 fuzzy set theory, 571
 generative ensembles, 568, 573
 hierarchical, 571–572
 input decimation approach, 575
 mixture of experts, 576
 multiple learners, 567
 nongenerative ensembles, 568, 570

OC methods, 576
 random forests, 578, 579
 random subspace methods, 575
 randomized, 577–578
 resampling methods, 573–575
 rotation forests, 576
 supervised, 579
 taxonomies of, 567–570
 theories on, 565
 tree, 105
Ensemble-based classifier learning, 606
Ensembles, 242, 509, 564
 of CTs, 263
 generative, 569
 of heterogeneous classifiers, 573
 nongenerative, 569
 RFs, 578
 selection methods, 572–573
 trainable and nontrainable, 567
Entropy, 240n, 305
 entropic prior, 304
 for LAT analysis, 245
 in MEM filter, 309
Epoch, 517
 armchair astronomer, 91
 displacement on sky, 125
 precision cosmology, 134
 ROC curves, 93
Equidistant cylindrical projection (ECP), 136, 139
EREW model. *See* Exclusive read, exclusive write
   model (EREW model)
EROs. *See* Extremely red objects (EROs)
ERP. *See* Edit distance with real penalty (ERP)
Error correcting output coding (ECOC), 577
ESA. *See* European Space Agency (ESA)
Euclidean distance, 465, 629. *See also* Distance
 alignment for, 630
 drawback for, 629
 hierarchical clustering, 636–637
 in Kernel density estimation problem, 474
 minimization, 398
 1-NN classification, 627–628
Euclidean Minimum Spanning Tree (EMST), 464
 Borůvka's algorithm, 469
 DUALTREEBORUVKA, 469–471
 graph-based MST algorithms, 468–469
European Space Agency (ESA), 420
Exclusive read, exclusive write model (EREW model),
   599
Exhaustive algorithm. *See* Naïve algorithm
EXP. *See* Exponential family (EXP)

Expectation–maximization algorithm (EM algorithm), 5, 558, 624
Expectation–maximization clustering (EM clustering), 549. *See also* Data clustering
    algorithm, 550
    on artificial data set, 551
    cluster model update step, 551
    generative approach, 549–550
    probabilistic cluster models, 549
Expert-based classification, 96
Exponential family (EXP), 525
Extinction, 413
Extremely Large Telescopes (ELTs), 162
Extremely red objects (EROs), 453
Extrinsic geometry, 409n

F

Faint Images of the Radio Sky at Twenty-cm survey (FIRST survey), 169, 579
Faraday effect, 413
Far-Infrared Absolute Spectrophotometer (FIRAS), 56
Fastest Fourier transform in the west code (FFTW code), 146
FastICA technique, 68–69
Feature-based classification, 99. *See also* Domain-based classification
    context-specific features, 103
    feature creation, 100
    frequency-domain features, 100, 101
    light-curve distance measures, 102
    methods, 99–100
    supervised approaches, 103–105, 106, 107
    time-ordered features, 102
    unsupervised and semisupervised approaches, 105
Fermi Gamma-ray Space Telescope, 184
Fermi LAT mission, 41, 238
    CT usage, 263
    features, 42
    light curves, 43, 44
FFTW code. *See* Fastest Fourier transform in the west code (FFTW code)
Field-of-view (FOV), 42, 254
    a priori, 254
    star observation, 356
Filter design, 332
FindComponentNeighbors, 469, 470
FIRAS. *See* Far-Infrared Absolute Spectrophotometer (FIRAS)
First overtone to fundamental mode (FO/FU), 386

FIRST survey. *See* Faint Images of the Radio Sky at Twenty-cm survey (FIRST survey)
Fisher matrix, 72
Fitting model principle, 270
Flare Advocates, 53
Flatland analogy, 409
FM. *See* Frequency-modulation (FM)
F-Meas. *See* F-Measure
F-Measure, 244, 245
    calculation, 262–263
    training bias, 245
    for training sample, 249
FO/FU. *See* First overtone to fundamental mode (FO/FU)
Forests, 242
FOV. *See* Field-of-view (FOV)
Free–free emission, 62
Frequency-domain features, 100, 101
Frequency-modulation (FM), 490
Friends-of-friends algorithm, 4, 341
Full width half maximum (FWHM), 135
FWHM. *See* Full width half maximum (FWHM)

G

GAIA
    mission, 168
    science alerts, 399
Galactic diffuse emission, 197
Galaxy, 214n, 343, 344. *See also* Very luminous infrared galaxies (VLIRGs)
    active, 222–223
    automated classification scheme, 216
    cosmological populations, 453
    distance information, 298
    elliptical, 456
    expert classifier approach, 216
    GIM2D, 216–217
    for GZ1 project, 219
    local galaxy density, 457
    merging, 222
    morphology history, 214, 215, 216
    plane galaxy parameters, 459
    solving classification problem, 217
    types, 214, 215
Galaxy clusters, 316, 337, 343, 348. *See also* Halo
    Abell 1882 system, 346
    catalog, 344
    color clustering, 344
    E/S0 ridgeline, 338, 345, 346
    finding techniques, 343
    geometric clustering, 347
    high-peak model, 338

Galaxy clusters (*continued*)
   matched filter method, 344
   in SDSS spectroscopic redshift catalog, 345
   uses of, 338
   vs. simulated halos, 340
Galaxy Evolution Explorer (GALEX), 114, 115
Galaxy Image 2D (GIM2D), 216
Galaxy Zoo. *See also* Galaxy Zoo 1 (GZ1); Galaxy Zoo
   2 (GZ2)
   in citizen science project context,
      232–233
   citizen scientists, 224, 225
   evolution, 223
   Galaxy Zoo forum, 225, 226
   Hubble tuning fork, 214, 215
   Internet forum, 218–219
   launching, 218
   MOSES, 217
   object's characteristics, 225
   Stardust@Home Project, 217–218
   Supernova Project, 228
Galaxy Zoo 1 (GZ1), 219
   active galaxies, 222–223
   classification biases, 220
   classification comparisons, 220–221
   from clicks to classifications, 219–220
   color and morphology, 222
   merging galaxies, 222
   peer-reviewed paper list, 221
   rare and unusual objects, 223
   spiral arm directions, 222
Galaxy Zoo 2 (GZ2), 223–224, 228
GALEX. *See* Galaxy Evolution Explorer
   (GALEX)
GAM. *See* Gamma ray (GAM)
Gamma ray (GAM), 240
   point sources, 184
Gamma ray bursts (GRBs), 42n, 43. *See also* Blazars
   in LAT data, 45
   search for, 47–50
Gamma-ray Burst Monitor (GBM), 45
Gamma-ray bursts (GRBs), 42n, 618
   all-sky counts map, 51
   blind search for, 47–50
   finding flaring sources, 50
   HEALPix-based localization, 51–52
   in LAT data, 45–47
   ROIs, 52
   wavelet analysis, 50, 51
Gamma-ray Large Area Space Telescope (GLAST),
   186, 238
Garbage in—garbage out (GIGO), 250n

Gaussian denoising. *See also* Poisson denoising
   application to, 187, 188
   multichannel, 206–207
Gaussian filter, 303
Gaussian mixture classifier, 104
Gaussian mixture modeling, 106, 549, 558.
      See also Expectation-maximization
      clustering (EM clustering)
Gaussian noise. *See also* Poisson noise
   multichannel Gaussian denoising, 206–207
   2D–1D decomposition/reconstruction,
      205–206
   2D–1D wavelet transform, 202, 203, 205
Gauss–Legendre sky pixelization package (GLESP
      package), 153
   code structure, 153–154
   data format, 156
   operation types, 154–156
   programs, 156
   test and precision, 157
Gauss–Legendre sky pixelization scheme (GLESP
      scheme), 134, 142. *See also* Hierarchical
      Equal Area iso-Latitude Pixelisation
      (HEALPix)
   data formats, 156
   package, 153, 154
   pixel shapes and distribution comparison, 145
   pixel window function, 151–153
   pixelization scheme comparison, 145
   position of pixel centers, 144
   properties, 146–147
   repixelization problem, 148–149
   temperature variation decomposition, 143
   test and precision, 157
   trapezoidal pixels, 144
   weighting coefficients, 143, 144
GBM. *See* Gamma-ray Burst Monitor (GBM)
GCN. *See* GRBs Coordinates Network (GCN)
GCVS. *See General Catalog of the Variable Stars*
   (GCVS)
Gelatin-coated dry plate process, 20
General Catalog of the Variable Stars (GCVS), 385
General relativity. *See also* Astronomy; Universe
   Bayesian statistics, 18–19
   data mining, 20
   gravitational deflection of light, 20–21
   issue of light rays, 290
   light speed in vacuum, 290
   Newtonian deflection, 21
   reverse data mining, 22
Generalized least squares (GLS), 135

Generalized linear models, 524. *See also* Chunks; Spherical approximation
   maximum likelihood estimation, 526–527
   notation for observations, 524
   probability models, 525
   statistical models, 525–526
Generalized Morphological Component Analysis (GMCA), 71
GI alignment. *See* Gravitational–Intrinsic alignment (GI alignment)
Gibbs sampling theory, 73
GIGO. *See* Garbage In—Garbage out (GIGO)
GIM2D. *See* Galaxy Image 2D (GIM2D)
Gini Index, 240n
GLAST. *See* Gamma ray Large Area Space Telescope (GLAST)
GLESP package. *See* Gauss–Legendre sky pixelization package (GLESP package)
GLESP scheme. *See* Gauss–Legendre sky pixelization scheme (GLESP scheme)
GLS. *See* Generalized least squares (GLS)
GMCA. *See* Generalized Morphological Component Analysis (GMCA)
GPUs. *See* Graphical processing units (GPUs)
Gradient component. *See* Electric component (E component)
Graphical processing units (GPUs), 129, 130, 529
Gravitational lensing, 289, 290. *See also* Weak gravitational lensing
Gravitational radiation. *See* Gravitational wave
Gravitational shear, 293, 294
   challenges, 298–299
   cosmic shear amplitude, 295
   direct deconvolution, 296
   galaxy distance information, 298
   galaxy ellipticity, 295
   GI alignment, 297, 298
   instrumental and atmospheric bias, 295
   intrinsic alignments and correlations, 297
   KSB method, 296
   lensfit method, 296
Gravitational wave, 408, 409–411. *See also* Black holes
   affected by, 413
   astronomy, 425
   of binary coalescence, 415
   chirp signals, 425
   connection to source, 411–413
   coupling between matter and gravity, 413
   detection, 431. *See* Laser interferometry; Pulsar timing array (PTA); Spacecraft doppler tracking
   detectors, 424
   flatland analogy, 409
   γ-ray bursts, 414–415
   gravitational wave bursts, 425
   gravity, 409
   interaction advantages, 414
   leading-order contribution, 412
   monopole moment of gravitational charge distribution, 412
   periodic signals, 425
   propagation impact on, 413
   quadrupole formula, 412–413
   space–time, 410–411
   stochastic gravitational wave, 425
Gravitational wave astronomy data analysis, 424
   detection from inspiraling compact binaries, 427–428
   noise characterization, 426
   periodic gravitational wave detection, 427
   problem characterization, 425–426
   stochastic gravitational wave detection, 426–427
   wave burst detection, 428
Gravitational wave astronomy machine learning, 428
   future work, 429
   glitch identification and classification, 429
   inspiraling massive black-hole binaries, 430
   noise nonstationarity, 428
   response function estimation, 429–430
   source population statistics, 430–431
Gravitational wave bursts, 425
   frequency and character, 420
   LISA's sensitivity effect, 430
Gravitational–intrinsic alignment (GI alignment), 297, 298
Gravity, 409
   influence of, 408
   relativistic theory, 20
   spherical collapse, 341
GRBs. *See* Gamma ray bursts (GRBs)
GRBs Coordinates Network (GCN), 46
Grid computing, 609–610
Grid data mining for astronomy project (GRIST project), 609
GRIST project. *See* Grid data mining for astronomy project (GRIST project)
GZ"Mergers" project, 228–229
GZ1. *See* Galaxy Zoo 1 (GZ1)
GZ2. *See* Galaxy Zoo 2 (GZ2)

H

Halo, 241, 348. *See also* Galaxy clusters
   concept, 338–340
   dark matter particle halo, 340
   finding algorithm, 341

Halo (*continued*)
    finding technique, 340
    galaxy clusters *vs.*, 340
    subhalo identification, 341–343
Harmonic mode–mode coupling, 74
Hat matrix, 658
HEALPix. *See* Hierarchical Equal Area iso-Latitude
        Pixelisation (HEALPix)
Hertzsprung–Russell diagrams (H–R diagrams),
        96
    fractional variability of stars, 97
    red variables, 387
    of simulated stars, 401
Hierarchical Equal Area iso-Latitude Pixelisation
        (HEALPix), 137, 140. *See also*
        Gauss–Legendre sky pixelization scheme
        (GLESP scheme)
    MS-VSTS comparison, 192
    orthographic view, 142
    requirements, 141
    for spherical data, 186, 187
    tessellation, 141
Hierarchical triangular mesh (HTM), 121, 137
High-energy gamma-ray sky, 238
High-l codes. *See also* Low-l codes
    MASTER, 73–75
    SPICE, 76
    Xpol, 75–76
High-peak model, 338
HIGH-performance computing power (HPC power),
        447
HPC power. *See* HIGH-performance computing
        power (HPC power)
H–R diagrams. *See* Hertzsprung–Russell diagrams
        (H–R diagrams)
HSD. *See* Hybrid Steepest Descent (HSD)
HST. *See* Hubble Space Telescope (HST)
HTM. *See* Hierarchical triangular mesh (HTM)
HTM sphere pixelization scheme, 137
Hubble, Edwin, 214
Hubble Space Telescope (HST), 224
    photometric redshift accuracy, 332
    Quasar, 453
    template repair methods, 331
Hubble tuning fork, 214, 215
Hubble Zoo, 224, 228
Hubble's law, 22
Human
    classification. *See* Expert-based classification
    eye–brain system, 28
    genome, 650
Hume's skepticism, 16

Hybrid Steepest Descent (HSD), 193
HyLIRG. *See* Hyperluminous infrared galaxy
        (HyLIRG)
Hyperluminous infrared galaxy (HyLIRG), 453
Hyperplane, 517–519

I

i.i.d. *See* Independently and identically distributed
        (i.i.d.)
IACT. *See* Imaging Air Cherenkov Telescope (IACT)
ICA. *See* Independent component analysis (ICA)
ICF. *See* Intrinsic correlation functions (ICF)
Icosahedron pixelization, 138–139
ICs. *See* Independent components (ICs)
Igloo pixelization, 139–140
II correlation. *See* Intrinsic–intrinsic correlation (II
        correlation)
i-kMeans, 624
ILC. *See* Internal linear combination (ILC)
Image differencing method, 92, 93
Image Resolution (IR) Knob, 257
Imaging Air Cherenkov Telescope (IACT), 240, 263
Importance sampling probabilities, 649n
Independent component analysis (ICA), 66. *See also*
        Spectral matching independent
        component analysis (SMICA)
    ensemble learning for, 581
    FastICA technique, 68–69
    solving BSS problem, 68
Independent components (ICs), 581
Independently and identically distributed (i.i.d.), 506
Information technology research (ITR), 451
Infrared astronomical satellite (IRAS), 273
Inpainting techniques, 301
    sparse, 79, 301
    weak lensing, 302
Instrument Science Operations Center (ISOC), 41
    Fermi ISOC pipeline, 52
Integrated Sachs Wolfe effect (ISW effect), 58
Interior point algorithm (IPM), 601
Internal linear combination (ILC), 67–68
Interstellar medium (ISM), 413, 618
Interval estimation procedure, 14
Intrinsic correlation functions (ICF), 305
Intrinsic geometry, 409n
Intrinsic–intrinsic correlation (II correlation), 297,
        298
Inverse Voronoi volume. *See* Local galaxy density
IPM. *See* Interior point algorithm (IPM)
IR. *See* Image resolution (IR)
IRAS. *See* Infrared astronomical satellite (IRAS)
Island Universes, 338

ISM. *See* Interstellar medium (ISM)
ISOC. *See* Instrument Science Operations Center (ISOC)
Isotopologues, 268–269
Isotropic undecimated wavelet transform (IUWT), 186, 187
    for curvelets and poisson noise, 192–193
    spherical, 205
    synthesis operator, 194
ISW effect. *See* Integrated Sachs Wolfe effect (ISW effect)
Iteratively reweighted least squares, 537. *See also* Spherical approximation
ITR. *See* Information technology research (ITR)
IUWT. *See* Isotropic undecimated wavelet transform (IUWT)

J

JL lemma. *See* Johnson–Lindenstrauss lemma (JL lemma)
Johnson–Lindenstrauss lemma (JL lemma), 655, 656

K

Kaiser–Squires–Broadhurst method (KSB method), 296
    galaxy's ellipticity, 298
    shapelets, 299
KAIT. *See* Katzman automatic imaging telescope (KAIT)
Karhunen–Loeve transform, 309
Katzman automatic imaging telescope (KAIT), 93
KDA. *See* Kernel discriminant analysis (KDA)
KDD. *See* Knowledge discovery in databases (KDD)
KDE. *See* Kernel density estimation (KDE)
Kepler magnitude, 357n
    Gaussian prior, 373
    photometric precision as, 368
Kepler mission, 167–168, 377. *See also* PLATO Mission
    Kepler magnitude, 357n
    launching, 356
    SOC functions, 357–358
Kepler SOC pipeline, 358, 377. *See also* Systematic error corrections
    data flow diagram, 358
    data validation, 366–368
    LDE, 360
    photometric analysis, 360–361
    pixel-level CAL, 359–360
    transient events, 360n
    transiting planet search, 362
    wavelet-based matched filter, 362–366

Kepler's exploitation, 13
Kernel clustering, 548, 558
Kernel density estimation (KDE), 103, 464, 605
    approximations, 475
    centroid approximation, 475
    conditional density estimation, 327
    far-field expansion, 475
    Gaussian kernel sum series expansion, 476
    guaranteeing error in dual-tree algorithms, 474–475
    KDE problem, 474
    local expansion, 477
Kernel discriminant analysis (KDA), 464, 478. *See also* Kernel density estimation (KDE)
    algorithm, 480
    avoiding full summation, 479
    Bayes' rule in classification, 479
Kernel function, 473, 558
    Gaussian, 388
    RBF, 520, 548
    in SVMs, 388
    uses, 621–622
Kernel regression, 464
    LPR and LLR, 478
    NWR, 477, 478
Kernel trick, 492
    in decision function, 497
    representer theorem, 492–493
k-means algorithm, 546, 599, 623. *See also* Expectation–maximization algorithm (EM algorithm)
    advantage, 599
    i-kMeans, 624
    limitations, 548
k-means clustering, 545, 558, 605. *See also* Data clustering
    algorithm, 546
    on artificial data set, 546
    compression via clustering, 547–548
    in eigenspace, 554, 555
    kernel *k*-means, 548
    k-means limitations, 548–549
    PCAD algorithm uses, 556
    variance, 545
K-means with Soft Constraints (KSC), 581
K-nearest neighbor classifier (KNN classifier), 104
KNN classifier. *See* K-nearest neighbor classifier (KNN classifier)
Knowledge discovery in databases (KDD), 177, 449
Known events/known classification rules, 454
Known events/unknown rules, 454
Kruskal's algorithm, 469

KSB method. *See* Kaiser–Squires–Broadhurst method (KSB method)
KSC. *See* K-means with Soft Constraints (KSC)

L

Lagrange multiplier method, 68
LAMOST. *See* Large Sky Area Multi-Object Fibre Spectroscopic Telescope (LAMOST)
Large Area Telescope (LAT), 41, 184, 239. *See also* Automated science processing (ASP); Classification tree (CT)
  ACD background rejection analysis, 255, 257
  analytic model for energy resolution, 253
  beam tests, 251
  BKG events, 257, 258
  contour generation, 261
  cross-subsystem correlations, 259, 260
  CT-generated probability distributions, 259, 260
  cut effect on IR, 253, 256
  direction reconstruction accuracy, 253
  energy events, 251, 252
  energy range issue, 251
  energy resolution knob, 254
  entropy, 241
  event analysis, 252
  event topology, 258
  good-energy probability, 253, 254
  McEnergySigma, 253
  PSF, 253, 254, 255
  recording information, 251
  shortcomings, 261–263
  subsystem, 250
  TKR and CAL subsystems, 258
  trigger, 239
Large Magellanic Cloud, 420
Large Sky Area Multi-Object Fibre Spectroscopic Telescope (LAMOST), 6
Large Synoptic Survey Telescope (LSST), 6, 166, 167, 384, 596, 638. *See also* Sky surveys
  dark energy, 172–173
  data mining research, 178
  Deep drilling subsurvey, 170
  inventory of solar system, 171–172
  mapping Milky Way, 172
  petascale data-to-knowledge challenge, 179–180
  science database, 618
  uses, 230, 231, 449
Laser Interferometer Gravitational-Wave Observatory (LIGO), 415
  detectors, 427
  PSD from, 418

Laser Interferometer Space Antenna (LISA), 419, 420
  data analysis, 424
  data character, 421
  detecting black-hole binaries, 430
  gravitational wave observation, 430–431
  LISA Pathfinder, 421
  noise characterization, 426
  recovery from antenna pointing, 429–430
Laser interferometry, 414
  antenna pattern of detector, 417
  data character, 417
  data quality, 418–419
  detector network, 417
  detector technology, 415–416
  displacement sensitivity, 418n
  10 Hz–1 kHz band, 414
  science data noise, 418
  working principle, 416
LAT. *See* Large Area Telescope (LAT)
Latent feature linearity (LFL), 525
LBVs. *See* Luminous blue variables (LBVs)
LC interval. *See* Long cadence interval (LC interval)
LCGT construction, 416
LCSS. *See* Longest Common Sub Sequence (LCSS)
LDA. *See* Linear discriminant analysis (LDA)
LDE. *See* Local detector electronics (LDE)
Least-squares approximation (LS approximation), 657. *See also* Randomized matrix algorithms
  basic structural result, 659–660
  hat matrix, 658
  perspectives, 657–658
  in practice, 661–662
  simple algorithm for, 658–659
  structural condition, 660
  in theory, 660–661
Least-squares estimates, 14
Least-squares regression (LS regression), 648
LFL. *See* Latent feature linearity (LFL)
Light
  gravitational deflection of, 20–21
  rays propagation, 290
  travel time disturbances, 416
Light-curve distance measures, 102
LIGO. *See* Laser Interferometer Gravitational-Wave Observatory (LIGO)
LIGO Scientific Collaboration (LSC), 419
Likelihood of β, 526
Line lists, 279
Linear discriminant analysis (LDA), 6, 104
LISA. *See* Laser Interferometer Space Antenna (LISA)
LLR. *See* Local linear regression (LLR)

Local detector electronics (LDE), 360
Local galaxy density, 457, 458
Local linear regression (LLR), 478
Local thermodynamic equilibrium (LTE), 272
LOFAR. *See* LOw Frequency Array (LOFAR)
Logistic model
    intercept estimation for, 538
    random chunk effect, 528
LogitBoost algorithm, 574
Log-likelihood function, 72, 230, 377, 526
Long cadence interval (LC interval), 356
Long γ-ray bursts progenitors, 414–415
Longest Common Sub Sequence (LCSS), 629
    advantages, 631
    algorithm, 630
    variation comparison, 633
Long-period variables (LPV), 385
Loss-less map, 136
LOw Frequency Array (LOFAR), 94, 230
Low-*l* codes. *See also* High-*l* codes
    commander algorithm, 73
    MADCAP, 72
Low-rank matrix approximation, 662. *See also*
        Randomized matrix algorithms
    additive-error bounds, 662
    importance sampling distribution, 663
    random projection algorithms, 666–668
    random sampling algorithm, 662–666
    structural condition, 665
LPV. *See* Long-period variables (LPV)
LS approximation. *See* Least-squares approximation
        (LS approximation)
LS regression. *See* Least-squares regression (LS
        regression)
LSC. *See* LIGO Scientific Collaboration (LSC)
LSST. *See* Large Synoptic Survey Telescope (LSST)
LTE. *See* Local thermodynamic equilibrium (LTE)
Luminous blue variables (LBVs), 618

M

Machine learning algorithms, 402. *See also* Data
        mining
    distinguishing Mira variables, 386
    to learning classification rules, 454
    periodic variability classification, 384–385
    production-scale, 540
Machine-learning (ML), 90, 240
    assessing causality, 30–31
    correlation *vs.* dependence, 31–32
    cosmic variance, 34–35
    detecting patterns, 28–30
    probability in science, 32–33

MACHO survey, 170
    feature space for variability classification, 395
    template light curves, 394
MAD. *See* Median absolute deviation (MAD);
        Median average deviation (MAD)
MADAM. *See* Map making through destriping for
        anisotropy measurments (MADAM)
MADCAP, 72
MAGIC analysis, 263–264
Magnetic component (B component), 300
Majority voting rule, 570
MANOVA tests. *See* Multivariate analysis of variance
        tests (MANOVA tests)
MAP approach. *See* Maximum a posteriori approach
        (MAP approach)
Map making through destriping for anisotropy
        measurments (MADAM), 82
Map-making, 61
    code, 82
    E–B-mode separation, 80
    linear, 308
Map-making problem, 134
    GLS solution, 136
    loss-less map, 136
    map estimation, 135–136
    TOD, 135
MapReduce framework, 598, 602
    data mining algorithms, 602–603
    least squares technique, 603
    machine learning algorithms, 604
    map and reduce functions, 602
Marginal regression, 537–538. *See also* Spherical
        approximation
Markov chain Monte Carlo method (MCMC
        method), 64
    for cosmological parameters, 76–77
    Metropolis-within-Gibbs, 82
Mars Orbiter Camera (MOC), 232
Mass dipole, 412
Massachusetts Institute of Technology (MIT), 378
MASTER. *See* Monte Carlo Apodized Spherical
        Transform Estimator (MASTER)
Matched filter (MF), 63, 344
    in TPS module, 359
    wavelet-based, 362, 363
Matrices, 649. *See also* Randomized matrix
        algorithms
    confusion, 389
    Hadamard transform, 654–655
    in large-scale scientific data analysis, 649–650
    $n \times l$ random projection matrix, 654
    object-feature matrices, 649

Matrices (*continued*)
    QR decomposition, 650
    random sampling and random projections, 652
    randomization, 652, 655–656
    retrospective and prospective, 656–657
    scientific applications, 650
maxBCG algorithm. *See* Brightest cluster galaxy
    algorithm (maxBCG algorithm)
Maximum a posteriori approach (MAP approach),
    371, 509
    estimator, 378
    implementation considerations, 377
    influence on light curves, 375
Maximum entropy method (MEM), 69, 304
    prior distribution, 70
    filter, 309
    implementation, 70
Maximum likelihood estimation (MLE), 46, 526–527.
    See also Generalized linear models
    likelihood of β, 526
    log-likelihood, 526
    for unchunked data structures, 537
Maximum likelihood method, 304
MC. *See* Monte Carlo (MC)
MCA. *See* Morphological component analysis (MCA)
McEnergy, 253
McEnergySigma, 253
MCMC method. *See* Markov chain Monte Carlo
    method (MCMC method)
MCS. *See* Multiple classifier systems (MCS)
MDD. *See* Mining of distributed data (MDD)
MDS. *See* Multidimensional scaling (MDS)
Median absolute deviation (MAD), 360
Median average deviation (MAD), 366
MEM. *See* Maximum entropy method (MEM)
Memory-based reasoning methods, 621. *See also*
    Pattern recognition
MergerWars, 229
Message passing interface (MPI), 598, 608–609
Mexican hat wavelet (MHW), 63
MF. *See* Matched filter (MF)
MHW. *See* Mexican hat wavelet (MHW)
Milky Way galaxy, 41, 172, 223
    mapping, 172
    motions of stars, 618
    simulated background model, 203
Millisecond pulsars (MSPs), 422
Minimum spanning tree (MST), 464, 553
Mining of distributed data (MDD), 451, 452
    distributed database querying, 454
    VLIRGs, 453
    VO-enabled science search, 453

Mira. *See* Asymptotic gaint branch stars (AGB
    stars)—star O Cet
Miras identification with SVM, 385. *See also* Red
    variables
    carbon star identification, 389–391
    confusion matrix, 387
    feature space projections, 386
    initial classification, 388
    machine learning classification, 384, 387
    result validation and misclassifications, 391
    support vector machines, 388
    three-class classification problem, 388–389
    two-class classification problem, 389
MIT. *See* Massachusetts Institute of Technology
    (MIT)
Mixture of experts methods, 568, 576. *See also*
    Ensemble methods
ML. *See* Machine-learning (ML)
MLE. *See* Maximum likelihood estimation (MLE)
MLP. *See* Multi layer perceptron (MLP)
MMDS. *See* Modern massive data set (MMDS)
MOC. *See* Mars Orbiter Camera (MOC)
Model family, 508
Modern massive data set (MMDS), 649
Molecular infrared spectroscopy, 269–270
Monopole moment of charge distribution, 412
Monte Carlo (MC), 73, 238
Monte Carlo Apodized Spherical Transform
    Estimator (MASTER), 73
Moore's law, 163, 448
Morphological component analysis (MCA), 79, 186,
    302
MOrphologically Selected Ellipticals in SDSS
    (MOSES), 217
MOSES. *See* MOrphologically Selected Ellipticals in
    SDSS (MOSES)
Motion-based time series, 618
MPI. *See* Message passing interface (MPI)
MRLens. *See* Multiresolution for weak lensing
    (MRLens)
MSE. *See* Multiscale entropy (MSE)
MSPs. *See* Millisecond pulsars (MSPs)
MST. *See* Minimum spanning tree (MST)
MS-VST. *See* Multiscale variance stabilizing
    transform (MS-VST)
MS-VST + IUWT + Inpainting method, 199, 200
MS-VST on sphere (MS-VSTS), 186, 189
    application to Gaussian denoising, 187, 188
    comparison, 192
    curvelet transform, 187, 188
    curvelets and Poisson noise, 192–193
    HEALPix pixelization, 186, 187

inpainting algorithm, 197, 200
IUWT on sphere, 186, 187
in Poisson denoising, 193
principle, 189
simulated Fermi data, 197, 203
VST of Poisson process, 189
wavelets and Poisson noise, 190–192
WMAP data, 187, 188
MS-VSTS. *See* MS-VST on sphere (MS-VSTS)
MS-VSTS + curvelets, 195–196
MS-VSTS + IUWT denoising, 190, 193, 195
convex constrained minimization problem, 194
multiresolution support adaptation, 193, 195
projection on multiresolution support, 194
MS-VSTS + IUWT denoising + background
extraction, 201–202
MS-VSTS-inpainting algorithm, 197, 199, 200
convex constrained minimization problem, 199
MS-VST + IUWT + inpainting method, 199, 200
MS-VSTS + IUWT denoising + background
extraction, 201–202
sensitivity to model errors, 202
simulated background model, 203
single HEALPix face, 204
source detection method, 199, 200
Multi layer perceptron (MLP), 576, 634
feedforward perceptron, 515
neuron, 515
sigmoid function, 515–516
Multichannel MEM-ICF method, 305
Multidimensional scaling (MDS), 498, 499, 500
Multiple classifier systems (MCS), 564
Multiple-stage models, 538–540. *See also* Spherical
approximation
Multiresolution for weak lensing (MRLens), 306.
See also Multiscale Bayesian filter
Multiresolution support adaptation, 194, 195
Multiscale Bayesian filter, 305, 306
Multiscale entropy (MSE), 305
Multiscale variance stabilizing transform (MS-VST),
185, 207. *See also* MS-VST on sphere
(MS-VSTS)
Multitree algorithms for large-scale astrostatistics,
463
all nearest neighbors, 463, 465, 480
dual-tree algorithm, 467
Euclidean minimum spanning trees, 468–471
kd-tree, 466
kernel density estimation, 473–477
kernel discriminant analysis, 478–480
kernel regression, 477–478
n-point correlation functions, 471–473, 480

pruning parts of computation, 466
single-tree algorithm, 467
space-partitioning trees, 467
Multivariate analysis of variance tests (MANOVA
tests), 5
Multivariate normal distributions (MVN
distributions), 5
Must-link constraint, 556, 558
MVN distributions. *See* Multivariate normal
distributions (MVN distributions)

N

Nadaraya–Watson regression (NWR), 477–478
Naïve algorithm, 465
Naïve Bayes
classifier, 513–514
decision rule, 570
learning algorithm, 513
NANOGrav. *See* North American Nanohertz
Observatory for Gravitational Waves
(NANOGrav)
NASA. *See* National Aeronautics and Space
Administration (NASA)
National Aeronautics and Space Administration
(NASA), 272
DSN, 357
Fermi Gamma-ray Space Telescope, 184
Kepler mission, 167–168
launching WMAP, 57
WFIRST, 168
WISE, 167
National Center for Supercomputing Applications
(NCSA), 448
National Science Foundation (NSF), 416, 609
data management plan, 419
US NVO, 610
National Virtual Observatory (NVO), 451, 609, 610
NCSA. *See* National Center for Supercomputing
Applications (NCSA)
Nearby Supernova Factory, 93
Near-earth object (NEO), 172
Nearest neighbor classification algorithm (NN
classification algorithm), 621
Nearest-neighbor identification, 455
Nebulous objects, 3
NEO. *See* Near-Earth object (NEO)
Neural nets. *See* Artificial Neural Networks (ANN)
Neural networks, 514–517. *See also* Support vector
machines (SVMs); Self-organizing maps
(SOMs)
feature space, 395
inputs to, 394

Neural networks (*continued*)
light curves for, 394, 397
output error, 396–397
pattern memory, 395
rate of false of negatives and positives, 396
transient variability classification, 393
Newton's First Law, 409n
"Newtonian" deflection, 20, 21
Newton–Raphson iteration (NR iteration), 72
NLA. *See* Numerical linear algebra (NLA)
NN classification algorithm. *See* Nearest neighbor classification algorithm (NN classification algorithm)
NNLS. *See* Nonnegative least-squares (NNLS)
NOISE-MSE, 305
Nondeterministic polynomial-complete (NP-complete), 577
Non-Gaussian statistics, 314
convergence bispectrum, 314
convergence three-point correlation function, 315
fourth-order moment of convergence, 315–316
peak counting, 316
skewness and kurtosis of aperture mass, 316
third-order moment of convergence, 315
Non-LTE situation, 275
Nonnegative least-squares (NNLS), 271
Nonsupervised classification approach, 495
density estimation problem avoidance, 495
one-class classification, 496
spherical MDS, 498
stationarity testing, 496–498, 499–500
support vector data description algorithm, 497
Normalized associated Legendre polynomials, 150–151
North American Nanohertz Observatory for Gravitational Waves (NANOGrav), 422
Northern Sky Variability Survey (NSVS), 385, 388
confusion matrix, 387
machine learning classifier comparison, 387
pulsation instability, 387
red variables in, 385, 386
NP-complete. *See* Nondeterministic polynomial-complete (NP-complete)
n-point correlation functions, 463, 471–472
counting pairs of points, 473
estimation, 472, 793
NR iteration. *See* Newton–Raphson iteration (NR iteration)
NSF. *See* United States National Science Foundation (NSF)
NSVS. *See* Northern Sky Variability Survey (NSVS)
Nuclide, 267

Nuisance effects, 527–528, 540
Numerical linear algebra (NLA), 648
NVO. *See* National Virtual Observatory (NVO)
NWR. *See* Nadaraya–Watson regression (NWR)

O

Object cross-correlation, 455
Object cross-identification, 455
OC. *See* Output coding (OC)
Oct-trees, 467
OGLE. *See* Optical gravitational lensing experiment (OGLE)
One-dimensional minimization algorithm, 64
Optical gravitational lensing experiment (OGLE), 101, 391
classification and novelty detection, 398–399
SOMs, 397–398
survey, 170
Orthogonal ICA method, 68–69
Output coding (OC), 565
ECOC, 577
methods, 576
PWC decomposition, 577

P

P2P. *See* Peer-to-peer (P2P)
PA module. *See* Photometric analysis module (PA module)
PAA. *See* Piecewise aggregate approximation (PAA)
PAHs. *See* Polycyclic aromatic hydrocarbons (PAHs)
Pairwise coupling decomposition (PWC decomposition), 577
Palomar Transient Factory (PTF), 5, 92n, 93, 170
Panoramic Survey Telescope and Rapid Response System (Pan-STARRS), 5, 166, 618
Pan-STARRS. *See* Panoramic Survey Telescope and Rapid Response System (Pan-STARRS)
Parallel computing framework, 596, 597. *See also* Data mining; Distributed computing framework
Parallel spectral clustering algorithm (PSC algorithm), 600
Parallel support vector machine (PSVM), 601. *See also* Support vector machines (SVMs)
Parameters for impatient cosmologist (PICO), 77
Pattern recognition, 617, 621, 639
algorithms, 621–622
approaches, 620
astronomical applications, 637–639
BOP, 635–637
CDM, 634–635
EM algorithm, 624

k-Means algorithm, 623–624
1-NN classification results, 628
SAX, 626–628
semi-supervised learning, 622–623
shape-based similarity, 629–631
similarity measure, 628, 632–633
specialized clustering algorithms, 624
structural similarity, 633–635
time-series data, 617–620. *See also* Time-series
    databases
time-series representations, 625
unsupervised learning, 623–624
PB. *See* Petabytes (PB)
PC. *See* Peak counting (PC)
PCA. *See* Principal components analysis (PCA)
PCAD algorithm. *See* Periodic curve anomaly
    detection algorithm (PCAD algorithm)
PDC module. *See* Presearch data conditioning
    module (PDC module)
Peak counting (PC), 316
Peer-to-peer (P2P), 598
Perceptron, 6. *See also* Artificial neural networks
    (ANN)
    hybrid, 576
    multilayer, 514, 515
Percolation algorithm, 4
    clustering scheme, 5
Periodic curve anomaly detection algorithm (PCAD
    algorithm), 556
Periodic signals, 425, 427
Periodic trigger (PT), 262
Periodic variability classification algorithms, 384
Petabytes (PB), 450
Peta-scale astronomical survey, 384
Petascale data-to-knowledge challenge, 179–180
pgwave, 51
PHAs. *See* Potentially Hazardous Objects (PHAs)
Photometric analysis module (PA module), 359, 360
Photometric redshifts, 298n, 323, 324
    broadband photometry, 326
    conditional density, 331
    determination, 324
    dropout techniques, 330
    empirical approaches, 324–325, 329
    limitations, 333
    mapping measurements with errors, 327–328
    optimal filter design, 332
    photometric errors, 330–331
    photometric measurements, 324
    photometric passbands, 324
    photometry accuracy, 332
    physical properties vs. observables, 326–327

piecewise fitting, 325
priors for models, 331
spatial orientation, 333
spectral energy distribution, 325
spectral lines, 332–333
template fitting, 325–326, 328–329
training set, 329–330
Photometric techniques, 338n
Pi GHz Sky Survey (PiGSS), 169
PICO. *See* Parameters for impatient cosmologist
    (PICO)
Piecewise aggregate approximation (PAA),
    626
Piecewise fitting, 325
Piecewise linear aggregate (PLA), 625
PiGSS. *See* Pi GHz Sky Survey (PiGSS)
Pixel domain, component separation in
    CCA, 69
    FastICA technique, 68
    ILC, 67
    template fitting, 66–67
Pixel response functions (PRFs), 361
Pixelated imaging, 92
    image differencing method, 92, 93
    image source extraction, 92
    Nearby Supernova Factory, 93
    PTF, 93
    ROC curve for image-differenced candidates,
        93–94
PLA. *See* Piecewise linear aggregate (PLA)
Planck Surveyor, 57–58
Planet detection, 355. *See also* Kepler SOC pipeline;
    Systematic error corrections
    Kepler mission, 356–358, 377
    photometric precision, 368
    planet hunting, early, 17–18
Planetary theory from data mining, 12–13
PLATO mission, 378
Point spread function (PSF), 43
    correction, 296
    debiasing, 295
    resolution, 51
Poisson denoising. *See also* Gaussian denoising;
    Poisson noise
    experiments, 196, 197
    Fermi simulated map without noise, 197
    MS-VSTS + curvelets, 195–196
    MS-VSTS + IUWT, 193–194, 195
    multiresolution support adaptation, 194,
        195
    single HEALPix face, 198

Poisson noise. *See also* Gaussian noise; Poisson
        denoising; MS-VST on sphere (MS-VSTS)
    constrained sparsity-promoting minimization
        problem, 208
    curvelets and, 192–193
    detection–reconstruction, 207–208
    Fermi mission, 185
    intensity estimation techniques, 185
    MS-VST, 207
    multiscale variance stabilizing transform, 207
    spherical MSVST 2D–1D wavelet denoising
        algorithm, 209
    wavelets and, 190–192
PolEMICA, 82
Polycyclic aromatic hydrocarbons (PAHs), 270, 274.
        See also Astronomical spectra analyses
    database and model, 274–277
    excitation and de-excitation, 275
    in space, 273
Position-dependent weighting function, 74
Post facto statistical analysis, 28, 29
Potentially Hazardous Objects (PHAs), 172
PowellSnake, 63–64
Power spectral density (PSD), 363, 488
    detector noise, 418, 426
    to measure fluctuation of, 501
Power spectrum estimation, 71–72. *See also* Cosmic
        microwave background (CMB)
    in CMB data analysis pipeline, 60
    CMB map inpainting, 79
    fromweak lensing data, 312
    Gibbs sampling method, 82
    high-l codes, 73–76
    low-l codes, 72–73
    for model spectra likelihood function evaluation,
        72
    pseudo-power spectrum, 74
Presearch data conditioning module (PDC module),
        359. *See also* Systematic error corrections
PRFs. *See* Pixel response functions (PRFs)
Principal components analysis (PCA), 102
    based classification, 217
    based clustering technique, 605
    to define optimal spectral band, 581
    distributed algorithm, 610
    DMD algorithms and, 456
    experiments with SDSS–2MASS data, 457
    galaxy study, 216
    NLA and, 651
    outlier identification, 556
    for spectral classification, 652
Priors for models, 331

Probability models, 525
Pruning, 479, 512
    ensemble, 573
    parts of the computation, 466
    rule, 473, 478
PSC algorithm. *See* Parallel spectral clustering
        algorithm (PSC algorithm)
PSD. *See* Power spectral density (PSD)
Pseudo-power spectrum, 74
PSF. *See* Point Spread Function (PSF)
PSVM. *See* Parallel support vector machine (PSVM)
PT. *See* Periodic trigger (PT)
PTA. *See* Pulsar timing array (PTA)
PTF. *See* Palomar Transient Factory (PTF)
Ptolemy's theory, 12–13
Pulsar timing array (PTA), 421, 422
    data character, 424
    detector network, 423–424
    detector technology, 422
    working principle, 423
PWC decomposition. *See* Pairwise coupling
        decomposition (PWC decomposition)

Q

QR decomposition, 650n. *See also* Matrices
QSO. *See* Quasi-stellar object (QSO)
Quadratic discriminant analysis classifiers.
        See Gaussian mixture classifier
Quadrilateralized Sky Cube Projection, 134
Quad-trees, 467
Quasi-stellar object (QSO), 98
    variability selection of, 99

R

RA. *See* Reconstruction analysis (RA)
Radial basis function kernel (RBF kernel), 548, 554
    Gaussian RBF, 621–622
    MLP/RBF, 576
Radial Wiener filter, 309
Radio interferometry, 94
Radio-frequency interference (RFI), 94
RAM. *See* Random access memory (RAM)
Random access memory (RAM), 481
Random forest (RF), 93, 578, 579. *See also* Ensemble
        methods
Random forest classifier, 93, 107
    cross-validation error rates, 106
    and high variance, 105
    in MAGIC analysis, 263–264
    performance, 104
    ROC curves from, 104
Randomized Hadamard transform, 654–655

Randomized matrix algorithms, 648, 668
  applications, 650
  developments, 648
  for low-rank matrix approximation, 662–668
  LS approximation, 657–662
  matrices, 649–652
  matrix problems, 652–657
  structured random projection, 654
  in TCS, 648
RBF kernel. *See* Radial basis function kernel (RBF kernel)
R–C optics. *See* Ritchey–Chrétien optics (R–C optics)
RC3. *See* Third Reference Catalogue of Bright Galaxies (RC3)
Receiver operating characteristic curve (ROC curve), 93
  for image-differenced candidates, 94
  for supernovae selection, 103, 104
Recon. *See* Reconstruction program (Recon)
ReconEnergy, 253
Reconstruction analysis (RA), 238
Reconstruction program (Recon), 251
  ReconEnergy, 253
Red variables, 386, 387
  confusion matrix for, 389
  feature space for, 390
  of NS *vs.* catalog, 385
  SVM classification of, 390
Reddening, 413
Redshift. *See also* Photometric redshifts; Sachs–Wolfe effect
  catalog, 325, 345
  cluster, 338
  convergence map and, 307
  cosmic distance ladder, 420
  dark energy and, 172
  determination, 324
  and distance, 337
  estimate, 329, 332
  extragalactic events and, 103
  of extragalactic object, 324
  of galaxy, 340, 343
  GZ1 classifications, 220
  low redshift Universe, 314
  photometric, 298n, 323, 324, 326, 338n
  possibilities, 326
  spectroscopic, 298n, 338n
  in systematic error reduction, 298
Refractive index in gravitational field, 290, 291
Regions-of-interest (ROI), 52
Regression tree (RT), 621
Regularization, 537

Repixelization, 148
Representer theorem, 493
  time–frequency distribution interpretation, 494
Reproducing kernel Hilbert space (RKHS), 492
RF. *See* Random forest (RF)
RFI. *See* Radio-frequency interference (RFI)
Ritchey–Chrétien optics (R–C optics), 166
RKHS. *See* Reproducing kernel Hilbert space (RKHS)
ROC curve. *See* Receiver operating characteristic curve (ROC curve)
ROentger SATellite mission (ROSAT mission), 580, 604
ROI. *See* Regions-of-interest (ROI)
ROSAT mission. *See* ROentger SATellite mission (ROSAT mission)
ro-vibrational spectrum, 268–269
RT. *See* Regression tree (RT)

S

SAA. *See* South Atlantic Anomaly (SAA)
Sachs–Wolfe effect, 59, 413
SAX. *See* Symbolic Aggregate approXimation (SAX)
SC. *See* Short Cadence (SC)
Science Operations Center (SOC), 357
  data validation, 366–368
  Kepler, 358
  photometric analysis, 360–361
  pixel-level CAL, 359–360
  systematic error corrections, 361–362
  transiting planet search, 362
  wavelet-based matched filter, 362–366
Scientific knowledge discovery, 177, 178–179. *See also* Sky surveys—digital
SDSS. *See* Sloan Digital Sky Survey (SDSS)
Second-order statistics, 311
  CCA, 69
  measure the Gaussian properties, 314
  to separate E and B modes, 313
SEDs. *See* Spectral energy distributions (SEDs)
Self-Organizing Maps (SOMs), 106, 107, 397. *See also* Neural networks
  cartography experiments, 398
  classification accuracy, 404
  Gaia science alerts, 399
  H–R diagram, 401
  nodes in, 397–398
  quantization error, 399
  on simulated Gaia spectra, 400
  spectral type mapping, 402
  stellar parameter mapping, 403
  temporal variability classification, 398
  U-matrix, 400

Semisupervised clustering, 558

Semi-supervised learning, 107, 622–623. *See also*
        Supervised learning; Unsupervised
        learning

  astronomical applications, 637–639

SExtractor software, 7, 64

Shear field, 294, 295. *See also* Convergence field

Short cadence (SC), 356

Short γ-ray burst progenitors, 414

SHT. *See* Spherical harmonic transform (SHT)

Signal–noise ratio (SNR), 202

  millisecond pulsars timing, 422

  redshift accuracy, 332

  TCEs, 359

Silicon strip detectors (SSDs), 238

Single nucleotide polymorphism (SNP), 649

Single trial statistical method, 29, 30

Single-linkage clustering, 5

Single-pass broadband filters, 332

Single-tree algorithm, 467

Singular value decomposition (SVD), 361, 373, 649,
        652. *See* Karhunen–Loeve transform

SKA. *See* Square Kilometer Array (SKA)

Sky surveys, 383, 450, 596. *See also* Virtual
        observatories

  DES, 168

  digital, 164

  étendue comparison, 169–170

  GAIA, 168

  Kepler, 167–168

  large-scale, 163, 604

  LSST, 167, 618, 638

  next generation, 596

  Pan-STARRS, 166

  SkyMapper, 165–166, 169

  VISTA, 166

  WFIRST, 168

  wide-area radio surveys, 168–169

  WISE, 167

Sloan Digital Sky Survey (SDSS), 5. *See also* Sky
        surveys

  color images, 215

  étendue comparison, 169

  fitting formula, 331

  Main Galaxy sample, 216

  RRLyrae stars, 97

  SDSS SVM classifier validation, 8

  SDSS-C4 Cluster Catalog, 344

  sources, 114, 115

  SVM classification of SDSS point sources, 7

SMICA. *See* Spectral matching independent
        component analysis (SMICA)

SNe. *See* Supernovae (SNe)

SNP. *See* Single nucleotide polymorphism (SNP)

SNR. *See* Signal–noise ratio (SNR)

SOC. *See* Science Operations Center (SOC)

Society of Photo-Optical Instrumentation Engineers
        (SPIE), 364, 448n

SOHO. *See* Solar and Heliospheric Observatory
        (SOHO)

Solar and Heliospheric Observatory (SOHO), 369

Solar system, 58

  emission, 61

  inventory of, 171–172

Solid-state recorder (SSR), 356, 357

SOMs. *See* Self-organizing maps (SOMs)

South Atlantic Anomaly (SAA), 49n, 50

Space Telescope Science Institute (STScI), 359

Spacecraft Doppler tracking, 419–421

  data character, 421

  detector network, 421

  detector technology, 420

  0.01mHz–1 Hz band, 419

  radiation bursts, 420

  working principle, 421

Spacecraft's ascending node, 49n

Space-partitioning trees, 467

Space–time, 410–411

  curvature, 410, 411

  gravitational wave and, 416

  as nonintersecting spatial cross sections, 411

  structure and gravity, 409

Sparse representations of signals, 305

Spatially Inhomogeneous Correlation Estimator
        (SPICE), 76

Spectral clustering, 545, 553, 558, 599–600. *See also*
        Data clustering

  algorithm, 554

  on artificial data set, 555

  constraint incorporation, 556

  parallelization, 557

Spectral energy distributions (SEDs), 123

  Bayes factor and, 131

  redshifts, 323

Spectral lines, 332–333

Spectral matching independent component analysis
        (SMICA), 66, 70

  PolEMICA, 82

Spectroscopy, 267

  cluster redshift determinations, 338

  discriminatory power of, 275

  molecular, 268–269

  redshifts, 298n

  techniques, 338n

Spherical 2D–1D undecimated wavelet transform, 205–206, 208

Spherical approximation, 530, 540. *See also* Chunks; Generalized linear models
- assessment, 535–536
- benzene electrostatic potential, 531
- chunk-specific correlation coefficient, 534–535
- normal equations, 534
- numerator–denominator plot, 534, 535
- related methods, 537
- size 3 chunks, 530
- spherical likelihood gradient, 532
- spherical throttling ratio, 532–533
- three-element vector permutations, 531
- von Mises–Fisher distribution, 531–532

Spherical harmonic transform (SHT), 73, 79

Spherical MSVST 2D–1D wavelet denoising algorithm, 209

Spherical overdensity halo finding algorithm, 341

SPICE. *See* Spatially Inhomogeneous Correlation Estimator (SPICE)

SPIE. *See* Society of Photo-Optical Instrumentation Engineers (SPIE)

Spin ±2 spherical harmonic coefficients, 80

Spline interpolation approach, 148–149

SQL. *See* Structured query language (SQL)

Square Kilometer Array (SKA), 168, 423

SSC. *See* Super Conducting Super Collider (SSC)

SSDs. *See* Silicon strip detectors (SSDs)

SSR. *See* Solid–state recorder (SSR)

Stacking methods, 571

Stardust@Home Project, 217–218. *See also* Galaxy Zoo

Stars, 3, 337, 345. *See also* Black holes; Galaxy; Kepler mission; Milky Way galaxy
- carbon, 389–391
- classes of, 384–385
- correlation coefficient matrix, 371, 375
- density, 124
- fractional variability of, 97
- H–R diagram, 401
- infrared observations of, 277
- infrared spectra of, 278, 281, 282
- light curves for, 376
- modern classification, 4
- neutron stars, 414–415
- SDSS SVM classifier validation, 8
- time-series data for, 620
- with unknown proper motion, 124–127
- variable, 107, 387
- white dwarf stars, 420

Statistical clustering. *See* Unsupervised classification methods

Statistical leverage scores, 648–649

Statistical malfeasance, 34–35

Statistical models, 525–526. *See also* Generalized linear models

Stellar transits, 527

Stochastic gravitational wave, 425

Stripe 82, 223

Strong gravitational lensing, 289, 292. *See also* Weak gravitational lensing

Structured query language (SQL), 91, 129

Structured random projection, 654

STScI. *See* Space Telescope Science Institute (STScI)

Subhalo identification, 341–343

Sunyaev-Zel'dovich effect (SZ effect), 58
- cluster extraction, 64–65

Super Conducting Super Collider (SSC), 238

Supernovae (SNe), 4, 618
- classification significance, 638
- cosmographic probes, 96
- dark energy, 90
- discoveries, 93
- domain-specific classification, 98
- explosion, 414
- ROC curves for the selection of, 104
- SOM and, 402
- type Ia supernovae, 96n

Supervised classification methods, 100. *See also* Supervised learning; Unsupervised classification methods
- ANN, 105
- Bayesian network classification, 104
- classification trees, 105
- class-wise KDE, 103, 104
- cross-validation error rate distribution, 106
- Gaussian mixture classifiers, 104
- hierarchical classification, 105
- KDE classification, 103
- KNN, 104–105
- SVMs, 104
- tree ensemble methods, 105
- variable star classification hierarchy, 107

Supervised learning, 505, 619, 621. *See also* Support vector machines (SVMs); Supervised classification methods
- astronomical applications, 637–639
- output of, 506

Support vector machines (SVMs), 6, 388, 517, 621. *See also* Classification learning algorithm; Neural networks
- accuracy, 385

Support vector machines (SVMs) (*continued*)
  applications, 6, 7, 8
  classification of red variables, 390
  confusion matrix for, 387
  disadvantages, 520–521
  hyperplane, 517–519
  and Mira variables, 386
  one-class SVM, 496–498
  PSVM, 601
  SDSS point sources, 7
  SDSS SVM classifier validation, 8
  support vectors, 520
  variable stars classification, 104
Surprise detection, 454, 456
Survey astronomer, 164
Survey telescopes, 163
SVD. *See* Singular value decomposition (SVD)
SVMs. *See* Support vector machines (SVMs)
Symbolic Aggregate approXimation (SAX), 626–628, 639
  SAX-sequence matrix, 635
Synchrotron emission, 61–62
Systematic data exploration, 455–456
Systematic error corrections, 361–362, 363. *See also* Kepler SOC pipeline
  empirical MAP implementation considerations, 377
  instrumental signatures, 370
  intrinsic stellar variability, 371
  light curve for star, 369, 370, 371, 376
  MAP application to Kepler data, 373–377
  MAP approach, 371–373
  PDC problems, 368–369
  singular components for light curves, 373
  singular values for light curves, 374
SZ effect. *See* Sunyaev-Zel'dovich effect (SZ effect)

T

TB. *See* Terabytes (TB)
TCEs. *See* Threshold crossing events (TCEs)
TCP. *See* Transients Classification Pipeline (TCP)
TCS. *See* Theoretical computer science (TCS)
TEASING, 73
Template fitting, 66–67, 325–326, 328–329
Template map, 67
Terabytes (TB), 450
TESS. *See* Transiting Exoplanet Survey Satellite (TESS)
TF. *See* Time–frequency (TF)
TFA. *See* Trend filtering algorithm (TFA)
TH. *See* Thresholding operator (TH)

Theoretical computer science (TCS), 648
  relative-error bounds, 662
Thermal bremsstrahlung, 62
Thin lens approximation, 291
Third Reference Catalogue of Bright Galaxies (RC3), 216
Third-order statistics, 314
Threshold crossing events (TCEs), 359
Thresholded plurality vote, 570
Thresholding operator (TH), 206, 207
Throttling ratio, 532–533
Time dimension, 170. *See also* Sky surveys—digital
  characterization, 170–171
  classification accuracy, 171
Time-domain
  astronomy, 105, 106, 451
  features, 100
  surveys, 171, 617–618
Time-domain classification, 96. *See also*
      Feature-based classification; Time-domain
      discovery
  domain-based classification, 96–99
  future challenges, 107–108
  variable star taxonomy, 105, 107
Time-domain discovery, 91–92
  for fast-moving sources, 92n
  pixelated imaging, 92–94
  radio interferometry, 94
  variability detection and analysis, 94–95
Time–frequency (TF), 489
  analysis, 488
  FM signal spectrogram, 490
  nonstationary processes, 488
  stationarization via surrogates, 489–491
Time–frequency learning machines, 491, 493–494, 500. *See also* Nonsupervised classification
      approach
  kernel reproduction, 492
  kernel trick, 492–493
  Wigner distribution *vs.* spectrogram, 495
Time-ordered data (TOD), 60, 61
  for polarization data, 80
Time-series databases, 617, 618. *See also* Pattern
      recognition
  astronomical applications, 637–639
  data mining framework, 625
  discretization, 626
  flux-based time series, 618
  memory-based reasoning methods, 621
  motion-based time series, 618
  research areas, 619
  semi-supervised classification technique, 622

similarity measure, 628–635
specialized clustering algorithms, 624
SVMs, 621–622
time-series discords, 623
for variable stars, 620
Time-warp edit distance (TWED), 631
error rates, 632
for multivariate time series, 633
Timing residuals, 423
TKR. *See* Tracker (TKR)
TOD. *See* Time-ordered data (TOD)
TPS. *See* Transiting planet search (TPS)
Tracker (TKR), 250, 258
Training sets, 5
Transient variability classification, 393–397
Transients Classification Pipeline (TCP), 618
Transiting Exoplanet Survey Satellite (TESS), 378
Transiting planet search (TPS), 357, 362
in Kepler science pipeline, 358
signal flow diagram for, 364
wavelet-based matched filter, 362–366
Transverse Wiener filter, 309
Tree ensemble methods, 105
Trend filtering algorithm (TFA), 369
TWED. *See* Time-warp edit distance (TWED)
2D–1D wavelet transform, 202
ideal wavelet function, 203
spatial and temporal wavelets, 203
wavelet coefficients, 203, 205
2MASS. *See* 2-Micron All Sky Survey (2MASS)
2-Micron All Sky Survey (2MASS), 114, 115, 163, 164, 387, 596

U

UIR. *See* Unidentified Infrared (UIR)
ULIRGs. *See* Ultra-luminous IR galaxies (ULIRGs)
Ultra-luminous IR galaxies (ULIRGs), 453
Unanimity vote rule, 570
Unidentified Infrared (UIR), 272
United States National Science Foundation (NSF), 416
GRIST, 609
open data, 418–419
U.S. NVO, 610
United States Naval Observatory (USNO), 579
United States Virtual Observatory (VO), 451. *See also* Data mining
data availability, 452
Universal approximators. *See* Neural networks
Universe, 22, 34, 56, 162. *See also* Dark matter 2D mapping; Galaxy clusters
CMB measurements and, 58, 78

color, 323
composition, 288
distance ladder measurements, 96–97
facts of, 214
island, 338
luminosity function, 343
period and luminosity relationship, 89–90
sky surveys, 163
Unknown events/known rules, 454
Unknown events/unknown rules, 454
Unsupervised classification methods, 100, 105. *See also* Supervised classification methods; Unsupervised learning
autoclass method, 106
Gaussian mixture modeling, 106
SOM, 106, 107
Unsupervised clustering, 4–5
Unsupervised learning, 402, 558, 619, 623–624. *See also* Self-organizing maps (SOMs); Supervised learning; Semi-supervised learning; Unsupervised classification
algorithm, 177
applications, 638
astronomical applications, 637–639
USNO. *See* United States Naval Observatory (USNO)

V

Vacuum energy density. *See* Dark energy
VAO. *See* Virtual Astronomical Observatory (VAO)
Variability of solar irradiance and gravity oscillations (VIRGO), 369n
Variable optical sky, 170
Variable star classification hierarchy, 107
Variance stabilizing transform (VST), 185. *See also* MS-VST on sphere (MS-VSTS)
Vector field, 300
Vector quantization, 547, 559
Very luminous infrared galaxies (VLIRGs), 453
VIM. *See* Visual integration and mining (VIM)
VIRGO. *See* Variability of solar irradiance and gravity oscillations (VIRGO)
Virgo cluster, 323, 337
Virtual Astronomical Observatory (VAO), 451
Virtual observatories, 450. *See also* Distributed data mining (DDM); Distributed mining of data (DMD); United States Virtual Observatory (VO)
cyberinfrastructure, 447
data volumes, 448
NVOs, 609
VISTA, 165, 166

Visual integration and mining (VIM), 452
VLIRGs. *See* Very luminous infrared galaxies (VLIRGs)
VO. *See* United States Virtual Observatory (VO)
VO-Event messaging protocol, 451
Von Mises–Fisher distribution, 531–532, 540
Voronoi cells, 347
Voronoi Tessellation, 347. *See also* Kernel density estimation
    local galaxy density, 457
VST. *See* Variance stabilizing transform (VST)

W

Wagging, 574. *See also* Bootstrap aggregating (Bagging)
Wavelet domain, component separation in
    Gaussian noise removal, 185
    GMCA, 71
    toward wavelet-based methods, 70–71
Wavelet-based matched filter, 362–366
Weak gravitational lensing, 288, 292, 294, 310, 316.
        See also Cosmological model constraints; Weak lensing theory
    challenges, 298–299
    convergence, 299
    cosmic shear amplitude, 295
    dark matter mapping, 299–309
    direct deconvolution, 296
    E mode, 301
    filling masked regions, 301
    galaxy distance information, 298
    galaxy ellipticity, 295
    GI alignment, 297, 298
    instrumental and atmospheric bias, 295
    intrinsic alignments and correlations, 297
    KSB method, 296
    lensfit method, 296
    mass reconstruction, 299
    MCA, 302
    multiresolution representations, 301–302
    power spectrum measurement, 301
    two-point statistics of the cosmic shear field, 301
Weak lensing theory, 289
    convergence, 293–294
    deflection angle, 290–291
    distortion matrix, 292–293
    gravitational shear, 293
    lens equation, 291–292

light speed, 290
    refractive index in gravitational field, 290, 291
Weighting method. *See* Kernel regression
WFIRST. *See* Wide Field Infrared Survey Telescope (WFIRST)
WGN. *See* White Gaussian noise (WGN)
White dwarf stars, 420
White Gaussian noise (WGN), 363
Wide Field Infrared Survey Telescope (WFIRST), 165, 168
Wide-area radio surveys, 168–169
Wide-field Infrared Survey Explorer (WISE), 165, 167
Wiener filter, 303, 306, 309, 427n
Wiener filtered map, generalized, 73
Wigner distribution, 493–494
    vs. spectrogram, 495
Wigner–Ville spectrum (WVS), 488
Wilkinson microwave anisotropy probe (WMAP), 57–58, 481
    CMBdata map, 78
    dark energy validation, 481
    data, 187
    ILC for template fitting, 67
    IUWT on WMAP data, 187
    pixels, 61
    WMAP data and wavelet transform, 188
Window function, 72, 120
    GLESP pixel, 151–153
WISE. *See* Wide-field Infrared Survey Explorer (WISE)
WMAP. *See* Wilkinson microwave anisotropy probe (WMAP)
WVS. *See* Wigner–Ville spectrum (WVS)

X

XFaster, 75
X-informatics. *See* Discovery informatics

Z

Zodiacal light, 18
Zooniverse, 214, 219n, 226–227, 233. *See also* Galaxy Zoo
    citizen science projects, 230–232
    codebase, 228
    data mining, 230
    development, 228
    from Galaxy Zoo to, 227
    suitable tasks for, 228–229

Printed and bound by CPI Group (UK) Ltd, Croydon, CR0 4YY

21/10/2024

01777040-0018